Ecology of Phytoplankton

Phytoplankton communities dominate the pelagic ecosystems that cover 70% of the world's surface area. In this marvellous new book Colin Reynolds deals with the adaptations, physiology and population dynamics of the phytoplankton communities of lakes and rivers, of seas and the great oceans. The book will serve both as a text and a major work of reference, providing basic information on composition, morphology and physiology of the main phyletic groups represented in marine and freshwater systems. In addition Reynolds reviews recent advances in community ecology, developing an appreciation of assembly processes, coexistence and competition, disturbance and diversity. Aimed primarily at students of the plankton, it develops many concepts relevant to ecology in the widest sense, and as such will appeal to a wide readership among students of ecology, limnology and oceanography.

Born in London, Colin completed his formal education at Sir John Cass College, University of London. He worked briefly with the Metropolitan Water Board and as a tutor with the Field Studies Council. In 1970, he joined the staff at the Windermere Laboratory of the Freshwater Biological Association. He studied the phytoplankton of eutrophic meres, then on the renowned 'Lund Tubes', the large limnetic enclosures in Blelham Tarn, before turning his attention to the phytoplankton of rivers. During the 1990s, working with Dr Tony Irish and, later, also Dr Alex Elliott, he helped to develop a family of models based on, the dynamic responses of phytoplankton populations that are now widely used by managers. He has published two books, edited a dozen others and has published over 220 scientific papers as well as about 150 reports for clients. He has given advanced courses in UK, Germany, Argentina, Australia and Uruguay. He was the winner of the 1994 Limnetic Ecology Prize; he was awarded a coveted Naumann–Thienemann Medal of SIL and was honoured by Her Majesty the Queen as a Member of the British Empire. Colin also served on his municipal authority for 18 years and was elected mayor of Kendal in 1992–93.

ECOLOGY, BIODIVERSITY, AND CONSERVATION

The world's biological diversity faces unprecedented threats. The urgent challenge facing the concerned biologist is to understand ecological processes well enough to maintain their functioning in the face of the pressures resulting from human population growth. Those concerned with the conservation of biodiversity and with restoration also need to be acquainted with the political, social, historical, economic and legal frameworks within which ecological and conservation practice must be developed. This series will present balanced, comprehensive, up-to-date and critical reviews of selected topics within the sciences of ecology and conservation biology, both botanical and zoological, and both 'pure' and 'applied'. It is aimed at advanced (final-year undergraduates, graduate students, researchers and university teachers, as well as ecologists and conservationists in industry, government and the voluntary sectors. The series encompasses a wide range of approaches and scales (spatial, temporal, and taxonomic), including quantitative, theoretical, population, community, ecosystem, landscape, historical, experimental, behavioural and evolutionary studies. The emphasis is on science related to the real world of plants and animals, rather than on purely theoretical abstractions and mathematical models. Books in this series will, wherever possible, consider issues from a broad perspective. Some books will challenge existing paradigms and present new ecological concepts, empirical or theoretical models, and testable hypotheses. Other books will explore new approaches and present syntheses on topics of ecological importance.

The Ecology of Phytoplankton

C. S. Reynolds

CAMBRIDGE UNIVERSITY PRESS
Cambridge, New York, Melbourne, Madrid, Cape Town, Singapore, São Paulo, Delhi

Cambridge University Press
The Edinburgh Building, Cambridge CB2 8RU, UK

Published in the United States of America by Cambridge University Press, New York

www.cambridge.org
Information on this title: www.cambridge.org/9780521605199

First published 2006
Reprinted 2007

A catalogue record for this publication is available from the British Library

ISBN 978-0-521-84413-0 hardback
ISBN 978-0-521-60519-9 paperback

Transferred to digital printing (with corrections) 2009

This book is dedicated to
my wife, JEAN, to whom its writing
represented an intrusion into
domestic life, and to Charles Sinker,
John Lund and Ramón Margalef. Each is
a constant source of inspiration to me.

Contents

Preface   *page* ix
Acknowledgements   xii

Chapter 1. Phytoplankton   1
1.1 Definitions and terminology   1
1.2 Historical context of phytoplankton studies   3
1.3 The diversification of phytoplankton   4
1.4 General features of phytoplankton   15
1.5 The construction and composition of freshwater
    phytoplankton   24
1.6 Marine phytoplankton   34
1.7 Summary   36

Chapter 2. Entrainment and distribution in the pelagic   38
2.1 Introduction   38
2.2 Motion in aquatic environments   39
2.3 Turbulence   42
2.4 Phytoplankton sinking and floating   49
2.5 Adaptive and evolutionary mechanisms for
    regulating w_s 53
2.6 Sinking and entrainment in natural turbulence   67
2.7 The spatial distribution of phytoplankton   77
2.8 Summary   90

Chapter 3. Photosynthesis and carbon acquisition in
phytoplankton   93
3.1 Introduction   93
3.2 Essential biochemistry of photosynthesis   94
3.3 Light-dependent environmental sensitivity of
    photosynthesis   101
3.4 Sensitivity of aquatic photosynthesis to carbon
    sources   124
3.5 Capacity, achievement and fate of primary
    production at the ecosystem scale   131
3.6 Summary   143

Chapter 4. Nutrient uptake and assimilation in
phytoplankton   145
4.1 Introduction   145
4.2 Cell uptake and intracellular transport of
    nutrients   146
4.3 Phosphorus: requirements, uptake, deployment in
    phytoplankton   151

4.4 Nitrogen: requirements, sources, uptake and
  metabolism in phytoplankton 161
4.5 The role of micronutrients 166
4.6 Major ions 171
4.7 Silicon: requirements, uptake, deployment in
  phytoplankton 173
4.8 Summary 175

Chapter 5. Growth and replication of phytoplankton 178
5.1 Introduction: characterising growth 178
5.2 The mechanics and control of growth 179
5.3 The dynamics of phytoplankton growth and
  replication in controlled conditions 183
5.4 Replication rates under sub-ideal conditions 189
5.5 Growth of phytoplankton in natural
  environments 217
5.6 Summary 236

Chapter 6. Mortality and loss processes in phytoplankton 239
6.1 Introduction 239
6.2 Wash-out and dilution 240
6.3 Sedimentation 243
6.4 Consumption by herbivores 250
6.5 Susceptibility to pathogens and parasites 292
6.6 Death and decomposition 296
6.7 Aggregated impacts of loss processes on
  phytoplankton composition 297
6.8 Summary 300

Chapter 7. Community assembly in the plankton: pattern,
process and dynamics 302
7.1 Introduction 302
7.2 Patterns of species composition and temporal
  change in phytoplankton assemblages 302
7.3 Assembly processes in the phytoplankton 350
7.4 Summary 385

Chapter 8. Phytoplankton ecology and aquatic ecosystems:
mechanisms and management 387
8.1 Introduction 387
8.2 Material transfers and energy flow in pelagic
  systems 387
8.3 Anthropogenic change in pelagic environments 395
8.4 Summary 432
8.5 A last word 435

Glossary 437
Units, symbols and abbreviations 440

References 447

Index to lakes, rivers and seas 508

Index to genera and species of
  phytoplankton 511

Index to genera and species of other
  organisms 520

General index 524

Preface

This is the third book I have written on the subject of phytoplankton ecology. When I finished the first, *The Ecology of Freshwater Phytoplankton* (Reynolds, 1984a), I vowed that it would also be my last. I felt better about it once it was published but, as I recognised that science was moving on, I became increasingly frustrated about the growing datedness of its information. When an opportunity was presented to me, in the form of the 1994 Ecology Institute Prize, to write my second book on the ecology of plankton, *Vegetation Processes in the Pelagic* (Reynolds, 1997a), I was able to draw on the enormous strides that were being made towards understanding the part played by the biochemistry, physiology and population dynamics of plankton in the overall functioning of the great aquatic ecosystems. Any feeling of satisfaction that that exercise brought to me has also been overtaken by events of the last decade, which have seen new tools deployed to the greater amplification of knowledge and new facts uncovered to be threaded into the web of understanding of how the world works.

Of course, this is the way of science. There is no scientific text that can be closed with a sigh, 'So that's it, then'. There are always more questions. I actually have rather more now than I had at the same stage of finishing the 1984 volume. No, the best that can be expected, or even hoped for, is a periodic stocktake: 'This is what we have learned, this is how we think we can explain things and this is where it fits into what we thought we knew already; this will stand until we learn something else.' This is truly the way of science. Taking observations, verifying them by experimentation, moving from hypothesis to fact, we are able to formulate progressively closer approximations to the truth.

In fact, the second violation of my 1984 vow has a more powerful and less high-principled driver. It is just that the progress in plankton ecology since 1984 has been astounding, turning almost each one of the first book's basic assumptions on its head. Besides widening the scope of the present volume to address more overtly the marine phytoplankton, I have set out to construct a new perspective on the expanded knowledge base. I have to say at once that the omission of 'freshwater' from the new title does not imply that the book covers the ecology of marine plankton in equivalent detail. It does, however, signify a genuine attempt to bridge the deep but wholly artificial chasm that exists between marine and freshwater science, which political organisation and science funding have perpetuated.

At a personal level, this wider view is a satisfying thing to develop, being almost a plea for absolution – 'I am sorry for getting it wrong before, this is what I should have said!' At a wider level, I am conscious that many people still use and frequently cite my 1984 book; I would like them to know that I no longer believe everything, or even very much, of what I wrote then. As if to emphasise this, I have adopted a very similar approach to the subject, again using eight chapters (albeit with altered titles). These are developed according to a similar sequence of topics, through morphology, suspension, ecophysiology and dynamics to the structuring of communities and their functions within ecosystems. This arrangement allows me to contrast directly the new knowledge and the understanding it has rendered redundant.

So just what are these mould-breaking findings? In truth, they impinge upon the subject matter in each of the chapters. Advances in microscopy have allowed ultrastructural details of planktic organisms to be revealed for the first time. The advances in molecular biology, in particular the introduction of techniques for isolating chromosomes and ribosomes, fragmenting them by restriction enzymes and reading genetic sequences, have totally altered perceptions about phyletic relationships among planktic taxa and suppositions about their evolution. The classification of organisms is undergoing change of revolutionary proportions, while morphological variation among (supposedly) homogeneous genotypes questions the very concept of putting names to individual organisms. At the scale of cells, the whole concept of how they are moved in the water has been addressed mathematically. It is now appreciated that planktic cells experience critical physical forces that are very different from those affecting (say) fish: viscosity and small-scale turbulence determine the immediate environment of microorganisms; surface tension is a lethal and inescapable spectre; while shear forces dominate dispersion and the spatial distributions of populations. These discoveries flow from the giant leaps in quantification and measurements made by physical limnologists and oceanographers since the early 1980s. These have also impinged on the revision of how sinking and settlement of phytoplankton are viewed and they have helped to consolidate a robust theory of filter-feeding by zooplankton.

The way in which nutrients are sequestered from dilute and dispersed sources in the water and then deployed in the assembly and replication of new generations of phytoplankton has been intensively investigated by physiologists. Recent findings have greatly modified perceptions about what is meant by 'limiting nutrients' and what happens when one or other is in short supply. As Sommer (1996) commented, past suppositions about the repercussions on community structure have had to be revised, both through the direct implications for interspecific competition for resources and, indirectly, through the effects of variable nutritional value of potential foods to the web of dependent consumers.

Arguably, the greatest shift in understanding concerns the way in which the pelagic ecosystem works. Although the abundance of planktic bacteria and the relatively vast reserve of dissolved organic carbon (DOC) had long been recognised, the microorganismic turnover of carbon has only been investigated intensively during the last two decades. It was soon recognised that the metazoan food web of the open oceans is linked to the producer network via the turnover of the microbes and that this statement applies to many larger freshwater systems as well. The metabolism of the variety of substances embraced by 'DOC' varies with source and chain length but a labile fraction originates from phytoplankton photosynthesis that is leaked or actively discharged into the water. Far from holding to the traditional view of the pelagic food chain – algae, zooplankton, fish – plankton ecologists now have to acknowledge that marine food webs are regulated 'by a sea of microbes' (Karl, 1999), through the muliple interactions of organic and inorganic resources and by the lock of protistan predators and acellular pathogens (Smetacek, 2002). Even in lakes, where the case for the top–down control of phytoplankton by herbivorous grazers is championed, the otherwise dominant microbially mediated supply of resources to higher trophic levels is demonstrably subsidised by components from the littoral (Schindler et al., 1996; Vadeboncoeur et al., 2002).

There have been many other revolutions. One more to mention here is the progress in ecosystem ecology, or more particularly, the bridge between the organismic and population ecology and the behaviour of entire systems. How ecosystems behave, how their structure is maintained and what is critical to that maintenance, what the biogeochemical consequences might be and how they respond to human exploitation and management, have all become quantifiable. The linking threads are based upon thermodynamic rules of energy capture, exergy storage and structural emergence, applied through to the systems level (Link, 2002; Odum, 2002).

In the later chapters in this volume, I attempt to apply these concepts to phytoplankton-based systems, where the opportunity is again taken to emphasise the value to the science of ecology of studying the dynamics of microorganisms in the pursuit of high-order pattern and assembly rules (Reynolds, 1997, 2002b). The dual challenge remains, to convince students of forests and other terrestrial ecosystems that microbial systems do conform to analogous rules, albeit at very truncated real-time scales, and to persuade microbiologists to look up from the microscope for long enough to see how their knowledge might be applied to ecological issues.

I am proud to acknowledge the many people who have influenced or contributed to the subject matter of this book. I thank Charles Sinker for inspiring a deep appreciation of ecology and its mechanisms. I am grateful to John Lund, CBE,

FRS for the opportunity to work on phytoplankton as a postgraduate and for the constant inspiration and access to his knowledge that he has given me. Of the many practising theoretical ecologists whose works I have read, I have felt the greatest affinity to the ideas and logic of Ramón Margalef; I greatly enjoyed the opportunities to discuss these with him and regret that there will be no more of them.

I gratefully acknowledge the various scientists whose work has profoundly influenced particular parts of this book and my thinking generally. They include (in alphabetical order) Sallie Chisholm, Paul Falkowski, Maciej Gliwicz, Phil Grime, Alan Hildrew, G. E. Hutchinson, Jörg Imberger, Petur Jónasson, Sven-Erik Jørgensen, Dave Karl, Winfried Lampert, John Lawton, John Raven, Marten Scheffer, Ted Smayda, Milan Straškraba, Reinhold Tüxen, Anthony Walsby and Thomas Weisse. I have also been most fortunate in having been able, at various times, to work with and discuss many ideas with colleagues who include Keith Beven, Sylvia Bonilla, Odécio Cáceres, Paul Carling, Jean-Pierre Descy, Mónica Diaz, Graham Harris, Vera Huszar, Dieter Imboden, Kana Ishikawa, Medina Kadiri, Susan Kilham, Michio Kumagai, Bill Li, Vivian Montecino, Mohi Munawar, Masami Nakanishi, Shin-Ichi Nakano, Luigi Naselli-Flores, Pat Neale, Søren Nielsen, Judit Padisák, Fernando Pedrozo, Victor Smetaček, Ulrich Sommer, José Tundisi and Peter Tyler. I am especially grateful to Catherine Legrand who generously allowed me to use and interpret her experimental data on *Alexandrium*. Nearer to home, I have similarly benefited from long and helpful discussions with such erstwhile Windermere colleagues as Hilda Canter-Lund, Bill Davison, Malcolm Elliott, Bland Finlay, Glen George, Ivan Heaney, Stephen Maberly, Jack Talling and Ed Tipping.

During my years at The Ferry House, I was ably and closely supported by several co-workers, among whom special thanks are due to Tony Irish, Sheila Wiseman, George Jaworski and Brian Godfrey. Peter Allen, Christine Butterwick, Julie Corry (later Parker), Mitzi De Ville, Joy Elsworth, Alastair Ferguson, Mark Glaister, David Gouldney, Matthew Rogers, Stephen Thackeray and Julie Thompson also worked with me at particular times. Throughout this period, I was privileged to work in a 'well-found' laboratory with abundant technical and practical support. I freely acknowledge use of the world's finest collection of the freshwater literature and the assistance provided at various times by John Horne, Ian Pettman, Ian McCullough, Olive Jolly and Marilyn Moore. Secretarial assistance has come from Margaret Thompson, Elisabeth Evans and Joyce Hawksworth. Trevor Furnass has provided abundant reprographic assistance over many years. I am forever in the debt of Hilda Canter-Lund, FRPS for the use of her internationally renowned photomicrographs.

A special word is due to the doctoral students whom I have supervised. The thirst for knowledge and understanding of a good pupil generally provide a foil and focus in the other direction. I owe much to the diligent curiosity of Chris van Vlymen, Helena Cmiech, Karen Saxby (now Rouen), Siân Davies, Alex Elliott, Carla Kruk and Phil Davis.

My final word of appreciation is reserved for acknowledgement of the tolerance and forbearance of my wife and family. I cheered through many juvenile football matches and dutifully attended a host of ballet and choir performances and, yes, it was quite fun to relive three more school curricula. Nevertheless, my children had less of my time than they were entitled to expect. Jean has generously shared with my science the full focus of my attention. Yet, in 35 years of marriage, she has never once complained, nor done less than encourage the pursuit of my work. I am proud to dedicate this book to her.

Acknowledgements

Except where stated, the illustrations in this book are reproduced, redrawn or otherwise slightly modified from sources noted in the individual captions. The author and the publisher are grateful to the various copyright holders, listed below, who have given permission to use copyright material in this volume. While every effort has been made to clear permissions as appropriate, the publisher would appreciate notification of any omission.

Figures 1.1 to 1.8, 1.10, 2.8 to 2.13, 2.17, 2.20 to 2.31, 3.3 to 3.9, 3.16 and 3.17, 5.20, 6.2, 6.7, 6.11. 7.6 and 7.18 are already copyrighted to Cambridge University Press.

Figure 1.9 is redrawn by permission of Oxford University Press.

Figure 1.11 is the copyright of the American Society of Limnology and Oceanography.

Figures 2.1 and 2.2, 2.5 to 2.7, 2.15 and 2.16, 2.18 and 2.19, 3.12, 3.14, 3.19, 4.1, 4.3 to 4.5, 5.1 to 5.5, 5.8, 5.10, 5.12 and 5.13, 5.20 and 5.21, 6.1, 6.2, 6.4, 6.14, 7.8, 7.10 and 7.11, 7.14, 7.16 and 7.17, 7.20 and 7.22 are redrawn by permission of The Ecology Institute, Oldendorf.

Figures 2.3 and 4.7 are redrawn from the source noted in the captions, with acknowledgement to Artemis Press.

Figures 2.4, 3.18, 5.11, 5.18, 7.5, 7.15, 8.2 and 8.3 are redrawn from the various sources noted in the respective captions and with acknowledgement to Elsevier Science, B.V.

Figure 2.14 is redrawn from the *British Phycological Journal* by permission of Taylor & Francis Ltd (http://www.tandf.co.uk/journals).

Figure 3.1 is redrawn by permission of Nature Publishing Group.

Figures 3.2, 3.11, 3.13, 4.2, 5.6, 6.4, 6.6, 6.9, 6.10 and 6.13 come from various titles that are the copyright of Blackwell Science (the specific sources are noted in the figure captions) and are redrawn by permission.

Figures 3.7, 3.15, 4.6 and 7.2.3 (or parts thereof) are redrawn from *Freshwater Biology* by permission of Blackwell Science.

Figure 3.7 incorporates items redrawn from *Biological Reviews* with acknowledgement to the Cambridge Philosophical Society.

Figure 5.9 is redrawn by permission of John Wiley & Sons Ltd.

Figure 5.14 is redrawn by permission of Springer-Verlag GmbH.

Figures 5.15 to 5.17, 5.19, 6.8 and 6.9 are redrawn by permission of SpringerScience+Business BV.

Figures 6.12, 6.15, 7.1 to 7.4, 7.9 and 8.6 are reproduced from *Journal of Plankton Research* by permission of Oxford University Press. Dr K. Bruning also gave permission to produce Fig. 6.12.

Figure 7.7 is redrawn by permission of the Director, Marine Biological Association.

Figures 7.12 to 7.14, 7.24 and 7.25 are redrawn from *Verhandlungen der internationale Vereinigung für theoretische und angewandte Limnologie* by permission of Dr E. Nägele (Publisher) (http://www.schwezerbart.de).

Figure 7.19 is redrawn with acknowledgement to the Athlone Press of the University of London.

Figure 7.21 is redrawn from *Aquatic Ecosystems Health and Management* by permission of Taylor & Francis, Inc. (http://www.taylorandfrancis.com).

Figure 8.1 is redrawn from *Scientia Maritima* by permission of Institut de Ciències del Mar.

Figures 8.5, 8.7 and 8.8 are redrawn by permission of the Chief Executive, Freshwater Biological Association.

Chapter 1

Phytoplankton

1.1 | Definitions and terminology

The correct place to begin any exposition of a major component in biospheric functioning is with precise definitions and crisp discrimination. This should be a relatively simple exercise but for the need to satisfy a consensus of understanding and usage. Particularly among the biological sciences, scientific knowledge is evolving rapidly and, as it does so, it often modifies and outgrows the constraints of the previously acceptable terminology. I recognised this problem for plankton science in an earlier monograph (Reynolds, 1984a). Since then, the difficulty has worsened and it impinges on many sections of the present book. The best means of dealing with it is to accept the issue as a symptom of the good health and dynamism of the science and to avoid constraining future philosophical development by a redundant terminological framework.

The need for definitions is not subverted, however, but it transforms to an insistence that those that are ventured are provisional and, thus, open to challenge and change. To be able to reveal something also of the historical context of the usage is to give some indication of the limitations of the terminology and of the areas of conjecture impinging upon it.

So it is with 'plankton'. The general understanding of this term is that it refers to the collective of *organisms* that are *adapted* to spend part or all of their lives in apparent *suspension* in the *open water* of the sea, of lakes, ponds and rivers. The italicised words are crucial to the concept and are not necessarily contested. Thus, 'plankton' excludes other suspensoids that are either non-living, such as clay particles and precipitated chemicals, or are fragments or cadavers derived from biogenic sources. Despite the existence of the now largely redundant subdivision *tychoplankton* (see Box 1.1), 'plankton' normally comprises those living organisms that are only fortuitously and temporarily present, imported from adjacent habitats but which neither grew in this habitat nor are suitably adapted to survive in the truly open water, ostensibly independent of shore and bottom. Such locations support distinct suites of surface-adhering organisms with their own distinctive survival adaptations.

'Suspension' has been more problematic, having quite rigid physical qualifications of density and movement relative to water. As will be rehearsed in Chapter 2, only rarely can plankton be *isopycnic* (having the same density) with the medium and will have a tendency to *float* upwards or *sink* downwards relative to it. The rate of movement is also size dependent, so that 'apparent suspension' is most consistently achieved by organisms of small (<1 mm) size. Crucially, this feature is mirrored in the fact that the intrinsic movements of small organisms are frequently too feeble to overcome the velocity and direction of much of the spectrum of water movements. The inability to control horizontal position or to swim against significant currents in open waters separates 'plankton' from the 'nekton' of active swimmers, which include adult fish, large cephalopods, aquatic reptiles, birds and mammals.

Box 1.1 | Some definitions used in the literaure on plankton

seston	the totality of particulate matter in water; all material not in solution
tripton	non-living seston
plankton	living seston, adapted for a life spent wholly or partly in quasi-suspension in open water; and whose powers of motility do not exceed turbulent entrainment (see Chapter 2)
nekton	animals adapted to living all or part of their lives in open water but whose intrinsic movements are almost independent of turbulence
euplankton	redundant term to distinguish fully adapted, truly planktic organisms from other living organisms fortuitously present in the water
tychoplankton	non-adapted organisms from adjacent habitats and present in the water mainly by chance
meroplankton	planktic organisms passing a major part of the life history out of the plankton (e.g. on the bottom sediments)
limnoplankton	plankton of lakes
heleoplankton	plankton of ponds
potamoplankton	plankton of rivers
phytoplankton	planktic photoautotrophs and major producer of the pelagic
bacterioplankton	planktic prokaryotes
mycoplankton	planktic fungi
zooplankton	planktic metazoa and heterotrophic protistans

Some more, now redundant, terms

The terms *nannoplankton, ultraplankton, μ-algae* are older names for various smaller size categories of phytoplankton, eclipsed by the classification of Sieburth *et al.* (1978) (see Box 1.2).

In this way, plankton comprises organisms that range in size from that of viruses (a few tens of nanometres) to those of large jellyfish (a metre or more). Representative organisms include bacteria, protistans, fungi and metazoans. In the past, it has seemed relatively straightforward to separate the organisms of the plankton, both into broad phyletic categories (e.g. bacterioplankton, mycoplankton) or into similarly broad functional categories (photosynthetic algae of the phytoplankton, phagotrophic animals of the zooplankton). Again, as knowledge of the organisms, their phyletic affinities and physiological capabilities has expanded, it has become clear that the divisions used hitherto do not precisely coincide: there are photosynthetic bacteria, phagotrophic algae and flagellates that take up organic carbon from solution. Here, as in general, precision will be considered relevant and important in the context of organismic properties (their names, phylogenies, their morphological and physiological characteristics). On the other hand, the generic contributions to systems (at the habitat or ecosystem scales) of the photosynthetic primary producers, phagotrophic consumers and heterotrophic decomposers may be attributed reasonably but imprecisely to phytoplankton, zooplankton and bacterioplankton.

The definiton of phytoplankton adopted for this book is the collective of photosynthetic microorganisms, adapted to live partly or continuously in open water. As such, it is the photoautotrophic part of the plankton and a major primary producer of organic carbon in the pelagic of the seas and of inland waters. The distinction of phytoplankton from other categories of plankton and suspended matter are listed in Box 1.1.

It may be added that it is correct to refer to phytoplankton as a singular term ('phytoplankton is' rather than 'phytoplankton are'). A single organism is a *phytoplanktont* or (more ususally) *phytoplankter*. Incidentally, the adjective 'planktic' is etymologically preferable to the more commonly used 'planktonic'.

1.2 | Historical context of phytoplankton studies

The first use of the term 'plankton' is attributed in several texts (Ruttner, 1953; Hutchinson, 1967) to Viktor Hensen, who, in the latter half of the nineteenth century, began to apply quantitative methods to gauge the distribution, abundance and productivity of the microscopic organisms of the open sea. The monograph that is usually cited (Hensen, 1887) is, in fact, rather obscure and probably not well read in recent times but Smetacek *et al.* (2002) have provided a probing and engaging review of the original, within the context of early development of plankton science. Most of the present section is based on their article.

The existence of a planktic community of organisms in open water had been demonstrated many years previously by Johannes Müller. Knowledge of some of the organisms themselves stretches further back, to the earliest days of microscopy. From the 1840s, Müller would demonstrate net collections to his students, using the word *Auftrieb* to characterise the community (Smetacek *et al.*, 2002). The literal transla-tion to English is 'up drive', approximately 'buoyancy' or 'flotation', a clear reference to Müller's assumption that the material floated up to the surface waters – like so much oceanic dirt! It took one of Müller's students, Ernst Haeckel, to champion the beauty of planktic protistans and metazoans. His monograph on the Radiolaria was also one of the first to embrace Darwin's (1859) evolutionary theory in order to show structural affinities and divergences. Haeckel, of course, became best known for his work on morphology, ontogeny and phylogeny. According to Smetacek *et al.* (2002), his interest and skills as a draughtsman advanced scientific awareness of the range of planktic form (most significantly, Haeckel, 1904) but to the detriment of any real progress in understanding of functional differentiation. Until the late 1880s, it was not appreciated that the organisms of the *Auftrieb*, even the algae among them, could contribute much to the nutrition of the larger animals of the sea. Instead, it seems to have been supposed that organic matter in the fluvial discharge from the land was the major nutritive input. It is thus rather interesting to note that, a century or so later, this possibility has enjoyed something of a revival (see Chapters 3 and 8).

If Haeckel had conveyed the beauty of the pelagic protistans, it was certainly Viktor Hensen who had been more concerned about their role in a functional ecosystem. Hensen was a physiologist who brought a degree of empiricism to his study of the perplexing fluctuations in North Sea fish stocks. He had reasoned that fish stocks and yields were related to the production and distribution of the juvenile stages. Through devising techniques for sampling, quantification and assessing distribution patterns, always carefully verified by microscopic examination, Hensen recognised both the ubiquity of phytoplankton and its superior abundance and quality over coastal inputs of terrestrial detritus. He saw the connection between phytoplankton and the light in the near-surface layer, the nutritive resource it provided to copepods and other small animals, and the value of these as a food source to fish.

Thus, in addition to bequeathing a new name for the basal biotic component in pelagic ecosystems, Hensen may be regarded justifiably as the first quantitative plankton ecologist and as the person who established a formal methodology for its study. Deducing the relative contributions of Hensen and Haeckel to the foundation of modern plankton science, Smetacek et al. (2002) concluded that it is the work of the latter that has been the more influential. This is an opinion with which not everyone will agree but this is of little consequence. However, Smetacek et al. (2002) offered a most profound and resonant observation in suggesting that Hensen's general understanding of the role of plankton ('the big picture') was essentially correct but erroneous in its details, whereas in Haeckel's case, it was the other way round. Nevertheless, both have good claim to fatherhood of plankton science!

1.3 | The diversification of phytoplankton

Current estimates suggest that between 4000 and 5000 legitimate species of marine phytoplankton have been described (Sournia et al. 1991; Tett and Barton, 1995). I have not seen a comparable estimate for the number of species in inland waters, beyond the extrapolation I made (Reynolds, 1996a) that the number is unlikely to be substantially smaller. In both lists, there is not just a large number of mutually distinct taxa of photosynthetic microorganisms but there is a wide variety of shape, size and phylogenetic affinity. As has also been pointed out before (Reynolds, 1994a), the morphological range is comparable to the one spanning forest trees and the herbs that grow at their base. The phyletic divergence of the representatives is yet wider. It would be surprising if the species of the phytoplankton were uniform in their requirements, dynamics and susceptibilities to loss processes. Once again, there is a strong case for attempting to categorise the phytoplankton both on the phylogeny of organisms and on the functional basis of their roles in aquatic ecosystems. Both objectives are adopted for the writing of this volume. Whereas the former is addressed only in the present chapter, the

latter quest occupies most of the rest of the book. However, it is not giving away too much to anticipate that systematics provides an important foundation for species-specific physiology and which is itself part-related to morphology. Accordingly, great attention is paid here to the differentiation of individualistic properties of representative species of phytoplankton.

However, there is value in being able simultaneously to distinguish among functional categories (trees from herbs!). The scaling system and nomenclature proposed by Sieburth et al. (1978) has been widely adopted in phytoplankton ecology to distinguish functional separations within the phytoplankton. It has also eclipsed the use of such terms as *μ-algae* and *ultraplankton* to separate the lower size range of planktic organisms from those (*netplankton*) large enough to be retained by the meshes of a standard phytoplankton net. The scheme of prefixes has been applied to size categories of zooplankton, with equal success. The size-based categories are set out in Box 1.2.

At the level of phyla, the classification of the phytoplankton is based on long-standing criteria, distinguished by microscopists and biochemists over the last 150 years or so, from which there is little dissent. In contrast, subdivision within classes, orders etc., and the tracing of intraphyletic relationships, affinities within and among families, even the validity of supposedly well-characterised species, has become subject to massive reappraisal. The new factor that has come into play is the powerful armoury of the molecular biologists, including the methods for reading gene sequences and for the statistical matching of these to measure the closeness to other species.

Of course, the potential outcome is a much more robust, genetically verified family tree of authentic species of phytoplankton. This may be some years away. For the present, it seems pointless to reproduce a detailed classification of the phytoplankton that will soon be made redundant. Even the evolutionary connectivities among the phyla and their relationship to the geochemical development of the planetary structures are undergoing deep re-evaluation (Delwiche, 2000; Falkowski, 2002). For these reasons, the

Box 1.2	The classification of phytoplankton according to the scaling nomenclature of Sieburth *et al.* (1978)

Maximum linear dimension	Name[a]
0.2–2 µm	picophytoplankton
2–20 µm	nanophytoplankton
20–200 µm	microphytoplankton
200 µm–2 mm	mesophytoplankton
>2 mm	macrophytoplankton

[a] The prefixes denote the same size categories when used with '-zooplankton', '-algae', '-cyanobacteria', 'flagellates', etc.

taxonomic listings in Table 1.1 are deliberately conservative.

Although the life forms of the plankton include acellular microorganisms (viruses) and a range of well-characterised Archaea (the halobacteria, methanogens and sulphur-reducing bacteria, formerly comprising the Archaebacteria), the most basic photosynthetic organisms of the phytoplankton belong to the Bacteria (formerly, Eubacteria). The separation of the ancestral bacteria from the archaeans (distinguished by the possession of membranes formed of branched hydrocarbons and ether linkages, as opposed to the straight-chain fatty acids and ester linkages found in the membranes of all other organisms: Atlas and Bartha, 1993) occurred early in microbial evolution (Woese, 1987; Woese *et al.*, 1990).

The appearance of phototrophic forms, distinguished by their crucial ability to use light energy in order to synthesise adenosine triphosphate (ATP) (see Chapter 3), was also an ancient event that took place some 3000 million years ago (3 Ga .BP (before present)). Some of these organisms were photoheterotrophs, requiring organic precursors for the synthesis of their own cells. Modern forms include green flexibacteria (Chloroflexaceae) and purple non-sulphur bacteria (Rhodospirillaceae), which contain pigments similar to chlorophyll (bacteriochlorophyll *a*, *b* or *c*). Others were true photoautotrophs, capable of reducing carbon dioxide as a source of cell carbon (photosynthesis). Light energy is used to strip electrons from a donor substance. In most modern plants, water is the source of reductant electrons and oxygen is liberated as a by-product (oxygenic photosynthesis). Despite their phyletic proximity to the photoheterotrophs and sharing a similar complement of bacteriochlorophylls (Béjà *et al.*, 2002), the Anoxyphotobacteria use alternative sources of electrons and, in consequence, generate oxidation products other than oxygen (anoxygenic photosynthesis). Their modern-day representatives are the purple and green sulphur bacteria of anoxic sediments. Some of these are planktic in the sense that they inhabit anoxic, intensively stratified layers deep in small and suitably stable lakes. The trait might be seen as a legacy of having evolved in a wholly anoxic world. However, aerobic, anoxygenic phototrophic bacteria, containing bacterichlorophyll *a*, have been isolated from oxic marine environments (Shiba *et al.*, 1979); it has also become clear that their contribution to the oceanic carbon cycle is not necessarily insignificant (Kolber *et al.*, 2001; Goericke, 2002).

Nevertheless, the oxygenic photosynthesis pioneered by the Cyanobacteria from about 2.8 Ga before present has proved to be a crucial step in the evolution of life in water and, subsequently, on land. Moreover, the composition of the atmosphere was eventually changed through the biological oxidation of water and the simultaneous removal and burial of carbon in marine sediments (Falkowski, 2002). Cyanobacterial photosynthesis is mediated primarily by chlorophyll *a*, borne on thylakoid membranes. Accessory

Table 1.1 | Survey of the organisms in the phytoplankton

Domain: BACTERIA
    Division: **Cyanobacteria** (blue-green algae)
        Unicellular and colonial bacteria, lacking membrane bound plastids. Primary photosynthetic pigment is chlorophyll a, with accessory phycobilins (phycocyanin, phycoerythrin). Assimilation products, glycogen, cyanophycin. Four main sub-groups, of which three have planktic representatives.
        Order: CHROOCOCCALES
        Unicellular or coenobial Cyanobacteria but never filamentous. Most planktic genera form mucilaginous colonies, and these are mainly in fresh water. Picophytoplanktic forms abundant in the oceans.
            Includes: *Aphanocapsa, Aphanothece, Chroococcus, Cyanodictyon, Gomphosphaeria, Merismopedia, Microcystis, Snowella, Synechococcus, Synechocystis, Woronichinia*
        Order: OSCILLATORIALES
        Uniseriate–filamentous Cyanobacteria whose cells all undergo division in the same plane. Marine and freshwater genera.
            Includes: *Arthrospira, Limnothrix, Lyngbya, Planktothrix, Pseudanabaena, Spirulina, Trichodesmium, Tychonema*
        Order: NOSTOCALES
        Unbranched–filamentous Cyanobacteria whose cells all undergo division in the same plane and certain of which may be facultatively differentiated into heterocysts. In the plankton of fresh waters and dilute seas.
            Includes: *Anabaena, Anabaenopsis, Aphanizomenon, Cylindrospermopsis, Gloeotrichia, Nodularia*
    Exempt Division: **Prochlorobacteria**
        Order: PROCHLORALES
        Unicellular and colonial bacteria, lacking membrane-bound plastids. Photosynthetic pigments are chlorophyll *a* and *b*, but lack phycobilins.
            Includes: *Prochloroccus, Prochloron, Prochlorothrix*
    Division: **Anoxyphotobacteria**
        Mostly unicellular bacteria whose (anaerobic) photosynthesis depends upon an electron donor other than water and so do not generate oxygen. Inhabit anaerobic sediments and (where appropriate) water layers where light penetrates sufficiently. Two main groups:
            Family: Chromatiaceae (purple sulphur bacteria) Cells able to photosynthesise with sulphide as sole electron donor. Cells contain bacteriochlorophyll *a*, *b* or *c*.
                Includes: *Chromatium, Thiocystis, Thiopedia.*
            Family: Chlorobiaceae (green sulphur bacteria) Cells able to photosynthesise with sulphide as sole electron donor. Cells contain bacteriochlorophyll *a*, *b* or *c*.
                Includes: *Chlorobium, Clathrocystis, Pelodictyon.*
Domain: EUCARYA
    Phylum: **Glaucophyta**
        Cyanelle-bearing organisms, with freshwater planktic representatives.
            Includes: *Cyanophora, Glaucocystis.*
    Phylum: **Prasinophyta**
        Unicellular, mostly motile green algae with 1–16 laterally or apically placed flagella, cell walls covered with fine scales and plastids containing chlorophyll *a* and *b*. Assimilatory products mannitol, starch.

(cont.)

Table 1.1 (cont.)

CLASS: Pedinophyceae
Order: PEDINOMONADALES
 Small cells, with single lateral flagellum.
  Includes: *Pedinomonas*
CLASS: Prasinophyceae
Order: CHLORODENDRALES
 Flattened, 4-flagellated cells.
  Includes: *Nephroselmis*, *Scherffelia* (freshwater); *Mantoniella*, *Micromonas*
  (marine)
Order: PYRAMIMONADALES
 Cells with 4 or 8 (rarely 16) flagella arising from an anterior depression. Marine
 and freshwater.
  Includes: *Pyramimonas*
Order: SCOURFIELDIALES
 Cells with two, sometimes unequal, flagella. Known from freshwater ponds.
  Includes: *Scourfieldia*

Phylum: **Chlorophyta** (green algae)
 Green-pigmented, unicellular, colonial, filamentous, siphonaceous and thalloid
 algae. One or more chloroplasts containing chlorophyll *a* and *b*. Assimilation
 product, starch (rarely, lipid).
CLASS: Chlorophyceae
 Several orders of which the following have planktic representatives:
Order: TETRASPORALES
 Non-flagellate cells embedded in mucilaginous or palmelloid colonies, but with
 motile propagules.
  Includes: *Paulschulzia*, *Pseudosphaerocystis*
Order: VOLVOCALES
 Unicellular or colonial biflagellates, cells with cup-shaped chloroplasts.
  Includes: *Chlamydomonas*, *Eudorina*, *Pandorina*, *Phacotus*, *Volvox* (in fresh
  waters); *Dunaliella*, *Nannochloris* (marine)
Order: CHLOROCOCCALES
 Non-flagellate, unicellular or coenobial (sometimes mucilaginous) algae, with
 many planktic genera.
  Includes: *Ankistrodesmus*, *Ankyra*, *Botryococcus*, *Chlorella*,
  Coelastrum, *Coenochloris*, *Crucigena*, *Choricystis*, *Dictyosphaerium*,
  Elakatothrix, *Kirchneriella*, *Monorophidium*, *Oocystis*, *Pediastrum*,
  Scenedesmus, *Tetrastrum*
Order: ULOTRICHALES
 Unicellular or mostly unbranched filamentous with band-shaped chloroplasts.
  Includes: *Geminella*, *Koliella*, *Stichococcus*
Order: ZYGNEMATALES
 Unicellular or filamentous green algae, reproducing isogamously by conjugation.
 Planktic genera are mostly members of the Desmidaceae, mostly unicellular or
 (rarely) filmentous coenobia with cells more or less constricted into two
 semi-cells linked by an interconnecting isthmus. Exclusively freshwater genera.
  Includes: *Arthrodesmus*, *Closterium*, *Cosmarium*, *Euastrum*, *Spondylosium*,
  Staurastrum, *Staurodesmus*, *Xanthidium*

(cont.)

Table 1.1 (cont.)

Phylum: **Euglenophyta**

Green-pigmented unicellular biflagellates. Plastids numerous and irregular, containing chlorophyll *a* and *b*. Reproduction by longitudinal fission. Assimilation product, paramylon, oil. One Class, Euglenophyceae, with two orders.

Order: EUTREPTIALES

Cells having two emergent flagella, of approximately equal length. Marine and freshwater species.
Includes: *Eutreptia*

Order: EUGLENALES

Cells having two flagella, one very short, one long and emergent.
Includes: *Euglena, Lepocinclis, Phacus, Trachelmonas*

Phylum: **Cryptophyta**

Order: CRYPTOMONADALES

Naked, unequally biflagellates with one or two large plastids, containing chlorophyll *a* and c_2 (but not chlorophyll *b*); accessory phycobiliproteins or other pigments colour cells brown, blue, blue-green or red; assimilatory product, starch. Freshwater and marine species.
Includes: *Chilomonas, Chroomonas, Cryptomonas, Plagioselmis, Pyrenomonas, Rhodomonas*

Phylum: **Raphidophyta**

Order: RAPHIDOMONADALES (syn. CHLOROMONADALES)

Biflagellate, cellulose-walled cells; two or more plastids containing chlorophyll *a*; cells yellow-green due to predominant accessory pigment, diatoxanthin; assimilatory product, lipid. Freshwater.
Includes: *Gonyostomum*

Phylum: **Xanthophyta** (yellow-green algae)

Unicellular, colonial, filamentous and coenocytic algae. Motile species generally subapically and unequally biflagellated; two or many more discoid plastids per cell containing chlorophyll *a*. Cells mostly yellow-green due to predominant accessory pigment, diatoxanthin; assimilation product, lipid. Several orders, two with freshwater planktic representatives.

Order: MISCHOCOCCALES

Rigid-walled, unicellular, sometimes colonial xanthophytes.
Includes: *Goniochloris, Nephrodiella, Ophiocytium*

Order: TRIBONEMATALES

Simple or branched uniseriate filamentous xanthophytes.
Includes: *Tribonema*

Phylum: **Eustigmatophyta**

Coccoid unicellular, flagellated or unequally biflagellated yellow-green algae with masking of chlorophyll *a* by accessory pigment violaxanthin. Assimilation product, probably lipid.
Includes: *Chlorobotrys, Monodus*

Phylum: **Chrysophyta** (golden algae)

Unicellular, colonial and filamentous. often uniflagellate, or unequally biflagellate algae. Contain chlorophyll *a*, c_1 and c_2, generally masked by abundant accessory pigment, fucoxanthin, imparting distinctive golden colour to cells. Cells sometimes naked or or enclosed in an urn-shaped lorica, sometimes with siliceous scales. Assimilation products, lipid, leucosin. Much reclassified group, has several classes and orders in the plankton.

(cont.)

Table 1.1 (cont.)

CLASS: Chrysophyceae
Order: CHROMULINALES
Mostly planktic, unicellular or colony-forming flagellates with one or two unequal flagella, occasionally naked, often in a hyaline lorica or gelatinous envelope.
Includes: *Chromulina, Chrysococcus, Chrysolykos, Chrysosphaerella, Dinobryon, Kephyrion, Ochromonas, Uroglena*
Order: HIBBERDIALES
Unicellular or colony-forming epiphytic gold algae but some planktic representatives.
Includes: *Bitrichia*
CLASS: Dictyochophyceae
Order: PEDINELLALES
Radially symmetrical, very unequally biflagellate unicells or coenobia.
Includes: *Pedinella* (freshwater); *Apedinella, Pelagococcus, Pelagomonas, Pseudopedinella* (marine)
CLASS: Synurophyceae
Order: SYNURALES
Unicellular or colony-forming flagellates, bearing distinctive siliceous scales.
Includes: *Mallomonas, Synura*
Phylum: **Bacillariophyta** (diatoms)
Unicellular and coenobial yellow-brown, non-motile algae with numerous discoid plastids, containing chlorophyll a, c_1 and c_2, masked by accessory pigment, fucoxanthin. Cell walls pectinaceous, in two distinct and overlapping halves, and impregnated with cryptocrystalline silica. Assimilatory products, chrysose, lipids. Two large orders, both conspicuously represented in the marine and freshwater phytoplankton.
CLASS: Bacillariophyceae
Order: BIDDULPHIALES (centric diatoms)
Diatoms with cylindrical halves, sometimes well separated by girdle bands. Some species form (pseudo-)filaments by adhesion of cells at their valve ends.
Includes: *Aulacoseira, Cyclotella, Stephanodiscus, Urosolenia* (freshwater); *Cerataulina, Chaetoceros, Detonula, Rhizosolenia, Skeletonema, Thalassiosira* (marine)
Order: BACILLARIALES (pennate diatoms)
Diatoms with boat-like halves, no girdle bands. Some species form coenobia by adhesion of cells on their girdle edges.
Includes: *Asterionella, Diatoma, Fragilaria, Synedra, Tabellaria* (freshwater); *Achnanthes, Fragilariopsis, Nitzschia* (marine)
Phylum: **Haptophyta**
CLASS: Haptophyceae
Gold or yellow-brown algae, usually unicellular, with two subequal flagella and a coiled haptonema, but with amoeboid, coccoid or palmelloid stages. Pigments, chlorophyll a, c_1 and c_2, masked by accessory pigment (usually fucoxanthin). Assimilatory product, chrysolaminarin. Cell walls with scales, sometimes more or less calcified.
Order: PAVLOVALES
Cells with haired flagella and small haptonema. Marine and freshwater species.
Includes: *Diacronema, Pavlova*

(cont.)

Table 1.1 (cont.)

Order: PRYMNESIALES

Cells with smooth flagella, haptonema usually small. Mainly marine or brackish but some common in freshwater plankton.

Includes: *Chrysochromulina, Isochrysis, Phaeocystis, Prymnesium*

Order: COCCOLITHOPHORIDALES

Cell suface covered by small, often complex, flat calcified scales (coccoliths). Exclusively marine.

Include: *Coccolithus, Emiliana, Florisphaera, Gephyrocapsa, Umbellosphaera*

Phylum: **Dinophyta**

Mostly unicellular, sometimes colonial, algae with two flagella of unequal length and orientation. Complex plastids containing chlorophyll a, c_1 and c_2, generally masked by accessory pigments. Cell walls firm, or reinforced with polygonal plates. Assimilation products: starch, oil. Conspicuously represented in marine and freshwater plankton. Two classes and (according to some authorities) up to 11 orders.

CLASS: Dinophyceae

Biflagellates, with one transverse flagellum encircling the cell, the other directed posteriorly.

Order: GYMNODINIALES

Free-living, free-swimming with flagella located in well-developed transverse and sulcal grooves, without thecal plates. Mostly marine.

Includes: *Amphidinium, Gymnodinium, Woloszynskia*

Order: GONYAULACALES

Armoured, plated, free-living unicells, the apical plates being asymmetrical. Marine and freshwater.

Includes: *Ceratium, Lingulodinium*

Order: PERIDINIALES

Armoured, plated, free-living unicells, with symmetrical apical plates. Marine and freshwater.

Includes: *Glenodinium, Gyrodinium, Peridinium*

Order: PHYTODINIALES

Coccoid dinoflagellates with thick cell walls but lacking thecal plates. Many epiphytic for part of life history. Some in plankton of humic fresh waters.

Includes: *Hemidinium*

CLASS: Adinophyceae

Order: PROROCENTRALES

Naked or cellulose-covered cells comprising two watchglass-shaped halves. Marine and freshwater species.

Includes: *Exuviella, Prorocentrum*

---

pigments, called phycobilins, are associated with these membranes, where they are carried in granular phycobilisomes. Life forms among the Cyanobacteria have diversified from simple coccoids and rods into loose mucilaginous colonies, called coenobia, into filamentous and to pseudotissued forms. Four main evolutionary lines are recognised, three of which (the chroococcalean, the oscillatorialean and the nostocalean; the stigonematalean line is the exception) have major planktic representatives that have diversified greatly among marine and freshwater systems. The most ancient group of the surviving groups of photosynthetic organisms is, in terms of individuals, the most abundant on the planet.

Links to eukaryotic protists, plants and animals from the Cyanobacteria had been supposed explicitly and sought implicitly. The discovery of a prokaryote containing chlorophyll *a* and *b* but lacking phycobilins, thus resembling the pigmentation of green plants, seemed to fit the bill (Lewin, 1981). *Prochloron*, a symbiont of salps, is not itself planktic but is recoverable in collections of marine plankton. The first description of *Prochlorothrix* from the freshwater phytoplankton in the Netherlands (Burger-Wiersma *et al.*, 1989) helped to consolidate the impression of an evolutionary 'missing link' of chlorophyll-*a*- and -*b*-containing bacteria. Then came another remarkable finding: the most abundant picoplankter in the low-latitude ocean was not a *Synechococcus*, as had been thitherto supposed, but another oxyphototrophic prokaryote containing divinyl chlorophyll-*a* and -*b* pigments but no bilins (Chisholm *et al.*, 1988, 1992); it was named *Prochlorococcus*. The elucidation of a biospheric role of a previously unrecognised organism is achievement enough by itself (Pinevich *et al.*, 2000); for the organisms apparently to occupy this transitional position in the evolution of plant life doubles the sense of scientific satisfaction. Nevertheless, subsequent investigations of the phylogenetic relationships of the newly defined Prochlorobacteria, using immunological and molecular techniques, failed to group *Prochlorococcus* with the other Prochlorales or even to separate it distinctly from *Synechococcus* (Moore *et al.*, 1998; Urbach *et al.*, 1998). The present view is that it is expedient to regard the Prochlorales as aberrent Cyanobacteria (Lewin, 2002).

The common root of all eukaryotic algae and higher plants is now understood to be based upon original primary endosymbioses involving early eukaryote protistans and Cyanobacteria (Margulis, 1970, 1981). As more is learned about the genomes and gene sequences of microorganisms, so the role of 'lateral' gene transfers in shaping them is increasingly appreciated (Doolittle *et al.*, 2003). For instance, in terms of ultrastructure, the similarity of 16S rRNA sequences, several common genes and the identical photosynthetic proteins, all point to cyanobacterial origin of eukaruote plastids (Bhattacharya and Medlin, 1998; Douglas and Raven, 2003). Pragmatically, we may judge this to have been a highly successful combination. There may well have been others of which nothing is known, apart from the small group of glaucophytes that carry cyanelles rather than plastids. The cyanelles are supposed to be an evolutionary intermediate between cyanobacterial cells and chloroplasts (admittedly, much closer to the latter). Neither cyanelles nor plastids can grow independently of the eukaryote host and they are apportioned among daughters when the host cell divides. There is no evidence that the handful of genera ascribed to this phylum are closely related to each other, so it may well be an artificial grouping. *Cyanophora* is known from the plankton of shallow, productive calcareous lakes (Whitton in John *et al.*, 2002).

Molecular investigation has revealed that the seemingly disparate algal phyla conform to one or other of two main lineages. The 'green line' of eukaryotes with endosymbiotic Cyanobacteria reflects the development of the chlorophyte and euglenophyte phyla and to the important offshoots to the bryophytes and the vascular plant phyla. The 'red line', with its secondary and even tertiary endosymbioses, embraces the evolution of the rhodophytes, the chrysophytes and the haptophytes, is of equal or perhaps greater fascination to the plankton ecologist interested in diversity.

A key distinguishing feature of the algae of the green line is the inclusion of chlorophyll *b* among the photosynthetic pigments and, typically, the accumulation of glucose polymers (such as starch, paramylon) as the main product of carbon assimilation. The subdivision of the green algae between the prasinophyte and the chlorophyte phyla reflects the evolutionary development and anatomic diversification within the line, although both are believed to have a long history on the planet (~1.5 Ga). Both are also well represented by modern genera, in water generally and in the freshwater phytoplankton in particular. Of the modern prasinophyte orders, the Pedinomonadales, the Chlorodendrales and the Pyramimonadales each have significant planktic representation, in the sense of producing populations of common occurrence and forming 'blooms' on occasions. Several modern chlorophyte orders (including Oedogoniales, Chaetophorales, Cladophorales, Coleochaetales, Prasiolales, Charales, Ulvales a.o.) are without modern planktic representation. In contrast, there are large numbers of volvocalean, chlorococcalean and zygnematalean species in lakes and ponds and the Tetrasporales and Ulotrichales are also well represented. These show a very wide span of cell size and organisation, with flagellated and non-motile cells, unicells and filamentous or ball-like coenobia, with varying degrees of mucilaginous investment and of varying consistency. The highest level of colonial development is arguably in *Volvox*, in which hundreds of networked biflagellate cells are coordinated to bring about the controlled movement of the whole. Colonies also reproduce by the budding off and release of near-fully formed daughter colonies. The desmid members of the Zygnematales are amongst the best-studied green plankters. Mostly unicellular, the often elaborate and beautiful architecture of the semi-cells invite the gaze and curiosity of the microscopist.

The euglenoids are unicellular flagellates. A majority of the 800 or so known species are colourless heterotrophs or phagotrophs and are placed by zoologists in the protist order Euglenida. Molecular investigations reveal them to be a single, if disparate group, some of which acquired the phototrophic capability through secondary symbioses. It appears that even the phototrophic euglenoids are capable of absorbing and assimilating particular simple organic solutes. Many of the extant species are associated with organically rich habitats (ponds and lagoons, lake margins, sediments).

The 'red line' of eukaryotic evolution is based on rhodophyte plastids that contain phycobilins and chlorophyll *a*, and whose single thylakoids lie separately and regularly spaced in the plastid stroma (see, e.g., Kirk, 1994). The modern phylum Rhodophyta is well represented in marine (especially; mainly as red seaweeds) and freshwater habitats but no modern or extinct planktic forms are known. However, among the interesting derivative groups that are believed to owe to secondary endosymbioses of rhodophyte

cells, there is a striking variety of planktic forms.

Closest to the ancestral root are the cryptophytes. These contain chlorophyll c_2, as well as chlorophyll *a* and phycobilins, in plastid thylakoids that are usually paired. Living cells are generally green but with characteristic, species-specific tendencies to be bluish, reddish or olive-tinged. The modern planktic representatives are exclusively unicellular; they remain poorly known, partly because thay are not easy to identify by conventional means. However, about 100 species each have been named for marine and fresh waters, where, collectively, they occur widely in terms of latitude, trophic state and season.

Next comes the small group of single-celled flagellates which, despite showing similarities with the cryptophytes, dinoflagellates and euglenophytes, are presently distinguished in the phylum Raphidophyta. One genus, *Gonyostomum*, is cosmopolitan and is found, sometimes in abundance, in acidic, humic lakes. The green colour imparted to these algae by chlorophyll *a* is, to some extent, masked by a xanthophyll (in this case, diatoxanthin) to yield the rather yellowish pigmentation. This statement applies even more to the yellow-green algae making up the phyla Xanthophyta and Eustigmatophyta. The xanthophytes are varied in form and habit with a number of familiar unicellular non-flagellate or biflagellate genera in the freshwater plankton, as well as the filamentous *Tribonema* of hard-water lakes. The eustigmatophytes are unicellular coccoid flagellates of uncertain affinities that take their name from the prominent orange eye-spots.

The golden algae (Chrysophyta) represent a further recombination along the red line, giving rise to a diverse selection of modern unicellular, colonial or filamentous algae. With a distinctive blend of chlorophyll *a*, c_1 and c_2, and the major presence of the xanthophyll fucoxanthin, the chrysophytes are presumed to be close to the Phaeophyta, which includes all the macrophytic brown seaweeds but no planktic vegetative forms. Most of the chrysophytes have, in contrast, remained microphytic, with numerous planktic genera. A majority of these come from fresh water, where they are traditionally supposed to indicate low nutrient status and productivity (but see Section 3.4.3: they may simply be unable to use carbon sources other than carbon dioxide). Mostly unicellular or coenobial flagellates, many species are enclosed in smooth protective loricae, or they may be beset with numerous delicate siliceous scales. The group has been subject to considerable taxonomic revision and reinterpretation of its phylogenies in recent years. The choanoflagellates (formerly Craspedophyceae, Order Monosigales) are no longer considered to be allied to the Chrysophytes.

The last three phyla named in Table 1.1, each conspicuously represented in both limnetic and marine plankton – indeed, they are the main pelagic eukaryotes in the oceans – are also remarkable in having relatively recent origins, in the mesozoic period. The Bacillariophyta (the diatoms) is a highly distinctive phylum of single cells, filaments and coenobia. The characteristics are the possession of golden-brown plastids containing the chlorophylls a, c_1 and c_2 and the accessory pigment fucoxanthin, and the well-known presence of a siliceous frustule or exoskeleton. Generally, the latter takes the form of a sort of lidded glass box, with one of two valves fitting in to the other, and bound by one or more girdle bands. The valves are often patterned with grooves, perforations and callosities in ways that greatly facilitate identification. Species are ascribed to one or other of the two main diatom classes. In the Biddulphiales, or centric diatoms, the valves are usually cylindrical, making a frustule resembling a traditional pill box; in the Bacillariales, or pennate diatoms, the valves are elongate but the girdles are short, having the appearance of the halves of a date box. While much is known and has been written on their morphology and evolution (see, for instance, Round et al., 1990), the origin of the siliceous frustule remains obscure.

The Haptophyta are typically unicellular gold or yellow-brown algae, though having amoeboid, coccoid or palmelloid stages in some cases. The pigment blend of chlorophylls a, c_1 and c_2, with accessory fucoxanthin, resembles that of other gold-brown phyla. The haptophytes are distinguished by the possession of a haptonema, located between the flagella. In some species it is a prominent thread, as long as the cell; in others it is smaller or even vestigial but, in most instances, can be bent or coiled. Most of the known extant haptophyte species are marine; some genera, such as Chrysochromulina, are represented by species that are relatively frequent members of the plankton of continental shelves and of mesotrophic lakes. Phaeocystis is another haptophyte common in enriched coastal waters, where it may impart a visible yellow-green colour to the water at times, and give a notoriously slimy texture to the water (Hardy, 1964).

The coccolithophorids are exclusively marine haptophytes and among the most distinctive microorganisms of the sea. They have a characteristic surface covering of coccoliths – flattened, often delicately fenestrated, scales impregnated with calcium carbonate. They fossilise particularly well and it is their accumulation which mainly gave rise to the massive deposits of chalk that gave its name to the Cretaceous (from Greek kreta, chalk) period, 120–65 Ma BP. Modern coccolithophorids still occur locally in sufficient profusion to generate 'white water' events. One of the best-studied of the modern coccolithophorids is Emiliana.

The final group in this brief survey is the dinoflagellates. These are mostly unicellular, rarely colonial biflagellated cells; some are relatively large (up to 200 to 300 μm across) and have complex morphology. Pigmentation generally, but not wholly, reflects a red-line ancestry, the complex plastids containing chlorophyll a, c_1 and c_2 and either fucoxanthin or peridinin as accessory pigments, possibly testifying to tertiary endosymbioses (Delwiche, 2000). The group shows an impressive degree of adaptive radiation, with naked gymnodinioid nanoplankters through to large, migratory gonyaulacoid swimmers armoured with sculpted plates and to deepwater shade forms with smooth cellulose walls such as Pyrocystis. Some genera are non-planktic and even pass part of the life cycle as epiphytes. Freshwater species of Ceratium and larger species of Peridinium are conspicuous in the plankton of certain types of lakes during summer stratification, while smaller species of Peridinium and other genera (e.g. Glenodinium) are associated with mixed water columns of shallow ponds.

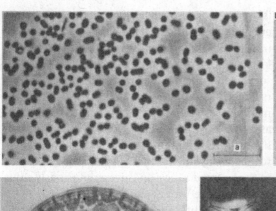

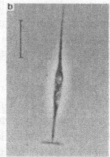

Figure 1.1 Non-motile unicellular phytoplankters. (a) *Synechococcus* sp.; (b) *Ankyra judayi*; (c) *Stephanodiscus rotula*; (d) *Closterium* cf. *acutum*. Scale bar, 10 μm. Original photomicrographs by Dr H. M. Canter-Lund, reproduced from Reynolds (1984a).

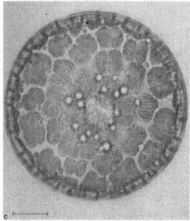

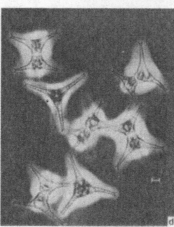

The relatively recent appearance of diatoms, coccolithophorids and dinoflagellates in the fossil record provides a clear illustration of how evolutionary diversification comes about. Although it cannot be certain that any of these three groups did not exist beforehand, there is no doubt about their extraordinary rise during the Mesozoic. The trigger may well have been the massive extinctions towards the end of the Permian period about 250 Ma BP, when a huge release of volcanic lava, ash and shrouding dust from what is now northern Siberia brought about a world-wide cooling. The trend was quickly reversed by accumulating atmospheric carbon dioxide and a period of severe global warming (which, with positive feedback of methane mobilisation from marine sediments, raised ambient temperatures by as much as 10–11 °C). Life on Earth suffered a severe setback, perhaps as close as it has ever come to total eradication. In a period of less than 0.1 Ma, many species fell extinct and the survivors were severely curtailed. As the planet cooled over the next 20 or so million years, the rump biota, on land as in water, were able to expand and radiate into habitats and niches that were otherwise unoccupied (Falkowski, 2002).

Dinoflagellate fossils are found in the early Triassic, the coccolithophorids from the late Triassic (around 180 Ma BP). Together with the diatoms, many new species appeared in the Jurassic and Cretaceous periods. In the sea, these three groups assumed a dominance over most other forms, the picocyanobacteria excluded, which persists to the present day.

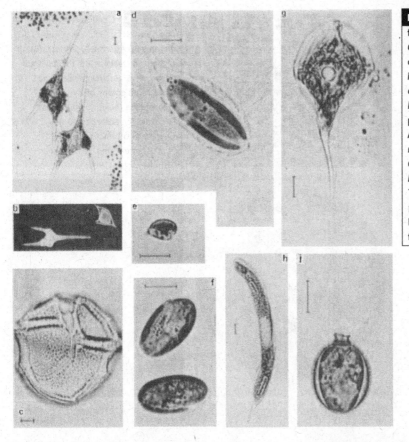

Figure 1.2 Planktic unicellular flagellates. (a) Two variants of *Ceratium hirundinella*; (b) overwintering cyst of *Ceratium hirundinella*, with vegetative cell for comparison; (c) empty case of *Peridinium willei* to show exoskeletal plates and flagellar grooves; (d) *Mallomonas caudata*; (e) *Plagioselmis nannoplanctica*; (f) two cells of *Cryptomonas ovata*; (g) *Phacus longicauda*; (h) *Euglena* sp.; (j) *Trachelomonas hispida*. Scale bar, 10 μm. Original photomicrographs by Dr H. M. Canter-Lund, reproduced from Reynolds (1984a).

1.4 | General features of phytoplankton

Despite being drawn from a diverse range of what appear to be distantly related phylogenetic groups (Table 1.1), there are features that phytoplankton share in common. In an earlier book (Reynolds, 1984a), I suggested that these features reflected powerful convergent forces in evolution, implying that the adaptive requirements for a planktic existence had risen independently within each of the major phyla represented. This *may* have been a correct deduction, although there is no compelling evidence that it *is* so. On the other hand, for small, unicellular microorganisms to live freely in suspension in water is an ancient trait, while the transition to a full planktic existence is seen to be a relatively short step. It remains an open question whether the supposed endosymbiotic recombinations could have occurred in the plankton, or whether they occurred among other precursors that subsequently established new lines of planktic invaders.

It is not a problem that can yet be answered satisfactorily. However, it does not detract from the fact that to function and survive in the plankton does require some specialised adaptations. It is worth emphasising again that just as phytoplankton comprises organisms other than algae, so not all algae (or even very many of them) are necessarily planktic. Moreover, neither the shortness of the supposed step to a planktic existence nor the generally low level of structural complexity of planktic unicells and coenobia should deceive us that they are necessarily simple organisms. Indeed, much of this book deals with the problems of life conducted in a fluid environment, often in complete isolation from solid boundaries, and the often sophisticated means by which planktic organisms overcome them. Thus, in spite of the diversity of phylogeny (Table 1.1), even a cursory consideration of the range of planktic algae (see Figs. 1.1–1.5)

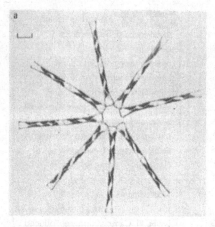

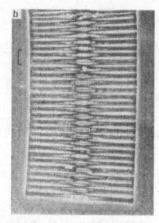

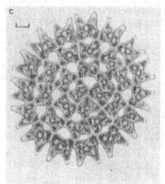

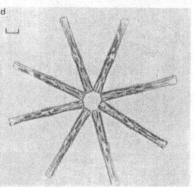

Figure 1.3 Coenobial phytoplankters. Colonies of the diatoms (a) *Asterionella formosa*, (b) *Fragilaria crotonensis* and (d) *Tabellaria flocculosa* var. *asterionelloides*. The fenestrated colony of the chlorophyte *Pediastrum duplex* is shown in (c). Scale bar, 10 μm. Original photomicrographs by Dr H. M. Canter-Lund, reproduced from Reynolds (1984a).

reveals a commensurate diversity of form, function and adaptive strategies.

What features, then, are characteristic and common to phytoplankton, and how have they been selected? The overriding requirements of any organism are to increase and multiply its kind and for a sufficient number of the progeny to survive for long enough to be able to invest in the next generation. For the photoautotroph, this translates to being able to fix sufficient carbon and build sufficient biomass to form the next generation, before it is lost to consumers or to any of the several other potential fates that await it. For the photoautotroph living in water, the important advantages of archimedean support and the temperature buffering afforded by the high specific heat of water (for more, see Chapter 2) must be balanced against the difficuties of absorbing sufficient nutriment from often very dilute solution (the subject of Chapter 4) and of intercepting sufficient light energy to sustain photosynthetic carbon fixation in excess of immediate respiratory needs (Chapter 3). However, radiant energy of suitable wavelengths (*photosynthetically active radiation*, or PAR) is neither universally or uniformly available in water but is sharply and hyperbolically attenuated with depth, through its absorption by the water and scattering by particulate matter (to be discussed in Chapter 3). The consequence is that for a given phytoplankter at anything more than a few meters in depth, there is likely to be a critical depth (the *compensation point*) below which net photosynthetic accumulation is impossible. It follows that the survival of the phytoplankter depends upon its ability to enter or remain in the upper, insolated part of the water mass for at least part of its life.

This much is well understood and the point has been emphasised in many other texts. These have also proffered the view that the essential characteristic of a planktic photoautotroph is to minimise its rate of sinking. This might be literally true if the water was static (in which case,

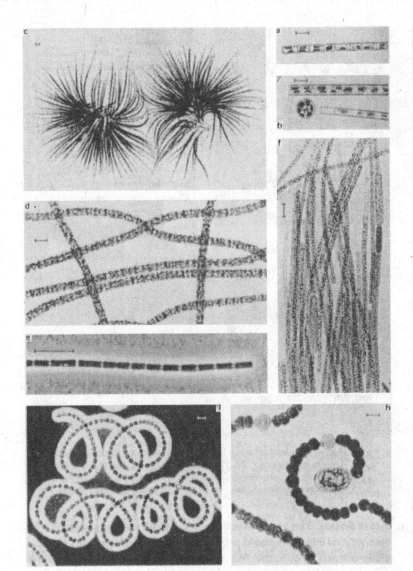

Figure 1.4 Filamentous phytoplankters. Filamentous coenobia of the diatom *Aulacoseira subarctica* (a, b; b also shows a spherical auxospore) and of the Cyanobacteria (c) *Gloeotrichia echinulata*, (d) *Planktothrix mougeotii*, (e) *Limnothrix redekei* (note polar gas vacuoles), (f) *Aphanizomenon flos-aquae* (with one akinete formed and another differentiating) and *Anabaena flos-aquae* (g) in India ink, to show the extent of mucilage, and (h) enlarged, to show two heterocysts and one akinete. Scale bar, 10 μm. Original photomicrographs by Dr H. M. Canter-Lund, reproduced from Reynolds (1984a).

neutral buoyancy would provide the only ideal adaptation). However, natural water bodies are almost never still. Movement is generated as a consequence of the water being warmed or cooling, causing convection with vertical and horizontal displacements. It is enhanced or modified by gravitation, by wind stress on the water surface and by the inertia due to the Earth's rotation (*Coriolis' force*). Major flows are compensated by return currents at depth and by a wide spectrum of intermediate eddies of diminishing size and of progressively smaller scales of turbulent diffusivity, culminating in molecular viscosity (these motions are characterised in Chapter 2).

To a greater or lesser degree, these movements of the medium overwhelm the sinking trajectories of phytoplankton. The traditional view of planktic adaptations as mechanisms to slow sinking rate needs to be adjusted. The essential requirement of phytoplankton is to *maximise the opportunities for suspension* in the various parts of the eddy spectrum. In many instances, the adaptations manifestly enhance the *entrainability* of planktic organisms by turbulent eddies. These include small size and low excess density (i.e. organismic density is close to that of water, $\sim$1000 kg m^{-3}), which features do contribute to a slow rate of sinking. They also include

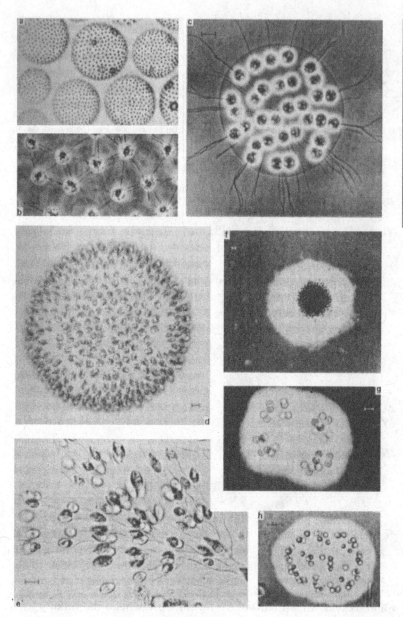

Figure 1.5 Colonial phytoplankters. Motile colonies of (a) *Volvox aureus*, with (b) detail of cells, (c) *Eudorina elegans*, (d) *Uroglena* sp. and (e) *Dinobryon divergens*; and non-motile colonies, all mounted in India ink to show the extent of mucilage, of (f) *Microcystis aeruginosa*, (g) *Pseudosphaerocystis lacustris* and (h) *Dictyosphaerium pulchellum*. Scale bar, 10 μm. Original photomicrographs by Dr H. M. Canter-Lund, reproduced from Reynolds (1984a).

mechanisms for increasing frictional resistance with the water, independently of size and density. At the same time, other phytoplankters show adaptations that favour disentrainment, at least from weak turbulence, coupled with relatively large size (often achieved by colony formation), streamlining and an ability to propel themselves rapidly through water. Such organisms exploit a different part of the eddy spectrum from the first group. The principle extends to the larger organisms of the nekton, – cephalopods, fish, reptiles and mammals – which are able to direct their own movements to overcome a still broader range of the pelagic eddy spectrum.

All these aspects of turbulent entrainment and disentrainment are explored more deeply and more empirically in Chapter 2. For the moment, it is important to understand how they impinge upon phytoplankton morphology in a general sense.

1.4.1 Size and shape

Apart from the issue of suspension, there is a further set of constraints that resists large size among phytoplankters. Autotrophy implies a requirement for inorganic nutrients that must be absorbed from the surrounding medium. These are generally so dilute and so much much less concentrated than they have to be inside the plankter's cell that uptake is generally against a very steep concentration gradient that requires the expenditure of energy to counter it. Once inside the cell, the nutrient must be translocated to the site of its deployment, invoking diffusion and transport along internal molecular pathways. Together, these twin constraints place a high premium on short internal distances: cells that are absolutely small or, otherwise, have one or two linear dimensions truncated (so that cells are flattened or are slender) benefit from this adaptation. Conversely, simply increasing the diameter (d) of a spherical cell is to increase the constraint for, though the surface area increases in proportion to d^2, the volume increases with d^3. However, distortion from the spherical form, together with surface convolution, provides a way of increasing surface in closer proportion to increasing volume, so that the latter is enclosed by relatively more surface than the geometrical minimum required to bound the same volume (a sphere). In this respect, the adaptive requirements for maximising entrainability and for enhancing the assimilation of nutrients taken up across the surface coincide.

It is worth adding, however, that nutrient uptake from the dilute solution is enhanced if the medium flows over the cell surface, displacing that which may have already become depleted. Movement of the cell relative to the adjacent water achieves the similar effect, with measurable benefit to uptake rate (Pasciak and Gavis, 1974; but see discussion in Section 4.2.1). It may be hypothesised that it is advantageous for the planker not to achieve isopycnic suspension in the water but to retain an ability to sink or float relative to the immediate surroundings, regardless of the rate and direction of travel of the latter, just to improve the sequestration of nutrients.

These traits are represented and sometimes blended in the morphological adaptations of specific plankters. They can be best illustrated by the plankters themselves and by examining how they influence their lives and ecologies. The wide ranges of form, size, volume and surface area are illustrated by the data for freshwater plankton presented in Table 1.2. The list is an edited, simplified and updated version of a similar table in Reynolds (1984a) which drew on the author's own measurements but quoted from other compilations (Pavoni, 1963; Nalewajko, 1966; Besch *et al.*, 1972; Bellinger, 1974; Findenegg, Nauwerck in Vollenweider, 1974; Willén, 1976; Bailey-Watts, 1978; Trevisan, 1978). The sizes are not precise and are often variable within an order of magnitude. However, the listing spans nearly eight orders, from the smallest cyanobacterial unicells of ~1 μm^3 or less, the composite structures of multicellular coenobia and filaments with volumes ranging between 10^3 and 10^5 μm^3, through to units of $>10^6$ μm^3 in which cells are embedded within a mucilaginous matrix. Indeed, the list is conservative in so far as colonies of *Microcystis* of >1 mm in diameter have been observed in nature (author's observations; i.e. up to 10^9 μm^3 in volume). Because all phytoplankters are 'small' in human terms, requiring good microscopes to see them, it is not always appreciated that the nine or more orders of magnitude over which their sizes range is comparable to that spanning forest trees to the herbs growing at their bases. Like the example, the biologies and ecologies of the individual organisms vary considerably through the spectrum of sizes.

1.4.2 Regulating surface-to-volume ratio

Dwelling on the issue of size and shape, we will find, as already hinted above, that a good deal of plankton physiology is correlated to the ratio of the surface area of a unit (s) to its volume (v). 'Unit' in this context refers to the live habit of the planker: where the vegetative form is unicellular (exemplified by the species listed in Table 1.2A), it is only the single cell that interacts with its environment and is, plainly, synonymous with the 'unit'. If cells are joined together to comprise a larger single structure, for whatever advantage, then the individual cells are no longer

Table 1.2 | Nominal mean maximum linear dimensions (MLD), approximate volumes (v) and surface areas (s) of some freshwater phytoplankton

Species	Shape	MLD (μm)	v (μm³)	s (μm²)	s/v (μm⁻¹)
(A) Unicells					
Synechoccoccus	ell	≤ 4	18 (1–20)	35	1.94
Ankyra judayi	bicon	16	24 (3–67)	60	2.50
Monoraphidium griffithsii	cyl	35	30	110	3.67
Chlorella pyrenoidosa	sph	4	33 (8–40)	50	1.52
Kephyrion littorale	sph	5	65	78	1.20
Plagioselmis nannoplanctica	ell	11	72 (39–134)	108	1.50
Chrysochromulina parva	cyl	6	85	113	1.33
Monodus sp.	ell	8	105	113	1.09
Chromulina sp.	ell	15	440	315	0.716
Chrysococcus sp.	sph	10	520	315	0.596
Stephanodiscus hantzschii	cyl	11	600 (180–1 200)	404	0.673
Cyclotella praeterissima	cyl	10	760 (540–980)	460	0.605
Cyclotella meneghiniana	cyl	15	1 600	780	0.488
Cryptomonas ovata	ell	21	2710 (1 950–3 750)	1 030	0.381
Mallomonas caudata	ell	40	4 200 (3 420–10 000)	3 490	0.831
Closterium aciculare	bicon	360	4 520	4 550	1.01
Stephanodiscus rotula	cyl	26	5 930 (2 220–18 870)	1 980	0.334
Cosmarium depressum	(a)	24	7 780 (400–30 000)	2 770	0.356
Synedra ulna	bicon	110	7 900	4 100	0.519
Staurastrum pingue	(b)	90	9 450 (4 920–16 020)	6 150	0.651
Ceratium hirundinella	(c)	201	43 740 (19 080–62 670)	9 600	0.219
Peridinium cinctum	ell	55	65 500 (33 500–73 100)	7 070	0.108
(B) Coenobia					
Dictyosphaerium pulchellum (40 cells)	(d)	40	900	1 540	1.71
Scenedesmus quadricauda (4 cells)	(e)	80	1 000	908	0.908
Asterionella formosa (8 cells)	(f)	130	5 160 (4 430–6 000)	6 690	1.30

(cont.)

Table 1.2 (cont.)

Species	Shape	MLD (μm)	v (μm³)	s (μm²)	s/v (μm⁻¹)
Fragilaria crotonensis (10 cells)	(g)	70	6 230 (4 970–7 490)	9 190	1.48
Dinobryon divergens (10 cells)	(h)	145	7 000 (6 000–8 500)	5 350	0.764
Tabellaria flocculosa var. asterionelloides (8 cells)	(f)	96	13 800 (6 520–13 600)	9 800	0.710
Pediastrum boryanum (32 cells)	(j)	100	16 000	18 200	1.14
(C) Filaments					
Aulacoseira subarctica (10 cells)	cyl (k)	240	5 930 (4 740–7 310)	4 350	0.734
Planktothrix mougeotii (1 mm length)	cyl (k)	1000	46 600	24 300	0.521
Anabaena circinalis (20 cells)	(m) (n)	60 75	2 040 29 000	2 110 6 200	1.03 0.214
Aphanizomenon flos-aquae (50 cells)	(p) (q)	125 125	610 15 400	990 5 200	1.62 0.338
(D) Mucilaginous colonies					
Coenochloris fottii (cells 80–1200 μm³)	sph	46	51×10^3	6.65×10^3	0.13
Eudorina unicocca (cells 120–1200 μm³)	sph	130	1.15×10^6	53.1×10^3	0.046
Uroglena lindii (cells 100 μm³)	sph	160	2.2×10^6	81×10^3	0.037
Microcystis aeruginosa (cells 30–100 μm³)	sph	200	4.2×10^6	126×10^3	0.030
Volvox globator (cells ~ 60 μm³)	sph (r) (s)	450 450	47.7×10^6 6.4×10^6	636×10^3 636×10^3	0.013 0.099

Notes: The volumes and surface areas are necessarily approximate. The values cited are those adopted and presented in Reynolds (1984a); some later additions taken from Reynolds (1993a), mostly based on his own measurements. The volumes given in brackets cover the ranges quoted elsewhere in the literature (see text). Note that the volumes and surface areas are calculated by analogy to the nearest geometrical shape. Surface sculpturing is mostly ignored. Shapes considered include: sph (for a sphere), cyl (cylinder), ell (ellipsoid), bicon (two cones fused at their bases, area of contact ignored from surface area calculation). Other adjustments noted as follows:

[a] Cell visualised as two adjacent ellipsoids, area of contact ignored.
[b] Cell visualised as two prisms and six cuboidal arms, area of contact ignored.
[c] Cell visualised as two frusta on elliptical bases, two cylindrical (apical) and two conical (lateral) horns.
[d] Coenobium envisaged as 40 contiguous spheres, area of contact ignored.
[e] Coenobium envisaged as four adjacent cuboids, volume of spines ignored
[f] Coenobium envisaged as eight cuboids, area of contact ignored.

Notes to Table 1.2 (cont.)

[g] Each cell visualised as four trapezoids; area of contact between cells ignored.

[h] Coenobium envisaged as a seies of cones, area of contact ignored

[j] Coenobium envisaged as a discus-shaped sphaeroid.

[k] Coenobium envisaged as a chain of cylinders, area of contact between cells ignored.

[l] Coenobium envisaged as a single cylinder, terminal taper ignored.

[m] Filament visualised as a chain of spheres, area of contact between them ignored

[n] Filament visualised as it appears in life, enveloped in mucilage, turned into a complete 'doughnut' ring, with a cross-sectional diameter of 21 μm.

[p] Filament visualised as a chain of ovoids, area of contact between them ignored.

[q] For the typical 'raft' habit of this plankter, a bundle of filaments is envisaged, having an overall length of 125 μm and a diameter of 12.5 μm.

[r] The volume calculation is based on the external dimensions.

[s] In fact the cells in the vegetative stage are located exclusively on the wall of a hollow sphere. This second volume calculation supposes an average wall thickness of 10 μm and subtracts the hollow volume.

independent but rather constitute a multicellular 'unit' whose behaviour and experienced environment is simultaneously shared by all the others in the unit. Such larger structures may deploy cells either in a plate- or ball-like *coenobium* (exemplified by the species listed in Table 1.2B) or, end-to-end, to make a uniseriate *filament* (Table 1.2C). Generally, added complexity brings increased size but, as volume increases as the cube but surface as the square of the linear dimension, there is natural tendency to sacrifice a high surface-to-volume ratio.

However, a counteractive tendency is found among the coenobial and filamentous units (and also among larger unicells), in which increased size is accompanied by increased departure from the spherical form. This means, in large-volume units, more surface bounds the volume than the strict geometrical minimum provided by the sphere. In addition to distortion, surface folding, the development of protuberances, lobes and horns all contribute to providing more surface for not much more volume. The trend is shown in Fig. 1.6, in which the surface areas of the species listed in Tables 1.2A, B and C are plotted against the corresponding, central-value volumes, in log/log format. The smaller spherical and ovoid plankters are seen to lie close to the (lower) slope representing the geometric minimum of surface on volume, $s \propto v$, but progressively larger units drift away from it. The second regression, fitted to the plotted data, has a steeper gradient of $v^{0.82}$. The individual values of

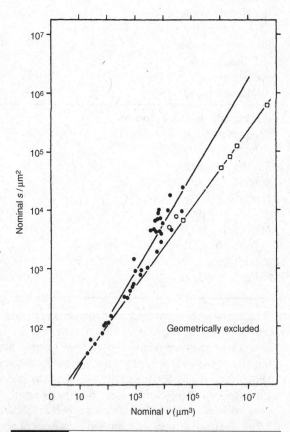

Figure 1.6 Log/log relationship between the surface areas (s) and volumes (v) of selected freshwater phytoplankters shown in Table 1.2. The lower line is fitted to the points (□) referring to species forming quasi-spherical mucilaginous coenobia (log s = 0.67 log v + 0.7); the upper line is the regression of coenobia (log s = 0.82 log v + 0.49) fitted to all other points (●). Redrawn from Reynolds (1984a).

s/v entered in Table 1.2B and C show several examples of algal units having volumes of between 10^3 and 10^5 μm^3 but maintaining surface-to-volume ratios of >1. In the case of *Asterionella*, the slender individual cuboidal cells are distorted in one plane and are attached to their immediate coenobial neighbours by terminal pads representing a very small part of the total unit surface. The resultant pseudostellate coenobium is actually a flattened spiral. In *Fragilaria crotonensis*, a still greater *s/v* ratio is achieved. Though superficially similar to those of *Asterionella*, its cells are widest in their mid-region and where they link, mid-valve to mid-valve, to their neighbours to form the 'double comb' appearance that distinguishes this species from others (mostly non-planktic) of the same genus. Among the filamentous forms, there is a widespread tendency for cells to be elongated in one plane and to be attached to their neighbours at their polar ends so that the long axes are cumulated (e.g. in *Aulacoseira*, Fig. 1.4).

Again confining the argument to the planktic forms covered in Tables 2.2A, B and C, whose volumes, together, cover five orders of magnitude, these show surface-to-volume ratios that fall in scarcely more than one-and-a-half orders (3.6 to 0.1). This evident conservatism of the surface-to-volume ratio among phytoplankton was noted in a memorable paper by Lewis (1976). He argued that this is not a geometric coincidence but an evolutionary outcome of natural selection of the adaptations for a planktic existence. Thus, however strong is the selective pressure favouring increased size and complexity, the necessity to maintain high *s/v*, whether for entrainment or nutrient exchange or both, remains of overriding importance. In other words, the relatively rigid constraints imposed by maintenance of an optimum surface-to-volume ratio constitute the most influential single factor governing the shape of planktic algae.

Lewis (1976) developed this hypothesis through an empirical analysis of phytoplankton shapes. To a greater or lesser extent, departure from the spherical form through the provision of additional surface area is achieved by shape attenuation in one or, perhaps, two planes, respectively resulting in slender, needle-like forms or flattened, plate-like structures. Lewis

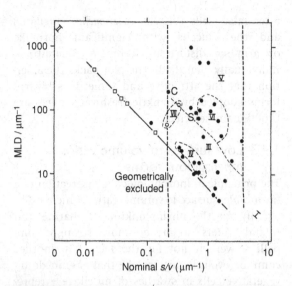

Figure 1.7 The shapes of phytoplankters: log/log plot of maximum linear dimension (MLD) versus nominal *s/v* of individual phytoplankters (data in Table 1.2). Similar morphologies are grouped together (I, spherical cells; II, spherical colonies; III, squat ellipsoids and cylinders, of which IV are exclusively centric diatoms; V, attenuate cells and filaments; VI, coenobia; VII, bundles of filaments of *Anabaena* or *Aphanizomenon*); C (for *Ceratium*) and S (for *Staurastrum*) identify shapes with large protuberances. The vertical dotted lines define the distributions of most marine phytoplankters according to Lewis (1976). Redrawn from Reynolds (1984a).

represented these modifications by plotting the maximum linear dimension (MLD) of the unit against its surface-to-volume ratio. His approach is followed in the construction of Fig. 1.7, in which the relevant data from Table 1.2 are plotted. The diagonal line is a geometric boundary, representing the diminution of *s/v* of spheres against the increment in diameter and, indeed, upon which the spherical unicells (marked I) and colonies (II) are located. All other shapes fall above this line, the further above it being the more distorted with respect to the sphere of the same MLD. The broken dotted lines bound the *s/v* ratios of non-spherical forms. All the ellipsoid shapes (III, such as *Mallomonas*, *Rhodomonas*), squat cylinders (IV, including *Cyclotella* and *Stephanodiscus* spp.), attenuated needle-like cells and filaments (V: *Monoraphidium*, *Closterium*, *Aulacoseira*, *Planktothrix*) fall within this area. So do the coenobial forms comprising individual attenuated cells (VI, e.g. *Asterionella*, *Fragilaria*) and the unicells with significant horn-like or arm-like distortions (*Ceratium*, *Staurastrum*, individually identified). The plot backs the assertion that the attractive and sometimes bizarre forms adopted by planktic freshwater algae are functionally selected.

1.4.3 Low surface-to-volume ratio: mucilaginous forms

The principle of morphological conservation of a favourable surface-to-volume ratio, which holds equally for the phytoplankton of marine and inland waters, might be more strongly compelling were it not for the fact that another common evolutionary trend – that of embedding vegetative cells in swathes of mucilage – represents a total antithesis. The formation of globular colonies is prevalent among the freshwater Cyanobacteria, Chlorophyta and the Chrysophyta. It is also observed in the vegetative life-history stages of the haptophyte, *Phaeocystis*, though, generally, the trait is not common among the marine phytoplankton. In many instances, the secondary structures are predominantly mucilaginous and the live cells may occupy as little as 2% of the total unit volume in *Coenochloris* and *Uroglena* and scarcely exceeds 20% in *Microcystis* or *Eudorina*. It was originally supposed to provide a low-density buoyancy aid but it has since been shown that any advantage is quickly lost to increased size (see Chapter 2). In some instances, the individual cells are flagellate (as in *Uroglena* and *Eudorina*) and the flagella pass through the mucilage to the exterior, where their coordinated beating propels the whole colony through the medium. Because the surface offers little friction, the mucilage is said to be helpful in assisting rapid passage and migration through weakly turbulent water. Certainly, in the case of the colonial gas-vacuolate Cyanobacteria that are able to regulate their buoyancy (e.g. *Microcystis*, *Snowella*, *Woronichinia*), larger colonies float more rapidly than smaller ones of the same density (Reynolds, 1987a). Merely adjusting buoyancy then becomes a potentialy effective means of recovering or controlling vertical position in the water (Ganf, 1974a). It is interesting that, in the two buoyancy-regulating filamentous species of

Cyanobacteria included in Table 1.2 (*Anabaena circinalis*, *Aphanizomenon flos-aquae*), the supposed advantage of the filamentous habit is sacrificed through a combination of aggregation, coiling and mucilage production to the attainment of rapid rates of migration (Booker and Walsby, 1979).

Provided colonies have a simultaneous capacity for controlled motility, there are good teleological grounds for deducing circumstances when massive provision of mucilage represents a discrete and alternative adaptation to a planktic existence. However, the idea that streamlining is more than a fortuitous benefit is challenged by the many non-motile species that exist as mucilaginous colonies. There are other demonstrable benefits from a mucilaginous exterior, including defence against fungal attack, grazers, digestion or metal toxicity, and there are circumstances in which it might assist in the sequestration or storage of nutrients or in protecting cells from an excessively oxidative environment (see Box 6.1, p. 271). Even mucilage itself, essentially a matrix of hygroscopic carbohydrate polymers immobilising relatively large amounts of water, is highly variable in its consistency, intraspecifically as well as interspecifically.

Thus, doubts persist about the true function of mucilage investment. However, a consistent geometric consequence of mucilage investment is that the planktic unit is left with an exceptionally low surface-to-volume ratio (i.e. area II, towards the left in Fig. 1.7).

1.5 | The construction and composition of freshwater phytoplankton

The architecture of the cells of planktic algae conforms to a basic model, common to the majority of eukaryotic plants. A series of differentiated protoplasmic structures are enclosed within a vital membrane, the *plasmalemma*. This membrane is complex, comprising three or four distinct layers. In a majority of algae there is a further, non-living *cell wall*, made of cellulose or other, relatively pure, condensed carbohydrate polymer, such as pecten. Among some algal groups, the wall may be more or less impregnated with inorganic deposits of calcium carbonate or silica. High-power scanning electron microscopy reveals that these deposits can form a more or less continuous but variably thickened and fenestrated surface (as in the siliceous frustule wall of diatoms) or can comprise an investment of individual scales (like those of the Synurophyceae, made of silica, or of the coccolithophorids, made of carbonate). These *exoskeletons* are distinctive and species-diagnostic. Some algae lack a polymer wall and are described as 'naked'. Both naked and walled cells may carry an additional layer of secreted mucilage.

The intracellular protoplasm (cytoplasm) is generally a viscous, gel-like suspension in which the nucleus, one or more plastids and various other organelles, including the endoplasmic reticulum and the mitochondria, and some condensed storage products are maintained. The plastids vary hugely and interspecifically in shape – from a solitary axial cup (as in the Volvocales), numerous discoids (typical of centric diatoms), one or two broad parietal or axial plates (as in Cryptophytes) or more complex shapes (many desmids). All take on the intense coloration of the dominant photosynthetic pigments they contain – chlorophyll *a* and β-carotene and, variously, other chlorophylls and/or accessory xanthophylls. The stored condensates of anabolism are also conspicuously variable among the algae: starch in the chlorophytes and cryptophytes, other carbohydrates in the euglenoids (paramylon) and the Chrysophyceae, oils in the Xanthophyceae). Many also store protein in the cytoplasm. The quantities of all storage products vary with metabolism and environmental circumstances.

Intracellular vacuoles are to be found in most planktic algae but the large sap-filled spaces characteristic of higher-plant cells are relatively rare, other than in the diatoms. Osmoregulatory *contractile vacuoles* occur widely, though not universally, among the planktic algae, varying in number and distribution among the phylogenetic groups.

The prokaryotic cell of a planktic Cyanobacterium is also bounded by a plasmalemma. It,

too, is multilayered but has the distinctive bacterial configuration. The cells lack a membrane-bound nucleus and plastids, the genetic material and photosynthetic thylakoid membranes being unconfined through the main body (stroma) of the cell. The pigments, chlorophyll and accessory phycobilins, colour the whole cell. Glycogen is the principal photosynthetic condensate and proteinaceous structured granules may also be accumulated. Many planktic genera contain, potentially or actually, specialised intracellular proteinaceous gas-filled vacuoles which may impart buoyancy to the cell.

From an ecological point of view, the ultrastructural properties of phytoplankton cells assume considerable relevance to the resource requirements of their assembly, as well as to adaptive behaviour, productivity and dynamics of populations. Thus, it is important to establish a number of empirical criteria of cell composition that impinge upon the fitness of individual plankters and the stress thresholds relative to light, temperature and nutrient availability. These include methods for assessing the biomass of phytoplankton populations and the environmental capacity to support them.

1.5.1 Dry weight

Characteristically, the major constituent of the live plankter is water. If the organism is air-dried to remove all uncombined water, the residue will comprise both organic (mainly protoplasm and storage condensates) and inorganic (such as the carbonate or silica impregnated into the cell walls) fractions. Oxidation of the organic fraction, by further heating in air to ~500 °C, yields an ash approximating to the original inorganic constituents. The relative masses of the ash and ash-free (i.e. organic) fractions of the original material may then be back-calculated.

The dry weights and the ash contents of a selection of freshwater phytoplankton are presented in Table 1.3. Generally, cell dry mass (W_c) increases with increasing cell volume (v), as shown in Fig. 1.8. The regression, fitted to all data points, has the equation $W_c = 0.47 v^{0.99}$, with a high coefficient of correlation (0.97). At first sight, the relationship yields the useful general prediction that the dry weights of live

Table 1.3 | Air-dry weights, ash-free free dry weights, chlorophyll content and volume of individual cells[a] from natural populations (all values are means of collected data having considerable ranges of variability)

Species	Dry weight (pg cell^{-1})	Ash-free dry weight (pg cell^{-1})	Ash % dry	Chla (pg cell^{-1})	Volume (μm^3 cell^{-1})
Cyanobacteria					
Anabaena circinalis[b]	45	–	–	0.72	99
Aphanizomenon flos-aquae[b]	3.9	–	–	0.04	8.2
Microcystis aeruginosa[b]	32	–	–	0.36	73
Planktothrix mougeotii[a,b]	28000	–	–	243	46 600
Chlorophytes					
Ankyra judayi[b]	–	–	–	0.45	24
Chlorella pyrenoidosa[b]	15	–	–	0.15	33
Chlorella pyrenoidosa[c]	5.1–6.4	4.5–5.7	11.4	–	20
Closterium aciculare[b]	–	–	–	89	4 520
Eudorina elegans[b]	273	–	–	5.5	320
Eudorina elegans[c]	251–292	233–268	7.9	–	320
Eudorina unicocca[b]	–	–	–	9.5	586
Monoraphidium contortum[c]	5.2–5.7	4.7–5.0	10.4	–	30
Scenedesmus quadricauda[c]	99–104	91–95	8.5	–	200
Staurastrum pingue[b]	–	–	–	57	9 450
Staurastrum sp.[c]	4680–4940	4480–4620	5.3	–	20 500
Volvox aureus[b]	99	–	–	1.1	60
Diatoms					
Asterionella formosa[b]	292–349	–	–	1.8	554–736
Asterionella formosa[c]	243–291	104–136	55	–	650
Asterionella formosa[d]	318	171	46	1.7	645
Aulacoseira binderana[c]	247–281	137–159	44	–	1 380
Aulacoseira granulata[b]	519	–	–	4–5	847
Fragilaria capucina[c]	197–215	99–109	50	–	350
Fragilaria crotonensis[b]	272	–	–	2	623
Stephanodiscus sp.[c]	115–122	62–66	47	–	310
Stephanodiscus hantzschii[b]	58	–	–	0.9	600
Stephanodiscus rotula[b]	2770	–	–	41	5 930
Tabellaria flocculosa v. asterionelloides[b]	525	279	47	2.5	1 725
Tabellaria flocculosa v. asterionelloides[c]	383–407	205–210	47	–	820
Cryptophytes					
Cryptomonas ovata[b]	2090	–	–	33	2 710
Dinoflagellates					
Ceratium hirundinella[b]	18790	–	–	237	43 740

[a] Data for Planktothrix is per 1 mm length of filament.
[b] From compilation of previously unpublished measurements of field-collected material, as specified in Reynolds (1984a).
[c] From Nalewajko (1966).
[d] From later data of Reynolds (see 1997a).
Source: List assembled by Reynolds (1984a) from thitherto unpublished field data, and from data of Nalewajko (1966).

planktic cells will be equivalent to between 0.41 and 0.47 pg μm^{-3}. In fact, the data suggest an order-of-magnitude range, from 0.10 to 1.65 pg μm^{-3}. In part, this reveals an inherent danger in the interpretation of a wide spread of raw data through log/log representations. Caution is required in interpolating species-specific derivations from a general statistical relationship. However, the additional difficulty must also be recognised that the regression is fitted to phytoplankton of considerable structural variation (vacuole space, carbonate or silica impregnation of walls). Accumulated dry mass is also influenced by the physiological state of cells and the environmental conditions obtaining immediately prior to the harvesting of the analysed material. It is, perhaps, surprising that the data used to construct Table 1.3 and Fig. 1.8, based on analyses of material drawn largely from wild populations, should show any consistency at all. Thus, the patterns detected are worthy of slightly deeper investigation.

For instance, the variation in the percentage ash content (where known) appears to be considerable. Nalewajko (1966; see also Table 1.3) found that ash accounted for between 5.3% and 19.9% (mean 10.2%) of the dry weights of 16 species of planktic chlorophytes but for between 27% and 55% (mean 41.4%) of those of 11 silicified diatoms.

1.5.2 Skeletal silica

Much of the high percentage ash content of the diatoms is attributable to the extent of silicification of the cell walls. Besides their value to taxonomic diagnosis, the ornate and often delicate exoskeletal structures, celebrated in such photographic collections as that of Round *et al.* (1990), command wonder at the evolutionary trait and at the genetic control of frustule assembly. There are ecological ramifications, too, arising from the relatively high density of the deposited silicon polymers, seemingly quite opposite to the adaptive requirements for a planktic existence. Moreover, the amounts of silicon consumed in the development of each individual diatom cell have to be met by uptake of silicon dissolved in the medium. Whatever may have been the situation at the time of their evolution, the present-day concentrations of dissolved

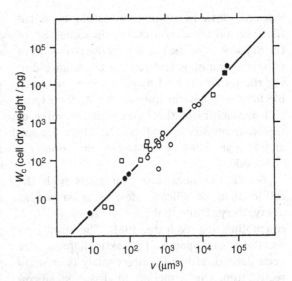

Figure 1.8 Log/log plot of cell dry mass (W_c) against cell volume (v) for various freshwater phytoplankers (data in Table 3: ●, Cyanobacteria; □, diatoms; ○, chlorophytes; others). The equation of the regression is $W_c = 0.47 \, v^{0.99}$. Redrawn from Reynolds (1984a).

silicon in lakes and oceans are frequently inadequate to meet the potential demands of unfettered diatom development. On the other hand, the requirement is obligate and within relatively narrow, species-specific ranges and uptake is subject to physiologically definable levels. The biological availability of silicon, its consumption and deployment, as well as the fate of its biogenic polymers, are of special relevance to planktic ecology, as they may well determine the environmental carrying capacity of new diatom production. Thus, they have some selective value for particular types of diatom, or for other types of non-siliceous planker, when external supplies are substantially deficient.

The cells of all living organisms have a requirement for the small amounts of silicon involved in the synthesis of nucleic acids and proteins (generally <0.1% of dry mass: Sullivan and Volcani, 1981). However, it is the demands of those groups of protistans and poriferans that characteristically employ silicon in skeletal structures – notably diatoms, other chrysophytes radiolarians and sponges – that impinge most on the geochemical cycling of silicon (Simpson and

Volcani, 1981). In passing, it should be noted that skeletal silica also makes up some 10% of the dry weight of the grasses, whose co-evolution with the mammals and relative abundance during the tertiary period may have been responsible for the long-term fluctuations in the export and availability of the main soluble source of silicon (monosilicic acid: Siever, 1962; Stumm and Morgan, 1996) in the aquatic environments (Falkowski, 2002).

For the moment, our concern is with the cell content of silicon. Most is deposited as a cryptocrystalline polymer of silica $((SiO_2)_n)$, resembling opal (Volcani, 1981). The silica contents of several species of planktic diatoms have been derived, either by direct analysis or, indirectly, from the depletion of dissolved silicon by a known specific recruitment of cells by growth. Unlike other elements critical to their survival, diatoms take up scarcely more silicon than is immediately required to form the frustules of the next generation (Paasche, 1980; Sullivan and Volcani, 1981). As already indicated, the amounts deposited are generally quite species-specific (Reynolds, 1984a; see also Table 1.4), at least when variability in cell size is taken into account (Lund, 1965; Jaworski et al., 1988). Cell-specific silicon requirements differ considerably among planktic species, reportedly ranging between 0.5% (in the marine Phaeodactylum tricornutum: Lewin et al., 1958) and 37% of dry weight (in some freshwater Aulacoseira spp.: Lund, 1965; Sicko-Goad et al., 1984).

In terms of mass of silicon per cell, broad relationships with the mean volume and with the mean surface area are demonstrable (Fig. 1.9). The regressions reflect the increasing silicon deployment with increasing cell size but their slopes, within the interpretative limits of log/log relationship, suggest that increased cell size is accompanied by a decreasing ratio of silicon : enclosed volume and an increasing ratio of silicon : area. This is in accord with the expectation of Einsele and Grim (1938), among the earliest investigators of the silicon requirements of diatoms, that interspecific variations in deployment are related to differences in shape (surface area-to-volume effects), together with the relative investment in such species-specific features

as fenestration, strengthening ribs, bracing struts and spines. Later data on some of the species of diatom considered by Einsele and Grim (1938) suggest that the area-specific silicon content is particularly responsive to increasing size (see Table 1.5).

1.5.3 Organic composition

Discounting the typical 5–12% ash (possibly up to 80% in the case of some diatoms) content of air-dried plankters, the balancing mass is supposed to be the organic components of the cell, derived from the living protoplasm. In comprising mainly proteins, lipids and condensed carbohydrates, albeit in variable proportions, the elemental composition of the ash-free dry material is dominated by carbon (C), hydrogen (H), oxygen (O) and nitrogen (N), together with smaller amounts of phosphorus (P) and sulphur (S). At least 14 other elements (Ca, Mg, Na, Cl, K, Si, Fe, Mn, Mo, Cu, Co, Zn, B, Va) are consistently recoverable if sufficient analytical rigour is applied (Lund, 1965). It is impressive that, whilst alive, every planktic cell had not only the capability of taking up these elements from extremely dilute media but also the success in doing so. This should be borne in mind when interpreting the relative quantities in which these elements occur in the dry matter of cells, because not all were necessarily equally available relative to demand. Besides falling deficient in one element that is relatively scarce, others that are relatively abundant may tend to accumulate in the cell. In this way, the ratios in which the elements make up the ash-free air-dried algal tissue often give a reliable reflection of the conditions of nutrient availability in the growth medium. Compounded by the special mechanisms that cells may have for taking up and retaining elements from unreliable environmental sources (to be explored in Chapter 4), the absolute quantities of several of the component elements are liable to wide variation.

Not surprisingly, the elemental ratios in natural phytoplankton plankton have for long aroused the interest of physiological ecologists and of biogeochemists and they have been much studied and reported. Absolute quantities do vary substantially, as do the ratios among them. Yet, remarkably, the same data indicate

Table 1.4 The silicon content of some freshwater planktic diatoms

Species	Si (pg cell^{-1})	Si (% dry wt)	nSiO$_2$ (% dry wt)	References[a]
Asterionella formosa	65.2 (45.5–80.3)	–	–	Einsele and Grim (1938)
	64.3 (46.9–82.1)	21	45	Lund (1950, 1965)
	61.0 (52.1–69.9)	–	–	Reynolds and Wiseman (1982)
	64.3	20	43	Reynolds (1997a), given as Si
	0.239W$^{1.003}$	24	51	Regression of Jaworski et al. (1988) fitted to measurements made on a cultured clone of diminishing size, data given as Si
Fragilaria crotonensis	88.7 (79.8–100.9)	–	–	Einsele and Grim (1938)
	88.7 (88.2–89.2)	22	46	Lund (1965)
	117.8 (98.0–142.7)	–	–	Reynolds and Wiseman (1982)
	49.7 (41.3–55.9)	–	–	Reynolds (1973a)
Aulacoseira granulata	61.0	–	–	Einsele and Grim (1938)
	291.0	25	54	Prowse and Talling (1958)
	138.0	27	57	Thitherto unpublished records cited in Reynolds (1984a), calculated indirectly from SiO$_2$ uptake
Aulacoseira subarctica	111.2 (77.4–128.6)	30	63	Lund (1965)
Stephanodiscus rotula	3989	–	–	Einsele and Grim (1938)
	1942 (1704–2182)	–	–	Thitherto unpublished records cited in Reynolds (1984a), calculated indirectly from SiO$_2$ uptake
	1075	–	–	Gibson et al. (1971)
	978	32	69	Thitherto unpublished records cited in Reynolds (1984a), calculated indirectly from SiO$_2$ uptake
	751	27	58	Thitherto unpublished records cited in Reynolds (1984a), calculated indirectly from SiO$_2$ uptake
Stephanodiscus hantzschii	19.2 (15.0–22.1)	26	55	Lund (1965), Swale (1963)
	16.4	28	60	Thitherto unpublished records cited in Reynolds (1984a), calculated indirectly from SiO$_2$ uptake
Tabellaria flocculosa var. flocculosa	185.4 (173.6–197.1)	–	–	Einsele and Grim (1938)
	145.5 (117.3–197.1)	25	53	Lund (1965)

[a] Original citations quote content of SiO$_2$, except where stated otherwise.

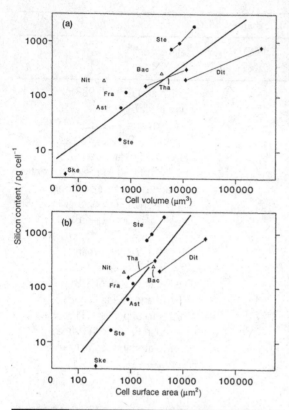

Figure 1.9 The silicon content of selected diatoms from the freshwater (●) or marine (■) phytoplankton or other aquatic habitat (△), plotted on log/log scales against (a) cell volume and (b) surface area. Ast refers to *Asterionella*, Fra to *Fragilaria* and Ste to species of *Stephanodiscus*; Bac refers to *Bacillaria*, Dit to *Ditylum*, Nit to *Nitzschia*, Ske to *Skeletonema* and Tha to *Thalassosira*. The equations of the least-squares regression fitted to the data in (a) is log [Si] = 0.707 log v − 0.263 (r = 0.85); that for (b) is log [Si] = 1.197 log s − 1.634 (r = 0.83). Redrawn, with permission, from Reynolds (1986a).

dry weight (Ketchum and Redfield, 1949). A slightly lower range (45–51%) was derived by Round (1965) and Fogg (1975) from measurements on freshwater phytoplankton. However, the same sources of data showed extremes of about 35%, in cells deprived of light or a supply of inorganic carbon, and 70%, if deficiencies of other elements impeded the opportunities for growth.

The importance of carbon assimilation by photoautotrophs to system dynamics has encouraged interest in being able to make direct estimates of organismic carbon content as a function of biovolume. It will be obvious, from the recognition of the variability in the absolute contents of carbon, its proportion of wet or dry biomass, and the relative fractions of ash and vacuolar space, that any general relationship must be subject to a generous margin of error. For instance, Mullin *et al.* (1966) derived an order-of-magnitude range of 0.012–0.26 pg C μm^{-3} for a selection of 14 marine phytoplankters that included large and small diatoms. Reynolds' (1984a) analysis of data, pertaining exclusively to freshwater forms, adopted simultaneous approaches to diatoms and non-diatoms. The relatively low ash content and absence of large vacuoles among the latter permitted a much narrower relationship between carbon and biovolume (averaging 0.21–0.24 pg C μm^{-3}). Supposing carbon makes up a little under half of the ash-free dry mass and that dry mass averages 0.47 pg μm^{-3} (Fig. 1.8), this figure is highly plausible. For diatoms, there seemed little alternative but to calculate carbon as a function of the silica-free dry mass.

This approach does not satisfy the quest for a volume-to-carbon conversion for mixed diatom-dominated assemblages, which continues to tax ecosystem ecologists. A recent re-exploration by Gosselain *et al.* (2000) confirms the wisdom of separating diatoms from other plankters. It provides an evaluation of several of the available formulaic methods for estimating the carbon contents of various diatoms.

Of the other elements comprising biomass, nitrogen accounts for some 4–9% of the ash-free dry mass of freshwater phytoplankters, depending on growth conditions (Ketchum and Redfield,

collectively that the quantities of the components vary within generally consistent limits and, though they do fluctuate, the ratios with other constituents do not vary by more than can be reasonably explained in these terms.

For instance, carbon generally makes up about half the dry organic mass of organic cells. The normal content of phytoplankton strains cultured under ideal laboratory conditions of constant saturating illumination, constant temperature and an adequate supply of all nutrients, was found to be 51–56% of the ash-free

Table 1.5 | The silicon content of some planktic diatoms relative to cell volume and surface area

Species	Si (pg cell^{-1})	v (μm^3)	s (μm^2)	Si (pg μm^{-3})	Si (pg μm^{-2})	References[a]
Asterionella formosa	61.0	630	860	0.097	0.071	Reynolds and Wiseman (1982)
Fragilaria crotonensis	117.8	780	1080	0.151	0.109	Reynolds and Wiseman (1982)
Stephanodiscus rotula	1075	8600	2574	0.125	0.418	Gibson et al. (1971)
	1942	15980	4390	0.122	0.442	Thitherto unpublished records cited in Reynolds (1984a), calculated indirectly from SiO$_2$ uptake
	978	8300	2580	0.118	0.379	Thitherto unpublished records cited in Reynolds (1984a), calculated indirectly from SiO$_2$ uptake
	751	5930	1980	0.127	0.379	Thitherto unpublished records cited in Reynolds (1984a), calculated indirectly from SiO$_2$ uptake
Stephanodiscus hantzschii	16.4	600	404	0.027	0.041	Thitherto unpublished records cited in Reynolds (1984a), calculated indirectly from SiO$_2$ uptake

[a] All citations converted from the original published data quoted content in terms of SiO$_2$, by multiplying by 0.4693.

1949; Lund, 1965, 1970; Round, 1965). Maximum growth rates are sustained by cells containing nitrogen equivalent to some 7–8.5% of ash-free dry mass. Among freshwater algae, at least, the phosphorus content is yet more variable, although again, maximum growth rate is attained in cells containing phosphorus equivalent to around 1–1.2% of ash-free dry mass (Lund, 1965; Round, 1965). Growth is undoubtedly possible at rather lower cell concentrations than this but further cell divisions cannot be sustained when the internal phosphorus content is too small to divide among daughters and cannot be replaced by uptake. This concept of a *minimum cell quota* (Droop, 1973) has been much used in the understanding the dynamics of nutrient limitation and algal growth: for phytoplankton, the threshold minimum seems to fall within the range 0.2–0.4% of ash-free dry mass. The investigation of Mackereth (1953) of the phosphorus contents of the diatom *Asterionella formosa*, which reported a range of 0.06 to 1.42 pg P per cell, is much cited to illustrate how low the cell quota may fall. The lower value, which is, incidentally, corroborated by data in earlier works (Rodhe, 1948; Lund, 1950), corresponds to $\sim$0.03% of ash-free dry mass. On the other hand, cell phosphorus quotas may be considerably higher than the minimum (certainly up to 3% of ash-free dry mass is possible: Reynolds, 1992a), especially when uptake rates exceed those of deployment and cells retain more than their immediate needs (so-called *luxury uptake*). Uptake and retention of phosphorus when carbon or nitrogen supplies are limiting uptake (cell C or N quotas low) may also result in high quotas of cell phosphorus.

Analogous arguments apply to the minimal quota of all the other cell components. However, it is the variability in the carbon, nitrogen and phosphorus contents that is most used by plankton ecologists to determine the physiological state of phytoplankton. Taking the ideal quotas relative to the ash-free dry mass of healthy, growing cells as being 50% carbon, 8.5% nitrogen and 1.2% phosphorus, these elements occur in the approximate mutual relation 41C : 7N : 1P (note, C : N ~6). Division by the respective atomic weights of the elements (~12, 14, 31) and normalising to phosphorus yields a defining molecular ratio for healthy biomass, 106C : 16 N : 1P.

This ratio set is well known and is generally referred to as the *Redfield ratio*. As a young marine scientist, A. C. Redfield had noted that the composition of particulate matter in the sea was stable and uniform in a statistical sense (Redfield, 1934) and, as he later made clear, 'reflected . . . the chemistry of the water from which materials are withdrawn and to which they are returned' (Redfield, 1958). The notion of a constant chemical condition was clearly intended to apply on a geochemical scale but the less-quoted investigations of Fleming (1940) and Corner and Davies (1971) confirm the generality of the ratio to living plankton.

It is, of course, very close to the approximate ratio in which the same elements occur in the protoplasm of growing bacteria, higher plants and animals (Margalef, 1997). Stumm and Morgan (1981; see also 1996) extended the ideal stoichiometric representation of protoplasmic composition to the other major components (those comprising >1% of ash-free dry mass – hydrogen, oxygen and sulphur) or some of those that frequently limit phytoplankton growth in nature (silicon, iron). The top row of Table 1.6 shows the information by atoms and the second by mass, both relative to P. The third line is recalculated from the second but related to sulphur. Unlike carbon, nitrogen or phosphorus, sulphur is usually superabundant relative to phytoplankton requirements and plankters have no special sulphur-storage facility. Following Cuhel and Lean (1987a, b), sulphur is a far more stable base reference and deserving of wider use than it receives. Unfortunately, few studies have adopted the recommendation. The fourth line relates the Redfield ratio to the base of carbon (= 100, for convenience), while further entries give elemental ratios for specific algae, reported in the literature but cast relative to carbon. *Chlorella* is a freshwater chlorophyte and *Asterionella* (*formosa*) is a freshwater diatom, having a siliceous frustule. The 'peridinians' are marine. Approximations of the order of typical elemental concentrations in lake water are included for reference. They are sufficiently coarse to pass as being applicable to the seas as well. The important point is that plankters are faced with the problem of gathering some of these essential components from extremely dilute and often vulnerable sources.

As applied to phytoplankton, the Redfield ratio is not diagnostic but an approximation to a normal ideal. However, departures are real enough and they give a strong indication that the cell is deficient in one of the three components. Extreme molecular ratios of 1300 C : P and 115 N : P in cells of the marine haptophyte, *Pavlova lutheri*, cultured to phosphorus exhaustion, and of 35 C : P and 5 N : P in nitrogen-deficient strains of the chlorophyte *Dunaliella* (from Goldman et al., 1979), illustrate the range and sensitivity of the C : N : P relationship to nutrient limitation.

Because the normal (Redfield) ratio is indicative of the health and vigour that underpin rapid cell growth and replication, and given that departures from the normal ratio •result from the exacting conditions of specific nutrient deficiencies, it is tempting to suppose that cells to which the normal ratios apply are not so constrained and must therefore be growing rapidly (Goldman, 1980). It would follow that, given the stability of the ratio in the sea, natural populations having close-to-Redfield composition are not only not nutrient-limited but may be growing at maximal rates. This may be sometimes true but there is a possibility that biomass production in oceanic phytoplankton is less constrained by N or P than was once thought (see Chapter 4). However, there are other constraints on growth rate and upon nutrient assimilation into new biomass, which may tend to uncouple growth rate from nutrient uptake rate (see Chapter 5). Tett *et al.* (1985) provided examples – of phytoplankton in continuous culture, of natural populations of Cyanobacteria

Table 1.6 Ideal chemical composition of phytoplankton tissue and relative abundance of major components by mass

	C	H	O	N	P	S	Si	Fe	References
Redfield atomic ratio (atomic stoichiometry rel to P)	106	263	110	16	1	0.7	trace	0.05	Stumm and Morgan (1981)
Redfield ratio by mass (stoichiometry rel to P)	41	8.5	57	7	1	0.7	trace	0.1	Stumm and Morgan (1981)
Redfield ratio by mass (stoichiometry rel to S)	60	12	81	10	1.4	1			Stumm and Morgan (1981)
Redfield ratio by mass (stoichiometry rel to C)	100			16.6	2.4				Stumm and Morgan (1981)
Chlorella (dry weight rel to C)	100			15	2.5	1.6	trace		Round (1965)
Peridinians (dry weight rel to C)	100			13.8	1.7		6.6	3.4	Sverdrup et al. (1942)
Asterionella (dry weight rel to C)	100			14	1.7		76		Lund (1965)
Medium (mol L^{-1})	10^{-3}	10^{2}	10^{2}	10^{-4}	10^{-6}	10^{-3}	10^{-2}	$<10^{-5}$	Author's approximation but omitting dissolved nitrogen gas

stratified deep in the light gradient, and of spring blooms of the diatom *Skeletonema costatum* in a Scottish sea loch – where growth rates were kept very low but the cell contents of carbon, nitrogen and phosphorus stayed close to the Redfield ideal in each instance. It is even possible that natural cells do not drift as far from the ideal as it possible to force them under laboratory conditions. How cells balance availability, uptake, storage and self-replication over the period of a single generation will be explored in Chapter 4.

1.5.4 Chlorophyll content

Besides being a distinguishing constituent of phytoplankton and having a universal distribution among all the photoautotrophic algae and cyanobacteria, the photosynthetic pigment chlorophyll *a* is also widely used as a convenient index of phytoplankton biomass. This makes its contribution to cellular composition extremely important to extrapolating to phytoplankton abundance and to its use as a base for estimating phytoplankton productivity. Although there have been many values published alluding to the absolute chlorophyll *a* contents of freshwater phytoplankton (some reviewed in Reynolds, 1984a), these quantities are now known to be so variable that they have little value by themselves. The variability is most obviously linked to the cell's requirement for carbon and the light energy available to drive its fixation. In broad

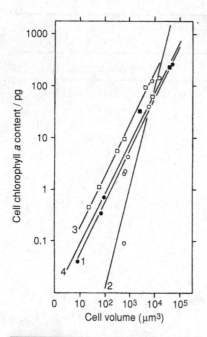

Figure 1.10 Log/log plot of cell chlorophyll-a content against cell volume of various freshwater phytoplankters (data in Table 1.3): ●, Cyanobacteria; ○, diatoms; □, chlorophytes; ■, others). Regression equations are fitted to the data for Cyanobacteria (1, log chl = 1.00 log v – 2.26), diatoms (2, log chl = 1.45 log v – 3.77) and chlorophytes (3, log chl = 0.88 log v – 1.51) and to *all* points (4, log chl = 0.98 log v – 2.07). Redrawn from Reynolds (1984a).

dry weight) between its deep-stratified and free-mixed phases (Reynolds 1997a).

On the other hand and analogously with Redfield stoichiometry, the probabilistic relationships derived from mixed populations over periods of time point to 0.003–0.007 pg μm^{-3}, or roughly 0.013–0.031 pg chl (pg cell C)$^{-1}$, or 0.7–1.6% of the ash-free dry mass as each being typical. For many purposes, the common approximations that chlorophyll accounts for 1% of the dry mass and about 2% of the value of the cell carbon quota are not at all unreasonable average estimates. Indeed, field chlorophyll measurements are commonly converted to approximate producer biomass, expressed as active cell carbon by the application of a ratio 50:1 by weight. A margin of variation, from 30:1 to 70:1, should nevertheless be allowed. It should be borne in mind too that the amount of chlorophyll *a* is proportionate to cell volume and not cell number, bigger cells carrying proportionately more chlorophyll than small ones, and that there may be systematic interphyletic differences in the typical cell-specific contents. The plot of data taken from Reynolds (1984a) emphasises both points (Fig. 1.10).

terms, the weaker is the photon flux, and the greater is the probability of limitation of growth rate by light, then the greater is the need for the light-harvesting centres of which the chlorophyll is an essential component. Where appropriate, this behaviour may be accompanied by the production of additional quotas of accessory pigments (for a fuller discussion, see Chapter 3). Synthesis of chlorophyll *a* is also sensitive to nutrient supply and deployment, directly or as a consequence of altered internal resource allocation. The measurements of biomass-specific estimates of chlorophyll *a* presented in Reynolds (1984a) range over an order of magnitude, between 0.0015 and 0.0197 pg μm^{-3} of live cell volume, which corresponds to 3 to 39 mg g^{-1} of dry mass (0.3% to 3.9%). The true range is probably wider: later data showed that the chlorophyll-*a* content of just one species of cyanobacterium, *Planktothrix* cf. *mougeotii*, may vary nearly ninefold (0.45–3.9%

1.6 | Marine phytoplankton

The information on the size, morphology and elemental composition of phytoplankton presented in this chapter has been dominated by relationships detected among freshwater species. This is partly attributable to the interests and experiences of the author and not to any lack of corresponding data for the sea; useful data compilations are to be found in, for instance, Mullin *et al.* (1966), Strathmann (1967), Sournia (1978) and Verity *et al.* (1992). However, it seemed more interesting to take the data and derivations of Montagnes *et al.* (1994), corrected for live cell volumes as opposed to those of material shrunk by preservatives, and to compare their findings with the patterns presently discerned among freshwater species.

Montagnes *et al.* (1994) compiled a thorough compilation of the dimensions and volumes of

Table 1.7 Some cell measurements of cultured marine phytoplankton

Species	shape[a]	MLD (μm)	v (μm³)	C (pg cell⁻¹)	N (pg cell⁻¹)	Chla
Prasinophytes						
Micromonas pusilla	ps	2	2	0.8	0.1	0.01
Mantoniella squamata	sph	4	25	3.6	0.6	0.15
Chlorophytes						
Nannochloris ocelata	sph	3	4	1.4	0.2	0.05
Dunaliella tertiolecta	sph	8	201	41.7	7.9	1.8
Chlamydomonas sp.	sph	19	3 300	969.7	129.6	11.3
Diatoms						
Thalassiosira pseudonana	cyl	4	20	5.9	0.94	0.2
Thalassiosira weissflogii	syl	19	286	64.4	11.1	1.5
Detonula pumila	cyl	36	4 697	355.2	61.4	7.4
Chrysophytes						
Pelagococcus sp.	sph	5	18	2.8	0.7	0.1
Pseudopedinella pyriformis	ps	9	80	18.0	2.9	0.7
Apedinella spinifera	sph	10	222	47.6	9.9	2.3
Haptophytes						
Emiliana huxleyi	sph	5	25	4.6	1.0	0.1
Pavlova lutheri	ps	6	25	8.5	1.2	0.2
Isochrysis galbana	sph	6	38	7.0	1.2	0.2
Phaeocystis pouchetii	ps	6	45	5.5	1.3	0.1
Chrysochromulina herdlensis	sph	7	74	8.2	1.9	0.2
Prymnesium parvum	ps	8	79	15.1	2.0	0.3
Coccolithus pelagicus	sph	12	620	65.5	14.5	1.9
Cryptophytes						
Rhodomonas lens	ps	10	203	40.7	11.4	0.7
Chroomonas salina	ps	11	167	32.4	7.9	0.9
Pyrenomonas salina	ps	12	181	32.4	7.0	1.4
Cryptomonas profunda	ps	17	765	104.7	21.0	2.6
Dinophytes						
Gymnodinium simplex	ps	11	224	38.0	8.3	0.7
Gymnodinium vitiligo	ps	14	683	113.8	22.6	2.3
Gyrodinium uncatenatum	ps	32	11 246	2275.3	441.0	37.3
Gymnodinium sanguineum	ps	56	31 761	2913.2	688.4	57.4
Karenia mikimotoi	ps	24	3 399	513.9	93.8	12.9
Raphidophytes						
Heterosigma carterae	ps	16	362	87.8	17.7	3.8

[a] Shapes: cyl, cylindrical; ps, prolate spheroid; sph, spherical.
Source: After Montagnes et al. (1994).

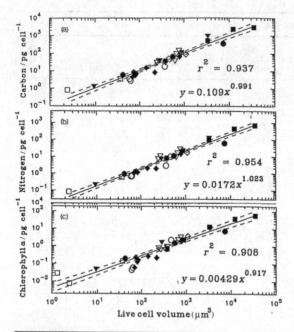

Figure 1.11 Log/log relationships between (a) cell carbon, (b) cell nitrogen and (c) chlorophyll-*a* content and cell volume in the marine phytoplankters listed in Table 1.7. Redrawn, with permission, from Montagnes *et al.* (1994).

30 or so species of phytoplankton, representative of various phyla and covering a good range of cell sizes, together with measured cell quotas of carbon, nitrogen, protein and chlorophyll *a*. The material reproduced in Table 1.7 represents but a fragment of the original. Live cell volumes cover a similar series of magnitudes as the freshwater species listed in Table 1.2. The carbon content of cells varied between 0.08 and 0.4 pg μm^{-3}, with a mean value of 0.20. The ratio of carbon-to-nitrogen varied between 3.6 and 7.6 by mass, with a mean of 5.43. Chlorophyll *a* fell within the range 0.001–0.009 pg μm^{-3}.

Some log/log relationships from the work of Montagnes *et al.* (1994) are plotted in Fig. 1.11. Carbon, nitrogen and chlorophyll are each closely correlated to live cell volume and in a way which is similar to the corresponding relationships among the freshwater phytoplankton. Thus (and, again, as is true for the freshwater phytoplankton), despite a remarkable diversity of phylogeny and morphology, as well as a 5-orders-of-magnitude range of cell volumes, there is an equally striking pattern of cell composition

and a statistically predictable pattern of the relative quantities of the various constituents. This simple fact contributes to the fascination for students of phytoplankton ecology as the subject embraces the observable wonder of the seasonal replacement of one dominant among many species by another among others, as well as the opportunity to express the dynamics of production and attrition and of population wax and wane in empirical terms interlinked by powerful and predictable statistical relationships.

The scene is set for the subsequent chapters.

1.7 | Summary

The chapter provides an introduction to phytoplankton. The phytoplankton is defined as a collective of photosynthetic microorganisms, adapted to live partly or continuously in the open of the seas, of lakes (including reservoirs), ponds and river waters, where they contribute part or most of the organic carbon available to pelagic food webs. Although their taxonomy is currently undergoing major revision and even the phylogenies are questioned, it is difficult to be categorical about the species representation and phyletic make-up of phytoplankton. It is reasonable to point to the description of some 4000 to 5000 species from the sea and, probably, a similar order of species from inland waters. The species belong to what appear to be 14 legitimate phyla, coming from both bacterial and eukaryotic protist domains. In both the marine and the freshwater phytoplankton, there is a wide diversity of size, morphology, colony formation. Though generally microscopic, phytoplankton covers a range of organism sizes comparable to that spanning forest trees and the herbs that grow at their bases.

The early history of phytoplankton studies is recapped. Although a knowledge of some of the organisms goes back to the invention of the microscope, and many genera were well known to nineteenth-century microscopists, their role in supporting the aquatic food webs of open water, culminating in commercially exploitable fish populations, was not realised until the 1870s. The early work by Müller, Haeckel and Hensen (who invented the name 'plankton') is briefly described. Some of the terms used in plankton science are noted with their meanings, while those that appear still to be conceptually useful are singled out for retention.

Despite variation of several orders of magnitude in the sizes of plankters, there is a powerful trend towards conservatism of the surface-to-volume ratio, which is achieved through distortion and departure from the spherical form among the larger species. This aids exchange of gases, nutrients and other solutes across the cell surface and it also has some role in prolongation of suspension. In an apparently diametrically opposite trend, some algae form mucilaginous coenobia that have very low surface-to-volume ratios. When it is combined with some other power of motility, the streamlining effect allows the colony to move relatively quickly through water and to move to a more favourable position in the water column.

The construction and composition of plankton are critically reviewed. Apart from a variety of scales, exoskeleta, plastid type and pigment composition, the ultrastructural components and architecture of the living protoplasm are comparable among the phytoplankton. Similarly, the elemental make-up of the protoplast is similar among all groups of phytoplankton, ideally occurring in approximately stable relative proportions. Discounting the ash from the

mineral-reinforced walls, carbon accounts for about 50% of the dry mass, nitrogen about 8–9% and phophorus between 1% and 1.5%. Relative to phosphorus, these amounts correspond to a probabilistic atomic ratio of 106 C : 16 N : 1 P, close to the so-called Redfield ratio for particulate matter in the ocean. It is also similar to the composition of most living protoplasm. The amounts are related also to hydrogen, oxygen, silicon, sulphur and iron. Up to 12 other elements are regularly present in phytoplankton in trace proportions. Departures from the ratio are rarely systematic, merely indicative of one of the highly variable components falling to the minimum cell quota.

The amount of chlorophyll a is also highly variable according to growth conditions but nevertheless tends to average about 1% of the ash-free dry mass of the cell and to represent about 2% of the elemental carbon. A carbon:chlorophyll value of 50 : 1 is considered typical but it may vary routinely between about 70 : 1 (cells in high light) to 30 : 1 or lower (in cells exposed to consistently low light).

Despite the extreme diversity of phylogeny, morphology and size, both the marine and the freshwater phytoplankton are characterised by a striking and statistically predictable blend of elemental constituents. This proves very helpful in quantifying production and attrition processes contributing to the dynamics of natural, functioning assemblages of plankton.

Chapter 2

Entrainment and distribution in the pelagic

2.1 | Introduction

The aims of this chapter are to develop an appreciation of the adaptive requirements of phytoplankton for pelagic life and to demonstrate the consequences of its embedding in the movements of the suspending water mass. The exploration begins by dismissing the simplistic notion that the essential requirement of plankton is to prevent or minimise the rate of sinking, in the sense that this will prolong its residence in the upper part of the water column. This would be a clear nonsense, were there no counteractive mechanism to ensure that organisms start out at the top of the water in the first instance. Moreover, slow sinking from the upper layers is of illusory respite if the downward passage to depths beyond the adequacy of penetrating light, whether that is 50 cm or 50 m beneath the water surface, is inevitable, unless there is some mechanism for the organism's return. Manifestly, it is not enough just to reduce the rate of irreversible sinking to qualify as a phytoplankter.

Prolonged residence in the upper insolated layers of the open water of lakes and seas (the photic zone) is, without doubt, a primary requirement of the individual phytoplankter, if it is to synthesise sufficient organic carbon to build the tissue of the next generation. The survival of the genetic stock and the seed population capable of providing the base of subsequent generations may also depend upon the survival of a relatively small number of extant individuals. It is not a condition either for the individual or for the persistence of the clone that residence is continuous, only that individuals of any given generation spend sufficient of their life in the photic zone to make the net autotrophic gains in synthesised carbon, over the burden of respiration, to be able to sustain the next cell replication. The point here is that the essential adaptation is to maximise the exposure to adequate light, by any appropriate mechanism.

The mechanisms for this are not self-evident, unless the behaviour of the water itself is taken into account. For this is the feature that the classical explanation of phytoplankton adaptations rather omits – that, at every scale, the water is never a passive component. Under the influence of its warming and cooling, of the influence of gravity, the pull of the Moon, of the work of wind and even of the rotation of the Earth, water is in motion. Some of these inputs are continuously variable, and their various interactions with the internal viscous forces contribute to a spectrum of motion that is characteristically variable, in both time and space.

Thus, there is an explicit, inescapable and variable velocity component to the medium. Moreover, the movement is, almost always, turbulent, so that flow tends to be in billowing eddies rather than along direct trajectories. Such movements are capable of alternately enhancing or counteracting the intrinsic velocity of the vertical tendency of the settling planker and, within the finite bounds of the water mass, may force its lateral displacement or even push it upwards.

These possibilities are the basis of the principle of entrainment of phytoplankton in

Table 2.1 | Comparison of the physical properties of air, pure water and sea water

	Air	Pure water	Sea water
Maximum density, ρ_w (kg m^{-3})	1.2	1×10^3	$\sim 1.03 \times 10^3$
Absolute viscosity, η (kg m^{-1} s^{-1})	1.8×10^{-5}	1×10^{-3}	$\sim 1.1 \times 10^{-3}$

the motion of natural water bodies. Empirical description of entrainment relies on the analogous relativity between the intrinsic sinking velocity of the plankter (w_s) and the turbulent velocity of the motion (u^*) This immediately introduces an anomaly, which must be addressed at once. The idea is that the smaller is w_s relative to u^* then the more complete is the entrainment of the particle in the motion. This is another way of saying that the best way to ensure entrainment is to minimise the rate of sinking. Isn't this just the idea that was so summarily dismissed in the first paragraph? No, for the comment was directed to the redundant, notional context of a slow settlement of plankters through a static water column. What is really needed, if full entrainment is the goal, is a slow rate of settlement relative to the water immediately adjacent to the organism and its instantaneous trajectory and velocity.

The approach adopted in developing this chapter is to first consider the nature, scale and variability of water movements and the estimation of u^*. Then the question of settling velocities, buoyant velocities and swimming rates (w_s) is reviewed, before the consequences on spatial and temporal distributions are considered at the end.

2.2 | Motion in aquatic environments

The aquatic environment is the greatest habitat to be continuously exploited by organisms. Liquid water presently covers about 71% of planet Earth, the sea alone occupying 361.3×10^6 km^2 of it. The estimated volume of the sea ($\sim 1\,350\,000\,000$ km^3) accounts for 97.4% of all the water on the planet. Taking off the volume stored in the polar ice caps (27.8×10^6 km^3) and the amount stored in the ground ($\sim 8 \times 10^6$ km^3), the balance, shared between lakes, rivers and the atmosphere, is less than 0.02%. However, even this fraction, totalling c. 225 000 km^3, is overwhelmingly dominated by the volume of standing inland waters: the 13 largest lakes in the world (by volume) alone hold 160 000 km^3 (Herdendorf, 1990). At any moment of time, most of this volume is actually so inhospitable to primary producers that it is not conducive to phytoplankton development but, because it is fluid and in persistent motion, all the volume is potentially available, sooner or later. The global rate of the hydrological renewal, in the cycle of precipitation, flow and evaporation, results in an estimated annual loss from the ocean of 353 000 km^3, made good by direct precipitation (roughly, 324 000 km^3) and net river run-off from the land masses ($\sim 29\,000$ km^3). The theoretical replacement time for the ocean is thus around 3800 years.

2.2.1 Physical properties of water
How this vast and enduring body of water reacts to the forces placed upon it is related to the somewhat anomalous physical properties of water itself. Given its low molecular weight (18 daltons), water is a relatively dense, viscous and barely compressible fluid (see Table 2.1 for reference), with relatively high melting and boiling points. This behaviour is due to the asymmetry of the water molecule and to the fact that the two hydrogen atoms, each sharing its electron with the oxygen atom, are held at a relatively narrow angle on one side of the molecule. In turn, this gives a polarity to the molecule, one side (the 'hydrogen side') having a net positive charge and the other (the 'oxygen side') a net negative one. The molecules then have a mutual attraction, giving rise to the formation of *aquo polymers*. It is the complexation into larger molecules which raises the melting point of what is otherwise a low-molecular-weight compound into

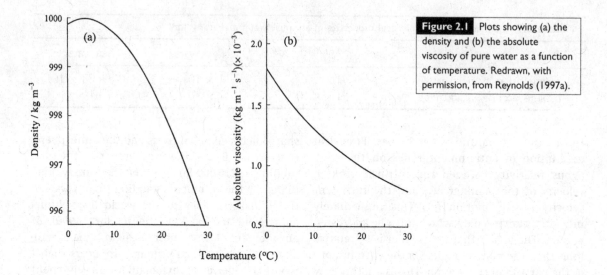

Figure 2.1 Plots showing (a) the density and (b) the absolute viscosity of pure water as a function of temperature. Redrawn, with permission, from Reynolds (1997a).

the range we perceive as being normal. As temperature is raised and the motion of molecules is increased, individual molecules break from the complexes. In most liquids, the molecules come to occupy more space, that is the liquid expands and the density decreases. In water, this effect is countered by the fact that the liberated molecules fall within the complexes, so that the same number of molecules occupies *less* space, leading to *increased* density. In pure water, the latter effect dominates up to 3.98 °C; above this temperature, the separation of molecules becomes, progressively, the dominant effect and the liquid expands accordingly (see Fig. 2.1).

The molecular behaviour explains not only why fresh water achieves its greatest density at close to 4 °C but also why, under appropriate conditions, ice forms at the surface of a lake (where, incidentally, it insulates the deeper water against further heat loss to the atmosphere), and why, with every degree step above 4 °C, the *difference* in density also becomes greater. Limnologists are well aware of the effect this has in enhancing the mechanical-energy requirement to mix increasingly warmed surface waters with the dense layers below; the limnetic ecologist is familiar with the impact of both processes on the environments of phytoplankton.

The same principles apply in the sea and in salt lakes, except that the higher concentrations of dissolved ionic salts, their separation into constituent charged ions and their attraction to the opposite-charged poles of the water molecules all contribute further modifications to the polymerisation. Individual ions become surrounded by water molecules in a hydrated layer, disrupting their structure and altering the properties of the liquid. The salinity of sea water ranges from a trace (in some estuaries and adjacent to melting glaciers) to a maximum of about 40 g kg^{-1} (the Red Sea; note that this is greatly exceeded in some inland lakes). In most of the open ocean, salinity is generally about 35 ($\pm$3) g kg^{-1}, having a density of 27 ($\pm$2) kg m^{-3} greater than pure water of the same temperature. The presence of salt depresses not just the freezing point but also the temperature of maximum density. When the salt content is about 25 g kg^{-1}, these temperatures coincide, at -1.3 °C. Thus, in most of the sea, the density of water increases with lowering temperatures right down to freezing. Sea ice does not form at the surface, as does lake ice, simply as a consequence of cooling of the water. Normally, some other component (dilution by rain and or terrestrial run-off) is necessary to decrease the density of the topmost water.

Molecular behaviour influences the temperature-dependent viscous properties of water. Viscosity, manifest as the resistance provided to one water layer to the slippage of another across it, decreases rapidly with rising temperature (Fig. 2.1b); according to the standard definition of viscosity, this means there is a decreasing resistance of one water layer to the slippage of another across it, for the same given difference in temperature. Viscosity is greater in sea water than in pure water: an increment about 0.1×10^{-3} kg m^{-1} s^{-1} applies to water containing 35 g kg^{-1} over a normal temperature range. High viscosity, like large differences in density, is an effective deterrent to physical mixing and mechanical heat transfer.

2.2.2 Generating oceanic circulation

It is also relevant to the generation of major flows that water has a high specific heat, which in essence means that it takes a lot of heating to raise its temperature (4186 J to raise 1 kg by 1 °C, or by 1 kelvin). However, it is just as slow to lose it again, save that evaporation, turning water liquid into vapour, is very consumptive of accumulated heat (2.243×10^6 J kg^{-1}). Nevertheless, the exchange of incoming and outgoing heat is a major component in the physical behaviour of the oceans. In fact, the patterns of motion are subject to a complex of drivers and the outcome is usually complicated. Empirical description of motion can only be probabilistic and, in any case, far beyond the scope of this book. In essence, the energy to drive the circulation comes from the Sun. Because of its relevance also to local variability in heat exchanges and its obvious links to the energy fluxes used in photosynthesis, a deeper consideration is given later to the solar irradiance fluxes. For the moment, it is necessary to accept that the proportion of the solar energy flux that penetrates the atmosphere to heat the surface of the sea or lake is first a function of the *solar constant*. This is the energy income to a notional surface held perpendicular to the solar electromagnetic flux, before there is any reflection, absorption or consumption in the Earth's atmosphere. Confusingly, it is not constant, as the elliptical orbit of the Earth around the Sun varies around the mean distance (149.6×10^6 km) fluctuates during the year within $\pm 2.5 \times 10^6$ km. Besides, the heat radiated from the Sun also fluctuates. Nevertheless, there is a valuable reference ($\sim$1.36 kW m^{-2}) against which the absorption, reflection and backscatter by dust, water vapour and other gases and, especially, clouds can be scaled. Even before those losses are deducted (see Fig. 2.2a), however, the heat flux per unit area

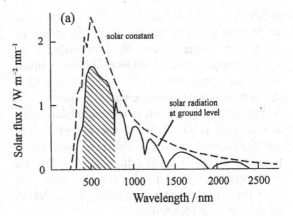

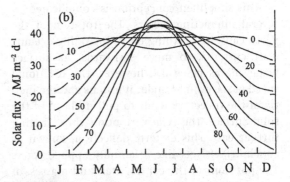

Figure 2.2 (a) The spectrum of the solar flux at ground level compared to that of the 'solar constant' at the top of the atmosphere; the visible wavelengths (light) are shown hatched. (b) Daily integrals of undepleted solar radiation at the top of the atmosphere, shown as a function of latitude (degrees) and time of year in the northern hemisphere. The approximate match for the southern hemisphere is gained by displacing the horizontal scale by 6 months. Redrawn, with permission, from Reynolds (1997a).

diminishes with latitude and not even the combination of the tilt of the Earth's axis and its annual excursion round the sun even this out. The plots in Fig. 2.2b show the annual variation in the undepleted daily flux at each of the selected northern-hemisphere latitudes.

Although the highest potential daily heat flux is everywhere quite similar, sustained heating through the year is always likely to be greatest in the tropics but never for such long diurnal periods as occur at high latitudes in summer. On a rotating but homogeneously water-covered Earth having a continuously clear atmosphere, there would be considerable latitudinal differences in the heat flux directed to the surface (other things being equal, sufficient at the equator to raise the temperature of the top metre by ~1 °C every daylight hour, for 12 months of the year). Ignoring night-time losses and any convectional heat penetration, the expectation is that the now less dense surface water, heated in the tropics, would spread out to the higher latitudes, at least until it had cooled to the temperature of the high-latitude water. In compensation, water must be drawn from the higher latitudes, via a deeper return flow. In this way, we may visualise the initiation of a convectional circulation of hemispheric proportions.

This simplified conception is complicated by several interacting factors. The rotation of the Earth causes everything on it, including oceanic drift currents, to move eastwards. As surface water moves poleward, however, the rotational speed of the ground under it lessens and the inertia of the trajectory tends to pull it ahead of the solid surface, the relative motion thus drifting further east. This easterly deflection, known as *Coriolis' effect*, acts like a laterally applied force. The positions of the continental land masses, of course, obstruct the free development of these motions while the irregularity of their distribution gives rise to compensatory latitudinal flows among the major oceans (especially in the southern hemisphere). The variable depth of the ocean floor also interferes with the passage of deep return currents which, locally, may be forced to deflect upwards and to 'short-circuit' the potential hemispheric circulation.

Also superimposed upon the circulatory pattern are the tidal cycles exerted by the variable gravitational pull on the water exerted by the rotation of the Moon around the Earth, having frequencies of ~25 hours and ~28 days. The effect of tides on the pattern of circulation may not be large in the open ocean but may dominate inshore circulations near blocking landforms that may trap tidal surges (the Bay of Fundy, between Nova Scotia and New Brunswick, experiences the greatest tidal extremes in the world – over 13 m – and some of the most aggressive tidal mixing).

Surface currents, especially in lakes, are proximally influenced by wind. Wind is the motion of air in the adjacent fluid environment, the atmosphere; its movements are subject to analogous global-scale forces. It is its much lower mass, density and viscosity that gives the impression of a different behaviour. In fact, there is close coupling between them, in the sense that strong winds generate waves, drive surface drift currents and force the transfer of some of mechanical energy to the water column. At the same time, differences in inertia and in specific heat bring differential rates of warming over land and water, leading to differences in air pressure and the superimposition of prevalent wind conditions.

The predictability of wind action on individual water bodies is generally difficult (as are most aspects of weather forecasting), save in probabilistic terms, based on the statistics of experience and pattern recognition. However, the linkage between wind effects and the motion of water in which phytoplankton is resident has been deeply explored. Broad flow patterns of surface currents in the oceans have been discerned and described by mariners over a period of centuries and since committed to oceanographers' maps (for an overview, see Fig. 2.3). Patterns of circulation in certain large lakes have been described over a rather shorter period of time (e.g. Mortimer, 1974; Csanady, 1978) and those of many smaller lakes have been added in recent years; the example in Fig. 2.4 is just one such instance.

2.3 | Turbulence

2.3.1 Generation of turbulence

Despite the rather self-evident relationship that plankton mostly goes where the water takes it, the large-scale motion of water bodies tells us frustratingly little about what the conditions of life are like at the spatial scales appropriate to individual species of phytoplankton (generally <2 mm), or about the trajectories followed by the individual phytoplankter whose survival depends on its passing a reasonable fraction of its life in the insolated upper reaches of the water column. Although it has long been appreciated that the energy of the major circulations is dissipated through cascades of smaller and smaller gyratory structures, now called the Kolmogorov eddy

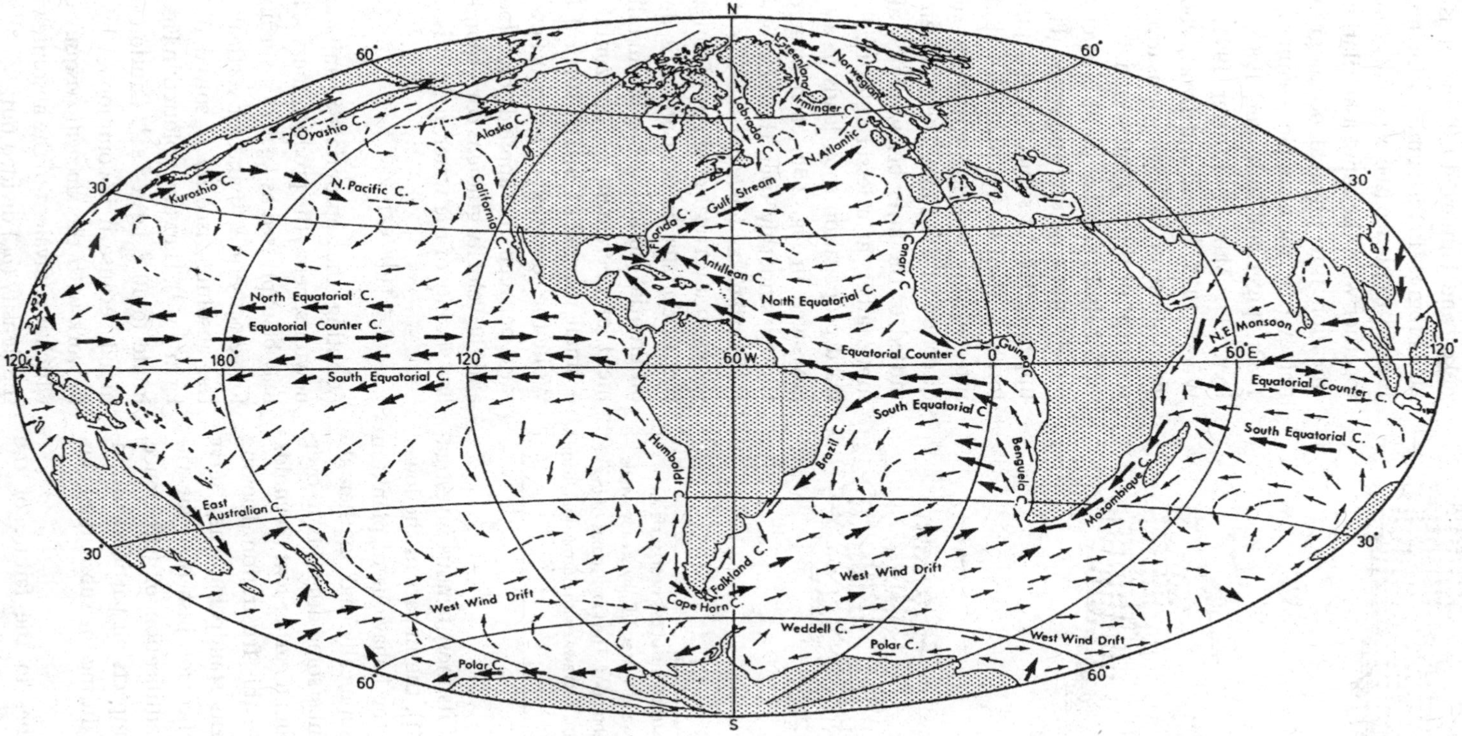

Figure 2.3 The currents at the surface of the world's oceans in the northern winter. Redrawn, with permission, from Harvey (1976).

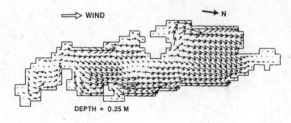

(a) Wind direction = 172°

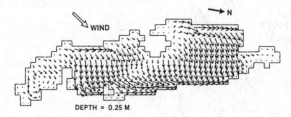

(b) Wind direction = 217°

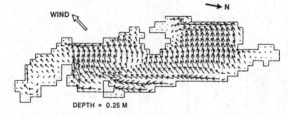

(c) Wind direction = 37°

Figure 2.4 Model reconstructions of the near-surface currents generated in Esthwaite Water, UK, by steady winds of 5 m s^{-1} and various orientations. The longest axis of the lake is approximately 2.2 km. Redrawn, with permission, from Falconer *et al.* (1991).

spectrum after one of its most famous investigators (Kolmogorov, 1941), the impact of the lower end of the series on the behaviour of plankton had remained philosophically and mathematically obscure. My cumbersome attempts to overcome this deficiency in an earlier text (Reynolds, 1984a) only emphasise this frustration. Looking back from the present standpoint, they serve as a point of reference for just how far the appreciation and quantification of turbulence, together with their impacts on particle entrainment, have moved on during the subsequent couple of decades.

'Turbulence' testifies to the failure of the molecular structure of a fluid to accommodate introduced mechanical energy. The mysteries

of turbulence have gradually given way to a developing turbulence theory (Levich, 1962; Tennekes and Lumley, 1972) but it is the remarkable progress in instrumentation and direct sensing of turbulence (see especially Imberger, 1998), together with the rapid assimilation of its quantification into physical oceanography and limnology (see, for example, Denman and Gargett, 1983; Spigel and Imberger, 1987; Imberger and Ivey, 1991; Mann and Lazier, 1991; Imboden and Wüest, 1995; Wüest and Lorke, 2003) that have given most to the characterisation of the environment of the phytoplankton.

A helpful starting point is to envisage a completely static column of water (Fig. 2.5a). If its upper surface is subjected to a mild horizontal force, τ, then water molecules at the air–water interface are dragged across the surface in the direction of the force. Their movement is transmitted to the layer below, which also begins to move, albeit at a lesser velocity. Further downward propagation soon leads to a configuration envisaged in Fig. 2.5b, each layer of molecules sliding smoothly over the one below, in what is described as *laminar flow*. The structure conforms to a vertical gradient of horizontal velocities, u, the steepness of which is defined by the differential notation, du/dz (literally the increment or decrement of horizontal velocity for a small increment in the vertical direction, z). While the condition of laminar flow persists, the ratio between the applied force (per unit area) and the velocity gradient corresponds to the *absolute viscosity* of the water, η. That is,

$$\eta = \tau(du/dz)^{-1} \qquad (2.1)$$

Adopting the appropriate SI units for force (N = newtons, being the product, mass × acceleration, may be expressed as kg m s^{-2}) per unit area (m^2), for velocity (m s^{-1}) and for vertical distance (m), the absolute viscosity is solved in poises (P = kg m^{-1} s^{-1}). The values plotted in Fig. 2.1b approximate to 10^{-3} kg m^{-1} s^{-1}. Caution over units is urged because it is common in hydrodynamics to work with the *kinematic viscosity* of a fluid, ν, which is equivalent to the absolute viscosity with the density (ρ_w) divided out:

$$\nu = \eta(\rho_w)^{-1} = \tau(\rho_w \, du/dz)^{-1} \quad \text{m}^2 \text{s}^{-1} \qquad (2.2)$$

(a) 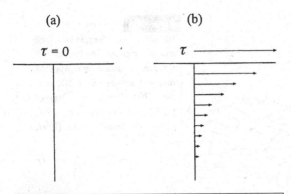 (b) (c)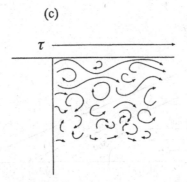

$\tau = 0$ τ ———————→ τ ———————→

Figure 2.5 The generation of turbulence by shear forces. In (a), the water beneath the horizontal surface is unstressed and at rest. In (b), a mild force, τ, is applied which that drags water molecules at the surface in the direction of the force; (b) their movement serves to drag those immediately below, and so on, giving rise to an ordered structure of laminar flow. In (c), the transmitted energy of the intensified force can no longer be dissipated through the velocity gradient which breaks down chaotically into turbulence. Redrawn, with permission, from Reynolds (1997a).

Given the density of water of $\sim 10^3$ kg m^{-3} (Fig. 2.1a), its kinematic viscosity approximates to 10^{-6} m^2 s^{-1}.

If the applied force is now increased sufficiently, it begins to shear molecules from the upper surface of the water column. Thus, the smooth, ordered velocity gradient fails to accommodate the applied energy; the structure breaks up into a complex series of swirling, recoiling eddies (Fig. 2.5c). A new, *turbulent* motion is superimposed upon the original direction of flow. The layer now assumes a net mean velocity in the same direction ($\bar{u}$ m s^{-1}). Now, at any given point within the turbulent flow, there would be detected a series of *velocity fluctuations*, accelerating to ($\bar{u} + u'$) and decelerating to ($\bar{u} - u'$) m s^{-1}. Simultaneously, the displacements in the vertical (z) direction introduce a velocity component which fluctuates between ($0 + w'$) and ($0 - w'$) m s^{-1}. This pattern is maintained for so long as the appropriate level of forcing persists. The driving energy is, as it were, extracted into the largest eddies, is progressively dissipated through smaller and smaller eddies of the Kolmogorov spectrum and is finally discharged as heat, as the smallest eddies are overwhelmed by viscosity.

The transition between ordered and turbulent flow patterns has long been supposed to depend upon the ratio between the driving and viscous forces; this ratio is expressed by the dimensionless Reynolds number, Re:

$$Re = (\rho_w u l_a)\eta^{-1} = u l_a \nu^{-1} \qquad (2.3)$$

where l_a is the length dimension available to the dissipation of the energy, usually the depth of the flow. Turbulence will develop wherever there is a sufficient depth of flow with sufficient horizontal velocity. Solving Eq. (2.3) for a notional small stream travelling at 0.1 m s^{-1} in a channel 0.1 m deep, $Re \approx 10^4$; for a 10-mm layer in a well-established thermocline in a small lake subjected to a horizontal drift of 10 mm s^{-1}, $Re \approx 10^2$. The former is manifestly turbulent but the latter maintains its laminations. There is no unique point at which turbulence develops or subsides; rather there is a transitional range which, for water, is equivalent to Reynolds numbers between 500 and 2000. The depth–velocity dependence of turbulence is sketched in Fig. 2.6.

The point in the spectrum where the eddies are overwhelmed by molecular forces and collapse into viscosity is more difficult to predict without information on their velocities. As suggested above, it is now possible to measure the velocity fluctuations directly with the aid of sophisticated accoustic sensors but the quantities still need to be interpreted within a theoretical context. More significantly, the theoretical framework can be used estimate the intensity of the turbulence from properties of the flow which are measurable with relative ease.

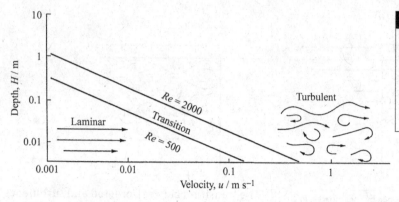

Figure 2.6 The onset of turbulence as a function of velocity and water depth. The boundary is not precise, occurring at Reynolds numbers between $Re = 500$ and $Re = 2000$. Plot based on Reynolds (1992b) and redrawn, with permission, from Reynolds (1997a).

2.3.2 Quantifying turbulence

The key empirical quantification of turbulence is based upon the time-averaged velocity fluctuations in the horizontal and vertical directions ($\pm u'$, $\pm w'$). Because the summation of positive and negative measurements must tend quickly to zero, the positive roots of their squares, $[(\pm u')^2]^{1/2}$, $[(\pm w')^2]^{1/2}$, are cumulated instead. The *turbulent intensity*, $(u^*)^2$, comes from the product of their root mean squares:

$$(u^*)^2 = [(\pm u')^2]^{1/2}[(\pm w')^2]^{1/2} \quad \mathrm{m^2\,s^{-2}} \quad (2.4)$$

The square root (u^*) has the dimensions of velocity and is known variously as the *turbulent velocity*, the *friction velocity* or the *shear velocity*. In natural systems, both (u^*) and $(u^*)^2$ are extremely variable, in time and in space, depending upon the energy of the mechanical forcing and the speed at which it is dissipated through the eddy spectrum. By considering the effects of forcing in contrasted situations, relevant ranges of values for (u^*) may be nominated.

Thus, the simplest model that may be proposed applies to a body of open water, of infinite depth and horizontal expanse and lacking any gradient in temperature. We subject it to a (wind) stress of constant velocity and direction. The momentum transferred across the surface must balance the force applied. So we may propose the following equalities:

$$\tau = \rho_a c_d U^2 = \rho_w (u^*)^2 \quad \mathrm{kg\,m^{-2}\,s^{-2}} \quad (2.5)$$

where ρ_a is the density of air (~ 1.2 kg m^{-3}), U is the wind velocity (properly, measured 10 m above the water surface) and c_d is a dimensionless

coefficient of frictional drag on the surface ($\sim 1.3 \times 10^{-3}$). The equation is imprecise for a number of reasons, one being the interference in transfer caused by surface waves. Nevertheless, the implied linear relationship, $(u^*) \approx U/800$, is sufficiently robust in the wind speed range 5–20 m s^{-1} ($u^* \sim 6 \times 10^{-3}$ to 2.5×10^{-2} m s^{-1}) for it to stand as a good rule-of-thumb quantity.

A general derivation for water flowing down a channel in response to gravity relates (u^*) to $\bar{u}$, as has been developed, *inter alia*, by Smith (1975):

$$(u^*) = \bar{u}[2.5\ln(12H/r_p)]^{-1} \quad \mathrm{m\,s^{-1}} \quad (2.6)$$

where H is the depth of the flow and r_p is the roughness of the bed, as defined by the heights of projections from the bottom. The ratio between them is such that (u^*) is generally 1/30 to 1/10 the value of $\bar{u}$. A general relationship relating (u^*) to channel form is:

$$(u^*) \cdot [g(A_x/p)s_b]^{1/2} \quad \mathrm{m\,s^{-1}} \quad (2.7)$$

where g is gravitational acceleration (in m s^{-2}), s_b is the gradient of the bed (in m m^{-1}), A_x is the cross-sectional area of the channel (in m^2) and p is its wetted perimeter (in m). In wide channels A_x/p approximates to the mean depth, H. Thus,

$$(u^*) \cdot [gHs_b]^{1/2} \quad \mathrm{m\,s^{-1}} \quad (2.8)$$

The turbulent velocity in a river 5 m deep and falling 0.2 m in every km, approximates to $(u^*) = 10^{-2}$ m s^{-1}. Turbulent intensity increases in rivers with increasing relative roughness, with increasing depth and increasing gradient. Theoretical contours of (u^*) are mapped in Fig. 2.7 in terms of gradient and water depth and links them to

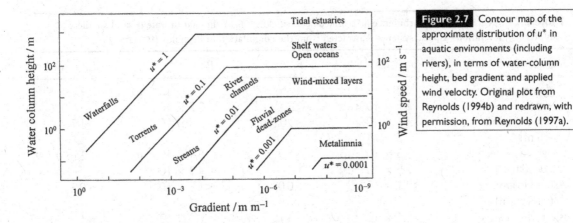

Figure 2.7 Contour map of the approximate distribution of u^* in aquatic environments (including rivers), in terms of water-column height, bed gradient and applied wind velocity. Original plot from Reynolds (1994b) and redrawn, with permission, from Reynolds (1997a).

those driven by surface wind stress: the dog-legs thus represent the 'switch points', where atmospheric forcing overtakes gravitational flow as the main source of turbulent energy in the water. Major aquatic habitats are noted on the map.

2.3.3 Turbulent dissipation

Before proceeding to the comparison of (u^*) with sinking velocities of plankters (u_s), it is helpful to grasp how the turbulent energy runs down through the eddy spectrum and how this, in turn, sets the environmental *grain*. In truly open turbulence, the largest eddies generated should propagate smoothly into smaller ones, having progressively lesser velocities as well as lesser dimensions. Momentum is lost until the residual inertia is finally overcome once again by viscosity and order returns. In the absence of any constraining solid surfaces (shores or bottom) or density gradients, it is possible to envisage a structure in which the largest eddies are adjacent to the source of their mechanical forcing (such as wind stress on the water surface) and a layer of active, propagating turbulence (in this case, from the surface downwards), until the turbulence is finally overwhelmed some distance away (in this case, its lower base). The entire structure might then be regarded as a single boundary layer, separating the energy source from the non-energised water. The mechanical properties of the boundary layer then relate to the sizes of the dimensions of the largest eddies (l_e) and the gradient with which their velocities are dimin-

ished. For the wind-stirred boundary layer, with a vertical velocity gradient, $(\mathrm{d}u/\mathrm{d}z)$,

$$(u^*) \approx l_e(\mathrm{d}u/\mathrm{d}z) \quad \mathrm{m\,s^{-1}} \tag{2.9}$$

Even in the open ocean, the wind-mixed layer rarely extends more than 200–250 m from the surface (Nixon, 1988; Mann and Lazier, 1991). It is clear that so long as the inputs remain steady, the boundary-layer structure serves to dissipate the input of kinetic energy through the spectrum of subsidiary eddies. The *rate* of energy dissipation, E, also turns out to be an important quantity in plankton ecology. Dimensionally, it is equivalent to the product of the turbulent intensity and the velocity gradient. Thus,

$$E = (u^*)^2\,(\mathrm{d}u/\mathrm{d}z) \quad \mathrm{m^2\,s^{-2}\,m\,s^{-1}\,m^{-1}} \tag{2.10}$$

By rearranging Eq. (2.9) for $(\mathrm{d}u/\mathrm{d}z)$ and substituting for it in Eq. (2.10), it follows that:

$$E \approx (u^*)^3 l_e^{-1} \quad \mathrm{m^2\,s^{-3}} \tag{2.11}$$

Where the vertical dimension is constrained, however, either because the basin is considerably less deep than 250 m in depth or because density gradients resist the downward eddy propagation, the smaller surface mixed layer must still dissipate the turbulent kinetic energy within the space available. Were this not so, the motion would have to spill out of the containing structure in, for instance, breaking waves or some rapid erosion of the perimeter shoreline. What happens is that the residual energy reaches into smaller eddies before it is overcome by viscosity. Thus it is that the most relevant feature of the

Table 2.2 Shear velocity of turbulence (u^*), mixed-layer thickness (h_m), dissipation rate (E) and smallest eddy size (l_m) for various kinetic systems. Abstracted from the compilation of Reynolds (1994a).

Site	(u^*) (m s^{-1})	h_m (m)	E (m^2 s^{-3})	l_m (mm)
Bodensee winter winds, 5–20 m s^{-1}	6.2×10^{-3} to 2.5×10^{-2}	44 to 177	1.4×10^{-8} to 2.2×10^{-7}	2.9 to 1.5
summer winds, <8 m s^{-1}	$\leq 1 \times 10^{-3}$	≤ 20	$\leq 1.3 \times 10^{-7}$	≥ 1.7
Lough Neagh winds, 10–20 m s^{-1}	1.3×10^{-2} to 2.5×10^{-2}	8.9	5.4×10^{-7} to 4.3×10^{-6}	1.2 to 0.7
Ashes Hollow (hill stream)	$\leq 3.1 \times 10^{-1}$	0.05	$\leq 1.5 \times 10^{-6}$	~0.9
River Thames (Reading)	$\leq 9.4 \times 10^{-3}$	4.1	5.1×10^{-7}	~1.2
Open ocean	$\leq 3.3 \times 10^{-2}$	≤ 233	$\leq 3.8 \times 10^{-7}$	1.27
Shelf water (Irish Sea)	$\leq 1.2 \times 10^{-1}$	≤ 100	$\leq 4 \times 10^{-5}$	>0.4
Tidal estuary (Severn)				
spring	$\leq 1.3 \times 10^{-1}$	≥ 10	$\leq 5.5 \times 10^{-4}$	≥ 0.20
neap	$\geq 4.3 \times 10^{-2}$	≤ 40	$\geq 5.0 \times 10^{-6}$	≤ 0.67

physical environment of plankton is determined *not* by the intensity of mechanical energy introduced but by the rate of its dissipation and the sizes of the smallest eddies that it can sustain. Simply, the greater is the rate of dissipation, the finer is the structural grain.

In much the same way, we can deduce that the size of the smallest eddies (l_m) in a structure is independent of the forcing but depends only on the rate of energy dissipation (work) per unit mass (E, in J kg^{-1} s^{-1}, which cancels to m^2 s^{-3}; see glossary of units, symbols and abbreviations) and the kinematic viscosity (ν, in m^2 s^{-1}):

$$l_m = (\nu^3/E)^{1/4} \quad \text{m} \qquad (2.12)$$

Solutions of Eq. (2.12) range from the order of millimetres in mixed layers, extending to metres in stratified layers (Spigel and Imberger, 1987). According to the data on well-mixed systems compiled by Reynolds (1994a), some of which are reproduced here as Table 2.2, the smallest eddy sizes calculated to be experienced in oceans and deeper lakes are hardly smaller than 1.3 mm. In rivers and shallow lakes, the smallest eddies may be are only half as large for the same input of kinetic energy. Tidal mixing of estuaries and coastal embayments powers some of the fastest

rates of dissipation, and here the smallest eddies may be in the range 200–400 μm.

2.3.4 Turbulent embedding of phytoplankton

These considerations are directly comparable with the dimensions of phytoplankton (Box 1.2, Tables 1.2, 1.7). Microplanktic algae are smaller, by one or more orders of magnitude, than the smallest eddy sizes in what are arguably among the most aggressively mixed, fastest dissipating turbulence fields that they might inhabit. There are some observations and the evidence of some experiments (Bykovskiy, 1978) that together suggest larger species of phytoplankton do not tolerate eddy diminution and intensified shear implicit in enhanced, fine-grained turbulence fields but are, instead, readily fragmented. It is an unverified hypothesis to argue that phytoplankton have evolved along lines that exploited the viscous range of the aquatic eddy spectrum, rather than to have invested in the mechanical tissue necessary to resist the collapse and fragmentation of larger structures (Reynolds, 1997a).

If the dominant vegetation of pelagic environments is truly selected by its ability to escape the smallest scales of turbulence, then the corollary is that the individual organisms are, in effect, *embedded* deep within the turbulence structures to which the water has frequently to accommodate. This is worth emphasising: planktic algae live most of their lives in an immediate environment that is wholly viscous but, at a slightly larger scale, one that is simultaneously liable to be transported far and rapidly through the turbulence field, and with varying intensity and frequency. The pelagic world of phytoplankton might be analogised to one of little viscous packets being moved rapidly in any of three dimensions. In reality, the packets have no enduring integrity but it is the behaviour of phytoplankton relative to the immediate water *and* to the transport of the water within the mixed layer that determines the suspension and settling characteristics of the whole population.

The consequences of living in a viscous medium have been graphically recounted in a much-celebrated paper by Purcell (1977). For instance, it is not possible for a planktic alga or bacterium to 'swim' through the medium as does (say) a water beetle (3–20 mm), by means of a reciprocating, rowing movement of paddle-like limbs, any more than can a man floundering in a vat of treacle. The alternative options for forward progression that are exploited by microplankters and smaller organisms include the serial deformation of the protoplast (amoeboid movement), the spiral rotation of the body (as do many ciliates and euglenoids) and the rotating of a flagellum like a corkscrew (as in the bacterium *Escherichia*). The speed of self-propulsion relative to the medium (u_s) of (say) a *Chlamydomonas* cell, 10 μm in diameter (d), is, at about 10 μm s^{-1}, trivial in absolute terms though nevertheless impressive in body-lengths covered per second. The Reynolds number of its motion per second, solved by analogy to Eq. (2.3), confirms that the alga moves smoothly through the water, its motion creating no turbulence:

$$Re = (\rho_w u_s d)\eta^{-1}$$
$$\approx 10^{-4} \quad (2.13)$$

The power required to maintain this momentum, $\sim$0.5 W kg^{-1}, is also quite trivial when compared to the power generation of its saturated rate of photosynthesis ($\sim$1.4 kW kg^{-1}). If it stops operating its flagella, however, the alga comes to a complete rest in $\sim$1 μs, having travelled no more than another 10 nm (10^{-8} m) in relation to the adjacent medium (Purcell, 1977).

Embedding is directly relevant to the issues of entrainment and distribution of phytoplankton, insofar as the behaviour of the plankter relative to a body of water in motion is strongly influenced by the behaviour of the plankter within its immediate viscous environment. To progress this exploration requires us to account for buoyancy and gravitation behaviour in relation to suspension.

2.4 | Phytoplankton sinking and floating

The buoyant properties of non-motile plankters, having rigid walls but lacking flagella or cilia, moving through a column of water, in response to gravity ($g = 9.8081$ m s^{-2}), are subject to the same forces that govern the settlement of inert particles in viscous fluids, which were quantified over a century and a half ago (Stokes, 1851). As the body moves, it displaces some of the fluid. Provided the movement of the displaced fluid over the particle is laminar, thus causing no turbulent drag, then its velocity (w_s) is related to its size (diameter, d) and the difference between its density from that of the water (ρ_w). For a spherical particle of uniform density (ρ_c),

$$w_s = gd^2(\rho_c - \rho_w)(18\,\eta)^{-1} \quad \text{m s}^{-1} \quad (2.14)$$

This is the well-known Stokes equation. Note that for a buoyant particle ($\rho_c < \rho_w$), Eq. 2.14 has a negative solution, representing a rate of flotation upwards. An empirical verification by McNown and Malaika (1950), who measured the sinking rates of machined metal shapes in viscous oils, is also frequently cited in the literature on phytoplankton. The Stokes equation is implicitly taken as a valid base for predicting the sinking behaviour of phytoplankton but it is necessary also to test all its assumptions and components if we are to grasp the many mechanisms that distort the relationship when applied to living organisms of other than spherical shapes.

2.4.1 Planktic movement and laminar flow

Let us take the condition of laminar flow. McNown and Malaika (1950) showed good adherence to the Stokes formulation whilst $Re < 0.1$ and that the error was $<10\%$ for $Re < 0.5$. For comparison, Walsby and Reynolds (1980) applied published data for phytoplankton to solve Eq. (2.13) for various phytoplankton, approximating ρ_w as 10^3 kg m^{-3} and η as 10^{-3} kg m^{-1} s^{-1} in each case. For the large marine centric diatom *Coscinodiscus wailseii* ($d \sim 150 \times 10^{-6}$ m), and substituting $w_s = 0.1 \times 10^{-3}$ m s^{-1} for u_s (from Smayda, 1970), Eq. (2.13) was balanced by $Re = 0.015$. Similarly, using measurements from Reynolds (1973a) for a freshwater centric diatom, *Stephanodiscus rotula* ($d \sim 50 \times 10^{-6}$ m, $w_s = 25 \times 10^{-6}$ m s^{-1}), $Re = 0.00125$. The deduction that the movements of most phytoplankton comfortably conform to the laminar flow condition is, however, challenged by very large plankters. According to Smayda's (1970) data, the sinking of the extraordinary *Ethmodiscus rex*, one of the largest known diatoms ($d \sim 1$ mm, $w_s = 6 \times 10^{-3}$ m s^{-1}), generates an $Re \sim 6$. Working with a size range of colonies of the Cyanobacterium *Microcystis*, Reynolds (1987a) showed that the Stokes equation (2.14) predicted velocities well in colonies of known densities up to ($d =$) 200×10^{-6} m in diameter, but in larger colonies (d up to 4 mm), velocities became significantly overpredicted especially when $Re > 1$.

2.4.2 Departure from spherical shape: form resistance

If the laminar-flow condition may thus be assumed to apply to the movements of microphytoplankton, probably at all times, it is not at all clear that the Stokes equation can apply other than to spherical organisms, coenobia or colonies, less than 200 μm in diameter. In fact, for the majority of phytoplankters that are not spherical, the shape distortion has a significant impact on the rate of settling. Distortion from the sphere inevitably results in a greater surface area to an unchanged volume and density and, hence, a greater volume-specific frictional surface,

otherwise known as *form resistance*. The effect of departure from spherical form was clearly demonstrated by the experiments of McNown and Malaika (1950). In most instances (the 'teardrop' shape being a notable exception), subspherical metal shapes sank through oil more slowly than the sphere of the same volume. It is difficult to account quantitatively for these results, as mathematical theory is not so well developed that the effects of distortion can be readily calculated. However, McNown and Malaika (1950) also published their findings on the sinking of spheroids, both oblate (flattened in one axis, like a medicinal pill, or what British readers will understand as 'Smartie'-shaped) or prolate (shortened in two axes, towards the shape of classic airships). These have provided plankton scientists with an important foundation on the impacts of form on algal sinking. Some interesting theoretical or experimental investigations have been pursued using cylinders (Hutchinson, 1967; Komar, 1980), chains of spheres (Davey and Walsby, 1985) and, more recently, some ingenious alga-like shapes fashioned in polyvinyl chloride (PVC) or malleable 'Plasticine' (Padisák et al., 2003a). These have helped to amplify an understanding of the importance of shape in the behaviour of phytoplankton.

In the case of spheroids, the reduction in sinking is related to the ratio of the vertical axis (say, a) to the square root of the product of the other two ($\sqrt{bc}$). The fastest-sinking spheroid is one in which $a \approx 2b$ and $b = c$: the horizontal cross-sectional area is smaller than that of the sphere of the same volume but with most of the volume in the vertical where it offers less drag, the spheroid actually sinks faster than the sphere. As the ratio is increased [$a/(\sqrt{bc})$ to >3], drag increases and velocity falls below that of the equivalent sphere. Analogously, making $a < b$ and $a/(\sqrt{bc}) < 1$, drag increases more than the horizontal cross-sectional area and to >3. Spheroids with the most disparate diameters, that is, the narrower or the flatter they are with respect to the sphere of identical volume and density, offer increased form resistance and up to a twofold reduction in sinking rate.

In the case of cylinders, increasing the length but keeping the cross-sectional diameter constant increases sinking velocity, although this approaches a maximum when length exceeds diameter about five times. Cylinders at this critical length have about the same sinking speed as spheres of diameter 3.5 times the cylinder section (d_c). As a cylinder having a length of $7(d_c)$ has the same volume as such a sphere, we may deduce that cylinders relatively longer than this will sink more slowly than the equivalent sphere, so long as all other Stokesian conditions are fulfilled.

It has been suggested in several earlier studies that distortions in shape have another role in orienting the cell, that (for instance) the cylindrical form makes it turn normal to the direction of sinking and that this may be advantageous in presenting the maximum photosynthetic area to the penetrating light. Walsby and Reynolds (1980) argued strongly for the counterview. In a truly turbulence-free viscous medium, a shape should proceed to sink at any angle at which it is set. There is an exception to this, of course, which will apply if the mass distribution is significantly non-uniform: the 'teardrop' reorientates so that it sinks 'heavy' end first. In the experiments of Padisák et al. (2003a), some of the models of Tetrastrum were made deliberately unstable by providing spines on one side only: these, too, reoriented themselves on release but then remained in the new position throughout the subsequent descent. Their observations on Staurastrum models that go on reorienting recalls the observations of Duthie (1965) on real algae of this genus, which reorientated persistently during descent to the extent that they rotated and veered away from a vertical path. At the time, the influence of convection on the sinking behaviour of Staurastrum could not be certainly excluded. The results of Padisák et al. (2003a) suggest that the form of the cells engenders the behaviour as it reproduced in a viscous medium.

For the present, the impact on sinking of distortions as complex as those of Fragilaria or Pediastrum coenobia requires more prosaic methods of assessment. The most widely followed of these is to calculate a coefficient of form resistance (φ_r) by determining all the variables in the Stokes equation (2.14) for a sphere of equivalent volume (having the diameter, d_s) and comparing the calculated rate of sinking (w_s calc) with the observed rate of sinking by direct measurement, w_s. Thus,

$$\varphi_r = w_s \text{ calc}/w_s \qquad (2.15)$$

The Stokes equation should also be modified in respect of phytoplankton (Eq. 2.16) by including a term for form resistance, accepting that the value of φ_r may be so close to 1 that the estimate provided by Eq. (2.14) would have been acceptable.

$$w_s = g(d_s)^2(\rho_c - \rho_w)(18\,\eta\,\varphi_r)^{-1} \quad \text{m s}^{-1} \quad (2.16)$$

This approach allows systematic variability in the coefficient to be investigated as a feature of phytoplankton morphology. Some of the interesting findings that impinge on the evolutionary ecology of phytoplankton are considered in the next section but it is important first to mention the practical difficulties that have been encountered in estimating form resistance in live phytoplankton and the ways in which they have been solved. As observed elsewhere (Chapter 1) precise estimates of the volume of a plankter (whence d_s is calculable) are difficult to determine if the shape is less than geometrically regular. Making a concentrated suspension and determining the volume of liquid it displaces offers an alternative to careful serial measurements of individuals. Walsby and Xypolyta (1977) gave details of a procedure using ^{14}C-labelled dextran to estimate the unoccupied space in a concentrated suspension. The usefulness of the approach nevertheless depends upon a high uniformity among the organisms under consideration – cultured clones are more promising than wild material in this respect.

The densities of phytoplankton used to be difficult to determine precisely, having to rely on good measurements of mass as well as of volume. Now, it has become relatively easy to set up solution gradients of high-density solutes, introduce the test organisms then centrifuge them until they come to rest at the point of isopycny between the organism and solute (Walsby and Reynolds, 1980). Following Conway and Trainor (1972), Ficoll is frequently selected as an appropriate solute. Being physiologically inert and osmotically inactive improves its utility.

Even to get accurate measurements of sinking rate (w_s) is problematic. Here, the main issue is to be able to keep the water static, when almost any conventional observation system involving uninsulated light sources is beset by generated convection. Techniques have been developed or applied, using some combination of strict thermostatic control, thin observation cells (Wiseman and Reynolds, 1981) or non-heat-generating measuring systems based on fluorometry (see, for instance, Eppley *et al.*, 1967; Tilman and Kilham, 1976; Jaworski *et al.*, 1981). Another approach has been to make measurements in solutions of high viscosity: Davey and Walsby (1985) used glycerol. Alternatively, to measure the rate of loss across a boundary layer from an initially mixed suspension works with convection and yields acceptable results on field-collected material when sophisticated techniques may not be readily available (Reynolds *et al.*, 1986). This actually imitates, in part, the way that plankton settles from natural water columns (see Section 2.6).

Once confidence was gained in the measurement of sinking rates, another, more tantalising source of variability was detected. Several, quite independent investigations of the species-specific sinking rates of diatoms each yielded order-of-magnitude variability. For any given species, the sinking rates seemed least when the cells were healthy and physiologicly active but were as much as three to seven times faster in similar cells that were naturally moribund (Eppley *et al.*, 1967; Smayda, 1970; Reynolds, 1973a), or whose photosynthesis was experimentally inhibited or carbon-limited (Jaworski *et al.*, 1981), or which had been exposed to sublethal doses of algicide (Margalef, 1957; Smayda, 1974), or had been otherwise freshly killed (Wiseman and Reynolds, 1981). However, comparing the fastest rates from each of the studies that had made measurements on comparable material (eight-celled stellate coenobia of the freshwater diatom *Asterionella formosa*), some conformity among the various results became apparent (Jaworski *et al.*, 1988).

Wiseman *et al.* (1983) had previously established that the one sure way to get the consistent, inter-experimental results necessary to be able to investigate the morphological form resistance of

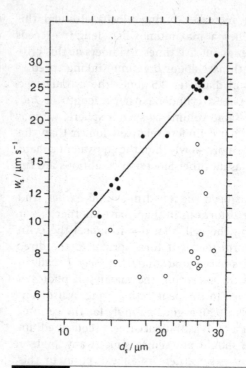

Figure 2.8 Log/log plot of the instantaneous intrinsic settling rates (w_s) of *Stephanodiscus rotula* cells, collected from the field and plotted against mean cell diameter, d_s (○). There is apparently no correlation. However, when corresponding samples are killed by heat prior to determination (●), a strong positive correlation is found. Redrawn from Reynolds (1984a).

phytoplankton was to first kill the diatoms under test. A case in point is shown in Fig. 2.8, where the mean sinking rates of *Stephanodiscus rotula* cells sampled during the increase and decrease phases of a natural lake population seem to vary randomly. However, the corresponding rates of cells killed by dipping in boiling water just before measurement were demonstrably correlated to size. It is now quite generally accepted that live, healthy diatoms have the capacity to *lower* their sinking rates below that of dead or moribund ones. The mechanism of change is not obviously contributed by variability in size or shape or even density. Sinking of live diatoms and, possibly, other algae is plainly influenced by the intervention of further, vital components that must be taken into account in any judgement on how phytoplankton regulate their sinking rates.

To go on now to review some of the analytical investigations into the sinking rates of phytoplankton and the role of form resistance provides the simultaneous opportunity to observe the cumulative influences of the biotic components of the modified Stokes equation (2.16).

2.5 | Adaptive and evolutionary mechanisms for regulating w_s

Of the six variables in the modified Stokes equation, three (g, ρ_w and η) are either constants or are independent variables. The other three (size, density and form resistance) are organismic properties and, as such, are open to adaptation and evolutionary modification through natural selection. It is possible that certain metabolites released into the water by organisms also affect the viscosity of the adjacent medium. The present section looks briefly at the influences of each on sinking behaviour and the extent to which a planktic existence selects for particular adaptive trends.

2.5.1 Size
The relatively small size of planktic algae has been alluded to in Chapter 1. For instance, the diameters of the spheres of equivalent volumes to species named in Table 1.2 cover a range from 1 to 450 μm. It has been suggested that this is itself an adaptive feature for living in fine-grained turbulence. Many species may present rather greater maximal dimensions if (presumably) the rate of turbulent energy dissipation allows. The effect of size on the settling velocities of centric diatoms was demonstrated empirically by Smayda (1970). A striking feature of his regression of the logarithm of velocity w_s on the average cell diameter (d) is that its slope lies closer to 1 than 2, as would have been expected from the Stokes equation (2.14). The regression fitted to the plot of sinking rates of killed *Stephanodiscus* cells against the diameter (in Fig. 2.8) also has a slope of ~1.1. The observations suggest that the larger size (and, hence, the larger internal space) is compensated by a lower overall unit density (cf. Section 1.5.2 and Fig. 1.9). The implication is that more of the overall density of the small diatom is explained

by a relatively greater silicon content. However, it is probable that the effect is enhanced by the fact that a relatively greater part of the internal space of larger diatoms is occupied by cell sap rather than cytoplasm (Walsby and Reynolds, 1980) and that many marine diatoms, at least, are known to be able to vary the sap density relative to that of sea water (Gross and Zeuthen, 1948; Anderson and Sweeney, 1978). This mechanism of density regulation is not available to freshwater algae (see Section 2.5.2) but, either way, density effects may be vital to the suspension of larger diatoms in the sea, if the reduction of sinking rates is the ultimate adaptive aim. According to Smayda's (1970) data, cells of *Coscinodiscus wailesii* should settle at a rate of 40 m d^{-1}, had they the same net density as the much smaller *Cyclotella nana*, rather than the observed 8–9 m d^{-1}.

2.5.2 Density
The cytoplasm of living cells comprises components that are considerably more dense than water (proteins, ~1300; carbohydrates, ~1500; nucleic acids ~1700 kg m^{-3}), so that the average density of live cells is rarely less than ~1050 kg m^{-3}. Inclusions such as polyphosphate bodies (~2500 kg m^{-3}) and exoskeletal structures of calcite and, especially, the opaline silica of diatom frustules (~2600 kg m^{-3}) increase the average density still further. Some of the excess density can be offset by the presence of oils and lipids that are lighter than water, the lightest having a density in the order of 860 kg m^{-3} (Sargent, 1976). However, these oils rarely account for more than 20% of the cell dry mass. Without adaptation, most freshwater phytoplankters are bound to be heavier than the medium and naturally sink! The corresponding deduction in respect of marine phytoplankton suspended in sea water (ρ_w generally ~1030 kg m^{-3}) is made with rather less confidence, where the scope for regulation of ρ_c can be more purposeful.

The list in Table 2.3 is an abbreviated version of one that was included in Reynolds (1984a). It is intended to illustrate the range of densities rather than the comprehensiveness of the data. In particular, it is easy to distinguish the gas-vacuolate Cyanobacteria which, in life, have densities of ≤1000 kg m^{-3} that enable them to float,

Table 2.3 | Some representative determinations of phytoplankton densities

Species	ρ_c (range, in kg m^{-3})	Method[a]	References
Cyanobacteria			
Anabaena flos-aquae	920–1030[b]	1	Reynolds (1987a)
Microcystis aeruginosa (colonies)	985–1005[b]	1	Reynolds (1987a)
Planktothrix agardhii	985–1085[b]	1	Reynolds (1987a)
Chlorophyta			
Chlorococcum sp.	1020–1140	2	Oliver et al. (1981)
Chlorella vulgaris	1088–1102	2	Oliver et al. (1981)
Bacillariophyta			
Stephanodiscus rotula	1078–1104	3	Reynolds (1984a)
Synedra acus (culture)	1092–1138	2	Reynolds (1984a)
Thalassiosira weissflogii	1121	3	Walsby and Xypolyta (1977)
Tabellaria flocculosa	1128–1156	2	Reynolds (1984a)
Asterionella formosa	1151–1215	2	Wiseman et al. (1983)
Fragilaria crotonensis	1183–1209	2	Reynolds (1984a)
Aulacoseira subarctica[c]	1155–1183	2	Reynolds (1984a)
Aulacoseira subarctica[d]	1237–1263	2	Reynolds (1984a)

[a] Methods: (1) calculations based on experimental changes in velocity and gas-vesicle content (see text); (2) deteminations by centrifugation through artificial density gradient to isopycny; (3) gravimetric determinations of mass and volume.
[b] Range covers colonies or filaments with maximum known gas vacuolation to colonies or filaments after subjection to pressure collapse (see text).
[c] Post-auxosporal (wide filaments).
[d] Pre-auxosporal (narrow filaments).
All measurements on wild material, unless otherwise stated.

but which are heavier than fresh water if the vesicles are collapsed by pressure treatment (see below). Two, small unarmoured chlorophytes are included (data of Oliver et al., 1981); how typical they are of non-siliceous algae is not known. The silica-clad diatoms have densities generally ≥1100 kg m^{-3}, although there is a good deal of variability. Interestingly, it seems likely that density varies inversely to internal volume (Asterionella vs. Stephanodiscus, slender pre-auxosporal Aulacoseira vs wide post-auxosporal cells).

Apart from these generalisations, average densities of many planktic algae are plainly influenced by a number of discrete mechanisms. These include lipid accumulation, ionic regulation, mucilage production and, in Cyanobacteria, the regulation of gas-filled space.

Lipid accumulation
Fats and oils normally account for some 2–20% of the ash-free dry mass of phytoplankton cells, perhaps increasing to 40% in some instances of cellular senescence (Smayda, 1970; Fogg and Thake, 1987). Most lipids are lighter than water and, inevitably, their presence counters the normal excess density to some limited extent. Oil accumulation is responsible for the ability of colonies of the green alga Botryococcus to float to the surface in small lakes and at certain times of population senescence (Belcher, 1968). However, it is improbable that oil or lipid storage could reverse the tendency of diatoms to sink. Reynolds (1984a) calculated that were the entire internal volume of an Asterionella cell to be completely filled with the lightest known oil, its overall

density ($\sim$1005 $kg\,m^{-3}$) would still not be enough to make it float. Walsby and Reynolds (1980) concluded that the reduction in density consequential upon intracellular lipid accumulation would, unquestionably, contribute to a reduced rate of sinking but they doubted any primary adaptive significance, neither was there evidence of its use as a buoyancy-regulating mechanism.

Ionic regulation

Inherent differences in the densities of equimolar solutions of organic ions raise the possibility that selective retention of 'light' ions at the expense of heavier ones could enable organisms to lower their overall densities. In a classical paper, Gross and Zeuthen (1948) calculated the density of the cell sap of the marine diatom *Ditylum brightwellii* to be $\sim$1020 $kg\,m^{-3}$, that is, significantly lower than the density of the suspending sea water and actually sufficient to bring overall density of the live diatom close to neutral buoyancy. The density difference between sap and sea water was explicable on the basis of a substantial replacement of the divalent ions (Ca^{2+}, Mg^{2+}) by monovalent ones (Na^+, K^+) with respect to their concentrations in sea water. Some years later, Anderson and Sweeney (1978) were able to follow changes in the ionic composition of cell sap of *Ditylum* cells grown under alternating light–dark periods. They were able to show that density may, indeed, be varied by up to $\pm$15 $kg\,m^{-3}$, through the selective accumulation of sodium or potassium ions, though interestingly, not sufficiently to overcome the net negative buoyancy of the cells. Elsewhere, Kahn and Swift (1978) were investigating the relevance of ionic regulation to the buoyancy of the dinoflagellate *Pyrocystis noctiluca*; they showed that by selective adjustment of the content of Ca^{2+}, Mg^{2+} and $(SO_4)^{2-}$, the alga could become positively buoyant.

The effectiveness of this mechanism is not to be doubted but its generality must be regarded with caution. For it to be effective does depend upon maintaining a relatively large sap volume. The scope of density reduction is limited, in so far as the dominant cations in sea water are the lighter ones and the lowest sap density is the isotonic solution of the lightest ions available. The scope for ionic regulation in phytoplankton of freshwater ($\rho_w < 1002$ $kg\,m^{-3}$) is too narrow to be advantageously exploited.

Mucilage

The mucilaginous investment that is such a common feature of phytoplankton, especially of freshwater Cyanobacteria, chlorophytes and chrysophytes, has long been supposed to function as a buoyancy aid. Again, that mucilage does reduce overall density is, generally, indisputable. Whether this is a primary function is less certain (see Box 6.1, p. 271) and it is mathematically demonstrable that the presence of a mucilaginous sheath does not always reduce sinking rate; in fact it may positively enhance it.

The presence, relative abundance and consistency of mucilage is highly variable among phytoplankton. Mucilages are gels formed of loose networks of hydrophilic polysaccharides which, though of high density ($\sim$1500 $kg\,m^{-3}$) themselves, are able to hold such large volumes of water that their average density (ρ_m) approaches isopycny. Reynolds *et al.* (1981) estimated the density of the mucilage of *Microcystis* to average $\rho_w + 0.7$ $kg\,m^{-3}$. The presence of mucilage cannot make the organism less dense than the suspending water but it can bring the the average density of the cell or colony maintaining it much closer to that of the medium (ρ_c). However, the clear advantage that this might bring to reducing w_s in, for instance, Eq. (2.16) must be set against a compensatory increase in overall size (d_s is increased). Thus, for mucilage to be effective in depressing sinking rate, the density advantage must outweigh the disadvantage of increased size.

This relationship was investigated in detail by Hutchinson (1967). If a spherical cell of density, ρ_c, is enclosed in mucilage of density, ρ_m, such that its overall diameter is increased by a factor a, then its sinking rate will be less than that of the uninvested cell, *provided*:

$$(\rho_c - \rho_m)/(\rho_m - \rho_w) > a\,(a+1) \qquad (2.17)$$

Because a is always >1, the density difference between cell and mucilage must be at least that between mucilage and water. Supposing ρ_c to be $\sim$1016 $kg\,m^{-3}$, ρ_w to be 999 and ρ_m to be 999.7,

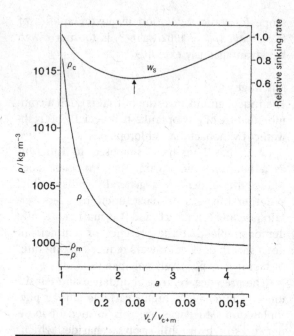

Figure 2.9 The effect of mucilage thickness (a, as a multiple of diameter, d_s) on the average density ($\rho = \rho_c + \rho_m$) of a spherical alga of constant diameter and density ($\rho_c = 1016\,kg\,m^{-3}$). The arrow on the relative velocity plot indicates the point of maximum advantage of mucilage investment in the context of sinking-velocity reduction. Redrawn from Reynolds (1984a).

Eq. (2.17) can be solved as $23.3 > a\,(a+1)$, or that a must be <4.3, if the presence of mucilage is to reduce the sinking rate. The maximum advantage can be solved graphically (see Fig. 2.9); with the nominated values, the greatest advantage occurs when $a \sim 2.3$. Of course, the precise optimal value of a varies from alga to alga, depending partly upon the nature of the mucilage but mainly upon the cell density. For a diatom with a cell density of $1200\,kg\,m^{-3}$, the greatest value of a that would bring a net reduction in its sinking rate could be as high as 16.4, with maximum advantage at $a \sim 8.7$.

In order to compare the mucilage provision among planktic algae, which are not all spherical, it is useful first to express cell volume as a proportion of the total unit volume (v_c/v_{c+m}, as included in Fig. 2.9). Some examples are given in Table 2.4. In each instance, the range of values of a is calculated as the ratio of the diameter of the sphere equivalent to the full unit volume and the diameter of the sphere equivalent to the total volume of cells enclosed. The data presented appear to fall within the range of benefit (in terms of sinking rate), although in the case of *Pseudosphaerocystis* and some *Coenochloris* colonies, the ratios appear to unfavourable, at least if the assumptions about the component densities apply in all these cases.

Gas vacuoles

The surest way of lowering average density is to maintain gas-filled space within the protoplast. This is precisely what some of the planktic Cyanobacteria do and, as is well known, the organisms become buoyant at times, accumulating at the surface as a scum, constituting what was originally called a 'water bloom'. The term 'bloom' has since been applied to almost any planktic population (not even necessarily algal) significantly above the norm: it is another of those words that has been misused to the point of being rendered unhelpful. However, the biology of scum-forming Cyanobacteria is a fascinating topic, and not only because of the almost universal contempt in which most environmental and water-supply managers hold their unsightliness and potential toxicity (Bartram and Chorus, 1999; see also Section 8.3.2). Part of the remarkable account of the functional morphology and population dynamics concerns the ability of these Cyanobacteria to regulate the buoyancy provided by their gas vacuoles (Reynolds *et al.* 1987). Nested within this is the unfolding appreciation of the structure and function of the buoyancy provision itself. Much of the progress over the last 30 years has been spearheaded by A. E. Walsby and his co-workers. Walsby's (1994) review is one of the most comprehensive, and it is this to which the reader is referred for all details.

Here, it is sufficient to emphasise that, from the time their existence was first established (Klebahn, 1895), gas vacuoles have been assumed to have the function of providing buoyancy. Although this may have been neither their original nor their only function (Porter and Jost, 1973, 1976), these uniquely prokaryotic organelles certainly do reduce the average density of the cell in which they occur. They are not bubbles – surface tension is much too powerful to permit

Table 2.4 Relative volumes of cell material (v_c) as a proportion of the full unit volume (v_{c+m}) of named mucilage-producing phytoplankters, expressed in terms of Hutchinson's (1967) factor a (see text)

Plankter	v_c/v_{c+m}	a	References
Microcystis aeruginosa	0.03–0.05	1.59–3.22	Reynolds et al. (1981)
Anabaena circinalis	0.045–0.124	2.00–2.81	Previously unpublished measurements reported in Reynolds (1984a)
Chlamydocapsa planctonica	0.126	1.99	Previously unpublished measurements reported in Reynolds (1984a)
Coenochloris fottii	0.005–0.532	1.24–9.52	Previously unpublished measurements reported in Reynolds (1984a)
Eudorina unicocca	0.055–0.262	1.64–2.63	Previously unpublished measurements reported in Reynolds (1984a)
Pseudosphaerocystis lacustris	0.008–0.013	4.30–4.92	Previously unpublished measurements reported in Reynolds (1984a)
Staurastrum brevispinum	–	2.2–2.3	From direct measurements taken from Fig. 27 of Ruttner (1953)
Fragilaria crotonensis	0.032–0.047	2.78–3.14	From direct measurements of linear dimensions taken from Plates 1c and 1d of Canter and Jaworski (1978), with calculation of a

their existence at the scale of micrometres – but rigid stacks of proteinaceous cylindrical or prismatic envelopes called *gas vesicles* (Bowen and Jensen, 1965). In the Cyanobacteria, they generally measure between 200 and 800 nm in length. The diameters of isolated gas vesicles vary interspecifically between 50 and 120 nm, but are reasonably constant within any given species. Each molecule of the specialised gas-vesicle protein has a hydrophobic end and they are aligned in ribs in the vesicle wall so that the entry of liquid water into the internal space is prevented. However, the vesicle wall is fully permeable to gases and in no sense do the vesicles hold gas under anything but ambient pressure. The gas inside the vesicles is usually dominated by nitrogen with certain metabolic by-products but it is clear that the gas composition is of much less significance than is the gas-filled space, which is created as the vesicle

assembles. The structures are vulnerable to external pressure, including the internal turgor pressure of the cell. They have a certain strength but, once a critical pressure has been exceeded, they collapse by implosion. They cannot be reinflated; they can only be built *de novo*, although the gas-vacuole protein is believed to be recyclable.

The critical pressures of isolated gas vesicles are inversely correlated to their diameters (Hayes and Walsby, 1986). The higher is the critical pressure of the vesicles, the greater the hydrostatic pressure and, thus, the greater water depth they can withstand. Intriguingly, vesicle size, like organism size and shape, co-varies with the principal ecological ranges in which individual species occur and, arguably, the habitats to which they are best adapted. In *Anabaena flos-aquae*, a common scum-forming species in small eutrophic lakes, vesicles measuring about 85 nm in diameter have a critical pressure of 0.3 to 0.7 MPa (Dinsdale and Walsby, 1972). In *Microcystis aeruginosa*, a species sometimes found in larger and more physically variable lakes, vesicles averaging 70 nm in diameter have critical pressures in the order of 0.6–1.1 MPa (Reynolds, 1973b; Thomas and Walsby, 1985). The species of *Planktothrix* (formerly *Oscillatoria*) of deep, glaciated lakes in mountainous regions have vesicles measuring 60–65 nm that withstand pressures of between 0.7 and 1.2 MPa (Walsby and Klemer, 1974; Walsby *et al.*, 1983; Utkilen *et al.*, 1985a). *Anabaena lemmermanni*, a species more usually found distributed in the deep mixed layers of larger temperate lakes, also has much stronger vesicles than most other *Anabaena* species (0.93 MPa: Walsby *et al.*, 1991). Vesicles from the oceanic *Trichodesmium thiebautii* were found to tolerate up to 3.7 MPa (Walsby, 1978).

It is now recognised that gas-vesicle size is subject to very strong selective pressure. Narrow gas vesicles are less efficient at providing buoyancy and, for a given yield of gas-filled space, they are assembled at greater energetic cost. Narrower ones should only be selected if the extra strength is required (Walsby and Bleything, 1988). Now that the genes controlling gas-vesicle assembly can be identified relatively easily (Beard *et al.*, 1999, 2000), the selection by hydrographic events (for instance, incidences of deep mixing of *Planktothrix* populations) for the survival of relatively more of the stronger or relatively more of the weaker kind is one of the most elegant demonstrations of gene-based natural selection to have been contrived (Bright and Walsby, 1999; Davis *et al.*, 2003).

The buoyancy-providing role of the gas vesicles has been studied for over 30 years. By preparing very thick suspensions of Cyanobacteria, placing them in specific-gravity bottles and then subjecting them to pressures sufficient to collapse the vesicles, the volume of gas displaced can be measured very accurately. Expressed relative to the cell volume, the percentage of gas-filled space is readily calculated (Walsby, 1971). Reynolds (1972) used this method to collect the data used to construct Fig. 2.10, which is included to show that there is, for any given alga, a poten-

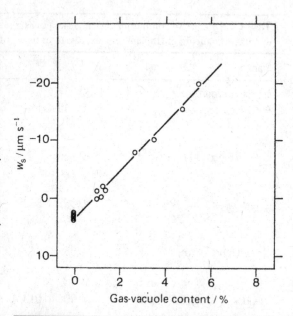

Figure 2.10 The relationship between sinking velocity (w_s) or floating velocity ($-w_s$) of cells of *Anabaena circinalis* and their gas-vacuole content as a proportion of cell volume. After Reynolds (1972) and redrawn from Reynolds (1984a).

tially linear relationship between density (and buoyant velocity) and the gas-vacuole content. It is 'potential' in so far as other items in the complement of cell materials affect the density, and the velocity is sensitive to size and form resistance. Measurements of the gas-vesicle content required to gain neutral buoyancy vary between 0.7% and 2.3% of the cell volume (Reynolds and Walsby, 1975).

To conclude this very brief overview of the buoyancy provision that gas vacuoles impart, it is relevant to the ecology of these organisms to refer to the mechanisms of *buoyancy regulation*. To be continuously buoyant is arguably advantageous in deep, continuously mixed water layers. In small, possibly sheltered lake basins and in larger ones at low latitudes, where the variability in convective mixing is highly responsive to diurnal heat income and net nocturnal heat loss, it is biologically useful to be able to alter or even reverse buoyancy. There are at least three ways in which the planktic Cyanobacteria do this. The relative content of gas vesicles is, in the first instance, the outcome of the balance between their assembly and their collapse. As cells simultaneously grow and divide, gas vesicles will be 'diluted out by growth' unless the cellular resource allocation to their assembly keeps pace. There is plenty of evidence (reviewed in Reynolds, 1987a) that the processes are not closely coupled and that the relative content of vesicles increases during slow (especially light-limited) growth and decreases during rapid growth. It is also apparent that for the species with the strongest vesicles, this is the main mechanism of control and, of course, it operates at the scale of generation times. For the species with weaker vesicles that are vulnerable to collapse in the face of rising turgor generated by low-molecular-weight carbohydrates, there is a rather more responsive mechanism of reversing buoyancy. Cells floating into higher light intensities photosynthesise more rapidly, raise cell turgor, collapse vesicles, lose buoyancy, sink back where the cycle can start again. The cycle of buoyant adjustments can operate on a diel basis and bring about daily migratory cycles over depth ranges of 2–4 m, cells accumulating near the surface by night and at greater depths by the end of the daylight period (Reynolds, 1975; Konopka et al., 1978). Such behaviour may be invoked to explain the diel migratory cycles of Anabaena spp. (reported by Talling, 1957a; Pushkar', 1975; Ganf and Oliver, 1982) and Aphanizomenon (Sirenko et al., 1968; Horne, 1979), or imitated in laboratory mesocosm (Booker et al., 1976; Booker and Walsby, 1981).

The third mechanism can also result in fairly fine control of buoyancy trimming in Cyanobacteria with gas vesicles of intermediate strength, such as those of Microcystis, beyond the scope of the turgor-collapse mechanism. Here, the buoyancy provided by a coarsely variable complement of robust gas vesicles is countered by a finely variable accumulation of photosynthetic polymers (chiefly glycogen) of high molecular weight (Kromkamp and Mur, 1984; Thomas and Walsby, 1985). So long as approximate neutral buoyancy is maintained, relatively small differences in the glycogen content take the average density of the colonies either side of neutral buoyancy, in response to insolation and photosyntetic rate. The large size attained by colonies magnifies the

small differences in density (as predicted in the modified Stokes equation, 2.16), allowing Microcystis colonies to migrate on a diel basis in stable water columns and to recover vertical position very rapidly after disruptive storms. Apart from the first detailed descriptions of this behaviour in shallow, tropical Lake George (Ganf, 1974a), similar adjustments, the ability of Microcystis to attain this control on its buoyancy, is apparent from the studies of Okino (1973), Reynolds (1973b, 1989a), Reynolds et al. (1981) and Okada and Aiba (1983).

No less striking is the formation of persistent plate-like layers in the stable metalimnia of certain relatively deep lakes: the plankters may be almost lacking from the water column but for a band of 1 m or rather less, where they remain poised, often at quite low light intensities. The behaviour has been known for many years from alpine lakes in central Europe (Findenegg, 1947; Thomas, 1949, 1950; Ravera and Vollenweider, 1968; Zimmermann, 1969; Utkilen et al., 1985a) and generally involves the solitary filaments of Planktothrix of the rubescens–prolifica–mougeotii group but it is also known from small, stratifying continental lakes elsewhere (Juday, 1934; Atkin, 1949; Eberley, 1959, 1964, Lund, 1959; Brook et al., 1971, Gorlenko and Kuznetsov, 1972; Walsby and Klemer, 1974) and to involve other genera (Lyngbya, or Planktolyngbya, Spirulina: Reynolds et al., 1983a; Hino et al., 1986). The ability to maintain station is attributable to close regulation of gas-vesicle content but the very low light intensities suggest that this is regulated by allocation. Zimmermann's (1969) study showed that, through the season, P. rubescens moves up and down in the water column of Vierwaldstättersee, mainly in response to changes in the downwelling irradiance. The cells are able to maintain biomass or even to grow slowly in situ and the behaviour has been interpreted as a sort of 'aestivation', to escape the period of minimal resource supply. However, the recent season-long investigation by Bright and Walsby (2000) of the P. rubescens stratified in the Zürichsee, points to a sophisticated set of adaptations to gain positive growth in the only region of the lake where a small nutrient base and a low light income are simultaneously available. The ability to control organismic density is crucial to the exploitation of the opportunity.

2.5.3 Form resistance

Many plankters are markedly subspherical in shape and theory suggests that, in a majority of instances, the departure results in the organism having a slower passive rate of sinking (or floating) than the sphere of equivalent volume and overall density. As has been suggested earlier, there have really been few attempts to verify that this is true for a majority of species, and then mainly through resort to empirical evaluation of the coefficient of form resistance, φ_r. Even where significant form resistance is established experimentally (see entries in Table 2.5), it does not prove distortion to be necessarily adaptive in the context of floating and sinking. Nevertheless, the experimental demonstrations of the impact on sinking rate made by the presence of horns or spines, cell elongation in one (or possibly two) axes and the creation of secondary shapes by coenobial formations of chains, filaments and spirals make a fascinating study. In the end, they may provide the key to how larger plankters actually do maximise their suspension opportunities.

Protuberances and spines

The value of distortions to staying in suspension goes back a long way in planktology, certainly to Gran's (1912) interpretation, quoted by Hardy (1964), of a verifiable tendency for *Ceratium* species of less viscous tropical seas to have longer and, often, more branched horns than the species typical of colder, high-latitude seas. Yet it is only relatively recently that effects were quantitatively demonstrated. Smayda and Boleyn (1966a) investigated several aspects of the variability in sinking rate in the marine diatom *Rhizosolenia setigera*, including the fact that spineless pre-auxospore cells settle significantly faster than the spined vegetative cells that follow auxospore 'germination' (see p. 64). The spines that occur on the end cells of four-celled coenobia of the freshwater chlorophyte *Scenedesmus quadricauda* are said to reduce the sinking rate relative to a spineless

strain, although there is a possibility that there was a density difference between the two forms (Conway and Trainor, 1972). However, the investigations of the role of the 70-nm chitin fibres that adorn the frustules of *Thalassiosira weissflogii* (formerly *T. fluviatilis*) in slowing the sinking speed of cells have been carefully evaluated by Walsby and Xypolyta (1977). Cells from which the fibres had been removed with a fungal chitinase sank almost twice as fast as those not so treated, even though the density of the fibres (1495 kg m^{-3}) was rather greater than that of the fibreless cells. The overall volume of the untreated cells was also larger (1.9-fold) than that of the fibreless cells but the surface area was 2.8 times greater. Only the increased form resistance could have been responsible for the reduced sinking rate.

Chain formation

Joining two or more cells together obviously increases the volume of the settling particle in the same ratio. It also increases the surface area, but for the area of mutual contact between individual cells in the chain. Theory dictates that the chain must sink faster than the individual component cells – were sinking rate the only criterion, joining cells together could not be claimed to be an adaptation to suspension. On the other hand, as pointed out by Walsby and Reynolds (1980), if there is another constraint favouring larger size (say, resistance to grazing), it is equally clear that the linear arrangement preserves much more surface drag than a sphere of the same volume of the aggregate of cells. Hutchinson (1967) invoked the results of some experiments by Kunkel (1948), who had measured sinking rates of identical glass beads, either singly or cemented together in linear chains of one, two, three, four or eight. Hutchinson (1967) calculated the relative form resistance and fitted a linear plot against chain length with the equation:

$$\varphi_r = 0.837 + 0.163b \qquad (2.18)$$

where b is the number of beads. Superficially, this supported observations of Smayda and Boleyn (1965, 1966a, b) in *Thalassiosira*, *Chaetoceros* and other chain-forming marine diatoms that sinking

Table 2.5 Comparison of measured sinking rates (w_s) of various freshwater plankters and the rates (w_s calc) calculated from Stokes' equation for spheres of identical volume and density

Plankter[a]	Dynamic shape	w_s ($\mu m\ s^{-1}$)	(w_s) calc	φ_r	References
Chlorella vulgaris	± spherical	–	–	0.98–1.07	Oliver et al. (1981)
Chlorococcum sp.	± spherical	–	–	1.02–1.04	Oliver et al. (1981)
Cyclotella meneghiniana	Squat cylinder	–	–	1.03	Oliver et al. (1981)
*Stephanodiscus rotula	Squat cylinder				
(d_s = 12–14 μm)		11.5 ± 0.8	13 ± 1	1.06	From data plotted in Fig. 2.8
(d_s = 24–28 μm)		27.6 ± 2.6	26 ± 2	0.94	From data plotted in Fig. 2.8
*Synedra acus	Attenuate cylinder (MLD = 17d)	7.3 ± 1.2	29.8	4.1	Reynolds (1984a)
*Aulacoseira subarctica					
(1–2 cells)	Cylinder	7.4 ± 2.8	17.2	2.3	Reynolds (1984a)
(7–8 cells)	Attenuate cylinder	11.4 ± 4.1	50.1	4.4	Reynolds (1984a)
*Tabellaria flocculosa var asterionelloides	Stellate, 8-armed colony	10.3 ± 1.0	54.1	5.5	Reynolds (1984a)
*Asterionella formosa					
(4 cells)	Stellate colony	5.8 ± 0.2	18.2	3.2	Reynolds (1984a)
(8 cells)	Stellate colony	7.3 ± 0.6	28.9	3.9	From data plotted in Fig. 2.14
(16 cells)	Stellate colony	10.7 ± 1.2	45.9	4.3	Reynolds (1984a)
(8 very short cells)	Chain of ovoids	4.0 ± 0.6	7.4–8.0	1.9	Jaworski et al. (1988)
*Fragilaria crotonensis					
(single cell)	Cylinder	3.9 ± 0.2	10.6	2.8	Reynolds (1984a)
(11–12 cells)	Plate	11.2 ± 0.5	54.1	4.8	Reynolds (1984a)

[a]Asterisks indicate experiments on diatones killed prior to measurement of sinking rates, to overcome vital interference (see Section 2.5.4).

rates increase with increasing chain length. However, Walsby and Reynolds (1980) argued that the linear expression is misleading as increasing chain length should tend to a finite maximum. The sinking velocity of the chain (say w_c) is linked to the sinking rate of the sphere of similar volume (w_s), through $w_c = w_s/\varphi_r$. According to Stokes' equation (2.16), $w_s \propto (d_s/2)^2$, where d_s is the diameter of the equivalent sphere. However, $d_s/2 \propto b^{0.33}$, so $w_s \propto b^{0.67}$, i.e. w_s does not increase linearly, as Hutchinson's equation predicts, neither does it fall asymptotically to zero as implied.

Cylinder elongation and cylindrical filaments

This theory is upheld by the data of another set of observations presented by Reynolds (1984a) for filaments of *Melosira italica* (now *Aulacoseira subarctica*). This freshwater diatom comprises cylindrical cells joined together by the valve ends, effectively lengthening the cylinder in a linear way. In Reynolds' experiments, the mean length of the cells was $h_c = 19.0$ μm, the mean diameter (d_c) was 6.3 μm. The external volume of the cylindical cell was calculated from $\pi(d_c/2)^2 h_c \sim 592$ μm^3. The ratio h_c/d_c, a sort of index of cylindricity, is 3.0 and the diameter of the sphere (d_s) of the same volume is ~ 10.4 μm. Adding another cell doubles the length, volume and cylindricity, but the area of mutual contact between them means that the surface is not quite doubled but the diameter of the equivalent sphere is increased by about a third, to 13.1 μm. The sinking rates (w_s) of individual filaments of a killed suspension of an otherwise healthy, late-exponential strain of *Aulacoseira* (excess density ~ 251 kg m^{-3}) were measured directly and grouped according to the number of cells in the filament. The measurements were compared directly with the rates calculated for spheres of equivalent volume and excess density (w_s calc) in Fig. 2.11; equivalent values of φ_r (as w_s calc/w_s) are also included.

The plot is instructive in several ways. Lengthening increases size as it does w_s. Filament formation does not decrease sinking rate with respect to single cells. On the other hand, the increments become smaller with each cell added, with an asymptote (in this instance) of about 15 μm s^{-1}, reached by a filament of 11–12 cells in length

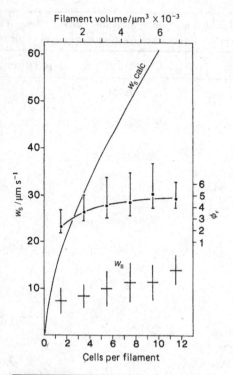

Figure 2.11 Plot of sinking rates (w_s) against length (as cells per filament) of *Aulacoseira subarctica* filaments compared with the sinking rates calculated for a spheres of the same volume and density (w_s calc). The ratio, $\phi_r = (w_s$ calc)/w_s, is also shown against the horizontal axis. Redrawn from Reynolds (1984a)

(200–220 μm, $h_c/d_c \sim 30$, $\varphi_r \sim 4.5$). This velocity is, moreover, about the same as that calculated for a sphere of similar density and of diameter equivalent to only 1.2 cells. Thus, the sacrifice of extra sinking speed is small in relation to the gain in size and where the loss of surface area is probably insignificant.

It is, of course, a feature of many pennate diatoms in the plankton to have finely cylindrical cells. On the basis of a small number of measurements on a species of *Synedra* and treating the essentially cuboidal cells as cylinders ($h_c = 128\pm11$ μm, greatest $d_c \sim 8.9$ μm, h_c/d_c 13–16), Reynolds (1984a) solved φ_r (as w_s calc/w_s) ~ 4.1. For the shorter individual cuboidal cells of *Fragilaria crotonensis*, whose length (~ 70 μm) exceeded mean width (3.4 μm) by a factor of over 20, φ_r was determined to be about 2.75. In the experiments with single cells of *Asterionella*

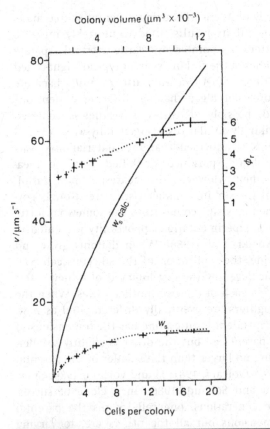

Figure 2.12 Plot of sinking rates (w_s) against the number of cells of killed *Fragilaria crotonensis* colonies compared with the sinking rates calculated for a sphere of the same volume and density (w_s calc). The ratio, $\phi_r = (w_s$ calc$)/w_s$, is also shown against the horizontal axis. Redrawn from Reynolds (1984a).

formosa (length 66 μm, mean width 3.5 μm) the form resistance was found to fall in the range 2.3 to 2.8. Compared with the squat cylindrical shapes of the centric diatoms *Cyclotella* and *Stephanodiscus* species for which measurements are available, the impact of attenuation on form resistance is plainly evident (see Table 2.5).

Colony formation

Of course, both *Fragilaria* and *Asterionella* are more familiarly recognised as coenobial algae, the cells in either case remaining tenuously attached on the valve surfaces, in the central region in *Fragilaria* (to form a sort of double-sided comb) and at the flared, distal end in *Asterionella*.

These distinctive new shapes generate some interesting sinking properties. In Fig. 2.12, mea-sured sinking rates of killed coenobia of *Fragilaria crotonensis* are plotted against the numbers of cells in the colonies and compared with the curve of (w_s calc) predicted for spheres of the same volumes. Whereas, once again, Stokes' equation predicts a continuous increase in sinking rate against size, coenobial lengthening tends to stability at 16–20 cells, not far from the point, indeed, at which the lateral expanse exceeds the lengths of the component cells and the coenobium starts to become ribbon-like. It is not unusual to encounter ribbons of *Fragilaria* in nature exceeding 150 or 200 cells (i.e. some 500–700 μm in length) but it is perhaps initially surprising to find that they sink no faster than filaments one-tenth their length! A further tendency that is only evident in these long chains is that the ribbons are not always flat but are sometimes twisted, with a frequency of between 140 and 200 cells per complete spiral. The extent to which this secondary structure influences sinking behaviour is not known.

The secondary shape of *Asterionella* coenobia is reminiscent of spokes in a rimless wheel, set generally at nearly 45° to each other. When there are eight such cells, they present a handsome star shape (alluded to in the generic name), although the mutual attachments determine that the colony is not flat but is like a shallow spiral staircase. Cell no. 9 starts a second layer; in cultures and in fast-growing natural populations, coenobia of more than eight cells are frequently observed (although, in the author's experience, chains of over 20–24 cells are in the 'rare' category). Following a similar approach to that used for *Aulacoseira* and *Fragilaria*, Reynolds (1984a) plotted measured sinking rates of killed coenobia against the numbers of cells in the colony and, again, compared them with corresponding curve (w_s calc) for spheres of the same volumes (see Fig. 2.13). Now, although the observed sinking rates, w_s, suggest the same tendency to be asymptotic up to a point where there are 6–10 cells in the coenobium, it is equally clear that higher numbers of cells make the colony sink faster, with no further gain in φ_r after ~4.0. Reynolds (1984a) interpreted this result as demonstrating the advantage to form resistance of creating a new shape, from a cylinder to a spoked disc,

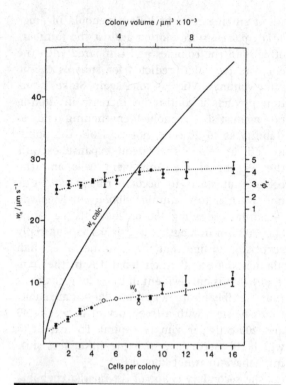

Figure 2.13 Plot of sinking rates (w_s) of killed stellate colonies of *Asterionella formosa* of comprising varying numbers of cells, compared with the sinking rates calculated for a spheres of the same volume and density (w_s calc). The ratio, $\phi_r = (w_s \text{ calc})/w_s$ is also shown against the horizontal axis. Redrawn from Reynolds (1984a).

which however is lost if space between cells is progressively plugged by more, dense cells that add nothing to the hydrodynamic resistance.

It is a satisfying piece of teleology that the maximum advantage should seem to be achieved at the most typical coenobial size. Indeed, there is further observational and experimental evidence that the shape determines more of the sinking behaviour of stellate colonies than any of the Stokesian components. First, the argument would need to hold for *Tabellaria flocculosa* var. *asterionelloides*, whose robust cells form eight-armed stellate colonies that contrast with the more typical habits adopted by this genus (see Knudsen, 1953). Frequently, up to 16 cells comprise the colonies but they remain in adherent pairs, preserving a regular eight-radiate form.

It is also of significance that some of the entries in Fig. 2.13 were derived from another

study of *Asterionella*, using different source material, but its results fitted comfortably into the pattern. As pointed out earlier, several separate studies of the sinking rates of typical, eight-celled colonies of *Asterionella*, using quite disparate sources of algae, having differing dimensions and, possibly, densities, nevertheless achieved strikingly similar results (generally $w_s = 5.5–11.0$ μm s^{-1}), provided that the material used was moribund, poisoned or killed prior to measurement. However, the clearest demonstration that the sinking characteristics are strongly governed by shape comes from the somewhat fortuitous experimentation opportunity presented to Jaworski *et al.* (1988). When diatoms grow and divide, the replication of the siliceous cell wall is achieved by the development of a 'new' valve inside each of the two mother valves. When the daughters are eventually differentiated as new, independent entities, one has the dimensions of its parent cell but the other is slightly smaller, being no larger than the smaller of the parental valves (Volcani, 1981; Li and Volcani, 1984; Crawford and Schmid, 1986). In a clone of successive generations, one cell retains the parental dimensions but all the others are, to varying extents, smaller. Average size must diminish in proportion to the number of divisions. This is clearly not a process without limit, as the species-specific sizes of diatoms normally remain within stable and predictable limits. Size is recovered periodically through distinctive *auxospore* stages, which give rise to vegetative cells, with relatively large overall dimensions and large sap vacuoles. This process is observed in culture as well as in nature. When it was noticed, however, that subcultured cells of *Asterionella* clone L354, one of those isolated from wild types and maintained at the Ferry House Laboratory of the Freshwater Biological Association, were becoming unusually small, it was decided to include it in the sinking studies that were then being conducted in the laboratory. These observations were commenced in 1981 and were continued until 1986 (Jaworski *et al.*, 1988). The clone never did recover size but the rate of growth became extremely slow towards the end. By the time of the last measurements, late in 1985, the cells had shrunk from long cuboids, 65 μm in length, to stubby ovoids,

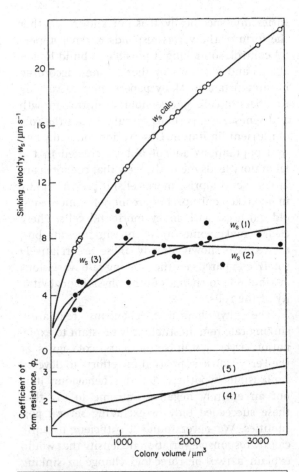

Figure 2.14 Plot of sinking rates of killed, stellate eight-celled colonies of *Asterionella* cells of diminutive length and volume, compared with the sinking rates calculated for a sphere of the same volume and density (w_s calc). Three regressions are fitted to the observed values: (1) applies to all points, (2) to those colonies > 1000 μm^3 in volume and (3) to those < 1000 μm^3 in volume. The ratio, $\phi_r = (w_s \text{ calc})/w_s$ is shown against the horizontal axis with respect to the estimates of w_s derived from regression equation (2.1) (curve 4) and those combining equations (2.2) and (2.3) (curve 5). Redrawn with permission, from Jaworski et al. (1988).

measuring about 5.5 μm with a basal width of 4.2 μm and a height of 2.1 μm. Significantly, the cells maintained mutual connections and there remained plenty of eight-armed colonies in the culture. The dry mass of cells and the silica content of the walls diminished with size but overall density increased (from ~1100 to 1200 kg m^{-3}).

The sinking rates (w_s) of killed cells measured over the five year period are plotted in Fig. 2.14

against colony volume and compared with w_s calc of equivalent spheres having the same densities. As the size of the cells diminished over the first 2 years (during which time the mean colony volume reduced by nearly two-thirds and the mean density difference with the medium almost doubled), sinking rates actually increased slightly. The rising mean of about 7.2 to one of 8.0 μm s^{-1} does not violate the hypothesis that the stellate cell arrangement dominates the sinking rate. Thereafter, with the aggregate volumes of colonies below 1200 μm^3, sinking rate diminished. The curve fitted to all sinking rate determinations (curve 1 in Fig. 2.14) reveals less of what is happening than do the two separate fitted curves referring to colonies >1200 μm^3 (curve 2) and those <1200 μm^3 (curve 3). By taking each of these regression lines and using them to solve form resistance as the calculated sinking rate for a sphere divided by the regression-predicted actual sinking rate, the relationship is further amplified. Curve (4) for all points (corresponding to regression 1) shows the typical form-resistance outline, albeit rather flattened. The right-hand side of curve 5, corresponding to regression 2, quickly takes $\varphi_r > 2$ and > 3, towards the 'plateau' level shown in Fig. 2.13. To the left of curve 5 (corresponding to regression 3), sinking rate varies more closely with volume (though not so steeply as w_s calc); φ_r is in the range 1.6–2.4. The change in behaviour has a pivot point when the colony falls below 1200 μm^3, when individual cell volumes are <150 μm^3 and the cells are 18–20 μm in length. The likely significance of this is that the hydromechanical characteristics of the star shape are eventually overtaken by those of a gyre of ovoids, more reminiscent of *Anabaena* than *Asterionella*.

Some of these measurements are included in the summaries of calculated form resistance owing to shape distortions.

2.5.4 Vital regulation of sinking rate

We may return, briefly, to the ability of live phytoplankton, especially diatoms, to exercise this further control on their own sinking rates below that expected from a modified Stokes equation, with all components properly evaluated. It has to be confirmed first that we use a correct interpretation of facts. Smayda (1970) referred to the acceleration of sinking rates in moribund diatoms, and the several mechanisms by which this might come about. These include the aggregation of dying cells and, through their involvement with other planktic detritus (zooplankton exuviae and faecal fragments, colloidal organics, fragments of plant remains), their formation in to larger floccular particles, collectively known as 'marine snow' (Alldredge and Silver, 1988). In these aggregates, particles may sink faster than might be predicted if they were separated. However, the consistency and reproducibility of behaviour of killed cells having similar form resistance should prompt us to regard this as the 'normal' sinking performance and to ask how it might be that live, healthy cells reduce their sinking rates below those that Stokes' law would predict.

The scale of these reductions is impressive, the 'live' rate being up to one order of magnitude, and frequently two- to four-fold less than the 'killed' rate. As well as the case of *Stephanodiscus rotula* illustrated in Fig. 2.8, there is an abundance of data to show the sinking rates of healthy, eight-celled *Asterionella formosa* colonies to be typically 2–3 μm s^{-1} (about 0.2–0.3 m d^{-1}) rather than the 7–8 μm s^{-1} explained by the modified Stokes' equation (Smayda, 1974; Tilman and Kilham, 1976; Jaworski *et al.*, 1981; Wiseman and Reynolds, 1981). Similarly, variations in the sinking rate of *Fragilaria crotonensis* may be <0.4 m d^{-1} for long periods but quite quickly increase to up to 1.1 m d^{-1} ($\leq$13 μm s^{-1}) when cells are nutrient limited or have been exposed to excessive sunlight (Reynolds, 1973a, 1983a). Indeed, these studies have suggested that it is a useful biological adaptation for an otherwise non-motile organism to be able to increase sinking rate spontaneously and to 'accelerate out of danger' from excessive irradiance, especially in stabilising water columns (Reynolds *et al.*, 1986).

It might appear that regulation of sinking rate is under the control of the diatom. Certainly, variability in the number of cells per colony may be, to an extent, self-regulated as the intercellular links vary in structure, some being much more amenable to separation than others. The frequency of 'linking valves' and 'separation valves'

varies interspecifically. It is not known whether the form of the valves responds to environmental control, so making it possible to build longer chains and filaments or shorter ones, according to circumstances. Many papers refer to varying numbers of cells per coenobium during growth and senescence, perhaps speculating on the role of nutrient limitation. Observations on many natural populations and of isolates treated in the laboratory leads me to the view that coenobia are larger (i.e. comprise more cells) if grown rapidly in unshaken cultures but are more fragmented in old cultures, with many moribund cells. These clearly make some impact on sinking rate but, as we have shown already, these are relatively small compared to the sinking-rate variations attributable to changes in the physiological vitality of the cells.

The physiological mechanisms regulating sinking rate remain stubbornly resistant to explanation. Over a number of years, colleagues at the Ferry House supported my efforts to develop a plausible hypothesis for this behaviour. It is not an entirely negative outcome to say that these succeeded only in excluding several possibilities. We never found a sufficient or sufficiently responsive variation in density that would explain a two- or three-fold change in sinking rate. While we were able to bring pressure to bear on planktic Cyanobacteria to collapse gas vesicles, to subject diatoms to similar treatment – up to about 12 bars, anyway – produced no response at all. Yet if sufficient of the specific photosynthetic inhibitor DCMU [3-(3,4-dichlorophenyl)-1,1-dimethyl urea] is added to a healthy, slow-sinking suspension of *Asterionella* cells to block their photosynthesis, sinking rate rose quickly to the rates of killed or moribund examples. In time, both effects are reversible (photosynthetic capacity, sinking-rate control are recovered). Contemporaneous work in our laboratory on the susceptibility of *Asterionella formosa* to attack by parasitic fungi (see especially Canter and Jaworski, 1981) had just revealed that, under conditions of low light or darkness, infective chytrid zoospores are not attracted to *Asterionella* cells as they are in the light. We deduced that actively photosynthesising cells either broadcast a signal to the adjacent medium advertising their presence or that they create a local change in the medium that is exploited or avoided by the infective spores. The lateral thinking that arose from our discussions led to the hypothesis that photosynthesising cells were immobilising water around their periphery. Thus, the particle acquires a new identity and the new dimensions of alga + water, in much the similar way that an investment of mucilage provides (see Section 2.5.2).

Considering that a sufficient swathe of mucilage has not been observed in these algae, the candidate mechanism that we proposed was that the surface charge on the cells was variable and that this might affect the amount of water thus immobilised. This was not an original inspiration but an echo of an earlier hypothesis, put forward by Margalef (1957). Based on his own observations of differing polarity and electrokinetic ('zeta') potential of *Scenedesmus* cells, he developed a theory of 'structural viscosity', where algae regulated the viscosity of their immediate surroundings through the electrical charge on the outer cell wall. It must be emphasised that, like any other small particles dispersed in an electrolyte (albeit, a very weak one), algae carry a surface charge in any case. This is, in part, determined by the ionic strength of the medium. Moreover, several publications detailing direct measurements of surface charge using electrophoretic procedures were available (Ives, 1956; Grünberg, 1968; Hegewald, 1972; Zhuravleva and Matsekevich, 1974). It became our objective to demonstrate that variable sinking rates are related to physiologically mediated changes in surface charge. We used an electrophoresis microscope to determine simultaneously the sinking rates and electrophoretic mobility of *Asterionella* colonies, incubated under varying laboratory conditions (Wiseman and Reynolds, 1981). The outcome was quite clear, insofar as large changes in sinking rate could not be correlated with relatively small variations in surface charge. The experiments succeeded only in rejecting another hypothesis about sinking-rate regulation and in establishing a nice method for the direct measurement of sinking rates.

The (as yet) unexplored alternative hypothesis we put forward (Wiseman and Reynolds, 1981) referred to a quite different role for mucilage, that it might be trailed in threads from cells, like a parachute or in the manner of the chitin hairs of *Thalassiosira*. Unlike chitin, threads of mucilage require active maintenance by healthy cells but would be sufficiently frail to be shed quickly, when they become a liability or too costly to maintain. Traditional algal anatomists would concur in conversation that such mucilaginous threads and trails exist but there seems very little published to uphold a compelling case. Even the use of Indian-ink irrigation, a popular technique for revealing mucilaginous structures, has proved unhelpful to the argument. However, a recent description of mucilaginous protuberances radiating from the marginal cells of *Pediastrum duplex* colonies (Krienitz, 1990) has been confirmed in the photomicrographs of Padisák *et al.* (2003a).

Back in the 1980s, we had proposed a number of approaches to investigate the hypothesis, including the possibility of using WETSTEM electron microscopy, for the observation of living materials at high magnification, which was then just becoming available. However, this was also the time when the sponsorship of science was moving rapidly from academic, curiosity-led problems such as this. Purchase or lease of suitable apparatus was less the problem than was the continued support to sustain an active group of personnel. Resolution of the mechanism of vital regulation of sinking rate by diatoms remains open to future research.

2.6 | Sinking and entrainment in natural turbulence

2.6.1 Sinking, floating and entrainment

Preceding sections of this chapter have reviewed the scales of the quantities of the two key components of plankton entrainability – the velocities of the intrinsic tendency of plankton to sink, swim or float and the velocities of motion in the medium. Both typically cover several orders of magnitude. The sinking rates of diatoms span something like 1 μm s^{-1} to 6 mm s^{-1}. The flotation rates of buoyant colonies of the Cyanobacteria such as *Anabaena* and *Aphanizomenon* may reach 40–60 μm s^{-1}, typical colonies

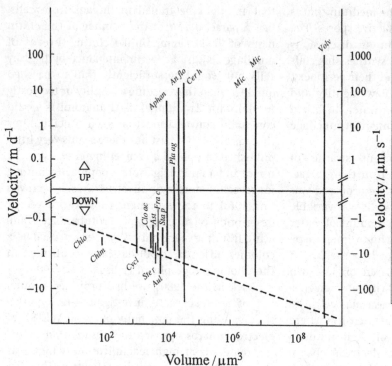

Figure 2.15 Ranges of sinking ('DOWN') and floating ('UP') velocities of freshwater phytoplankters, or, where appropriate of vertical swimming rates of motile species, plotted against unit volume. The algae are: *An flo, Anabaena flos-aquae; Aphan, Aphanizomenon flos-aquae; Ast, Asterionella formosa; Aul, Aulacoseira subarctica; Cer h, Ceratium hirundinella; Chlm, Chlorococcum; Chlo, Chlorella; Clo ac, Closterium aciculare; Cycl, Cyclotella meneghiniana; Fra c, Fragilaria crotonensis; Mic; Microcystis aeruginosa; Pla ag, Planktothrix agardhii; Sta p, Staurastrum pingue; Ste r, Stephanodiscus rotula; Volv, Volvox aureus.* Redrawn with permission from Reynolds (1997a).

of *Gloeotrichia* and *Microcystis* may achieve 100–300 μm s^{-1}, while some of the largest aggregations achieve 3–4 mm s^{-1} (Reynolds *et al.*, 1987; Oliver, 1994). Among motile organisms, reported 'swimming speeds' range between 3–30 μm s^{-1} for the nanoplanktic flagellates to 200–500 μm s^{-1} for the larger dinoflagellates, such as freshwater *Ceratium* and *Peridinium* (Talling, 1971; Pollingher, 1988) and marine *Gymnodinium catenatum* and *Lingulodinium* spp. (see Smayda, 2002). Large colonies of *Volvox* can attain almost 1 mm s^{-1} (Sommer and Gliwicz, 1986) while the ciliate *Mesodinium* is reported to have a maximum swimming speed of 8 mm s^{-1} (Crawford and Lindholm, 1997).

At the other end of the motility spectrum, solitary bacteria and picoplankters probably sink no faster than 0.01–0.02 μm s^{-1} (data collected in Reynolds, 1987a). The data plotted in Fig. 2.15 show that, despite the compounding of several factors in the modified Stokes equation (2.16), the intrinsic rates at which phytoplankton move (or potentially move) in relation to the adjacent medium are powerfully related to their sizes. Smaller algae sink or 'swim' so slowly that the motion of the water is supposed to keep them in

suspension. Larger species potentially move faster or farther but they need to be either flagellate or to govern their own buoyancy to counter the tendency to sink. Indeed, there is a strong indication that their ability to overcome elimination from the plankton, at least in the extant vegetative stages of their life cycles, depends upon the amplification of motility that large size confers. In essence, phytoplankton motility can be differentiated among those that can do little to stop themselves from sinking (mostly diatoms), the very large, which self-regulate their movements, and the very small for which it seems to matter rather little.

Even so, when the comparison is made, the range of intrinsic rates of sinking (w_s), floating ($-w_s$) and flagellar self-propulsion (u_s) represented in Fig. 2.15 (mostly $\leq 10^{-3}$ m s^{-1}) are 1–6 orders of magnitude smaller than the sample turbulent velocities cited in Table 2.2 (mostly $\geq 10^{-2}$ m s^{-1}). Generally speaking, the deduction is that $w_s \ll u^*$. This does not mean that the sinking potential (or the floating or migratory potential) is overcome, just that gravitating plankters are constantly being redistributed. What really matters to sinking particles is the relative magnitudes of w_s and the upward thrusts of the turbulent eddies w' (see Section 2.3.1). If $w_s > w'$, nothing prevents the particle from sinking. While, however, $w_s < w'$ some particles can be transported upwards faster than they gravitate downwards – and their sinking trajectories are reinitiated at a higher point in the turbulence field. Of course, there are, other things being equal, downward eddy thrusts which add to rate of vertical descent of the particles. Given that the upward and downward values of w' are self-cancelling, w_s is not affected. However, the greater is the magnitude of w' relative to w_s, the more dominant is the redistribution and the more delayed is the descent of the particles.

In this way, the ability of fluid turbulence to maintain sinking particles in apparent suspension depends on the ratio of sinking speed to the vertical turbulent velocity fluctuations. Empirical judgement suggested that this entrainment threshold occurs at 1–2 orders of magnitude greater than the intrinsic motion of the particle. In a detailed consideration of this relationship, Humphries and Imberger (1982) introduced a quotient (herein referred to as Ψ) to represent the boundary between behaviour dominated by turbulent diffusivity of the medium and behaviour dominated by particle buoyancy:

$$\Psi = w_s/15[(w')^2]^{1/2} \qquad (2.19)$$

Noting that, in open turbulence, the magnitude of u^* is not dissimilar from $[(w')^2]^{1/2}$ (see Eq. (2.4)), Reynolds (1994a) proposed that substitution of u^* in Eq. (2.19) gave a useful approximation to the value of Ψ. The main line drawn into Fig. 2.16 ($\Psi = 1$) against axes representing sinking rate (w_s) and turbulent velocity (u^*), is proposed as the boundary between effective entrainment (diffusivity dominates distribution) and effective disentrainment (particle properties dominate distribution).

The adjective 'effective' is important, because entrainment is never complete while w_s has finite value; neither is disentrainment total while there is any possibility that motion in the water can deflect the particle from its intrinsic vertical trajectory. However, the main point requiring emphasis at this point is that, in lakes, rivers

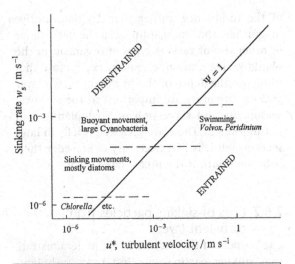

Figure 2.16 The entrainment criterion, as expressed in Eq. (2.19). In essence, the larger is the alga and the greater its intrinsic settling (or flotation) velocity, then the greater is the turbulent intensity required to entrain it. Redrawn with permission from Reynolds (1997a).

and oceans, u^* is a highly variable quantity (see, e.g., Table 2.2), with the variability being often expressed over high temporal frequencies (from the order of a few minutes) and, sometimes, over quite short spatial scales. Whilst in near-surface layers even the lower values u^* may often still be an order of magnitude greater than the intrinsic particle properties, the entrainment condition is not necessarily continuous in the vertical direction. Just taking the example of the transfer of the momentum of wind stress on the water surface and the propagation of turbulent eddies in the water column below (Eq. 2.5), it is clear that the loss of velocity through the spectrum of diminishing eddies will continue downwards into the water to the point where the residual energy is overcome by viscosity. As the penetrating turbulence decays with depth, the entraining capacity steadily weakens towards a point where neither u^* nor $[(w')^2]^{1/2}$ can any longer satisfy the particle-entrainment condition. In other words, the turbulence field is finite in extent and is open to the loss of sedimenting particles (and, equally, to the recruitment of buoyant ones).

This leads us to a significant deduction about the suspension and continued entrainment of phytoplankton in lakes and the sea. It is the *extent* of the turbulence, rather than its quantitative magnitude, that most influences the persistence or otherwise of various types of organism in the plankton. For the same reason, the factors that influence the depth of the turbulent mixed layer and its variability are important in the survival, seasonality and succession of phytoplankton in natural waters. These are reviewed briefly in later sections but it is first necessary to consider their behaviour within the mixed layer itself.

2.6.2 Loss of sinking particles from turbulent layers

The purpose of this section is not to demonstrate that sinking particles are lost from turbulent, surface-mixed layers but to provide the basis of estimating the rate of loss. The converse, how slowly they are lost, is the essence of adaptation to planktic survival. The development here is rather briefer than that in Reynolds (1984a), as its principles are now broadly accepted by plankton scientists. Its physical basis is rather older, owing to Dobbins (1944) and Cordoba-Molina et al. (1978). Smith (1982) considered its application to plankton. Interestingly, empirical validation of the theory comes from using plankton algae in laboratory-scale measurements.

Let us first take the example of a completely static water column, of height h_w (in m), open at the surface with a smooth bottom, to which small inert, uniform particles are added at the top. Supposing their density exceeds that of the water, that they satisfy the laminar-flow condition of the Stokes equation and sink through the water column at a predictable velocity, w_s (in m s^{-1}), then the time they take to settle out from the column is $t' = h_w/w_s$ (in s). If a large number (N_0, m^{-3}) of such particles are initially distributed uniformly through the water column, after which its static condition is immediately restored, they would settle out at the same rates but, depending on the distance to be travelled, in times ranging from zero to t'. The last particle will not settle in a time significantly less than t', which continues to represent the minimum period in which the column is cleared of particles. At any intermediate time, t, the proportion of particles settled is given by $N_0 w_s t/h_w$. The

number remaining in the column (N_t) is approximately

$$N_t = N_0(1 - w_s t/h_w) \qquad (2.20)$$

Let us suppose that the column is now instantaneously and homogeneously mixed, such that the particles still in the column are redistributed throughout the column but those that have already settled into the basal boundary layer are not resuspended. This action reintroduces particles (albeit now more dilute) to the top of the column and they recommence their downward trajectory. Obviously, the time to complete settling is now longer than t' (though not longer than $2t'$).

The process could be repeated, each time leaving the settled particles undisturbed but redistributing the unsettled particles on each occasion. If, within the original period, t', m such mixings are accommodated at regular intervals, separating quiescent periods each t'/m in duration. The general formula for the population remaining in suspension after the first short period is derived from Eq. (2.20):

$$N_{t/m} = N_0(1 - w_s t'/mh_w) \qquad (2.21)$$

After the second, it will be

$$N_0(1 - w_s t'/mh_w)(1 - w_s t'/mh_w)$$

and after the mth,

$$N_{t'} = N_0(1 - w_s t'/mh_w)^m \qquad (2.22)$$

Because $t' = h_w/w_s$, Eq. (2.22) simplifies to

$$N_{t'} = N_0(1 - 1/m)^m \qquad (2.23)$$

As m becomes large, the series tends to an exponential decay curve

$$N_{t'} = N_0(1/e) \qquad (2.24)$$

where e is the base of natural logarithms ($\sim$2.72). Solving empirically,

$$N_{t'} = 0.368 N_0 \qquad (2.25)$$

This derivation is instructive in several respects. The literal interpretation of Eq. (2.25) is that repeated (i.e. continuous) mixing of a layer should be expected still to retain 36.8% of an initial population of sinking particles at the end of a period during which particles would have

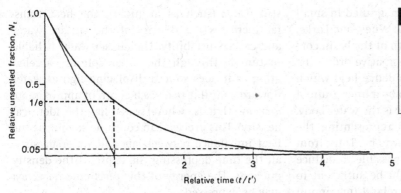

Figure 2.17 The number of particles retained in a continuously mixed supension compared to the retention of the same particles in a static water column of identical height. Redrawn from Reynolds (1984a).

left the same layer had it been unmixed. Moreover, the time to achieve total elimination (t_e) is an asymptote to infinity but we may deduce that the time to achieve 95% or 99% elimination is (respectively) calculable from

$$t_e/t' = \log_e 0.05/\log 0.368 = 3.0$$

or

$$t_e/t' = \log_e 0.01/\log 0.368 = 4.6$$

Using this approach, the longevity of suspension can be plotted (Fig. 2.17). It takes three times longer for 95% of particles to escape a mixed layer than were the same depth of water left unmixed. It may also be noted that the number of mixings does not have to be vast to achieve this effect. Substituting in Eq. (2.23), if $m = 2$, $N_{t'} = 0.25\,N_0$; if $m = 5$, $N_t = 0.33\,N_0$; if $m = 20$, $N_t = 0.36\,N_0$.

The formulation does not predict the value of the base time, t'. However, it is abundantly evident from the rest of this chapter that, the smaller is the specific sinking rate, w_s, the longer will be t' for any given column of length, h_m. Further, the greater is the mixed depth, then the longer is the period of maintenance. Any tendency towards truncation of the mixed depth, h_m, will accelerate the loss rate of any species that does not have, or can effect, an absolutely slow rate of sinking.

This confirms the adaptive significance of a slow intrinsic sinking rate. It has much less to do with delaying settlement directly but, rather, through the extension it confers to average residence time within an actively-mixed, entraining water layer. The relationship also provides a major variable in the population ecology of phytoplankton, because mixed layers may range from hundreds of metres in depth down to few millimetres and, in any given water body, the variability in mixed-layer depth may occur on timescales of as little as minutes to hours.

2.6.3 Mixed depth variability in natural water columns

Turbulent extent defines the vertical and horizontal displacement of particles that fulfil the entrainment criterion (Fig. 2.16). The vertical extent of turbulent boundary layers, unconstrained by the basin morphometry or by the presence of density gradients (open turbulence), is related primarily to the kinetic energy transferred: rearranging Eq. (2.9),

$$l_e \approx (u^*)(du/dz)^{-1} \quad \text{m} \qquad (2.26)$$

where (du/dz) is the vertical gradient of horizontal velocities (in $\text{m s}^{-1}\,\text{m}^{-1}$) and l_e is the dimension of the largest eddies. Entries in Table 2.2 pertaining to the upper layers of the open ocean and also of a moderately large lake like the Bodensee (Germany/Switzerland) imply an increase in mixing depth of about 9 m for each increment in wind forcing of $1\,\text{m s}^{-1}$. Of course, even this relationship applies only under a constant wind: an increase in wind speed necessarily invokes a restructuring, which may take many minutes to organise (see below). Similarly, the contraction of the thickness of the mixed layer following a weakening of the wind stress is gradual, pending the dissipation of inertia. The structure of turbulence under a variable, gusting wind is extremely complex!

The complexity is further magnified in small basins (Imberger and Ivey, 1991; Wüest and Lorke, 2003). Where the physical depth of the basin constrains even this degree of dissipative order, the water is fully mixed by a turbulence field which is, as already argued, made up by a finer grain of eddies. Moreover, the shallower is the water body, the lower is the wind speed representing the onset of full basin mixing likely to be. Thus, from the entries in Table 2.2, it is possible to deduce that a wind of 3.5 m s^{-1} might be sufficient to mix Lough Neagh, Northern Ireland (maximum depth 31 m, mean depth 8.9 m). In fact, the lake is usually well mixed by wind and is often quite turbid with particles entrained by direct shear stress on the sediment.

Density gradients, especially those due to the thermal expansion of the near-surface water subject to solar heating, also provide a significant barrier to the vertical dissipation of the kinetic energy of mixing. Although there is some outward conduction of heat from the interior of the Earth and some heat is released in the dissipation of mechanical energy, over 99% of the heat received by most water bodies comes directly from the Sun. The solar flux influences the ecology of phytoplankton in a number of ways but, in the present context, our concern is solely the direct role of surface heat exchanges upon the vertical extent of the surface boundary layer.

Starting with the case when there is no wind and solar heating brings expansion and decreasing water density (i.e. its temperature is >4 °C), a positive heat flux is attenuated beneath the water surface so that the heating is confined to a narrow near-surface layer. The heat reaching a depth, z, is expressed:

$$Q_z = Q_T^* e^{-kz} \quad W m^{-2} \qquad (2.27)$$

where e is the base of natural logarithms and k is an exponential coefficient of heat absorption. Q_T^* is that fraction of the net heat flux, Q_T which penetrates beyond the top millimeter or so. Roughly, Q_T^* averages about half the incoming short-wave radiation, Q_S. The effect of acquired buoyancy suppresses the downward transport of heat, save by conduction.

If the water temperature is <4 °C, or if it is >4 °C but the heat flux is away from the still water (such as at night), the heat transfer increases the density of the surface water and causes instability. The denser water is liable to tumble through the water column, accelerating as it does so and displacing lighter water upwards, until it reaches a depth of approximate *isopycny* (that is, where water has the identical density). This process can continue so long as the heat imbalance between air and water persists, all the time depressing the depth of the density gradient. The energy of this *penetrative convection* may be expressed:

$$B_Q = (g\gamma\, Q_T^*)(\rho_w\, \sigma)^{-1} \quad W m^2 s^{-2} J^{-1} (= m^2 s^{-3}) \qquad (2.28)$$

At all other times, the buoyancy acquired by the warmer water resists its downward transport through propagating eddies, whether generated by internal convection or externally, such as through the work of wind. They are instead confined to a layer of lesser thickness. Its depth, h_b, tends to a point at which the kinetic energy (J_k) and buoyancy (J_b) forces are balanced. Its instantaneous value, also known as the *Monin–Obukhov* length, may be calculated, considering that the kinetic energy flux, in W m^{-2}, is given by:

$$J_k = \tau(u^*) = \rho_w(u^*)^3 \quad kg\, s^{-3} \qquad (2.29)$$

while the buoyancy flux is the product of the expansion due to the net heat flux to the water (Q_T^*), also in W m^{-2},

$$J_b = 1/2 g h_b\, \gamma\, Q_T^* \cdot \sigma^{-1} \quad kg\, s^{-3} \qquad (2.30)$$

where γ is the temperature-dependent coefficient of thermal expansion of water, σ is its specific heat (4186 J kg^{-1} K^{-1}) and g is gravitational acceleration (9.8081 m s^{-2}).

Then, when $J_b = J_k$,

$$h_b = 2\sigma\rho_w(u^*)^3(g\gamma\, Q_T^*)^{-1} \quad m \qquad (2.31)$$

Owing to the organisational lags and the variability in the opposing energy sources, Eq. (2.29) should be considered more illustrative than predictive. Nevertheless, simulations that recognise the complexity of the heat exchanges across the surface can give close approximations to actual events, both over the day (Imberger, 1985) and over seasons (Marti and Imboden, 1986).

The effect of wind is to distribute the heat evenly throughout its depth, h_m. If the heat flux across its lower boundary is due only to conduction and negligible, the rate of temperature change of the whole mixed layer can be approximated from the net heat flux,

$$d\theta/dt = Q_T^*(h_m\rho_w\sigma)^{-1} \quad Ks^{-1} \tag{2.32}$$

The extent of the turbulent mixed boundary layer may be then viewed as the outcome of a continuous 'war' between the buoyancy generating forces and the dissipative forces. The battles favour one or other of the opponents, depending mostly upon the heat income, Q_T^*, and the kinetic energy input, τ, as encapsulated in the Monin–Obukhov equation (2.31). Note that if the instantaneous calculation of h_b is less than the lagged, observable h_m, buoyancy forces are dominant and the system will become more stable. If $h_b > h_m$, turbulence is dominant and the mixed layer should be expected to deepen. It becomes easy to appreciate how, at least in warm climates, when the water temperatures are above 20 °C for sustained periods, diurnal stratification and shrinkage of the mixed depth occurs during the morning and net cooling leads to its breakdown and extension of the mixed depth during the afternoon or evening. The example in Fig. 2.18 shows the outcome of diel fluctuations in heat- and mechanical-energy fluxes to the density structure of an Australian reservoir.

2.6.4 Vertical structure in the pelagic

Over periods of days of strong heating and/or weak winds, during which convective energy is insufficient to bring about complete overnight mixing, there will develop a residual density difference between the surface mixed layer and the water beneath it, leading to the formation of a more enduring density gradient, or thermocline. Its resistance to mixing is acquired during preceding buoyancy phases. This resistance is expressed by the (dimensionless) bulk Richardson number, Ri_b, expressing the ratio between the two sets of forces:

$$Ri_b = [\Delta\rho_w g\, h_m]\,[\rho_w(u^*)^2]^{-1} \tag{2.33}$$

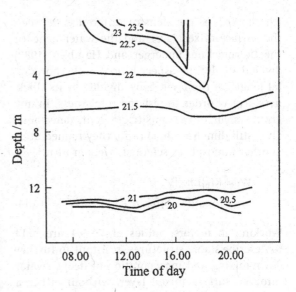

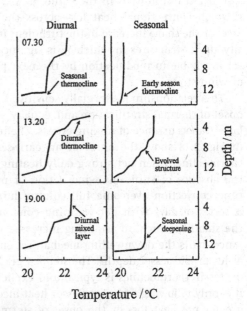

Figure 2.18 Diel variability in the mixed depth of a subtropical reservoir (Wellington Reservoir, Western Australia), reflecting the net heat exchanges with the atmosphere. The top panel shows the depth distribution of isotherms through a single day. The left-hand column of smaller graphs shows features of the evolving temperature structure (based on Imberger, 1985). The right-hand column proposes stages in the season-long development of enduring stratification. Redrawn with permission from Reynolds (1997a).

where $\Delta\rho_w$ is the density difference between the surface mixed layer and the water beneath the thermocline. Imberger and Hamblin (1982) divided Ri_b by the aspect ratio (i.e. the horizontal length of the layer, L, by, divided by its thickness, h_m) in order to test the robustness, in any given system, of the density gradients detectable. This (still dimensionless) ratio, they named after another atmosphere scientist, Wedderburn.

$$W = Ri_b[L\,h_m]^{-1}$$
$$= [\Delta\rho_w g(h_m)^2][\rho_w(u^*)^2 L]^{-1} \qquad (2.34)$$

Working in meters, values of $W > 1$ are held to describe stable structures, resistant to further down-mixing and incorporation of deeper water into the surface mixed layer, without either a significant diminution in the value of $\Delta\rho_w$ (e.g. through convectional heat loss across the surface) or the sharp increase in the turbulent intensity, (u^{*2}). Structures in which W is significantly <1 are liable to modification by the next phase of wind stress.

This relationship is especially sensitive to the onset of thermal stratification and, equally, simulates the occurrence of mixing events. The insets in Fig. 2.18 show the onset of an early-season thermocline, when net strong daily heating and the absence of sufficient wind action or nighttime convection overcome full column mixing. A series of days with net warming compounds the stability which, in acquiring increased resistance, halts the downwelling mechanical energy at lesser and lesser depths. The stepped gradient of 'fossil' thermoclines is typical and explicable. It is only following a change (lesser heat income, greater net heat loss or the onset of storms, W diminishing) that deeper penetration by turbulence eats into the colder water and sharpens the thermocline at the base of the mixed layer.

This is the basic mechanism for the onset and eventual breakdown in temperate lakes and seas. It also serves to track the seasonal behaviour of many more kinds of system other than those of middle- to high-latitude lakes. It applies to very deep lakes and seas, which may remain incompletely mixed (*meromictic*) for years on end. It also covers circumstances of water bodies too shallow or too wind-exposed to stratify for more

than a few hours or days at a time (*polymictic*). It explains the patterns of seasonal stratification in tropical lakes wherein stable density structures are precipitated by relatively small gains in heat content but are correspondingly liable to major mixing episodes for a relatively small drop in suface temperature (*atelomictic* lakes). It can be used to test the contribution of ionic strength (e.g. salt content in reinforcing density gradients).

Examples of all these kinds of stratification, classified by Lewis (1983), may equally be viewed from the opposite standpoint as a series describing variability in the extent and duration of turbulent mixing. The intriguing consequence is that the depth of the turbulent mixed layer (h_m) may remain nearly constant, when it is the full depth of the water (H) in a shallow, wind-exposed site. In a large, deep lake, it may fluctuate between <1 m and >100 m, in some instances, within a matter of a few hours.

The Wedderburn formulation equation has also been used to determine whether lakes will stratify at all. Putting $W = 1$ and interpolating the observed summer-thermocline depths of a series of temperate lakes, Reynolds (1992c) rearranged Eq. (2.34) to determine the density difference, $\Delta\rho_w$, between the waters separated by the seasonal thermocline. In most instances, the outcome was not less than 0.7 to 0.9 kg m^{-3}. At the depths of the respective thermoclines, the density difference would resist erosion by surface-layer circulations generated by winds up to $\sim$20 m s^{-1}. Winds much stronger than these would cause deepening of the mixed layer and depression of the thermocline. Interpolating the corresponding values for $\Delta\rho_w$ and u^*, Eq. (2.34) was solved for h_m against various nominated values for L. The resultant slope separated almost perfectly the dataset of stratifying and non-stratifying lakes assembled by Gorham and Boyce (1989) (Fig. 2.19).

This outcome is a satisfying vindication of theoretical modelling. Its principal virtue in plankton biology is to empiricise the relationships by which familiar environmental components govern the entrainment and transport of plankton-sized particles and how often the various conditions might apply.

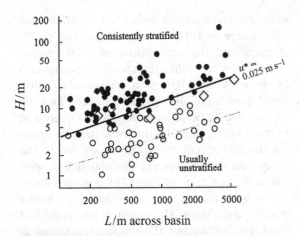

Figure 2.19 Depth, *H*, plotted against *L*, the length across various lakes of routinely stratifying (•) and generally unstratified lakes (○) considered by Gorham and Boyce (1989). The line corresponds to Reynolds' (1992c) prediction of the wind stress required to overcome a density difference of 0.7 kg m⁻³ (equivalent to $u^* = 0.025$ m s⁻¹). The diamond symbols refer to lakes said to stratify in some years but not in others Redrawn with permission from Reynolds (1997a).

This is a suitable point at which to emphasise an important distinction between the thickness of the mixed layer and the depth of the summer thermocline. As demonstrated here, the latter really represents the transition between the upper parts of the water column (in lakes, the *epilimnion*) that are liable to frequent wind-mixing events and the lower part that is isolated from the atmosphere and the effects of direct wind stress (the *hypolimnion*). The thickness of the intermediate layer (in lakes, the *metalimnion*) is defined by the steepness of the main vertical gradient of temperature (the *thermocline*) or density (*pycnocline*) between the upper and lower layers, though neither layer is necessarily uniform itself. The top of the thermocline may represent the point to which wind mixing and/or convection last penetrated. Otherwise, the mixed layer is entirely dynamic, its depth and structure always relating to the current or very recent (the previous hour) balance between J_b and J_k. The mixed layer can be well within the epilimnion or its full extent. Any tendency to exceed it, however, results in the simultaneous deepening of the epilimnion, the surface circulation shearing off and entraining erstwhile metalimnetic water

and simultaneously lowering the depth of the thermocline.

2.6.5 Mixing times

The intensity of turbulence required to entrain phytoplankton covers almost 2 orders of magnitude: non-motile algae with sinking rates of ~40 μm s⁻¹ are effectively dispersed through turbulence fields where $u^* \leq 600$ μm s⁻¹ (Eq. 2.19), whereas $u^* > 50$ mm s⁻¹ is sufficient to disperse the least entrainable buoyant plankters ($w_s \geq 1$ mm s⁻¹) (see Section 2.6.1). As has already been suggested, the intensity of mixing is often less important to the alga than is the vertical depth through which it is mixed. The depth of water through which phytoplankton is randomised can be approximated from the Wedderburn equation (2.34). Putting $W = 1$ and $u^* \geq 0.6$ m s⁻¹, the numerator is equivalent to ≥ 0.36 for each 1000 m of horizontal distance, *L*. Dividing out gravity, the product, $\Delta\rho_w (h_m)^2$ solves at ~0.037 kg m⁻¹. This is equivalent to an average density gradient of ≥ 0.04 kg m⁻³ m⁻¹ per km across a lake for a 1-m mixed layer, ~0.01 for a 2-m layer, ~0.0045 for a 3-m layer, and so on. The weaker is the average density gradient, then the greater is the depth of entrainment likely to be. The limiting condition is the maximum penetration of turbulent dissipation, unimpeded by density constraints. Where a density difference blocks the free passage of entraining turbulence, the effective floor of the layer of entrainment is defined by a significant local steepening of the vertical density gradient. Reynolds' (1984a) consideration of the entrainment of diatoms, mostly having sinking rates, w_s, substantially less than 40 μm s⁻¹, indicated that the formation of local density gradients of >0.02 kg m⁻³ m⁻¹ probably coincided with the extent of full entrainment, that is, in substantial agreement with the above averages. For the highly buoyant cyanobacterial colonies, however, disentrainment will occur in much stronger levels of turbulence and from mixed layers bounded by much weaker gradients.

Assumption of homogeneous dispersion of particles fully entrained in the actively mixed layer allows us to approximate the average velocity of their transport and, hence, the average time of their passage through the mixed layer. In their elegant development of this topic, Denman and Gargett (1983) showed that the average time of travel (t_m) through a mixed layer unconstrained by physical boundaries or density gradients corresponds to:

$$t_m = 0.2 \, h_m (u^*)^{-1} \quad s \tag{2.35}$$

Because, in this instance, both h_m and u^* are directly scaled to the wind speed, U (Eqs. 2.5, 2.11), t_m is theoretically constant. Interpolation into Eq. (2.35) of entries in Table 2.2 in respect of Bodensee permit its solution at 1416 s. The probability of a complete mixing cycle is thus 2×1416 s (≈ 47 minutes).

In the case of the wind-mixed layer of a small or shallow basin, or one bounded by a density gradient, the timescale through the layer is inversely proportional to the flux of turbulent kinetic energy:

$$t_m = h_m (2u^*)^{-1} \quad s \tag{2.36}$$

Following this logic, a wind of $8 \, m \, s^{-1}$ may be expected to mix a 20-m epilimnion in 2000 s (33 minutes) but a 2-m layer in just 200 s (3.3 minutes). A wind speed of $4 \, m \, s^{-1}$ would take twice as long in either case.

These approximations are among the most important recent derivations pertaining to the environment of phytoplankton. They have a profound relevance to the harvest of light energy and the adaptations of species to maximise the opportunities provided by turbulent transport (see Chapter 3).

2.6.6 Particle settling from variable mixed layers: an experiment

As part of an effort to improve the empirical description of the sedimentary losses of phytoplankton from suspension, Reynolds employed several approaches to measuring the sedimentary flux in the large limnetic enclosures in Blelham Tarn, UK. These cylindrical vessels, 45 m in diameter, anchored in 11–12 m of water contained sufficient water ($\sim$18 000 m^3) to behave like natural water columns. Their hydraulic isolation ensured all populations husbanded therein were captive and virtually free from external contamination (Lack and Lund, 1974; Lund and Reynolds, 1982; see also Section 5.5.1). The deployment of

sediment traps was to have been a part of this programme and it was decided that the choice of traps and the authenticity of their catches could be tested within the special environment of the enclosures. The plan was to add a measured quantity of alien particles to the water column (actually, *Lycopodium* spores, well steeped in wetting agent and preservative), then to monitor the subsequent loss from suspension in the water and to compare the calculated flux with the sediment trap catches. Three such experiments were carried out, under differing hydrographic conditions. The results were published (Reynolds, 1979a) but the unexpected bonus of the experiments was the contrasting rates of loss from suspension of ostensibly identical spores under the varying conditions.

In the first experiment, carried out in winter, the spores were dispersed over the enclosure surface, during windy conditions which intensified in the subsequent few days. A near-uniform distribution with depth was quickly established (see Fig. 2.20). The spores ($d = 32.80 \pm 3.18$; $\rho_c = 1049 \, kg \, m^{-3}$; $\varphi_r \sim 2.2$) had a measured sinking rate (w_s) of 15.75 $\mu m \, s^{-1}$ at 17 °C, which, adjusted for the density and viscosity of the water at the 4–5 °C obtaining in the field, predicted an *in-situ* intrinsic sinking rate of 0.96 m d^{-1}. The theoretical time for spores to eliminate the enclosure (at the time, $H \sim 11.8$ m) was thus calculated to be ($t' = $) 12.3 days. In fact, the elimination proceeded smoothly, always from a near-uniformly distributed residual population at an average exponential rate of -0.10 m d^{-1}, which value corresponds to a 95% removal in ($t_e = $) 30 days. The ratio t_e/t' is lower than predicted in Section 2.6.2 (2.44 against 3.0). This may be explained by probable violation of the initial assumption of full mixing of the water column throughout the experiment. Although no significant density gradient developed, continuous and complete vertical mixing of the enclosure cannot be verified. Nevertheless, the outcome is sufficiently close to the model (Fig. 2.20) solution for us not to reject the hypothesis that entrained particles are lost from suspension at an exponential rate close to $-(w_s/h_m)$.

In the second experiment, commenced in June, spores were dispersed at the top of the

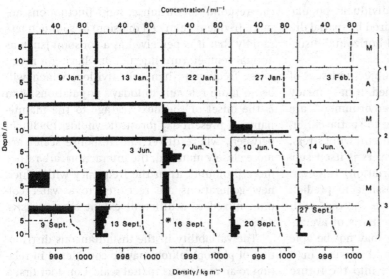

Figure 2.20 Modelled (M) and actual (A) depth–time distributions of preserved *Lycopodium* spores (of predetermined sinking characteristics) introduced at the water surface of one of the Blelham enclosures on each of three occasions (1, 9 January; 2, 3 June; 3, 9 September) during 1976, under sharply differing conditions of thermal stability. *Lycopodium* concentrations plotted as cylindrical curves; density gradients plotted as dashed lines. * – indicates no field data are available. Redrawn from Reynolds (1984a).

stratified enclosure during relatively calm conditions. Sampling within 30 minutes showed a good dispersion but still restricted to the top 1 m only. However, 4 days later, spores were found at all depths but the bulk of the original addition was accounted for in a 'cloud' of spores located at a depth of 5–7 m. After a further 7 days, measurable concentrations were detected only in the bottom 2 m of the column, meaning that, effectively, the addition had cleared 10 m in 11 days, at a rate not less than 0.91 m d^{-1}. Adjusted for the density and viscosity of the water at the top of the water column, the predicted sinking rate was 1.42 m d^{-1}. Thus, overall, the value of t' for the first 10 m (= 7 days) was exceeded by the observed t_e (= 11 days) by a factor of only 1.57. Part of the explanation is that sinking spores would have sunk more slowly than 1.42 m d^{-1} in the colder hypolimnion. However, the model explanation envisages a daily export of the population from the upper mixed layer (varying between 0.5 and 4 m during the course of the experiment), calculated as $N \exp -(w_s/h_m)$, whence it continues to settle unentrained at the rate w_s m d^{-1}. To judge from Fig. 2.20, this is an oversimplification but the prediction of the elimination is reasonable.

The same model was applied to predict the distribution and settlement of *Lycopodium* spores in the third experiment, conducted during the autumnal period of weakening stratification and mixed-layer deepening. Variability in wind forcing was quite high and a certain degree of re-entrainment is known to have occurred but the time taken to achieve 95% elimination from the upper 9 m of the water column (t_e = 18.0 days) at the calculated *in-situ* sinking rate (w_s) of 1.32 m d^{-1} exceeded the equivalent t' value (9/1.32 = 6.82) by a factor of 2.6.

The three results are held to confirm that the depth of entrainment by mixing is the major constraint on elimination of non-motile plankters heavier than water, that the eventual elimination is however delayed rather than avoided, and that prolongation of the period of suspension is in proportion to the depth of the mixed layer, wherein $u^* \geq 15 \, (w_s)$.

2.7 | The spatial distribution of phytoplankton

The focus of this chapter, the conditions of entrainment and embedding of phytoplankton in the constant movement of natural water masses, is now extended to the conditions where water movements are either insufficiently strong or insufficiently extensive to randomise the spatial distribution of phytoplankton. This section is concerned with the circumstances of plankters becoming disentrained and the consequences of weakening entrainment for individuals, populations and communities, as augured by spatial differentiation in the vertical and horizontal distributions of natural assemblages.

Distributional variation is subject to issues of scaling which need to be clarified. It has already been made plain that aquatic environments are manifestly heterogenous, owing to spatial differences in temperature, solute content, wind stress, etc., and that each of these drivers is itself subject to almost continuous variation. However, while precise values are impossible to predict, the range of variability may be forecast with some confidence, either on the basis of averaging or experience, or both. We may not be able to predict the intensity of wind mixing in a lake some three weeks or more into the future but we may estimate from the knowledge base the probability with which a given wind intensity will prevail. The changes in temperature, insolation, hydraulic exchanges and the delivery of essential nutrients affecting a given stretch of water also occur on simultaneously differing scales of temporal oscillation – over minutes to hours, night–day alternations, with changing season, interannually and over much broader scales of climatic change. The nesting of the smaller temporal scales within the larger scales holds consequences for phytoplankters in the other direction, too, towards the probabilities of being ingested by filter-feeders, of the adequacy of light at the depths to which entrained cells may be circulated, even to the probability that the energy of the next photon hitting the photosynthetic apparatus will be captured. The point is that the reactions of individual organelles, cells, populations and assemblages are now generally predictable, but the impacts can only be judged at the relevant temporal scales. These responses and their outcomes are considered in later chapters in the context of the relevant processes (photosynthesis, assimilation, growth and population dynamics). However, the interrelation of scales makes for fascinating study (see, for instance, Reynolds, 1999a, 2002a): in the end, the distinction is determined by the reactivity of the response. This means that critical variations alter more rapidly than the process of interest

can respond (for instance, light fluctuations are more frequent than cell division) or so much less rapidly that it is perceived as a constant (such as annual temperature fluctuations having a much lower frequency than cell division) which will be no more relevant to today's populations than is the onset of the next ice age to the maintenance of present-day forests (Reynolds, 1993b). In between, where driver and response scales are more closely matched, the interactions are rather more profound, as in the frequency with which new generations are recruited to a water column mixed to a different extent on successive days.

The variability in the instantaneous distribution of phytoplankton may be considered in relation to an analogous spatial scale. Consider first a randomised suspension of unicellular flagellates, such as *Chlamydomonas* or *Dunaliella*. Viewed at the 1–10 μm scale, distribution appears highly patchy, resolvable on the basis of presence or absence. In the range 10–1000 μm, the same distribution is increasingly perceived to be near-uniform but, in the turbulence field of a wind-mixed layer, variability over the 1–10 mm scale may attest to the interaction of algal movements with water at the viscous scale (Reynolds *et al.*, 1993a). In the range 10–100 mm and, perhaps, 10–1000 mm, the distribution may again appear uniform. Beyond that, the increasing tendency for there to be variations in the intensity of mixing leads to the separation of water masses in the vertical (at the scale of tens to hundreds of metres) and in the horizontal (hundreds of metres to hundreds of kilometres), at least to the extent that they represent quite isolated and coexisting environments, each having quite distinct conditions for the survival of the flagellates and the rate of their recruitment by growth. This is but one example of the principle that the relative uniformity or heterogeneity within an ecological system depends mainly upon the spatial and temporal scale at which it is observed (Juhász-Nagy, 1992).

Uniformity and randomisation, on the one hand, and differentiation ('patchiness'), on the other, may thus be detected simultaneously within a single, often quite small system.

Moreover, the biological differentiation of individual patches may well increase the longer their mutual isolation persists. Thus it remains important to make clear the spatial scale that is under consideration, whether in the context of vertical or horizontal distribution.

2.7.1 Vertical distribution of phytoplankton

Against this background, it seems appropriate to emphasise that the expectation of the vertical distributions of phytoplankton is that they should conform to the vertical differentiation of the water column, in terms of its current (or, at least, very recent) kinetic structure. The latter may comprise a wind-mixed convective layer overlying a typically less energetic layer of turbulence that is supposed to diminish with increasing depth, towards a benthic boundary layer in which turbulence is overcome by friction with the solid surface. However, in the seas as in lakes, the horizontal drift of the convective layer has to be compensated by counterflows, which movement promotes internal eddies and provides some turbulent kinetic energy from below. The formation of vertical density gradients may allow the confinement of the horizontal circulation to the upper part of the water column, leaving a distinct and kinetically rather inert water mass of the pycnocline, with very weak vertical motion. Density gradients form at depths in large, deep lakes and in the sea but do not contain basin-wide circulations; even though the gradients may persist, they may be rhythmically or chaotically displaced through the interplay of the gravitational 'sloshing' movements of deep water masses and the convective movements of the surface layer.

The vertical extent of the convective layer is, as we have seen, highly variable and subject to change at high frequency. It can vary from a few millimetres to tens of metres over a period of hours and between tens to hundreds of metres over a few days. The presence of density gradients reduces the entrainability of the deeper water into the surface flow. There is often great complexity at the interface (note, even if the top layer of the deeper water is sheared off and incorporated into the convective circulation of the upper, a gradient between the layers persists, albeit by now at a greater physical depth). Internal waves may form as a consequence of differential velocities, mirroring those formed at the water surface by the drag of high-velocity winds; and, just as surface waves break when the velocity differences can no longer be contained, so internal waves become unstable and collapse (*Kelvin–Helmholtz instabilities*: see, e.g., Imberger, 1985, for details), releasing more bottom water into the upper convection.

In shallow water columns, the extension of the convective layer is confined to the physical water depth, in which all energy-dissipative interactions and return flows must be accommodated. This necessarily results in very complex and aggressive mixing processes, often extending to the bottom boundary layer and, on occasions, penetrating it to the extent of entraining, and resuspending, the unconsolidated sedimented material.

Supposing wind-driven convective layers everywhere to be characterised by $u^* \geq 5 \times 10^{-3} \, \mathrm{m\,s^{-1}}$ (i.e. $5 \, \mathrm{mm\,s^{-1}}$), they should be capable of fulfilling the entrainment criterion for plankton with sinking, floating or swimming speeds of $\leq 250 \times 10^{-6} \, \mathrm{m\,s^{-1}}$. Comparison with Section 2.6.1 supports the deduction that the distribution of almost all phytoplankters must be quickly randomised through the vertical extent of convective layers. However, if the driving energy weakens, so that the convective layer contracts (in line with the Monin–Obukhov prediction [Eq. 2.31] or, without simultaneous heat gain, because u^* diminishes), plankters that were entrained towards the bottom of the layer become increasingly liable to disentrainment *in situ*, where, *ergo*, their own intrinsic movements begin to be expressed.

Four examples of distributional responses to physical structure are sufficient to demonstrate the basic behaviours of phytoplankters that are heavier than water ($\rho_c > \rho_w$), those that are frequently lighter than water ($\rho_c < \rho_w$), those non-motile species that are more nearly isopycnic ($\rho_c \sim \rho_w$) and those that are sufficiently motile for any density difference to be at times surmountable. The selection also employs some of the

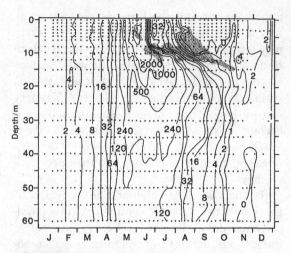

Figure 2.21 Depth–time plot of the vertical distribution of *Asterionella formosa* in the North Basin of Windermere through 1947. Isopleths in live cells mL^{-1}. The shaded area represents the extent of the metalimnion. Original from Lund *et al.* (1963), and redrawn from Reynolds (1984a).

differing ways that distributional data may be represented.

Non-motile, negatively buoyant plankters ($\rho_c > \rho_w$)

For the first case, the classical study of Lund *et al.* (1963) on the season-long distribution of *Asterionella* cells in the North Basin of Windermere in 1947 is illustrated (Fig. 2.21). The densities of diatoms mostly exceed, sometimes considerably, that of the surrounding water in lakes and seas. Those bound to be negatively buoyant (they have positive sinking rates) are destined to be lost progressively from suspension, at variable rates that are due to the relationship between the (variable) intrinsic particle sinking rate and the (variable) depth of penetration of sufficient kinetic energy to fulfil the species-specific entrainment criterion. Even before the critical quantities were known, numerous studies had demonstrated the sensitivity of diatom distribution to water movements and to the onset of thermal stratification in particular, both in lakes (Ruttner, 1938; Findenegg, 1943; Lund, 1959; Nauwerck, 1963) and in the sea (Margalef, 1958, 1978; Parsons and Takahashi, 1973; Smayda, 1973, 1980; Holligan and Harbour, 1977).

While numerous other factors intervene in the seasonal population dynamics and periodicity of diatoms, the importance of the depth of the surface mixed layer as another ecological threshold that must be satisfied for successful maintenance and recruitment of diatoms has been demonstrated many times (Reynolds and Wiseman, 1982; Reynolds *et al.*, 1983b; Sommer, 1988a; Huisman *et al.*, 1999; Padisák *et al.*, 2003b). If the rate of recruitment through cell growth and replication fails to make good the aggregate rate of all losses, including to settlement, then the standing population must go into decline. The extent of the mixed depth is already implicated in the ability of a seed population to remain in suspension. As the *Lycopodium* experiments demonstrate (see Section 2.6.6), the nearer this contracts to the surface, the faster is the rate of loss from the diminishing mixed layer itself and from the enlarging, stagnating layer beneath it. This occurs independently of the chemical capacity of the water to support growth, although it is often influenced by spontaneous changes in the intrinsic sinking rate (Reynolds and Wiseman, 1982; Neale *et al.*, 1991b).

The plot of Lund *et al.* (1963) (Fig. 2.21) shows the strong tendency in the first 3 months of the year towards vertical similarity in the concentration of *Asterionella* (with the isopleths, in cells mL^{-1}, themselves vertically arranged), as their numbers slowly rise during the spring increase. With the progressive increase in day length and potential intensity of solar irradiance, the lake starts to stratify, with a pycnocline (represented in Fig. 2.21 by the fine stippling) developing at a depth of between 5 and 10 m from the surface. The contours reflect the segregating response of the vertical distribution, with an initial near-surface acceleration in recruitment but followed soon afterwards by rapid decline in numbers, as sinking losses by dilution from the truncated mixed layer overtake recruitment. The distribution of contours beneath the pycnocline acquire a diagonal trend (reflecting algal settlement) while the near-horizontal lines in the pycnocline itself confirm the heterogeneity of numbers in the vertical direction and the strong vertical gradient in algal concentration in the region of the pycnocline. It is not until the final breakdown of thermal stratification (usually in December in Windermere) that approximate homogeneity in the vertical is recovered.

Positively buoyant plankters ($\rho_c < \rho_w$)

The vertical distribution of buoyant organisms, which include many of the planktic, gas-vacuolate Cyanobacteria during at least stages of their development, is similarly responsive to variability in the diffusive strength of vertical convection, save that algae float, rather than sink, through the more stable layers. A further difference is that the population of the upper mixed layer potentially experiences concentration by net recruitment from upward-moving organisms rather than dilution as downward-moving organisms are shed from it.

It has long been appreciated that the formation of surface scums of buoyant Cyanobacteria (variously known, colloquially and in many languages, as water blooms, flowering of the waters, etc.: Reynolds and Walsby, 1975), and involving such genera as *Anabaena, Anabaenopsis, Aphanizomenon, Gloeotrichia, Gomphosphaeria, Woronichinia* and (especially) *Microcystis* are prone to form in still, windless conditions (Griffiths, 1939). Buoyant *Trichodesmium* filaments also form locally dense surface patches in warm tropical seas under calm conditions (Ramamurthy, 1970): the little flakes of filaments also merited the sailors' colloquialism of 'sea sawdust'.

The mechanisms of scum formation are not straightforward but, rather, require the coincidence of three preconditions: a pre-existing population, a significant proportion of this being rendered positively buoyant on the balance of its gas-vesicle content, and the hydrographic conditions being such as to allow their disentrainment (Reynolds and Walsby, 1975). The present discussion assumes that the first two criteria are satisfied, scum formation now depending upon the relatively short-term onset of lowered diffusivity to the water surface, so that the magnitude of the flotation velocity ($-w_s$) is no longer overwhelmed by the turbulent velocity (u^*). The example presented in Fig. 2.22 traces the changing distribution of colonies of *Microcystis aeruginosa*, in relation to the thermal structure in a small temperate lake during one 24-h period of

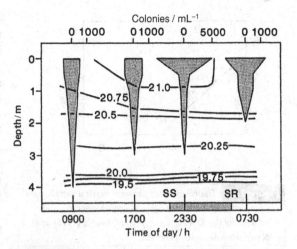

Figure 2.22 Changes in the vertical distribution of *Microcystis aeruginosa* colonies in a small lake (shown as cylindrical curves) in relation to temperature (isopleths in °C), during 28/29 July, 1971 (SS, sunset; SR, sunrise). Data of Reynolds (1973b) and redrawn from Reynolds (1984a).

anticyclonic weather in July 1971. The colonies were, on average, buoyant throughout, having a mean flotation rate of $\sim 9\,\mu m\,s^{-1}$ during the first day that increased almost twofold during the hours of darkness (Reynolds, 1973b). Note that an established temperature gradient, extending downwards from a depth of about 3.5 m, already contained the buoyant population and the changing vertical distribution of *Microcystis* occurred in relation to the *secondary microstratification* that developed during the course of the day (density gradient 0.1 kg m^{-3} m^{-1}). A light breeze occurred in the early evening of 28 July, before windless conditions resumed. Some convectional cooling also occurred during the night, sufficient to redistribute the population to a small extent but not to dissipate the surface scum that had formed, which, from an average concentration of ~ 160 colonies mL^{-1}, increased 37-fold (to 5940 mL^{-1} at 23.30).

The method of plotting, greatly favoured by the early plankton ecologists, invokes the use of 'cylindrical curves'. These are drawn as laterally viewed solid cones or more complex 'table-legs', the cross-sectional diameter at any given point being proportional to the cube root of the concentration. These shapes capture well the discontinuities in a given vertical distribution but may

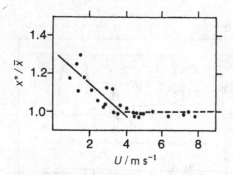

Figure 2.23 The relationship between vertical patchiness of *Microcystis* in the shallow Eglwys Nynydd reservoir and wind velocity (U). Data of George and Edwards (1976) and redrawn from Reynolds (1984a).

be less helpful than contoured depth–time plots (such as used in Fig. 2.21) over long periods of time. Neither do they necessarily convey the generalism between vertical discontinuity and the physical heterogeneity. In consideration of a 2-year series of *Microcystis* depth profiles in a small, shallow reservoir, (Eglwys Nynydd: area 1.01 km^2, mean depth 3.5 m), George and Edwards (1976) calculated a crowding statistic (x^*, owing to Lloyd, 1967) and its ratio to the mean concentration over the full depth ($\bar{x}$) to demonstrate the susceptibility of vertical distribution to the wind-forced energy. Putting $x^* = [\bar{x} + s^2/\bar{x} - 1]$, where s^2 is the variance between the individual samples in each vertical series, they showed that the relative crowding in the vertical, $(x^*/\bar{x})$, occurred only at low wind speeds ($U < 4\,\mathrm{m\,s^{-1}}$) and in approximate proportion to U), but wind speeds over $4\,\mathrm{m\,s^{-1}}$ were always sufficient to randomise *Microcystis* through the full 3.5-m depth of the reservoir (Fig. 2.23).

Neutrally buoyant plankters ($\rho_c \sim \rho_w$)

In this instance, 'neutral' implies 'approximately neutral'. As already discussed above (Section 2.5), it is not possible, nor particularly desirable, for plankters to be continuously isopycnic with the medium. Nevertheless, many species of phytoplankton that are non-motile and are unencumbered by skeletal ballast (or the gas-filled spaces to offset it) survive through maintaining a state in which they do not travel far after disentrainment. This state is achieved among very small unicells of the nanoplankton and picoplankton

(where, regardless of density, small size determines a low Stokesian velocity) and among rather larger, mucilage-invested phytoplankters, which are genuinely able to 'dilute' the excess mass of cell protoplasm and structures in the relatively large volume of water that the mucilaginous sheath immobilises. In either case, the adaptations to pelagic survival take the organisms clos-, est to the ideal condition for suspension. These are, literally, the most readily entrainable plankters according to the definition (Section 2.6.1, Eq. 2.19).

In Fig. 2.24, sequences in the vertical distributions of two non-motile green algae – the nanoplanktic chlorococcal *Ankyra* and the microplanktic, coenobial palmelloid *Coenochloris* (formerly ascribed to *Sphaerocystis*) – are depicted in relation to stratification in the large limnetic enclosures in Blelham Tarn. The original data refer to average concentrations in metre-thick layers (sampled by means of a 1-m long Friedinger trap: Irish, 1980) or in multiples thereof. The data are plotted as stacks of individual cylinders, the diameters of which correspond to the respective cube roots of the concentrations. In both cases, the algae are dispersed approximately uniformly through the epilimnion, while numbers in the hypolimnion remain low, owing to weak recruitment either by growth or by sedimentation ($w_s < 0.1\ \mathrm{m\ d^{-1}}$). The effect of grazers has not been excluded but simultaneous studies on loss rates suggest that the impact may have been small, while neither species was well-represented in simultaneous sediment-trap collections or samples from the surface deposits (Reynolds *et al.*, 1982a). On the other hand, deepening of the mixed layer and depression of the thermocline result in the immediate randomisation of approximately neutrally buoyant algae throughout the newly expanded layer. Such species are thus regarded as being always likely to become freely distributed within water layers subject to turbulent mixing, then to settle from them only very slowly and, of course, to be unable to recover a former distribution when mixing weakens (Happey-Wood, 1988).

Motile plankters ($u_s > u^*$)

Flagellates and ciliates are capable of directed movements that, actually as well as potentially,

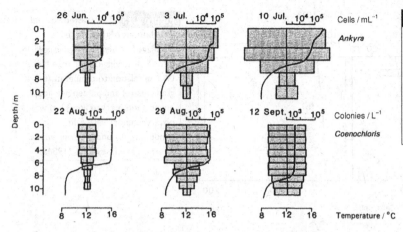

Figure 2.24 Instances in the vertical distribution of non-motile phytoplankton in a Lund Enclosure during the summer of 1978, shown as cylindrical curves. *Ankyra* is a unicellular nanoplankter; Coenochloris occurs as palmelloid colonies. Redrawn from Reynolds (1984a).

may result in discontinuous vertical distributions in lakes and seas. This ability is compounded by the capacity for self-propulsion of the alga (the word 'swim' is studiously avoided – see Section 2.3.4). In terms of body-lengths per unit time, the rates of progression may impress the microscopist but, in reality, rarely exceed the order of $0.1–1\,\mathrm{mm\,s^{-1}}$.

In general, the rates of progress that are possible in natural water columns are related to size and to the attendant ability to disentrain from the scale of water movements (Sommer, 1988b). Moreover, the detectable impacts on vertical distribution also depend upon some directionality in the movements or some common set of responses being simultaneously expressed: if all movements are random, fast rates of movement scarcely lead to any predictable pattern of distribution. For instance, the impressive vertical migrations of populations of large dinoflagellates are powerfully and self-evidently responsive to the movements of individual cells within environmental gradients of light and nutrient availability. This applies even more impressively to the colonial volvocalean migrations (Section 2.6.1) where all the flagellar beating of all the cells in the colony have to be under simultaneous control. The point needs emphasis as the flagellar movements of, for instance, the colonial chrysophytes (including the large and superficially *Volvox*-like *Uroglena*), seem less well coordinated: they neither 'swim' so fast, nor do their movements produce such readily interpretable distributions as *Volvox* (Sandgren, 1988b).

On the other hand, there is a large number of published field studies attesting to the vertical heterogeneity of flagellate distribution in ponds, lakes and coastal embayments. Reference to no more than a few investigative studies is needed (Nauwerck, 1963; Moss, 1967; Reynolds, 1976a; Cloern, 1977; Moll and Stoermer, 1982; Donato-Rondon, 2001). Works detailing behaviour of particular phylogenetic groups include Ichimura *et al.* (1968) and Klaveness (1988) on cryptomonads; Pick *et al.* (1984) and Sandgren (1988b) on chrysophytes; Croome and Tyler (1984) and Häder (1986) on euglenoids. Note also that not all flagellate movements are directed towards the surface. There are many instances of conspicuous surface avoidance (Heaney and Furnass, 1980; Heaney and Talling, 1980a; Gálvez *et al.*, 1988; Kamykowski *et al.*, 1992) and of assembling deep-water 'depth maxima' of flagellates, analogous to those of *Planktothrix* and photosynthetic bacteria (Vicente and Miracle, 1988; Gasol *et al.*, 1992).

The example illustrated in Fig. 2.25 shows the contrasted distribution of *Ceratium hirundinella*. This freshwater dinoflagellate is known for its strong motility (up to $0.3\,\mathrm{mm\,s^{-1}}$) (see Section 2.6.1) and its well-studied capacity for vertical migration under suitable hydrographic conditions (Talling, 1971; Reynolds, 1976b; Harris *et al.*, 1979; Heaney and Talling, 1980a, b; Frempong, 1984; Pollingher, 1988; James *et al.*, 1992). These properties enable it quickly to take up advantageous distribution with respect to light gradients, when diffusivity permits ($u^* < 10^{-4}\,\mathrm{m\,s^{-1}}$). The right-hand profile in Fig. 2.25 shows the vertical distribution of *Ceratium* during windy weather in a small, eutrophic temperate lake; the left-hand profile shows a distribution under

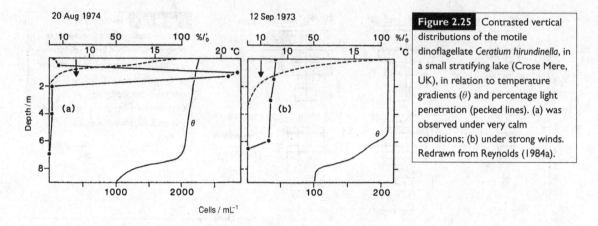

Figure 2.25 Contrasted vertical distributions of the motile dinoflagellate *Ceratium hirundinella*, in a small stratifying lake (Crose Mere, UK), in relation to temperature gradients (θ) and percentage light penetration (pecked lines). (a) was observed under very calm conditions; (b) under strong winds. Redrawn from Reynolds (1984a).

only weakly stratified conditions but one that is strongly allied to the light gradient.

2.7.2 Horizontal distribution of phytoplankton

A considerable literature on the horizontal variability in plankton distribution has grown up, seemingly aimed, in part, towards invalidating any preconceived assumption of homogeneity. It is difficult to determine just where this assumption might have arisen, as investigations readily demonstrate that distributions are often far from homogeneous. It may be that a predominance of papers in the mid-twentieth century focused on the population dynamics and vertical distributions of phytoplankton in small lakes paying insufficient attention to simultaneous horizontal heterogeneity. Such assumptions, real or supposed, have no place in modern plankton science. However, even now, it is important to present a perspective on just how much variation might be expected, over what sort of horizontal distances and how it might reflect the contributory physical processes.

Small-scale patchiness

Omitting the very smallest scales (see preamble to Section 2.7), phytoplankton is generally well-randomised within freshly collected water samples (typical volumes in the range 0.5–5 litres, roughly corresponding to a linear scale of 50–200 mm). Thus, there is normally a low coefficient of variation between the concentrations of plankton in successive samples taken at the same place. The first focus of this section is the horizontal distance separating similar samples show-

ing significant or systematic differences in concentration.

There have been few systematic attempts to resolve this question directly. In a rarely cited study, Nasev *et al.* (1978) analysed the confidence interval about phytoplankton counting by partitioning the variance attaching to each step in the estimation – from sampling through to counting. Provided adequate steps were taken to suppress the errors of subsampling and counting (Javorničky, 1958; Lund *et al.*, 1958; Willén, 1976), systematic differences in the numbers present in the original samples could be detected at scales of a few tens of metres but, on other occasions, not for hundreds. Irish and Clarke (1984) analysed the estimates of specific algal populations of algae in similar samples collected from within the confines of a single Blelham enclosure (area 1641 m², diameter, 45.7 m) at locations nominated on a stratified-random grid. They found that the coefficients of variation varied among different species of plankton, from about 5%, in the case of non-motile, neutrally buoyant algae, to up to 22% for some larger, buoyancy-regulating Cyanobacteria. In another, unrelated study, Stephenson *et al.* (1984), showed that spatial variability increased with increasing enclosure size.

A general conclusion is that sampling designs underpinning *in-situ* studies of phytoplankton population dynamics must not fail to take notice of the horizontal dimension. However, the size of the basin under investigation is also important. For instance, a coefficient of variation of even 22% is small compared with the outcome of growth and cell division, where a population doubling represents a variation of 100% per

---

Box 2.1 | Langmuir circulations

Langmuir circulations are elongated, wind-induced convection cells that form at the surface of lakes and of the sea, having characters first formalised by Langmuir (1938). They take the form of parallel rotations, that spiral approximately in the direction of the wind, in the general manner sketched in Fig. 2.26. Their structure is more clearly understood than is their mechanics but it is plain that the cells arise through the interaction of the horizontal drag currents and the gravitational resistance of deeper water to entrainment. Thus, they provide the additional means of spatially confined energy dissipation at the upper end of the eddy spectrum (Leibovich, 1983). In this way, they represent a fairly aggressive mixing process at the mesoscale but the ordered structure of the convection cells does lead naturally to a surprising level of microstructural differentiation. Adjacent spirals have interfaces where both are either upwelling simultaneously or downwelling simultaneously. In the former case, there is a divergence at the surface; in the latter there is a convergence. This gives rise to the striking formation of surface windrows or streaks that comprise bubbles and such buoyant particles as seaweed fragments, leaves and plant remains, insect exuviae and animal products as they are disentrained at the convergences of downwelling water.

The dynamics and dimensions of Langmuir circulation cells are now fairly well known. The circumstances of their formation never arise at all at low wind speeds ($U < 3$–4 m s^{-1}: Scott et al., 1969; Assaf et al., 1971). Spacing of streaks may be as little as 3–6 m apart at these lower wind speeds, when there is an rough correlation between the downwelling depth and the width of the cell (ratio 2.0–2.8). In the open water of large lakes and the sea, where there is little impediment to Langmuir circulation, the distance between the larger streaks (50–100 m) maintains this approximate dimensional proportionality, being comparable with that of the mixed depth (Harris and Lott, 1973; Boyce, 1974). The velocity of downwelling ($w > 2.5 \times 10^{-2}$ m s^{-1}) is said to be proportional to the wind speed ($\sim 0.8 \times 10^{-2}$ U): Scott et al., 1969; Faller, 1971), but the average velocities of the upwellings and cross-currents are typically less.

Consequences for microalgae have been considered (notably by Smayda, 1970, and George and Edwards, 1973) and are reviewed in the main text.

---

generation time (Reynolds, 1986b). Moreover, a spatial difference within a closed area of water only 45.7 m across is unlikely to persist, as the forcing of the gradient is hardly likely to be stable. A change in wind intensity and direction is likely to redistribute the same population within the same limited space.

We may follow this progression of thinking to the wider confines of an entire small lake, or to the relatively unconfined areas of the open sea. Before that, however, it is opportune to draw attention to a relatively better-known horizontal sorting of phytoplankton at the scale of a few metres and, curiously perhaps, is dependent upon significant wind forcing on the lake surface.

The mechanism concerns the Langmuir circulations, which are consequent upon a strong wind acting on a shallow surface layer, when accelerated dissipation from a spatially constrained volume generates ordered structures. These are manifest as stripe-like 'windrows' of foam bubbles on the water surface. Even now, the formation of Langmuir cells is imperfectly understood but their main properties are fairly well described (see Box 2.1).

Although the characteristic current velocities prevalent within Langmuir circulations (>10–20 mm s^{-1}) would be well sufficient to entrain phytoplankton around the spiral trajectories, the cells do have identifiable relative dead

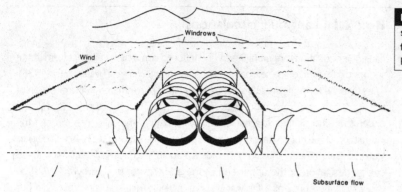

Figure 2.26 Diagrammatic section across wind-induced surface flow to show Langmuir circulations. Redrawn from Reynolds (1984a).

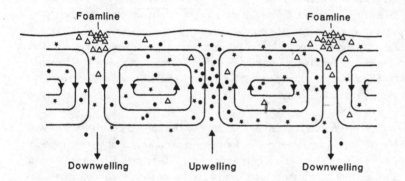

Figure 2.27 Schematic section through Langmuir rotations to show the likely distributions of non-buoyant (●), positively buoyant (△) and neutrally buoyant, fully entrained (*) organisms. Based on an original in George (1981) and redrawn from Reynolds (1984a).

spots, towards the centre of the spiral, at the base of the upwelling and, especially, at the top of the convergent downwellings, marked by the foamlines (see Fig. 2.26). Smayda (1970) predicted the distributions of planktic algae, categorised by their intrinsic settling velocities, within a cross-section adjacent to Langmuir spirals. Independent observations by George and Edwards (1973) and Harris and Lott (1973) on the distributions of real (*Daphnia*) and artificial (paper) markers in the field lent support for Smayda's predictions. Although mostly well-entrained, sinking particles ($\rho_c > \rho_w$) take longer to clear the upwellings and accumulate selectively there, buoyant particles ($\rho_c < \rho_w$) will similarly take longer to clear the downwellings and those entering the foamline will tend to be retained. A schematic, based on figures in Smayda (1970) and George (1981), is included as Fig. 2.27.

Such distributions of algae are not easy to verify by traditional sampling–counting methods, because the behaviour depends not only on the match of the necessary physical conditions – the circulating velocity, the width and penetration of the rotations are all wind-influenced – but their

persistence (Evans and Taylor, 1980). Whereas it may take some minutes to organise and generate the circulation, a wind of fluctuating speed and direction will be constantly initiating new patterns and superimposing them on previous ones. This behaviour does not suppress the fact that larger, more motile plankters remain liable to crude sorting, on the basis of their individual buoyant properties, into a horizontal patchiness at the relatively small scales of a few metres to a few tens of metres.

Patchiness in small lake basins

With or without superimposed Langmuir spirals, the horizontal drift is likely, at least in lakes, to be interrupted by shallows, margins or islands, where the flow is subject to new constraints. Supposing that little of the drifting water escapes the basin, most is returned upwind in subsurface countercurrents (see Imberger and Spigel, 1987). In small basins, there is a clear horizontal circulation, which George and Edwards (1976) analogised to a conveyor belt. While this process seems destined towards the basin-scale horizontal integration of populations, the movements of

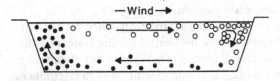

Figure 2.28 Whole-lake 'conveyor belt' model of non-buoyant (•) and positively buoyant (○) phytoplankters, proposed by George and Edwards (1976). Redrawn from Reynolds (1984a).

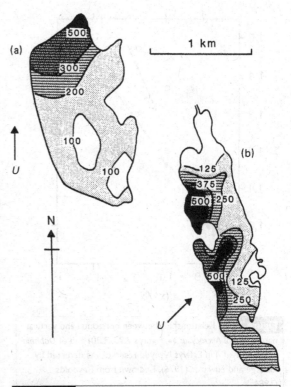

Figure 2.29 Advective horizontal patchiness of phytoplankters in relation to wind direction: (a) positively buoyant *Microcystis* in Eglwys Nynydd reservoir (after George and Edwards, 1976), isopleths in μg chlorophyll a L^{-1}; (b) surface-avoiding, motile *Ceratium* in Esthwaite Water (after Heaney, 1976), isopleths in cells mL^{-1}. Redrawn from Reynolds (1984a).

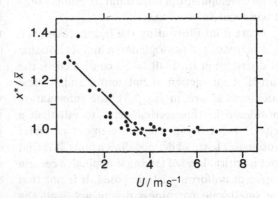

Figure 2.30 The relationship between horizontal patchiness of *Microcystis* in the shallow Eglwys Nynydd reservoir and wind velocity (U). Data of George and Edwards (1976) and redrawn from Reynolds (1984a).

plankters in the vertical plane may well super-impose a distinct *advective* patchiness in the horizontal plane. The mechanism is analogous to the behavioural segregation in the Langmuir circulation, though on a larger scale. Put simply, the upward movement of buoyant organisms is enhanced in upwind upwellings but resisted in downwind downwellings; conversely, sinking organisms accelerate in downwellings but accumulate in the upcurrents. Positively buoyant organisms accumulate on downwind (lee) shores; negatively buoyant organisms are relatively more abundant to windward (Fig. 2.28).

Such distributions of zooplankton have been observed, with concentrations of downward-swimming crustaceans collecting upwind (Cole-brook, 1960; George and Edwards, 1976). Similar patterns have been described for downward-migrating dinoflagellates (Heaney, 1976; George and Heaney, 1978); on the other hand, the down-wind accumulation of buoyant *Microcystis* has been verified graphically by George and Edwards (1976). Representative maps of these contrasting outcomes are shown in Fig. 2.29.

The representation in Fig. 2.29a is one of a number of such 'snapshots' of variable patchiness during a long period of *Microcystis* dominance in the Eglwys Nynydd reservoir. The field data allowed George and Edwards (1976) to calculate a crowding index of horizontal patchiness (x^*), analogous to that solved for the vertical dimension (See Section 2.7.1), and to show that its relationship to the mean population ($\bar{x}$) was a close correlative of the accumulated wind effect. Their data are redrawn here (Fig. 2.30) but with the horizontal axis rescaled as an equivalent steady wind speed. Patchiness is strongest

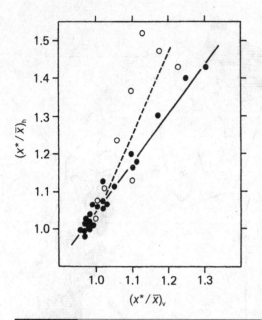

Figure 2.31 Relationships between horizontal and vertical patchiness of *Microcystis* (•, Figures 2.23, 2.30) and of *Daphnia* populations (○) in Eglwys Nynydd reservoir, as detected by George and Edwards (1976). Redrawn from Reynolds (1984a).

Large-scale patchiness

Nevertheless, the relationship does have a time dimension, the horizontal mixing time, and this may accommodate other sources of change. For instance, if a constant wind of $4\ m\,s^{-1}$ induces a surface drift of the order of $400\ m\,d^{-1}$ across a 1-km basin, the probable mixing time is 5 d. If, in the same 5 d, a patchy population is recruited through one or more successive doublings, then the same probability of its achieving uniformity requires stronger forcing or a shorter mixing time. This relationship between transport and recruitment becomes increasingly prevalent in larger lake basins where the role of the return current in establishing uniformity is progressively diminished: maintenance of large-scale patches (kilometers, days) needs persistent spatial differences in recruitment rate. The latter might be due directly to a local enhancement in organismic replication (because of warm or shallow water, or a point source of nutrient) or to consistently enhanced removal rates by local aggregations of herbivorous animals. However, to be evident at all, the patch must give way to the concentrations in a surrounding larger stretch of water, through diffusion and erosion by hydraulic exchanges at the periphery. Several publications have considered this relationship. Two of these, in particular (Skellam, 1951; Kierstead and Slobodkin, 1953), have given us the so-called KISS explicative model, relating the critical size of the patch to the interplay between the rates of reproductive recruitment and of horizontal diffusivity. Specifically, Kierstead and Slobodkin (1953) predicted the radius of a critical patch (r_c) as:

$$r_c = 2.4048(D_x/k_n)^2 \qquad (2.37)$$

where D_x is the horizontal diffusivity and k_n is the net rate of population increase or decrease. Interpolating values for k_n appropriate to the generation times of phytoplankton (the order of 0.1 to 1.0 doublings per day) and for typical wind-driven diffusivities ($D_x \sim 5 \times 10^{-3}$ to 2×10^{-6} $cm^2\,s^{-1}$: Okubo, 1971), critical radii of 60 m to 32 km may be derived. This 3-order range spans the general cases of large-scale phytoplankton patches in the open ocean considered by (for instance) Steele (1976), and Okubo (1978), with

when winds are light but it weakens as winds start to exceed $3\ m\,s^{-1}$, disappearing altogether at $U > 5\ m\,s^{-1}$. The work on *Ceratium* in Esthwaite Water (Heaney, 1976; George and Heaney, 1978; Heaney and Talling, 1980a, b) points consistently to the development of horizontal patchiness only at wind speeds $< \sim 4\ m\,s^{-1}$.

Apart from illustrating the link between vertical behaviour of phytoplankton and its horizontal distribution in small lakes, confirmed in the statistical interaction of horizontal and vertical patchiness shown in Fig. 2.31, the information considered in this section helps to establish a general point about the confinement of water motion to a basin of defined dimensions. It is that once a critical level of forcing is applied, a certain degree of uniformity is reimposed. It is not that the small-scale patchiness disappears – all the causes of its creation remain intact – so much as that the variance at the small scales becomes very similar at larger ones: small-scale heterogeneity collapses into large-scale homogeneity.

the most probable cases having a critical minimum of $\sim$1 km (review of Platt and Denman, 1980; see also Therriault and Platt, 1981).

Even under the most favourable conditions of low diffusivities and localised rapid growth, patches smaller than 1 km are liable to rapid dispersion. Moreover, wind-driven diffusivity may be considerably enhanced by other horizontal transport mechanisms, including by flow in river channels (see Smith, 1975), tidal mixing in estuaries (data of Lucas et al., 1999) and, in stratified small-to-medium lakes, by internal waves (Stocker and Imberger, 2003; Wüest and Lorke, 2003). In spite of this, some instances of small patch persistence are on record. Reynolds et al. (1993a) reported a set of observations on an intensely localised explosive growth of Dinobryon in Lake Balaton, following a mass germination of spores disturbed by dredging operations. The increase in cell concentration within the widening patch was overtaken after a week or two, partly through dispersal in the circulation of the eastern basin of the lake and into that of the western end but, ultimately, because the rates of Dinobryon growth and recruitment soon ran down.

At the other end of the scale, satellite-sensed distributions of phytoplankton in the ocean reveal consistent areas of relatively high biomass, covering tens to hundreds of kilometres in some cases – usually shallow shelf waters, well supplied by riverine outflows, or along oceanic fronts and at deep-water upwellings (see review of Falkowski et al., 1998). The size and long-term stability of these structures are due to the geographical persistence of the favourable conditions that maintain production (shallow water, enriched nutrient supply) relative to the rates of horizontal diffusivity in these unconfined locations.

Such behaviour is observable in larger lakes, especially where there are persistent gradients (chiefly in the supply of nutrients) that survive seiching. Enduring patchiness was memorably demonstrated by Watson and Kalff (1981) along a persistent nutrient gradient in the ribbon-like glacial Lake Memphrémagog (Canada/USA). Persistent gradients of phytoplankton concentration are evident from long-term surveys of the North American Great Lakes (Munawar and Munawar, 1996, 2000). Of additional interest in

large, northern continental lakes is the vernal patchiness of phytoplankton attributable to the early-season growth in the inshore waters that are retained by horizontal temperature gradients associated with the centripetal seasonal warming – the so-called 'thermal bar'. This phenomenon, first described in detail by Munawar and Munawar (1975) in the context of diatom growth in Lake Ontario, has been reported from other large lakes: Issyk-kul (Shaboonin, 1982), Ladozhskoye, Onezhskoye (Petrova, 1986) and Baykal (Shimarev et al., 1993, Likhoshway et al., 1996).

In general, it is fair to say that the KISS model is illustrative rather than deterministic, and it is only imprecisely applicable to a majority of small lakes subject to internal circulation and advection. Here, the predictive utility of the later general model derived by Joseph and Sendner (1958) is sometimes preferred. The fitted equation is used to predict critical patch radius as a function of the advective velocity, u_s:

$$r_c = 3.67(u_s/k_n) \qquad (2.38)$$

If k_n is one division per day and $u_s = 5 \times 10^{-3}$ m s^{-1} (roughly what is generated by a wind force of 4 m s^{-1}), $r_c \sim 1.6$ km. At five times the rate of horizontal advection ($u_s = 25 \times 10^{-3}$ m s^{-1}), the critical radius is increased to $\sim$8 km. Again, the actual values probably have less relevance than does the principle that patchiness in phytoplankton developing in lake basins less than 10 km in diameter is likely to be temporally transient and not systematically persistent.

Relevance of patchiness

Many of the mysteries of patchiness that concerned plankton scientists in the third quarter of the twentieth century may have been cleared up, but the issue remains an important one, for two main reasons. One, self-evidently, lies in the design of sampling strategies. If the purpose is merely to characterise the community structure, much information may be yielded from infrequent samples collected at a single location (Kadiri and Reynolds, 1993) but, as soon as the exercise concerns the quantification of plankton populations and the dynamics of their change, it is essential to intensify the sampling in both time and space. Sampling design is covered in many methodological manuals (see, for instance, Sournia, 1978) but it is often useful to follow specific case studies where temporal and spatial variability needed to be resolved statistically (Moll and Rohlf, 1981).

The second reason is the perspective that is required for the ecological interpretation of information on the structure and distribution of planktic communities in nature. This is crucial to the ideas to be developed in subsequent chapters of this book. It is not just a case of defining the confidence intervals of quantitative deductions about organisms whose distribution has long been regarded as non-random, and over a wide range of scales (Cassie, 1959; McAlice, 1970). In order to sort out the multiple constraints on the selection, succession and sequencing of natural phytoplankton populations, it is always necessary to distinguish the dynamic driver from the starting base. The assemblage that is observed in a give parcel of water at given place and at a given time is unrelated to the present conditions but is the outcome of a myriad of processing constraints applying to a finite inoculum of individual organisms of historic and probably inexplicable provenance.

In this chapter, the focus has spanned the dissipation of a fortuitous and localised recruitment of algae in a relatively small, shallow lake through to the relative uniformity of the plankton composition of an oceanic basin. The one may depend upon the rapidity of growth in relation to diffusivity (Reynolds et al., 1993a); the other upon the extent of a single and possibly severe growth constraint over an extensive area of open ocean (Denman and Platt, 1975). Thus, it is important to emphasise that although the entraining motions and horizontal diffusivity of pelagic water masses influence profoundly the distribution of phytoplankton, they do not confine organisms to a fixed position in relation to the motion. Transport in the constrained circulation of a small lake or passage in an open ocean current each sets a background for the dynamics of change and variations in composition. The various outcomes arising from differing relative contributions of the same basic entraining processes are remarkably disparate. On the one hand, we can explain the seasonal formation of cyanobacterial plates in deep alpine lakes (Bright and Walsby, 2000). On the other, there is no conceptual objection to the inability of McGowan and Walker (1985) to demonstrate any significant variation in the rank-order of species abundances in the North Atlantic at spatial scales up to 800 km, despite strong small-scale spatial and interseasonal and interannual heterogeneity. This is attributable to the large-scale coherences in the basin-scale forcing functions (direct measurements of the mid-depth circulation of this part of the Atlantic Ocean have been given by Bower et al., 2002). Analogous relationships among planktic components have been shown to be widespread through extensive circulation provinces of the tropical and subtropical Atlantic Ocean (Finenko et al., 2003), despite significant intracompartmental variability in abundance and considerable intercompartmetal structural differences. The observations demonstrate the nature of the interaction of population dynamics with the distinguishable movements of the water masses in shaping the species structures of the pelagic.

2.8 | Summary

The chapter explores the nature of the relationship that phytoplankters have with the physical properties of their environment. Water is a dense, non-compressible, relatively viscous fluid, having aberrent, non-linear tendencies to expand and contract. The water masses of lakes and seas are subject to convection generated by solar heating and, more especially, by cooling heat losses from the surface. These motions are enhanced in the surface layers interfacing with the atmosphere, where frictional stresses impart mechanical energy to the water through boundary-layer wave generation and frictional drag. Diel cycles of insolation, geographical variations in heating and cooling, atmospheric pressure and the amplifying inertia caused by the rotation of the Earth represent continuous but variable drivers of motion in aquatic environments. These can rarely be regarded as still: water is continuously in motion. However, the viscous resistance of the water determines that the introduced motion is damped and dissipated through a spectrum of turbulent eddies of diminishing size, until molecular forces overwhelm the residual kinetic energy. Instrumentation confirms emerging turbulence theory about the extent of water layers subject to turbulent mixing and the sizes of the smallest eddies (generally around 1 mm), which, together, most characterise the medium in which all pelagic organisms, and phytoplankton in particular, have to function.

The most striking general conclusion is that most phytoplankters experience an immediate environment that is characteristically viscous. Yet the physical scale is such that individuals of most categories of plankter (those less than 0.2 mm in size) and their adjacent media are liable to be transported wherever the characteristic motion determines. From the standpoint of the plankter, the important criteria of the turbulent layer are its vertical extent (h_m) and the rate at which it dissipates its turbulent kinetic energy (E). Both are related to the intensity of the turbulence (u^{*2}) and, thus, to the turbulent velocity (u^*).

The traditional supposition that the survival strategy of phytoplankton centres on an ability to minimise sinking is carefully updated in the context of pelagic motion. Extended residence in the upper water layers remains the central requirement at most times. This is attained, in many instances, by maximising the entrainability of the plankter within the motion. Viewed at a slightly larger scale, many phytoplankters optimise their 'embedding' within the surface mixed layer. Criteria for plankter entrainment are considered – for it can only be complete if the plankter has precisely the same density as (or is isopycnic with) the suspending aqueous medium. Even were this always desirable, it would be difficult to attain. Not only does the water vary in density with temperature and solute-content but the components of phytoplankton cells are rather more dense than water (typically amounting to 1020–1263 kg m^{-3}, compared to <1000; Table 2.3). Following Humphries and Imberger (1982), relative entrainment (Ψ) is instead suggested to be governed by the relationship between particle buoyancy and turbulent diffusivity. Effectively, in order to achieve turbulent entrainment, an alga's sinking rate, w_s (or its flotation rate, $-w_s$) must be exceeded by u^* by a factor of ≥ 15. Thus, the best descriptor of algal entrainability turns out to be its sinking rate and, the greater is the adaptive ability to minimise it, the better able is the alga to contribute to its persistence in an adequately mixed water column.

The adaptive mechanisms for lowering the sinking velocity are reviewed in the context of the Stokes equation and its various derivatives. Of the equation components, only particle size, particle density and particle form resistance are considered subject to evolutionary or behavioural adaptation. Examples of each adaptation are quantified. Adaptations to control or offset density and the beneficial effects of distortion from the spherical form are demonstrated. The consequences of chain formation and cylindrical elongation (into filaments) on sinking rate are explored and the effects of cell aggregation to form the distinctive coenobia of *Asterionella* and *Fragilaria* are evaluated. In relation to the presumed vital regulatory component in sinking rate, some possible mechanisms are discussed. Some of the explanations offered are eliminated but there remain others that await careful investigation.

Some larger, motile organisms are successful plankters by virtue of adaptations that are antithetical to increasing entrainability. Large, motile species of *Microcystis*, *Volvox*, *Ceratium* and *Peridinium* combine relatively large size, motility and shape-streamlining to be able to escape moderate-to-low turbulent intensities in order to perform controlled migrations, at rates of several metres per day. Reducing sinking rate is far from being a unique or universal adaptation qualifying microorganisms for a planktic existence.

In the later sections of the chapter, various types of behaviour are illustrated through specific examples of the vertical distributions of planktic algae in relation to the increased differentiation of density structure in the water column. The impacts are extrapolated to horizontal distribution and to the instances of small-scale patchiness and advective patchiness in small lakes, resolving in terms of algal migratory speeds in relation to the velocity of advective currents. The viability and persistence of phytoplankton patches in expansive, large-scale systems, where return currents are extremely remote, relate to the comparative rates of recruitment within the patch and of the erosion at the patch periphery. Some case studies are presented to show some very contrasted large-scale outcomes, distinguishing enduring community similarities over 800 km in the horizontal from sharply localised patches in systems able to support ongoing or persistent rapid recuitment of organisms from point sources.

Many different distributional outcomes can be explained by the behaviour and dispersiveness of particular species in given systems, although the subjugated deployment of the same processes elsewhere may contribute to the formation of quite different patterns.

Chapter 3

Photosynthesis and carbon acquisition in phytoplankton

3.1 | Introduction

The first aim of this chapter is to summarise the biochemical basis of photosynthesis in planktic algae and to review the physiological sensitivities of carbon fixation and assimilation under the environmental conditions experienced by natural populations of phytoplankton. These fundamental aspects of autotrophy are plainly relevant to the dynamics and population ecology of individual algal species, functioning within the constraints set by temperature and by the natural fluxes of light energy and inorganic carbon. They are also relevant to the function of entire pelagic systems as, frequently, they furnish the major source of energy, in the form of reduced carbon, to heterotrophic consumers. The yields of fish, birds and mammals in aquatic systems are ultimately related to the harvestable and assimilable sources of carbon bonds. In turn, the energy and resource fluxes through the entire biosphere are greatly influenced by pelagic primary producers, impinging on the gaseous composition of the atmosphere and the heat balance of the whole planet.

Here, we shall be concerned with events at the population, community and ecosystem levels. However, it is necessary to emphasise at the start of the chapter that recent advances in understanding of planetary carbon stores and fluxes assist our appreciation of the relative global importance of aquatic photosynthesis. To those biologists of my generation brought up with the exclusive axiom that animals derive their energy by respiring (oxidising) the carbohydrates and proteins manufactured (reduced) by photosynthesising plants, the presently perceived realities of aquatic-reductant fluxes may seem quite counter-intuitive. The original postulate is not in error: it succeeds in describing how a section of the trophic relationships of the pelagic is conducted. It is just that it is far from being the whole story. For instance, the photosynthetic reduction of carbon is antedated by several hundreds of millions of years (≥ 0.4 Ga: Falkowski, 2002) by the chemosynthesis by Archaeans of reduced carbon. This continues to be maintained in deep-ocean hydrothermal vents, where there is no sunlight and only minimal supplies of organic nutrients (Karl et al., 1980; Jannasch and Mottl, 1985). Even in the upper, illuminated waters of lakes and seas, most (perhaps 60–95%) of the organic carbon present is not organismic but in solution (Sugimura and Suzuki, 1988; Wetzel, 1995; Thomas, 1997). A large proportion of this is humic in character and, thus, supposed to be derived from terrestrial soils and ecosystems. True, much of this carbon would have been reduced orginally through terrestrial photosynthesis but the extent of its contribution to the assembly of marine biomass is still not fully clear. Setting this aside, the direct phagotrophic transfer of photosynthetic primary products from phytoplankton to zooplanktic consumers is not universally achieved in the pelagic but is, in fact, commonly mediated by the activities of free-living microbes. Thus, the dynamic relationships among phytoplankton and their potential phagotrophic consumers acquire a new interpretative significance, which is to be addressed in this and later chapters.

The present chapter prepares some of the ground necessary to understanding the relation of planktic photoautotrophy to the dynamics of phytoplankton populations. After considering the biochemical and physiological basis of photosynthetic production, the chapter compares the various limitations on the assembly of photoautotrophic biomass in natural lakes and seas, and it considers the implications for species selection and assemblage composition.

3.2 | Essential biochemistry of photosynthesis

It has been stated or implied several times already that the paramount requirement of photoautotrophic plankton to prolong residence in, or gain frequent access to, the upper, illuminated layers of the pelagic is consequential upon the requirement for light. The need to capture solar energy in order to drive photosynthetic carbon fixation and anabolic growth is no different from that experienced by any other chlorophyll-containing photoautotroph inhabiting the surface of the Earth. Indeed, the mechanisms and ultrastructural provisions for bringing this about constitutes one of the most universally conserved processes amongst all photoautotrophic organisms. On the other hand, to achieve, within the bounds of an effectively opaque and fluid environment, a net excess of energy harvested over the energy consumed in metabolism requires certain features of photosynthetic production that are peculiar to the plankton. Thus, our approach should be to rehearse the fundamental requirements and sensitivities of photosynthetic production and then seek to review the aspects of the pelagic lifestyle that constrain their adaptation and govern their yields.

Enormous strides in photosynthetic chemistry have been made, especially over the last 30 years or so, especially at the molecular and submolecular levels (Barber and Anderson, 2002). This progress is not likely to stop so that, undoubtedly, whatever is written here will have soon

been overtaken by new information. At the same time, it is possible to predict that future progress will concern the biochemical and biophysical intricacies of control and regulation more than the broad principles of process and order-of-magnitude yields, which are generally accepted by physiological ecologists. Thus, the contemporary biochemical basis for assessing phytoplankton production will continue to be valid for some time to come.

Photosynthesis comprises a series of reactions that involve the absorption of light quanta (photons); the deployment of power to the reduction of water molecules and the release of oxygen; and the capture of the liberated electrons in the synthesis of energy-conserving compounds, which are used ultimately in the Calvin cycle of carbon-dioxide carboxylation to form hexose (Falkowski and Raven, 1997; Geider and MacIntyre, 2002). The aggregate of these reactions may be summarised:

$$H_2O + CO_2 + photons = 1/6[C_6H_{12}O_6] + O_2 \quad (3.1)$$

As with most summaries, Eq. (3.1) omits not merely detail but several important intermediate feedback switches, involving carbon, oxygen and reductant, all of which have a bearing upon the output products and their physiological allocation in active phytoplankters. These are best appreciated against the background of the supposed 'normal pathway' of photosynthetic electron transport. The latter was famously proposed by Hill and Bendall (1960). Their z-model of two, linked redox gradients (photosystems) has been well substantiated, biochemically and ultrastructurally. In the first of these (perversely, still referred to as photosystem II, or PSII), electrons are stripped, ultimately from water, and transported to a reductant pool. In the second (photosystem I, or PSI), photon energy is used to re-elevate the electrochemical potential sufficiently to transfer electrons to carbon dioxide, through the reduction of nicotinamide adenine dinuceotide phosphate (NADP to NADPH).

The (Calvin cycle) carbon reduction is based on the carboxylation reaction. Catalysed by ribulose 1,5-biphosphate carboxylase (RUBISCO), one molecule each of carbon dioxide, water and ribulose 1,5-biphosphate (RuBP) react to yield two molecules of the initial fixation product, glycerate 3-phosphate (G3P). This latter reacts with ATP and NADPH to form the sugar precursor, glyceraldehyde 3-phosphate (GA3P), which now incorporates the high energy phosphate bond. In the remaining steps of the Calvin cycle, GA3P is further metabolised, first to triose, then to hexose, and RuBP is regenerated.

At the molecular level, photosynthetic reactivity is plainly sensitive to the supply of carbon and water, the photon harvesting and, like all other biochemical processes, to the ambient temperature. Measurement of photosynthesis may invoke a yield of fixed carbon, the quantum efficiency of its synthesis (yield per photon), or the amount of oxygen liberated. None of these is any longer difficult to quantify but the difficulty is still the correct interpretation of the bulk results. It is still necessary to consider carefully the regulatory role of the ultrastructural and biochemical components that govern the photosynthesis of phytoplankton. Special attention is directed to the issues of photon harvesting, the internal electron transfer, carbon uptake, RUBISCO activity and the behaviour of the regulatory safeguards that phytoplankters invoke in order to function in highly variable environments.

3.2.1 Light harvesting, excitation and electron capture

Light is the visible part of the spectrum of electromagnetic radiation emanating from the sun. Electromagnetic energy occurs in indivisible units, called quanta, that travel along sinusoidal trajectories, at a velocity (in air) of $c \sim 3 \times 10^8$ m s^{-1}. The wavelengths of the quanta define their properties – those with wavelengths (λ) between 400 and 700 nm (400 – 700 × 10^{-9} m) correspond with the visible wavelengths we call light (and within which waveband the quanta are called *photons*). The waveband of *photosynthetically active radiation* (PAR) coincides almost exactly with that of light. The white light of the visible spectrum is the aggregate of the flux of photons of differing wavelengths, ranging from the shorter (blue) to the longer (red) parts of the spectrum.

Relative to the solar constant (see Section 2.2.2), the PAR waveband represents some 46–48% of the total quantum flux. The corresponding photon flux density averages 1.77×10^{21} m^{-2} s^{-1}. Division by the Avogadro number (1 mol = 6.023×10^{23} photons) expresses the maximum flux in the more customary units, einsteins or mols, 2.94 mmol photon m^{-2} s^{-1}. The energy of a single photon, $\acute{\varepsilon}$, varies with the wavelength,

$$\acute{\varepsilon} = h'c/\lambda \qquad (3.2)$$

where h' is Planck's constant, having the value 6.63×10^{-34} J s (e.g. Kirk, 1994). Photons at the red end of the PAR spectrum each contain about 2.84×10^{-19} J, about 57% of the content of blue-light photons (4.97×10^{-19} J).

While a given radiation flux of light of a single wavelength can be readily expressed in J s^{-1} (and vice versa), precise conversion across a spectral band does not apply. The approximate relationship proposed by Morel and Smith (1974) for the interconversion of solar radiation in the 400–700 nm band of 2.77×10^{18} quanta s^{-1} W^{-1} (equivalent to 3.62×10^{-19} J per photon, or 218 kJ per mol photon) has general applicability (Kirk, 1994).

Photosynthesis depends upon the interception and absorption of photons. Both photosystems involve the photosynthetic pigment chlorophyll *a* (and, where applicable, other chlorophylls), which is characteristically complexed with particular proteins, and certain other pigments in many instances. These are accommodated within structures known as light-harvesting complexes (LHC) and it is these that act as antennae in picking up incoming photons. For instance, the light-harvesting complex of the eukaryotic photochemical system II (LHCII) typically comprises some 200–300 chlorophyll molecules (mostly of chlorophyll *a*; up to 30% may be of chlorophyll *b*), the specific chlorophyll-binding proteins and a variable number of xanthophyll and carotene molecules, to a combined molecular mass of 300–400 kDa (Dau, 1994; Goussias *et al.*, 2002). The prokaryotic Cyanobacteria lack chlorophyll *b* and the light-harvesting chlorophyll-proteins of PSII. They rely instead on the phycobiliproteins, assembled in bodies known as phycobilisomes (Grossman *et al.*, 1993; Rüdiger, 1994).

At the heart of the eukaryote LHCII is the antennal chlorophyll-protein known as P_{680}. It is here that the reactions of PSII are initiated, when the complex is exposed to light. The energy of a single photon is sufficient to raise a P_{680} electron from its ground-state to its excited-state orbital. Next to the P_{680} is the phaeophytin acceptor molecule (usually referred to as 'Phaeo') and the two further acceptor quinones (Q_A and Q_B) that comprise the PSII reaction centre. In sequence, this acceptor chain passes the electrons to PSI. The reaction ($P_{680} \rightarrow P_{680}^+$) is one of the most powerful biological oxidations known to science; the electrons are readily captured by the Phaeo acceptor. In its now-reduced state, Phaeo$^-$ in turn activates the Q_A acceptor: its reduction to Q_A^- stimulates acceptance of the electron by Q_B.

In this way, the electrons are serially transported towards PSI. Once it has accepted two electrons, Q_B dissociates to enter a pool of reduced plastoquinone ('PQ'). Molecules of PQH_2 are eventually oxidised by the cytochrome known as b_6/f, which carries the electrons to PSI.

The plastoquinone pool functions as a system capacitor, like a sort of surge tank of reductant (D. Walker, 1992; Kolber and Falkowski, 1993), whose activity can be viewed in the context of PSII light harvesting. At quiescence, the entire reaction centre is said to be 'open': P_{680} is in its reduced state, Phaeo and Q_A are oxidised. Then, photon excitation of the P_{680} initiates a flow of electrons to the plastoquinine pool, whence they may be removed as rapidly as PSI can accept them. At the same time, the otherwise uncomplemented positive charge of excited P_{680}^+ is balanced by the stripping of electrons from water (that is, P_{680}^+ is reduced back to P_{680}). Note that four photochemical reactions are necessary to generate one dioxygen molecule from two molecules of water ($2H_2O \rightarrow 4H^+ + 4e + O_2$). It is now understood that P_{680}^+ is actually reduced through the action of manganese ions, via a redox-active tyrosine (Barber and Nield, 2002). However, until the P_{680}^+ molecule is re-reduced, the reaction centre is unable to accept further electrons and it is said to be 'closed'. It remains so until Q_A is reoxidised.

The light-harvesting complex and reaction centre of PSI are built around an analogous chlorophyll-protein complex (known as P_{700}) and acceptor (usually denoted A). Again, photons excite the equivalent number of P_{700} electrons to the point where they can be accepted by A. Next in the electron transfer pathway is ferredoxin, transfer of electrons to which reoxidises A$^-$ while donation of the equivalent number from the plastoquinone pool re-reduces P_{700} molecules. The electrons may be passed from ferredoxin, and beyond PSI, to bring about the reduction of NADP to NADPH that provides power to drive Calvin-cycle carboxylation. The subsequent reactions of NADPH with carbon dioxide are not directly dependent upon the photon flux and can continue in darkness (see Section 3.2.3).

The PSI generation of the carbon-reducing power nevertheless also requires the photosynthetic transfer of four electrons per atom of carbon. Under ideal conditions, the light reactions in photosynthesis may be summarised:

$$2NADP + 3ADP + 3P + 2H_2O + 8e$$
$$\rightarrow 2NADPH + 3ATP + 3P + 2H^+ + O_2 \quad (3.3)$$

3.2.2 Photosystem architecture

The electron transfer that this equation represents is readily facilitated by the physical arrangement of the two main components (PSII, PSI) and the intercoupling plastoquinone pool, the b_6/f cytochrome complex and, in most algae and plants, the soluble electron carrier plastocyanin (in Cyanobacteria, cytochrome c may substitute). The basic architecture and the location of the biochemical functions of the photosynthetic units seems to be extremely well conserved among eukaryotic algae, plants and their ancestral cyanobacterial lines. The best-known features were revealed long ago, through light microscopy and early transmission electron microscopy. The granule-like units, comprising LHC antennae and the reaction centres, are strung on proteinaceous membranes, called thylakoids. In the cells of eukaryotes, stacks of thylakoids are contained within one or more separate membrane-bound envelopes, the chromophores (also called plastids or, where they occur in chlorophyte algae and all higher plants, chloroplasts) whose shape and arrangement is often taxon-specific. Cyanobacteria lack separate chromophores; the thylakoids

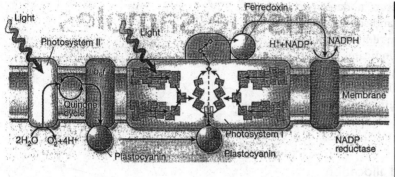

Figure 3.1 Diagram of the configuration of the structure and the flow of excitation energy through the photsystems. Electrons are extracted from water in photosystem II and transported through the quinone cycle and released to photosystem I. Electrons are accepted by ferredoxin, to bring about the reduction of NADP to NADPH that enables the cell to synthesise its molecular components. Redrawn, with permission from Kühlbrandt (2001).

are rather loosely dispersed through the body of the cell. Apart from anchoring the various transmembrane structures (including, in the case of the Cyanobacteria, the phycobilisomes), the thylakoid also maintains a regulatory charge gradient, down which the electrons are passed.

The molecular structure of the energy-harvesting apparatus has become clearer as a result of the recent application of electron crystallography. Since Kühlbrandt and Wang (1991) published the three-dimensional structure of a light-harvesting complex, other investigative studies have followed, showing, at increasingly fine resolution, the organisation and interlinkages of the major sub-units of PSII in plants and Cyanobacteria (McDermott *et al.*, 1995; Zouni *et al.*, 2001; Barber and Nield, 2002) and also of PSI (Jordan *et al.*, 2001). The recent overview and model of Fromme *et al.* (2002) upholds that proposed by Kühlbrandt *et al.* (1994) and updates it in several respects.

The cited literature should be consulted for more of the fascinating details of the structures and organisational patterns of light harvesting and the electron-transport chain. Here, we should emphasise the generalised configuration and functional dynamics of the various subunits involved in photon absorption and electron capture, for it is these which impinge upon their physiological performance and their adaptability to operation under sub-ideal conditions. As will be seen (in Section 3.3), the relevant outputs of an adequate carbon-reducing capacity relate to system performance under ambient light fluxes, how it behaves in poor light (low photon fluxes)

and, equally, its reactions to damagingly high light levels. The relevant ultrastructural and biochemical input parameters concern how much light-harvesting capacity there is present in an alga and how much reductant it can deliver per unit time.

The arrangement and linkage of the photosystems are schematised in Fig. 3.1. The size of the LHCII structures studied by Kühlbrandt *et al.* (1994) averaged 13 nm in area and 4.8 nm in thickness. The PSII complexes from *Synechococcus* measured roughly 19×10 nm across and 12 nm thick (Zouni *et al.*, 2001). The LHCI complexes from *Synechococcus* revealed by Jordan *et al.* (2001) are apparently of similar size. On the basis of there being 200–300 chlorophyll molecules in a typical LHC, Reynolds (1997a) calculated that 1 g chlorophyll could be organised into 2.2 to 3.4×10^{18} LHCs. Because the area that 1 g of chlorophyll subtends in the light field can be as great as $20\,m^2$ (see Section 3.3.3), each LHC contributes an average photon absorption of up to $10 \times 10^{-18}\,m^{-2}$ (i.e. $10\,nm^2$).

It was also supposed that the photon absorption is in inverse proportion to the product of the area of the LHC and the aggregate time for the electron transport chain to accept photons and clear electrons, ready for the next photon. Kolber and Falkowski (1993) approximated the aggregate time of reactions linking initial excitation (occupying less than 100 fs, or 10^{-13}: Knox, 1977) to the re-oxidation of Q_A^- to be 0.6 ms. The principal rate-limiting step is the onward passage of electrons from the plastoquinone pool, which, depending upon temperature, needed between 2 and 15 ms.

Thus, a single pathway might accommodate up to 66 reactions per second at 0 °C and some 500 s^{-1} at 30 °C, with a matching carbon-reducing power.

As a rough approximation indicates that, at 30 °C, 1 g chlorophyll containing >2 × 10^{18} active LHCs has the capacity to deliver >10^{21} electrons every second and a theoretical potential to reduce more than 1.25 × 10^{20} atoms of carbon [i.e. > ∼200 μmol C (g chla)$^{-1}$ s^{-1}].

3.2.3 Carbon reduction and allocation

As noted above, the fixation of carbon dioxide occurs downstream of the energy capture, where the reducing power inherent in NADPH is deployed in the synthesis of carbohydrate. The flow of reductant drives the Calvin cycle of RuBP consumption and regeneration, during which carbon dioxide is drawn in and glucose is discharged. The cycle is summarised in Fig. 3.2. In the algae and in many higher plants, RUBISCO-mediated carboxylation of RuBP yields the first stable product of so-called C$_3$ photosynthetic carbon fixation, the 3-carbon glycerate 3-phosphate (G3P). (Note that in this, the process differs from those terrestrial C$_4$ fixers that synthesise four-carbon malate or aspartate.)

After the further NADPH-reduction of G3P to glyceraldehyde 3-phosphate (GA3P), the metabolism proceeds through a series of sugar-phosphate intermediates to yield a hexose (usually glucose). In this way, one molecule of hexose may be exported from the Calvin cycle for every five of GA3P returned to the cycle of RuBP regeneration and, ideally, one for every six molecules of carbon dioxide imported. In this case of steady-state photosynthesis, the following equation summarises the mass balance through the Calvin cycle:

$$CO_2 + 2NADPH + 3ATP + 2H^+$$
$$\rightarrow 1/6C_6H_{12}O_6 + H_2O + 2NADP$$
$$+ 3ADP + 3P \qquad (3.4)$$

According to demand, the glucose may be respired immediately to fuel the energy demands of metabolism, or it may be submitted to the amination reactions leading to protein synthesis. Excesses may be polymerised into polysaccharides (glycogen, starch, paramylon). In this

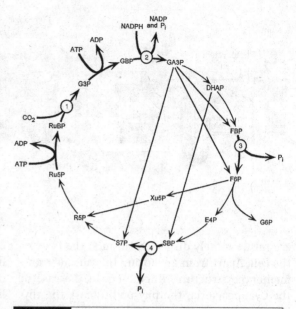

Figure 3.2 The Calvin cycle. Carboxylation by RUBISCO of RuBP at 1 is driven by ATP and NADPH generated by the light reactions of photosynthesis, and results ultimately in the synthesis of sugar precursors and the renewed availability of RuBP substrate, thus maintaining the cycle. The cycle is regulated at the numbered reactions, where it may be short-circuited as shown. Abbreviations: DHAP, dihydroxyaceton phosphate; E4P, erythrose 4-phosphate; FBP, fructose 1,5-biphosphate; F6P, fructose 6-phosphate; GA3P, glyceraldehyde 3-phosphate; GBP, glycerate 1,3-biphosphate; G3P, glycerate 3-phosphate; G6P, glucose 6-phosphate; P$_i$, inorganic phosphate; RuBP, ribulose 1,5-biphosphate; Ru5P, ribulose 5-phosphate; R5P, ribose 5-phospate; SBP, sedoheptulose 1,7-biphosphate; S7P, sedoheptulose 7-phosphate; Xu5P, xylulose 5-phosphate. Redrawn with permission from Geider and MacIntyre (2002).

way, the overall photosynthetic Eq. (3.1) is balanced, at the minimal energy cost of eight photons per atom of carbon fixed. Thus, the theoretical maximum *quantum yield of photosynthesis* (φ) is 0.125 mol C (mol photon)$^{-1}$ (D. Walker, 1992).

However, neither the cycle nor its fixed-carbon yield is immutable but it is subject to deviation and to autoregulation, according to circumstances. As stated at the outset, these have reverberations at successive levels of cell growth, community composition, ecosystem function and the geochemistry of the biosphere. The variability may owe to imbalances in the light harvest and carbon capture, or to difficulties in allocating the carbon fixed. To evaluate these resourcing impacts requires us to look again at the sensitivity of the Calvin-cycle reactions, beginning with the initial carboxylation and the action of the RUBISCO enzyme.

RUBISCO, the catalyst of the CO_2–RuBP conjunction, is a most highly conserved enzyme, occurring, with little variation, throughout the photosynthetic carbon-fixers (Geider and MacIntyre, 2002). From the bacteria, through the 'red line' and the 'green line' of algae (see Section 1.3), to the seed-bearing angiosperms, present-day photosynthetic organisms have to contend with acknowledged catalytical weaknesses of RUBISCO. These are due, in part, to the fact that carbon-dioxide-based photosynthesis evolved under different atmospheric conditions from those that presently obtain. In particular, the progressive decline in the partial pressure of CO_2 exposes the rather weak affinity of RUBISCO for CO_2 (Tortell, 2000). According to Raven (1997), the maximum reported rates of carboxylation (80 mol CO_2 (mol RUBISCO)$^{-1}$s^{-1}: Geider and MacIntyre, 2002) are low compared to those mediated by other carboxylases. Even these levels of activity are dependent upon a significant concentration of carbon dioxide at the reaction site (with reported half-saturation constants of 12–60 µM among eukaryotic algae: Badger et al., 1998). Supposing that the cell-specific rate of carbon fixation could be raised by elevating the amount of active RUBISCO available, the investment in its large molecule (~560 kDa) is relatively expensive. RUBISCO may account for 1–10% of cell carbon and 2–10% of its protein (Geider and MacIntyre, 2002).

Having more RUBISCO capacity is not necessarily helpful either, owing to the susceptibility of RUBISCO to oxygen inhibition: at low CO_2 concentrations (<10 µM) and high O_2 concentrations (>400 µM), RUBISCO functions as an oxidase, in initiating an alternative reaction that leads to the formation of glycerate 3-phosphate and phosphoglycolate. In the steady-state Calvin-cycle operation, the activity of RUBISCO serves to maintain the balance between NADPH generation and the output of carbohydrates. For a given supply of reductant from PSI, the rate of carbon fixation may be seen to depend upon an adequate intracellular carbon supply and upon the RUBISCO capacity or, at least, upon that proportion of RUBISCO capacity that is actually 'active'. To be catalytcally competent, the active site of RUBISCO has also to be carbamylated by the binding of a magnesium ion and a non-substrate CO_2 molecule. Under low light and/or low carbon availability, RUBISCO is inactivated (decarbamylated), by the reversible action of an enzyme (appropriately known as RUBISCO inactivase), to match the slower rate of RuBP regeneration. The resultant down-cycle sequestration of phosphate ions and lower ATP regeneration brings about an increase in ADP:ATP ratio and, thus, a decrease in RUBISCO activity (for further details of Calvin-cycle self-regulation, refer to Geider and MacIntyre, 2002).

The action of RUBISCO inactivase is itself sensitive to the ADP:ATP ratio and to the redox state of PSI. Thus, RUBISCO activity responds positively to a cue of a light-stimulated acceleration in photosynthetic electron flow. With conditions of high-light-driven reductant fluxes and high CO_2 availability at the sites of carboxylation, the limitation of photosynthetic rate switches to the rate of RuBP complexation and renewal, both of which become subject to the overriding constraint of the RUBISCO capacity (Tortell, 2000). However, the kinetics of RUBISCO activity impose a heavy demand in terms of the delivery of carbon dioxide to the carboxylation sites. Although many phytoplankters invoke biophysical mechanisms for concentrating carbon dioxide (see Section 3.4), the relatively high levels needed to saturate the carboxylation function of RUBISCO may frequently be overtaken. Circumstances that combine low CO_2 with the high rates of reductant and oxygen generation possible in strong light are liable to effect the competitive switch to the oxygenase function of RUBISCO and the inception of photorespiration.

Photorespiration is a term introduced in the physiology of vascular plants to refer to the sequence of reactions that commence with the formation of phosphoglycolate from the oxygenation of RuBP by RUBISCO (Osmond, 1981). In the present context, the term covers the metabolism of reductant power and controlling photosynthesis at low CO_2 concentrations. The manufacture of phosphglycolate carries a significant energetic cost through the altered ATP balance (see Raven et al., 2000), though this is partly recouped in the continued (albeit smaller) RUBISCO-mediated contribution of G3P to the Calvin cycle. Meanwhile, the phosphoglycolate is itself dephosphorylated (by phosphoglycolate phosphatase) to form glycolic acid. In the 'green line' of algae (including the prasinophytes, chlorophytes and euglenophytes) and higher plants, this glycolate can be further oxidised, to glyoxalate and thence to G3P. The full sequence of reactions has been called the 'photosynthetic carbon oxidation cycle' (PCOC) (Raven, 1997). In the Cyanobacteria and in the 'red line' of algae, this capacity seems to be generally lacking. When experiencing oxidative stress at high irradiance levels, these organisms cells will excrete glycolate into the medium.

Excreted glycolate is sufficiently conspicuous outside affected cells for its production to have been studied for many years as a principal 'extracellular product' of phytoplankton photosynthesis (Fogg, 1971). It is now known that not only glycolate but also other photosynthetic intermediates and soluble anabolic products are released from cells into the medium. This apparent squandering of costly, autogenic products seemed to be an unlikely activity in which 'healthy' cells might engage (cf. Sharp, 1977). However, it is now appreciated that, far from being a consequence of ill health, the venting of unusable dissolved organic carbon (DOC) into the medium constitutes a vital aspect of the cell's homeostatic maintenance (Reynolds, 1997a). It is especially important, for example, when the producer cells are unable to match other growth-sustaining materials to the synthesis of the carbohydrate base. In natural environments, the DOC compounds thus released – glycolate, monosaccharides, carboxylic acid, amino acids (Sorokin, 1999, p. 64; see also Grover and Chrzanowski, 2000; Søndergaard et al., 2000) – are readily taken up and metabolised by pelagic microorganisms. The far-reaching ecological consequences of this behaviour are explored in later sections of this book (Sections 3.5.4, 8.2.1).

So far as the biochemistry of photosynthesis is concerned, these alternative sinks for primary product make it less easy to be precise about the yields and the energetic efficiency of photosynthetic carbon fixation. The basic equation (3.1) indicates equimolecular exchanges between carbon dioxide consumed and oxygen released (the photosynthetic quotient, PQ, mol O_2 evolved/mol CO_2 assimilated, is 1). In fact, both components are subject to partially independent variation. Oxygen cycling may occur within the photosynthetis electron transfer chain (the Mehler reaction), independently of the amount of carbon delivered through the system. The 'competition' between the carboxylation and oxidation activity of RUBISCO are swayed in favour of oxygen production, photorespiration and glycolate metabolism (Geider and MacIntyre, 2002). The PQ may move from close to 1.0 in normally photosynthesising cells (actually, it is generally measured to be 1.1 to 1.2: Kirk, 1994) to the range 1.2 to 1.8 under high rates of carbon-limited photosynthesis. Low photosynthetic rates under high partial pressures of oxygen may force $PQ < 1$ (Burris, 1981).

The effects on energy efficiency are also sensitive to biochemical flexibility. Taking glucose as an example, the energy stored and released in the complete oxidation of its molecule is equivalent to 2.821 kJ mol^{-1}, or ~470 kJ per mol carbon synthesised. The electron stoichiometry of the synthesis cannot be less than 8 mol photon (mol $C)^{-1}$ but, energetically, the photon efficiency is weaker. The interconversion of Morel and Smith (1974; see above; 1 mol photon ~218 kJ) implies an average investment of the energy of 12.94 photons mol^{-1}. This coincides more closely to the highest quantum yields determined experimentally (0.07–0.09 mol C per mol photon: Bannister and Weidemann, 1984; D. Walker, 1992).

Clearly, even these yields are subject to the variability in the fate of primary photosynthate. Moreover, the alternative allocations of the fixed carbon (whether polymerised and stored, respired, allocated to protein synthesis or excreted) need to be borne in mind. It is well accepted that about half the photosynthate in actively growing, nutrient-replete cells is invested in protein synthesis and in the replication of cell material (Li and Platt, 1982; Reynolds et al., 1985). However, this proportion is very susceptible to the physiological stresses experienced by plankters in their natural environments as a consequence of low light incomes, carbon deficiencies or severe nutrient depletion. These effects are explored in subsequent sections.

3.3 | Light-dependent environmental sensitivity of photosynthesis

In this section, the focus moves towards the physiology of photosynthetic behaviour of phytoplankton in natural lakes and seas, especially its relationship with underwater light availability. According to a recently compiled history of phytoplankton productivity (Barber and Hilting, 2002), quantification of pelagic photosynthesis developed through a series of sharp conceptual and (especially) methodological jumps. After a rapid series of discoveries in the late eighteenth century, establishing that plants need light and carbon dioxide to produce oxygen and organic matter from carbon dioxide, there was much slower progress in estimating the rates and magnitude of the exchanges. This is especially true for aquatic primary production, until the idea that it could have much bearing on the trophodynamics of the sea became a matter of serious debate (see Section 1.2). It was not until the beginning of the twentieth century, as concerns came to focus increasingly on the rates of reproduction and consumption of planktic food plants, that the pressing need for quantitative measures of plankton production was identified (Gran, 1912). Building on the techniques and observations of Whipple (1899), who had shown a light dependence of the growth of phytoplankton in closed bottles suspended at various depths of water, and using the Winkler (1888) back-titration method for estimating dissolved oxygen concentration, Gran and colleagues devised a method of measuring the photosynthetic evolution of oxygen in sealed bottles of natural phytoplankton, within measured time periods. Darkened bottles were set up to provide controls for respirational consumption. They published (in 1918) the results of a study of photosynthesis carried out in the Christiania Fjord (now Oslofjord). According to Barber and Hilting (2002), the method was not widely

adopted until it had been described in English (Gaarder and Gran, 1927). Although it was possible at that time to estimate carbon dioxide uptake in similar bottles, essentially through the use of pH-sensitive indicator dyes, the measurement of photosynthetic rate through changes in oxygen concentration in light and dark bottles was soon adopted as a standard method in biological limnology and oceanography.

3.3.1 Measurement of light-dependent photosynthetic oxygen production

Numerous studies based on oxygen generation in light and darkened bottles were published in the 50 years between 1935 and 1985. Many of the findings were substantially covered in a thorough synthesis by Harris (1978). Since then, the method has been displaced by more direct and more sensitive techniques. Nevertheless, the experiments based on measurements of photosynthetic oxygen production in closed bottles suspended at selected depths in the water column yielded consistent generalised results and have bequeathed to plankton science many of the conceptual aspects and quantitative descriptors of productive capacity. The set of sample results illustrated in Fig. 3.3 depicts a typical depth profile of photosynthetic (4-h) exposures of unmodified lake plankton in relation to the underwater light field in an unstratified temperate lake in winter (temperature ~5 °C). The features of general interest include the following:

- The plot of gross photosynthetic oxygen generation is shown as a function of depth in Fig. 3.3a. The curve is fitted by eye. However, it is plain that photosynthesis over the 4 hours peaks a little way beneath the surface, with slower rates being detected at depth. In this instance, as in a large number of other similar experiments, there is an apparent depression in photosynthetic rate towards the surface.
- The 'gross photosynthesis' is the measured aggregate oxygen production at a given depth averaged over the exposure period. It is calculated as the observed increase in oxygen concentration in the 'light' bottle (usually a mean of duplicates) plus the observed decrease in concentration in the corresponding 'dark'

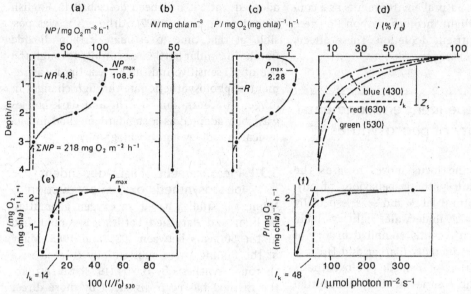

Figure 3.3 Specimen depth distributions of (a) total gross photosynthetic rate (NP) and total gross respiration rate (NR); (b) the photosynthetic population (N), in terms of chlorophyll a; (c) chlorophyll-specific photosynthetic rate (P = NP/N) and respiration rate (R = NR/N); (d) underwater irradiance (I) in each of three spectral blocks, expressed as a percentage of the irradiance I_0' obtaining immediately beneath the surface. P is replotted against I, either (e) as a percentage of I_0' in th green spectral block (peak: 530 nm), or as the reworked estimate of the intensity of the visible light above the water (in μmol photon m^{-2} s^{-1}) averaged through the exposure period. Original data of the author, redrawn from Reynolds (1984a).

control, it being supposed that the respiration in the dark applies equally to the similar material in the light. The mean gross photosynthetic rate at the given depth (here, expressed in mg O_2 m^{-3} h^{-1}) represents NP, the product of a biomass-specific rate of photosynthesis (P, here in mg O_2 (mg chla)$^{-1}$ h^{-1}) and the biomass present (N, shown in Fig 3.3b and expressed as the amount of chlorophyll a in the enclosed plankton, in (mg chla) m^{-3}). At this stage, it is the biomass specific behaviour that is of first interest: P (=NP/N) is plotted in Fig. 3.3c. The shape of the curve of P is scarcely distorted from that of NP in this instance, owing to the uniformity of N with depth. Discontinuities in the depth distribution of N do affect the shape of the NP plot (see below and Fig. 3.4).

• It was neither essential nor common to calculate the respiratory uptake, as the depletion of oxygen concentration in the dark controls. All that was necessary was the additional measurement of the initial oxygen concentration at the start of the experiment. The change in the dark bottles over the course of the experiment, normalised to the base period, is the equivalent to the respiration of the enclosed organisms in the dark, NR (here, expressed in mg O_2 m^{-3} h^{-1}); R (=NR/N) is thus supposed to be the biomass-specific respiration rate (in mg O_2 consumed (mg chla)$^{-1}$ h^{-1}) inserted in Fig. 3.3c. However, the extrapolation must be applied cautiously. Whereas NP, as determined, is attributable to photosynthesis, the separate determination of NR does not exclude respiration by non-photosynthetic organisms, including bacteria and zooplankton. Moreover, it cannot be assumed that even basal respiration rate of phytoplankton is identical in light and darkness: according to Geider and MacIntyre (2002), oxygen consumption in photosynthesising microflagellates is 3- to 20- (mean: 7-) fold greater than dark respiration rate. Accelerated metabolism and excretion of photosynthate in starved or stressed phytoplankton may, conceivably, account for all the materials fixed in photosynthesis: (P−R) is small but dark R need not be large either.

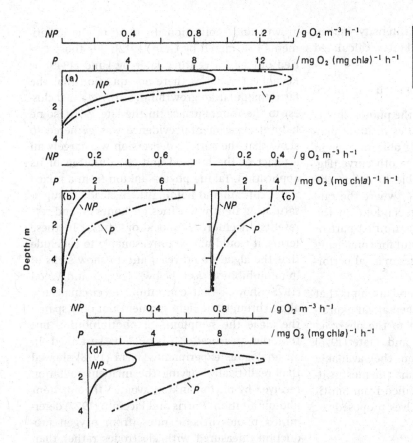

Figure 3.4 Some variations in the basic form of depth profiles of gross (NP) and chlorophyll-specific rates (P), explained in the text.

- The decline in biomass-specific P below the subsurface maximum (P_{max}) is supposed to be a function of the underwater extinction. In the present example, the underwater light attenuation measured photometrically in each of three spectral bands of visible light (blue, absorption peaking at 430 nm; green, peaking at 530 nm; and red, peaking at 630 nm) is included as Fig. 3.3d. Note that the spectral balance changes with depth (blue light being absorbed faster than red or green light in this case).
- Replotting the biomass-specific rates of P against I_z, the level of the residual light penetrating to each of the depths of measurement (expressed as a percentage of the surface measurement), gives the new curve shown in Fig. 3.3e. This emphasises the sensitivity of P to I_z, at least at low percentage residual light penetration, and a much more plateau-like feature around P_{max}.
- Finally, the P vs. I curve is replotted in Fig. 3.3f in terms of the depth-dependent irradiance across the spectrum, approximated from $I_z =$ $(I_{430} + I_{530} + I_{630})_z/3$, and taking the immediate subsurface irradiance (I_0) as the average over the experiment (in this instance, $I_0 = 800$ µmol photons m^{-2} s^{-1}).

This last plot, Fig. 3.3f, conforms to the format of the generic P vs. I curve. It has a steeply rising portion, in which P increases in direct proportion to I_z; the slope of this line, generally notated as α ($=P/I_z$), is a measure of photosynthetic efficiency at low light intensities. Expressing P in mg O$_2$ (mg chl$a$)$^{-1}$ h^{-1} and I_z in mol photons m^{-2} h^{-1}, the slope, α, has the dimensions of photosynthetic rate per unit of irradiance, mg O$_2$ (mg chl$a$)$^{-1}$ (mol photon)$^{-1}$. As the incident photon flux is increased, P becomes increasingly less light dependent and, so, increasingly saturated by the light available. The irradiance level representing the *onset of light saturation* is judged to occur at the point of intersection between the extrapolation of the linear light-dependent part of the $P - I_z$ curve with the back projection of P_{max}, being the fastest, light-independent rate of photosynthesis measured. This intensity, known as I_k, can be judged by eye or, better, calculated as:

$$I_k = P_{max}/\alpha \ \text{mol photons m}^{-2} \text{h}^{-1} \quad (3.5)$$

It is usual (because that is how the photon flux is measured) to give light intensities in μmol photons m^{-2} s^{-1}. In terms of being able to explain the shape of the original P vs. depth curve (Fig. 3.3a), it is also useful to be able to define the point in the water column, $z_{(I_k)}$, where the rate of photosynthesis is directly dependent on the photon flux. The left-hand, light-limited part of the P vs. I_z curve is the most useful for comparing the interexperimental differences in algal performances.

Kirk (1994) described the several attempts that have been made to find a mathematical expression that gives a reasonable fit to the observed relationship of P to I. Jassby and Platt (1976) tested several different expressions then available against their own data. They found the most suitable expressions to be one modified from Smith (1936) and the one they themselves proposed:

$$P = P_{max} [1 - \exp(\alpha - I_z/P_{max})]$$
$$\text{(Smith, 1936)} \quad (3.6)$$
$$P = P_{max} \tanh(\alpha - I_z/P_{max})$$
$$\text{(Jassby \& Platt, 1976)} \quad (3.7)$$

In contrast, the significance and mathematical treatment of the right-hand part of the P vs. I_z curve, corresponding to the apparent near-surface depression of photosynthesis in field experiments, was, for a long time, a puzzling feature. It was not necessarily a feature of all P vs. z curves: some of the possible variations are exemplified in Fig. 3.4. Anomalies in the depth distribution of NP, caused by phytoplankton abundance (Fig. 3.4a), low water temperature (Fig 3.4c) or by surface scumming of dominant Cyanobacteria (Fig. 3.4d) fail to obscure the incidence of subsurface biomass-specific photosynthetic rates. However, they are not seen when dull skies ensure that, even at the water surface, photosynthetic rates are not light-saturated [$z_{(Ik)}$ < 0 m!].

This near-surface depression in measured photosynthertic rate is not, however, reflected in growth and replication. In none of the experiments considered by Gran (1912), nor those carried out half a century later by Lund (1949) or Reynolds (1973a), is there any question that the fastest population growth rates are obtained closest to the water surface. In the late 1970s, some helpful experimental evidence was gathered to show that the surface depression was largely an artefact of the method and the duration of its application. Taking phytoplankton from a mixed water column and holding them steady at supersaturating light intensities for hours on end represents an enforced 'shock' or 'stress'. In these terms, it would not be unreasonable to conclude that the algae would react and to show signs of 'photoinhibition' (see below). Jewson and Wood (1975) showed that continuing to circulate the algae through the light gradient not only spared the algae the symptoms of photinhibition but that the measured P_{max} could be sustained. In an analogous experiment, Marra (1978) showed that realistically varying the incident radiation received by phytoplankton avoided the apparent photoinhibition. Harris and Piccinin (1977) determined photosynthetic rates (from oxygen production, measured with electrodes rather than by titration) in bottled suspensions of *Oocystis* exposed to high light intensities (>1300 μmol photons m^{-2} s^{-1}) and temperatures (>20 °C) for varying lengths of time. Their results suggested that an elevated photosynthetic rate was maintained for 10 minutes or so but then declined steeply with prolonged exposure. Either the algae were photoinhibited or damaged or they had reacted to prevent such damage (photoprotection) in some way that would enable to retain their vitality to maintain growth (see Section 3.3.4).

For these reasons, determinations of depth-integrated photosynthesis (ΣNP, in mg O_2 m^{-2} h^{-1}) need no longer need depend on the planimetric measurement of the area enclosed by the measurements of photosynthetic rates against depth. They are estimable, for instance, from the P vs. I_z curve in Fig. 3.3f, as the area of a trapezium equivalent to $P_{max} \times 2 [(I_0 - I_k) + (I_0 - I_c)$, where I_0 is the light intensity at the water surface and I_c is the intercept of zero photosynthesis. In terms of P vs. depth, this simplifies to the product P_{max} × (the depth from the surface to the point in the water column where the light will half-saturate it); i.e.

$$\Sigma NP \approx P_{max} \times z_{(0.5 I_k)} \qquad (3.8)$$

If the dark respiration rate, NR, is uniform with depth, then the integral is simply the product of the full depth range over which it applies (the height of the full water column, H) depth

$$\Sigma NR \approx R \times H \qquad (3.9)$$

Based upon the numerous published records, several compendia of the key indices of photosynthetic oxygen production of phytoplankton in closed bottles have been assembled (Harris, 1978; Kirk, 1994; Padisák, 2003). Clearly, the gross rate of photosynthesis (NP_{max}) and the depth integral (ΣNP) respond to two variables, which are, within limits, highly variable. Other factors notwithstanding, observed P_{max} should always be the light-saturated maximum rate at the given temperature. Among the fastest reported examples are 30–32 mg O_2 (mg chl$a$)$^{-1}$ h^{-1} (noted in warm tropical lakes in Africa, by Talling, 1965; Talling et al., 1973; Ganf, 1975), whereas the specific photosynthetic rates in temperate lakes rarely exceed the 20 mg O_2 (mg chl$a$)$^{-1}$ h^{-1} found by Bindloss (1974). Thus, as might be expected, examples of high community rates of oxygen production come from warm lakes supporting large populations of phytoplankton algal chlorophyll and when there are also ample reserves of exploitable CO_2: integrals in the range 6–18 g O_2 m^{-3} h^{-1} have been noted in Lac Tchad, Chad (Lévêque et al., 1972), Red Rock Tarn, Australia (Hammer et al., 1973) and Lake George, Uganda (Ganf, 1975). The onset of light saturation, I_k, is also temperature influenced but is generally <300 μmol photons m^{-2} s^{-1}. There are many citations of much lower I_k determinations, 15–50 μmol photons m^{-2} s^{-1}, generally at temperatures <10 °C (Kirk, 1994). The most consistent values seem likely to relate to the slope, α, which, at low light levels, is less dependent on temperature and more dependent upon light-harvesting efficiency. Taking the plot in Fig. 3.3f, the slope α (i.e. P on I) is calcu-

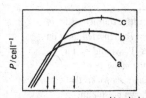 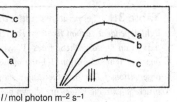

I / mol photon m^{-2} s^{-1}

Figure 3.5 Hypothetical P vs. I plots to contrast seasonal variations in temperature on photosynthetic behaviour, with special reference to changes in P_{max} (tagged) and I_k (arrowed). In either plot, the sequence a→b→c is one of increasing temperature. In the left-hand plot, the dependence upon light is constant; in the right-hand plot, P/I varies to keep I_k constant. Redrawn from Reynolds (1984a).

lated by rearranging Eq. (3.5):

$$\alpha = P_{max}/I_k = 2.28 \, mgO_2 \, (mg \, chl a)^{-1} h^{-1}$$
$$/48 \mu mol \, photons \, m^{-2} s^{-1}$$
$$= 0.0475 \, mgO_2 \, (mg \, chl a)^{-1} h^{-1}$$
$$(\mu mol \, photons \, m^{-2} s^{-1})^{-1}$$
$$= 13.2 \, mg \, O_2 \, (mg \, chl a)^{-1} (mol \, photon)^{-1} m^2$$

Comparisons among differing P vs. I plots often differentiate patterns of photosynthetic behaviour. In the left-hand box of Fig. 3.5, the slopes (α) show similar light dependence of photosynthesis at low irradiances but the sequence of increasing P_{max} values could be the response of the same alga to increasing temperatures. In the right-hand plot, data for three different algae are shown, all saturating at similar levels but with differing photosynthetic efficiencies. A higher α enables a faster rate of photosynthesis to be maintained when light intensity is low.

3.3.2 Measurement of light-dependent photosynthetic carbon fixation

On the basis of an equimolecular photosynthetic quotient ($PQ \sim 1$), the light efficiency of photosynthesis could be calculated from the oxygen evolution data to be ~5.0 mg C (mg chl$a$)$^{-1}$ (mol photon)$^{-1}$ m^2. This is actually below the average of a large number of extrapolated values, which fall mainly in the range 6–18 mg C (mg chl$a$)$^{-1}$ (mol photon)$^{-1}$ m^2 (Harris, 1978). However, for over 50 years now, it has been possible to work directly in the currency of carbon, applying the so-called ^{14}C method of Steemann Nielsen (1952)

Table 3.1 | *Temperature-sensitive characteristics of light-dependent carbon fixation. For reported data on maximum photosynthetic rates (P_{max}) and on the onset of light saturation (I_k), the majority of observations fall within the ranges shown in the brackets; extreme values are shown outside the brackets. Under conditions of light limitation ($I_z < I_k$), temperature dependence of photosynthesis is weaker and the single range of photosynthetic efficiencies applies to the available data*

P_{max}
　in mg C (mg chl$a$)$^{-1}$ h^{-1}
　　θ 2–5 °C　　　　　　　　0.5 (1.0–2.6) 3.0
　　θ 17–20 °C　　　　　　　2.5 (3.3–8.6) 9.7
　　θ 27–30 °C　　　　　　　7–20
I_k
　in μmol photon m^{-2} s^{-1}
　　θ 2–5 °C　　　　　　　　17–30
　　θ 17–20 °C　　　　　　　20 (90–150) 320
　　θ 27–30 °C　　　　　　　60 (180–250) 360
α
　in mg C (mg chl$a$)$^{-1}$　　　2 (6–18) 37
　(mol photon)$^{-1}$
　m^2

Source: Generalised values synthesised from the literature (Harris, 1978; Reynolds, 1984a, 1990, 1994a; Padisák 2003).

for measuring the photosynthetic incorporation of carbon dioxide labelled with the radioactive isotope. Still using darkened and undarkened bottles suspended in the light field, the essential principle is that the natural carbon source is augmented by a dose of radio-labelled NaH^{14}CO$_3$ in solution, which carbon source is exploited and taken up and fixed into the photosynthetic algae. At the end of the exposure, the flask contents are filtered and the residues are submitted to Geiger counting (for a recent methodological guide, see Howarth and Michaels, 2000), and the quantity thus assimilated is calculated. It is supposed that ^{14}C will be fixed in photosynthesis in the same proportion to ^{12}C that is available in the pool at the start of the exposure. Then,

$$^{12}CO_2 \text{ uptake}/^{12}CO_2 \text{ available}$$
$$= {}^{14}CO_2 \text{ uptake}/^{14}CO_2 \text{ available} \quad (3.10)$$

Since its introduction, the method has been improved in detail (the liquid-scintillation and gas-phase counting technique is nowadays preferred) but, in essence, the original method has survived intact (Søndergaard, 2002). Providing proper licensing and handling protocols are followed meticulously, the method is easy to apply and yields reproducible results. Comparisons with simultaneous measurements of oxygen evolution normally give tolerable agreement, if allowance is made for a PQ of ~1.15 (Kirk, 1994). A large number of results have been published in the literature and these have been the subject of a series of syntheses (including Harris, 1978, 1986; Fogg and Thake, 1987; Kirk, 1994; Padisák, 2003). It is sufficient in the present context simply to summarise the key characteristics that have been reported (maximum measured chlorophyll-specific rates of light-saturated carbon incorporation and the chlorophyll-specific photosynthetic efficiencies under sub-saturating light intensities (see Table 3.1).

Two comments are important to make, however. One is the positive linkage of light saturation with temperature, at least within the range of the majority of observations – between 5 and 25 °C. As with many other cellular processes, activity increases with increasing temperatures, up to maximal levels varying between 25 and 40 °C. There is a plain dependence for the photosynthetic rates to accelerate with higher temperatures and, so, for there to be a higher threshold flux of photons necessary to saturate it. With no change in the strongly light-limited rates of photosynthesis, α may vary little (i.e. light-limited photosynthesis is not temperature-constrained; left-hand plot in Fig. 3.5). Thus, I_k increases with temperature, in broadly similar proportion to the increase in P_{max}.

As a function of customary temperature, P_{max} increases non-linearly, roughly doubling with each 10 °C rise in temperature. This multiple, formalised as the Q_{10} factor, is now used infrequently as a physiological index: preference is now accorded to the slope of reactivity on the Arrhenius scale, which expresses absolute temperature (in kelvins) as a reciprocal scale, 1/(temperature in K). As an example, the measured temperature sensitivity of light-saturated

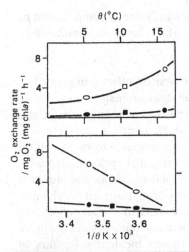

photosynthesis and dark respiration of laboratory isolates of *Asterionella formosa* are shown in Fig. 3.6. The normalised, Arrhenius coefficient is -18.88×10^{-3} per reciprocal kelvin. In the more familiar, if incorrect, terms, Q_{10} is 2.18. Almost all quoted values, applying both to named species and to phytoplankton in general, fall in the range 1.8–2.25 (Eppley, 1972; Harris, 1978). Some variation is probable, not least because photosynthesis is a complex of many individual reactions. However, its maximum, light-saturated rate is primarily a function of temperature (Morel, 1991), even though the several descriptive equations fitted to experimental data differ mutually (Eppley, 1972; Megard, 1972) and have been shown to be underestimates against the maximum growth rates that have been observed (Brush *et al.*, 2002). Whereas photon capture has a Q_{10} close to 2 (see Section 3.2.1), protein assembly and internal transport have greater temperature sensitivity (Tamiya *et al.*, 1953; see also Section 5.3.2 for the influence of temperature on growth). The Q_{10} of steady, dark respiration rates of healthy phytoplankters is similarly close to 2.

The second comment is that all these deductions are subject to uncertainties about the precision of the radiocarbon method. Interpretational difficulties were recognised early on (includ-

ing by Steemann Nielsen himself) and, perplexingly, these persist to the present day. The most important concerns the metabolic exchanges and cycling of carbon, in which the labelled carbon participates relatively freely. At first, labelled carbon moves in only one direction, from solution to photosynthate; it is a manifestation of gross photosynthesis. As the experiment proceeds, some of the ^{14}C-labelled carbohydrate may be assimilated but it may just as easily be used in basal respiration, or it may well be subject to photorespiration or excretion (see Section 3.2.3). This means that, as the incubation proceeds, the method is ostensibly measuring something closer to net photosynthesis. Long incubations may determine only net photosynthetic ^{14}C incorporation (Steemann Nielsen, 1955; Dring and Jewson, 1979). Comparing net ^{14}C assimilation with net oxygen production over 24-h incubations, by which time respired [14]CO$_2$ is being refixed, takes PQ closer to 1.4 (see Marra, 2002).

The switches towards this more balanced state of exchanges will be approached at different rates, depending upon temperature and the irradiance to which the incubating material is exposed and on the physiological condition of the alga at the outset. The behaviour has been expressed through various descriptive equations (notably those formulated by Hobson *et al.*, 1976; Dring and Jewson, 1982; Marra *et al.*, 1988; Williams and Lefevre, 1996). A probabilistic outcome is that the largest proportion of the gross uptake of ^{14}C is assimilated into new protein and biomass in healthy cells when they are simultaneously exposed to sub-saturating light intensities. Conversely, with approaching light saturation of the growth-assimilatory demand for photosynthate (recognising that this may be constrained by factors other than the supply of photosynthate), then more of the excess is vented or metabolised in other ways. Thus, the ratio of net photosynthetic carbon fixation to gross photosynthetic carbon fixation ($P_n : P_g$) is in the proportion of the fraction of the gross photosynthate that can be assimilated, i.e. $(P_g - R)/P_g$ (Marra, 2002). Here, R may represent not just the basal, autotrophic respiration (R_a) but, in addition, all metabolic elimination of excess photosynthate (R_h). Even under the optimal conditions envisaged, R_a never disappears but is always finite, being, at least, about 4% of P_{max} (or the maximum sustainable P_g at the same temperature), and typically 7–10% (Talling, 1957b, 1971; Reynolds, 1984a). Reynolds' (1997a) best predictor of basal respiration in a number of named freshwater phytoplankters at 20 °C is related to the surface-to-volume ratio $[R_{a20} = 0.079 \, (s/v)^{0.325}]$. On the other hand, the sum of physiological losses in light-saturated and/or nutrient-deficient cells, including that vented as DOC, can be extremely high, approaching 100% of the fixation rate: the quotient $(P_g - R - R_h)/P_g$ and the ratio $P_n : P_g$ both fall toward zero.

This is a plausible way of explaining difficulties experienced in accounting for the fate of the carbon fixed in photosynthesis (Talling, 1984; Tilzer, 1984; Reynolds et al., 1985) and the sometimes very large gaps between primary carbon fixation and net biomass accumulation (Jassby and Goldman, 1974a; Forsberg, 1985). Thus, beyond gaining a broad perspective on the constraints acting on chlorophyll-specific photosynthetic rates, it is necessary also for the physiological ecologist to grasp the manner in which the environmental conditions mould the deployment of fixed carbon into the population dynamics of phytoplankton.

3.3.3 Photosynthetic production at sub-saturating light intensities

In this section, we focus on the adaptive mechanisms which phytoplankters use to optimise their carbon fixation under irradiance fluxes that markedly undersaturate the capacity of the individual light-harvesting centres. To do this, it is necessary to relate the light-harvesting capacity to the cell rather than to chlorophyll per se, as we invoke the number of centres in existence within the cell and their intracellular distribution as adaptive variables, as well as the role played by accessory pigments. It is also necessary to adopt a more quantitative appreciation of the diminution of harvestable light as a function of increasing water depth and the contribution to vertical light attenuation that the plankton makes itself. Finally, water movements entrain and transport phytoplankton through part of this gradient, frequently mixing them to depths beyond the point

where photosynthesis is entirely compensated by respirational loss. This theme is continued in Section 3.3.4.

Characterising the photosynthetic impact of the underwater light intensity

Taking first the condition of phytoplankton at a fixed or relatively stable depth in the water column, the light energy available to them will, in any case, fluctuate on a diel cycle, in step with daytime changes in solar radiation, and also on a less predictable basis brought about by changes in cloud cover and changes in surface reflectance owing to surface wind action. Besides variance in I_0, the instantaneous incident solar flux of PAR on the water surface, there is variance in the irradiance in the flux passing to just beneath the air–water interface, I_0'. Reflectance of incoming light is least (about 5%) at high angles of incidence but the proportion reflected increases steeply at lower angles of incidence, especially <30°. Wind-induced surface waves modify the reflectance, both reducing and amplifying penetration into the water at the scale of milliseconds. Near sunrise and sunset, waves are important to maintaining a flux of light across the water surface (Larkum and Barrett, 1983).

Variability in the light income to a water body is represented in the examples in Fig. 3.7. Measured daily integrals of the solar insolation input across the surface of Esthwaite Water, UK (a temperate-region lake, experiencing a predominantly oceanic climate) through a period of just over one year are shown in the main plot. Besides showing a 100-fold variation between the highest (56.4 mol photons m^{-2} d^{-1}) and lowest light inputs (0.5 mol photons m^{-2} d^{-1}), the plot reveals that part of this owes to the substantial annual fluctuation in the maximum possible insolation (based on the solar constant, its latitudinal correction, the deduction of non-PAR and with allowance for albedo scatter in the atmosphere) due to the location of the lake (54 °N). Superimposed upon that are the near chaotic fluctuations that are due to day-to-day variations in the extent and thickness of cloud cover and which cut the light income by anything between 2% or as much as 94% of the theoretical maximum (Davis et al., 2003). The inset in Fig. 3.7 presents two day-long

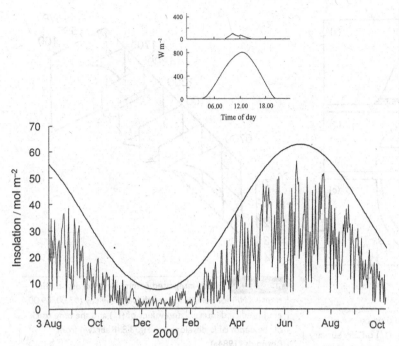

Figure 3.7 Main plot: Measured daily insolation at the surface of Blelham Tarn (August 1999 to October 2000), compared to the theoretical maximum insolation, as calculated from the latitude and assuming a clear, dry atmosphere (redrawn with permission from the data of Davis et al., 2003). The insets show the diel course of solar radiation intensity, measured at the same location on an overcast January day and a cloudless day close to the summer solstice. Based on data presented originally by Talling (1973) and redrawn with permission from Reynolds (1997a).

time courses of insolation, measured at the same latitude by Talling (1971). Despite coming from quite different data sets collected at quite different times, these two extremes amplify the variation shown in the main plot of Fig. 3.7. The first corresponds to a rainy, overcast winter day on which the aggregate insolation is below 1 mol photons m^{-2} d^{-1}; the second is a cloudless near-solstice day when the maximum instantaneous insolation reached over 1.6 mmol photons m^{-2} s^{-1}.

Downwelling radiation is subject to absorption and scatter beneath the water surface, where it is more or less attenuated steeply and exponentially with depth, as shown in Fig. 3.3d. This attenuation can usefully be expressed by the coefficient of vertical light extinction, ε. On the basis of the spectral integration used to translate the P vs. depth to P vs. I, the coefficient may be estimated on the basis of the equation

$$I_z = I_0' \cdot e^{-\varepsilon z} \qquad (3.11)$$

where e is the natural logarithmic base, and whence

$$\varepsilon = -\ln(I_z/I_0')/h(0 - z)$$

or

$$\varepsilon = (\ln I_0' - \ln I_z)/h(0 - z) \qquad (3.12)$$

where $h(0 - z)$ is the vertical distance from the surface to depth z.

From Fig. 3.3d, $\varepsilon(0.5\,\text{m}$ to $1.5\,\text{m})$ is solved from $(\ln I_{0.5} - \ln I_{1.5})$ as $\varepsilon = 1.516\,\text{m}^{-1}$. We may now apply this spectral integral of the attenuating light to characterise the daytime changes in the underwater light field (see Fig. 3.8). The plot in Fig 3.8a shows the reconstructed time course of the irradiance at the top of the water column of Crose Mere on 25 February 1971. Beneath it are marked the values of I_k and $0.5I_k$ derived in the original experiment (Fig. 3.3). For clarity, the same information is plotted on to Fig. 3.8b, now against a natural logarithmic scale. Assuming no change in any component save the incoming radiation through the day, Fig. 3.8c represents the diurnal time track between sunrise and sunset of the depths of I_k and $0.5I_k$. Other factors being equal (including the coefficient of vertical extinction of light, ε), the depth at which chlorophyll-specific photosynthesis can be saturated increases to a maximum at around the diurnal solar zenith, as a function of the insolation

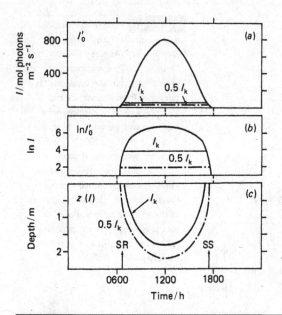

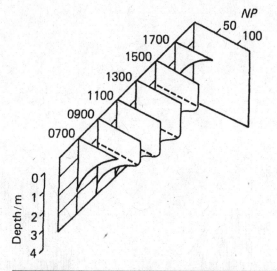

Figure 3.9 Depth-integrated community photosynthetic rates (ΣNP) for selected times through the day (07.00, 09.00, etc.) predicted interpolated values of (I_0') and the time track of the depth of I_k developed in Fig. 3.8. Redrawn from Reynolds (1984a).

Figure 3.8 (a) Hypothetical plot of the time-course of immediate subsurface irradiance intensity (I_0') on 25 February 1971 (the date of the measurements presented in Fig. 3.3); (b) the same shown on a semilogarithmic plot. In (a) and (b), the contemporaneous determinations of I_k and 0.5 I_k are inserted. In (c), the time courses of the water depths reached by irradiance intensities respectively equivalent to I_k and 0.5 I_k are plotted SR, sunrise; SS, sunset. Redrawn from Reynolds (1984a).

(I_0'). Outside the I_k perimeter, chlorophyll-specific fixation rates are light-limited, as predicted by the contemporaneous P vs. I curve. For instance, extrapolation of relevant data from the experiment shown in Figure 3.3 allows us to fit selected reconstructions of P vs. z curves applying at various times of the day (see Fig. 3.9).

In reality, matters are more complex than that, especially if cloud cover (and, hence, the insolation, I_0'), or the extinction coefficient is altered during the course of the day. The greatest depth of I_k need not be maximal in the middle of the day. Modern *in-situ* recording and telemetry of continuous radiation make it possible to track minute-to-minute variability in the underwater light conditions. Moreover, it is now relatively simple to translate light measurements to instantaneous photosynthetic rates, from the P vs. I curve, and to integrate them through depth (ΣNP, in mg C m^{-2} h^{-1}) and through time

($\Sigma\Sigma NP$, in mg C m^{-2} d^{-1}). A. E. Walsby and colleagues have been particularly successful in developing this approach (using Microsoft Excel 7 software), details of which they make available on the World Wide Web (Walsby, 2001; see also Bright and Walsby, 2000; Davis *et al.*, 2003). The technique is helpful in establishing the environmental requirements and limitations on algal growth.

The impact of optical properties of water on the underwater light spectrum

As Kirk's manual (1994) and several other of his contributions (see, for instance, Kirk, 2003) have powerfully emphasised, the attraction and usefulness of a single average vertical attenuation coefficient (ε) remains an approximation of the complexities of the underwater dissipation of light energy. First, as already established, light is not absorbed equally across the visible spectrum, even in pure water (see Fig. 3.10). Photons travelling with wavelengths of about 400–480 nm are least likely to be captured by water molecules; those with wavelengths closer to 700 nm are 30 times more likely to be absorbed. For this reason, water cannot be regarded as being colourless.

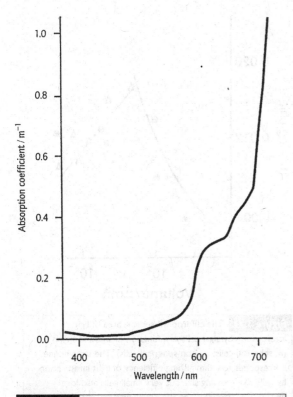

Figure 3.10 The absorption of visible light (375–725 nm) by pure water. Drawn from data in Kirk (1994).

The selective absorption in the red leaves oceanic water of high clarity distinctly blue-green in colour. This effect is strongly evident at increasing water depths beneath the surface, where the most penetrative wavelengths come increasingly to dominate the diminishing light field. In lakes, there is a tendency towards higher solute concentrations, including, significantly, of plant or humic derivatives. These absorb wavelengths in the blue end of the spectrum, to leave a yellow or brownish tinge to the water, as the older names ('Gelbstoff', 'gilvin') might imply. Under these circumstances, the averaged extinction coefficient (distinguished here as ε_{av}) becomes less steep with depth and cannot strictly be normalised just by logarithmic expression. ε_{av} or ε as used generally, calculated as in Eq. (3.12) from the slope of I_z on z, ε_{av} really has only a local value, applying to relatively restricted depth bands.

To express attenuation of light of a given wavelength or within a narrow waveband overcomes part of the difficulty but, especially in

waters of high clarity, a second source of error is encountered. This relates to another fundamental of attenuation: besides being absorbed and scattered, according to the properties of the water, attenuation is modified perceptibly by the angular distribution of the light, which becomes increasingly diffuse with increasing depth. Kirk (2003) proposed the use of an irradiance weighting of the wavelength-specific gradient, by integrating light measurements over the entire water column, so that

$$^{w}\varepsilon_{av} = \left[\int_{\infty}^{0} \varepsilon_z I_z \, dz \right] \bigg/ \left[\int_{\infty}^{0} I_z \, dz \right] \quad (3.13)$$

where $^{w}\varepsilon_{av}$ is the weighted average attenuation coefficient and ε_z and I_z are the attenuation coefficient and residual light values at each depth increment.

The use of this integral improves resolution of the subsaturating light levels in clear waters but the simpler, less precise attenuation coefficient derived in Eq. (3.12) remains adequate for describing the underwater light field in most lakes and many coastal waters, where there may be more humic material in solution and there may be more particulate material in suspension. Moreover, greater concentrations of algal chlorophyll also contribute to the rapid relative attenuation of light with depth. The average attenuation coefficient comprises:

$$\varepsilon = \varepsilon_w + \varepsilon_p + N\varepsilon_a \quad (3.14)$$

ε_w is the attenuation coefficient due to the water. As already indicated, there is no unique and meaningful value that applies across the visible spectrum. The minimum absorption in pure water, equivalent to 0.0145 m^{-1}, occurs at a wavelength of ~440 nm. The clearest (least absorbing) natural waters are in the open ocean (Sargasso, Gulf Stream off Bahamas), where the coefficients of vertical attenuation at 440 nm ($\varepsilon_{w\,440}$) of ambient visible insolation (I'_0) are <0.01 m^{-1} (various sources tabulated in Kirk, 1994). Values quoted for the open Atlantic, Indian and Pacific Oceans range between 0.02 and 0.05 m^{-1}. Among the clearest lakes for which data are to hand, Crater Lake, Oregon has an exceptionally low coefficient of incident light attenuation ($\varepsilon \sim 0.06$ m^{-1}: Tyler, in Kirk, 1994). High-altitude lakes in the Andes range return values of 0.12 to 0.2 m^{-1} (Reynolds, 1987b, and unpublished observations). Elsewhere, the effects of solutes contribute to higher average extinction coefficients, in coastal seas ($\varepsilon_w > 0.15$ m^{-1}) and fresh waters generally ($\varepsilon_w > 0.2$ m^{-1}). At the other extreme, coloration due to humic staining can be intense; Kirk (1994) tabulated some representative values from estuaries and coastal seas receiving drainage from extensive peatland catchments (Gulf of Bothnia, Baltic Sea; Clyde, Scotland) of 0.4 to 0.65 m^{-1}, from some humic lakes and reservoirs in Australia (1 to 3.5 m^{-1}) and from peat-bog ponds and streams in Ireland (2 to 20 m^{-1}).

ε_p is the attenuation coefficient due to the solid particulates in suspension in the water. The particles may emanate from eroded soils or resuspended silts or biogenic detritus. Fine precipitates, for instance, of calcium carbonate, also contribute to particulate turbidity. The most persistent are clay particles (typically <5 μm), eroded from unconsolidated deposits, which, by absorbing and backscattering the photon flux, can impart high turbidity. In many lakes and seas, the contribution of background turbidity to attenuation may be trivial (<0.05 m^{-1}) but Kirk (1994) cites many examples where it is anything but trivial; canals, rivers, meltwater streams flowing into lakes or reservoirs can impart attenuation coefficients of 1 to 4 m^{-1}, with extreme values in the range 10–20 m^{-1}. Investigating the turbidity of the tidally mixed estuary of the Severn, UK, Reynolds and West (unpublished data, 1988) found a correlation between ε_p and the mass of suspended clay and fine silt, such that an attenuation coefficient of 20 m^{-1} corresponds approximately to 1 kg suspended clay m^{-3}, whence $\varepsilon_p \sim 20$ m^2 kg^{-1}.

$N\varepsilon_a$ is the attenuation due to phytoplankton. Evolved to be able to intercept light energy, phytoplankton can be, in aggregate, the main component to vertical attenuation. The extent is proportional to the mass of phytoplankton present per unit volume (N, in mg chl$a$ m^{-3}); however, the chlorophyll-specific attenuation coefficient, ε_a, varies considerably. Most general accounts attribute attenuation coefficients of between 0.01 to 0.02 m^{-1} (mg chl$a$ m^{-3})$^{-1}$, or, more simply, 0.01 to 0.02 m^2 (mg chl$a$)$^{-1}$. Variations are, in

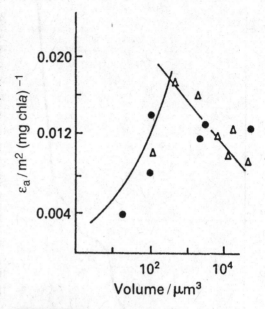

Figure 3.11 The chlorophyll-specific area of light interception by algal cells as a function of size and shape (●, spherical cells; △, non-spherical cells). The straight line corresponds to a diminishing efficiency of light interception by cells of increasing size but very small cells also lose efficiency at sizes close to the wavelength of light. Redrawn with permission from Reynolds (1987b).

part, due to the size and shape of the algae (see Fig. 3.11). Among microplankters, larger quasi-spherical shapes, having larger volumes, hold absolutely greater amounts of chlorophyll than smaller shapes. Relative to the cross-section presented to the light field, however, the chlorophyll is more compactly arranged in larger units and more light passes between them than in smaller units carrying the same amount of chlorophyll. This approximates to the expectations of the 'packet effect' (or 'sieve effect'), first studied by Duysens (1956) in the context of solutions and suspensions but developed for phytoplankton by Kirk (1975a, b, 1976). However, among the free-living nanoplanktic and picoplanktic cells (<100 μm^3), the chlorophyll-specific area of projection is diminished, to as low as 0.004 m^2 (mg chl$a$)$^{-1}$. This may have something to do with the size of plastids being similar to the wavelength of light. Alternatively, it might be explained by the second anticipated source of variation, the chlorophyll content of

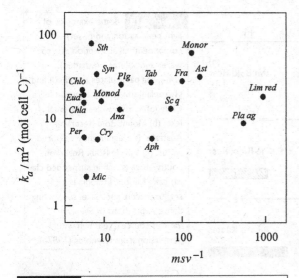

Figure 3.12 Carbon-specific area of freshwater planktic algae (k_a) plotted against the dimensionless shape index, msv^{-1}. Near-spherical algae line up close to $msv^{-1} = 6$, smaller species tending to have greater interception properties than larger ones. Shape distortions increase msv^{-1} without sacrifice of k_a. The algae are *Ana, Anabaena flos-aquae; Aphan, Aphanizomenon flos-aquae; Ast, Asterionella formosa; Chla, Chlamydomonas; Chlo, Chlorella; Cry, Cryptomonas ovata; Eud, Eudorina unicocca; Fra, Fragilaria crotonensis; Lim red, Limnothrix redekei; Mic, Microcystis aeruginosa; Monod, Monodus; Monor, Monoraphidium contortum; Per, Peridinium cinctum; Pla ag, Planktothrix agardhii; Plg, Plagioselmis; Sc q, Scenedesmus quadricauda; Tab, Tabellaria flocculosa.* Redrawn with permission from Reynolds (1997a).

the cells, relative to (say) cell carbon. Plotting k_a, the area projected per mol of cell carbon, against an index of shape (msv^{-1} is the product of the maximum dimension and the surface-to-volume ratio), shows the package effect to be upheld for quasi-spherical units (from *Chlorella* to *Microcystis*; $msv^{-1} = [d4\pi(d/2)^2/4/3\pi(d/2)^3] = 6$), while distortion from spherical form usually enhances the area that the equivalent sphere might projection (Reynolds, 1993a) (see also Fig. 3.12). Note that the individual carbon-based projections are less liable to variation than the chlorophyll-based derivations.

The attenuation components self-compound (Eq. 3.14) to influence the diminution of the depth to which incident radiation of given wavelength penetrates (Eq. 3.11). In some extreme instances of attenuation with depth, one of the components may well dominate over the others. In Fig. 3.13, instances of high attenuation due to a relatively high ε_w (in the humic Mt Bold Reservoir, Australia), ε_p (in the P. K. le Roux Reservoir, South Africa) and $N\varepsilon_a$ (in an Ethiopian soda lake) are compared with the transparency of a mountain lake (2800 m a.s.l.) in the middle Andes. In each of these relatively deep-water examples, the gradient in I_z is exclusively related to the depth and to the coefficient of attenuation. The same principles apply in shallow waters but, because light can reach the bottom and be reflected back into the water column, the gradient of I_z may not be so steep or so smooth as that shown in Fig. 3.13 (discussion in Ackleson, 2003). Nevertheless, the present examples give a feel for the column depth in which diurnal photosynthesis can be light-saturated ($I_z > I_k >$ (say) 0.1 mmol photons m^{-2} s^{-1}) and, more to the point, the vertical extent of the water column where phytoplankters need to adapt their light-harvesting potential to be able to match their capabilities for carbon fixation.

Phytoplankton adaptations to sub-saturating irradiances

Several mechanisms exist for enhancing cell-specific photosynthetic potential at low levels of irradiance. One of these is simply to increase the cell-specific light-harvesting capacity, by adding to the number of LHCs in individual cells. This may be manifest at the anatomical level in the synthesis of more chlorophyll and deploying it in more (or more extensive) plastids placed to intercept more of the available photon flux falling on the cell. Most phytoplankton are able to adjust their chlorophyll content within a range of ±50% of average and to do so within the timescale of one or two cell generations. For example, Reynolds (1984a) made reference to measurements of the chlorophyll content of *Asterionella formosa* cells taken at various stages of the seasonal growth cycle in a natural lake, during which the cell-specific quota fluctuated between 1.3 and 2.3 mg chl*a* (10^9 cells)$^{-1}$, i.e. 1.3–2.3 pg chl*a* (cell)$^{-1}$. Supposing the capacity of the *Asterionella* to fix carbon to have been as measured in Fig. 3.3, equivalent to ~0.8 mg C (mg chl*a*)$^{-1}$ h^{-1},

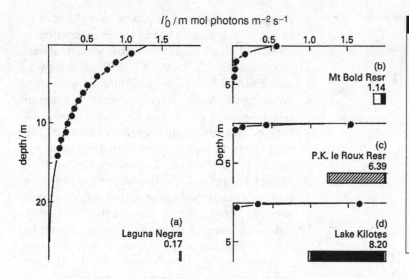

Figure 3.13 Some examples of light penetration and (inset) components of absorption due to water and solutes (unshaded), non-living particulates (hatched) and phytoplankton (solid). Laguna Negra, Chile (a), is a very clear mountain lake; (b) Mount Bold Reservoir, Australia, is a significantly coloured water; (c) P. K. le Roux Reservoir, South Africa is rich in suspended clay; (d) Lake Kilotes, Ethiopia is a shallow, fertile soda lake, supporting dense populations of *Spirulina*. Various sources; redrawn with permission from Reynolds (1987b).

then the cells with the lower chlorophyll *a* complement would have fixed 1.04 mg C $(10^9$ cells$)^{-1}$ h^{-1}, or 0.012 mg C (mg cell C) h^{-1}. Other things being equal, the cells with the higher chlorophyll complement might have been capable of fixing 1.84 mg C $(10^9$ cells$)^{-1}$ h^{-1}, or 0.022 mg C (mg cell C) h^{-1}. The main point, however, is that the cells with the higher chlorophyll content are capable of fixing the same amount of carbon as those with the lower complement but at a lower photon flux density: in this instance, the high-chlorophyll cells achieve 1.04 mg C $(10^9$ cells$)^{-1}$ h^{-1} not at ≥ 48 μmol photons m^{-2} s^{-1} but at ≥ 27 μmol photons m^{-2} s^{-1}. The extra chlorophyll *a* increases the steepness of biomass-specific P on I (the slope α).

The measurements of biomass-specific chlorophyll *a* referred to in Section 1.5.4 range over an order of magnitude, between 0.0015 and 0.0197 pg μm^{-3} of live cell volume. This corresponds approximately to 3 to 39 mg g^{-1} of dry weight (0.3 to 3.9%) or, relative to cell carbon, 6.5 to 87 mg chl*a* (g C)$^{-1}$. Many of the lowest values come from marine phytoplankters in culture (Cloern *et al.*, 1995); most of the highest come from cultured or natural material, but grown under persistent low light intensities (Reynolds, 1992a, 1997a). The data show that the frequently adopted ratio of cell carbon to chlorophyll content (50:1, or 20 mg chl*a* (gC)$^{-1}$] must be applied with caution, though it remains a good approxi-

mation for cells exposed to other than very low light intensities.

It appears that, over successive generations, phytoplankton vary the amount of chlorophyll, both upwards (in response to poor photon fluxes) and downwards (when similar cells of the same species are exposed to saturating light fluxes). This represents an ability to optimise the allocation of cellular resources in response to the particular internal rate-limiting function, bearing in mind that the synthesis and maintenance of the light-harvesting apparatus carries a significant energetic cost and no more of it will be sponsored than a given steady insolation state may require (Raven, 1984). Moreover, under persistently low average irradiances, there is plainly a limit to the extra light-harvesting capacity that can be installed before the returns in cell-specific photon capture diminish to zero. If all the photons falling on the cell are being intercepted, more harvesting centres will not improve the energy income. On the other hand, this logic indicates the advantage (preadaptation?) of having a relatively large carbon-specific area of projection. Most of the species indicated towards the top of Fig. 3.12 are able to operate under relatively low photon fluxes, in part, through enhancement of the chlorophyll deployment across the light field available. Many of the slender or filamentous diatoms, as well as the solitary filamentous Cyanobacteria (*Limnothrix, Planktothrix*), which project perhaps 10–30 m^2 (mol cell C)$^{-1}$, perform well in this respect. The quasi-spherical colonies of Microcystis show the opposite extreme (k_a: 2–3 m^2 (mol cell C)$^{-1}$).

Another physiological adaptation to persistent light limitation is to increase the complement of accessory photosynthetic pigments. In general, this assists photon capture by widening the wavebands of high absorbance, effectively plugging the gaps in the activity spectrum of chlorophyll a. In particular, the phycobiliproteins (phycocyanins, phycoerythrins of the Cyanobacteria and Cryptophyta) and the various xanthophylls (of the Chrysophyta, Bacillariophyta and Haptophyta; see Table 1.1) increase the light harvesting in the middle parts of the visible spectrum. The close association of accessory pigments with the LHCs facilitates the transfer of excitation energy to chlorophyll a. The corollary of widening the spectral bands of absorption is a colour shift, the chlorophyll green becoming masked with blues, browns or purples. This is acknowledged in the term 'chromatic adaptation' (Tandeau de Marsac, 1977). Some of its best-known instances involve Cyanobacteria formerly ascribed to the genus Oscillatoria. Post et al. (1985) described the photosynthetic performance of a two- to three-fold increase in the chlorophyll content and a three- to four-fold increase in c-phycocyanin pigment in low-light grown cultures of Planktothrix agardhii. Photosynthetic attributes of pink-coloured, deep-stratified populations of P. agardhii were investigated by Utkilen et al. (1985b). A remarkable case of chromatic adjustment in a non-buoyant population of Tychonema bourrelleyi, as it slowly sank through the full light gradient in the water column of Windermere, is given in Ganf et al. (1991). Chromatic adaptation reaches an extreme claret-colour in populations of Planktothrix rubescens, which stratify deep in the metalimnetic light gradient of alpine and glacial ribbon lakes (Meffert, 1971; Bright and Walsby, 2000). The early-twentieth-century appearance of Planktothrix at the surface of Murtensee, Switzerland, was popularly supposed to have come from the bodies of the army of the ruling dukes of Burgundy, defeated and slain in a battle at Murten in the fifteenth century. The connotation 'Burgundy blood alga' celebrates the colour of the alga as much as the independence that was won.

Reynolds et al. (1983a) described an analogous chromatic adaptation of a Cyanobacterium, now ascribed to Planktolyngbya, stratified in the metalimnion of a tropical forest lake (Lagoa Carioca) in eastern Brazil. In each of those cases where measurement has been made, chromatic adaptation increased the chlorophyll-specific photosynthetic yield and the cell-specific photosynthetic efficiency (α). In the P. agardhii strain studied by Post et al. (1985), the efficiency (α) was ~7 times steeper in cultures grown at 20 °C on a 16 h : 8h light–dark cycle and a photon flux of 7 μmol photons m^{-2} s^{-1} than in material in similarly treated cultures exposed to >60 μmol photons m^{-2} s^{-1} (0.78 vs. 0.11 mg O$_2$ (mg dry weight)$^{-1}$ (mol photon)$^{-1}$ m^2; or, in terms of carbon, approximately 0.54 vs. 0.08 mol C (mol cell C)$^{-1}$ (mol photon)$^{-1}$ m^2). Supposing a basal (dark) rate of respiration for Planktothrix at 20 °C (derived from [$R_{a20} = 0.079 \, (s/v)^{0.325}$]; see Sections 3.3.2 and 5.4.1) of 0.064 mol C (mol cell C)$^{-1}$ d^{-1}, or 0.74 × 10^{-6} mol C (mol cell C)$^{-1}$ s^{-1}, it is possible to deduce that compensation is literally achievable at ~1.4 μmol photons m^{-2} s^{-1}. Realistically, allowing for faster respiration during photosynthesis and for the dark period (8 h out of 24 h), net photosynthetic gain is possible over about 3 to 4 μmol photons m^{-2} s^{-1}.

Even this performance may be considerably improved on by stratified bacterial photolithotrophs, with net growth being sustained by as little as 4–10 nmol photons m^{-2} s^{-1} (review of Raven et al., 2000). Chromatic photoadaptation also sustains net photoautotrophic production in cryptomonad-dominated layers in karstic dolines (solution hollows) at ambient photon flux densities of <2 μmol photons m^{-2} s^{-1} (Vicente and Miracle, 1988). This seems to be acceptable as a reasonable threshold for photoautotrophy in phytoplankton. Deep chlorophyll maxima dominated by chromatically adapted cryptomonads have also been observed in somewhat larger lakes and reservoirs (Moll and Stoermer, 1982), sometimes close to the oxycline, from which short diel migrations, either upwards to higher light or downwards (to more abundant nutrient resources) are possible (Knapp et al., 2003).

Photosynthetic limits in lakes and seas

It is often convenient to subdivide the water column on the basis of its ability to sustain net photosynthesis or otherwise. The foregoing sections demonstrate three functional subdivisions based on the criterion of light availability. In the first (the uppermost), light is able to saturate photosynthesis ($I_z > I_k$); in the second, light is a constraint, being limiting to chlorophyll-specific photosynthesis ($I_z < I_k$), but whose effects may be photoadaptively offset in order to optimise the rate of biomass-specific photosynthesis. In the third, even biomass-specific photosynthesis is incapable of compensating the biomass-specific demands of respiration and maintenance ($I_z < I_{P=R}$). The actual water depths for these irradiance thresholds are notionally simple to calculate from the I vs. z curve but, of course, they are not fixed in any sense. The immediate subsurface intensity through the solar day and it is subject to superimposed variability in cloud cover and atmospheric albedo, as well in the fluctuating surface reflectance and subsurface scattering by particulates, induced by wind action. Irradiance thresholds translate to given depths only on an instantaneous basis.

Despite the self-evident weakness of any depth–light threshold relationship, it is still valuable to intercompare various underwater light environments by reference to the impacts of their light attenuation properties. A commonly cited index used in connection with the ability of a water column to support phytoplankton growth is the average depth 'reached by 1% of surface irradiance'; this has also been used to define the depth of the so-called *euphotic zone*. Bearing in mind that the PAR flux at the surface at mid-day varies within at least an order of magnitude (200–2000 µmol photons m^{-2} s^{-1}), the 1% irradiance boundary is approximated no more closely than 2–20 µmol photons m^{-2} s^{-1}. Besides the temporal variability in its precise location in the water column, the quantity also suffers from its conceptual coarseness. Irradiances within this range could saturate the requirements of some species while simultaneously failing to compensate the respiration of others. The depth of the euphotic zone (h_p) is not a general property of the underwater environment, although it remains valid as a species-specific statement of an individual plankter's position in the light gradient relative to its requirement to be able to compensate its respiratory costs.

From Eq. (3.12), we can propose:

$$h_p = \ln(I_0'/0.5I_k)\varepsilon^{-1} \qquad (3.15)$$

recognising that I_k is a property of the species of phytoplankton present and that its cell-carbon specificity is attributable to its carbon-specific photosynthetic efficiency; i.e.

$$I_k = P/\alpha \qquad (3.16)$$

where P is the carbon-specific rate of photosynthesis (in mol C (mol cell C) s^{-1}) and α is the efficiency (in mol C (mol cell C)$^{-1}$ (mol photon)$^{-1}$ m^2).

This development also infers the value of the attenuation coefficient, ε, as a basis for intercomparing aquatic environments. It has the advantage of being a property of the environment (albeit a transitory one) although care is necessary in citing the waveband being used (ε_{440}, ε_{530}, ε_{av}, etc.). Talling (1960, and many later publications) demonstrated a predictive robustness in the approximation of euphotic depth from the minimum attenuation coefficient as the quotient, $3.7/\varepsilon_{min}$. In his examples, the least attenuation was in the green wavebands ($\lambda \sim 530$ nm) but, as a rough guide to the depth in which photosynthesis is possible in the sea, the relationship holds quite well for other wavebands. Table 3.2 is included to contrast the photosynthetic limits of the clearest oceans and some of the most turbid estuarine waters, on the strength of the approximation that, for many phytoplankters and for much of the day-light period, positive net photosynthesis ($P_g \geq R_a$) is possible in the water column defined by $h_p = 3.7/\varepsilon_{min}$. It is emphasised that almost all the net primary photoautotrophic production in the sea occurs within the top 100 m or so and, in lakes, within the top 60 m. In both cases, it is usually much more constrained than this.

Another, much more convenient measure of relative transparency of natural waters is available, the *Secchi disk*. A weighted circular plate, painted all-white or with alternate black and white quadrants, is lowered into the water and the depth beneath the surface that it just

Table 3.2 Comparison of the depth of water likely to be capable of supporting net photosynthetic production (h_p) in some representative lakes and seas, supposing $h_p = 3.7/\varepsilon_{min}$ (cf. Talling, 1960). Values of ε_{min}, the minimum coefficient of attenuation across the visible spectrum are taken (1) from Kirk (1994) or (2) from sources quoted in Reynolds (1987b)

Water	ε_{min} m^{-1}	h_p m	Source
Oceans			
Sargasso Sea	0.03	123	(1)
Pacific, 100 km off Mexico	0.11	34	(1)
Shelf waters	0.15–0.18	20–25	(1)
Lakes and reservoirs			
Crater Lake	0.06	62	(1)
Lake Superior	0.10–0.20	18.5–37.0	(1)
Windermere (North Basin)	0.28–0.72	5.1–13.2	(2)
Crose Mere	0.32–4.20	0.9–11.6	(2)
Lake Kilotes	8.20	0.45	(2)
Mt Bold Reservoir	1.14	3.25	(2)
P. K. le Roux Reservoir	6.39	0.58	(2)
Severn Estuary	10–20	0.2–0.4	(2)

disappears from view is noted as the *Secchi depth* or *Secchi-disk depth*. It is easy to use, has no working parts to malfunction and, in the hands of a single operater, it can give fairly consistent results. However, these attractions are countered by difficulties of quantitative interpretation of its measurements (see Box 3.1). However, documented records of Secchi-disk depth (z_s) span a wide range, from 0.2 m to 77 m (Berman *et al.*, 1985) and are adequate to separate clear waters ($z_s \geq 10$ m) from the turbid ($z_s \leq 3$ m) and to be sensitive to temporal changes in the clarity of any one of them.

Photoadaptation to vertical mixing

This section considers the adaptive responses to turbulent entrainment and vertical transport beyond the column compensation point. Vertical mixing per se is not necessarily problematic for a microplanktic photoautotroph and, in entraining plankters to and from the high solar irradiances that may obtain near the top of the water column, may help to avoid the photoinhibition response observed in phytoplankton captured in static bottles (Jewson and Wood, 1975). Entrainment resists the development of other restricting gradients (for instance, of diminishing dissolved carbon dioxide or accumulating oxygen).

It is likely to be wholly beneficial in maintaining photosynthetic vigour, just so long as the vertical extent of entrainment is within the depth range offering irradiances between I_0' to I_k.

With relatively deeper mixing, however (either as a result of stronger physical forcing or of greater underwater light attenuation), entrained plankters are carried beyond the depth of I_k and, in many circumstances, beyond the productive compensation point. During a period of time (probabilistically, the mixing time, t_m: see Section 2.6.5), the individual plankter may be successively exposed to light intensities that are saturating, sub-saturating or altogether inadequate to support net photosynthetic production. These effects are represented in Fig. 3.14, where a notional 'Lagrangian' path of a single alga, moved randomly in the vertical axis by turbulent entrainment through an equally notional light gradient (a), is exposed to a predictably fluctuant photon flux (b). From the prediction of light-dependent photosynthesis (c), the instantaneous rate of photosynthesis of the phytoplankter is also now predictable (d). Integrating through time, it is clear that the net photosynthetic gain is impaired below the potential of P_{max}. Moreover, the deeper is the mixed depth (h_m) with respect to the depth of the column in which net

Box 3.1 | The Secchi disk

Many variants of the Secchi disk have been employed in limnology and plankton science but the recommended standard is made of aluminium and painted white, is 300 mm in diameter and suspended by three cords attached to a single rope about 30 cm above. It is lowered carefully into the water until the observer just loses sight of it. It is often said to measure transparency or light penetration but these are not literally accurate. The image is lost to the observer through scattering of light rays. However, it is a sufficiently useful and serviceable instrument for there to have been several attempts to relate measurements of Secchi-disk depth (z_s) to more formal photometric light determinations (Vollenweider, 1974; Preisendorfer, 1976; Stewart, 1976). The quest is not aided by differences between observers or, for a single observer, by differences between sun and cloud or between calm and waves. On the basis of simultaneous measurements, Poole and Atkins (1929) deduced that z_s and ε are in approximate inverse proportion and, thus, that the product $z_s \times \varepsilon_{min}$ should be roughly constant. Their evaluation of this constant (1.44) is only just representative of the later estimates (1.4 to 3.0, with a mean of about 2.2: Vollenweider, 1974). Then the light intensity remaining at z_s is between 5% and 24% of I_0 (mean ~15%), whence the compensation depth is perhaps 1.2 to 2.7 × z_s.

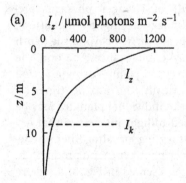
(a)

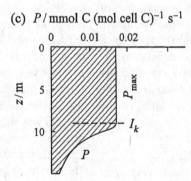
(c)

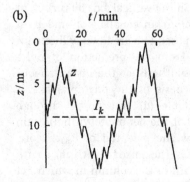
(b)

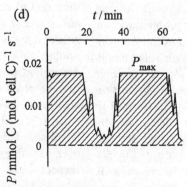
(d)

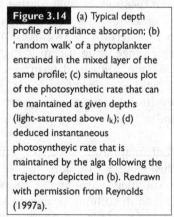

Figure 3.14 (a) Typical depth profile of irradiance absorption; (b) 'random walk' of a phytoplankter entrained in the mixed layer of the same profile; (c) simultaneous plot of the photosynthetic rate that can be maintained at given depths (light-saturated above I_k); (d) deduced instantaneous photosyntheyic rate that is maintained by the alga following the trajectory depicted in (b). Redrawn with permission from Reynolds (1997a).

photosynthesis is possible (h_p), the more restrictive are the mixing conditions on the prospects of photosynthetic gain.

By extension of this argument, the smaller is h_p in relation to h_m, then the more difficult it is to sustain any net photosynthesis at all. It was long a tenet of biological oceanography that the major mechanism permitting phytoplankton recruitment through growth depended upon the depth of mechanical mixing relaxing sufficiently relative to light penetration for net photosynthesis to be sustainable. This relationship was considered by Sverdrup et al. (1942), although it is the eventual mathematical formulation of what is still known as Sverdrup's (1953) 'critical depth model' to which reference is most frequently made. In particular, the idea that the spring bloom in temperate waters, in lakes and rivers as well as open seas and continental shelves, is dependent upon the onset of thermal stratification, at least when it is compounded by a seasonal increase in the day length, remains a broadly plausible concept. However, it is lacking in precision, is open to too literal an interpretation and is not amenable to simulation in models. The problem is due partly to the perception of stratification (as pointed out previously, lack of a pronounced temperature or density gradient is not, by itself, evidence of active vertical mixing). Compounding this is the issue of short-term variability and the likelihood of incomplete mixing within a layer defined by a 'fossil structure' (as defined in Section 2.6.4).

These are important statements regarding an important paradigm, so care is needed to emphasise their essence. It is perfectly true, for instance, that mixing to depth does not only homogenise probable, time-averaged integrals of insolation but 'dilutes' it as well. I used an integral, I^* (Reynolds, 1987c), to estimate light concentration in homogeneously mixed layers, based on the difference between the light availability at the surface and at the bottom, as extrapolated from the attenuation coefficient.

$$\ln I^* = (\ln I'_0 + \ln I_m)/2 \qquad (3.17)$$

where I_m is the extrapolated irradiance at the base of the contemporary mixed layer. Solved by Eq. (3.11), as $I_m = I'_0 \cdot e^{-\varepsilon z m}$, deep mixing can

be shown to be profound. Supposing I'_0 is 800 $\mu mol\, m^{-2}\, s^{-1}$ and ε is 1.0 m^{-1}, then the light reaching the bottom of an 8-m mixed later would be 0.27 $\mu mol\, m^{-2}\, s^{-1}$ and I^* for the whole 8-m layer is just under 15 $\mu mol\, m^{-2}\, s^{-1}$. Doubling the mixed depth to 16 m, means that the light reaching the bottom would be $9 \times 10^{-5}\, \mu mol\, m^{-2}\, s^{-1}$ and the integral for the 16-m layer would be <0.3 $\mu mol\, m^{-2}\, s^{-1}$. A factor of two in the depth of mixing changes a light dose expected to sustain significant net photosynthetic gain to one which will not even satisfy respiration.

Furthermore, starting on the basis of areal integrations of measured photosynthesis and respiration rates versus depth (ΣNP and ΣNR), extrapolations of net photosynthesis over 24 h may be approximated, as calculated as $\Sigma\Sigma NP - \Sigma\Sigma NR$. For short enough days (temperate winters!), high enough attenuation coefficients and verifiable mixing limits, it is probable that low or zero rates of observed phytoplankton increase are correctly attributed to inadequate energy income. To illustrate this, Reynolds (1997b) reworked some earlier data (Reynolds, 1978a; Reynolds and Bellinger, 1992) on the year-round observations on the phytoplankton dynamics in the turbid ($\varepsilon > 0.5$ m), 30-m deep, eutrophic Rostherne Mere, UK. He showed that, in the period January–March, net photosynthetic gain would have been possible only when the mixed depth was <4 m and that the development of any significant spring diatom bloom was normally delayed until April. In a classical paper on the whole-column photosynthetic integrals in Windermere, Talling (1957c) showed that the observed population increase of *Asterionella* in Windermere in the first 3 to 4 months of the year would certainly account for a very high proportion (around 96%) of the extrapolated integral photosynthesis (and, hence, a simultaneously very low biomass-specific respiration rate) to be able to sustain it.

Later attempts to model the population growth from first principles underestimated the oft-observed growth performance in Windermere, unless some allowance was made for a diminishing Monin–Obukhov length (see Section 2.6.3) through the spring period (Reynolds, 1990; Neale et al., 1991a). In other words, the actual

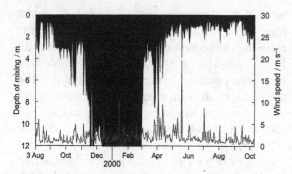

Figure 3.15 Measured daily mean wind speeds at Blelham Tarn, UK, between August 1999 and October 2000 (continuous line), and the mixed depth, as calculated from wind speed and temperature of the water column. Redrawn with permission from Davis et al. (2003).

spring growth, averaged over a number of consecutive years, is really a response to the weakening 'dilution' of the incoming light in a diminishing mixed depth, even though no conventional thermal stratification is yet established. In reality, this is not at all a particularly smooth temporal progression. The day-to-day variability in cloud cover and wind forcing continues but the frequency of the sunnier, less windy days does accelerate, just as the days are lengthening significantly. It follows that, as the year advances, there are more days on which photosynthetic gain, growth and recruitment in the upper water layers is possible.

Another decade of improvements in monitoring approaches and in simulation modelling techniques permits us to derive a greater resolution on the variability in the underwater photosynthetic environment. In Fig. 3.15, the fluctuations in the mixed depth of Blelham Tarn (between 0.5 and 12 m) through a winter–spring period reveal that there is neither a smooth nor single abrupt switch between fully mixed and stably stratified conditions, more a changing frequency of alternation. The contemporaneous compensation depth varied between 2 and 8 m (Davis et al., 2003), net growth being possible when compensation depth exceeds mixing depth.

Two other modelling approaches anticipated similarly modified views of the Sverdrup critical depth model. Woods and Onken (1983) constructed a 'Lagrangian' ensemble, in which the responses of each of a number (at least 20) of plankters, being simultaneously 'walked randomly' through a simulated light gradient, were summated to derive an aggregate for the population. Huisman et al. (1999) used Okubo's (1980) turbulent-diffusion model and a concept of residual light at the base of the turbulent layer (devised by Huisman and Weissing, 1994) to demonstrate the importance of a critical light threshold and, incidentally, to show up the shortcomings of the literal 'critical depth' model. Though this view has not passed unchallenged, it has been supported independently in the theoretical consideration of Szeligiwicz (1998). He also verifies the point that a critical depth is not the same for all species simultaneously, that those with a lower critical light compensation will perform better in deeply mixed layers and that their own adaptive behaviours may modify the critical light and critical depth while the environmental conditions persist.

The nature of these adaptive responses to low aggregate light doses in mixed layers is, in many ways, similar to those arising from the low aggregate light exposure of plankters residing deep in the light gradient. However, mixed-layer entrainment offers short bursts (a few minutes) of exposure to relatively high light intensities separated by probabilistically relatively long periods (of the order of 30–40 minutes) in effective darkness. It would seem important for these organisms to undertake as much photosynthesis as possible in the exposure 'windows' to non-limiting irradiance fluxes, which requires an enhanced light-harvesting capacity rather than a wide spectrum of absorption. Phytoplankton reputed to grow relatively well under deep-mixed conditions include the diatoms (especially those with attenuate cells or that form filaments) and certain Cyanobacteria and chlorophyte genera with analogous morphological adaptations (high msv^{-1}: Reynolds, 1988a); all project substantial carbon-specific areas (10–100 m^2 per mol cell C) but they vary their contents of chlorophyll a more than accessory pigments.

Mostly these effects have been detected through population growth and recruitment in culture. Turning off the light for a part of the day soon brings growth limitation into regulation

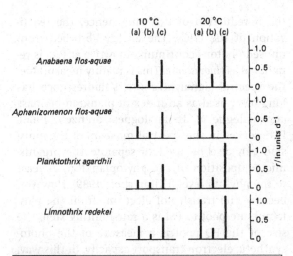

Figure 3.16 Growth-rate responses of Cyanobacteria in culture at two different temperatures (10 °C, 20 °C) to various and two photoperiods: (a) continuous light at 27 μmol photons m^{-2} s^{-1} and (b) under a 6 h : 18 h light–dark alternation. (c) shows the daily growth rate in (b) extrapolated to 24 h, to show the improved efficiency of energy use. Data of Foy et al., (1976), redrawn from Reynolds (1984a).

by light-dependent photosynthesis more than light-independent assimilation – exposing algae to saturating light for only 12 or 6 h out of 24 h always results in a reduced daily growth rate. However, photoadaptive responses to shortened photoperiod raise the rate of biomass-specific energy harvest to the extent that growth normalised per light hour is raised. This principle was memorably demonstrated by Foy et al. (1976) (Fig. 3.16).

Litchman (2000, 2003) has taken this approach further, exploring the effect of shorter experimental photoperiods and their discrimination among the performances of the test algae. Fluctuation periods were varied between 1 and 24 h and intensities were varied between 5 and 240 μmol photons m^{-2} s^{-1}. Photoperiod evoked little photoadaptation at the higher intensities but, at lower intensities, differences in species-specific responses became evident. In general, the effect of fluctuating light tended to be greater when irradiance fluctuated between levels alternately limiting and saturating growth requirements.

Experimentally imposed fluctuations in light-exposure levels on periodicities of days to weeks also affect the composition and diversity of phytoplankton assemblages (Flöder et al., 2002). These operate through the replication, recruitment and attrition of successive generations to populations and will be considered in a later chapter (see Sections 5.4.1, 5.4.2). However, to simulate in the laboratory some of the extreme behaviours observed in the field required observations relating to photoperiods rather shorter than an hour. Robarts and Howard-Williams (1989) described the response of a low-light-adapted Anabaena species in a turbid, mixed lake (Rotongaio, New Zealand) whose rate of photosynthesis could accommodate exposure to light at the water surface for 6 minutes but was slowed abruptly under further exposure. In this instance, the productive advantages were to be gained only in the photoperiods of less than 6 minutes, to which the organism had clearly adapted. These observations are considered in the context of photoprotection and photoinhibition, in the next section.

3.3.4 Photoprotection, photoinhibition and photooxidation

Against the background of environmental variability, there may be superimposed variations in the contemporary ambient range of fluctuations, subjecting hitherto supposedly acclimated plankters to additional demands of accommodation. Among the most crucial of these is a weakening of the mechanical forcing, either as a result of a sharp reduction in the wind speed or of sharp increase in the photon flux (perhaps as the cloud clears) or, as is often the case, the coincidence of both events. In all these instances, the abrupt shortening of the Monin–Obukhov length is, far from being the net beneficial influence cited above (in Section 3.3.3), potentially highly dangerous. Part of the hitherto entrained population becomes disentrained deep in the water column, where the irradiance is markedly sub-saturating. Another is retained within a new, much shallower, surface circulation, exposed to a much elevated I^* value and to a probable excess of radiation in the harmful, high-energy ultraviolet wavelengths. The greater the previous adaptation to low average insolation and the more enhanced is their light-harvesting capacity then, clearly, the greater is the danger of damage to the cells affected. Analogous risks confront plankters near the surface of lakes becalmed overnight and subject to rapid post-dawn increases in insolation. In a lesser way, perhaps, even short bursts of strong light on a mixed layer or lulls in the wind intensity acting on turbid water under bright sunshine will result in potentially sharp increases in the photon flux experienced by individual algae. Moreover, this is the fate of isolates of wild populations, sampled from the water column, subsampled into glass bottles and then held captive at the top of the water column; it is little wonder that their performance becomes impaired (Harris and Piccinin, 1977) (see Section 3.3.1).

In fact, photoautotrophic plankters are equipped with a battery of defences for coping with and surviving exposure to excessive solar radiation levels. As has already been said, some of these have the effect of cutting photosynthetic rate and the response was formerly interpreted as 'photoinhibition'. Strong light certainly can inhibit photosynthesis and do a lot of physical damage to the photosynthetic apparatus. However, many of the observed responses are photoprotective and serve to avoid serious damage occurring to the cell. These are reviewed briefly below; the sequence is more or less the one in which live cells, suddenly confronted by supersaturating photon fluxes, invoke them in response. Some excellent, detailed reviews of this topic include Neale (1987), Demmig-Adams and Adams (1992) and Long et al. (1994).

Fluorescence

In simply moving upwards from a sub-saturating light to a depth where the photon flux density supersaturates not just the demand for growth but also the carbon-fixing ability of PSI, the entrained cell will experience two almost concurrent effects, evoking two compounding reactions. The greater bombardment of the LHCs by photons means that some of these now arrive at reaction centres that are still closed, pending reoxidation of the acceptor quinone, Q_A (see Section 3.2.1). At the same time, the accelerated accumulation of PQH_2 in the plastoquinone pool slows

the re-reduction of P_{680} and, hence, the reactivation of the LHCs. The energy absorbed from unused photons continuing to arrive at P_{680} is re-radiated as *fluorescence*. This is readily measurable: the spectral signal of emitted fluorescence has long been used as an index of plankton biomass (an analogue of an analogue: Lorenzen, 1966). Differences in the spectral make-up of the emission can also be used to separate the organismic composition of the phytoplankton, at least to the phylum level (Hilton et al., 1989). However, because the transfer of electrons from the plastoquinone pool to PSI is a rate-limiting step, the size of the PQ pool is a measure of the photosynthetic electron transport capacity. In this way, PSII fluorescence may also be exploited as a sensitive analogue of photosynthetic activity (Kolber and Falkowski, 1993). Light-stimulated in vivo fluorescence from cells exposed to a flash of weak light in the dark (F_0, when all centres are open) is compared with the fluorescence following a subsequent saturating flash (F_m, corresponding to their total closure). The presence of open centres quenches the fluorescence signal proportionately, so the difference, ($F_v = F_m - F_0$), becomes a direct measure of the photosynthetic electron-transport capacity available and the extent of the reduction in the quantum yield of photosynthesis caused by exposure to high light intensities.

As a relatively short-term response, the chlorophyll-a fluorescence yield alters as the plankters are moved up and down through the mixed layer. The measurement of fluorescence to investigate the transport and the speed of photoadaptive and photoprotective reactions of phytoplankton to variable underwater light climates is one of the exciting new areas of applied plankton physiology (Oliver and Whittington, 1998).

Avoidance reactions

Provided they are adequately disentrained and their intrinsic movements are adequately effective, motile organisms migrate downwards from high irradiance levels. Avoidance reactions have been observed especially among the larger motile dinoflagellates (see, especially, Heaney and Furnass, 1980a; Heaney and Talling, 1980a) and larger, buoyancy-regulating Cyanobacteria (Reynolds, 1975, 1978b; Reynolds et al., 1987). For non-motile diatoms, a rapid sinking rate may provide an essential escape from near-surface 'stranding' through disentrainment, especially in low-latitude lakes. The relatively high sinking rates in (especially) Aulacoseira granulata may be a factor in the frequency of its role as dominant diatom in many tropical lakes where there is a diel variation in mixed water depth (see Reynolds et al., 1986). The effect may be significantly enhanced by spontaneous acceleration of the sinking rate of cells coping with an abrupt increase in insolation (Reynolds and Wiseman, 1982; Neale et al., 1991b), perhaps as a result of the withdrawal of the alleged mechanism of vital regulation (see Section 2.5.4).

Plastid orientation and contraction

Planktic cells are generally too small for plastid relocation to have the significance it does in the cells of higher plants (Long et al., 1994) but, over periods of minutes to hours of exposure to high light intensities, contraction of the chromophores of planktic diatoms lowers the cross-sectional areas projected by the cell chlorophyll (Neale, 1987).

Protective pigmentation

In cells exposed to frequent or continuing high light intensities over a generation time or more, over-excitation of the PSII LHCs is avoided by changes affecting the xanthophylls. These oxygenated carotenoids are subject to a series of light-dependent reactions, which, among the chlorophytes (as among green higher plants), results in the accumulation of zeaxanthin under excess light conditions and its reconversion to violoxanthin on the return of normal light conditions. Among the dinoflagellates and the chrysophyte orders (sensu lato, here including the diatoms), an analagous reaction involves the conversion of diadinoxanthin to diatoxanthin, when light is excessive, with oxidation back to diadinoxanthin in darkness. The reaction is said to be about 10 times faster than the analogous reaction in higher plants (Long et al., 1994, quoting the work of M. Olaizola). The principal function of the xanthophyll cycle in protecting PSII from excessive photon flux density operates by siphoning off a good part of the energy as heat. Many details of the cycle and its fine-tuning are considered by Demmig-Adams and Adams (1992). Here, it is important to emphasise how these adaptations of phytoplankton to high light assist in maintaining photosynthetic productivity. Carotenoids are especially effective in protecting cells against short-wave radiation and the risk of photooxidative stress.

Compounds specific to the absorption of ultraviolet wavelengths, previously known from the sheaths of epilithic mat-forming Cyanobacteria of hot springs (Garcia-Pichel and Castenholz, 1991), where they screen cells from damaging wavelengths of radiation (max absorption ~370 nm), have been found recently in the natural phytoplankton of high mountain lakes. Laurion et al. (2002) suggested that, together with the carotenoids, these mycosporine-like amino acids may occur widely among limnetic phytoplankton species, especially in response to exposure to ultraviolet wavelengths. Ibelings et al. (1994) demonstrated just this sort of acclimation of planktic species, especially in Microcystis, where the sustained presence of zeaxanthin contributes to an ongoing ability to dissipate excess excitation energy as heat. As originally proposed by Paerl et al. (1983), the mechanism substantially protects cells from overexposure of surface blooms to high light.

Excretion

In nutrient-limited cells, photosynthate is scarcely consumed in growth. Even under quite modest light levels, simultaneous accumulation of fixed carbon and free oxidant in the cell risk serious photooxidative damage to the cell. This is countered principally through the production of antioxidants, such as ascorbate and glutathione. High oxygen levels may trigger the Mehler reaction in PSI in which oxygen is reduced to water (Section 3.2.3). Moreover, high O_2 concentrations (>400 μM) induce the oxidase reaction of RUBISCO, and the photorespiration of RuBP to phosphoglycolic acid. Release of glycolate and other photosynthetic intermediates into the water is one of the 'healthy' (cf. Sharp, 1977) ways in which cells of other algal groups regulate the internal environment by venting unusable DOC into the medium. This behaviour carries important consequences for the structure and function of pelagic communities (see Section 3.5.4).

If these fail . . .
Nevertheless, prolonged exposure of phytoplankton cells to high light intensities over periods of days to weeks usually results in pigment loss, loss of enzyme activity, photooxidation of proteins and, ultimately, death. Such dire consequences to the photosystems and cell structures certainly do enter the realm of severe photoinhibition and photodamage. Floating scums of buoyant Cyanobacteria are especially vulnerable to photodamage; death sequences have been graphically reported by Abeliovich and Shilo (1972). In a more recent account, Ibelings and Maberly (1998) described the loss of photosynthetic capacity in response to excessive insolation and carbon depletion in laboratory simulations of the conditions experienced in surface blooms.

At lesser extremes, the resilience of cells and opportunities for repair may allow recovery of physiological vigour. Thus, the many effects of environmental variability that can lead to a fall in the net planktic production of photosynthate, once universally labelled as 'photoinhibition', should properly be viewed as a suite of homeostatic protective mechanisms. They enable phytoplankton to survive a large part of the full range of environmental extremes that may be encountered as a consequence of pelagic embedding (see also the discussion in Long *et al.*, 1994).

3.4 | Sensitivity of aquatic photosynthesis to carbon sources

Besides light, a supply of carbon dioxide is essential to normal photosynthetic production. However, while the necessity of an instantaneous light source to the fixation of carbon is self-evidently axiomatic, the potential limitation of photosynthetic production by inorganic carbon has been a surprisingly contentious issue. Kuentzel (1969) considered the carbon supply to be one of several factors crucial to the development of algal blooms in response to lake enrichment. Shapiro's (1973) experimental demonstrations of the ability of bloom-forming Cyanobacteria to grow at high pH levels (indicative of deficiencies in the reserves of CO_2 in solution) provided very strong support for this view. On the other hand, Schindler's (1971) whole-lake manipulations in the Experimental Lakes Area of Canada pointed to the direct linkage of production responses to added phosphorus and nitrogen. His data were persuasive. Moreover, the intuitive supposition of an adjacent, effectively infinite reserve of atmospheric carbon dioxide, readily soluble in water, was enough to allay most doubts that carbon availability is a significant constraint upon the yields of aquatic biomass.

However, matters did not rest there and much important research has ensued. Crucially, the first point that must be recognised is that these arguments are not, in fact, directly opposed, nor are they mutually exclusive. The aquatic sources of inorganic carbon are, indeed, ultimately plentiful and renewable and do not constitute a *biomass-limiting* constraint. At the same time, it has to be accepted that the ambient concentration of dissolved carbon dioxide is highly variable, that it is inextricably linked to the pH-dependent bicarbonate system and that algal production is, anyway, a principal driver in the transformations. The potential role of the carbon supply as a *rate-limiting* constraint on photosynthetic behaviour is plainly indicated.

3.4.1 Sources and fluxes of available inorganic carbon

The sea-level, air-equilibrated concentration of carbon dioxide in water at $0\,^{\circ}C$ is $\sim 23\,\mu M$, falling to $\sim 13\,\mu M$ at $20\,^{\circ}C$ (say, 0.5–$1.0\,mg\ CO_2\ L^{-1}$, or ~ 0.15–$0.3\,mg\,C\ L^{-1}$). Besides being sensitive to temperature, the equilibrium concentration depends upon the atmospheric partial pressure of CO_2, which is, in turn, affected by altitude. In many natural waters, carbonic acid is the only free acid present and, thus, the concentrations of alkalinity (base ions), carbon dioxide

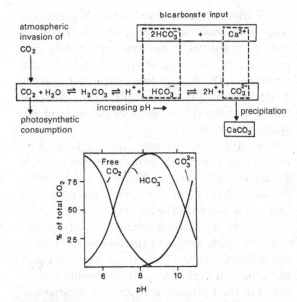

Figure 3.17 The pH–carbon dioxide–carbonate system in natural waters. The relative quantities of the three components, CO_2, HCO_3- and CO_3^{2-}, determine the pH of the water, as shown in the inset. Changes in the concentration of one component shifts the equilibrium. Photosynthetic withdrawal of CO_2 can raise pH to the point where CO_3^{2-} is precipitated as a calcium salt. For a more complete explanation, see Stumm and Morgan (1996). Redrawn from Reynolds (1984a).

and pH (acidity) are continuously interrelated. In this range, bicarbonate is generally the dominant anion (0 to $\sim$3.5 meq L^{-1}; higher alkalinities may be associated with alternative solute composition, sodium or potassium being the dominant cation, rather than calcium). Potentially, bicarbonate may dissociate to release free CO_2, according to the reversible reactions shown in Fig. 3.17. In this way, bicarbonate in solution represents an exploitable store of dissolved inorganic carbon (DIC) of up to 42 mg CO_2 L^{-1}. For much of the time it serves to buffer the water to the mildly alkaline side of neutrality (pH $\sim$8.3), potentially to the point of calcium carbonate precipitation (see Fig. 3.17). In base-poor lakes, this buffering capacity is proportionately weaker. In extremely acidic sodium-sulphate waters (pH < 4.5), there is (by definition) no alkalinity at all and DIC is present only as dissolved carbon dioxide or carbonic acid. Here, as in the first case considered,

the equilibrated mass of gas in solution (again, generally $\sim$0.3 mg C L^{-1}) complies with Henry's law and does not exceed the proportionality of the partial pressure of the gas in contact with the liquid.

At face value, the instantaneous carbon capacity of natural waters to support phytoplankton is unlikely to exceed 0.3 mg C L^{-1} (or about 0.02 mol C m^{-3}). Where present, bicarbonate raises the DIC reserve up to 2 orders of magnitude greater. Supposing a C : Chl$a$ of 50, these capacities are equivalent to the supportive capacity for 6 to 600 μg chl$a$ l^{-1}. Plainly, carbon limitation of the phytoplankton supportive capacity is hardly likely to arise among the many water bodies in the world in which biomass is severely restricted by 1 or 2 orders of magnitude ($\sim$0.6 to 6 μg chl$a$ L^{-1}). Neither is the standing biomass of non-calcareous waters prevented from considerably exceeding 6 μg chl$a$ L^{-1}. The instantaneous carbon dioxide availability in even the soft, non-calcareous waters may significantly exceed the air-equilibrated concentration and some production may be maintained when the DIC reserve is exhausted. This inspires queries about the internal sources of carbon dioxide and the rates of their replenishment.

The proximal sources of 'new' carbon include the solution of CO_2 at the air–water interface, not just at the surface of the water body in question but in the rainfall leaching the atmosphere and falling directly or, indirectly, in the overland flow discharging into it from the surrounding watershed. The quantities transported to lakes can be considerable but the concentrations are still subject to equilibrium constraints. That fraction of the inflow made up from groundwater sources can become relatively enriched with CO_2 under pressure (to the extent that some may vaporise when normal air pressure is encountered). Direct vulcanism (through fumaroles) can provide additional sources of carbon dioxide to sea water in certain locations.

Usually, the major source of DIC is derived from chemical weathering of carbonate rocks and debris in soils, including terrestrially sequestered atmospheric CO_2, which is transported in run-off. Anthropogenically increased CO_2 levels and accelerated erosion have contributed to historic sharp increases in alkalinity in some major catchments, including that of the Mississippi River (Raymond and Cole, 2003). Significant additional sources of carbon in lakes may come from deliveries of readily oxidisable organic carbon, both particulate (POC) and in solution (DOC). Anthropogenic sources (sewage, acid deposition, mine discharge) may be of local importance.

These various carbon sources are available to primary producers and, thence, to assimilation in aquatic food webs. In lakes, some of this carbon may be removed as organisms, their wastes or cadavers, either to the sediments, or to downstream transport, eventually becoming part of the POC flux to the sea. What is usually rather a larger part of the biogenically assembled carbon is respired by the producers or metabolised and respired by their heterotrophic consumers (grazers and decomposers), mostly back to carbon dioxide. This gas can now be vented to the atmosphere by equilibration. In lakes, especially, and at times of low biological activity, carbon dioxide is present in solution at concentrations considerably over those predicted by Henry's law (Satake and Saijo, 1974). In deep, oligomictic and meromictic lakes, hydrostatic pressure adds to the level of carbon-dioxide supersaturation that is possible. Mechanical release of such reserves, such as occurred at Lake Nyos, Cameroon, in 1986, carries dire consequences for people and livestock in the adjacent hinterland (see Löffler, 1988).

It becomes clear that there is a wide range of carbon availabilities among lakes and seas, as there is in the principal carbon sources. The stores and supplies can often be adequate but, at times, demand is capable of exhausting them faster than they can be replenished. This can be especially true in individual lakes having high biomass-supportive capacities and making strong seasonal demands on the carbon flux. Maberly (1996) constructed a balance sheet of annual carbon dioxide exchanges in Esthwaite Water (Cumbria, UK), a small (1.0 km^2), stratifying (15 m), soft-water (alkalinity: 0.4 meq L^{-1}) but eutrophic lake. During the autumn, winter and spring, free-CO_2 concentrations of up to 120 μM (1.4 mg C L^{-1}), almost seven times the expected atmospheric

equilibrium, were observed. At such times, the lake would have been losing CO_2 to the atmosphere. In contrast, photosynthetic carbon consumption in the summer typically depletes the epilimnetic DIC to very low levels (Heaney et al., 1986), occasionally to zero (pH ∼10.3). At these times, the atmosphere becomes the main photosynthetic carbon source.

Over the year, this lake probably loses three to four times more CO_2 to the atmosphere (up to 2.8 mol m^{-2} a^{-1}) than it absorbs (Maberly, 1996). No more than 4% of the annual production of biomass was found to be attributable to CO_2 solution across the water surface. Most of the net resource influx arrives in the lake in solution in the inflow streams. As elsewhere, the main part of the annual load of free CO_2 is roughly proportional to the hydraulic load and the bicarbonate load is, approximately, the product of the mean bicarbonate alkalinity in the inflow and the number of annual hydraulic replacements.

The idea that smaller lakes and rivers are not necessarily net sinks for atmospheric CO_2 but, rather, may often be outgassing CO_2 to the atmosphere, is relatively new (Cole et al., 1994; Cole and Caraco, 1998). In the wet tropics, catchment sources of CO_2 can make an especially significant contribution to the dissolved content and to losses back to the atmosphere (Richey et al., 2002). In yet another study, Jones et al. (2001) calculated that net CO_2 efflux from temperate Loch Ness may represent around 6% of the net ecosystem production of the catchment.

Proportionately, the amount of carbon dioxide loaded hydraulically must diminish with increasing size of the water body, as (presumably) the proportion of the carbon dioxide influx contributed by net inward invasion across the water surface increases. Yet it is plain that, even in Esthwaite Water, there are times of high carbon demand and low resource-renewal rate, marked by high pH values (∼10) when the accelerated absorption of atmospheric carbon dioxide across the water surface must supplement the truncated terrestrial and internally recycled sources.

In water bodies much larger than Esthwaite Water, the hydraulic loads are relatively very much smaller and the oxidation of external POC may be equally diminished on a relative scale. In this case, the exchange of carbon dioxide is mediated mainly by respiration and the dynamics tend to be dominated by metabolic turnover, the loss to sedimentary depletion of particulate carbon and the supplement of 'new', invading atmospheric CO_2.

The direction and rate of gas-exchange flux (F_C) across the water surface is governed by the relationship

$$F_C = G_C \xi \Delta p_{CO_2} \qquad (3.18)$$

where ξ is the solubility coefficient (in mol m^{-3} atmosphere^{-1}), Δp_{CO2} is the difference in partial pressure of carbon dioxide between water and air, and G_C is the gas exchange coefficient, or linear migration rate (m s^{-1}). In fact, the magnitude of actual exchanges is difficult to establish. However, the work of Frankignoulle (1988), Upstill-Goddard et al. (1990), Watson et al. (1991) and Crusius and Wanninkhof (2003), who are among those who have attempted to determine gas-transfer rates by reference to models or to the movements of sulphur hexafluoride tracer (SF$_6$), provides important verifications. The seasonal variability in p$_{CO2}$ becomes a crucially powerful driver in those instances when, as in Esthwaite Water, photosynthetic withdrawal from the aquatic phase takes the air–water difference to its maximum (up to 9×10^{-4} atmosphere, i.e. the fastest consumption stimulates the most rapid invasion of the lake). However, the transfer velocity is accelerated as a function of wind speed and surface roughness, from $\sim 10^{-5}$ m s^{-1}, at wind velocities beneath the critical value of 3.5–3.7 m s^{-1}, to an order of magnitude greater, at 15 m s^{-1} (Watson et al., 1991; Crusius and Wanninkhof, 2003). Given high values of Δp_{CO2}, the corresponding invasion fluxes are calculated to be in the order of 3–30 $\times 10^{-8}$ mol m^{-2} s^{-1}, or between 31 and 310 mg C m^{-2} d^{-1}. Once again, using the approximate 50 : 1 conversion, this is theoretically sufficient to sponsor a productive increment of only 0.6 to 6 mg chlorophyll m^{-2} d^{-1}.

These calculations fully amplify the observation that large crops of algae, especially in lakes of low bicarbonate alkalinity, do not just deplete the store of CO_2 available, with a sharp rise in pH, but they create the conditions for carbon limitation of their own photosynthesis, at least until the demand falls or other sources of carbon can assuage it. In the open sea, where, it is alleged, atmospheric dissolution represents the major resource of new carbon, frequent strong winds may well fulfil one of the criteria of gaseous invasion. The typical low biomass represented by oceanic phytoplankton assemblages rarely raises the pH far above neutrality. Even so, the maximum rates of invasion under the conditions envisaged here can hardly be expected to supply much more than 100 g C m^{-2} a^{-1}.

3.4.2 Phytoplankton uptake of carbon dioxide

Given the relatively high half-saturation constants for RUBISCO carboxylation (12–60 μM) (see Section 3.2.3), photosynthetic carbon fixation is plainly vulnerable to rate limitation by the low aquatic concentrations to which CO_2 may be drawn (<10 μM) (see Section 3.4.1). Even with relatively plentiful supplies of carbon dioxide, the harvesting mechanisms of aquatic plants need to be well developed (Raven, 1991). In the first place, satisfaction of the principal requirements of planktic cells embedded in the viscous range is subject to Fick's laws of diffusion. The number of moles of a solute (n) that will diffuse across an area (a) in unit time, t, is a function of the gradient in solute concentration, C_0, (i.e. dC_0/dx) and the coefficient of molecular diffusion of the substance (m):

$$n = am(dC_0/dx)t \text{ mol m}^2 \text{ s}^{-1} \qquad (3.19)$$

Reynolds (1997a) used data for the single, spherical cell of Chlorella (diameter $\leq 4 \times 10^{-6}$ m, approximate surface area $\leq 50.3 \times 10^{-12}$ m^2) to illustrate the limits of diffusion dependence. Given (i) that, for an average small-sized solute molecule (such as carbon dioxide), $m \sim 10^{-9}$ m^2 s^{-1}, that (ii) the thickness of the adjacent water layer from which nutrients may be absorbed is equal to the cell radius and (iii) the concentration of carbon dioxide molecules beyond is at air equilibrium (11 μmol CO_2 L^{-1}, or 11×10^{-3} mol m^{-3}), then Eq. (3.19) is solved to deliver 275×10^{-18} mol s^{-1}. Now, let us assume the volume of the cell (v) is 33.5×10^{-18} m^3

and contains 0.63×10^{-12} mol carbon (Tables 1.2, 1.3). If we also assume every molecule of carbon dioxide so encountered is successfully taken into the cell, then the requirement to sustain the doubling of biomass without change in the internal carbon concentration is a further 0.63×10^{-12} mol C (cell C)$^{-1}$. While the concentration gradient is maintained, the diffusion rate calculated from Eq. (3.19) is capable of delivering the entire carbon requirement to the cell in $\sim$2300 s (i.e. just over 38 minutes).

For proportionately lower concentrations of carbon dioxide, Eq. (3.19) delivers smaller amounts per carbon per unit time and the time to accumulate the material for the next doubling is correspondingly extended. A concentration of $0.3\,\mu$mol CO_2 L^{-1} could not sustain a doubling in less than 1 day, when growth rate would be considered to be carbon limited. With pH already close to 8.3, the (uncatalysed) dissociation of bicarbonate would support a continuing supply of carbon dioxide. When that too became exhausted, pH drifts quickly upwards as carbonate becomes the dominant form of inorganic carbon.

Many planktic algae avoid (or at least delay) carbon-dioxide limitation and slow bicarbonate dissociation through resort to a carbon-concentration mechanism, or CCM. Although the kinetic characteristics of the key RUBISCO enzyme (especially its high half-saturation requirement for CO_2) cannot be modified, the CCM provides the means to maintain its activity by concentrating CO_2 at the sites of carboxylation. Since their function was first recognised (Badger et al., 1980, Allen and Spence, 1981; Lucas and Berry, 1985), the mechanisms assisting survival of low-CO_2 conditions have continued to be intensively investigated. Progress has been reported in several helpful reviews (Raven, 1991, 1997; Badger et al., 1998; Moroney and Chen, 1998). Working with the green flagellate *Chlamydomonas reinhardtii*, Sültemeyer et al. (1991) showed that the algal CCM involves a series of ATP-mediated cross-membrane transfers – at the cell wall, the plasma membrane and the chloroplast membrane – which transport and concentrate bicarbonate ions as well as carbon dioxide. Breakdown of the bicarbonate is accelerated through the action of carbonic anhydrase. The carbon dioxide thus available to the carboxylation of RuBP is effectively concentrated by a factor of 40.

The more intensively studied CCM of the Cyanobacterium *Synechococcus* also transports and accumulates carbon dioxide and bicarbonate ions, achieving concentration factors in the order of 4000-fold (Badger and Gallacher, 1987). Recent work has revealed the mechanism and genetic control of each of four separate uptake pathways in *Synechococcus* PCC7942 (Omata et al., 2002: Price et al., 2002). Two take up CO_2 at a relatively low affinity, one constitutive and the other inducible, involving thylakoid-based dehydrogenase complexes. There is a third, inducible, high-affinity bicarbonate transporter (known as BCT-1) that is activated by a cAMP receptor protein at times of carbon starvation (a fuller discussion is included in Section 5.2.1). The fourth mechanism is a constitutive Na$^+$-dependent bicarbonate transport system that is selectively activated (perhaps by phosphorylation).

CCMs represent a remarkable adaptation of some (but not all) photosynthetic microorganisms to the onset of carbon limitation of production rates. They are energetically expensive to operate and, not surprisingly, are invoked to assist survival and maintenance only under severe conditions of DIC depletion. The photon cost of fixation of CO_2 concentrated by the cell as opposed to the harvest at equilibration is roughly doubled (>16 mol photon per mol C fixed) and the compensation point is raised to around 10 μmol photons m^{-2} s^{-1} (Raven et al., 2000).

3.4.3 Species-specific sensitivities to low DIC concentrations

The differential ability of aquatic plants to utilise the inorganic carbon supply in fresh waters has been recognised as such for several decades. However, the association of particular types of phytoplankton with particular types of water stretches back almost 100 years, to the days of the Wests and the Pearsalls (West and West, 1909; Pearsall, 1924, 1932) and to the lake classification schemes based on biological metabolism devised by Thienemann (1918) and Naumann (1919). The importance of inorganic nutrients, nitrogen and phosphorus in particular, in governing aquatic metabolism was quickly and correctly appreciated. As the broad correlations detected among indicative types of freshwater phytoplankton and the metabolic state of lakes became developed, particular species or groups of species became classified as indicators of oligotrophic or of eutrophic conditions (Rodhe, 1948; Rawson, 1956). Many chrysophyte, desmid and certain diatom species were seen to be indicative of oligotrophic, phosphorus-deficient conditions (e.g. Findenegg, 1943). Rodhe (1948) went as far as suggesting that phosphorus levels $>20\,\mu g\,P\,L^{-1}$ may actually have been toxic to chrysophytes. On the other hand, Cyanobacteria, especially those species of *Anabaena*, *Aphanizomenon* and *Microcystis* that became abundant as a consequence of anthropogenic eutrophication, were believed to express a preference for high-phosphorus conditions.

Many of these differences can now be explained in terms of the chemistry of carbon rather than of other nutrients. There is no question that the levels of biomass of phytoplankton that may be sustained in a pelagic system are related to the resources available and that the amounts of accessible phosphorus or nitrogen or (in the oceans) iron may well be the biomass-limiting resource (see Chapter 4). Because carbon is unlikely ever to be a capacity-limiting resource and because a large body of literature projects a weight of experimental evidence for species-specific differentials in the uptake capabilities and requirements in respect of (especially) phosphorus and nitrogen, it is understandable that interpretations of species selection in terms of available nutrients should persist (Reynolds, 1998a, 2000a). In fact, the experimental evidence for interspecific differentiation among the dynamics of planktic algae on the basis of their differential abilities to exploit the supplies of carbon has been to hand for many years. Now, detailed biochemical and physiological explanations are available to support the critical role of carbon in distinguishing 'oligotrophic' and 'eutrophic' assemblages.

An indicative anomaly is the example of calcareous (marl) lakes set in karstic, limestone upland areas. Their waters are, by definition, rich in bicarbonate but usually deficient in nitrogen and, partly as a consequence of precipitation as hydroxyapatite (calcium phosphate), phosphorus. The modest phytoplanktic biomass they carry is, however, often dominated by the species of volvocalean green algae, diatoms, dinoflagellates and bloom-forming species of *Anabaena*, *Gloeotrichia* or other Cyanobacteria, that Rodhe (1948) had associated with nutrient-enriched systems. In describing the sparse phytoplankton of Malham Tarn, situated in the carboniferous limestone formations of northern England, Lund (1961) remarked that it was 'quantitatively typical of an oligotrophic lake but qualitatively representative of a eutrophic one'. In contrast, planktic elements most indicative of supposedly oligotrophic plankton (including diatoms of the genera *Cyclotella* and *Urosolenia*, such colonial green algae as *Coenochloris*, *Paulschulzia*, *Pseudosphaerocystis* and *Oocystis* of the *O. lacustris* group, desmid genera such as *Cosmarium*, *Staurastrum* and *Staurodesmus* and chrysophytes that might include species of *Chrysosphaerella*, *Dinobryon*, *Mallomonas* and *Uroglena*) were conspicuously lacking.

Moss (1972, 1973a, b, c) conducted an important series of experimental investigations of the factors influencing the distributions of algae associated with the eutrophication of erstwhile oligotrophic lakes. He found systematic differences in the dynamic responses of algae to variable pH and/or variable carbon sources to be more striking than those due to variation in the amounts of nutrient supplied (see especially Moss, 1973a), with clear separation between the characteristically eutrophic species that could maintain growth at relatively high pH and low concentrations of free carbon dioxide, and the oligotrophic, 'soft-water' species, which could not. Moss's (1973c) conclusion that the response of natural phytoplankton assemblages to nutrient enrichment ('eutrophication') is not dependent on the principal variant (more or less nutrient) but on the productivity demands on the totality of resources.

Talling's (1976) more detailed experiments on the capacity of freshwater phytoplankton to remove dissolved inorganic carbon from the water established a series (*Aulacoseira subarctica* → *Asterionella formosa* → *Fragilaria crotonensis* →

Ceratium hirudinella/*Microcystis aeruginosa*) of increasing tolerance of CO_2 depletion and an increasing capability of staging large population maxima under alkaline, CO_2-depleted conditions. The work of Shapiro (1990) confirmed the apparent high carbon affinities of several Cyanobacteria, especially of *Anabaena* and *Microcystis*, which could maintain slow net growth at pH > 10. That the supply of carbon, rather than any other factor, is limiting under such high-pH conditions is supported by the fact that bubbling with CO_2 will restore the growth rate of *Microcystis* (Qiu and Gao, 2002).

On the other hand, Saxby-Rouen *et al.* (1998; see also Saxby, 1990; Saxby-Rouen *et al.*, 1996) showed convincingly that the chrysophyte *Synura petersenii* is unable to use bicarbonate at all and gave strong indications that species of *Dinobryon* and *Mallomonas* probably also lack the capability. Ball (in Moroney, 2001) has presented evidence that a number of chrysophyte species, including *Synura petersenii* and *Mallomonas caudata*, lack any known kind of carbon-concentrating mechanism. Lehman (1976) had already shown that high phosphorus concentration was no bar to the growth of *Dinobryon*. Reynolds' (1986b) manipulations of phytoplankton composition in the large limnetic enclosures in Blelham Tarn (see Section 5.5.1), showed that phosphorus was as stimulatory to the growth of chrysophytes (*Dinobryon, Mallomonas, Uroglena*) as to any other kind of phytoplankter, provided that the pH did not exceed 8.5. To emphasise the point: neither phosphate nor bicarbonate interferes with the growth of these chrysophytes, so long as they have access to free CO_2.

It is now easy to interpret these various findings in the light of understanding about differential abilities to exploit the various available sources of DIC. Eutrophic phytoplankters, including colonial volvocaleans, many Cyanobacteria and several dinoflagellates, are those that tolerate the low free-CO_2 conditions of naturally high-alkalinity lakes. The species found in soft waters in which enrichment with nitrogen and phosphorus stimulates greater demands on the DIC reserves may well be selected by their ability to exploit bicarbonate directly and/or to focus carbon supplies on the sites of carboxylation. Olig-

otrophic species have no, or only modest, abilities in this direction. The species of *Aulacoseira* and *Anabaena* studied by Talling (1976) are intermediate on this scale. Talling's deduction that the CO_2 system in natural waters 'plays a large part in determining the qualitative composition as well as the photosynthetic activity of the freshwater phytoplankton' was prophetic.

Evidence is accumulating to suggest that the carbon dioxide system may be similarly selective in the sea. Normally, upward pH drift in the sea used to be considered unusual. With the exception of *Emiliana huxleyi*, the formation of whose coccoliths was investigated by Paasche (1964), evidence for the ability of marine phytoplankters to use bicarbonate was still lacking as recently as the mid-1980s (Riebesell and Wolf-Gladrow, 2002). The investigations of Riebesell *et al.* (1993) confirmed that certain species of marine diatom (*Ditylum brightwellii, Thalassiosira punctigera, Rhizosolenia alata*) appear to depend exclusively on the diffusive flux of dissolved CO_2. Growth rates became carbon limited at DIC concentrations below $10-20$ μM ($0.012-0.024$ mg C L^{-1}; pH > 8.1) and stalled completely at <5 μM. The dependence upon diffusive transport and non-catalysed conversion of bicarbonate becomes more problematic among those larger phytoplankters that have a relatively low ratio of surface area to volume, for the flux to the boundary layer and the natural dissociation of bicarbonate is just too slow to compensate the CO_2 deficit at the cell surface in the wake of a high photosynthetic demand. In contrast, however, some shelf-water species, such as *Skeletonema costatum* and *Thalassiosira weissflogii*, show no growth-rate dependence on free-CO_2 concentrations, even at pH levels (>8.5) requiring use of bicarbonate and/or some method of carbon concentration (Burkhardt *et al.*, 1999). Whether carbon dioxide or bicarbonate predominates as the proximal carbon source for the alga is not altogether clear. Bicarbonate may be taken up and converted to carbon dioxide by the action of the enzyme carbonic anhydrase, hydroxyl being excreted to balance the charge, so adding to the prevalence of bicarbonate. Carbonic-anhydrase activity is also detectable on the outer surface of these phytoplankters where the use of bicarbonate is accelerated, especially in response to reducing concentrations of free CO_2 (Nimer et al., 1997; Sültemeyer, 1998). However, even the benefits that this ability brings are finite, amounting, in effect, to an acceleration of the re-establishment of the carbonate system (Riebesell and Wolf-Gladrow, 2002). Carbonic anhydrase activity is said to reach its peak at CO_2 concentrations of ~ 1 µM, (Elzenga et al., 2000) when dissociation of bicarbonate is likely to yield not more than 10–20% of the carbon flux occurring at the air equilibrium in sea water. Carbon limitation of photosynthetic assimilation and potential growth rate of marine phytoplankton is certainly possible and may occur more frequently in high-production waters than has previously been acknowledged.

3.4.4 Other carbon sources

To be able to fix carbon in photosynthesis is the abiding property separating the majority of photoautotrophic organisms ('plants') from the majority of phagotrophic heterotrophs ('animals'). This division is by no means so obvious among the protists where nominally photosynthetic algae show capacities to ingest particulate matter and bacteria as a facultative or a quite typical feature of their lifestyles. This kind of *mixotrophy* is seen among the dinoflagellates and certain types of chrysophyte (mostly chromulines). Alternatively, the bacterium-like ability to absorb selected dissolved organic compounds across the cell surface ('*osmotrophy*') is possessed among some chlorophyte algae (of the Chlorococcales) and in certain Euglenophyta, and among cryptomonads (Lewitus and Kana, 1994).

Whereas osmotrophy clearly represents a means of sourcing assimilable carbon, without the requirement for its prior photosynthetic reduction, mixotrophy is generally regarded as a facultative ability to supplement nutrients other than carbon (chiefly N or P) under conditions of nutrient limitation of production (Riemann et al., 1995; Li et al., 2000). However, there is a valuable energetic subsidy to be derived too, although the exploitative opportunity varies (Geider and Mac-Intyre, 2002); some marine dinoflagellates are said to be 'voraciously heterotrophic', ingesting other protists. In some lakes, *Gymnodinium helveticum* seems to be predominantly phagotrophic,

to the point of abandoning chlorophyll pigmentation. Among the Chromulinales, the resort to phagotrophy seems very strongly associated with a shortage of nutrients; though normally pigmented, however, cells resorting to bacterivory are much paler and show reduced photosynthetic capacity above a threshold prey density (references in Geider and MacIntyre, 2002). On the other hand, some of these species (e.g. *Ochromonas*) are prominent nanoplanktic bacterivores and fulfil a key stage in the microbial loop (see Section 3.5).

An ability of Chlorococcales to take up glucose and other soluble sugars derivatives has been inferred or demonstrated on several occasions (Algéus, 1950: Berman et al., 1977; Lewitus and Kana, 1994). Their habitats, which frequently include organically rich ponds, provide the opportunities for assimilating organic solutes but the relationship appears not to be obligate. Though sometimes representing a major step in the carbon dynamics of ponds, the trait is far from being obligate; algal heterotrophy nevertheless can play an important role in the pelagic carbon cycle of large lakes.

3.5 | Capacity, achievement and fate of primary production at the ecosystem scale

This section is concerned with the estimation of the capacity of phytoplankton-based systems to fix carbon, how much of that capacity is realised in terms of primary product assembled and how much of that is, in turn, processed into the biomass of its consumers, including at higher trophic levels. This is no new challenge, the questions having been implicit in the earliest investigations of plankton biology. The stimulus to pursue them has varied perceptibly over this period, beginning with the objective of understanding the dynamics of biological systems thitherto appreciated only as steady states. From the 1970s, advances in satellite observation have greatly enhanced the means of detecting planetary behaviour and function, while, at the other end of the telescope (as it were), revolutionary changes in understanding how fixed carbon is transferred among ecosystem components have greatly enhanced the interpretation of the sophisticated techniques available for their remote sensing. In the 1990s, study of the fluxes of materials between atmosphere and oceanic systems and the means by which they are regulated has acquired a fresh urgency, born of the need to understand the nature and potential of the ocean as a geochemical sink for anthropogenically enhanced carbon dioxide levels.

These researches help to substantiate the following account, even though it is approached through a hierarchical sequence of integrals of capacity, from those of organelles to metapopulations of plankters stretching across oceans. Moreover, although the productive potential is shown to be vast, plankton biomass remains generally very dilute and broadly stable, and has remained so despite recent increases in atmospheric CO_2 partial pressures, and it is also argued that the current flux of carbon through this rarefied catabolic system is probably as rapid as it can be.

3.5.1 Primary production at the local scale

Subject only to methodological shortcomings and the free availability of exploitable inorganic carbon, the most useful indicator of the potential primary production, P_g (or P_n, *sensu* $P_g - R_a$) (see Section 3.3.2) comes from the areal integration of the instantaneous measurements of photosynthetic rate (ΣNP, in mg C fixed $m^{-2} h^{-1}$) (see Section 3.3.1). The productivity, *sensu* production per unit biomass, measured in mg C fixed (mg biomass C)$^{-1}$ h^{-1}, is a valuable comparator. However, for periods relating to the recruitment of new generations, it is helpful to measure (or to extrapolate) carbon uptake over longer periods, comparable, at least, with the generation time required for cell replication to occur. This generally means designing experimental exposures of 12 or 24 h. Such designs increase the risk of measurement error (through depletion of unreplenished carbon, possible oxygen poisoning and the increasing recycling of 'old' carbon; see Section 3.3.2). These problems may be overcome by mounting contiguous shorter experiments (see the notable example of Stadelmann *et al.*, 1974) but it is generally desirable to integrate results from a small number of short, representative field measurements (i.e. to extrapolate the values $\Sigma\Sigma NP$, $\Sigma\Sigma NR$). With alternative techniques for proxy estimates of biomass and production rates (remote variable chlorophyll fluorometry; Kolber and Falkowski (1993) and see Section 3.3.4), the usefulness of models to relate instantaneous estimates to water-column integrals over periods of hours to days is self-evident.

Integral solutions for calculating primary production over 24 h were devised some 50 years ago. Several, generically similar formulations are available, differing in the detail of the manner in which they overcome the difficult diurnal integration of the light field, especially the diel cycle of underwater irradiance in reponse to the day time variation in I_0' and its relationship to I_k. For instance, Vollenweider (1965) chose an empirical integral to the diel shift in I_0', employing the quotient, $\Gamma[(0.70 \pm 0.07)\, I_{0\,max}']$, where Γ is the length of the daylight period from sunrise to sunset. Extrapolation of $\Sigma\Sigma NP$, in respect of the measured NP_{max} (or, rather, an empirically fitted proportion, $[(0.75 \pm 0.08)\, NP_{max}]$, that compensates for the proportion of the day when $I_0' < I_k$) invokes the coefficient of underwater light extinction (ε). The Vollenweider solution is:

$$\Sigma\Sigma NP = (0.75 \pm 0.08)NP_{max} \times \Gamma$$
$$\times \ln([0.70 \pm 0.07]I_{0\,max}'/0.5I_k) \times 1/\varepsilon$$

$$(3.20)$$

Talling (1957c) had earlier tackled the integration problem by treating the daily light income as a derivative of the daytime mean intensity, I_0', applying over the whole day (Γ). He expressed I_0' in units of light divisions (LD), where LD $= \ln(I_{0\,max}'/0.5I_k)\,/\,\ln 2$, having the dimensions of time. The daily integral (LDH) is the product of LD and Γ, approximating to:

$$LDH = \Gamma[\ln(I_{0\,max}'/0.5I_k)/\ln 2] \qquad (3.21)$$

The completed Talling solution, equivalent to Eq. (3.20) is:

$$\Sigma\Sigma NP = \ln 2 \times NP_{max} \times \Gamma \times LDH \times 1/\varepsilon$$

$$(3.22)$$

Numerically, the two solutions are similar. Applied to the data shown in Fig. 3.3, for example, where $N = 47.6$ mg chl$a$ m^{-3}, $P_{max} = 2.28$ mg

O_2 (mg chl$a$)$^{-1}$ h^{-1}, $\Gamma = 10.8$ h, $I'_{0\,\text{max}} = 800$ and 0.5 $I_k = 24$ µmol photons m^{-2} s^{-1}; and supposing $\varepsilon = 1.33$ (ε_{min}) $= 1.33$ ($\varepsilon_w + \varepsilon_p + N\varepsilon_a$), where ($\varepsilon_w + \varepsilon_p$) $= 0.422$ m^{-1} and $\varepsilon_a = 0.0158$ m^2 (mg chl$a$)$^{-1}$, $\Sigma\Sigma NP$ is solved by Eq. (3.20) to between 1531 and 2020 mg O_2 m^{-2} d^{-1}. For the Talling solution, the daily mean integral irradiance (406 µmol photons m^{-2} s^{-1}) is used to predict $\Sigma\Sigma NP$ $= 2119$ mg O_2 m^{-2} d^{-1}. For comparison, interpolation of the profiles represented in Fig. 3.9 summates to approximately 2057 mg O_2 m^{-2} d^{-1}.

Most of the variations in the other integrative approaches relate to the description of the light field. Steel's (1972, 1973) models used a proportional factor to separate light-saturated and sub-saturated sections of the P vs. I curve. Taking advantage of advances in automated serial measurements of the underwater light field, A. E. Walsby and his colleagues have devised a means of estimating column photosynthesis at each iteration and then summing these to gain a direct estimate of $\Sigma\Sigma NP$ (Walsby, 2001) (see Section 3.3.3).

We may note, at this point, that all these estimates of gross production (P_g; on a daily basis, the equivalent of $\Sigma\Sigma NP$) need correction to yield a useful estimate of potential net production ($P_n = P_g - R_a$, as defined in Section 3.3.2). Again, taking the example from Fig. 3.3, we may approximate $\Sigma\Sigma NR$ as the product, 24 h $\times$ H $\times$ NR, where mean H $= 4.8$ m, $R = 0.101$ mgO_2 (mg chl$a$)$^{-1}$ h^{-1} and $N = 47.6$ mg chl$a$ m^{-3} to be $\Sigma\Sigma NR \sim 554$ mg O_2 m^{-2} d^{-1}. The difference with P_g gives the daily estimate of P_n (strictly, we should distinguish it as NP_n):

$$NP_n = \Sigma\Sigma NP - \Sigma\Sigma NR$$
$$= 2057 - 554 \text{ mg} O_2 \text{ m}^{-2} \text{ d}^{-1}$$
$$= 1.503 \text{ g} O_2 \text{ m}^{-2} \text{ d}^{-1}. \qquad (3.23)$$

Since many of these approaches were developed, it has become more common, and scarcely less convenient, to measure, or to estimate by analogy, photosynthetic production in terms of carbon. Supposing a photosynthetic quotient close to 1.0 (reasonable in view of the governing conditions – see Section 3.2.1), the net carbon fixation in the example considered might well have been about 0.564 g C m^{-2} d^{-1}. Relative to the producer population (4.8 m $\times$ 47.6 mg chl$a$ m^{-3}) this represents some 2.47 mg C (mg chl$a$)$^{-1}$ d^{-1}, or around 0.05 mg C (mg cell C)$^{-1}$ d^{-1}.

Such estimates of local production are the basis of determining the production of given habitats (P_n), the potential biomass- (B-)specific productive yields (P_n/B) and the organic carbon made available to aquatic food webs. The productive yield may sometimes seem relatively trivial in some instances. Marra (2002), reviewing experimental production measurements, showed daily assimilation rates in the subtropical gyre of the North Pacific in the order of 6 mg C m^{-3} d^{-1}. However, this rate was light saturated to a depth of 70 m. Positive light-limited photosynthetic rates were detected as deep as 120 m. Thus, area-integrated day rates of photosynthesis ($\sim$570 mg C m^{-2} d^{-1}) could be approximated that are comparable to those of a eutrophic lake. However, the concentration of phytoplankton chlorophyll ($\sim$0.08 mg chl$a$ m^{-3} through much of the upper water column reaching a 'maximum' of 0.25 mg chl$a$ m^{-3} near the depth of 0.5 I_k) indicates chlorophyll-specific fixation rates in the order of only 60 mg C (mg chl$a$)$^{-1}$ d^{-1}. Carbon-specific rates of $\sim$1.2 mg C (mg cell C)$^{-1}$ d^{-1} are indicated. This is more than enough to sustain a doubling of the cell carbon and, in theory, of the population of cells in the photic layer. It is curious, not to say confusing, that eutrophic lakes are often referred to as being 'productive' when ultraoligotrophic oceans and lakes are described as 'unproductive'. This may be justified in terms of biomass supported but, taking (P_n/B) as the index of productivity, then the usage is diametrically opposite to what is actually the case.

In areal terms, reports of directly measured productive yields in lakes generally range between $\sim$50 mg and 2.5 g C m^{-2} d^{-1} (review of Jónasson, 1978). These would seem to embrace directly measured rates in the sea, according to the tabulations in Raymont (1980). Estimates of annual primary production run from some 30–90 g C fixed m^{-2} a^{-1} (in very oligotrophic, high-latitude lakes and the open oceans that support producer biomass in the order of 1–5 mg C m^{-3} through a depth of 50–100 m; or, say, $\leq$500 mg C m^{-2}), to some 100–200 g C fixed

m^{-2} a^{-1} (in more mesotrophic systems and shelf waters able to support some 50–500 mg producer $C\,m^{-3}$ through depths between 15 and 30 m; i.e. ~$15\,g\,C\,m^{-2}$), and to 200–500 $g\,C\,m^{-2}$ a^{-1} among those relatively eutrophic systems sustaining the fixation of enriched lakes and upwellings, capable of supporting perhaps 1500–5000 mg C m^{-3} through depths of only 3–10 m i.e. ~$15\,g\,C\,m^{-2}$.

This series is rather imprecise but it serves to illustrate the fact that although the supportive capacity of pelagic habitats probably varies over 3 or 4 orders of magnitude, the annual production that is achieved varies over rather less than 2 (maximum annual fixation rates of up to 800–900 $g\,C\,m^{-2}$ are possible in some shallow lakes and estuaries). This is mainly because the exploitation of high supportive capacity results in a diminished euphotic depth. There is thus an asymptotic area-specific maximum capacity and, on many occasions, a diminishing productivity, *sensu* production rate over biomass (see also Margalef, 1997).

3.5.2 Primary production at the global scale

The rapid advances in, on the one hand, airborne and, especially, satellite-based remote-sensing techniques and, on the other, the techniques and resolution for analysing the signals thus detected, have verified and greatly amplified our appreciation of the scale of global net primary production (NPP). In barely 20 years, the capability has moved from qualitative observation, to remote quantification of biomass from the air (Hoge and Swift, 1983; Dekker *et al.*, 1995) and on to the detection of analogues of the rate of its assembly and dissembly (Behrenfeld *et al.*, 2002). The newest satellite techniques can provide the means to gather this information in a single overpass. For terrestrial systems, NPP is gauged from the light absorption by the plant canopy (APAR, absorbed photosynthetically active radiation) and an average efficiency of its utilisation (Field *et al.*, 1998). For aquatic systems, sensors are needed to derive the rate of underwater light attenuation (from which the magnitude of the photosynthetically active flux density at depth is estimable) and the rates of light

absorption and fluorescence attributable to the phytoplankton (Geider *et al.*, 2001). To do this with confidence requires methodological calibration and the application of interpolative production models (discussed in detail in Behrenfeld *et al.*, 2002). Most of the latter employ traditional functions (such as those reviewed in the previous section); calibration is painstaking and protracted but the remakable progress in interpreting photosynthetic properties of phytoplankton has led to synoptic mapping both of the distribution of phytoplankton at the basin mesoscale and of analogues of the rate of its carbon fixation (Behrenfield and Falkowski, 1997; Joint and Groom, 2000; Behrenfield *et al.*, 2002).

The beauty, the global generality and the simultaneous detail of such imagery are awesome. However, its scientific application has been first to confirm and to consolidate the previous generalised findings of biological oceanographers (e.g., Ryther, 1956; Raymont, 1980; Platt and Sathyendranath, 1988; Kyewalyanga *et al.*, 1992; see also Barber and Hilting, 2002, for review). In essence, the main oceans (Pacific, Atlantic, Indian) are deserts in terms of producer biomass (<50 mg chl*a* m^{-2}) while net primary production is assessed to be generally <200 $g\,C$ fixed m^{-2} a^{-1}. In the high latitudes, towards either pole, biomass and production tend to be more seasonal, with maximum production in the six summer months and least in winter. The greatest annual aggregates (200–500 $g\,C$ fixed m^{-2} a^{-1}) are detected mainly on the continental shelves. Production 'hotspots' (500–800 $g\,C$ fixed m^{-2} a^{-1}) are located in particularly shallow areas (e.g., the Baltic Sea, the Sea of Okhotsk), in shelf waters receiving nutrient-rich river outfalls (the Yellow Sea, the Gulf of St Lawrence) and in the upwellings of major cold currents (e.g., the Peru, around Galápagos; and the Benguela, Gulf of Guinea).

Satellite remote sensing has also helped to improve the resolution of global NPP aggregates and their relative contribution to the global carbon cycle. The estimates of total oceanic NPP, based on imagery, converge on values of around 45–50 Pg C a^{-1}. It is interesting that previous estimates, all based on summations and extrapolations of various *in-situ* measurements, are mostly

Table 3.3 | Annual net primary production (NPP) of various parts of the sea and of other major units of the biosphere

	NPP (Pg C)	Energy invested $(J \times 10^{-18})$
Marine domains		
Tropical/subtropical trades	13.0	509
Temperate westerlies	16.3	638
Polar	6.4	250
Coastal shelf	10.7	419
Salt marshes, estuarine	1.2	47
Coral reef	0.7	27
Total	48.3	1890
Terrestrial domains		
Tropical rainforests	17.8	697
Evergreen needleleaf forest	3.1	121
Deciduous broadleaf forest	1.5	58
Deciduous needleleaf forest	1.4	54
Mixed broad- and needleleaf forest	3.1	121
Savannah	16.8	658
Perennial grassland	2.4	94
Broadleaf scrub	1.0	39
Tundra	0.8	31
Desert	0.5	19
Cultivation	8.0	313
Total	56.4	2205

Source: Based on Geider et al. (2001), using data of Longhurst et al. (1995) and Field et al. (1998).

within about 50% of this (20–60 Pg C a^{-1}: see Barber and Hilting, 2002). Only Riley's (1944) estimate of 126 Pg C a^{-1}, based on oxygen exchanges, now seems exaggerated.

It is interesting to compare the estimates for various parts of the ocean and with other major biospheric units (see Table 3.3). Shelf waters contribute nearly a quarter of the total oceanic exchanges despite occupying less than about 1/20 of the area of the seas. Nevertheless, marine photosynthesis is responsible for just under half the global NPP of about 110 Pg (or 1.1 $\times 10^{17}$ g) C a^{-1}.

Much of this carbon is recycled in respiration and metabolism and reused within the year (see also Section 8.2.1). Net replenishment of atmospheric carbon dioxide would contribute a steady-state concentration (currently around 370 parts per million by volume, or 0.2 g C m^{-3}). Significant natural abiotic exchanges of carbon with the atmosphere include the removal due to carbonate solution and silcate weathering ($\sim$0.3 Pg C a^{-1}) but this is probably balanced by releases of CO_2 through calcite precipitation, carbonate metamorphism and vulcanism (Falkowski, 2002). As is well known, however, ambient atmospheric carbon dioxide concentration is currently increasing. This is generally attributed to the combustion of fossil fuels (presently around 5.5 $\pm$ 0.5 Pg C a^{-1} and rising) but the oxidation of terrestrial organic carbon as a consequence of land drainage and deforestation also makes significant contributions (some 1.6 $\pm$ 1.0 Pg C a^{-1}: data from Sarmiento and Wofsy, 1999, as quoted by Behrenfeld et al., 2002). In spite of this anthropogenic annual addition to the atmosphere of $\sim$7 Pg C, the present annual increment is said to be 'only' 3.3 ($\pm$0.2) Pg C a^{-1}; say $\sim$0.6% a^{-1} relative to an atmospheric pool of $\sim$500 Pg). The 'deficit' ($\sim$3.8 Pg C a^{-1}) is explained, in part, by a verified dissolution of atmospheric CO_2 into the sea and, in part, by transfers of organic and biotic components of unverified scale.

Considering that it seems more painful (in the short term) to cut anthropogenic carbon emissions, there is currently a great deal of interest in augmenting net annual flux of carbon to the oceanic store of dissolved carbon of around 40×10^{18} g C (Margalef, 1997).

3.5.3 Capacity

The achievement of even 48.3 Pg across some 360×10^6 km^2 of ocean (note, areal average $\sim$133 g C fixed m^{-2} a^{-1}) should be viewed against the typically small active photosynthetic standing-crop biomass (on average, perhaps just 1–2 g C m^{-2}, $P_n/B \sim 100$). The supportive potential of this conversion is certainly impressive but it represents poor productive yield against the potential yield were every photon of the solar flux captured and its energy successfully transferred to carbon reduction. It is possible that the rate is constrained; or that the rate is rapid but the producer biomass is constrained; or, again, that other processes impinge and the productive yield is constrained.

These are problems of capacity – how rapidly can production proceed before the metabolic losses might balance the productive gains or before the assembly of biomass might become, subject only to the supply of the material components. The problem will not be addressed fully before the physiology of resource gathering and growth have been addressed (Chapters 4 and 5). For the present, we will examine the constraints of insolation and carbon flux on the biomass of phtoplankton that may be present.

Commencing with the productive capacity of the solar quantum flux to the top of the atmosphere (see Fig. 2.2), the income of 25–45 MJ m^{-2} d^{-1} provides the basis for a theoretical annual income of 12–13 GJ m^{-2} annually. Taking only the PAR (46%) and subjecting the flux to absorption in the air by water vapour, reflec-

tion by cloud and backscattering by dust, it is unrealistic to expect much more than 30% of this to be available to photosynthetic organisms (say 3.6 GJ m^{-2} PAR). Using the relationships considered in Section 3.2.3, the potential maximum yield of fixed carbon is 3.6 GJ m^{-2}/470 kJ (mol C fixed)$^{-1}$, i.e. in the order of 7 kmol m^{-2} a^{-1}, or about 85 kg C m^{-2} a^{-1}. This is about 100 times greater than measured or inferred optima. Even if we start with something a little less optimistic, such as the PAR radiation measured at the temperate latitude of the North Sea (annual PAR flux $\sim$1.7 GJ m^{-2}), we still find that no more than about 1.5% of the energy available is trapped into the carbon harvest of primary production.

Looking at the issue in the opposite direction, the minimum energy investment in the carbon that is harvested (Table 3.3) by planktic primary producers is usually a small proportion of that which is available. To produce 50–800 g C m^{-2} a^{-1} requires the capture of a minimum of 9–144 mol photons m^{-2} a^{-1}, equivalent to 2–32 MJ m^{-2} a^{-1}, or less than 2% of the available light energy.

There are many reasons for this apparent inefficiency. The first consideration is the areal density of the LHCs needed to intercept every photon. This can be approximated from the area of the individual centre, the (temperature-dependent) length of time that the centres are closed following the preceding photon capture and, of course, the photon flux density. It was argued, in Section 3.2.1, that the area of a PSII LHC (σ_X) covers about 10 nm^2. Thus, in theory, it requires a minimum of 10^{17} LHCs to cover 1 m^2. It was also shown that 1 g chlorophyll could support between 2.2 and 3.4×10^{18} LHCs and, so, might project an area of $\sim$22 m^2 and harvest photons over its entire area. At very low flux densities, the intact LHC network (0.045 g chl$a$ m^{-2}) has some prospect – again, in theory – of intercepting all the bombarding photons but, as the flux density is increased, it is increasingly likely that photons will fall on closed centres and be lost. Saturation of the network (I_k) is, in fact predicted by:

$$I_k = (\sigma_X t_c)^{-1} \text{ photons m}^{-2} \text{ s}^{-1} \qquad (3.24)$$

where t_c is the limiting reaction time (in s photon^{-1}). At 20 °C, when $t_c \sim 4 \times 10^{-3}$ s (see Section 3.2.1), I_k is solved from Eq. (3.24) as (10 $\times$

10^{-18} m^2 × 4 × 10^{-3} s photon^{-1})$^{-1}$, i.e. about 2.5 × 10^{-19} photons m^{-2} s^{-1}, or ~42 µmol photon m^{-2} s^{-1}.

With supersaturation, more of the photons pass the network or bounce back. Increasing the areal density of LHCs increases the proportion of the total flux intercepted but, in the opposite way, the increasing overlap of LHC projection means that individual LHCs are activated less frequently.

There is no light constraint on the upper limit of light-intercepting biomass surface per se, but the cost of maintaining underemployed photosynthetic apparatus and associated organelles is significant. Ultimately, the productive capacity is set not by the maximum rate of photosynthesis but by the excess over respiration. Reynolds (1997b) developed a simple model of the capacity of the maximum photon flux to support the freshwater unicellular chlorophyte *Chlorella*. The alga was chosen for its well-characterised growth and photosynthetic properties. The 'maximum' set was 12.6 MJ m^{-2} d^{-1}, based upon the flux at the summer solstice at a latitude of 52 °N and supposing a dry, cloudless atmosphere throughout (equivalent daily photon flux, 57.6 mol). Higher daily incomes occur in lower latitudes but the small shortfall in the adopted value need not concern us here).

The available light energy is represented by the vertical axis in Fig. 3.18. The temperature is assumed to be invariable at 20 °C. A relatively small numbers of cells of *Chlorella* (diameter 4 µm, cross-sectional area 12.6 µm^2, carbon content 7.3 pg C or 0.61 × 10^{-12} mol C cell^{-1}, carbon specific projection 20.7 m^2 (mol cell C)$^{-1}$) are now introduced into the light field. They begin to harvest a small fraction of the flux of photons. At low concentrations, the cells have no difficulty in intercepting the light. At higher concentrations, cells near the surface will partly shade out those beneath them. Even if the whole population is gently mixed, the probability is that cells will experience increasingly suboptimal illumination and approach a condition where a larger aggregate portion of each day is passed by each cell in effective darkness. The onset of suboptimal absorption begins when there is a probabilistic occlusion by the *Chlorella* canopy, i.e. at

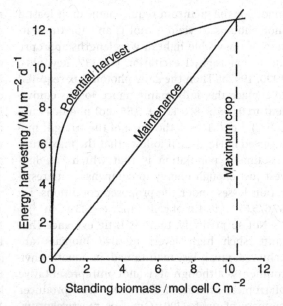

Figure 3.18 Comparison of the light-harvesting potential of phytoplankton as a function of biomass, compared with the energetic costs of respiration and maintenance. The maximum supportable biomass is that which continues to fix just enough carbon in photosynthesis to offset its maintenance requirements. Redrawn with permission from Reynolds (2002b).

an areal concentration of 1/(20.7 m^2 (mol cell C)$^{-1}$) = 0.0483 mol cell C m^{-2} (or ~0.58 g C, or about 11 mg chl*a* m^{-2}). Above this threshold, the cell-specific carbon yield decreases while the cell-specific maintenance remains constant. Thus, for increasingly large populations, total maintenance costs increase absolutely and directly with crop size; total C fixation also increases but with exponentially decreasing efficiency. While the daily photon flux to the lake surface remains unaltered, the asymptote is the standing crop which dissipates the entire 12.6 MJ m^{-2} d^{-1} without any increase in standing biomass (somewhere to the right in Fig. 3.18).

Taking the basal respiration rate of *Chlorella* at 20 °C to be ~1.1 × 10^{-6} mol C (mol cell C)$^{-1}$ s^{-1} (Reynolds, 1990), the energy that would be consumed in the maintenance of the biomass present is supposed to be not to be less than 0.095 mol C (mol cell C)$^{-1}$ d^{-1}. To calculate the maximum biomass sustainable on a photon flux of 57.6 mol photon m^{-2} d^{-1}, allowance must be made for the quantum requirement of at least 8 mol photons to yield 1 mol C and for the fraction of the visible light of wavelengths appropriate to chlorophyll excitation ($\sim$0.137: Reynolds, 1990, 1997b). Then the daily photon flux required to replace the daily maintenance loss is equivalent to $0.095 \times 8/0.137 = 5.55$ mol photons (mol cell C)$^{-1}$ d^{-1}. This is the slope of the straight line inserted in Fig. 3.18. It follows that the maximum sustainable population is that which can harvest just enough energy to compensate its respiration losses under the proposed conditions, i.e. $57.6/5.55 = 10.4$ mol cell C m^{-2} ($\sim$125 g C m^{-2}).

Not surprisingly, there is little evidence that quite such high levels of live biomass are achieved, much less maintained, in aquatic environments, although it is not unrepresentative of tropical rainforest. There the true producer biomass of up to 100 g C m^{-2} is massively augmented by 10–20 kg C m^{-2} of biogenic necromass (wood, sclerenchyma, etc.: Margalef, 1997). In enriched shallow lakes, phytoplankton biomass is frequently found to attain areal concentrations equivalent to 600–700 mg chl$a$ m^{-2} (Reynolds, 1986b, 2001a), more rarely 800–1000 mg chl$a$ m^{-2} (Talling et $al.$, 1973; i.e., 30–50 g C m^{-2}). This relative poverty is due, in part, to the markedly sub-ideal and fluctuating energy incomes that water bodies actualy experience. Moreover, frequent, weather-driven extensions of the mixed layer determine a lower carrying capacity for the entrained population (Section 3.3.3).

Light absorption by water is a powerful detraction from the potential areal carrying capacity that increases with the depth of entrainment. If we solve Eq. (3.17), for instance, against nominated values for $I'_0 = 1000$ and for $I_m = 1.225$ μmol photons m^{-2} s^{-1}, the mixed-layer integral is $I^* = 35$ μmol photons m^{-2} s^{-1}, which is sufficient to impose a significant energy constraint on further increase in biomass. From Eq. (3.11), we can work out that the coefficient of attenuation equivalent to diminish 1000 down to 1.225 μmol photons m^{-2} s^{-1} is equivalent to $\varepsilon z = -6.7$. Now, supposing that this extinction occurs in a mixed layer extending through just the top metre ($z = 1$), we may deduce from Eq. (3.14) that $\varepsilon = 6.7$ m$^{-1} = (\varepsilon_w + \varepsilon_p) + N\varepsilon_a$. Putting $(\varepsilon_w + \varepsilon_p) = 0.2$ m^{-1} and $\varepsilon_a = 0.01$ m^2 (mg chl$a$)$^{-1}$, the theo-

retical phytoplankton concentration that would generate these light conditions is demonstrably equivalent to 650 mg chl$a$ m^{-3} and that it would absorb some 97% of the incoming light. However, if the mixing extends to 10 m, $N\varepsilon_a = 4.7$, the chlorophyll a (470 mg m^{-2}) can now have a maximum mean concentration of only 47 mg m^{-3}, accounting for 70% of the light absorbed. If mixing extends 30 m, $N\varepsilon_a = 0.7$ and the chlorophyll a capacity (70 mg m^{-2}) can have a concentration of no more than 2.3 mg chl$a$ m^{-3}, which absorbs no more than 10% of the incoming light.

Finally, capacity may also be approximated from the integral equations for primary production and respiration. Taking the Vollenweider Eq. (3.20) for $\Sigma\Sigma NP$, for instance, capacity is deemed to be filled when it is precisely compensated by respiration, i.e. when $\Sigma\Sigma NP = \Sigma\Sigma NR$, and $\Sigma\Sigma NP/\Sigma\Sigma NR = 1$. Given that $\Sigma\Sigma NR$ is equivalent to the product, 24 h $\times H \times NR$, we may approximate that:

$$[0.75\,NP_{max} \times \Gamma \times \ln\,(0.70\,I'_{0\,max}/0.5I_k)$$
$$\times 1/\varepsilon]/[24\,\mathrm{h} \times H \times NR] = 1$$

Whence, supposing $\varepsilon = (\varepsilon_w + \varepsilon_p) + N\varepsilon_a$,

$$N = (1/\varepsilon_a)[0.75\,(P_{max}/R) \times (\Gamma/24)$$
$$\times \ln(0.70I'_{0\,max}/0.5I_k)$$
$$\times (1/H) - (\varepsilon_w + \varepsilon_p)] \qquad (3.25)$$

The general utility of this equation is realised in two ways. First, it can be used to gauge the annual variation in photosynthetic carrying capacity of a water body of known intrinsic light-absorbing properties ($\varepsilon_w + \varepsilon_p$) and known seasonal variation in mixed-layer depth (h_m in substitution of H). In the application and spreadsheet solution of Reynolds and Maberly (2002), (P_{max}/R) was set at 15 and ε_a at 0.01 m^2 (mg chl$a$)$^{-1}$.

Second, the equation has been used by Reynolds (1997a, 1998a) to illustrate the impact of the depth of convective mixing on phytoplankton carrying capacity (reproduced here as Fig. 3.19). Against axes of mixed depth (h_m) and background light extinction ($\varepsilon_w + \varepsilon_p$, with a minimum set, arbitrarily, for fresh waters at 0.2 m^{-1}) and supposing $\Gamma = 12$ h, Fig. 3.19 shows graphically how the maximum chlorophyll-carrying capacity is diluted by mixing, from $\sim$150 mg chl$a$ m^{-3} in

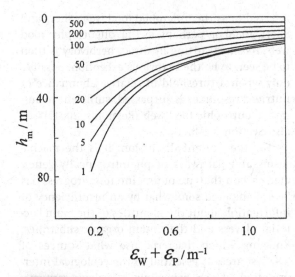

Figure 3.19 Chlorophyll-carrying capacity (as μg chl$a$ L^{-1}) of water columns as a function of the depth to which they are mixed (h_m) and the background coefficient of light extinction due to colour and suspended tripton ($\varepsilon_w + \varepsilon_p$). Solutions assume that $P_{max}/R = 15$ and day length is 12 h. Redrawn with permission from Reynolds (1997a).

a 10-m layer to ~20 mg chl$a$ m^{-3} in a 40-m mixed layer and to <1 mg chl$a$ m^{-3} in a layer mixed to 80 m. Steel and Duncan (1999) developed a similar model to emphasise the advantages of destratifying the eutrophic Thames Valley Reservoirs supplying London in order to lower their plankton carrying capacity below that of the nutrients.

3.5.4 Photosynthetic yield to the planktic food web

The purpose of this section is to comment on some aspects of the fate of photosynthetic products at the local and global level. The profound impact that they exert on carbon cycling in plankton-based aquatic systems is also addressed. The preceding sections show a wide variation in the eventual allocation of the carbon fixed in photosynthesis. Well corroborated in the literature, the range extends from some 92% to 95% investment in new biomass (Talling, 1957c; de Amezaga *et al.*, 1973; Knoechel and Kalff, 1978; Geider and Osborne, 1992) to disparately

low (Talling, 1966; Pollingher and Berman, 1977; Peterson, 1978; Hecky and Fee, 1981). Zero net gains in biomass relative to photosynthesis (i.e. carbon fixation is balanced or exceeded by net losses: Reynolds *et al.*, 1985) are observable when plankter growth is resisted by the total exhaustion of one or other of the essential nutrients. The relatively stable biomass of low-latitude oceanic phytoplankton, in spite of positive carbon-fixation rates (Karl *et al.*, 2002), conforms to the latter diagnosis. The general pattern is that the coupling of net growth to photosynthesis is closest in well mixed, light-limited and nutrient-replete (but possibly carbon-deficient) water columns and weakest under conditions characterised either by stratification, or light saturation to a substantial depth, or extreme nutrient limitations, or any combination of these.

That there should be a gap between the amounts of carbon fixed and those eventually constituting new biomass is not in itself surprising, neither is the relative magnitude of the difference. It was once supposed that the shortfall was explicable in terms of mortalities of producer biomass, chiefly to settlement and to consumers and pathogens (Jassby and Goldman, 1974a). Estimating loss rate of biomass was not then well-advanced but the magnitude of biomass losses necessary to explain productive shortfalls of this order is, on purely intuitive grounds, unrealistically large. It was another perceptive analysis (Forsberg, 1985) that pointed out that, in study after study, the alleged loss of biomass was so close to the measured photosynthetic gain that perhaps the 'lost biomass' had never been formed in the first place. Instead, the losses of fixed carbon are predominantly physiological (for instance, through enhanced respiration), as had indeed been suggested by both Talling (1984) and Tilzer (1984).

The fate and allocation of what, to the photosynthetic microorganism, is excess, unassimilable and mainly unstorable photosynthate, has taken a little longer to diagnose. Certainly, a proportion is respired directly, or is fully photorespired to carbon dioxide and water. It had already been clear for over a decade, however, that a proportion is excreted as DOC, especially when cells are stressed by high insolation or depleted nutrititive resource fluxes (Fogg, 1971; Sharp, 1977). The circumstances of gly-colate excretion, in particular, had been diag-nosed (Fogg, 1977), well before the circumstances of its regulated production through the acceler-ated oxygenase activity of the RUBISCO enzyme (see Section 3.2.3) and its beneficial role in pro-tecting against photooxidative stress had been elucidated (Geider and MacIntyre, 2002). Many other organic compounds are now known to be released by algae into the water, often in solu-tion but not necessarily all to do with metabolic homeostasis. They include monosaccharides, car-bohydrate polymers, carboxylic acid and amino acids (Sorokin, 1999; Grover and Chrzanowski, 2000; Søndergaard et al., 2000).

Though they may be unusable and unwanted by the primary producers, at least in the imme-diate short term, these organic solutes provide a ready and exploitable resource to pelagic bac-teria (Larsson and Hagström, 1979; Cole, 1982; Cole et al., 1982; Sell and Overbeck, 1992). The existence of bacteria in the plankton, both free-living and attached to small mineral and detri-tal particles in suspension, has for long been appreciated but, for many years, their role in doing much more than recycling organic detritus and liberating inorganic nutrients was scarcely appreciated. Interest in the ability of bacteria to assimilate the organic excretory products of pho-toautotrophs increased rapidly with the realisa-tion that a large part of the flux of photosyn-thetically fixed carbon is passed to the food web through a reservoir of DOC rather than through the direct phagotrophic activity of metazoans feeding on intact algal cells (Williams, 1970, 1981; Pomeroy, 1974). It soon emerged that the chain of consumption of bacterial carbon – nor-mally by phagotrophic nanoflagellates, then suc-cessively ciliates, crustacea and plantivorous fish – resulted in the transfer of carbon to the higher trophic levels. This alternative to the more tra-ditional view of the pelagic alga → zooplank-ton → fish food chain soon became known as the 'microbial loop' (Azam et al., 1983). Indeed, it now seems that the 'loop' is often the only viable means by which diffusely produced organic car-bon can be exploited efficiently by the fauna of resource-constrained pelagic systems. It is better

referred to as the 'microbial food web' (Sherr and Sherr, 1988) or, perhaps, as the 'oligotrophic food web'. Its short-cutting by direct herbivory is then to be seen to be the luxurious exception, possible only when a threshold of relative abundance of nutrient resources is surpassed, sufficient to sus-tain a 'eutrophic food web' (Reynolds, 2001a) (see also Section 8.2.4).

For the present discussion, it is the mecha-nisms and pathways of phototrophically gener-ated carbon that are of first interest. Progress has been hampered somewhat by an insufficiency of information about the identities of the main bac-terial players and their main organic substrates. Knowing which 'bacteria' use what sources of 'DOC' sources is essential to the ecological inter-pretation of the behaviour of pelagic systems. Until quite recently, the identification of bac-teria relied upon shape recognition, stain reac-tion and substrate assay. Now, microbiology is adopting powerful new methods for the isolation of nucleic acids (DNA and especially the 16S or 23S ribosomal RNA), their amplification through polymerase chain reaction (PCR) and their match-ing to primers specific to particular bacterial taxa. The approach is similar for both marine and freshwater bacterioplankton (good examples of each are given by Riemann and Winding, 2001; Gattuso et al., 2002). Methods of enumeration have advanced to routine automated counts by flow cytometry. The use of highly fluorescent nucleic acid stains makes for rapid and easy cyto-metric determination of microbe abundance and size distribution in simple bench-top apparatus (Gasol and del Giorgio, 2000). In consequence, the knowledge of the composition, abundance and dynamics of planktic microorganisms is now developing rapidly.

In both lakes and the seas, the free-living heterotrophic bacteria occur in the picoplanktic size range ($0.2–2\,\mu m$; $<4\,\mu m^3$), which they share with the photoautotrophic synechococcoids and eukaryote picophytoplankton (in lakes) and coc-coid prochlorophytes (in the open ocean). Other key participants in the oceanic microbial food webs (viruses, protists) are also prominently rep-resented in those of lakes. These structural sim-ilarities between marine and freshwater micro-bial communuties suggest that they both have

Table 3.4 Typical (upper) densities of bacterioplankton in lakes and seas and daily production rates.

	Standing population ($\times 10^{-6}$ mL^{-1})	Standing biomass (mg C m^{-3})	Production (mg C m^{-3} d^{-1})
Deep (>800 m) tropical oceanic waters	0.01–0.02	<0.4	<0.02
Surface tropical oceans	0.1–0.4	2–6	2–10
Oligotrophic lakes	0.5–0.8	8–15	4–12
Antarctic waters (summer)	1–2	20–60	?
Temperate ocean	1–2	20–70	10–50
Mesotrophic lakes	1–3	20–90	10–70
Inshore waters, estuaries	1.5–3	40–90	20–130
Oceanic coastal upwellings	2–5	90–500	40–150
Eutrophic lakes	3–8	150–500	70–150
Hypertrophic lakes, polluted lagoons	5–40	500–2500	120–700

Source: Based on Sorokin, 1998.

ancient and, possibly, common origins. The typical cell concentrations present in either broadly fit within 2 orders of magnitude (10^5–10^7 mL^{-1}), although pronounced seasonal variation is often detectable, depending upon temperature, the abundance of organic substrate and the intensity of bacterivorous grazing. It is clear that numbers generally reflect the trophic state and they are responsive to enhanced primary-producer activity (Nakano *et al.*, 1998), especially in the wake of phytoplankton 'bloom' periods (Coveney and Wetzel, 1995; Sorokin, 1999; Ducklow *et al.*, 2002) but there is no constant proportionality. Indeed, relative to algal biomass, bacterial numbers (10^4–$10^{7.6}$ mL^{-1}: Vadstein *et al.*, 1993; Sorokin, 1999) (see also Table 3.4) diminish with higher nutrient availability (Weisse and MacIsaac, 2000). Using relationships resolved by Lee and Fuhrman (1987) for marine bacterioplankton, heterotrophs and photoautotrophs may each account for (roundly) up to 0.1 mg C L^{-1} of the phytoplankton biomass of substantially oligotrophic systems. Here, bacterial activity may exceed that of algae (Biddanda *et al.*, 2001) and, thus, make relatively the greatest contribution to organic-matter cycling. In another recent study of the carbon flux through the oligotrophic microbial community of the Bay of Biscay ([chl*a*] < 0.7 mg m^{-3}), González *et al.* (2003) found that a gross primary production (GPP) rate of ~230 mg C m^{-2} d^{-1} was considerably exceeded by bacterial respiration rate (1740 mg C m^{-2} d^{-1}) but with little change either to the bacterial abundance or to the DOC pool (indicating almost no biomass accumulation and complete pelagic cycling of CO_2). Significantly, an upwelling event, with a consequent pulse of nitrogen, stimulated GPP (to over 900 mg C m^{-2} d^{-1}) and to the net recruitment through growth of phytoplankton, leading to an enhanced biomass of larger species 'exportable' as food or in the sedimentary flux. The bacterial biomass decreased, relatively and absolutely, presumably in response to the rerouting of photosynthetic carbon. Where conditions are typically more eutrophic, supporting biomasses of >2.5 mg algal C L^{-1}, bacterial mass may scarcely exceed 0.5 mg C L^{-1}.

The species structure of the bacterioplankton is highly varied and, collectively, includes species capable of oxidising substrates as varied as carbohydrates, various hydrocarbons, proteins and lipids (Perry, 2002). Presumably, the numbers of the particular types fluctuate in response to substrate supply and to grazing: distinct species 'successions', in time and in space, have been demonstrated as the community composition responds to the dissipation of substrate pulses moving through linear mainstem reservoirs (see especially Šimek et al., 1999). Interestingly, however, the clearest trends in species composition seem to respond – like the phytoplankton itself – to the availability of inorganic nutrients. Oligotrophic and mesotrophic assemblages in lakes (e.g. Šimek et al., 1999; Lindström, 2000; Riemann and Winding, 2001; Gattuso et al., 2002) and in the sea (Fuhrman et al., 2002; Cavicchioli et al., 2003; Kuuppo et al., 2003; Massana and Jürgens, 2003) are commonly dominated by species of the Cytophaga–Flavobacterium group and/or various genera of α- and γ-proteobacteria. It also seems likely that these are the main groups of bacteria colonising particulate organic detritus (Riemann and Winding, 2001), some of which anyway decomposes and disintegrates rapidly (Legendre and Rivkin, 2002a). The detailed studies of Cavicchioli et al. (2003) on the dynamics of the proteobacterium Sphingopyxis reveal the relevant properties of an oligotrophic heterotroph. Besides its small size and high surface-to-volume ratio, this obligately aerobic bacterium has a high affinity for nutrient uptake. Population growth rates are sensitive to availability of substrates (which include malate, acetate and amino acids) and to the supply of inorganic ions.

It is clear from this that, although planktic photoautotrophs and heterotrophs have quite independent carbon sources, they nevertheless have to compete for common sources of limiting inorganic nutrients. Moreover, it is likely that the bacteria are superior in this respect (Gurung and Urabe, 1999). Potentially, a mutualism develops between nutrient-deficient autotrophs and carbon-deficient heterotrophs. The elegant experiments of Gurung et al. (1999) on the plankton of the oligotrophic Biwa-Ko, Japan, illustrate how this balance might be maintained. Under low light, photosynthesis is low and bacterial growth is constrained by low organic carbon release. Increasing the light to nutrient-limited phytoplankton stimulates the supply of extracellular organic carbon and the growth of heterotrophs (and of their phagotrophic consumers). Raising the resources available to the photoautotrophs, however, interferes with the organic carbon release to the increasing limitation of heterotroph production. With increased nutrients,

the producers retain proportionately more photosynthate and invest it in the production of their own biomass.

In reality, planktic systems are rather more complex than this simple model might indicate. One major distorting factor is the complicating and paradoxical role played by other, usually much more abundant, sources of dissolved organic matter in pelagic environments. In particular, dissolved humic matter (DHM) is often, by far, the major component of the DOC content of natural waters. Indeed, in the open ocean, where there is a fairly invariable base concentration of ~ 1 mg L^{-1} of DOC (Williams, 1975; Sugimura and Suzuki, 1988), in lakes where concentrations typically fall in the range, 1–10 mg C L^{-1} (Thomas, 1997), and in brown, humic waters draining swamps and peatlands and in which humic matter accounts for 100–500 mg C L^{-1} (Gjessing, 1970; Freeman et al., 2001), DHM may represent some 50–90% of all the organic carbon (including organisms) in the pelagic (Wetzel, 1995; Thomas, 1997). The supposed origin of this varied material – decomposing terrestrial plant matter – is plainly self-evident in lake catchments, although neither the flux to the sea nor its persistence in the ocean has been fully verified.

DHM has the reputation of resistance, or recalcitrance, to degradation by bacteria. Humic material appears in water as substances, mainly phytogenic polymers, of relatively high molecular weight and complexed with various organic groups, which include acetates, formates, oxalates and labile amino acids. By the time they leach into water some decomposition has already taken place. The diversity of humic materials, already large, is increased further (Wershaw, 2000): to make any kind of general assessment of the availability of DHM to pelagic bacteria is still difficult, awaiting more research. However, Tranvik's (1998) thorough evaluation of the bacterial degradation of DOM in humic waters presents some well-considered analysis. Many humic compounds are amenable to bacterial decomposition but, generally, the yield of energy to bacteria is rather poorer than nonhumic DOM. Most is relatively refractory but the resultant pools are not unimportant as bacterial substrates, even though they turn over slowly. The rate of oxidation is influenced by the availability of other nutrients, their tendency to flocculate and their exposure to sunlight and photochemical cleavage. This last turns out to be crucial, as the photodegradation of organic macromolecules to more labile and more assimilable products is now known to occur under strong visible and ultraviolet irradiance (Bertilsson and Tranvik, 1998, 2000; Obernosterer *et al.*, 1999; Ziegler and Benner, 2000). As a result, many relatively simple, low-molecular-weight organic radicals may become available to microbes and may not necessarily be readily distinguishable from the DOM released by photoautotrophs (Tranvik and Bertilsson, 2001).

The emphasis may still be on the restricted nature of the photodegradation and its confinement to surface layers, for the general impression of slow decomposition of humic matter endures. It is likely that it is only in shallow-water systems where allochthonous inputs of DOM might sustain the predominantly hetertrophic activity that the relative abundance of organic carbon would lead us to expect. Elsewhere, it is mainly the non-humic, autochthonously produced DOC that seems likely to underpin heterotroph activity.

This last deduction fits most comfortably with the previously noted general coupling between bacterial mass and primary production: the supposition that, on average, around half the primary production of the oligotrophic pelagic passes through the DOC reservoir requires that this must also be the more dynamic source of carbon and this supports the more active part of bacterial respiration (Cole *et al.*, 1988; Ducklow, 2000). Of course, the relationship is approximate, it is difficult to predict precisely and is plainly subject to breakage. High rates of bacterivory, for instance, would cause one such mechanism. However, bacterial growth can become nutrient limited in very oligotrophic waters, to the extent of positive DOC accumulation (Williams, 1995; Obernosterer *et al.*, 2003), just as easily as it can be substrate limited in (say) nutrient-rich estuaries (Murrel, 2003). In this context, it is especially interesting to note the reports of oceanic microbiologists referring to atomic C : N ratios

in pelagic bacterial biomass of 4.5 to 7.1 (Kirchman, 1990) are interpreted as being indicative more of carbon than of nitrogen limitation of bacterial growth (Goldman and Dennett, 2000). Clearly, the relatively abundant forms of DOC in the oceanic pools often fail to satisfy the requirements of the most abundant planktic heterotrophs, which must therefore rely predominantly on the excretion of phototrophs, much as the Gurung *et al.* (1999) model suggests. Equally clearly, it is a relationship of high resilience (Laws, 2003).

3.6 | Summary

Pelagic primary production is the outcome of complex interplay among biochemical, physiological and ecological processes that include photosynthesis and the large-scale dynamics of various forms of carbon. Photosynthesis is the photochemical reduction of carbon dioxide to carbohydrate, drawing upon radiant energy to synthesise a store of potential chemical energy, pending its discharge when the carbohydrate (or its derivatives) is oxidised (respiration). As in other photoautotrophs, algae and photosynthetic bacteria employ two sequenced, chlorophyll-based photosystems. In the first, electrons are stripped from water and transported to a reductant pool. In the second, photon power re-elevates the electrochemical potential sufficiently to transfer electrons to carbon dioxide, through the reduction of nicotinamide adenine dinuceotide phosphate (NADP to NADPH). The carbon reduction process is built around the cyclical regeneration of ribulose 1,5-biphosphate (RuBP). RuBP is first combined with (carboxylated) carbon dioxide and water to form sugar precursors, under the control of the enzyme RUBISCO, and from which hexose is generated and RuBP is liberated (the Calvin cycle). The hexose may be polymerised (e.g. to starch or glycogen) or stored.

The theoretical photosynthetic quantum yield is 1 mol carbon for 8 mol photon, or 0.125 mol C (mol photon captured)$^{-1}$. Actual efficiency is closer to 0.08 mol C (mol photon)$^{-1}$, equivalent to 470 kJ (mol C fixed)$^{-1}$, or ~39 kJ (g C)$^{-1}$. The maximum rates of photosynthesis are related to the rate of electron clearance from the reductant pool (and which responds to the photon flux), as well as to an adequate supply of CO_2 to the RUBISCO reaction (if a concentration of >0.01 mM is not maintained, the enzyme acts as an oxygenase).

Physiologically, photosynthetic rate is sensitive to temperature, to light and carbon dioxide availability. Even at $30\,°C$, given saturating light and an adequate carbon supply, photosynthesis achieves $<20\,mg\,C\,(mg\,chl a)^{-1}\,h^{-1}$. Maximum photosynthetic rates are generally halved for each $10\,°C$ drop in temperature. Below saturation (usually <150 mol photons $m^{-2}\,s^{-1}$), photosynthetic rates fall in a light-dependent manner, in the proportion $6–18\,mg\,C\,(mg\,chl a)^{-1}\,(mol\,photon)^{-1}\,m^2$. Carbon dioxide concentrations below air saturation may also limit photosynthetic rates. Some algae are extremely efficient in adapting to photon harvesting under very low light fluxes or in the fluctuating light experienced by phytoplankton entrained in mixed water columns. Some algae are restricted to carbon dioxide as a carbon source and are sensitive to the very low concentrations experienced at pH >8. Others can use bicarbonate or employ energy-consuming carbon-concentrating mechanisms to focus the limited fluxes at the sites of synthesis. In this way, low light and low carbon availability select strongly for well-adapted species.

On a local basis, it is possible to calculate the carrying capacity of the environment and the rates of biomass assembly that might be sustainable. Down-mixing and light dilution place important limits on both. The carbon flux from the atmosphere is potentially – and, at times, is – a constraint on area-specific photosynthesis but is avoided in most lakes and at most times by inflowing CO_2-saturated and internal recycles. Indeed, most smaller lakes probably release more CO_2 to the atmosphere than they take from it. They are considered to be net heterotrophic. Only in very large, oligotrophic systems does the sedimentary export of carbon balance the atmospheric inorganic uptake flux (at some $50–90\,g\,C\,m^{-2}\,a^{-1}$).

Globally, pelagic photosynthesis accounts for around 45% of the planetary carbon fixation. In some circumstances, when photosynthesis is constrained (especially by light dilution) the carbon is invested in the growth of the photoautotroph. These organisms become potential food to pelagic grazers. In many other cases, light saturation or nutrient depletion result in carbon fixation in excess of contemporaneous growth requirements and photosynthate is either reoxidised or excreted as dissolved organic carbon (DOC). This augments an already relatively large pool of dissolved humic matter (DHM) but presents a much more amenable substrate for pelagic bacteria. Like those of photoautotrophs, concentrations of heterotrophic bacteria reflect the availability of inorganic nutrients and there is mutual competition. However, bacterial growth is often more carbon limited while the main producers are usually nutrient limited. Besides the mutualism that this situation engenders, the acquisition by bacteria of organic carbon products of the phytoplankton and the consumption of bacteria by microzooplankton represents the main route of pelagic photosynthate to the pelagic food web. This 'microbial loop' commonly dominates the first steps in the food chain, is certainly of great antiquity, and should no longer be regarded as a special exception to alga–herbivore–fish linkages. It is the latter that are the exception, being sustainable only in relatively resource-rich conditions.

Chapter 4

Nutrient uptake and assimilation in phytoplankton

4.1 | Introduction

This chapter addresses the resource requirements for the assembly of photoautotrophic biomass. In addition to light and carbon, growth of phytoplankton consumes 'nutrients' and, equally, may often be constrained by their availability and fluxes. Put at the most basic level, every replication of a phytoplankton cell roundly demands the uptake and assimilation of a quota of (usually) inorganic nutrients similar to that in the mother cell, if her daughters are to have the similar composition. Ignoring skeletal biominerals for the moment, we may recall from Section 1.5.3 that, in addition to carbon, the living protoplast comprises at least 19 other elements. Some are needed in considerable abundance (hydrogen, oxygen, nitrogen), others in rather smaller amounts (phosphorus, sulphur, potassium, sodium, calcium, magnesium and chlorine), for the assembly and production of the organic matter of protoplasm. Others occur as vital traces in support of cellular metabolism (silicon, iron, manganese, molybdenum, copper, cobalt, zinc, boron, vanadium). However, it is less the amounts in which these elements are required that constrains growth than does the ease or otherwise with which they are obtained. It is the demand (D) relative to the supply (S) that is ultimately critical, bearing in mind that a measurable presence is not a measure of availability if the element in question is not both soluble and diffusible and, so, assimilable by cells.

There is a huge literature on this topic. The purpose here is not to review the findings in detail or to give more than the sketchiest outline of the historical development of the understanding of nutrient limitation. Even the elements most often implicated in the constraint of phytoplankton growth (nitrogen, phosphorus, iron and one or two other trace elements, together with the well-known constraint on diatom growth set by its skeletal requirement for considerable quantities of silicon) are sufficiently well and clearly known not to require any long and detailed account of how this recognition came about. The approach that I have adopted is first to consider the mechanisms of nutrient uptake and the general constraints that govern the successful assimilation and anabolism of resources by phytoplankton. Then, mainly by reference to the key limiting elements (N, P, Fe, etc.), I seek to show how the abilities of phytoplankton to gather the resources necessary to support cell growth and replication might impinge upon the dynamics and ecology of populations.

Uptake and assimilation of these nutrients do need to be considered in rather more detail, as impairment to these processes, mostly through resource deficiencies, is frequently implicated in the comparative dynamics and relative abundance of phytoplankters. It is important to recognise that the first impediment to be overcome is that just about all of the nutrients to be drawn from the water occur in concentrations that, relative to their effective concentrations within the cell, are extremely dilute, or rarefied. How these steep gradients are overcome is an appropriate starting point for our consideration.

4.2 | Cell uptake and intracellular transport of nutrients

To describe adequately the main structures of a eukaryotic unicellular phytoplankter that are involved in the uptake, transport and assembly of inorganic components, it is helpful to refer to the simplified and stylised diagram in Fig. 4.1. Inside the multiple-layered plasmalemma (shown as a single line), there is a nucleus containing the genomic proteins (marked 'DNA'); the ribosomal centres of protein synthesis are represented by 'RNA' and part of the structure of the chloroplast and the thylakoid membranes are also sketched. Superimposed upon the cell is a series of arrows that provides a fragmentary indication of the key pathways located within the protoplast. The arrows refer, in part, to the dynamics of photosynthetic reduction of inorganic carbon dioxide and, in part, to the uptake and intracellular delivery of key nutrients to the

sites of their anabolism into proteins and, eventually, into organelles. The cell is ordered, with relative compositional homeostasis based on balanced resource deployment and controlled composition. Outside the cell, the external medium is chaotic: besides signalling irregular and rapid fluctuations in the photon flux, the solutes to which the cell is exposed are often patchily distributed, even at the scale of a few millimetres.

Some initial calculations illustrate the magnitude of the uptake requirement. Starting from the premise that the ash-free dry mass of the cytoplasm accounts for between 0.41 and 0.47 $pg\,\mu m^{-3}$ of live volume and that between 46% and 56% of the ash-free dry mass is carbon, then it follows that the carbon concentration in the replete, healthy, live cell is in the range 0.19 to 0.26 pg C μm^{-3}, or 225 ± 35 g C L^{-1}. This is equivalent to 18.8 mols C L^{-1}. Against the air-equilibrium concentration of carbon dioxide in water (0.5–1 mg L^{-1}, or between 11 and 23 μmol L^{-1}), the growing cell is literally accumulating carbon atoms against a concentration gradient in the order of 1 000 000 to 1. Moreover, in order to accomplish a doubling of cell material, it has to acquire another 1 mol carbon for every 1 mol of carbon in the newly isolated daughter tissue. The corresponding calculations for the average cell concentrations of nitrogen ($\sim$2.8 mols N L^{-1}) and phosphorus ($\sim$0.18 mols P L^{-1}) are of a similar magnitude greater than they might typically occur in natural waters (2–20 μmol N L^{-1}; 0.1–5 μmol P L^{-1}). In relation to the carbon requirement, each cell has to draw on the equivalent of $\sim$151 mmol N and 9.4 mmol P for each mol of C required to replicate the cell mass.

4.2.1 Supply of nutrients

Based on the example of carbon, the well-developed nutrient harvesting capabilities of algae have already been indicated (see, especially, Section 3.4.2). However, it is not simply a matter of engineering a high affinity for the carbon dioxide (or, indeed, other nutrient in the adjacent medium) as the mechanisms can only be effective over a short distance beyond the cell. The operational benefits are really restricted to within the boundary layer adjacent to the cell.

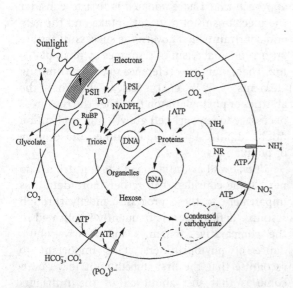

Figure 4.1 Diagram of a phytoplankton cell to show the essential pathways for the gathering and deployment of the key resources. Based on an illustration of Harris (1986) and reproduced with permission from Reynolds (1997a).

Here, the movement of solutes are subject to Fickian laws of diffusion (cf. Eq. 3.19). The renewal, or replenishment, of nutrients in this immediate microenvironment of the cell can also be critical and, hence, so is any attribute of the organism that enhances the rate of entry of essential solutes into that boundary layer. Such adaptations in this direction may raise directly the effectiveness of nutrient gathering by the cell.

The importance of the movement of water relative to the phytoplankter (or, as we now recognise, to the phytoplankter plus its boundary layer) was famously considered by Munk and Riley (1952). They were among the first to point out that the effect of motion – either active 'swimming' or passive sinking or flotation – is to increase the solute fluxes to the cell above those that would be experienced by one that is non-motile with respect to the adjacent medium. This seemingly axiomatic statement was verified through the experiments of Pasciak and Gavis (1974, 1975) and the interpolation of the results to the benefits to nutrient uptake kinetics of a diatom of sinking through nutrient-depleted water. In consideration of these data, Walsby and Reynolds (1980) determined the trade-offs between sinking and uptake rates in sinking diatoms: there was always a positive benefit in material delivery but at the ambient external concentrations critical to sufficiency, the compensatory sinking rates become unrealistically large. In other words, motion relative to the medium undoubtedly assists the renewal and delivery of nutrients to the immediate vicinity of the plankter but, ultimately, is no guarantee of satisfaction of the plankter's requirements at low concentrations.

A modern, empirical perspective on this topic has been pursued in the work of Wolf-Gladrow and Riebesell (1997; see also the review of Riebesell and Wolf-Gladrow, 2002). Starting from the perspective of the single spherical algal cell with an adjacent boundary layer of thickness a, the concentration (C) of a given nutrient in the immediate microenvironment is subject only to diffusive change, in conformity with the equation:

$$\delta C / \delta t = m(\Delta_N)^2 C \qquad (4.1)$$

where t is time, m is the coefficient of molecular diffusion of the solute (as in Eq. 3.19) and $\Delta_N = (\delta/\delta x,\ \delta/\delta y,\ \delta/\delta z)$ is an integral of the gradients in the x, y and z planes. Supposing steady state in a symmetrical sphere, this will reduce to:

$$d/dr_b(r_b^2 dC/dr_b) = 0 \qquad (4.2)$$

where r_b is the radial distance from the centre of the sphere. It may be solved for the space to the edge of the boundary, $C_{surface} = C(r_b = a)$, and beyond, $C_{bulk} = C(r_b \to \infty)$, so that:

$$C(r_b) = C_{bulk} - (C_{bulk} - C_{surface})a/r_b \qquad (4.3)$$

The flux (F_a) of the solute to the cell is calculable as:

$$F_a = 4\pi a^2 m dC/dr_b|_{r_{b=a}}$$
$$= 4\pi a m(C_{bulk} - C_{surface}) \qquad (4.4)$$

If the live cell now retains the inwardly diffusing solute molecules, $C_{surface}$ diminishes to zero and the flux increases towards a maximum:

$$F_{a\,max} = 4\pi a m C_{bulk} \qquad (4.5)$$

The effect of the motion of the cell, sinking, floating or 'swimming' through water is to increase the flux to the diffusive boundary layer at the same time as compressing its thickness (Lazier and Mann, 1989). The distribution of a nutrient solute next to the cell is modified with respect to Eq. (4.1) by the advection owing to the hydrodynamic flow velocity, u:

$$\delta C / \delta t + u\Delta_N C = m(\Delta_N)^2 C \qquad (4.6)$$

The advection–diffusion equation is not easily soluble. The approach of Riebesell and Wolf-Gladrow (2002) was to rewrite the problem in dimensionless Navier–Stokes terms, using particle Reynolds (Section 2.3.4 and Eq. 2.13), Péclet and Sherwood numbers. The Péclet number (Pe) compares the momentum of a moving particle to diffusive transport. For a phytoplankton cell whose movement satisfies the condition of non-turbulent, laminar flow ($Re < 0.1$; Section 2.4.1),

$$Pe = (u_s d) m^{-1} \qquad (4.7)$$

where u_s is the intrinsic velocity of a spherical cell of diameter d. In the present context, where the particle is introduced into the flow field u

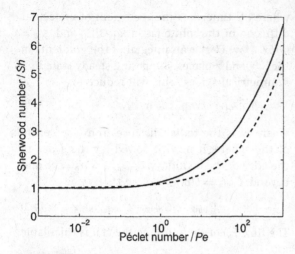

Figure 4.2 The Sherwood number as a function of the Péclet numbers for steady, uniform flow past an algal cell (solid line) and turbulent shear (dashed line). Figure redrawn with permission from an original in Riebesell and Wolf-Gladrow (2002).

(x, y, z, Re), the Péclet number also expresses the ratio of the scales of advective ($u \, \Delta_N \, C$) and diffusive solute transport ($m(\Delta_N)^2 C$). The Sherwood number (Sh) is the ratio between the total flux of a nutrient solute arriving at the surface of a cell in motion and the wholly diffusive flux. Riebesell and Wolf-Gladrow (2002) showed that, for particles in motion with very low Reynolds numbers, Sherwood numbers are non-linearly related to Péclet numbers but, in the range $0.01 \leq Pe \leq 10$ (which embraces the sinking motions of algae from *Chlorella* to *Stephanodiscus*; see Section 2.4.1), the relationship is adequately described by:

$$Sh = 1/2 + 1/2(1 + 2Pe)^{1/3} \qquad (4.8)$$

The relationship (sketched in Fig. 4.2) shows that, for small cells embedded deeply in the turbulence spectrum ($Pe < 1$), the benefit in terms of nutrient supply is marginal ($Sh \sim 1$). For larger units and motile forms generating $Re > 0.001$ and $Pe > 1$, the dependence on turbulence for the delivery of nutrients becomes increasingly significant ($Sh > 1$).

The conclusion fits comfortably with the demonstration of a direct relationship of algal size to Sherwood scaling, mediated through the

influence of the turbulent shear rate (Karp-Boss et al., 1996). They showed:

$$Sh = 1.014 + 0.15Pe^{1/2} \qquad (4.9)$$

when the Péclet number was derived from:

$$Pe = (d/2)^2 (E/\nu)^2 m^{-1} \qquad (4.10)$$

in which equation, d is the diameter of a spherical cell, E is the turbulent dissipation rate, in $m^2 \, s^{-3}$, ν is the kinematic viscosity of the water (in $m^2 \, s^{-1}$) and m is the coefficient of molecular diffusion of the solute (for further details, see Section 2.3.3).

A further deduction and reinterpretation of the comment of Walsby and Reynolds (1980; see above) is that relative motion does not in itself overcome rarefied nutrient resources. However, chronic and extensive resource deficiencies must exact a greater dependence of larger algae on turbulence to fulfil their absolute resources requirements to sustain growth requirements than they do of smaller ones. Conversely, smaller cells are less dependent upon turbulent diffusivity to deliver their nutrient requirements than are larger ones.

4.2.2 Moving nutrients into the cell

The transfer of nutrients from the enveloping boundary layer into and within the cells is biologically mediated, being effected principally through a series of substance-specific membrane transport systems. Modern molecular biology is providing the means to investigate both the mechanisms by which cells marshal and assemble components in cellular synthesis and how their operations are regulated. In the case of membrane transport systems, working against a concentration gradient, structure and function conform to a generalised arrangement common to most living cells. In essence, these accept specific target molecules and transfer them to the sites of deployment. These movements are generally not spontaneous and, so, require the expenditure of energy. Power, fuelled by ATP phosphorylation, is used to generate and maintain ion gradients and proton motive forces, through the coupling of energy-yielding reactions to energy-consuming steps (Simon, 1995).

DOMAINS   STRUCTURES

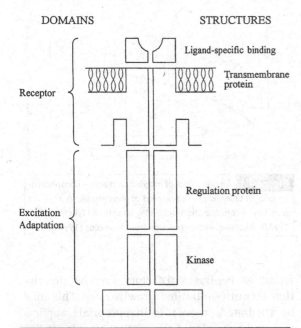

Figure 4.3 Basic structure of a receptor–excitation assembly, used to capture, bind and transport specific target molecules into and within the phytoplankton cell. Based on a figure in Simon (1995) and reproduced with permission from Reynolds (1997a).

In the specific case of nutrient uptake, the linkages involve sequences of protein–protein interactions in which the binding of a specific target ligand at a peripheral *receptor* stimulates an *excitation* of the transfer response. The basic structure of the transmembrane assembly is sketched in Fig. 4.3. The receptor region is periplasmic and constitutes the ligand-specific protein. Reaction with the target molecule stimulates a molecular transformation, which, in turn, becomes the excitant substrate to the proteins of the transmembrane region. The central reaction within this complex is to catalyse the *phosphorylation* of the substrate. This is, of course, the principal reaction through which cells regulate the transfer of redox power. The high-energy pyrophosphate bond between the second and third radicals of ATP is broken by a kinase, the conversion to the diphosphate releasing some 33 kJ $(mol)^{-1}$ of chemical energy.

The further proteins in the series react analogously and sequentially, the excitation of a receptor by the reaction with the target becoming the excitation of the next. The sequenced reactions of the transporter proteins provide a redox-gradient 'channel' along which the target molecule is passed. The whole functions rather like a line of people helping to douse a fire. The first lifts the filled bucket and passes it to the next, who, in turn passes it to a third. Only after the second has accepted a bucket from the first can the first turn to pick up another bucket. The second cannot accept another bucket until he has passed on the last and is once more receptive to the next.

In much the same way as the fire-fighters might be supplied with filled buckets more rapidly than they can be dispatched down the line, so the molecular sequence can become saturated and the fastest rate of uptake then fails to deplete the supply of target nutrients. At the other extreme, exhaustion of the immediate source of buckets or targets leaves the entire sequence idle but the full transport capacity remains 'open' and primed to react to the stimulation of the next arriving molecule.

The activity state of the transport system is communicated to the controlling genes. It is extremely important that the cell can react to the symptoms of shortages of supply (of, say, phosphorus) by regulating and closing down the assembly processes before the supply of components (or a particular component) is exhausted. What happens is that a second group of regulatory proteins associated with the transporters activate the transcription of particular genes called *operons*. While the uptake and transport mechanism is functioning normally, the operons repress the expression of further genes which regulate the reactions to nutrient starvation (Mann, 1995). For instance, an external shortage of a given nutrient (say, orthophosphate ions) will result in a diminishing frequency of receptor reactions and a weakening suppression of the genes that will activate the appropriate cellular response. The response may be to produce more phosphatases or to promote the metabolic close-down of the cell, including entry into a resting stage, before the cell starves to death (see also Sections 4.3.3, 5.2.1).

4.2.3 Empirical models of nutrient uptake

Many of the well-established paradigms relating to the uptake and deployment of nutrients and the ways in which these impinge upon the growth and dynamics of phytoplankton are founded on a welter of experimental observations. However, it is a small number of classic studies that have provided much of the insight and understanding, into which the majority of observations can be interpolated. For instance, the model of Dugdale (1967) recognised the saturable transport capacity (corresponding to the fully 'open' condition of the receptors) and the simultaneous dependence of the rate of uptake upon resource availability in the adjacent medium. Dugdale showed that the relationship between the rate of uptake of nutrients in starved cells and the concentrations in which the nutrient is proffered conforms to Michaelis–Menten enzyme kinetics, as expressed by the Monod equation. Dugdale's (1967) general equation stated:

$$V_U = V_{U_{max}} S / (K_U + S) \qquad (4.11)$$

The equation recognises that the rate of uptake of a nutrient by fully receptive cells is a function of the resource concentration, S, up to a saturable limit of $V_{U_{max}}$. K_U is the constant of half saturation (i.e. the concentration of nutrient satisfies half the maximum uptake capacity). There is no way to predict these values accurately save by experimental determination. Many measurements show close conformity to the predicted behaviour and, hence, to the generalised plot in Fig. 4.4, showing the uptake of phosphorus by *Chlorella*, as described by Nyholm (1977). Because the uptake rates and affinities are, however, very variable among individual phytoplankton species, they are most conveniently intercompared by reference to the magnitudes of alga- and nutrient-specific values of $V_{U_{max}}$ and K_U. For instance, a relatively high $V_{U_{max}}$ capacity combined with a low K_U is indicative of high uptake affinity for a given nutrient.

The Monod model has been widely applied and found to describe adequately the uptake by algae of essential micronutrients, under the starvation conditions described. It has been

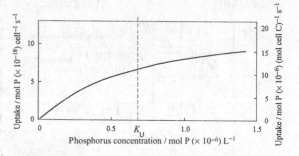

Figure 4.4 Uptake rates of molybdate-reactive phosphorus by cells of *Chlorella* sp., pre-starved of phosphorus, as a function of concentration, according to data of Nyholm (1977). Redrawn with permission from Reynolds (1997a).

found to be less satisfactory for the description of nutrient-limited growth rates. This may be attributed, in part, to inappropriate application and a misplaced assumption – analogous to the one about photosynthesis and growth – that growth rates are as rapid as the relevant materials can be assembled. In fact, growing cells may take up nutrients when they are abundant much more rapidly than they can deploy them, just as they can sustain growth at the expense of internal stores at times when the rate of uptake may be constrained by low external concentrations. Droop (1973, 1974) cleverly adapted the Dugdale model to include a variable internal store in order to represent the impact of the cell quota on the rate of growth. The impacts of nutrient deficiencies on cell replication are considered in Chapter 5 (see Sections 5.4.4, 5.4.5). However, it is useful to introduce here the concept of an internal store in the context of its influence on uptake and, indeed, its relevance to how we judge 'limitation' and its role in interspecific competition for resources.

It is, firstly, quite plain that the intracellular content of the cell starved of a given particular resource will not just be low (leaving the cell very responsive to new resource) but it will probably be close to the absolute minimum for the cell to stay alive. This is Droop's 'minimum cell quota' (q_0) and it is too small to be able to sustain any growth. Secondly, raising the actual internal content (q) above the minimal threshold (essentially through uptake) makes resource available to deployment and growth. At low but steady rates of supply, some proportionality between the rates of growth (r') and resource is expected to be evident. Interpolating ($q - q_0$) for S, r' for V_U and r'_{max} for $V_{U_{max}}$ in Eq. (4.11), the Droop equation states:

$$r' = r'_{max}(q - q_0)/(K_r + q - q_0) \qquad (4.12)$$

As it cannot be assumed that growth and uptake are half-saturated at the same concentration, we must also substitute a half-saturation constant of growth, (K_r). By analogy, we might also assume that, again at a low but steady rate of supply, r' and V_U and r'_{max} and $V_{U_{max}}$ are mutually interchangeable. Such equivalence is demonstrable, provided the steady-state condition holds (Goldman, 1977; Burmaster, 1979). However, when supply rates exceed deployment and the internal store increases, the uptake rate must slow down, even when a high external concentration obtains. It might then be proposed that nutrient uptake is better described by:

$$V_U = [(q_{max} - q)/(q_{max} - q_0)]$$
$$[V_{U max}S/(K_U + S)] \qquad (4.13)$$

where (q_{max}) is the replete cell quota. The larger is the instantaneous cell quota, q, the smaller will be the effective rate of uptake. The multiple allows the cell to accumulate even scarce resources (of, say, phosphorus) from low concentrations so long as the uptake of another (say, nitrogen) is controlling ('limiting') the rate of the deployment of both in the structure of new cell material. An independent increase in the supply of the second resource (nitrogen), however, with no simultaneous alteration in the availability of the first (phosphorus), might very quickly leave the rate of supply of the first as the limiting constraint, when the internal quota is likely to be drawn down. This principle underpins the use of intracellular nutrient ratios to indicate the nutrient status of cells and, thus, the identity of the instantaneously 'limiting factor'. Interspecific differences in the competitive abilities of algae to function at low resource availability are also held to influence the structure of communities.

The terms 'limitation' and 'competition' (in the context of satisfying resource requirements) have been used variously and inconsistently in the literature. In the following considerations, the usage is necessarily precise, adhering to the definitions shown in Box 4.1.

4.3 | Phosphorus: requirements, uptake, deployment in phytoplankton

The phosphorus relations of phytoplankton cells provide a good example of the ways in which the adaptations for gathering of an essential but frequently scarce resource impinge upon the dynamics of populations and the species structure of natural assemblages. As a component of nucleic acids governing protein synthesis and of the adenosine phosphate transformations that power intracellular transport, phosphorus is an essential requirement of living, functional planters. As observed earlier (Section 1.5.3), the phosphorus content of healthy, resource-replete, actively growing phytoplankton cells is generally close to 1–1.2% of ash-free dry mass (Round, 1965; Lund, 1965), with a molecular ratio to carbon of around 0.0094 (106 C:P). The minimum cell quota (q_0) may vary intespecifically, most probably, between 0.2% and 0.4% of ash-free dry mass (some 320–640 mol C:mol P) but, reportedly, almost an order lower in some species (Asterionella ~0.03% of ash-free dry mass: Rodhe, 1948; Mackereth, 1953), equivalent to molecular C:P ratios of ~4000). Conversely, intracellular storage capacity of phosphorus may allow q to rise in some species to ≥3% of dry mass (≤40 C:P). The interesting deduction is that, as a result of this so-called 'luxury uptake', the cell may contain 8–16 times the minimum quota and that, as a consequence, it is theoretically able to sustain three or possibly four cell doublings without taking up any more phosphorus.

4.3.1 The sources and biological availability of phosphorus in natural waters

The natural sources of phosphorus in water are the small amounts that occur in rainfall (generally 0.2 to 0.3 µM), augmented by phosphates derived from the weathering of phosphatic minerals, especially the crystalline apatites, such

Box 4.1 | Limitation and competition in the nutrient relations of phytoplankton

The word 'limitation' has been used, variously, to explain the control of phytoplankton growth dynamics, the poverty of plankton biomass and the dearth of the supportive nutrients. All non-toxic environments have a finite supportive capacity, which is generally based upon the notion that available resources are deployed in the assembly of biomass, ideally, in quasi-fixed quotas, up to a maximum, $B_{max} = K_i/q_0$, where $= K_i$ is the steady-state concentration of the ith resource and q_0 is the minimum cell quota in the biomass, supposing a uniform, Redfield-type composition, or the minimum cell quota in the biomass of the jth species. In this usage, the *limiting capacity* is the lowest of the individual supportive capacities, K_i. By implication, the ratios among the components of the cell will show supra-ideal values to the one that is sub-ideal and, thus, biomass limiting (Reynolds, 1992a; Reynolds and Maberly, 2002). The capacity limitation by the factor least available relative to demand is the expression of von Liebig's 'Law of the Minimum' (von Liebig, 1840). In practical terms, the identity of the capacity-limiting factor is revealed by the magnitude of the response to its augmentation. Gibson (1971) usefully deduced that a substance is *not* capacity-limiting if an increase in that factor produces no stimulation to the biomass that can be supported.

Growth dynamics may also limited, in the sense that the *rate* of biomass elaboration is determined by the rate of resource supply. Moreover, the *rate-limiting* factor is the one upon by whose rate of supply determines the rate of elaboration. Analysing data on the growth of the diatom *Asterionella* in Windermere over a period of 50 years, Reynolds and Irish (2000) were able to confirm Lund's (1950) view that the biomass capacity was set by the winter concentration of soluble reactive silicon. However, they also showed that the rate of its attainment had been phosphorus limited and that the timing of the silicon-limited maximum had advanced over the period of the documented enrichment of available phosphorus in this lake.

'Competition' is used inconsistently by biologists. However, the term 'competitor' is applied by aquatic ecologists, with great consistency, to refer to species that eventually rise, tortoise-like, to a steady-state dominance. Unfortunately, in the parlance of terrestrial plant ecologists, a good competitor is dynamic, fast-growing and applicable to Aesop's fabled hare. Mindful of the place that competition theory occupies in ecological and evolutionary theories, it seems important to have robust definitions. In this book, I use 'competition' in the sense of Keddy's (2001) definition as 'The negative effects that one organism has upon another by consuming, or controlling access to, a resource that is limiting in its availability.' Thus, a competitive outcome has only transpired if the activities of species 1 denies access to the resources required to nourish the activities of species 2. Being able to grow faster when fully resourced does not, by itself, make species 1 'more competitive' than species 2. It is merely more efficient in converting adequate resources into biomass. On the other hand, the behavioral or physiological flexibility of species 2 to better exploit a critically limiting resource affords a significant competitive advantage over species 1, at such times when that resource limitation is operative.

as fluorapatite and hydroxylapatite, and the amorphous phosphorite. All are forms of calcium phosphate, which has a low solubility in water at neutrality, and the bioavailability of phosphorus in drainage waters tends to be low (Emsley, 1980). Terrestrial plants and the ecosystems of which they are part share analogous problems of phosphorus sequestration. Not surprisingly, forested catchments, especially, remove and accumulate much of the modest quantities of inorganic phosphorus with which they are supplied, leaving little in the export to receiving waters save as organic derivatives of biogenic products. The losses of inorganic phosphorus to water can be greatly enhanced through anthropogenic activities (quarrying, agriculture and tillage and, especially, the treatment of sewage) but the general condition of natural waters draining from any but desert catchments and/or ones with an abundant occurrence of evaporite minerals is to be moderately or severely deficient in inorganic phosphorus (Reynolds and Davies, 2001).

In all its biologically available (or 'bioavailable') forms, phosphorus occurs in combination with oxygen in the ions of orthophosphoric acid, $OP(OH)_3$ (Emsley, 1980). Orthophosphoric acid itself is a weak tribasic acid and is freely water soluble. The relative proportions of the various anions (PO_4^{3-}, HPO_4^{2-} and $H_2PO_4^-$) vary with pH. The hydrogen radicals are all replaceable by metals. The orthophosphates of the alkali metals (except lithium) are also soluble but those of the alkaline earth metals and the transition elements are quite insoluble. Three of these – calcium, aluminium and iron – are especially relevant to the consideration of phosphorus availability and plankton behaviour. The precipitation of calcium phosphate effectively removes orthophosphate ions from solution, in stoichiometric proportions. The bioavailability of orthophosphate ions can be significantly affected through exchange with the hydroxyl ions that are otherwise immobilised, in large, non-stoichiometric numbers, on the surfaces of aluminium oxides; this sorption of the orthophosphate ions effectively renders them biologically unavailable. A similar behaviour characterises the reactions of phosphate with the precipitated hydroxides of iron (and man-

ganese), although there is the further complication of their redox sensitivities. At redox potentials below +200 mV, the higher-oxidised ion, Fe^{3+}, is reduced to the divalent Fe^{2+}. Whereas the hydrolysis of the trivalent ion leads to the precipitation of insoluble ferric hydroxide, divalent ferrous ions remain in solution. Raising the redox potential favours the opposite reaction (Fe^{2+} – e $\rightarrow$ Fe^{3+}, although it is usually enhanced by microbial oxidation): the floccular ferric hydroxide precipitate scavenges orthophosphate ions, again in exchange for hydroxyls. At close to neutrality, the orthophosphate ions are substantially immobilised ('occluded') to the extent that they are scarcely any longer available to algal or microbial uptake. Only a further change in redox or an increase in the ambient alkalinity of the medium alters this position. The phosphate ions that are released into solution are, potentially, fully bioavailable (Golterman et al., 1969).

Redox-mediated changes in phosphate solubility in sediment water and in limnetic hypolimnia were described over 60 years ago (Einsele, 1936; Mortimer, 1941, 1942). Since then, many of the fears about the impacts of phosphorus enrichment on aquatic ecosystems have continued to be dominated by the renewed bioavailability to phytoplankton of sediment phosphorus. Of course, there still needs to be phytoplankton access to these phosphorus sources. Though the 'release' of orthophosphate to the water seems just as likely an occurrence, most of this should be re-precipitated with ferric iron, once the redox is raised sufficiently ($\geq +200$ mV). Under severe reducing conditions (redox potential $\leq$ -200 mV), however, sulphate ions are reduced to sulphide ions. These readily precipitate with ferrous iron, thus scavenging the water of Fe^{2+} ions. The consequence then is that, on re-oxidation, the residual iron content will be diminished, less ferric hydroxide will precipitate and less phosphate may be scavenged. High phosphate levels in eutrophic systems may be more influenced by the redox transformations of sulphur than by those of iron.

Certainly, the solubility transformations at high pH and the behaviours of other elements at low redox can have profound effects on the bioavailability of phosphate in natural waters

Table 4.1 Phosphorus-containing fractions in water: nomenclature and availability

Phase	Abbreviation	Chemical sensitivity and bioavailability
Dissolved P	DP	Free orthophosphate ions, some in combination with organic derivatives. Assumed to be *freely bioavailable*
Soluble, molybdate-reactive P	MRP (or SRP)	DP + fine colloidal organic material. Demonstrably bioexhaustible and supposed to be *freely bioavailable*
Particulate P	PP	Phosphorus not in solution or in fine colloids but bound to suspended solids; fraction subdivisble as:
Water-extractable PP	IMRP	Phosphorus moves into solution in irrigating water; most frequently encountered in intact sediments, where it is mainly from the interstitial and is *conditionally bioavailable*
NH_4Cl-extractable PP	NH_4Cl-P	Particle-bound phosphorus, ion exchangeable and *conditionally bioavailable*
Citrate-dithionate-extractable PP	$Na_2S_2O_4$-P	Iron-bound phosphorus. *Scarcely bioavailable*, dependent upon low redox or high pH
NaOH-reactive P	NaOH-rP	Iron- and aluminium-bound phosphorus, sensitive to high pH. Otherwise *scarcely bioavailable*
Non-alkali-reactive PP	NaOH-nrP	PP that is not soluble in strong alkali; fraction includes:
HCl-reactive P	HCl-P	Phosphorus in compound with alkaline metals, esp. apatite. *Scarcely bioavailable*
non-HCl-reactive residue	resP	(Organic) PP soluble only in powerful oxidant (e.g. perchloric acid). *Not bioavailable*
$HClO_4$-digestible P	TP	Perchloric acid digestion releases all known combinations of phosphorus. Phosphorus quoted as 'TP' is only *partially bioavailable* as roughly determined by serial analysis of the above sequence

Source: Based on Table 1 of Reynolds and Davies (2001), compounded from various sources.

(Lijklema, 1977). On the other hand, the mechanisms favouring increased bioavailability of phosphorus are more active in environments that are already relatively enriched with respect to this and other nutrients. These are not exceptional or uncommon conditions among shallow, enriched lakes. However, the wide acceptance that a majority of lakes and many seas conform to a model of pristine conditions characterised by low phosphorus availability is well justified. Such habitats frequently carry a total-phosphorus concentration (TP, being the aggregate of all dissolved, mineral and biogenic particulate phosphorus in the water) generally under 1–2 μM. Moreover, only a small proportion of this TP may be in solution or be so readily soluble to be measurable by the standard molybdenum-blue method of Murphy and Riley (1962). Most of the balance will already be constituent in pelagic biomass or in non-bioavailable colloids and fine particles.

These various fractions are potentially separable by serial assays, each step using a progressively more aggressive chemical cleavage (see Table 4.1). Prior to these methods being developed, the understanding of phosphorus dynamics was poised between the detection of small amounts of molybdate-reactive phosphorus (MRP), with poor sensitivity, and the MRP content of companion samples after digestion with powerful oxidants, supposedly corresponding to the TP concentration. Neither necessarily affords a clear notion of the supportive capacity of the bioavailable forms (BAP): the MRP content, if reliably measurable at all, is an underestimate of BAP, with some or most of what is available having already been biologically assimilated. The TP determination is always likely to include fractions that are chemically immobilised and, in the short term, biologically inert. The whole issue of what is or is not bioavailable is complex and requires a different approach (see Section 4.3.2). Yet it is perfectly clear, from studies of systems as far apart as Windermere and Lake Michigan, that vernal 'blooms' of planktic diatoms, featuring significant increments in chlorophyll concentration and a $\geq$30-fold increase in the concentrations of cells in suspension take place against only small changes in MRP concentration. As revealed by conventional chemical analyses, these scarcely exceed 0.1 μM (i.e. ~3 mg P m^{-3}: Reynolds, 1992a). Some planktic algae and bacteria, at least, are sufficiently well adapted to gather phosphorus to fund several cell doublings despite chronically low ambient MRP concentrations.

4.3.2 MRP-uptake kinetics

Phosphorus uptake and transport in microorganisms are thought to depend on two separate uptake mechanisms. In the Enterobacteria, such as *Escherichia coli*, which can normally experience a much higher external concentration than a free-living phytoplankter, a low-affinity membrane transport system normally operates (Rao and Torriani, 1990). If external phosphorus concentration falls, however, to $\leq$20 μM (~0.6 mg P l^{-1}), the second, high-affinity system is activated. This one is ATP-driven and is linked directly to periplasmic phosphate-binding sites. Working with a cultured strain of the marine cyanobacterium, *Synechococcus* sp. WH7803, Scanlan *et al.* (1993) demonstrated the accelerated synthesis of several intracellular polypeptides as their cultures became increasingly depleted of soluble

phosphorus. One of these, a 32-kDa polypeptide, was localised in the cell wall, linked to an intracellular 100-kDa polypeptide. Together, these conform to the typical structure of a receptor transport system (cf. Fig. 4.3). These polypeptides showed 35% identity and 52% similarity with those of *E. coli*. They also showed that the encoding genes were almost identical to those isolated from other *Synechococcus* strains, which had already been linked to the induction of phosphatase activity in *Synechococcus* PCC7942 (Ray *et al.*, 1991). Alkaline phosphatases are a well-known group of zinc-based enzymes which break phosphate ions from organic polymers in the external medium close to the cell. These are also said to be produced only under conditions of declining external orthophosphate concentrations. Ihlenfeldt and Gibson (1975) noted phosphatase production in a freshwater *Synechococcus* at external concentrations of <4 μM P.

Eukaryotic phytoplankton has not been investigated to this level of biochemical detail. However, it seems likely that analogous mechanisms and similar sensitivities apply among the many phytoplankters that inhabit aquatic environments in which phosphate concentrations are frequently <1 μM P and, often, an order of magnitude less again (<0.1 μM P, i.e. <10^{-7} mol L^{-1}, <3 μg P L^{-1}).

Most of our present knowledge of the phosphorus-uptake kinetics of phytoplankton comes from the numerous laboratory studies on named species, carried out mainly in the middle years of the last century. Some of these have been used in the compilation of compendia and reviews (Reynolds, 1988a, 1993a; Padisák, 2003). In order to make valid interspecific comparisons (such as those in Fig. 4.5), it is necessary to convert often disparate measurements to appropriate common scales. Only volume- or carbon-specific uptake rates of planktic cells lend themselves to some generalisations. One of these is that the maximum phosphorus-uptake rates ($V_{U_{max}}$) of a range of freshwater phytoplankton are comparable, at least within 2 orders of magnitude, when expressed per unit area of algal unit surface (Fig. 4.5b: 0.5 to 35 × 10^{-19} mol P μm^{-2} s^{-1}). When normalised to cell carbon, the same data translate to maximum uptake rates

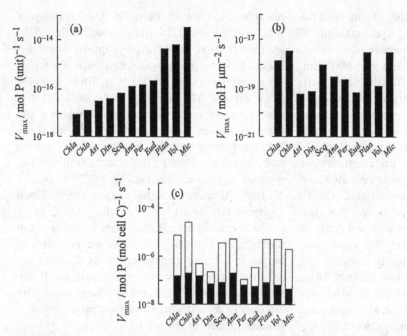

Figure 4.5 (a) Absolute maximal phosphorus uptake rates of phytoplankton cells and colonies as reported in the literature reviewed by Reynolds (1988a) and expressed on a common scale. (b) The same data normalised to the surface areas of the cells or colonies as appropriate. (c) The same data normalised to cell carbon (the shaded part of the histograms correspond to the fastest carbon-specific rate of P assimilation in growth; the balance represents spare capacity (note the logarithmic scales). The algae are *Ana*, *Anabaena flos-aquae*; *Ast*, *Asterionella formosa*; *Chla*, *Chlamydomonas* sp.; *Chlo*, *Chlorella*; *Din*, *Dinobryon divergens*; *Eud*, *Eudorina unicocca*; *Mic*, *Microcystis aeruginosa*; *Per*, *Peridinium cinctum*; *Plaa*, *Planktothrix agardhii*; *Scq*, *Scenedesmus quadricauda*: *Vol*, *Volvox aureus*. Data presented in Reynolds (1993a) and redrawn from Reynolds (1997a) with permission.

of between 0.1 and 21 × 10^{-6} mol P (mol cell C)$^{-1}$ s^{-1}. Against the theoretical requirement for phosphorus to sustain a doubling of the cell carbon (9.4 × 10^{-3} mol P (mol cell C)$^{-1}$), these maximal P-uptake rates ($V_{U_{max}}$) would be sufficient to meet the growth demand in from 440 to 94 000 s (~7 minutes to 26 h), supposing a saturating concentration and a constant rate of uptake. The steady-state phosphorus requirements of the same planktic species growing at their respective maximal cellular growth rates (from Chapter 5) are inserted in Fig. 4.5c to emphasise a second generalisation. It is that we need not be

too concerned about phosphorus-limited uptake rates that remain capable of saturating growth rates down to external concentrations (reading from Fig. 4.4) of the order of 0.1 × 10^{-6} mol P L^{-1} (see also Section 5.4.4).

Nevertheless, ambient concentrations of MRP in some natural waters may be chronically constrained to this order and, in many others, be frequently drawn down to such levels. We may readily accept that phytoplankton tolerant of such conditions must invoke high-affinity mechanisms for phosphorus uptake. Our interest should be sharply focused on the shape of the uptake curve at the extreme left-hand side of Fig. 4.4 and the beneficial distortion represented by a relatively low half-saturation concentration (K_U). Indeed, the lower is the concentration required to half-saturate the uptake of phosphorus, then the greater is the likely ability of the alga to fulfil its requirements at chronically low external concentrations. The faster is the uptake capacity at the low, markedly sub-saturating resource levels, then the greater is the alga's affinity for phosphorus and the greater is its ability to compete for scarce resources.

Pursuing this reasoning further, Sommer's (1984) experiments distinguished several differing adaptive strategies among freshwater phytoplankton for contending with variable supplies of

Table 4.2 Some species-specific values of maximum phosphorus uptake rate ($V_{U_{max}}$) at ~20 °C and the external concentration of MRP required to half-saturate the uptake rate (K_U, being the concentration required to sustain $0.5\ V_{U_{max}}$)

Species	$V_{U_{max}}$ μmol P (mol cell C)$^{-1}$ s^{-1}	K_U μmol P L^{-1}	References[a]
Chlamydomonas reinhardtii	7.35	0.59	Kennedy and Sandgren (unpublished, quoted by Reynolds, 1988a, with permission)
Chlorella pyrenoidosa	25.12	0.68	Nyholm (1977)
Asterionella formosa	0.51	1.9–2.8	Tilman and Kilham (1976)
Dinobryon sociale	0.21	0.39	Lehman (1976)
Scenedesmus quadricauda	3.4	1.2–4.0	Nalewajko and Lean (1978)
Anabaena flos-aquae	10.5	1.8–2.5	Nalewajko and Lean (1978)
Peridinium sp.	0.11	6.3	Lehman (1976)
Eudorina elegans	0.34	0.53	Kennedy and Sandgren (unpublished, quoted by Reynolds, 1988a, with permission)
Planktothrix agardhii	5.41	0.2–0.3	van Liere (1979), Ahlgren (1977, 1978)
Volvox aureus	5.24	1.62	Kennedy and Sandgren (unpublished, quoted by Reynolds, 1988a, with permission)
Microcystis aeruginosa	1.95	0.3	Holm and Armstrong (1981)

[a] The original data come from the works cited, as recalculated to a common scale of cell-carbon specificity by Reynolds (1988a).

phosphorus. Species might be relatively *velocity-adapted*, in which high rates of cellular growth and replication (r') are matched by suitably rapid rates of nutrient uptake ($V_{U_{max}}$), or else, they may be more *storage-adapted*, in which rapid, opportunistic uptake rates exceed relatively slow rates of deployment in growth, thereby permitting a net accumulation of an intracellular reserve of phosphorus. These adaptations are said to be distinguished by differences in the species-specific ratio ($V_{U_{max}}/r'_{max}$). Species may also show a tendency to be more or less *affinity-adapted* according to the species-specific ratio, $V_{U_{max}}/K_U$; as suggested, high affinity is imparted by a low K_U requirement.

Sommer's terminology is helpful but the derived measures are themselves subject to interexperimental variability, even for the same species. The values noted in Table 4.2 are those used in the construction of Fig. 4.5. To an extent, $V_{U_{max}}$ is necessarily greater than the rate of deployment of phosphorus in new cell material, supposing that this corresponds to r'_{max} (at 20 °C) and that the cell quota remains constant. From this, it may also be deduced that the uptake rate, V_U, has to be markedly undersaturated for a considerable time before r' can be said to be P-limited. The most helpful adaptation to enable algae to deal with chronically low external MRP concentrations is a very low half-saturation coefficient. However, it is clearly relevant for such algae to be able still to function on a relatively low internal phosphorus quota. Davies' (1997) recent investigations of the cell-phosphorus-related growth kinetics of natural *Asterionella* populations during the spring-bloom period in the English Lakes are illustrative. Plotting cell-increase rates against the corresponding cell P quotas at the time of their sampling, Davies (1997) was able to fit a single, statistically significant Michalis–Menten-type curve, that suggested growth rate was fully saturated by cell quotas of 5–10 pg P $(cell)^{-1}$ (or, roughly, 0.023–0.045 mol cell P $(mol\ cell\ C)^{-1}$) and half-saturated at about 0.7 pg P $(cell)^{-1}$ (i.e. ~0.003 mol cell P $(mol\ cell\ C)^{-1}$). Plotting cell phosphorus as a function of the MRP concentration in the lake water at the time of collection, she showed that maintenance of a quota of this magnitude was possible at an external concentration of round 0.75 µg P L^{-1} (0.024 µmol P L^{-1}). Plainly then, good growth is still possible in the face of external depletion so long as the cell quota is maintained. In contrast, at the very minimum cell quota (e.g. of Mackereth, 1953) corresponding to ~0.0003 mol cell P $(mol\ cell\ C)^{-1}$, growth is quite impossible.

4.3.3 Metabolic-rate limitation by phosphorus

How cells function in the face of low internal P resources and very low external P supplies has been investigated in recent years, using a variety of alternative techniques that overcome the problem of how to quantify chemically the small amounts of determinand present. For instance, Falkner *et al.* (1989) applied force-flow functions, derived by Thellier (1970), to demonstrate that the typical external concentrations of phosphate below which cells of Cyanobacteria fail to balance their minimal maintenance requirements indeed fall within the range 1–50 nmol L^{-1} (0.03–1.5 µg P L^{-1}). In the application of Aubriot *et al.* (2000), the importance of the affinity of uptake mechanisms and of the opportunism to invoke them in the face of erratic supplies was especially emphasised. Hudson *et al.* (2000) applied a radiobioassay technique which was also able to demonstrate that the amount of phosphorus in the medium supporting active phytoplankton populations can fall less than 10 nmol L^{-1} (i.e. $<10^{-8}$ M), without necessarily impairing productivity.

In another line of investigation, it has been shown that some strains of Cyanobacteria are able to maintain full growth down to external concentrations of 100 nmol P L^{-1} (~3 µg L^{-1}), without producing any of the regulator proteins that signal the activity state of the transport system to the controlling operons (Mann, 1995; Scanlan and Wilson, 1999). As suggested in Section 4.2.2, the presence of regulator proteins is indicative of incipient cell starvation, triggering the appropriate intracellular defensive reactions. Work on the bacterium *Vibrio* (Kjelleberg *et al.*, 1993) showed that symptoms include a sharp slowdown in cell growth, following an abrupt deceleration in the rates of protein synthesis. Assembly of macromolecules is halted by the action of synthesis inhibitors, followed by the reorganisation of the cell components and the adjustment of the fatty-acid content of the membranes to resist lysis. In turn, these actions are followed by a decline in the rate of respiration and other metabolic activity.

Central to these reactions are the transducing signals. Certain nucleotides are known to increase in response to falling nitrogen concentration and amino-acid synthesis. One of these, guanosine 3′,5′-bipyrophosphate (ppGpp), is generated in nitrogen-starved *E. coli* (Gentry *et al.*, 1993) and *Vibrio* (Kjelleberg *et al.*, 1993). Homologues to these are found in cyanobacterial cells experiencing a sharp reduction in photon flux (Mann, 1995). Incipient starvation and ribosomal stalling are thought to lead to ppGpp synthesis and, thence, to the communication of starvation. Mann's group was able to grow planktic cyanobacteria in media in which phosphorus concentrations fell to <0.1 µM (i.e. less than 3 µg P L^{-1}) before compounds like ppGpp began to appear in the cells. This is strongly suggestive of the probability that cells do not experience phosphorus shortages in media containing MRP concentrations greater than this.

Finally, in this context, the emerging technique of using fluorimetric labelling to detect the intracellular transients induced by incipient nutrient starvation (the so-called NIFT, nutrient-induced fluorescent transients) has been applied to microalgae grown under P-replete and P-deficient conditions to identify the reactivity of the cells. According to the experiments of Beardall *et al.* (2001), NIFT responses were wholly lacking in each of four species of freshwater microalgae in media containing ≥ 0.13 μM (4 μg P L^{-1}).

These various threads lead to a strong consensus that phosphorus availability does not limit phytoplankton activity and growth before the MRP concentration in the medium falls almost to the limits of conventional analytical detection. At this point, phytoplankton may draw on internal reserves such that activity is not immediately suppressed by lower external concentrations. Even at <0.1 μM, it is not the concentration of phosphorus that is critical so much as the capacity of the intracellular storage and the affinity of the biological uptake mechanism for the small amounts of bioavailable phosphorus being turned over in the system (Hudson *et al.*, 2000).

Two other mechanisms for contending with MRP limitation of metabolic activity are available to certain species of phytoplankton. The first involves the production of extracellular phosphatases. Many freshwater species, in fact, produce alkaline phosphatases which liberate phosphate from organic solutes that can then be absorbed by the alga (Cembella *et al.*, 1984). They are produced in response to external MRP deficiency, almost as soon as it develops (Healey, 1973). In the past, phosphatase activity has been considered to be indicative of phosphorus limitation (Rhee, 1973). There is little doubting the fact that phosphorus thus sequestered increases the resource availability to the cell. For phosphatase production to be able to offer any survival advantage, however, the phosphatase must be retained at or close to the cell surface (Turpin, 1988). Phosphatase activity might then raise significantly the ability of algae to tolerate chronically P-deficient conditions. There is little evidence to suggest that phosphatase production does much to enhance the growth dynamics of assemblages, or any component species, when inorganic phosphorus sources are effectively exhausted (Reynolds, 1992a).

The second mechanism involves the phagotrophic ingestion of organic particles, including especially other organisms such as bacteria. As indicated earlier (Section 3.4.4), those photosynthetic organisms capable of supplementing or, perhaps, fulfilling their requirements for nutrients and carbon by ingesting organic particulates are called *mixotrophs*. The best-known examples come from among the dinoflagellates (marine and freshwater Gymnodiniales and Gonyaulacales) and from among the Chromulinales, including *Ochromonas* (Riemann *et al.*, 1995; Geider and MacIntyre, 2002). Certain pigmented cryptomonads are reputedly mixotrophic (Porter *et al.*, 1985): this need not be surprising insofar as the phagotrophic abilities of the typically colourless cryptomonad genera (such as *Katablepharis* and *Cyathomonas*) have long been recognised (Klaveness, 1988). As a source of phosphorus, bacterivory and phagotrophy offer a rich alternative to scarce dissolved inorganic sources and, unlike phosphatase secretion, the available resource would seem to be less restricted. However, a low-phosphorus environment pervades all its trophic levels: bacterivory is not a sustainable alternative to deficient MRP if the bacteria are themselves simultaneously P-limited. Mixotrophy is particularly beneficial as a supplementary source of nutrient in those (generally smaller) water bodies that receive inputs of terrestrial organic matter) but are otherwise quite oligotrophic (Riemann *et al.*, 1995).

4.3.4 Capacity limitation and potential phosphorus yield

While it is clear that (probably all) phytoplankton can take up and assimilate the entire measurable MRP resource base, without first experiencing rate limitation, and that, thereafter, some species at least are extremely effective in maintaining their biomass, it is no less clear that the maximum supportable biomass, B_{max}, cannot exceed the capacity of the most scarce resource relative to demand (Box 4.1). It remains generally true that, in a large number of larger, deeper lakes in the higher latitudes, phosphorus is the nutrient that is exhausted first and, thus, the one that imposes the upper limit on the supportive capacity of the location. The generality is supported by the well-known 'Vollenweider model' and the impressive fit of the average phytoplankton biomass present in a selection lakes to the corresponding average phosphorus availability in the same lakes (Vollenweider, 1968; 1976; Organisation for Economic Co-operation and Development, 1982; see also Section 8.3.1). Several generically similar regression models from the same era (Sakamoto, 1966; Dillon and Rigler, 1974; Oglesby and Schaffner, 1975) provide analogous findings. It has to be recognised that the datasets are dominated by information from just the well-studied, northern-hemisphere oligotrophic lakes in which later understanding confirms the maximum phytoplankton carrying capacity is determined by the availability of phosphorus. It has equally to be recognised that the condition does not apply everywhere – it is less likely to apply to large, continental lakes at low latitudes, especially to lakes in arid regions, and to smaller, shallower lakes at all latitudes. It certainly does not apply to the open oceans, although its relevance to coastal waters should not be dismissed. Nevertheless, there remains a danger in supposing that the Vollenweider-type equations can be used to predict phytoplankton biomass in a given individual lake. Plainly, the criterion of capacity limitation by phosphorus must be demonstrable. Moreover, the equations are statistical and not predictive, indicating no more than an order-of-magnitude probability of average biomass that may be supported. The trite circumsciption of lakes or seas as being 'phosphorus-limited' (or 'nitrogen', or 'anything else' limited') is to be avoided completely.

Several authors have tried to express the *maximum* yield of biomass as a function of nutrient availability (Lund, 1978; Reynolds, 1978c). Lund regressed maximum summer chlorophyll against total phosphorus in a small lake (Blelham Tarn, UK) in each of 23 consecutive years during which the lake underwent considerable eutrophication. Reynolds (1978c) chose to regress the chlorophyll concentrations measured in several contrasted lakes in north-west England at the times of their vernal maxima against the corresponding MRP concentrations at the start of the spring growth. Despite certain obvious drawbacks to this approach (no allowance was made for intermediate hydraulic exchange and nutrient supply, neither were any other loss processes computed), the regression comes close to expressing the notion of a direct yield of algal chlorophyll for a known resource availability. This same regression equation (4.13) has been shown to be applicable to the prediction of the maximum algal concentration in other British lakes in which the MRP falls to analytically undetectable levels (Reynolds, 1992a; Reynolds and Davies, 2001), enhancing the supposition that the resource-limited yield is predictable from the available resource. It has also been used to estimate the chlorophyll-carrying capacity of the MRP resource in lakes where the maximum crop is susceptible to other limitations (Reynolds and Bellinger, 1992) and is now incorporated into the capacity-solving model of Reynolds and Maberly (2002).

Whereas, it was originally estimated that:

$$\log[\text{chl}a]_{max} = 0.585 \log[\text{MRP}]_{max} + 0.801$$

$$(4.14)$$

where $[\text{MRP}]_{max}$ is the highest observed concentration of MRP and $[\text{chl}a]_{max}$ is the predicted maximum chlorophyll (both units in $\mu g\ l^{-1}$ or $mg\ m^{-3}$), the later applications are used to predict an instantaneous yield against the supposed bioavailability of P. Thus,

$$[\text{chl}a]_{max} = 6.32[\text{BAP}]^{0.585} \qquad (4.15)$$

Estimating exactly what is bioavailable, without enormous analytical effort, remains problematic. However, on the assumption that phosphatase activity will raise the supportive capacity only negligibly and that mixotrophic enhancement rarely applies outside the habitats in which it is recognised (see Section 4.3.3), then the resource currently available to the phytoplankton is represented by the unused MRP in solution plus the intracellular phosphorus already in the algae. The *minimum* estimate of the resource in intracellular store can be gauged simply by reversing Eq. (4.15) to solve the BAP invested in the standing crop.

$$[\text{cell P}]_{min} = (0.158[\text{chl}a])^{1.709} \qquad (4.16)$$

BAP can be estimated by first solving Eq. (4.16), based on the existing chlorophyll concentration, then adding the equivalent intracellular cell P content thus predicted to the existing MRP concentration. Substituting this solution in Eq. (4.15) gives an instantaneous carrying capacity and potential chlorophyll yield.

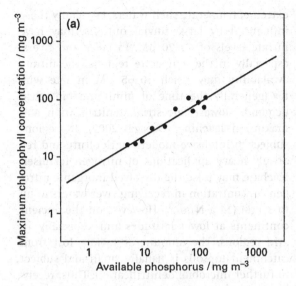

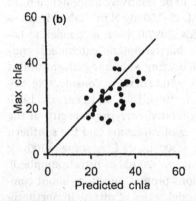

Figure 4.6 (a) Observed maximum chlorophyll concentrations in lakes in north-west England as a function of bioavailable phosphorus, as detected by Reynolds (1992a) and the regression originally proposed by Reynolds (1978c) on the basis of a study of just three lakes (see Eqs. 4.14, 4.15). (b) A later, larger dataset of observed chlorophyll maxima from UK lakes plotted against the predtion of the Reynolds regression. As expected, a majority of points lie below the predicted maximum but a number are above it, sometimes substantially so. Graphs redrawn from Reynolds and Davies (2001) and Reynolds and Maberly (2002).

Besides being broadly verifiable from observations (see Fig. 4.6b), Eq. (4.15) is consistent with the underpinning physiology. The slope of the equation as plotted in Fig. 4.6a predicts a higher return in chlorophyll for the BAP invested at low availabilities. Thus, 1 µg BAP L^{-1} is predicted to be capable of supporting up to 6.32 µg chla

L^{-1} (i.e. 6.32 µg chla (µg BAP)$^{-1}$); 10 µg BAP L^{-1} will support $\leq$24 µg chla L^{-1} (i.e. 2.4 µg chla (µg BAP)$^{-1}$), whereas the return on 100 µg BAP L^{-1} is $\leq$91.4 µg chla L^{-1} (i.e. 0.91 µg chla (µg BAP)$^{-1}$). A small, biomass-limiting BAP has to be used very efficiently but a larger base, one that perhaps challenges the next potential capacity of the biomass, is used with more 'luxury', at least before the external resource is exhausted. In terms of biomass, the effect may be even more striking, bearing in mind the tendency towards relatively lower biomass-specific chlorophyll contents of phytoplankters in sparse, light-saturated, nutrient-limited populations. Supposing a constant quota of 0.02 µg chla (µg cell C)$^{-1}$ (see Section 1.5.4), the phytoplankton carbon yields available from 1–100 µg BAP L^{-1} may be calculated to be from 316 down to 45.7 µg C (µg BAP)$^{-1}$. The corresponding range of phosphorus quotas, 0.0012–0.0085 mol P (mol cell C)$^{-1}$, neatly spans the condition of cells close to their minimum (q_0) to being close to the Redfield ideal (C : P ratio >800 to 118). This outcome is possibly more realistic than the direct solution of $B_{max} = K_i/q_0$ (as proposed in Box 4.1), which may greatly exaggerate outcomes extrapolated from abundant resource bases.

4.4 | Nitrogen: requirements, sources, uptake and metabolism in phytoplankton

Nitrogen is the second element whose relative scarcity impinges upon the ecology of phytoplankton. As a constituent of amino acids and, thus, all the proteins from which they are synthesised, nitrogen accounts for not less than 3% of the ash-free dry mass of living cells (about 0.05 mol N (mol C)$^{-1}$). This rises to around 7–8.5% in replete cells, capable of attaining rapid growth (0.12–0.15 mol N (mol C)$^{-1}$, i.e. 6.6–8.2 C : N) (see also Section 1.5.3), and to 10–12% in cells storing condensed proteins. However, molecular C : N ratios of <6 in vegetative cells are usually construed to be symptomatic of carbon deprivation (see Section 3.5.4). Relative to cell phosphorus, the nitrogen content of replete cells is generally in the range 13–19 mol N (mol P)$^{-1}$; higher molecular ratios (>30 N : P) are indicative of intracellular phosphorus deficiency; lower ratios (<10 N : P) are consistent with nitrogen shortages.

4.4.1 The sources and availability of nitrogen to phytoplankton

Despite the abundance of the element in the atmosphere, relative inertness of nitrogen gas rather restricts most photoautotrophic exploitation to nitrogen compounds. The element is also poorly represented in the Earth's crust: its occurrence is largely restricted to biogenic layers in sedimentary rocks. The principal forms of combined nitrogen available to photoautotrophs are the ions nitrate, nitrite and ammonium (NO_3^-, NO_2^- and NH_4^+), although this may not be exclusively true for all phytoplankton (see below). Very little of the available resource, either in lakes or in the sea, is due to direct atmosphere-to-water linkages: most of the sources of combined nitrogen in water are imported from terrestrial systems or are recycled within the aquatic system. The ready solubility of most inorganic nitrogen compounds, the rarity of their occurrence in secondary polymers and the redox sensitivity of their ionic configurations assist the frequency of transformations and relocation. The biogeochemical cycling of nitrogen is mediated mainly by organisms. Accordingly, its turnover is regulated predominantly at the physiological level, and so is extremely rapid when compared to the cycling of other elements such as phosphorus or silicon.

Of the three main sources of inorganic combined nitrogen, it is the highest-oxide form that occurs most widely in solution in lakes and seas. In the deep oceans, nitrate concentrations are generally in the range 20–40 μM (280–560 mg N m^{-3}) but, towards the surface (the upper 50–100 m or so), they may be drawn down severely as a result of algal and microbial uptake, to levels close to the limits of conventional analysis (~1 μM). Among the most nitrate-deficient waters are those of the North Pacific Gyre, the subtropical Atlantic (including the Sargasso) and Indian Oceans (McCarthy, 1980). At the other

extreme, temperate shelf waters, especially those influenced by large fluvial outfalls, may have nitrate levels of 60–70 μM. In lakes and rivers, especially in the temperate regions, the nitrate availability may reach 50–65 μM in late winter (generally the time of minimum biological demand, slowest terrestrial denitrification and maximum leaching: George, 2002). In regions subject to intensive modern agriculture and relatively heavy applications of nitrogen fertiliser, leachate may raise the dissolved inorganic nitrogen concentration in receiving river waters to up to 1 mM (14 g N m^{-3}). However, on the ancient continents at lower latitudes and, especially, in arid regions, the amount of nitrate lost from catchment topsoils is usually small and subject to further microbial denitrification. Thus, receiving waters tend to be relatively more deficient in nitrate (1–10 μM, 15–150 mg N m^{-3}) than in phosphorus (Reynolds, 1997a). Even at temperate latitudes, however, barren upland catchments may be capable of delivering only low concentrations of nitrate (≤15 μM). There is considerable evidence (Soto et al., 1994; Diaz and Pedrozo, 1996) of a nitrogen-regulated carrying capacity in the oligotrophic lakes of Patagonia and the southern Andes (total N <300 μg N L^{-1}, some <100 μg N L^{-1}; equivalent to 7–20 μM as nitrate supplied). These observations prompt questions about comparative current deliveries of nitrate in northern-hemisphere rainfall, which may have been relatively more augmented by industrial airfill than in the southern hemisphere.

Nitrate ions are sensitive to the low-redox conditions (<+300 mV) in sediments, the deep water of stratified, eutrophic lakes and seas and in other (usually polluted) waters experiencing high biochemical oxygen demand. Reduction to lower oxides (nitrite), to nitrogen gas and ammonia is accelerated through microbial oxidation of organic carbon and its requirement for alternative electron acceptors to the diminishing quantities of oxygen. Specifically, the activites of the denitrifying nitrate reducers like Thiobacillus denitrificans and various pseudomonad bacteria result in the venting of nitrogen gas to the atmosphere. Nitrate ammonification occurs through the agency of facultatively anaerobic bacteria, such as Aeromonas, Bacillus, Flavobacterium and

Vibrio, first reducing nitrate to nitrite. This may be excreted or, under appropriate conditions, some of these organisms reduce the nitrite further, to hydroxylamine (NH_2OH) and ammonium (Atlas and Bartha, 1993).

The ammonium ions are more soluble (so less volatile) than nitrogen, hence the reduction is more of a transformation within the pool of inorganic nitrogen, denoted by DIN (dissolved inorganic N), rather than a loss therefrom. Whereas nitrate may dominate the DIN fraction in the open water of seas and lakes, *in-situ* biologically mediated redox transformations may lead to the accumulation of comparable quantities of nitrite and, especially, ammonium (to >1 g N m^{-3}, 70 μM) in microaerophilous or anoxic environments. Ammonium is typically also present in oxic, unpolluted surface waters, though rarely in excess of ~150 mg N m^{-3} (or $\leq$10 μM) (Reynolds, 1984a).

The sources of nitrogen available to phytoplankton may be supplemented by certain dissolved and bioavailable organic nitrogen compounds (DON). These include urea (McCarthy, 1972), which is produced mainly as an excretory metabolite of animal protein metabolism, as well as through the bacterial degradation of purines and pyrimidines. McCarthy's (1980) compilation of urea concentrations recorded in the literature reveals concentrations under 3 μg-atoms N L^{-1} ($\leq$3 μM N) in the sea and up to ~9 μM in some North American rivers. Other sources of organic nitrogen directly available to phytoplankton include the small amounts (generally <1 μM N) of free amino acid present in lakes and seas (McCarthy, 1980). The relevant deaminases are said to be produced by microalgae only under conditions of DIN deficiency (Saubert, 1957; Turpin, 1988).

Of course, the size and dynamics of the DON pool is of additional indirect relevance to the pelagic function. Far from being refractory, DON is frequently the major source of nitrogen available to planktic microbes ($>80\%$ of the nitrogen available in oceanic surface waters is organic: Antia *et al.*, 1991) and some of it is evidently metabolised rapidly (in days rather than weeks; see the review of Berman and Bronk, 2003). Planktic algae and cyanobacteria may contribute to the DON pool as well as benefit from the microbial liberation of DIN.

4.4.2 Uptake of DIN by phytoplankton

Phytoplankton are generally capable of active uptake of DIN from external concentrations as low as 3–4 mg N m^{-3} (0.2–0.3 μM). Although nitrate is usually the most abundant of the DIN sources in the surface waters of lakes and seas, ammonium is taken up preferentially if concentrations exceed some 0.15–0.5 μM N (2–7 mg N m^{-3}). This is because the initial intracellular of assimilation of nitrogen proceeds via a reductive amination, forming glutamate, and a subsequent transamination to form other amino acids. The substrate is apparently always ammonium (Owens and Esaias, 1976). Thus, it is both probable and energetically preferable that the alga should use ammonium directly; nitrate and nitrite have to be reduced prior to assimilation in reactions catalysed by (respectively) nitrate reductase and nitrite reductase, so adding to the energetic cost of nitrogen metabolism. This difference in the energy requirement for the assimilation of nitrate and ammonium is reflected in the photosynthetic quotient, being about 1.1 mol O_2 (mol CO_2)$^{-1}$ when ammonium is assimilated and 1.4 when nitrate is the substrate (Geider and McIntyre, 2002, quoting Laws, 1991). Eppley *et al.* (1969a) devised an assay for nitrate-reductase activity in natural populations and which shows, consistently, that it is suppressed by ammonia concentrations exceeding 0.5–1.0 μg-atoms N L^{-1} (0.5–1.0 μM N) (see also McCarthy *et al.*, 1975). More recently, the genes encoding the kinases for bacterial nitrogen transport have been recognised (Stock *et al.*, 1989) and the action of ammonium in suppressing them has been similarly demonstrated (Vega-Palas *et al.*, 1992).

The kinetics of DIN uptake by marine phytoplankton have been studied extensively; those of freshwater species having received relatively less attention. Half-saturation concentrations for uptake (K_U) by named, small-celled oceanic species in culture (Eppley *et al.*, 1969b; Caperon and Meyer, 1972; Parsons and Takahashi, 1973) fall within the range 0.1–0.7 μM N (when nitrate is the substrate) and 0.1–0.5 μM N (with ammonium). Among neritic diatoms, the corresponding ranges are 0.4–5.1 μM $NO_3.N$ and 0.5–9.3 μM $NH_4.N$. Some half-saturation concentrations for nitrate uptake among freshwater plankters are available (Lehman et al., 1975; Reynolds, 1987a; Sommer, 1994), typically falling in the range 0.3–3.0 μM N. The maximum rates of DIN uptake at 20 °C (calculated to be generally equivalent to 0.6 to 35 μmol N (mol cell C)$^{-1}$ s^{-1}) are competent to saturate growth demand (D/S < 0.1 to 0.2: Riebesell and Wolf-Gladrow, 2002). As in the general case (see Section 4.2.3), uptake and consumption achieve parity at steady rates of growth, at external DIN concentrations generally $\leq$7 μmol N L^{-1}.

Conversely, nitrogen availability is unlikely to constrain phytoplankton activity and growth before the DIN concentration in the medium falls to below 7 μmol N L^{-1} ($\sim$100 mg N m^{-3}) in the case of large, low-affinity species or below $\sim$0.7 μmol N L^{-1} ($\sim$10 mg N m^{-3}) in the case of oceanic picoplankton). Activities become severely constrained once the cell nitrogen content falls below $\sim$0.07 mol N (mol C)$^{-1}$, when the cell reacts to its internal N deficiency by closing down non-essential processes. The minimum cell quota (q_0) of nitrogen in phytoplankton cells is said to be 0.02–0.05 mol N (mol C)$^{-1}$ (Sommer, 1994).

Applying the statistic (K/q_0), the ultimate yield or carrying capacity of the available inorganic combined nitrogen is around 20 mol C (mol N)$^{-1}$, with a possible extreme of $\sim$50 (17–42 g C : g N). Before the internal nitrogen becomes yield limiting, however, the equivalence is not likely to much exceed 10 and perhaps as little as 5 mol C (mol N)$^{-1}$ (say, 8.5 to 4.2 g C : g N). In terms of chlorophyll yield, the supportive capacity of 5–20 mol C (mol N)$^{-1}$ is equivalent to some 0.08–0.34 g chl$a$: g N. The factor used in the capacity-solving model of Reynolds and Maberly (2002), which is biassed by data from systems that are more likely to be P-deficient, is 0.11 g chl$a$: g N.

4.4.3 Nitrogen fixation
The ability to exploit the atmospheric reservoir of nitrogen gas (or, at least, that fraction dissolved in water: at sea-level air-equilibrium, $\sim$20 mg nitrogen L^{-1} at 0 °C, falling to $\sim$11 mg L^{-1} at 20 °C) as a source of nutrient is exclusively associated with prokaryotes. This ability to 'fix' (reduce) elemental dinitrogen to ammonia is a widespread trait among obligate heterotrophic chemolithotrophic bacteria, the photosynthetic bacteria and the Cyanobacteria. Certain of the latter, most especially, some of the nostocalean genera, are the only members of phytoplankton to have this capacity. Nitrogen fixation may have been a crucial step in the evolution of autotrophy in an increasingly oxygenic atmosphere, because of the relative volatility and extreme sparseness of nitrogen in the lithosphere. Ammonia was also rare owing to its photolysis in an atmosphere relatively undefended against ultraviolet radiation. The early emergence of biological fixation, through the production of the dinitrogen reductase enzyme, provided the first means of entry into ecosystems of large quantities of combined nitrogen (Falkowski, 2002). The enzyme catalyses the reduction of dinitrogen to ammonium using reductant produced via carbohydrate oxidation. Nitrogen fixation is a respiratory reaction:

$$2N_2 + 4H^+ + 3[CH_2O] + 3H_2O6\,4NH_4^+ + 3CO_2$$

$$(4.17)$$

Interestingly, dinitrogen reductases are based on iron–sulphur prosthetic groups that are redox sensitive: the enzymes operate only under strictly anaerobic conditions (as they did when they evolved). Nitrogen fixation is rapidly inactivated in the presence of oxygen (Yates, 1977). In order to fix nitrogen in an oxic ocean, the enzyme must be protected from poisoning by oxygen. As Paerl (1988) remarked, for compatibility between oxygenic photosynthesis and anoxic nitrogen fixation to have developed represents a remarkable evolutionary achievement for the Cyanobacteria.

Until the 1960s, the nitrogen-fixing capability of Cyanobacteria had only been suspected from nitrogen budgets (e.g. Dugdale et al., 1959). The introduction of the acetylene-reduction assay for nitrogenase activity (Stewart et al., 1967) made it possible to investigate which species fixed nitrogen, under what conditions and at which locations. Among the freshwater Nostocales, fixation is confined to the heterocysts (sometimes called heterocytes). These are specialised cells differentiated at intervals along the vegetative filaments (Fay et al., 1968). Their thick walls defend the intracellular anaerobic conditions necessary for the enzyme function. They are differentiated in life from normal vegetative cells responding to nitrogen deficiency. Besides the thickening of the wall, the cells lose their blue-coloured phycocyanin. However, they retain a chlorophyll-based light-harvesting capacity, attached to a functional PS I and ferredoxin transfer pathway to NADP reduction (cf. Section 3.2.1) but the oxygen-evolving PS II is, of course, defunct (Wolk, 1982; Paerl, 1988).

Heterocysts are not permanent features. Natural populations of *Anabaena, Aphanizomemon*, etc. can increase to significant levels of biomass without producing heterocysts. Differentiation is a facultative response to falling external DIN concentrations, to the extent that their relative frequency (heterocysts : vegetative cells) has been taken by ecologists to be a sign of incipient nitrogen limitation (Horne and Commins, 1987; Reynolds, 1987a). Their induction and distribution are regulated genetically by DNA promoters (see Mann, 1995, for details). Most of the relevant observations on their production (reviewed in Reynolds, 1987a) report the incidence of increased heterocyst frequency, from $<1 : 10\,000$ vegetative cells to as high as $1 : 10$, in populations of *Anabaena, Aphanizomenon* and *Nodularia*, coincident with DIN concentrations falling below $300-350$ mg m^{-3} ($19-25$ μM). The higher ratios are noted particularly at DIN concentrations ≤ 80 mg m^{-3} (≤ 6 μM N). Given that these concentrations will, ostensibly, at least half-saturate the maximum rates of uptake of combined nitrogen and completely saturate nitrogen demand (see above), the sensitivity of heterocyst production to rather higher DIN concentrations is puzzling. One possible explanation is that the heterocyst production and, indeed, the nitrogenase activity that they accommodate are actually sensitive to the external concentrations of ammonium, which may represent much the smaller fraction of the total DIN pool and is also the one that is the more rapidly drawn down. This would also have to imply that the nitrogen-fixation response is a preferential reaction to low external levels of NH$_4$.N (<0.5 μM N, or <7 mg N m^{-3}) and not of nitrate. Direct sensitivity of nitrogenase production in *Anabaena flos-aquae* to ammonium concen-

trations has been demonstrated in the laboratory (Ohmori and Hattori, 1974; see also Kerby *et al.*, 1987), albeit at higher levels. On the other hand, the isolates of non-nitrogen-fixing Cyanobacteria from nitrogen-deficient lakes (*Merismopedia, Microcystis, Synechococcus*) used in the experiments of Blomqvist *et al.* (1994), all responded much more positively to ammonium enrichment than they did to nitrate additions. They also responded less positively to nitrate additions than did planktic eukaryotes, including a *Peridinium*. Bearing in mind that the group evolved in an ammonium-scarce, nitrate-free, anoxic environment, a high affinity for ammonium nitrogen and a low-redox mechanism for its intracellular enhancement would appear to be useful adaptations. Both retain a relevance to survival and relative success of distinctive members of the group in modern environments that are extremely nitrogen-deficient or where relatively high phosphorus levels drive biomass accumulation of producers able to exploit extraneous sources of nitrogen.

Besides low external DIN concentrations and a low-redox, microaerophilous proximal environment, adequate nitrogen fixation remains dependent upon high electron transport energy as well as high rates of endogenous respiration (Paerl, 1988), driven (in this instance) by photosynthesis and good insolation. The low reactivity of N$_2$ requires that large amounts of ATP and reducing power are invested in the nitrogenase reaction (Postgate, 1987). Nitrogen fixation also requires phosphorus: Stewart and Alexander (1971) showed that nitrogenase activity was steadily lost in cultures of heterocystous *Anabaena* and other nostocalean species transferred to P-free medium and was not restored without the addition of phosphate to the medium to a concentration equivalent to 5 mg P m^{-3} (~ 0.16 μM P). The availability of molybdenum and/or vanadium/iron for the core of the nitrogenase enzyme is biochemically essential to nitrogen fixation (Postgate, 1987; Rueter and Petersen, 1987). It cannot be assumed that low ambient nitrogen levels are automatically compensated by nitrogen fixation and the successful exploitation of such conditions by N$_2$-fixing Cyanobacteria, without demonstrable satisfaction of the constraints imposed by light, phosphorus and micronutrient deficiencies.

Nitrogen fixation can occur, or has been induced, in other non-heterocystous genera of freshwater Cyanobacteria (*Plectonema*: Stewart and Lex, 1970; *Gloeocapsa*: Rippka *et al.*, 1971). The maintenance of an oxygen-free microenvironment remains a paramount precondition. One way in which this can be achieved is through the dense adpositioning of trichomes into dense bundles or rafts (Carpenter and Price, 1976; Paerl, 1988). The effect is further enhanced by bathing filaments in mucilage containing reducing sulphydryl groups (Sirenko *et al.*, 1968). Nitrogen fixation also occurs among mat-forming littoral species of *Oscillatoria* but only during darkness when there is no photosynthetic oxygen generation (Bautista and Paerl, 1985). For many other common freshwater genera (*Microcystis*, *Woronichinia*, *Gomphosphaeria*), no such facility has been demonstrated.

In the marine phytoplankton, nitrogen fixers are represented by the non-heterocystous marine species of oscillatorialean genus of *Trichodesmium*. Each of the three recognised species has adopted the life habit of forming large macroscopic rafts, or flakes, of uniseriate filaments. These bundles were sufficiently prominent in the very clear tropical and subtropical seas where they mainly occur for early mariners to have named them 'sea sawdust' (given the reddish-brown accessory pigmentation of the flakes, the name is apposite, elegantly conveying a good description of their appearance). Living in environments of the Atlantic Ocean maintaining very low levels of inorganic combined nitrogen (often <1 μM DIN), *Trichodesmium thiebautii* nevertheless fixes nitrogen, aerobically and whilst photosynthesising, sufficient to satisfy the bulk of its nitrogen requirements (Carpenter and McCarthy, 1975). This metabolism is energetically expensive and relatively slow but is adequate to support *Trichodesmium* dominance over almost all other, nitrogen-starved phytoplankters (Carpenter and Romans, 1991; Zehr, 1995).

This being so, it has long been puzzling to ecologists why there are not more genera of nitrogen fixers in the nitrogen-deficient oceans or even why *Trichodesmium* is not more abundant than it is and contributing a larger part to the oceanic turnover of nitrogen. Zehr's (1995) careful exploration of these questions confirmed the widespread occurrence among the Cyanobacteria of the nitrogenase-encoding DNA but that its expression in nitrogenase activity was as limited as previously circumscribed. It is not known to be expressed among the picoplanktic Cyanobacteria but fixers were sometimes incorporated in the microaerophilic zones of sinking particulate clusters of marine snow. Expression of nitrogenase activity among other members of the Oscillatoriales is confined to microaerophilous conditions, which *Trichodesmium* is uniquely able to contrive through its own growth habit. It is not relatively abundant where more nutrients (including DIN) deny to *Trichodesmium* its dynamic advantage. As to why it is not more abundant in the low-DIN oceans, it was supposed, for a time, that micronutrient deficiencies (in particular, of iron and molybdenum) so interfere with nitrogen fixation that the otherwise obvious potential advantage that nitrogen fixation might confer is suppressed. In fact, as is now well known, relatively high-nitrogen, low-chlorophyll regions of the great oceans augur that the biomass capacity is constrained by micronutrient availability per se (see Section 4.5), which access to alternative exploitable sources of nitrogen fails to alleviate. Not even the diatom *Hemiaulus*, with its nitrogen-fixing endosymbiont, *Richiella* (Heinbokel, 1986), is able to gain much advantage over other species of the tropical gyres. The problem is still to satisfy the simultaneous requirements of nitrogen fixation. The ability to fix nitrogen really provides an advantage only in those parts of the sea where DIN is truly limiting and where energy, phosphorus and adequate sources of iron, molybdenum and vanadium are simultaneously sufficient to support it.

4.5 | The role of micronutrients

After the six elements that each contribute $\geq 1\%$ of the ash-free dry mass of phytoplankton cells (in descending order of contribution by mass, C, O, H, N, P, S), all the others figure in relatively small fractions of the cytological structure or participate in its function. Some of these are used in small quantities, despite being normally relatively abundant in the solute content of lake and sea water, where they are constitute some of the major ions (Na, K, Ca, Mg, Cl). Most of the remainder used by plankton cells in small quantities also generally occur naturally at low concentrations. These used to be known as the 'trace elements' but are now more commonly referred to as *micronutrients*.

Much of the early knowledge of the important part played by micronutrients in the growth and physiological well-being of phytoplankton came not from analysis of lake or sea water but from the attempts to grow algae in prepared artificial media. The use of carefully formulated solutions, contrived and refined through the experimental pragmatism of such pioneers as Chu (e.g. 1942, 1943), Pringsheim (1946), Gerloff and co-workers (1952), Provasoli (see especially Provasoli *et al.*, 1957) and Gorham *et al.* (1964; see also Stein, 1973) progressively identified the additional 'ingredients' necessary to keep laboratory clones in a healthy, active, vegetative state. More recently, of course, chromatographic applications, atomic-absorption spectroscopy and X-ray microanalysis have helped to confirm and greatly amplify the elemental composition of planktic cells, in field samples as well as in laboratory cultures, even to the specific intracellular locations (Booth *et al.*, 1987; Krivtsov *et al.*, 1999). Yet more recently, a method for measuring the chemiluminescence emitted in reactions between metals and luminol appear to be both precise and sensitive at very low sample concentrations (Bowie *et al.*, 2002).

4.5.1 The toxic metals

The known micronutrients, as they are now understood, include several metals whose availability in natural waters may vary between deficiency and toxic concentrations. Some (barium, vanadium) are required in such trivial amounts that their specific inclusion in artificial culture media is considered unnecessary; their presence as impurities in other laboratory-grade chemicals or among the solutes that leach from the containing glassware suffices for most practical purposes. On the other hand, iron, manganese,

zinc, copper, molybdenum and cobalt are necessary additions to culture media (Huntsman and Sunda, 1980), even if the information about their deployment and effects is difficult to interpret. For instance, the cellular content of manganese (Mn) ranks next to that of iron. The finite requirement for its central role in re-reducing P_{680}^+ in photosynthesis (see Section 3.2.1) is usually fulfilled by the amounts present in lakes, which may be sufficiently abundant to bring about external deposition on the cell walls. In (unspecified) excesses of manganese ions are supposed to inhibit algal growth. Although there are occasional references to growth being stimulated by the addition of manganese (Goldman, 1964), there is little evidence to suggest that the metal is ever a significant growth-regulating factor. Similar conclusions apply to zinc, copper and cobalt, insofar as each participates vitally in one or more enzymic or cytochrome reactions. In solution at concentrations >10 nmol L^{-1} each is seriously toxic to a majority of algae. Copper sulphate is still widely used as an algicide (although, in many countries, its use in waters eventually supplied for drinking is banned) and is effective at concentrations of 0.3–1.0 mg CuSO$_4$ L^{-1} (2–6 μmol Cu L^{-1}). Toxicity varies interspecifically among algae and in relation to the organic content of the water (Huntsman and Sunda, 1980). Possible toxic effects of redox-sensitive metal species may be magnified in relevant habitats, including lakes subject to seasonal deep-water anoxia, where bioavailable species may be recycled (Achterberg *et al.*, 1997).

Clear incidences of regulation (or 'limitation') of algal activity through deficiency of these elements are scarce. In contrast, both molybdenum and (especially) iron are known to fulfil, on occasions, this key limiting role in pelagic systems. The case for molybdenum has been made on a number of occasions (Goldman, 1960; Dumont, 1972). In the best-known case, additions of a few micrograms of molybdenum per litre to water from Castle Lake (in an arid part of California) were sufficient to promote, quite strikingly, the growth and the attainment of a higher standing biomass of phytoplankton, where previously, despite the presence of adequate levels of bioavailable P and DIN, activities had been severely constrained (Goldman, 1960). In a later investigation of the same lake, molybdenum addition was shown to stimulate carbon fixation and nitrogen uptake rates, especially when nitrate dominated the nitrogen sources (Axler et al., 1980). Molybdenum is specifically involved in the nitrogen metabolism of the cell, participating as a co-factor in the action of nitrate reductase and (in Cyanobacteria) nitrogenase, and in the intracellular transport of nitrogen (Rueter and Peterson, 1987). The cell requirement is estimated to be about 1/50 000 of that for P ($\sim$0.2 µmol Mo (mol cell C)$^{-1}$), which can be accumulated from external concentrations of the order of 10^{-11} M. According to Steeg et al. (1986), Mo deficiency nevertheless results in symptoms of nitrogen limitation, including heterocyst formation among members of the Nostocales, even though rates of nitrogen fixation are themselves seriously impaired.

4.5.2 Iron

Iron, being, by weight, the most important of the trace components in algal cells, has long been implicated in the ecology of phytoplankton. The two most energy-demanding processes in the cell – photosynthetic carbon reduction and nitrogen reduction – involve the participation of iron-containing compounds deployed in electron transport (such as ferredoxin, nitrogenase) and pigment biosynthesis. Among the recognised direct symptoms of iron deficiency are reduced levels of cytochrome f (Glover, 1977) and the blockage of chlorophyll synthesis (and, where relevant, phycobilins: Spiller et al., 1982). Shortage of iron also impairs the structural assembly of thylakoid membranes (Guikema and Sherman, 1984). Thus, iron-deficient cells are able to harvest relatively fewer photons than iron-replete ones and photon energy is utilised less efficiently. Iron limitation results in poor photosynthetic yields of fixed carbon, lower reductive power and, hence, an impaired growth potential. Iron deficiencies also restrict directly the synthesis of nitrite reductase. For active dinitrogen fixers, the requirement for iron is relatively greater. The synthesis of nitrogenase and the electron power demand for the reduction of nitrogen draws upon the availability of up to 10 times more iron

than is needed by cells supplied with assimilable DIN to sustain the equivalent yield of cell carbon (Rueter and Peterson, 1987).

The iron content of phytoplankton cells is reckoned to be between 0.03% and 0.1% of ash-free dry mass, or about 0.1–0.4 mmol Fe : mol C. The problem that cells have in meeting even this modest requirement lies principally in the low solubility of the hydrous ferric oxides that precipitate in aerated, neutral waters. Thus, despite a relative abundance of total iron in fresh waters generally (10^{-7}–10^{-5} M Fe), most of it is present in flocs and particulates, extremely little ($\leq 10^{-15}$ M) being in true solution (other than at very low pH: Sigg and Xue, 1994). In the open oceans, the concentration of total iron is generally weaker (10^{-8}–10^{-7} M Fe). Particular interest has been directed towards the eastern equatorial Pacific, the sub-Arctic Pacific and the Southern Ocean (north of the Antarctic front) where concentrations are considerably lower still (perhaps $<10^{-11}$ M) and where iron limitation of photosynthesis and growth is demonstrable (Martin, 1992; Martin et al., 1994; and see below).

It has not always been clear just how phytoplankton cells satisfy their iron needs and at what point these become compromised by an inadequacy of availability. Supplying inorganic iron to cultures and maintaining it in solution requires the inclusion of chelating ligands, such as citrate (Gerloff et al., 1952). It was supposed that this role was fulfilled naturally by the humic or fulvic acids, and that it was suitably imitated by including in media formulations aqueous extracts of soil or by substitution of reproducible solutions of nitrilotriacetic acid (NTA) or trishydroxymethyl-aminomethane (tris). Procedures soon standardised on the use of ethylene diamine tetra-acetic acid (EDTA, usually as its sodium salt) in media in which cultures could be maintained over many generations. EDTA has proved very successful in this context.

It seems that the large molecules of EDTA are too large to be absorbed directly by algae. The role of the chelate is to maintain the source of iron in the medium, stably and accessibly, and which algae are then able to exploit directly, through a process of ligand exchange. The importance of providing an iron source was an incidental confirmation of the procedures for the bioassaying of natural-water samples that became popular in the 1960s and 1970s. The essence of the technique is to grow a test alga, under as near-reproducible laboratory conditions as possible, in water sampled from a given lake or sea under investigation, and in samples of the same water selectively enriched ('spiked') with the suspected regulatory nutrients, separately and in various combinations. Thus, the chemical component that most enhanced the yield of test organism relative to that in the unspiked water was deemed to be the capacity-regulating ('limiting') factor in the original sample (Skulberg, 1964; Maloney et al., 1973). The method would readily confirm previous suspicions about P or N deficiencies but frequently, the tests would point to a direct and previously unsuspected limitation by iron. Alternatively, the effects of N or P spikes were substantially enhanced when iron-EDTA and the relevant spike were added to the medium (Lund et al., 1975). This was true even for water samples from particular lakes previously and deliberately enriched with iron (Reynolds and Butterwick, 1979). The explanation for this behaviour lay almost wholly in the method and its requirement that lake water submitted to bioassay be first fine-filtered of all algal inocula and as many bacteria as possible, prior to the introduction of the test organism. Reynolds et al. (1981) used serial filtrations and intermediate analyses of total iron (TFe) to identify where the loss of iron fertility occurred. Even coarse filtration (50 μm) removed up to one-third of the TFe (as floccular material or finer precipitates on the algae) and glass-filter filtration (pore size 0.45 μm) removed over half of the remainder. From initial TFe concentrations close to 10^{-5} M (560 μg Fe L^{-1}), the passage of $\sim 10^{-6}$ M TFe iron in fine, near-colloidal suspension would nevertheless sustain the subsequent growth of test algae, at least to the point of exhaustion of the conventional N or P additions, without any further enhancement of the iron or the EDTA. On the basis of further experiments, reviewed in Reynolds (1997a), similar results were obtained with iron-starved algae reintroduced into artificial media containing $\geq 10^{-8}$ M TFe, whereas media containing $< 10^{-11}$ M were consistently too

dilute to support any growth of similarly-starved algae. Within this three-order span, results were erratic, either showing some or no growth but which could be stimulated by the addition of Fe-EDTA or, on occasions, by EDTA alone. These performances could not be explained satisfactorily. Some of the variability is attributable to the difficulties of manipulating such low concentrations, when even the impurities present interfere with the nominal interpretation. It would appear, however, that concentrations of residual TFe in the range 10^{-11}–10^{-10} M may well be exploitable by some algae, provided chelators continue to mediate their availability.

In all these cases, the maintenance of iron in solution by organic chelates is properly emphasised. However, equal emphasis is due to the existence of a mechanism for transferring chelated iron in the medium into the cell. It seems most likely that the uptake and assimilation of iron in the cell relies on reduction of the Fe-chelate at or near the cell surface. In turn, this presupposes that a redox enzyme is produced close to the cell membrane and whose action is to cleave iron from the organic chelates adjacent to the cell surface.

The minimum iron requirements of active nitrogen fixers must, fairly obviously, be relatively higher than those of facultative or obligate users of DIN. It has been suggested that dinitrogen-fixing Cyanobacteria need up to 10 times more iron than algae of the same species growing on DIN at the same rate (Rueter and Peterson, 1987). However, Kustka et al. (2003) have explored the complexities of iron-use efficiency in the diazotrophic Trichodesmium species and calculated the fixed-carbon and fixed-nitrogen quotas required to sustain daily growth rates of 0.1 d^{-1}. The iron use efficiency was such that 1 mol Fe sustained the elaboration of between 2900 and 7700 mol C d^{-1} (0.13–0.34 mmol Fe : mol C incorporated), thus requiring the supply of 27–48 μmol Fe (mol cell C) d^{-1}. To supply the iron demand of a population equivalent to 0.4–4 × 10^{-6} mol C L^{-1} ($\sim$0.1–1 μg chla L^{-1}) requires an iron source of 1–2 ×10^{-10} M TFe.

It is relevant to point out that many Cyanobacteria (though not just the dinitrogen fixers) are able to acquire and transport iron through the production of extracellular iron-binding compounds (called siderophores) that comprise part of their own high-affinity iron-transport systems (Simpson and Neilands, 1976). Production is induced under iron stress and repressed by its relative availability. This ability is said to confer some competitive advantage to Cyanobacteria over other algae (Murphy et al., 1976), though this would seem to apply only to nitrogen fixers under conditions of simultaneous nitrogen stress.

Piecing together the (mainly circumstantial) evidence, it seems scarcely likely that iron exerts any serious regulation on the activities of freshwater phytoplankton. Algae are exposed to relatively high concentrations of TFe supplied by terrestrial soils and that sufficient of the iron is normally maintained in dissolved or colloidal complexes with organic carbon (DOFe) for the carrying capacities of the nitrogen, phosphorus and light to be simultaneously iron-replete. In the sea, however, iron is much more dilute. Apart from fluvial inputs, concentrations are, in part, augmented directly by wet deposition of dust, derived from arid terrestrial (*aeolian*) sources (Karl, 2002), as well as from deep ocean vents. In much of the sea, organic ligands probably complex sufficient of this iron to ensure its availability to phytoplankton (again, as DOFe) but there remain areas of the ocean where there is just too little iron to avoid a deficiency of supply to autotrophs. It had been suggested that production in the oligotrophic ocean, long supposed to be regulated by nitrogen, would be stimulated by iron additions permitting more nitrogen (and thus carbon) to be fixed (Falkowski, 1997). In the relatively high-nutrient but iron-deficient low-chlorophyll areas of the southern Pacific and the circumpolar Southern Ocean, subject to the IRONEX fertilisation (Martin et al., 1994), it was photosynthesis that was first stimulated by iron addition (Kolber et al., 1994). In a later fertilisation experiment in the Southern Ocean (SOIREE; see Bowie et al., 2001), a pre-infusion concentration of 0.4 nM was raised to give a dissolved iron concentration of 2.7 nM. This was very rapidly depleted within the fertilised patch, to under 0.3 nM. Part of this was due to patch dilution but a distinct biological response confirmed that the biomass limitation was exclusively attributable to iron deficiency. A part of the enhanced iron pool (15–40%) supported the production of autotrophic diatoms and flagellates while the balance persisted (for over 40 days after fertilisation) within the tight cyclic linkages involving pelagic bacteria and microzooplanktic grazers.

We may deduce that, at least for these oceanic locations, the natural iron levels are simply too low ($\sim 10^{-10}$ M) to support any more autotrophic biomass than they do (i.e. iron availability is absolutely yield limiting and it is not just nitrogen fixation that is constrained). Moreover, the structure of typical iron-limited communities ensures that bioavailable iron is retained, as far as possible, in surface waters.

4.5.3 Organic micronutrients and vitamins

Besides the provision of a full spectrum of inorganic nutrients, successive generations of subcultured axenic (bacteria-free) clones of many species of microalgae are known to benefit from organic supplements at low concentrations. In particular, thiamine, biotin and cyanocobalamine (vitamin B_{12}) have been shown to be essential nutrients to some species of algae at least. The reviews of Provasoli and Carlucci (1974) and Swift (1980) highlight the widespread dependence upon organic micronutrients among algae of the 'red' evolutionary line. The centric diatoms and several pennate species have been shown to require a supply of vitamin B_{12}. Most dinoflagellates studied also require vitamin B_{12}, either alone or in combination with thiamine or biotin or both. Among the Haptophyceae, a majority of species tested have a requirement for thiamine, sometimes with B_{12} as well. Among the Chrysophyceae, most species require the supply of two or three vitamins.

The biochemical requirement for these substances is general: thiamine is a co-factor in the decarboxylation of pyruvic acid; biotin is a co-factor in the carboxylation and transcarboxylation reactions of photosynthesis. Vitamin B_{12} mediates reactions involving intramolecular recombinations involving C—C bond cleavages (Swift, 1980). The point is that many bacteria (including the Cyanobacteria), most green algae and higher plants are capable of synthesising these products themselves and are independent of an external supply. As a consequence of their metabolism, however, thiamine, biotin and cyanocobalamine are generally measurable components of the labile DOC fraction in the sea. Moreover, each is present in concentrations (equivalent to 0.5 to 5 ng C L^{-1}: Williams, 1975) sufficient to saturate the demands of individual plankters (said to be in the order 10^{-12}–10^{-10} mol L^{-1}: Swift, 1980). The demand and supply of organic trace substances seem not to exert any strong ecological outcome on the competitive potential of plankters in the wild.

4.6 | Major ions

Although the major ions in lake and sea water (including Ca, Mg, Na, K, Cl) are no less important to planktic cells than are P, N or the micronutrients, they are treated in less detail here because their ecological role in regulating species composition and abundance is either trivial or unclear.

4.6.1 Cations

Calcium belongs in the second of these categories. It does have an immediate and direct relevance to the species that deploy calcium salts (usually carbonate in the form of calcite) in skeletal structures. Among the most notable of these are the marine coccolith-producing haptophytes. For such organisms, there is a specific and finite requirement for calcium but the amounts dissolved in sea water are, globally, generally uniform. Of the mean solute content (35 ± 3 g kg^{-1}: Section 2.2.1), calcium accounts for just over 1% by mass (typical calcium content, ~0.4 g L^{-1}, i.e. ~10^{-2} M, or 20 meq L^{-1}) (see Fig. 4.7). Any preference of particular species for particular water masses is unlikely to be governed by differences in ambient calcium concentrations. Among fresh waters, however, calcium is frequently the dominant cation (≤120 mg L^{-1}, ≤6 meq L^{-1}), although concentrations (and specific 'hardnesses') in individual water bodies vary throughout the available range. The importance of calcium hardness resides in its relation to the anions; electrochemical balance in fresh waters is often contributed, substantially or in part, by bicarbonate ions. Being derived from the salt of

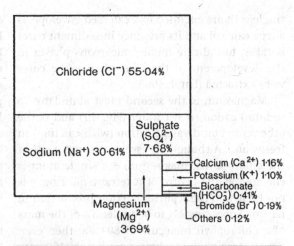

Figure 4.7 The major ionic constituents dissolved in sea water. Redrawn with permission from Harvey (1976).

a weak acid, they allow the strongly alkaline ions to press pH above neutrality, although this is resisted by the presence of free carbon dioxide in solution. Dissociation of the bicarbonate to release free CO_2 serves to buffer the water at a mildly alkaline level, as well providing an additional resource of photosynthetic DIC (Section 3.4.1; see Fig. 3.17). This crucial participation in the carbon-dioxide–bicarbonate system means that calcium can have a strong selective influence among phytoplankton that are variously sensitive to pH and carbon sources and, ultimately, those with an acknowledged capacity for carbon concentration (CCM; see Section 3.4.2). In this way, many chrysophyte species seem to be confined to soft-water (low-Ca) systems (unless there are local sources of CO_2 from organic fermentation), while many Cyanobacteria are supposed to have an affinity for calcareous waters. Indeed, a large number of cyanobacterial genera is reputed to be relatively intolerant of acid conditions (pH <6.0; Paerl, 1988; Shapiro, 1990). The mechanism underpinning the observation has not been satisfactorily explained. The generality about pH sensitivity in these organisms is doubtless confounded by the extreme tolerance of high pH by many planktic Cyanobacteria, yet some species (e.g. of *Merismopedia*, *Chroococcus*) are plainly able to function adequately in notably acidic waters (pH ≥5.5; author's observations). The green volvocalean alga *Phacotus*, whose single cells are enclosed in a calcified envelope, is a true calcicol and its presence in sediment cores is taken to indicate highly calcareous phases in the development of the lake whence the cores were extracted (Lund, 1965).

Magnesium is the second most abundant (to sodium) cation in sea water (Fig. 4.7) and is the other common divalent cation (with calcium) in freshwater. Although an essential component of the chlorophylls, magnesium – a single atom is chelated at the centre of a tetrapyrole ring – is not known to limit phytoplankton production in nature. Supposing Mg to be under 3% of the mass of a chlorophyll molecule of 894 da, then even a very large phytoplankton population (1000 mg chl$a$ m^{-3} is sustained by 30 mg Mg m^{-3}, or ~1.25 µM Mg) is unlikely to challenge the availability in the majority of natural waters.

Similarly, sodium and potassium are rarely considered to have much influence on algal composition, save through their impact on ionic strength and on the effects of ionic osmosis across the cell membrane. Marine algae equipped to deal with a medium containing some 10.7 g Na L^{-1} (i.e. ~0.47 M, or 470 meq L^{-1}) would simply burst if immersed in a soft, oligotrophic lake water containing between (say) 0.1 and 0.5 meq (or 2–10 mg Na) L^{-1}. Conversely, phytoplankters from dilute fresh waters shrivel and lyse in sea water. Neither would tolerate the extreme salinities of certain endorheic inland waters; the dissolved sodium content of the Dead Sea is ~1090 meq (25 g, or 1.1 mol) L^{-1}. The cationic strength is compounded by some 140 meq (5.5 g) L^{-1} potassium and 2540 meq (58 g) L^{-1} calcium. At the other extreme, minimum sodium concentrations required by algae and Cyanobacteria are to be found in the literature, the highest of these being 4–5 mg Na L^{-1} (say 0.2 meq L^{-1}) for a cultivated *Anabaena* (Kratz and Myers, 1955). However, many studies on other algae and Cyanobacteria report much lower thresholds than this (Reynolds and Walsby, 1975).

The requirements of phytoplankton for potassium are probably similarly non-controversial. This is despite its well-known importance as a constituent of agricultural fertilisers (acknowledging soil deficiencies) and, in both the sea and in many fresh waters, it being the least abundant of the four major cations, yet the most abun-

dant of the four in cytoplasm. A recently published experimental study (Jaworski *et al.*, 2003) has reappriased the situation. The authors were unable to show any dependence of the growth rate of the freshwater diatom *Asterionella* on potassium concentrations above 0.7 µM (27.5 µg K L^{-1}), while yields were only diminished in cultures in which the initial concentration was below this level. Another diatom (*Diatoma elongatum*) and a cryptomonad (*Plagioselmis nannoplanktica*) also showed no dependence on potassium initially supplied at >0.8 µM (31 µg K L^{-1}). We may conclude that the regulation of phytoplankton growth by potassium or sodium is unlikely to occur naturally.

It is relevant to remark briefly on another suggestion in the early ecological literature (Pearsall, 1922) that the species composition of the freshwater phytoplankton might be sensitive to the variable ratio of monovalent to divalent cations (M : D). Pearsall (1922) deduced that diatoms are more abundant in relatively calcareous waters (M : D < 1.5) whereas many desmids and some chrysophyceans occur in softer water. Talling and Talling (1965) showed desmids increased in major African lakes of high alkalinity (>2.5 meq L^{-1}) but where sodium rather than calcium is the dominant cation. The observations are, broadly, upheld by later work but the underpinning mechanisms can be explained by sensitivity to the carbon sources in the media concerned.

4.6.2 Anions

Apart from the crucial behaviour of bicarbonate, the major anions (chloride, sulphate) do not appear to limit algal production. Chloride is, by mass, the most abundant of the dissolved ions in sea water (~19.3 g L^{-1}, i.e. ~0.54 M, or 540 meq L^{-1}) (Fig. 4.7) and the principal agent in its salinity and halinity. Among softer fresh waters, it is generally the dominant anion (0.1–1 meq L^{-1}, 4–40 mg Cl L^{-1}) but, elsewhere, it may be less abundant than either bicarbonate or sulphate or both. The anions influence the medium and, in extreme, distinguish the properties of high-chloride, high-sulphate and carbonate-hydroxyl (soda) lakes. Sulphate also normally saturates the sulphur uptake requirements of algae down to concentrations of 0.1 meq L^{-1} (4.8 mg SO$_4^{2-}$, 0.05 mM).

In the context of sulphur biogeochemistry, this is an appropriate point to mention the biological production of dimethyl sulphide (DMS). This volatile compound evaporates from the sea to the air, where it constitutes the main natural, biogenic source of atmospheric sulphur. As the molecules in the air act as condensation nuclei, DMS production has consequences on the radiative flux to the ocean surface. In the 1980s, when the DMS fluxes were first recognised, excitement was engendered by the idea that the release by marine microalgae of such a substance might contribute to the regulation of global climate. It was cited as a practical demonstration of Lovelock's (1979) *Gaia principle*, with living systems creating and regulating the planetary conditions for their own survival. Since then, it has been recognised that the source of the DMS is a precursor osmolyte, dimethyl-sulphonioproponiate (DMSP), which is synthesised by marine microalgae and bacteria as a counter to excessive water loss. Measurements noted by Malin *et al.* (1993) showed correlations between DMS concentrations ranging between 1 and 94 nmol L^{-1} (mean 12) and those of the DMSP precursor and between the concentrations of DMSP and chlorophyll *a* during a summer bloom of coccolithophorids in the northeast Atlantic Ocean. The volatile DMS metabolite is released mainly as a consequence of the operation of the marine food web (Simo, 2001). Using ^{35}S-labelled DMSP, Kiene *et al.* (2000) have shown that DMSP supports a significant part of the carbon metabolism of the marine bacterioplankton and that it impinges upon the availability of chelated metals (see Section 4.5.2). The arguments for and against the tenancy of the Gaian hypothesis of supra-organismic regulation of the planetary biogeochemistry notwithstanding, it is plain that the DMSP–DMS metabolism plays a significant role in the ecological structuring of the oceans.

4.7 | Silicon: requirements, uptake, deployment in phytoplankton

All phytoplankton have a requirement for the small amounts of silicon involved in protein synthesis (<0.1% of dry mass; see Section 1.5.2).

In the context of the present chapter, however, the focus of interest is the extensive deployment of silicon polymers in the scales of the Synurophyceae and, especially, in the skeletal reinforcement of the pectinaceous cell walls of the Bacillariophyceae (diatoms). As the diatoms are among the most conspicuous and abundant groups represented in the phytoplankton of the sea and of fresh waters, the biological intervention in the movements of silicon are of profound ecological and biogeochemical significance. Several useful reviews of this topic (e.g. Werner, 1977; Paasche, 1980; Sullivan and Volcani, 1981; Reynolds, 1986a) appeared 20 to 30 years ago but more modern treatments are scarce. However, many aspects of the pelagic availability, uptake, deployment and dynamics of silicon have been well established for some time. In lots of ways, it is an ideal nutrient to study. Later work has considerably amplified, rather than revolutionised, the earlier findings.

Despite its chemical similarities to carbon, the second most common element in the Earth's crust is somewhat less reactive. It occurs almost invariably in combination with oxygen (as in the minerals quartz and cristobalite) and often also with aluminium, potassium or hydrogen (kaolinite, feldspars and micas – the so-called clay minerals). These are only sparingly soluble, but hydrolysis of aluminium silicates, aided by mechanical weathering, allows silicon into aqueous solution. Below pH ~9, the dissolved reactive silicon available that is exploitable by diatoms and other algae is the weak monosilicic acid (H_4SiO_4). Its upper concentration (at neutrality and 20 °C: ~$10^{-2.7}$ M, or ~56 mg Si L^{-1}) is regulated by the precipitation of amorphous silica (Siever, 1962; Stumm and Morgan, 1996). The concentrations of soluble reactive silicon (SRSi) that can be encountered in most open fresh waters are 1 or 2 orders lower (0.7–7 mg Si L^{-1}, 25–250 μM); the maximal levels found in oceanic upwellings (~3 mg Si L^{-1}) also fall within this range. In both habitats, the concentration may be drawn down substantially, as a consequence of uptake and growth by diatoms, by other algae, by radiolarian rhizopods and sponges.

Uptake and intracellular transport of H_4SiO_4 proceed by way of a membrane-bound carrier system that conforms to Michaelis–Menten kinetics (Azam and Chisholm, 1976; Paasche, 1980 Raven, 1983). Reported species-specific half-saturation constants (K_U) for the uptake of monosilicic acid by marine planktic diatoms are generally within the range 0.3–5 μmol L^{-1}, while those for freshwater species are slightly higher (Paasche, 1980). The next steps, leading to the precipitation of the opal-like cryptocrystalline silica polymer used in the skeletal elements and the species-specific morphogenesis and organisation into the punctuate plates, ribs, bracing spars, spines and other diagnostic features that both characterise the group and facilitate their identification, are closely regulated and coordinated by the genome of the living cell (Li and Volcani, 1984; Crawford and Schmid, 1986). The resultant structures, that may survive long after death and which can be isolated and purified by chemical treatment, inspire the deep interest of diatomists and taxonomists alike. Their forms are celebrated in compendia of scanning electromicrographs (for instance, Round *et al.*, 1990), published essentially as aids to diatom identification, although they generally project a powerful artistic appeal too. It is worth emphasising that the active uptake of silicic acid is necessary not simply to sustain the amounts of silica used in the skeleton but also to generate the saturated intracellular environment essential to aid its deposition in the wall (Raven, 1983). The mechanism is known to be extremely effective: external concentrations of SRSi can be lowered to barely detectable levels ($\leq$0.1 μmol L^{-1} in some instances: Hughes and Lund, 1962).

On the other hand, the cellular silicon requirement for a cell of a given species and size is quite predictable (see Section 1.5.2 and Fig. 1.9). The skeletal structure in diatoms comprises two interlocking *frustules*, or *valves*. At cell division (see Section 5.2.1), each of the daughter cells takes away one of the separating maternal frustules and elaborates a new one to just fit inside it. The amount of silicon taken up does not much differ from the amount required to form the new frustules and it is absorbed at the time of demand. This carries some negative implications for the new cell (see Section 5.2.2) but, because no more silicon is withdrawn than is required, the rate of its removal from solution provides a useful indirect measure of the rate of recruitment of new cells.

Nevertheless, interspecific differences in size, shape and vacuole size determine that the amount of silicon needed to complete the new cell varies in relation to the mass of cytoplasm (and, hence, its content of C, P and N). Among the selection of diatoms included in Tables 1.4 and 1.5, Si : C varies between 0.76 and 1.42 by mass (supposing the non-silica dry mass to equal ash-free dry mass and 50% of this to be carbon.). This fact is not to be confused with the differing concentrations at which various diatoms experience growth-rate limitation by silicon availability (see Section 5.4.4).

Diatoms make a major impact upon the abiotic geochemical silicon cycle. This is due partly to the global abundance of diatoms but it is compounded by their behaviour and propensity to redissolve. Owing to the high density of opaline silica ($\sim$2600 kg m^{-3}), most diatoms are significantly heavier than the water they displace and, hence, they sink continuously. Populations are correspondingly dependent upon turbulent entrainment for continued residence in surface waters and the loss rates through sinking remain sensitive to fluctuations in mixed depth (see Sections 2.7.1, 6.3.2). Much of the production is eventually destined to sediment out or to be eaten by pelagic consumers. Either way, there is a flux of diatomaceous silicon towards the abyss. During its passage through the water column, silicic acid is leached from the particulate material, at variable rates that relate to the size of particles, their degree of aggregation, pH and water temperature, though probably not much exceeding 0.1 $\times$ 10^{-9} mol m^{-2} s^{-1} ($\sim$3 pg Si m^{-2} s^{-1}) (Wollast, 1974; Werner, 1977; Raven, 1983). Matching these to the sinking rates of dead diatoms (many of which sink at <1 m d^{-1}), the time taken to dissolve completely (200–800 d) will have elapsed before they have settled through 1000 m (calculations of Reynolds, 1986a). Larger, faster-sinking centric diatoms may reach rather greater depths but the point is that very little silica reaches the deep ocean floor. Most will have been returned to solution in the water above, and its availability to future diatom growth is restored. It is easy to concur with Wollast's (1974) view that

45% of the mass of siliceous debris in the sea is redissolved between the surface and a depth of 1000 m. This uptake, incorporation, transport and resolution of silicon from diatoms plainly shortcuts the abiotic movements of silicon; Wollast (1974) estimated the annual consumption of silicon by diatoms (~12 Pg Si) reduced the residence time of oceanic silicon from ~13 000 to just a couple of hundred years.

In contrast, however, in the relatively truncated water columns of most lakes (between, say, 5 and 40 m in depth), a significant proportion of the spring 'bloom' of settling pelagic diatoms does reach the sediments intact (Reynolds and Wiseman, 1982; Reynolds *et al.*, 1982b). This does not prevent re-solution from continuing but, once part of the superficial lake sediment, the rate of solution quickly becomes subject to regulation by the (usually high) external concentration of silicic acid in the interstitial water and by re-precipitation on the frustule surfaces. Other substances, including organic remains, may interfere (Lewin, 1962; Berner, 1980). Thus, the pelagic diatoms become preserved in the accumulating sediment, providing a superb record of unfolding sedimentary events over a long period of time and environmental change. In this instance, the limnetic silicon flux enters a highly retarded phase of its potential biogeochemical cycle.

4.8 | Summary

The chapter deals with the components required to build algal cells and the means by which they are obtained, given the background of sometimes exceedingly dilute supplies. The materials are needed in differing quantities, some of which are in relatively plentiful supply (H, O, S), some are abundant in relation to relatively modest requirements (Ca, Mg, Na, K, Cl), while some occur as traces and are used as such (Mn, Zn, Cu, Co, Mo, Ba, Va). Four elements for which failure of supply to satiate demand has important ecological consequences are treated in some detail (P, N, Fe, Si). However, even the more abundant of these nutrients are orders of magnitude more dilute in the medium than in the living cell. They have to be drawn from the water against very steep concentration gradients. Plankters have sophisticated, ligand-specific membrane transport systems, comprising receptors and excitation responses, for the capture and internal assimilation of target nutrient molecules from within the vicinity of the cell. These work like pumps and they consume power supplied by ATP phosphorylation. However, cells are still reliant on external diffusivities to renew the water in their vicinities; quantitative expressions are available to demonstrate the importance of relative motion. The work of Wolf-Gladrow and Riebesell (1997) suggests that small cells may experience an advantage over larger cells in rarefied, nutrient-poor water, as they are less reliant upon turbulent motion to replenish their immediate environments.

The uptake of nutrients supplied to starved planktic cells conforms to the well-tested models based on Michaelis–Menten enzyme kinetics. Performances are characterised by reference to the maximum capacity to take up nutrient ($V_{U_{max}}$) and the external concentration (K_U) that will half-saturate this maximum rate of uptake (i.e. that that will sustain $0.5 \, V_{U_{max}}$). Clearly, a high biomass-specific uptake rate and/or an ability to half-saturate it at low concentrations represent advantageous adaptations. Actual performances are conditioned by what is already in the cell and its 'blocking' of the assimilation pathways (according to the Droop 'cell quota' concept). A new expression (Eq. 4.12) is ventured to show how the uptake rate is conditoned by the contemporaneous quota. These formulations are used to distinguish among uptake mechanisms that are variously 'velocity adapted', 'storage adapted' or 'affinity adapted'. The usage of the terms 'limitation' and 'competition' (in the context of satisfying resource requirements) is also rationalised (see Box 4.1).

Phosphorus generally accounts for 1–1.2% of the ash-free dry mass of healthy, active cells, in the approximate molecular ratio to carbon of 0.009. Minimum cell quota (q_0) may vary interspecifically, generally to 0.2–0.4% of ash-free dry mass but some species may survive on as little as 0.03%. Natural concentrations of bioavailable

P (usually less than the total concentration in the water but often rather more than the soluble, molybdate-reactive fraction (MRP, or SRP), are frequently around 0.2 to 0.3 μM. These are variously augmented by the weathering of phosphatic minerals, especially in desert catchments. Forested catchments may restrict even this supply but anthropogenic activities (quarrying, agriculture and tillage and, of course, the treatment of sewage) may significantly augment them. As phosphorus is often considered to be the biomass-limiting constraint in pelagic ecosystems, P enrichment can provide a significant stimulus to the sustainable biomass of phytoplankton. Many species can take up freely available phosphorus at very rapid rates, sufficient to sustain a doubling of cell mass in a matter of a few (≤ 7) minutes. The external concentrations required to saturate the rates of growth are generally under 0.13 μM P and the most affinity-adapted species can function at concentrations of between 10^{-8} and 10^{-7} M. In the presence of MRP concentrations >0.1 μM P, phytoplankton is scarcely 'phosphorus limited'. The conclusion is supported by each of four quite distinct approaches to determining whether cells are experiencing P regulation.

Persistent P deficiency can be countered in appropriately adapted species by the production of phosphatase (which cleaves P from certain organic binders) or by phagotrophy (consumption of P-containing organic particles or bacteria). Both rely on the sustained availability of these alternative sources of P.

Nitrogen accounts for 7–8.5% of the ash-free dry mass of healthy, active cells, in the approximate molecular ratio to carbon of 0.12–0.15. Minimum cell quota (q_0) may not be less than 3% ash-free dry mass of living cells. Nitrogen is relatively unreactive, organisms having to rely on sources of the element in inorganic combination (nitrate, ammonium) but which are extremely soluble. Aggregate concentrations of dissolved inorganic nitrogen (DIN) in the open sea are generally in the range 20–40 μM but are often depleted towards the surface. The most N-deficient waters are those of the North Pacific, the Sargasso and the Indian Ocean. Shelf waters may be relatively more replete, especially in temperate waters in late winter (concentrations 50–70 μM). Temperate lakes and rivers may offer similar levels of resource but, again, anthropogenic activities (especially the agricultural application of nitrogenous fertiliser) may augment these, up to 1 mM. On the continental masses at lower latitudes and, especially, in arid regions, DIN losses from catchment topsoils are small and subject to further microbial denitrification, so that receiving waters are often DIN-deficient in consequence (1–10 μM).

Nitrate is redox-sensitive. Ammonification is mediated by facultatively anaerobic bacteria. Ammonium is less volatile than elemental nitrogen, so DIN levels are not necessarily depleted as a consequence of anoxia.

Phytoplankton generally takes up DIN from external concentrations as low as 0.2–0.3 μM. Although nitrate is usually the most abundant of the DIN sources in surface waters, ammonium is taken up preferentially while concentrations exceed 0.15–0.5 μM N. Half-saturation of DIN uptake by small oceanic phytoplankters occurs at concentrations of 0.1–0.7 μM N with nitrate as substrate and 0.1–0.5 μM N with ammonium. Nitrogen availability is unlikely to constrain phytoplankton activity and growth before the DIN concentration in the medium falls to below 7 μmol N L^{-1} ($\sim$100 mg N m^{-3}), in the case of large, low-affinity species, or below $\sim$0.7 μmol N L^{-1}, in the case of oceanic picoplankton.

In the effective absence of DIN, phytoplankton exploits the pool of dissolved organic nitrogen, including urea. Certain groups of bacteria, including the Cyanobacteria, are additionally able to reduce ('fix') dissolved nitrogen gas. The relevant enzymes operate only under anaerobic conditions. In the Nostocales, usually the most effective dinitrogen fixers in the freshwater plankton, fixation is confined to specialised cells called heterocysts. They are produced facultatively under conditions of depleted DIN. Differentiation commences against a background of falling DIN, below 19–25 μM. It is likely that the reaction actually responds to depletion of ammonium nitrogen (to <0.5 μM $NH_4.N$). Successful fixation also depends upon threshold levels of light, phosphorus, iron and molybdenum being satisfied.

In parts of the Atlantic and Indian Oceans characterised by low DIN levels (often <1 μM DIN), nitrogen is fixed by *Trichodesmium* spp. The plankters succeed over non-fixers but dinitrogen fixers are not more widespread in DIN-deficient seas because the energy and micronutrient requirements are not simultaneouly satisfied.

Iron is a micronutrient, the availability of which is rarely problematic except in the large oceans. However, the amounts that occur in true solution are extremely small ($\sim 10^{-15}$ M) and the availability to algae depends upon its chelation by fine organic colloids. Some, at least, of this fraction is accessible to phytoplankton. A total-iron (TFe) content of 10^{-8} M seems adequate to support the needs of most species of phytoplankton, in which iron constitutes some 0.03% and 0.1% of ash-free dry mass (about 0.1–0.4 mmol Fe : mol C). On the other hand, media containing $<10^{-11}$ M are too dilute to support sustained growth of algae. It seems most likely that minimal productivity of the relatively high-nitrogen, low-chlorophyll areas of the Southern Ocean are absolutely iron deficient. The minimum iron requirements of active nitrogen fixers are suggested to be relatively higher, at about $1–2 \times 10^{-10}$ M TFe. Lack of nitrogen may preclude most other algae but lack of sufficient iron means that the nitrogen fixers are not free to exploit the situation.

Silicon plays a regulatory role in the plankton, not as a conventional nutrient but as a vital skeletal requirement of diatoms. The crypto-crystalline, opal-like silica polymer that makes up the structure of the diatom frustules is precipitated and organised within the cell from the dissolved reactive monosilicic acid that the cells take up from solution. The transformations between external solution, internal deposition and re-solution of the frustule after death are regulated, in part, by the solubilities of the silicic acid and of the silica polymer after death of the cell.

The amounts of silicon that are deposited in cell walls vary interspecifically and, intraspecifically, with size. Cell-specific silicon requirements range between 0.5% (in marine *Phaeodactylum*) and 37% of dry mass (in freshwater *Aulacoseira*); among well-studied diatoms, Si : C varies between 0.76 and 1.42 by mass. Populations draw down the concentrations present in natural waters, from a typical range, 25–250 μM, until depleted to half-saturation levels (K_U) of about 0.3–5 μM. However, uptake of Si scarcely exceeds the amounts deposited; silicon consumption provides an accurate guide to the numbers of diatoms produced.

Chapter 5

Growth and replication of phytoplankton

5.1 | Introduction: characterising growth

Whereas the previous two chapters have been directed towards the acquisition of resources (reduced carbon and the raw materials of biomass), the concern of the present one is the assembly of biomass and the dynamics of cell recruitment. Because most of the genera of phytoplankton either are unicellular or comprise relatively few-celled coenobia, the cell cycle occupies a central position in their ecology. Division of the cell, resulting in the replication of similar daughter cells, defines the generation. On the same basis, the completion of one full replication cycle, from the point of separation of one daughter from its parent to the time that it too divides into daughters, provides a fundamental time period, the *generation time*.

Moreover, provided that the daughters are, ultimately, sufficiently similar to the parent, the increase in numbers is a convenient analogue of the rate of growth in biomass. The rate of increase that is thus observed, in the field as in the laboratory, is very much the average of what is happening to all the cells present and is net of simultaneous failures and mortalities that may be occurring. The rate of increase in the natural population may well fall short of what most students understand to be its growth rate. It is, therefore, quite common for plankton biologists to emphasise 'true growth rates' and 'net growth rates'. In this work, '*growth rate*' (represented by r) will be used to refer to the rates

of intracellular processes leading to the completion of the cell cycle, net or otherwise of respiratory costs as specified. '*Increase rate*' (presented as r_n) will be used in the context of the accumulation of species-specific biomass, though frequently as detected by change in cell concentration. The potential increase in the biomass provided by the frequency of cell division and the intermediate net growth of the cells of each generation will be referred to as the biomass-specific population *replication rate*, signified by r'. In each instance, the dimensions relate the increment to the existing mass and are thus expressible, in specific growth units, using natural logarithms. In this way, the net rate of increase of an enlarging population, N, in a unit time, t, is equivalent to:

$$\delta N / \delta t = N_t / N_0 = N_0 e^{r_n^t} \tag{5.1}$$

where N_0 is a population at $t = 0$ and N_t is the population at time t; whence, the specific rate of increase is:

$$r_n = \ln(N_t / N_0) / t \tag{5.2}$$

Then, the replication rate is the rate of cell production before any rate of loss of finished cells to all mortalities (r_L), so

$$r' = r_n + r_L \tag{5.3}$$

Replication rate is net of all metabolic losses, not least those due to respiration rate (R). Where the computation of cytological anabolism (r) is before or after respiration and other metabolic losses will be stated in the text.

On the above basis (Eq. 5.2), one biomass doubling is expressed by the natural logarithm of 2 (ln 2 = 0.693) and its relation to time provides the rate. Growth rate may be expressed per second but growth of populations is sometimes more conveniently expressed in days. Thus, a doubling per day corresponds to a cell replication rate of 0.693 d^{-1}, which is sustained by an average specific growth rate of not less than 8.0225 × 10^{-6} s^{-1}, net of all metabolic and respirational losses.

The present chapter is essentially concerned with the growth of populations of planktic algae, the generation times they occupy and the factors that determine them. It establishes the fastest replication rates that may be sustained under ideal conditions and it rehearses their susceptibility to alteration by external constraints in subideal environments. In most instances, information is based upon the observable rates of change in plankton populations husbanded in experimental laboratory systems, or in some sort of field enclosures or, most frequently, on natural, multi-species assemblages in lakes or seas. They are expressed in the same natural logarithmic units. Their derivation necessarily relies upon the application of rigorous, representative and (so far as possible) non-destructive sampling techniques, to ensure that all errors due to selectivity or patchiness of distribution are minimised.

Estimating the numbers or the biomass of each species in each of the samples provides a further problem. To use such analogues as chlorophyll, fluorescence, light absorption or scatter is convenient but each compounds the errors of sampling through misplaced assumptions about the biomass equivalence. They also lose a lot of species-specific information. There really is no substitute for direct counting, using a good microscope and based on a pre-validated subsampling method, subject to known statistical confidence (Lund *et al.*, 1958). However, the original iodine-sedimentation/inverted microscope technique (of Utermöhl, 1931) has largely given way to the use of flat, haemocytometer-type cells (Youngman, 1971). With the advent and improvement of computer-assisted image analysis and recognition, algal counting is now much less of a chore. Properly used, the computerised aids yield results that can be as accurate as those of any human operator.

Results accumulated over a period of algal change can be processed to determine the mean rate of population change over that period. Often it is convenient to find the least-squares regression of the individual counts for each species ln(N_1), ln(N_2) ... ln(N_i), on the corresponding occasions (t_1, t_2 ... t_i). The slope of ln(N) on t is, manifestly, equivalent to r_n. Much of the information to be presented, in this and subsequent chapters, on the net rates of change in algal populations and the rates of growth and replication that may be inferred, is based on this approach. Modifications to this technique have been devised in respect of colony-forming algae, such as *Microcystis* and *Volvox* (Reynolds and Jaworski, 1978; Reynolds, 1983b).

5.2 | The mechanics and control of growth

5.2.1 The cell growth cycle

The 'cell cycle' refers to the progression from the newly separated daughter to the point where it itself separates into daughters, allocating its accumulated mass and structure between them. Of course, the process depends upon the adequate functioning of the cell's resource gathering and organisation (Chapters 3 and 4) but it also requires the accomplishment of several other key assimilatory steps. The correct proteins and lipids must be formed; they have to be arranged in the relevant cytological structures and, eventually, allocated between the pro-daughter cells. Meanwhile, the entire nucleic acid complement has to have been copied, the chromosomes segregated and (at least among eukaryotes) the nuclear separation (*karyokinesis*) initiated. All these processes take finite and significant periods of time to complete. Throughout, their organisation and control are orchestrated by the genome, including especially the RNA of the ribosomes.

The regulation of the life-cycle events among eukaryotes is remarkably conserved, many of their features having analogues in the bacteria (Vaulot, 1995). This much is well known, the process of nuclear division having been described by nineteenth-century microscopists studying

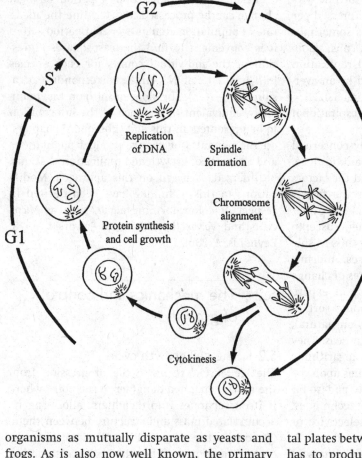

G2

S

G1

Replication
of DNA

Spindle
formation

Chromosome
alignment

Protein synthesis
and cell growth

Cytokinesis

M

Figure 5.1 The cell cycle. The sketch shows the vegetative growth (G) of the new cell and, at maturity, the break into mitotic division (M) and separation of the daughter cells. Completion depends crucially upon (S) the replication of the DNA which occurs only after the regulatory proteins have been 'satisfied' that the cell has all the resources necessary to sustain the daughters. Sketch based on figures in Murray and Kirschner (1991) and Vaulot (1995), and redrawn with permission from Reynolds (1997a).

organisms as mutually disparate as yeasts and frogs. As is also now well known, the primary alternation is, of course, between nuclear *interphase*, during which the cell increases in mass but the nucleus remains intact, and *mitosis*, during which the nucleus is replicated. Mitosis follows a strict sequence of steps, starting with the breakdown of the nuclear membrane (*prophase*); the duplicated chromosomes first align (*metaphase*) and then separate to the poles of the nuclear spindle (*anaphase*), before the propagated chromosomes become re-encapsulated in separate nuclei (*telophase*). The rest of the cell contents divide around the two daughter nuclei, the original maternal cytoplasm thus becoming divided to contribute the substance of each of the two new daughter cells, which are eventually excised from each other to complete the cell division. It is not yet wholly clear how elaborate organelles (such as flagella, vacuoles, eye-spots) are reallocated but the daughters soon copy what they are missing. Cell division in the markedly asymmetric dinoflagellates allocates the armoured exoskele-

tal plates between the separating daughters: each has to produce and assemble the replacement parts (for details, see Pfiester and Anderson, 1987). The cytokinetic bequest of one parental frustule requires each daughter diatom to produce anew the relevant complementary (internal) valve (see Crawford and Schmid, 1986, for more details). Similar issues of scale or coccolith replacement respectively confront dividing synurophyceaens (Leadbeater, 1990) and coccolithophorids (de Vrind-de Jong et al., 1994).

After separation, the next generation of daughters resumes growth during the next period of nuclear interphase. Following the discovery of Howard and Pelc (1951) that DNA synthesis is discontinuous and confined to a distinct segment of the cell development, the interphase is also subdivided into corresponding the periods. These are denoted by S, signifying synthesis, G1 for the preceding gap and G2 for the succeeding gap, terminated by the inception of mitosis prophase (see Fig. 5.1).

Assembling the growing cell has been analogised to a factory production line, with a highly sensitive quality supervision over each process (Reynolds, 1997a). Close coordination is required to sequence the events of interphase in the correct order, to check stocks and marshal the components and, if something is missing, to determine that production should be suspended. It is also needed to initiate mitosis in a way that keeps the size of the daughter cells so evidently similar to that of the parent cell prior to its own division. The 'supervision' is, in fact, achieved through the activity of a series of regulatory proteins, knowledge of which has developed relatively recently (Murray and Hunt, 1993; good summaries appear in Murray and Kirschner, 1991; Vaulot, 1995). A maturation-promoting factor (MPF) occurs at a low concentration in newly separated daughter cells but it increases steadily throughout interphase. Purification of MPF showed it to be made of two kinds of protein. One of these, cyclin B, is one of several cyclins produced and periodically destroyed through the cell cycle, each being specific to a particular part of the cycle. Cyclin B is the one that reaches its maximum concentration at the start of mitosis. The second part of the MPF is a kinase. It had been first discovered in a strain of mutant yeast, codified cdc. The mutation concerned the presence of a kinase-encoding gene, called cdc2, and the 34-kDa kinase protein was referred to as p34cdc2. Like all kinases, it is active only when phosphorylated (cf. Section 4.2.2). In this state, it triggers prophase spindle formation. After completion of the mitosis, the kinase is dephosphorylated and the cyclin B is rapidly degraded (by a cyclin protease), thus leading to the destruction of MPF. In the daughter cells, cyclin B is synthesised again and MPF begins to accumulate for the next division.

The cues are critical. In this instance, it is clearly the instruction to phosphorylate the kinase that triggers the mitosis. The commitment to nuclear division is, however, made earlier, when, following a sequence of signals processed through the preceding G1 period of interphase, the DNA is finally replicated (the S phase in Fig. 5.1). Activation depends upon satisfaction of resource adequacy, which is communicated by the operons informed by the intracellular-transport and protein-synthesis pathways (see Sections 3.4.2, 4.2.2). Transcription of the relevant operon genes is stimulated by the group of cAMP receptor proteins (or CRP) associated with the active pathways. Thus, marked resource undersaturation or slow delivery are reflected in weakened CRP flow and weakened operon activity. Much like the air-brakes on a train, active transport and synthesis act to suppress the cell's protective features but, as soon as normal functions begin to fail, the mechanisms for closing nonessential processes and conserving cell materials are immediately expressed. Incipient starvation and the stalling of the relevant ribosomes lead to the activation of the inhibitory nucleotides, such as ppGpp (see Section 4.3.3). These do not just arrest maturation but may induce the onset of a resting condition, with a substantial reduction in all metabolic activities, including RNA and protein synthesis and respiration. The machinery is said to remain in place for at least 200 h after such shutdowns (Mann, 1995). If renewed resources permit, however, renewed protein synthesis may be induced within minutes of their availability.

So long as the conditions remain benign, optimal functioning is supported and CRP generation is upheld. The cell recognises that it has sufficient in reserve to be able to fulfil the mitosis and so allow the DNA replication to proceed. Then, there really is no escape from the commitment. A further group of substances, called licensing factors, are bound to the chromosomes but are destroyed during DNA replication. Licensing factors cannot be re-formed while the nuclear membrane is intact, so that the DNA synthesis can occur only once per generation.

Though they lack a membrane-bound nucleus and any of the physical structure of a spindle upon which to sort the replicated chromosomes, the prokaryotes have no lesser need than eukaryotes for a closely controlled, phased cell cycle. At slow rates of growth, the DNA replication of E. coli occupies only a fraction of the generation time, giving a delineation analogous to the G1–S–G2 phasing represented in Fig. 5.1. In the fast-growing Synechococcus, DNA replication may occupy relatively more of the cell generation time. Completion of the DNA duplication is the cue for chromosome separation and cell division (Armbrust *et al.*, 1989). Cycle regulation by analogues of MPF is probable.

5.2.2 Cell division and population growth

All the eukaryotic species of phytoplankton that have been investigated conform to G1–S–G2 phasing (Vaulot, 1995). Curiously, the best-studied freshwater species belong to the Chlorococcales or to the Volvocales, which undergo a relatively prolonged growth period that is followed by a fairly rapid series of cell divisions, resulting in the formation of between four (as in genera of Chlorococcales such as *Chlorella* and *Scenedesmus*) to 16 or 32 (in *Eudorina*) or, perhaps, as many as 1000 (*Volvox* itself) daughter cells. Suppressed though the smooth transition between generations may be, it is also quite evident that, over two or three generations, the increase in specific biomass adheres closely to a smooth exponential rate (Reynolds and Rodgers, 1983). Reynolds' (1983b) deductions on the increase in biomass of a field population of *Volvox aureus* are also consistent with a smoothing of both cell growth and population growth with respect to the cell-division sequence. To treat the growth rate and the times of consecutive generations (t_G) in the same terms as simple binary fission times may be cautiously justified. Based on Eq. (5.2), we deduce:

$$r' = \ln 2 (t_G)^{-1} \tag{5.4}$$

On the other hand, every confidence can be accorded to deductions about the observable rate of population growth over consecutive generations of diatoms. There is a manifest similarity of size between parent and daughter cells, owing to the shared bequest of the parental frustules. The cellular requirements of each species are also remarkably constant, at least when cell size is taken into account (Lund, 1965; see Tables 1.4, 1.5). Cells take up little more monosilicic acid than they require to sustain the skeletal demands of the imminent division (Paasche, 1980), including that needed to maintain the necessary internal concentration (see Section 4.7). These characters lend themselves to accurate computation of biomass increase from the division of cells and its direct analogy to silicon uptake (Reynolds, 1986a). Despite the complicated kinetics of silicon uptake, the constraints on intracellular

transport and the intricacies of frustular morphogenesis in forming the two new frustules required to complete each cell division (see Section 4.8), the actual construction of the new silica structures is confined to a relatively short period. The latter extends from just after nuclear division to the point of eventual cell separation. Having passed G1 of interphase in its vegetative condition, the cell commences to form the new valve in a silica-deposition vesicle just beneath the plasmalemma (Drum and Pankratz, 1964). The origins of the silicon deposition vesicle and of the control of the highly species-specific patterning of the new valves are described in detail in Pickett-Heaps *et al.* (1990). The trigger for the process is DNA replication. Thus, no new wall forms without the initiation of the nuclear division. Equally, the commitment to division is taken *before* the parent cell has taken up sufficient soluble reactive silicon either to fulfil the skeletal demand of the cell division, or to be able to maintain the necessary internal concentration (Raven, 1983). This carries ecological consequences: if the requirement is pitched against low or falling external concentrations of monosilicic acid, it is possible that, in a growing population, cells begin division in a silica-replete medium but encounter deficiency before it is completed. Many cells may fail to complete the replication and die (Moed, 1973).

While external concentrations continue to satisfy uptake requirements (K_U: 0.3–5 µmol L^{-1}; Section 4.7), the demand is assuaged and the completion of the next generation can reasonably be anticipated. For all phytoplankton, the processes of gathering of raw materials equivalent to its current mass, of assembling them into species-specific proteins and lipids and, then, into the correct cytological structures and organelles, each occupy a finite period of time. Once accomplished, further time is required for the completion of the S, G2 and mitosis phases, prior to the occurrence of the final cytokinetic separation. As stated above, this event or, at least, the frequency with which it occurs, is of fundamental significance to the plankton ecologist. To predict, measure accurately or model the rate of growth of cells has long been an ambition of students of the phytoplankton. Most of the convenient, traditional determinations of the measured rates of change in numbers are, of course, surrogates of cell growth and they are always net of metabolic losses and, often, also net of mortalities of replicated cells. The rates of cell replication are rarely predictable from separate determinations of the capacities deduced from experimental measurements of photosynthetic rate or nutrient uptake rate (although these represent upper limits). It is easy to share the frustrations of all workers who have struggled with the problem of the determination of *in-situ* growth rates.

5.3 | The dynamics of phytoplankton growth and replication in controlled conditions

There is another way to address the potential rates of cell replication of phytoplankton and that is through the dynamics of isolates under and carefully controlled conditions in the laboratory. We might begin by assessing how well the cells of a given species perform under the most idealised conditions it is possible to devise. Once an optimum performance is established, further experiments may be devised to quantify the influences of each of the suspected controlling factors. Finally, the various impositions of sub-ideal growth conditions can be evaluated. The following sections apply this approach to a selection of freshwater species of phytoplankton.

5.3.1 Maximum replication rates as a function of algal morphology

The collective experience of culturing isolates of natural phytoplankton in the laboratory has been summarised by Fogg and Thake (1987). The fastest rates of species-specific increase are attained in prepared standard media, designed to saturate resource requirements, when exposed to constant, continuous light of an intensity sufficient to saturate photosynthesis, and at a steady, optimal temperature. Even then, maximal growth rates are not established instantaneously. There is usually a significant 'lag phase' during which the inoculated cells acclimatise to the ideal world into which they have been introduced. Within a day or two, however, the isolated population will

be increasing rapidly and, generally, will be doubling its mass at approximately regular intervals. It is early in this *exponential phase* that the maximal rate of replication is achieved, when r' is supposed to be equal to the observed net rate of increase, r_n, solved by Eq. (5.2). Later, as the resources in the medium become depleted or the density of cells in suspension begins to start self-shading, the rate of increase will slow down considerably (eventually, the *stationary phase*). Care is taken to discount the biomass increase in this phase from the computation of the exponent of the maximum specific growth rate, r'.

The fastest published rate of replication for any planktic photoautroph is still that claimed for a species of *Synechococcus* (at that time named *Anacystis nidulans*) by Kratz and Myers (1955). At a temperature of 41 °C, the Cyanobacterium increased its mass 2896-fold in a single day, through the equivalent of 11.5 doublings ($t_G = 2.09$ h), and sustaining a specific rate of exponential increase (r') of 7.97 d^{-1} (or 92.3 $\times$ 10^{-6} s^{-1}). Numerous other algal growth rates are recorded in the literature cover wide ranges of species, culture conditions and temperatures. Several notable attempts to rationalise and compile data for interspecific comparison include those by Hoogenhout and Amesz (1965), Reynolds (1984a, 1988a) and Padisák (2003). The selection of entries in Table 5.1 is hardly intended to be comprehensive but it does refer to standardised measurements, made at or extrapolated to 20 °C, on a diversity of organisms of contrasting sizes, shapes and habits. The temperature is critical only insofar as it is uniform and that it has been a popular standard among culturists of microorganisms. It is probably lower than that at which most individual species (though not all) achieve their best performances (see Section 5.3.2 below).

The entries in Table 5.1 show a significant range of variation (from $\sim$0.2 to nearly 2.0 d^{-1}). There is no immediately obvious pattern to the distribution of the quantities – certainly not one pertaining to the respective phylogenetic affinities of the organisms, nor to whether they are colonial or unicellular. The variations are not random either, for in instances where more than one authority has offered growth-rate determinations for the same species of alga, the mutual agreements between the studies has been

Table 5.1 | Maximum specific growth rates (r'_{20} d^{-1}) reported for some freshwater species of phytoplankton in laboratory cultures, under continuous energy and resource saturation at 20 °C

Phylum	Species	r'_{20} d^{-1}	References
Cyanobacteria	Synechococcus sp.	1.72[a]	Kratz and Myers (1955)
	Planktothrix agardhii	0.86	Van Liere (1979)
	Anabaena flos-aquae	0.78	Foy et al. (1976)
	Aphanizomenon flos-aquae	0.98	Foy et al. (1976)
	Microcystis aeruginosa[b]	1.11	Kappers (1984)
	Microcystis aeruginosa[c]	0.48	Reynolds et al. (1981)
Chlorophyta	Chlorella strain 221	1.84	Reynolds (1990)
	Ankistrodesmus braunii	1.59[a]	Hoogenhout and Amesz (1965)
	Eudorina unicocca	0.62	Reynolds and Rodgers (1983)
	Volvox aureus	0.46	Reynolds (1983b)
Cryptophyta	Cryptomonas ovata	0.81[a]	Cloern (1977)
Eustigmatophyta	Monodus subterraneus	0.64[a]	Hoogenhout and Amesz (1965)
Chrysophyta	Dinobryon divergens	1.00	Saxby (1990), Saxby-Rouen et al. (1997)
Bacillariophyta	Stephanodiscus hantzschii	1.18	Hoogenhout and Amesz (1965)
	Asterionella formosa	1.78	Lund (1949)
	Fragilaria crotonensis	1.37	Jaworski, in Reynolds (1983a)
	Tabellaria flocculosa var. asterionelloides	0.66	G. H. M. Jaworski (unpublished data)
Dinophyta	Ceratium hirundinella	0.21	G. H. M. Jaworski (unpublished data)

[a] Rate extrapolated to 20 °C.
[b] Unicellular culture.
[c] Colonial culture.

generally excellent (Reynolds, 1988a, 1997a). Where there has been a significant departure, as there is in the case of the two entries for *Microcystis*, it is attributable to the difference between working with a colonial strain and, as is common among laboratory cultures, one in which the colonial habit had been lost. This turns out to be an important observation, for it provides the clue to the robust pattern that accounts for a large part of the variability in organismic replication rate, which relates to organismic morphology. The apparent dependence of growth rate on algal size and shape, suggested by earlier analyses (Reynolds, 1984a) was convincingly confirmed when the replication rates (r'_{20}) were plotted against the surface-to-volume ratio (sv^{-1}) ratio *of the life-form*, regardless of whether it was a unicell, coenobium or mucilage-bound colony (Reynolds, 1989b). The relationship, between the

replication rates shown in Table 5.1 (save the entry for *Dinobryon* that was not available to the 1989 compilation) and the corresponding species-specific surface-to-volume ratios noted in Table 1.2, is reproduced in Fig. 5.2. The regression line fitted to points plotted in Fig. 5.2 is:

$$r'_{20} = 1.142(s/v)^{0.325} \mathrm{d}^{-1} \tag{5.5}$$

where s is the approximate area of the algal surface (in μm^2) and v is the corresponding volume (in μm^3). Both dimensions were estimated from microscope measurements and the compounding of relevant geometric shapes (Reynolds, 1984a). The coefficient of correlation is 0.72; thus, 52% of the variability in the original dataset is explained.

The outcome is instructive in several ways. First, it is satisfying that surface-to-volume

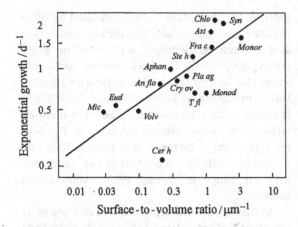

should provide such a strong allometric statement about the capacity of the alga to fulfil its
ultimate purpose. The empirical relation between
these two attributes is, of course, not constant,
even among individual spherical cells. Rather, it
diminishes with increasing diameter (d), with a
slope of exactly 2/3. The unit in which (sv^{-1}) is
expressed, ($\mu m^2/\mu m^3$ =) μm^{-1}, has the dimension of reciprocal length and conveys the idea
that the decline in assimilation and growth efficiency might be primarily a function of the
intracellular distance that metabolites must be
conducted within the cell. The implication is
that small spherical cells are metabolically more
active than large ones. If there is to be an advantage in being larger, it must either be at the
expense of the potential for rapid growth, or
it should invoke a simultaneous distortion in
shape. As the sphere is bounded by the least possible surface enclosing a given three-dimensional
space, any distortion from the spherical form
increases the surface area relative to the enclosed

volume. It was pointed out first by Lewis (1976)
that the morphologies of marine phytoplankton,
despite embracing a range of sizes covering 6 or
7 orders of magnitude, are such that many of the
larger ones are sufficiently non-spherical for the
corresponding surface-to-volume values to vary
only within 2. He deduced that this conservatism
of (sv^{-1}) was not incidental but a product of natural selection. When Reynolds (1984a) attempted
the analogous treatment for a selection of freshwater phytoplankton, nearly 3 orders of magnitude of variation in (sv^{-1}) were found, against
over 9 orders of variation in the corresponding
unit volumes (see 1.7). The cytological relationship of cell surface to cell mass is clearly a relevant factor in cell physiology.

The second interesting feature of Fig. 5.2 is the
slope of r'_{20} on (s/v): the exponent, 0.325 agrees
closely with Raven's (1982) theoretical argument
for growth conforming to a model relationship
of the type, r = a constant × (cell C content)$^{-0.33}$.
Raven's supposition invoked the slower increase
of surface (as the square of the diameter) than
the volume (as its cube) of larger-celled organisms. Assuming, for a moment, carbon content
to be a direct correlative of protoplasmic volume (as is argued in Section 4.2), and the surface area to be a 2/3 function of increasing volume, then the arithmetic of the indices to (s/v)
clearly sum to 1/3. Raven's (1982) derivation differs from the more frequently quoted deduction
of Banse (1976), namely r = a constant × (cell C
content)$^{-0.25}$, which had been based upon marine
diatoms. Reynolds' (1989b) equation (i.e. that in
Fig. 5.2) is not necessarily at odds with Banse's
findings, as the assumption of a constant relationship of C to external cell volume does not
hold for the (larger) diatom cells characterised
by large internal vacuolar spaces.

A further satisfaction about the relationship
comes from the fact that the scatter of points
shown in Fig. 5.2 becomes a cluster at the end
of the regression line fitted by Nielsen and Sand-
Jensen (1990) to the growth rates of higher plants
as a function of their surface-to-volume ratios.
The slopes of the two regressions are almost identical. Presumably, small size, the consequent relatively high surface-to-volume ratio, structural
simplicity and the exemption from having to

allocate resources to the production of mechanical and conducting tissues provide the main reasons for the high rates of specific biomass increase among planktic microalgae relative to those of littoral bryophytes and angiosperms. The ability of the plankter to exchange materials across and within its boundaries is a key determinant of its potential physiological performance. That ability is strongly conditioned by the ratio of its surface to its volume.

5.3.2 The effect of temperature

Sourcing from the various literature compilations mentioned in the previous section, Reynolds (1984a) deduced that, with the exception of acknowledged cold-water stenothermic and thermophilic species, most laboratory strains of planktic algae and cyanobacteria then tested achieved their maximal specific rates of replication in the temperature range 25–35 °C. A few (like the *Synechococcus* of Kratz and Myers, 1955) maintain an accelerated function beyond 40 °C but, exposed to their respective supra-optimal temperatures, the replication rates of most of the species considered here first stabilise and, sooner or later, fall away abruptly. From 0 °C to just below the temperature of the species-specific optimum, the replication rates of most plankters in culture appear, as expected, to increase exponentially as a function of temperature. However, the degree of temperature sensitivity of the division rate is evidently dissimilar among plankters. In some, growth rates vary by a factor of ~2 for each 10 °C step in temperature, as Lund (1965) recognised; for others, the temperature dependence of growth rate is more sensitive and the slope of r on temperature is steeper.

Seeking some general expression to describe the sensitivity of algal replication rates to temperature, Reynolds (1989b) used the same data compilations to identify sources of relevant information on growth performances. What was needed were the maximum rates of replication of named algal species maintained in culture under constant, photosynthesis-saturating light conditions and initially growth-saturating levels of nutrients but at two or (ideally) more constant temperatures. Data satisfying this criterion were found for 11 species. For the purpose of comparative

plotting (Fig. 5.3), data were normalised by relating the logarithm of daily specific replication rate at the given temperature, $\log(r'_\theta)$, to the given temperature (θ °C) rendered on an Arrhenius scale. The latter invokes the reciprocals of absolute temperature (in kelvins). Thus, 0 °C (or 273 K) is shown as its reciprocal, 0.003663. For manipulative convenience, the units shown in Fig. 5.3a are calculated in terms of $A = 1000 [1/(\theta K)]$. The thousand multiple simply brings the coefficient into the range of manageable, standard-form numbers.

In this format, the temperature-response plots reveal several interesting features. These include interspecific differences in the temperature of maximum performance, ranging from appearing at a little over 20 °C (a little under 3.42 A) in *Aphanizomenon flos-aquae* and *Planktothrix agardhii* (original data of Foy *et al.*, 1976) but somewhere >41 °C (<3.19 A) in the *Synechococcus* of Kratz and Myers (1955). The differences in the normalised parts of the species-specific slopes reflect interspecfic differences in sensitivity to variation in temperature. Thus, the slope fitted to the data for *Synechococcus*, for example, has the value $\beta = -3.50$ A^{-1}. Cast in the more traditional terms of the Q_{10} expression for the factor of rate acceleration over a 10- °C step in customary temperature (usually that from 10 to 20 °C), the normalised response has a Q_{10} of ~2.6. In contrast, the slope for colonial *Microcystis*, $\beta = -8.15$ A^{-1} ($Q_{10} \sim 9.6$), reveals a relatively greater temperature sensitivity. The slopes, of course, reflect interspecific differences in the sum of metabolic responses to temperature fluctuation, some of which are themselves differentially responsive to thermal influence. Whereas, for instance, photosynthetic electron transfer has a Q_{10} of approximately 2 over a 30 °C Range (see Section 3.2.1) and both light-saturated photosynthesis and dark respiration carry Q_{10} values in the range 1.8–2.25 (Section 3.3.2), protein assembly has a Q_{10} said to exceed 2.5 (Tamiya *et al.*, 1953). Whereas the rates of growth of the plankters whose (relativly gentler) slopes appear towards the top of Fig. 5.3a might reflect temperature constraints on individual assembly processes, the (steeper) slopes towards the bottom of the figure refer to larger, low (s/v) forms and coenobial Cyanobacteria whose responses

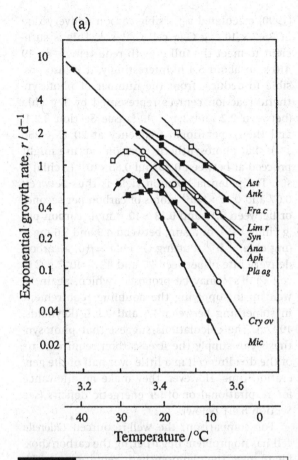

(a)

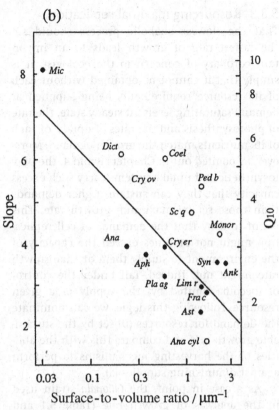

(b)

Figure 5.3 Temperature dependence of light-saturated growth in a selection of freshwater phytoplanktic species. (a) Datasets normalised against a temperature on an Arrhenius scale (10^3/K); (b) the slopes of the regressions plotted against the surface-to-volume ratios of the algae. The algae are: Ana, Anabaena flos-aquae; Ana cyl, Anabaena cylindrica; Ank, Ankyra judayi; Aph, Aphanizomenon; Ast, Asterionella formosa; Coel, Coelastrum microporum; Cry er, Cryptomonas erosa; Cry ov, Cryptomonas ovata; Dict, Dictyosphaerium pulchellum; Fra c, Fragilaria crotonensis; Lim r, Limnothrix redekei; Mic, Microcystis aeruginosa; Monor, Monoraphidium sp.; Ped b, Pediastrum boryanum; Pla ag, Planktothrix agardhii; Sc q, Scenedesmus quadricauda; Syn, Synechococcus. The least-squares regression inserted in (b) is fitted only to the solid points; its equation is $\beta = 3.378 - 2.505 \log (sv^{-1})$. Redrawn with permission from Reynolds (1997a).

are determined by the temperature sensitivity of the slowest processes of intracellular assimilation and relative rates of surface exchange (Foy *et al.*, 1976; Konopka and Brock, 1978).

This pattern of behaviour is emphasised in the plot (in Fig. 5.2b) of the species-specific slopes

of temperature sensitivity of replication (β, from Fig. 5.2a) against the corresponding organismic (sv^{-1}) value. β is a significant ($p < 0.05$) correlative of the surface-to-volume attributes of the eleven algae tested. The regression, $\beta = 3.378 - 2.505 \log (sv^{-1})$, has a coefficient of correlation of 0.84 and explains 70% of the variability in the data. The open circles entered in Fig 5.3a are not part of the generative dataset but come from a previously untraced paper of Dauta (1982) describing the growth responses of eight microalgae. They have been left as a verification of the predictive value of the regression.

The hypothesis that algal morphology also regulates the temperature sensitivity of the growth rate is not disproved. For the present, β can be invoked to predict the growth rate of an alga of known shape and size at a give temperature, θ, following Eq. (5.6):

$$\log(r'_\theta) = \log(r'_{20}) + \beta[1000/(273 + 20) - 1000/(273 + \theta)]\mathrm{d}^{-1} \qquad (5.6)$$

5.3.3 Resourcing maximal replication

That the slowest anabolic process should set the fastest rate of growth leads to an important corollary of concern to the ecologist. It is simply that it cannot be obtained without each of its resource requirements being supplied at demand-saturating levels. At steady state, the rate of photosynthesis and the rates of uptake of each of its nutrients match the growth demand. Moreover, as pointed out in Chapters 3 and 4, the photosynthetic and uptake systems carry such excess capacity that they can sustain higher demands than those set by maximum growth rate. This is not to deny that the demands of cell replication might not at times exceed the capacity of the environment to supply them or that growth rate might not, indeed, fall under the control of (become limited by) the supply of a given resource. Following this logic, we can nominate the demand for resources (D) set by the sustainable growth rate and compare this with the abilities of the harvesting mechanisms to perform against diminishing supplies (S).

As a case in point, the *Chlorella* strain used in the analysis of growth rate (Table 5.1 and Reynolds, 1990) achieved consistently a maximal rate of biomass increase of 1.84 d^{-1} at 20 °C. Given the size of its spherical cells ($d \sim 4$ μm; $s \sim 50$ μm^2; $v \sim 33$ μm^3), this is faster than is predicted by the Eq. (5.6) of the regression (Fig. 5.2), which gives $r'_{20} = 1.31$. Staying with the real data, one biomass doubling (taken as the equivalent of an orthodox cell cycle culminating in a single division) defines the generation time; by rearrangement of Eq. (5.3), $t_G = 9.05$ h. During this period of time, the alga will have taken up, assimilated and deployed 1 g of new carbon for every 1 g of cell carbon existing at the start of the cycle. It is not assumed that the increase in biomass is continuously smooth but the *average* exponential specific net growth rate over the generation time is (1.84/86 400 s =) 21.3 × 10^{-6} s^{-1}. This, in turn requires the assimilation of carbon fixed in photosynthesis at an instantaneous rate of 21.3 × 10^{-6} g C (g cell C)$^{-1}$ s^{-1}. From the maximum measured photosynthetic rate at $\sim$20 °C [17.15 mg O$_2$ (mg chl$a$)$^{-1}$ h^{-1}] and, assuming a photosynthetic quotient of 1 mol C : 1 mol O$_2$ (12 g C : 32 g O$_2$) and a C : chl$a$ of 50 by weight, Reynolds

(1990) calculated a possible carbon delivery rate of 35.7 × 10^{-6} g C (g cell C)$^{-1}$ s^{-1}. This is sufficient to meet the full growth requirement in 19 416 s, or about 5.4 h. Interestingly, it is also possible to deduce, from the number of photosynthetic reaction centres represented by 1 g chl$a$ (between 2.2 and 3.4 × 10^{18}) (see Section 3.2.1) and their operational frequency at 20 °C ($\sim$250 s^{-1}), that photosynthetic electron capture might proceed at between 0.55 and 0.85 × 10^{21} (g chl$a$)$^{-1}$ s^{-1}. The potential fixation yield is thus between 0.07 and 0.11 × 10^{21} atoms of carbon per second, or between 0.11 and 0.18 × 10^{-3} mols carbon per (g chl$a$)$^{-1}$ s^{-1}, or again, between 4.9 and 7.6 mg C (mg chl$a$)$^{-1}$ h^{-1}. Putting C : chl$a$ = 50, a carbon delivery rate of between 27 and 42 × 10^{-6} g C (g cell C)$^{-1}$ s^{-1} may be proposed, which, again, is well up to supplying the doubling requirement in something between 4.6 and 7.1 h (Reynolds, 1997a). The calculations suggest that photosynthesis can supply the fixed-carbon requirements of the dividing cell in a little over half of the generation time. However, they make no allowance for respirational or other energetic deficits (see Section 5.4.1 below).

For comparison, the well-resourced *Chlorella* cell has no problem in gathering the carbon dioxide to meet the photosynthetic requirement. The diffusion rate calculated from Eq. (3.19) in section 3.4.2 indicated that delivery of the entire doubling requirement of carbon in $\sim$2300 s (i.e. just over 38 minutes). In order to maintain steady internal Redfield stoichiometry, the growing cell must absorb 9.43 × 10^{-3} mol P (mol C incorporated)$^{-1}$ per generation (i.e. 1 mol/106 mol C). The phosphorus requirement for the doubling of the cell-carbon content of 0.63 × 10^{-12} mol is 5.9 × 10^{-15} mol P cell^{-1}, which, at its maximal rate of phosphorus uptake (13.5 × 10^{-18} mol P cell^{-1} s^{-1} (Fig. 4.5), the cell could, in theory, take up in only 0.44 × 10^3 s, that is, in just 7.3 minutes! Note, too, that an external concentration of 6.3 × 10^{-9} mol L^{-1} is sufficient to supply the entire phosphorus requirement over the full generation time of 9.05 h (see Fig. 4.6). Solving Eq. (4.12) for the supply of nitrogen to the same cell of *Chlorella*, a concentration of 7 μmol DIN L^{-1} should deliver $\sim$175 × 10^{-18} mol N s^{-1}, sufficient to fulfil the doubling requirement of

$(0.151 \times 0.63 \times 10^{-12} =) 0.095 \times 10^{-12}$ mol N in 540 s (9 minutes).

These calculations help us to judge that, for the rate of cell growth to qualify for the description 'limited', whether by light or by the availability of carbon, phosphorus or any other element, then it has to be demonstrable that the generation time between cell replications is prolonged. Moreover, it has to be shown that the additional time corresponds to that taken by the cells of the present generation to acquire sufficient of the 'rate-limiting' resource to complete the G1 stage and thus sustain the next division.

5.4 | Replication rates under sub-ideal conditions

The corollary of the previous section is that the achievement of maximal growth rates is not dependent upon maximal photosynthetic rates and nutrient uptake rates being achieved: the resource-gathering provisions have capacity in excess over the heaviest resource demands of growth. This luxury can be enjoyed only under ideal conditions of an abundant supply of resources, which scarcely obtain in the natural environments of phytoplankton: darkness alternates with periods of daylight, depth is equated with (at best) sub-saturating light levels and a balanced, abundant supply of nutrients is extremely rare. The question that arises relates to the levels at which resources start to impose restrictive 'limitations' on the dynamics of growth. The question is not wholly academic, as it impinges on the prevalent theories about competition for light and nutrients. Some, indeed, may need revision, while growth-simulation models founded upon resource harvesting are likely to be erroneous, except at very low nutrient levels.

5.4.1 The effect of truncated photoperiod

Perhaps the most obvious shortcoming that real habitats experience in comparison with idealised cultures is the alternation between light and dark: this *is* avoidable in nature but only at polar latitudes and, even then, for just a short period of the year. It also seems likely, following the

calculations in Section 5.3.3, that growth performance is most vulnerable to interruptions to net photosynthetic output and the supply of reduced-carbon skeletons to cell assembly. It is self-evident that photoperiod truncation by phases of real or effective darkness must eventually detract from the ability to sustain rapid growth. For small algae, even maintaining any adequate reserve of condensed photosynthate (such as starch, glycogen, paramylum, etc.) soon becomes problematic during extended periods of darkness, whereas all light-independent anabolism will become rapidly starved of newly fixed carbon.

Of the responses to light/dark alternation that might be anticipated, the most plausible is that the rate of cell replication becomes a direct function of the aggregate of the light periods. If the cell required a 24-h period of continuous light exposure in which to complete one generation, then, other things being equal, a day/night alternation of 12 h should determine that at least two photoperiods must be passed (not less than 36 h real time) before the cell can complete its replication. By extension of this logic, day/night alternations giving 6 h of light to 18 h of darkness, or 3 h to 21 h will each double the real time of replication. We may note that, on the basis of this logic, alternations of 6 h light to 6 h dark (or, for that matter, 1-minute alternations from light to dark) should not extend the generation time beyond the 12-h cycle of alternations.

Surprisingly, there is not a lot of experimental evidence to confirm or dismiss these conjectures. Although there are indications that the logic is neither ill-founded nor especially unrealistic, it does exclude two influential effects. One is the ongoing maintenance requirement of the cell – all those phosphorylations have to be sustained! The power demand, which persists through the hours of darkness and light, is met through the respirational reoxidation of carbohydrate. Unfortunately, it is still difficult to be certain about precisely how great that demand might be, at least partly because of historic difficulties in making good measurements and, in part, because sound testable hypotheses about maintenance have been lacking (see Sections 3.3.1, 3.3.2 and 3.5.1). Most of the available information about the respiration rates of planktic algae and Cyanobacteria comes from the 'dark controls' to experimental measurements of photosynthetic rate (of, for instance, Talling, 1957b; Steel, 1972; Jewson, 1976; Robarts and Zohary, 1987). Rates, appropriately expressed as a proportion of chlorophyll-specific P_{max}, typically range between 0.04 and 0.10, over a reasonable range of customary temperatures. Translating from the chlorophyll base to one of cell carbon, Reynolds (1990) deduced specific basal metabolic rates for *Chlorella*, *Asterionella* and *Microcystis* at 20 °C of (respectively) 1.3, 1.1 and 0.3 $\times$ 10^{-6} mol C (mol cell C)$^{-1}$ s^{-1}. These fit sufficiently well to a regression parallel to the one describing maximal replication rate (Eq. 5.4, Fig. 5.2), that it is possible to hypothesise that basal respiration at 20 °C (R_{20}) conforms to something close to:

$$R_{20} = 0.079(sv^{-1})^{0.325} \qquad (5.7)$$

In this case, R is roughly equivalent to 0.055 r'. Moreover, there is no *a priori* reason to suppose that basal respiration rate necessarily carries a temperature sensitivity different from other biochemical processes in the cell (that is, $Q_{10} \sim 2$). It is suggested, again provisionally, that the proportionality of Eq. (5.6) holds at other temperatures, i.e. $R_\theta \sim 0.055 r'_\theta$.

In the context of growth, however, the value is academic rather than deterministic, for it is by no means proven that the basal respiration applies equally in the light and in the dark. The implication of photosynthetic quotients (PQ) of 1.1–1.15 (Section 3.3.2) is that 10–15% more oxygen is generated in photosynthetic carbon fixation than is predicted by the stoichiometric fixed-carbon yield and that the 'missing' balance represents respirational losses in the light. Ganf (1980) observed that when *Microcystis* colonies were transferred from zero to saturating-light intensities, their respiration rate accelerated rapidly, from ~25 to ~50 µmol O_2 (mg chl*a*)$^{-1}$ h^{-1}, and did not fall back until after the colonies had been transferred back to the darkness. The time taken to return to base rate was proportional to the time spent at light saturation. Insofar as the same inability to store excess photosynthate requires some homeostatic defence reaction, such as accelerated respiration, glycolate

excretion, or photorespiration (see Section 3.3.4), the carbon losses from cells must be expected often to exceed basal metabolism. The experiments of Peterson (1978) show how easily the coupling between respiration and growth is broken and why the fractions of photosynthetic production lost to respiration, reported in the literature (reviewed by Tang and Peters, 1995), are so variable.

The second reservation about extrapolating growth over summated photoperiods concerns the photoadaptation of cells. Culturing cells under ideal irradiances actually tends to lead to a reduction in the cell-specific chlorophyll content, often to as little as 4 mg (g ash-free dry mass)$^{-1}$; Reynolds, 1987a; C : Chl ~100). On the other hand, populations exposed to low light intensities are able to adapt by increasing their pigment content (see Section 3.3.4). It is almost as if the cell's photosynthetic potential varies to match the growth requirement, rather than the opposite, as is generally presumed. Turning off the light for a part of the day provokes photoadaptative responses in respect of the shortened photoperiod to the extent that the energy available to invest in growth, normalised per light hour, is compensated, as shown by the data of Foy *et al.* (1976) (see Fig. 3.16).

There are limits to this argument, of course. On the other hand, the abilities of certain diatoms (Talling, 1957b; Reynolds, 1984a) and, especially, of some of the filamentous Cyanobacteria, such as *Planktothrix* (formerly *Oscillatoria*) *agardhii* (Jones, 1978; Foy and Gibson, 1982; Post *et al.*, 1985) to function on very low light doses has been well authenticated. The curves plotted in Fig. 5.4 represent a selection of experimentally derived fits of specific growth rates of named phytoplankters at 20 °C in cultures fully acclimatised to the daily photon fluxes noted. Far from being a linear function of light dose, except at very low average photon fluxes, the more adaptable species are able to increase biomass-specific photosynthetic efficiency so that the growth demand can continue to be saturated at significantly lowered light intensities.

How far the photosynthetic apparatus can be pushed to turn photons into new biomass is ultimately dependent upon the integrated

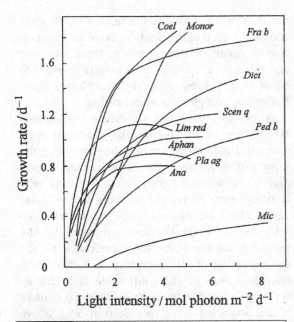

Figure 5.4 Light dependence of growth rate at 20 °C, as a function of intensity, in a selection of freshwater phytoplanktic species. The algae are: *Ana, Anabaena flos-aquae; Aphan, Aphanizomenon flos-aquae; Coel, Coelastrum microporum; Dict, Dictyosphaerium pulchellum; Fra b, Fragilaria bidens; Lim red, Limnothrix redekei; Mic, Microcystis aeruginosa; Monor, Monoraphidium sp.; Ped b, Pediastrum boryanum; Pla ag, Planktothrix agardhii; Scen q, Scenedesmus quadricauda.* Redrawn with permission from Reynolds (1997a).

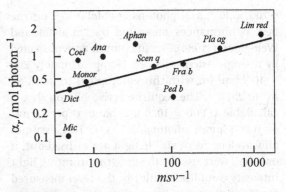

Figure 5.5 The initial slopes, α_r, of growth rate on light intensity from Fig. 5.4, plotted against the corresponding values of msv^{-1}, the product of maximum dimension and the surface-to-volume ratio, of each of the same selection of species of algae: *Ana, Anabaena flos-aquae; Aphan, Aphanizomenon flos-aquae; Coel, Coelastrum microporum; Dict, Dictyosphaerium pulchellum; Fra b, Fragilaria bidens; Lim red, Limnothrix redekei; Mic, Microcystis aeruginosa; Monor, Monoraphidium sp.; Ped b, Pediastrum boryanum; Pla ag, Planktothrix agardhii; Scen q, Scenedesmus quadricauda.* The least-squares regression fitted to the points is $\alpha_r = 0.257\,(msv^{-1})^{0.236}$. Redrawn with permission from Reynolds (1997a).

flux density, which does, eventually, force a dependence of r on I. The steeper is the slope of light-dependent growth (α_r), the more efficient is the dedication of harvested light energy. Values of α_r (expressed in units of specific replication rate (r) d^{-1} (mol photon)$^{-1}$ m^{-2} per d, which simplifies to (mol photon)$^{-1}$ m^2), are derived from the initial, light-dependent slopes in Fig. 5.4. Following Reynolds (1989b), they are plotted in Fig. 5.5 against the dimensionless product of the surface-to-volume ratio (sv^{-1}) and the maximum dimension of the alga (m). This had been found to provide the most satisfactory morphological descriptor of the interspecific variability in α_r. It also corresponded with the interpretation that the greatest flexibility of algae to enhance their light-dependent growth efficiency evidently resided with those having the greatest morphological attenuation of form. Ostensibly, slender and flattened shapes make the best

light antennae, at least when oriented correctly in the photon-flux field. The filamentous arrangement cells in the Oscillatoriales seems to be supremely efficient in this context, as both the numerous studies referred to above and the affinity the *Planktothrix* and *Limnothrix* species for turbid, well-mixed lakes (see Section 7.2.3) would indicate. Recalculating from the data of Post *et al.* (1985), acclimated *P. agardhii* may maintain a maximum growth rate of 9.8×10^{-6} mol C (mol cell C)$^{-1}$ s^{-1} at 20 °C, to as low as 18 μmol photons m^{-2} s^{-1}; $\alpha_r \approx 0.54$ mol C (mol cell C)$^{-1}$ (mol photon)$^{-1}$ m^2. The analogous experimental data for the diatom *Asterionella* have not been located but, piecing together information from field populations, Reynolds (1994a) showed it to rival the reputation of *Planktothrix* as a low-light adapting organism. From a specific growth rate, $r'_{20} = 20.6 \times 10^{-6}$ mol C (mol cell C)$^{-1}$ s^{-1} and a chlorophyll content of 2.3 pg cell^{-1}, that is ~0.324 g chl*a* (mol cell C)$^{-1}$, the sustaining chlorophyll-specific yield is 63.6×10^{-6} mol C (g chl*a*)$^{-1}$ s^{-1}. In turn, this requires a photon flux of, theoretically, not less than 509×10^{-6} mol photons (g chl*a*)$^{-1}$ s^{-1} and

$\sim 700 \times 10^{-6}$ mol photons (g chl$a$)$^{-1}$ s^{-1}, on the best performances measured by Bannister and Weidemann (1984). The maximum area projected by a single *Asterionella* cell is approximately 200×10^{-12} m^2 or, with this chlorophyll content, 87 m^2 (g chl$a$)$^{-1}$. The requisite active photon flux is calculable as 700×10^{-6} mol photons per 87 m^2, or just 8 μmol photons m^{-2} s^{-1}. This assumes all wavelengths of visible light are utilised but, if only half were usable, the growth-saturating light intensity would be similar to the level measured for *Planktothrix agardhii*.

It is not, at first sight, at all obvious that low-light adaptation should be related to algal morphology when it is functionally dependent upon appropriate enhancements in pigmentation. However, it is easily demonstrated that cell geometry and orientation raise the efficiency of light interception by the pigment complement (Kirk, 1975a, b, 1976). A spherical cell, d μm in diameter, has a volume, $v = 4/3\pi (d/2)^3$, while the area that it projects is that of the equivalent disc, $a = \pi(d/2)^2$. Because the carbon content is, primarily, a function of volume, the carbon-specific projection (k_a) of spherical algae diminishes with increasing diameter. For example, we may calculate that, for a single cell of *Chlorella* ($a = 12.6 \times 10^{-12}$ m^2; C content $= 0.61 \times 10^{-12}$ mol C), $k_a = 20.7$ m^2 (mol cell C)$^{-1}$; for a spherical *Microcystis* colony ($d = 200$ μm), comprising 12 000 cells, each containing 14 pg C (Reynolds and Jaworski, 1978), in which $a = 31.4 \times 10^{-9}$ m^2 for a content of 14×10^{-9} mol C, $k_a = 2.24$ m^2 (mol cell C)$^{-1}$.

In the case of non-spherical algae that are flattened in one, like those of *Pediastrum*, or in two planes, like those of *Closterium*, *Synedra* or *Asterionella*, the area projected depends upon orientation. The maximum area projected is when the two longest axes are perpendicular to a unidirectional photon source. The typical cell of *Asterionella* in a colony lying flat on a microscope slide is ~ 65 μm in length and shows a tapering valve with an average width of ~ 3.3 μm. In relation to its approximate carbon content of 85 pg (7.08×10^{-12} mol C; Reynolds, 1984a), the maximum value of k_a is ~ 28.2 m^2 (mol cell C)$^{-1}$. In other orientations relative to a single source of light, the area projected may greatly diminish. Kirk's (1976) calculations compensated for this

but these are not followed here. In a turbid environment, much of the light available for interception is already scattered and at these low average intensities, changes of orientation prove to be of little consequence. Well-distributed light-harvesting complexes are everything.

These considerations emphasise the influential nature of the relationship between algal shape in the interception of light energy and the impact of algal size in governing its metabolism. Morel and Bricaud (1981) recognised this relationship some years ago, referring to the 'packaging' effect on pigment deployment (cf. Duysens, 1956), where the area projected by the pigments assumed the same relevance as the concentration of LHC receptors. It is thus inextricably linked to the contestable size allometry of growth rate (with its $-1/3$ slope instead of the expected $-1/4$; see Section 5.3.1 above and Finkel, 2001). A general relationship between projection and morphometry is shown in Fig. 3.12. The independent variable is, again, the index msv^{-1}, the product of the maximum cell dimension (m) and the surface-to-volume ratio. Note that it is a dimensionless property, length always cancelling out. For spheres, $m = d$, and msv^{-1} is a constant $[d \times 4\pi(d/2)^2 \div 4/3\pi(d/2)^3 = 6]$. For any shape representing distortion from the spherical form, $msv^{-1} > 6$. Figure 3.12 also shows that the smaller algae generally project large carbon-specific areas but larger ones have to be significantly subspherical to match the k_a values of the smaller ones. It is especially interesting to observe that the algae that already project the greatest area in relation to their cell-carbon content are also mostly those with the maximum photoadaptive potential.

5.4.2 The effect of persistent low light intensities

Rather than experiencing alternations of dark interludes with windows of saturating light, phytoplankton of so-called *crepuscular habitats* are exposed to variability that offers only windows of gloom. The algae forming metalimnetic swarms and deep chlorophyll maxima (DCM) in stable layers in seas and lakes experience the same circadian alternations of night and day perceived by terrestrial and littoral plants but, because they are located so relatively deep in the light gradient, the daytime irradiances they experience are low.

The circumstances of a cell placed at a constant depth and receiving a dielly fluctuating but low-intensity insolation differ from those of one receiving short bursts of high illumination, even though the daily photon flux might be similar. However, the ultimate objective – to maximise absorption of the photons available – does invoke certain similarities of response. All the organisms that successfully exploit stable layers have to be capable of maintaining vertical position with respect to the light gradient. They are either motile (e.g. the flagellated chrysophytes that form layers in oligotrophic, softwater lakes and certain species of cryptophyte of slightly more enriched lakes), or they regulate buoyancy (as do some of the solitary filamentous Cyanobacteria, including, most familiarly, *Planktothrix rubescens* and other members of the *prolifica* group of species, and *Planktolyngbya limnetica*). Buoyancy-regulating sulphur bacteria of the Chromatiaceae and Chlorobiaceae may stratify in the oxygen gradient, provided this lies simultaneously within a stable density gradient and is also reached by a few downwelling photons (usually ≤ 5 μmol m^{-2} s^{-1}). Here, photoautotrophs function on inputs of light energy that are invariably low. The extent of photoadaptation demanded of them depends essentially upon the depth in the light gradient at which the organism is poised. This determines the quantity of penetrating irradiance and the wavelengths of the residual light least absorbed at lesser depths (the quality of the irradiance). Beneath relatively clear layers, with absorption predominantly in the blue wavelengths, the pigmentation may be expected to intensify but without obvious chromatic adaptation. The less is the residual light in the red wavelengths, however, the more advantageous is the facultative production of accessory pigments, such as phycoerythrin and phycocyanin, to the organism's ability to function phototrophically at depth.

As discussed in Section 3.3.3, the adaptation is most plainly observed in depth-zonated populations of Cyanobacteria (Reynolds *et al.*, 1983a) or in populations slowly sinking to greater depths

(Ganf *et al.*, 1991). Taking as the most extreme case of photosynthetic efficiency from Post *et al.* (1985) for *P. agardhii* as a proven example of the low-light adaptation that is possible at 20 °C (viz. 0.54 mol C (mol cell C)$^{-1}$ (mol photon)$^{-1}$ m^2; Section 5.4.1) and comparing it with the supposed basal rate of respiration of an alga of its dimensions, 0.079 $(s/v)^{0.325}$, i.e. $\sim$0.064 mol C (mol cell C) d^{-1}, or 0.74 $\times 10^{-6}$ mol C (mol cell C)$^{-1}$ s^{-1} at the same temperature, it is possible to deduce that compensation is achieved at $\sim$1.4 μmol photons m^{-2} s^{-1}, certainly in the order of 3–4 μmol photons m^{-2} s^{-1}, if allowance for the dark period is accommodated.

5.4.3 The effect of fluctuating light

Applying the results of laboratory experiments to the extrapolation of field conditions or to the interpetation of field data requires caution. In the context of the growth responses of phytoplankton entrained in mixed layer, insolation may change rapidly, either increasing or decreasing at random (Fig. 3.14). Depending on the light gradient and on the depth of turbulent entrainment, phytoplankters might experience anything from a probable period of 30–40 minutes in effective darkness with a few minutes of exposure to high light (deep mixing, steep light gradient), to a similar time period of fluctuating light levels that are nevertheless adequate to support net photosynthesis throughout (mixing within the photic zone). The generic nature of the adaptive responses available, discussed here and in Section 3.3.3, is clearly aimed towards optimising growth against highly erratic drivers. However, the probabilistic, Eulerian aggregation of the responses of the whole population does not take account of the photoprotective and recovery behaviour on the growth rates of individual cells to what are sometimes sharp, sudden and possibly crucial changes in insolation.

There are experimental data that lend perspective to this issue. Litchman (2000) designed experiments that brought irradiance fluctuations to each of four cultured microalgae, in each of three ranges (15–35, 15–85 and 65–135 μmol m^{-2} s^{-1}) and over three wavelengths of fluctuation (1, 8 and 24 h). Variations in the low-light range (15–35 μmol m^{-2} s^{-1}, wherein growth rate is expected to be proportional) were fairly neutral. The growth rate of the diatom *Nitzschia* was slightly increased, that of the green alga *Sphaerocystis schroeteri* was slightly depressed, compared to growth at a constant 20 μmol m^{-2} s^{-1}. In the saturating range (65–135 μmol m^{-2} s^{-1}), little effect was experienced, except that *Anabaena flos-aquae* was slightly increased over its performance at a steady 100 μmol m^{-2} s^{-1}. Over the wide range of fluctuations (15–85 μmol m^{-2} s^{-1}, spanning limitation to saturation), growth responses differed significantly from the behaviour under steady exposure. Growth of all species was maintained on the short (1-h) cycle; in each case, the relatively short exposure to high light remained within the capacity of the species-specific physiologies. On the longer cycles, however, growth rate was impaired in all species, though not all to the same extent. The reaction would once have been described as 'photoinhibition' but would now be better referred to photoprotection and the first steps toward photoadaptation (see Section 3.3.4). The experiments of Flöder *et al.* (2002) investigated the influence of fluctuating light intensities (range 20–100 μmol m^{-2} s^{-1}) on growth rates of natural phytoplankton assemblages collected from Biwa-Ko, imposed on cycles of 1, 3, 6 or 12 days. These ably illustrate the population responses consequential upon differential growth rates altering the composition of the assemblage.

5.4.4 The effects of nutrient deficiency

A cornerstone role has long been accorded to nutrients in the regulation of productive capacity in the plankton and in shaping the species composition. A very large literature on the nutrient limitation of phytoplankton and the interspecific competition for resources reflects the importance of their availability in pelagic ecology, even if these key processes seem, at times, to have been misrepresented. The point has been made earlier that the least-available resource, relative to the minimum requirements of organisms, sets a finite 'carrying capacity' for the habitat. To establish the limiting role of nutrients on growth rate is more difficult, for two reasons. As has been shown in Chapter 4 and again in Section 5.3.3, most phytoplankters are, under ideal

circumstances, able to take up nutrients far faster than they can deploy them. Moreover, they can continue to do so until very low resource concentrations are encountered. Even then, the luxury uptake in generations experiencing resource plenitude may support two or three generations born to resource deficiency.

At first sight, there is certainly a rapid transition from there being no competition for resources to there being little left over which to compete. There is a counter to this deduction, which refers back to the distinctions in the strategies of velocity, storage and affinity adaptation (Section 4.3.2). Naturally, fast-growing algae must be able to garner resources with equal velocity, from diminishing external concentrations. To be able to store resources in excess promises an advantage when external concentrations have been diminished, although it might be more beneficial to species that have the ability to migrate between the relative resource-richness of deep-water layers and resource-depleted but insolated surface waters. To survive, even to thrive, in waters chronically deficient in one (rarely more) major resource might well call for a superior competitive ability to win the scarce supplies and deny them to individuals of other species.

Evidence for the existence and implementation of these strategies is to be discussed later in this chapter. For the moment, the first task is to discern the resource levels that separate famine from bounty. It is convenient to consider these nutrient by nutrient.

Phosphorus deficiency

Satisfaction of the alga-specific P requirements for growth has been suggested to rest upon the ability to maintain a stoichiometric balance of assimilates approximating to 1 P atom to every 106 of carbon. This determines the generalised requirement that 9.4×10^{-3} mol P (mol C)$^{-1}$ is incorporated during each single replication time. As already indicated, the alga's uptake capacity is likely to be such that, under resource-rich conditions, it may achieve this in minutes rather than hours. It is not likely that any phytoplankter takes longer than its achievable generation time while external MRP concentrations exceed 0.13×10^{-6} M (4 μg P L^{-1}). For many species, indeed, this could be true at MRP concentrations as low as 10^{-8} M (0.3 μg P L^{-1}; see Sections 4.3.3, 5.3.3). Even then, the reserves accumulated during previous resource-replete generations may sustain one or two generations before the cell recognises impending shortages and perhaps three, or even four, before the cell quotas approach q_0 and exhaustion. Fast-growing species, having high sv^{-1} (>1.3 $μm^{-1}$, which, at about 30 °C, might permit specific growth rates of up to ~4 d^{-1}, or ~50 × 10^{-6} s^{-1}, to be attained), would be expected to reach exhaustion proportionately sooner. Yet averaging and normalising the total P requirements to the starting biomass carbon, the P demand of two generations (~0.5 × 10^{-6} mol P (mol initial cell C)$^{-1}$ s^{-1}) is still likely to be deliverable from a starting concentration of 0.03–0.13 × 10^{-6} M (1–4 μg P L^{-1}).

Data compilations that compare growth rates, nutrient-uptake rates and their half-saturation coefficients are available (see Padisák, 2003), from which it ought to be a straightforward exercise to verify, on the basis of hard, experimental results, some of the above conjectures. The trouble is that many of the half-saturation concentrations refer to uptake by starved cells, rather than to that needed to half-saturate the requirement to maintain growth (K_r). These quantities require the analysis of a lot of batch cultures, across a range of concentrations, or the application of a semi-continuous technique in which the algal culture is diluted with test medium at a rate adjusted to keep the algal concentration constant (that is, to balance the rate of growth). As a consequence, there are few good data to confirm the half-saturation constants of P-limited growth. Rhee (1973), using a *Scenedesmus*, Ahlgren (1985), with *Microcystis wesenbergii*, and Spijkerman and Coesel (1996a), using two desmids (*Cosmarium abbreviatum* and *Staurastrum pingue*), each found that the algae would grow at about one half the nutrient-saturated rate at 20 °C, in media supplying <6.0 μg P L^{-1} (0.19 μM). All but the *Scenedesmus* did so at <1.2 μg P L^{-1} (<0.04 μM). *Cosmarium* growth rate was half-saturated at 0.35 μg P L^{-1} (0.011 μM). Davies (1997), whose work with *Asterionella* has been highlighted earlier in the context of the interaction between P-uptake and cell quota

(Section 4.3.2), found that growth rate at 20 °C is half-saturated when the cell contains about 0.7 pg P (or about 0.003 mol cell P (mol cell C)$^{-1}$). The external MRP concentration required to balance this quota is about 0.75 μg P L^{-1} ($K_r = 0.024$ μM). The extensive work of Tilman and Kilham (1976) using semi-continuous cultures of diatoms pointed to the half saturation of growth in *Asterionella formosa* falling between 0.02 and 0.04 μmol P L^{-1} (0.6–1.2 μg P L^{-1}). However, the corresponding value for another diatom, *Cyclotella meneghiniana*, was substantially higher than this ($K_r = 0.25$ μmol P L^{-1}, or nearly 8 μg P L^{-1}). Finally, in this context, Nalewajko and Lean (1978) used ^{32}P-labelled phosphorus to track phosphorus uptake and turnover in cultures of *Anabaena flos-aquae* and *Scenedesmus quadricauda*, and remarked on the 'low' concentrations (~6 μg P L^{-1}; ~0.2 μmol P L^{-1}) at which full exchange activity is maintained.

Certainly, most of these data uphold the view that the onset of phosphorus 'famine', below which growth rate may be regulated by the supply of P, generally occurs at <0.1 μM P (~3 μg P L^{-1}). This may not apply to all species: *Scenedesmus* and the *Cyclotella* of Tilman and Kilham are possible exceptions. For most species, allusions to growth-rate limitation of phytoplankton by phosphorus when external MRP concentrations exceed 0.1 μM P may be doubted. It is possible that the higher species-specific thresholds are a consequence of adaptation to the relatively P-rich habitats in which they occur. Conversely, differential species-specific affinities uptake (K_r values spanning 0.01–0.2 μM P) can be said to influence where species may live and how competitive they might be for truly limiting supplies of phosphorus.

Nitrogen deficiency

Analogous calculations concerning algae and their nitrogen requirements are also available. Supposing the satisfaction of the alga-specific N requirements for growth to be similarly based upon the stoichiometric balancing against carbon in the atomic ratio 6.6 : 1, the generalised demand approximates to 151 × 10^{-3} mol N (mol C)$^{-1}$. Again the DIN-uptake capacity is well up to this under nitrogen-replete conditions. From the information on DIN uptake in *Chlorella* (Section 5.3.3), it may be deduced that the nitrogen complement needed to sustain a doubling of mass could proceed 60 times more slowly and still fulfil the maximum growth rate. Moreover, the external concentration needed to supply DIN at this rate would be about 0.12 μmol N L^{-1}, or 2 μg N L^{-1}). The half-saturation constants for nitrate and ammonium uptake (K_U) among small-celled oceanic phytoplankton are of similar order or only slightly higher (see Section 4.4.2) and are, thus, unlikely to experience symptoms of nitrogen-limited growth at external DIN concentrations >1–2 μg N L^{-1}. On the other hand, the half-saturation constants for nitrate and ammonium uptake (K_U) among larger diatoms of inshore waters may be up to an order greater again (0.5–5 μM N), so that problems of obtaining sufficient nitrogen to even half-saturate growth might be experienced at external DIN concentrations in the range 0.2–2 μM N (say 3–30 μg N L^{-1}).

The lower limits of the range of DIN concentrations able to support phytoplankton growth in inland waters are less well researched. It was once deduced (Reynolds, 1972) that, based on the assumption that the appearance of nitrogen-fixing organelles (nostocalean heterocysts) at DIN levels of 20–25 μM provided a simultaneous advantage to the fixers over non-fixers, the non-fixers were experiencing supply difficulties. As a result of later observations but without varying the essential logic, this critical range was revised downwards to ~6–7 μM N (80–100 μg N L^{-1}) (Reynolds, 1986b). As it now seems probable that heterocyst production responds to ammonia concentration and not DIN per se (discussion, section 4.4.3), it no longer follows that their appearance necessarily coincides with nitrogen shortage among the nitrate users. Current evidence indicates that they are competent to draw down DIN to concentrations to levels of 0.3–3 μM without growth-rate limitation.

Eventually, the ability of heterocystous cyanobacteria to fix nitrogen when DIN is simultaneously depleted influences the recruitment of to natural communities (Riddolls, 1985) but the concentration threshold favouring Cyanobacterial dominance, between 2 and 6 μM N, remains imprecise. In many of the nitrogen-deficient but simultaneously phosphorus-poor lakes of the Andean–Patagonian lakes investigated by Diaz and Pedrozo (1996), the phytoplankton biomass is demonstrably (by assays and by mathematical regession) constrained by nitrogen availability. The significant incidence of dinitrogen-fixing Cyanobacteria is relatively restricted to lakes, such as Bayley-Willis and Fonck, that have higher TP contents (and where, incidentally, the SRP is drawn down in summer to growth-limiting levels as identied above, viz. to <0.1 μM P). Elsewhere, the unsupplemented combined levels of nitrate-, nitrite- and ammonium- nitrogen levels stand at ≤0.8–1.0 μM (≤11–14 μg N L^{-1}), perhaps falling in summer to <0.2 μM (see Diaz *et al.*, 2000).

The revised verdict on the DIN levels representing the onset of nitrogen limitation among non-fixers is 0.1 μM (1.5 μg N L^{-1}) for oceanic picoplankton and 1–2 μM (15–30 μg N L^{-1}) for many microplankters in lakes and seas. If other conditions are satisfied (see Section 4.4.3), nitrogen fixers may avoid altogether the constraint of low DIN concentrations.

Silicon deficiency

Silicon limitation of growth in diatoms is absolute and well understood. Lund's (1949, 1950) classical studies on *Asterionella* in Windermere and its relationship with the availability of silicon helped to establish the importance of nutrients in planktic ecology. The clarity of the effects and the precision of the chemical thresholds were, and are still, impressive. That they were never emulated in the studies of other elemental requirements is attributable to the nature of the nutrients and the margin around the empiricism of the requirements that the organisms introduce through storage and luxury uptake. As stated (Section 4.6), the diatoms are the biggest planktic consumers of silicon; they take up no more than is required to build the silica valves of the frustules of the current generation, and the silicon polymer is laid down under close genetic supervision. All this leads to the silicon requirement of each new cell of a given species and size being readily predictable. The concentrations of silicic acid capable of delivering the silica requirement have been found through observation and experiment. The consequence of silicon 'limitation' is easy to detect as its consequence is that cells cannot complete the growth cycle. Moreover, failure of the putative cell to build its new frustular valve is fatal.

In the case of Lund's *Asterionella*, the average complement of 140 pg $(SiO_2)_n$ cell^{-1} could be satisfied from a concentration equivalent to ≥ 0.5 mg (SiO_2) L^{-1}. Translation of the units notwithstanding, both quantities have been abundantly verified. Solving the regressions of Jaworski *et al.* (1988), an *Asterionella* cell 65 µm in length has a probable Si content of 65 pg (2.3 pmol cell^{-1}). Using the experimental data of Tilman and Kilham (1976), the silicic acid concentration that will half-saturate the Si requirement is 3.9 µM (equivalent to 109 µg Si L^{-1}, or 0.23 mg $(SiO_2)_n$ L^{-1}).

Equivalent data from many other experiments, reviewed in Tilman *et al.* (1982) and in Sommer (1988a), show that growth rates of freshwater diatoms at 20 °C tend to be half-saturated at between 0.9 (in *Stephanodiscus minutus*) and 20 µM Si (in *Synedra filiformis*). That of *Cyclotella meneghiniana* is half-saturated at 1.4 µM Si (Tilman and Kilham, 1976). Among the investigated clones of marine *Thalassiosira pseudonana* and *T. nordenskioeldii*, half-saturation of the growth rate at 20 °C tends to occur in the range 0.2 to 1.5 µM (data of Paasche, 1973a, b).

The 100-fold range in uptake thresholds has ecological consequences, to be considered in the next section. For the moment, the deduction is that silicon concentrations begin to interfere with the growth of diatoms at concentrations below ~0.5 mg Si L^{-1} (say 1 mg L^{-1} as equivalent silica, or ~20 µM). In most lacustrine instances, growth-limiting concentrations are encountered mainly below about 0.1 mg Si L^{-1} (say <4 µM) and, in the sea, below 0.03 mg Si L^{-1} (<2 µM).

5.4.5 The effect of resource interactions: nutrients and light

Resource-based competition

The able demonstrations by Tilman and his co-workers of the interspecific differences in the capability of diatoms to take up the silicon and phosphorus required to sustain their growth from relatively low external concentrations also established an important conceptual theory of *resource-based competition*. Simply, if two species, A and B, having similar resource-saturated growth rates, are cultured together in a gradient of growth-limiting concentrations of a resource S_1, the one with the higher uptake rate (V_U) is clearly able to sustain a faster rate of growth at low concentrations of $[S_1]$ than the other. The theoretical growth performances of A and B are shown in Fig. 5.6a. Against a second resource, S_2, however, it is species B that performs better at low concentrations (Fig. 5.6b). Placed together in a medium deficient in both S_1 and S_2, it is possible, depending on the relative concentrations of either nutrient, for the species to be simultaneously limited by different nutrients. Plotting against the concentrations of both resources (Fig. 5.6c), the respective limitations can be used to predict competitive outcomes of variable resource combinations (Fig. 5.6d). Thus, at low concentrations of (S_1), an outcome will always be favoured in which A dominates; at low concentrations of ions of (S_2), it will be B that is favoured. At slightly higher concentrations of both resources, A and B may coexist successfully, while the one (B) remains S_1-limited and the other is still limited by S_2. This prediction was precisely the outcome of their investigations (Tilman and Kilham, 1976) of phosphorus- and silicon-limited growth between *Asterionella formosa* (K_r(P), ~0.03 µmol P L^{-1}, K_r (Si) 3.9 µmol Si L^{-1}) and *Cyclotella meneghiniana* (K_r(P), ~0.25 µmol P L^{-1}, K_r (Si) 1.4 µmol Si L^{-1}); *Cyclotella* dominated over *Asterionella* in mixed cultures at low Si : P ratios. The opposite was true at high Si : P ratios. On the basis of later investigations of other species, Tilman *et al.* (1982) emphasised the differential competitive abilities of diatoms to Si : P by plotting the experimentally solved K_r(P) against K_r(Si) for each (see Fig. 5.6e). The plot ably arranges species on the basis of Si : P preferences.

For most other plankton, silicon is a minor nutrient and, not surprisingly, they tolerate Si:P ratios very much lower than *Cyclotella*'s limit of 5.6 (Holm and Armstrong, 1981; Sommer, 1989). However, it is the relationships between the nitrogen and phosphorus requirements that have aroused enormous interest, especially in the context of a widely held belief that low

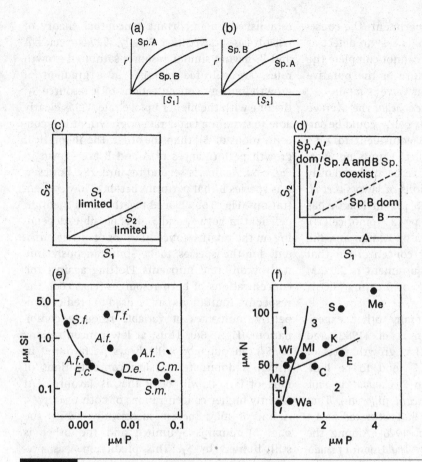

Figure 5.6 Resource competition and species interactions. Parts (a) and (b) compare the nutrient-limited growth rates two species of phytoplankter, Sp. A and Sp. B, against low, steady-state concentrations of resources S_1 and S_2. Growth of either may be limited (c) by the availability of either resource. Tilman's theory of resource-based interspecific competition acknowledges that the uptake constraints acting on Sp. A and Sp. B differ sufficiently for (d) Sp. A to dominate over Sp. B when $[S_1]$ is low and Sp. B to do so when $[S_1]$ is low, but Spp. A and B do not compete when limited by different resources. The relative competitive abilities of named diatoms for silicate and phosphate, as determined by Tilman *et al.* (1982), are shown in (e): *A.f.*, *Asterionella formosa*; *C.m.*, *Cyclotella meneghiniana*; *D.e.*, *Diatoma elongatum*; *F.c.*, *Fragilaria crotonensis*; *S.f.*, *Synedra filiformis*; *S.m.*, *Stephanodiscus minutulus*; *T.f.*, *Tabellaria flocculosa*. In (f), the effects on phytoplankton assemblages in a selection of natural lakes of differing N and P availabilities are represented: C, Crose Mere, and W; Windermere, are in UK; E is Esrum, Denmark; K, Kasumiga-Ura, and S, Sagami-Ko, Japan; Me, Mendota, T, Tahoe, and Wa, Washington, in USA; Mg is Maggiore, Italy/Swizerland and Ml, Mälaren, in Sweden. Area I applies to low-P lakes, dominated by diatoms and chrysophytes; area 2 covers lakes in which nitrogen-fixing Cyanobacteria are abundant through substantial parts of the year; area 3 lakes are are dominated by *Microcystis* for long periods. The composite combines figures redrawn from Reynolds (1984a) and Reynolds (1987b).

nitrogen-to-phosphorus ratios favour the (usually unwelcome) dominance of Cyanobacteria (Smith, 1983). Rhee (1978) suggested that the evident mutual competition along N : P gradients is influenced by differing N : P optima in the cells of various species. In a major programme of laboratory experimentation, Rhee and Gotham (1980) showed systematic differences in the ratios of species-specific optimal N and P quotas dur-

ing normal growth (when both storage effects and deficiencies should have been minimal). Outcomes ranged between 7, for the diatom *Stephanodiscus binderanus*, and 20–30, for three species of Chlorococcales. For the Cyanobacterium *Microcystis aeruginosa*, it was 9. These optima differ from the ideal ('Redfield') N : P stoichiometry centred at 16 molecular and from the suggested (Section 4.4) range of normality of 13–19. Given the range in physiological variability in N : C and P : C content of individual cells, extremes of N : P of 4 and 108 are theoretically tenable but values within the range 10 to 30 are scarcely indicative of nutrient stress.

Nevertheless, the findings of Rhee and Gotham (1980) do not violate any supposition about the importance of N : P ratios in favouring Cyanobacteria or otherwise. However, there are difficulties (see below) in applying resource ratios to either the interpretation or the prediction of the composition of natural communities. Besides, ratios of available resources change quite rapidly through time, without necessarily precipitating immediate changes in species composition. If the *total* nitrogen and phosphorus resources delivered to and present within a water body are considered, some broad compositional trends *are* discernible. The distribution of lakes shown in Fig. 5.6f against axes of total N and total P separates those that will support large populations of nitrogen-fixing Cyanobacteria as being low-N-high-P habitats, from those that support non-fixing *Microcystis* (high N, high P) and those that seem never to be dominated by bloom-forming genera (low-P lakes).

To be any more specific about predictions of differentiated growth or the composition of the population structures that it might yield needs much more information. One component obviously lacking from the previous paragraph is the carbon input and the solar-energy income that is essential to its compounding with nutrient resources into biomass. In developing a hypothesis about the deterministic importance of the *light : nutrient ratio* in lakes, Sterner et al. (1997) showed that the relationship between the mean mixed-layer light level (equivalent to the calculation of I^* in Section 3.3.3) in each of a number of lakes and its corresponding total P concentration is a good probabilistic predictor of the C : P content of the seston and of the efficiency of resource use by the system as a whole. They further hypothesised that the seston C : P ratio influenced the pathways of secondary production (essentially the food-web consumers of primary product) and thence, biassed the intensity of nutrient recycling, the strength of microbial processing and, indeed, the structuring of the entire ecosystem.

Stoichiometry

These plausible deductions invoked parallel developments in the appreciation of organismic stoichiometry, prompted, in part, by the work of Hessen and Lyche (1991) on the differential elemental make up (chiefly C:N:P) of zooplankton. Stoichiometric differences between the trophic components imply consequences for the system as a whole. For instance, animals with a relatively low N : P content feeding on algal foods having a high N:P make-up will retain P preferentially and, so, recycle wastes with a yet higher N:P content. This approach has been developed further by J. Elser and co-workers (usefully reviewed in Elser *et al.*, 2000). They were able to demonstrate striking connectivities among the molecular stoichiometries of growing cells of a wide range of algal species, their rates of growth and the evolutionary pressures underpinning their life-history traits. That is to say, the evolution of fast or slow growth rates and the allocation of the catalysing (P-rich) RNA molecules are inextricably interlinked. The matching of evolutionary traits, from their molecular bases to the environments in which whole organisms function and interact, now has a wide following. The recent book by Sterner and Elser (2002) on *Ecological Stoichiometry* conveys and nourishes much of this excitement. Besides providing an alternative perspective on ecosystem function and a persuasive argument for unifying concepts in its understanding, the book invokes pertinent explorations of just what goes on inside the living cell to yield recognisable structural stoichiometries in the first place and what activities lead to (relatively modest) departures therefrom.

To bring us back specifically to phytoplankton, if these attractive theories of resource ratios and biological stoichiometry are to be helpful to understanding how pelagic communities function, then it is important first to separate just what is cause and what is effect. It is necessary to emphasise the distinctions among the ratios of algal cell quotas, the ratios of resource availability and supply, and the competitive abilities of algae to take up elements at low concentrations. Taking cell quotas first, we have to accept that the ratio N : P = 16 is certainly not absolute, that it is subject to a margin of physiological variability, including in the complement of RNA, and that there may indeed be systematic, interphyletic differences in the optimal elemental balances. Nevertheless, the range of normality in the elemental composition of cells, from bacteria to elephants, is relatively quite narrow, reflecting general similarities in the cell complements of protein and nucleic acid (Geider and La Roche, 2000). From Chapter 4, it is plain that a factor of 50% variation in either the N or the P content is hardly exceptional, yet it yields a full range in N:P from 5 to 36. For the purpose of estimating stoichiometrically the phytoplankton-carrying capacity of the nutrients in a given habitat, less error attaches to the adoption of a mean complement of 16 to 1 than to the correct estimation of the base of bioavailable nutrients (Reynolds and Maberly, 2002). However, it is more straightforward (and more illuminating) to examine the support resource by resource. To be able to form a standing phytoplankton biomass equivalent to, say, 106 μmol C L^{-1} (1.27 mg C L^{-1}, $\sim$25 μg chl a L^{-1}) ostensibly requires the supply of 16 μmol N and 1 μmol P L^{-1} (i.e. 224 μg N, 31 μg P L^{-1}). If the water body can fulfil only 10 μmol N and 0.1 μmol P L^{-1} (note, N : P = 100), it is obvious that the growth demand will first exhaust the phosphorus, at a rather smaller chlorophyll yield than 25 μg chl a L^{-1}. Long before this maximum is reached, the biomass : P yield is stretched, according to Eq. (4.15), to 12.2 μg chl$a$ L^{-1}; the biomass has a probable content of carbon equivalent to $\sim$51 μmol C and 7.7 μmol N L^{-1}, but no more than the originally available 0.1 μmol P L^{-1}. We deduce a biomass N : P ratio of 77, correctly inferring the obviously severe phosphorus limitation on the biomass. The magnitude of the eventual P-limited quota constraint on the biomass is predictable on the basis of the initial phosphorus availability. That it was phosphorus, rather than nitrogen, that would impose the eventual limit is implicit in the starting resource ratio.

Now, if the water body could fulfil 1 μmol P but only 1.6 μmol N L^{-1} (i.e., 31 μg P, 22.4 μg N L^{-1}; N:P 1.6), growth would be expected to exhaust the nitrogen, for the production of 32 μmol biomass C L^{-1} (assuming the minimum C : N ratio quoted in Section 4.4) and which, stoichiometrically, would have a phosphorus content equivalent to not less than 0.3 μmol P L^{-1}

and a probable biomass N : P (5.3) indicative of N-limitation. The residual P concentration (arguably $\sim$0.7 μmol P L^{-1}) confirms that P is certainly not a constraint and it is also sufficient to support the activity of nitrogen-fixing Cyanobacteria. These could grow to the limit of the phosphorus capacity (47 μg chl$a$ L^{-1}), only to now be dominated by nitrogen-fixing Cyanobacteria having an intracellular N : P ratio of $\sim$30. Note that, provided it is not itself limited by some other factor, nitrogen fixation forces the nitrogen-deficient system to the capacity of its phosphorus supply (cf. Schindler, 1977).

Resource depletion and growth-rate regulation
Now let us consider the role of interspecific competition in the way different species might simultaneously satisfy their resource requirements to sustain their growth rates. Here we bring into sharp focus the uptake capabilities of the algae themselves and the sorts of threshold concentrations at which they fail to be able to take up specific nutrients as fast as maximum consumption would demand (1–2 μmol N, 0.1 μmol P L^{-1} (Section 5.4.4). The corollary of this is that if the concentrations of bioavailable N and P are significantly greater than these thresholds, neither imposes a rate-limiting constraint upon the growth of any alga. Several species could grow simultaneously and, while each satisfies its requirement, they are not in mutual direct competition (*sensu* Keddy, 2001; see Box 4.1). Each performs to its capacity (or to some independent regulation). *At this stage, the ratio of the available resources is quite irrelevant to the regulation of species-specific growth rates.*

This deduction arouses persistent controversy. Yet it is readily verifiable in the laboratory through the measurement of the early exponential increase of a test alga in prepared media offering nutrients in differing mutual ratios but at initially saturating concentrations. I am unaware of any publication that draws attention to this behaviour. However, I am most grateful to Dr Catherine Legrand, of the University of Kalmar, Sweden, for her permission to reproduce a graph that she presented at the 1999 Meeting of the American Society of Limnology and

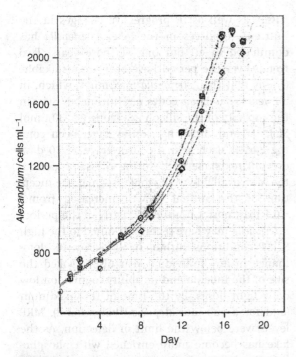

Figure 5.7 Initial increase in cell populations of *Alexandrium tamarense* KAC01 in laboratory cultures artificial media with widely differing ratios of N : P: ν – 160; – 16; λ – 1.6. Original data of Dr C. Legrand, replotted with permission. See text for further details.

Oceanography, held in Santa Fé, NM, USA. Her paper was concerned with the development of toxicity in a laboratory strain (KAC01) of the dinoflagellate *Alexandrium tamarense* in relation to the nitrogen and phosphorus resources supplied. Replicate treatments were grown in artificial media, at $17 \pm 1\,^{\circ}$C and subject to a 16-h light : 8-h dark cycle, differing only in the concentrations at which nitrogen and phosphorus were supplied. Media contained either 20 μM N and 0.75 μM P (i.e., N : P = 160), or 120 μM N and 7.5 μM (N : P = 16) or 12 μM N and 7.5 μM P (N : P = 1.6). The results, summarised in Fig. 5.7, refer to the increases in cell concentration of algae grown in the three media. They show clearly that initial growth performances in the three media were indistinguishable, though, not surprisingly, they diverged as the experiment progressed. (Note: When Dr Legrand gave her permission to use her data, it was in her full knowledge

of the point I wished to illustrate; however, she does not necessarily share my interpretation.)

In nature, phytoplankters inhabit dynamic environments, far removed from the contrived and quasi-steady states of the laboratory, and algal consumption will, in many instances, take one or another of the free resources to below the concentration representing its competitive thresholds. It is within this region of potential growth-rate limitation, that Tilman-type resource-based competition is expected to be strongly expressed. Thus, with external concentrations of (say) MRP falling to below 0.1 μmol P L^{-1}, *Asterionella* may maintain a rate of replication close to the maximum that the temperature and photoperiod may allow, yet *Cyclotella*, once its reserves are depleted, is able to maintain only a fraction of its resource-saturated growth rate. Given initial parity, its abundance relative to *Asterionella* is set to fall quickly behind. As the MRP concentration falls yet further, *Cyclotella* may cease to increase at all, while *Asterionella* is still absorbing phosphorus that is now effectively denied to the 'outcompeted' *Cyclotella*. Very soon, of course, the MRP is too depleted to be able to support further growth of *Asterionella* either, and the performance failure is still more abrupt. All this is predicted by the species-specific growth-rate dependence on P concentration (Fig. 5.6).

Against gradients of falling silicon concentration, *Cyclotella* can continue to absorb silicic acid from concentrations already severely constraining *Asterionella*. The outcome of interspecific competition betwen other pairs of diatoms may be verifiably predicted according to the model of Tilman *et al.* (1982). Another, quite separate, illustration of resource competition was described by Spijkerman and Coesel (1996b). Of three species of planktic desmid grown in continuous-flow cultures under stringent P-limiting conditions, one – *Cosmarium abbreviatum* – showed superior affinity for phosphorus at low concentrations over the other two (*Staurastrum pingue*, *S. chaetoceras*), even though these have faster maximum P-uptake rates. The outcome of competition in mixed chemostat cultures conformed to the prediction that, at concentrations of <0.02 μmol P L^{-1}, *Cosmarium* outcompeted the *Staurastra* but, at higher delivery rates, their faster rates of uptake and growth allowed the *Staurastra* to dominate the numbers of *Cosmarium*. If instead of low continuous supply, phosphorus was supplied in single daily doses of 0.7 μmol P L^{-1}, superior uptake rates enabled the *Staurastrum* species to sequester relatively more of the pulsed resource supply (velocity/storage then proving more advantageous than high affinity).

In none of these cases is the outcome attributable to anything other than the absolute characteristics of the critical resource supply to the needs and sequestration abilities of the organism(s) concerned. In none is the outcome the direct consequence of the ratio of resources available or of the rate at which critical resources are supplied. Thus, at this stage too, *it is not the ratio of available resources that determines the outcome*. The resource ratio is an interpretative convenience in identifying which of two scarce resources is likely to be, or to become, limiting and it aids the understanding of simultaneous limitation of coexisting species by different resources, provided both are below their respective critical thresholds. When the limiting concentrations of both resources are exceeded, the interspecific competition for those resources is correspondingly diminished, as both satisfy their immediate needs without interference to the other, and the likelihood of the one excluding the other is minimised. The explicit prediction of coexistence, inserted in the top right-hand corner of Fig. 5.6d, is correct but the explanation is different from that applying towards the bottom left-hand side.

There is a further perplexity over the general assumption of resource competition among species, which I have aired at length in certain earlier publications (Reynolds, 1997b, 1998a). From a comparison of the interannual variations in the dynamics and composition of the phytoplankton through successive spring phytoplankton blooms in Windermere since 1945, several dynamic characteristics have been recognised (Maberly *et al.*, 1994; Reynolds and Irish, 2000). Despite interannual differences in temperature, rainfall and stratification, in the size of the inocula and the rates of growth attained by each of several species of diatom (*Asterionella formosa*, *Aulacoseira subarctica* and *Cyclotella praeter-*

missa), as well as in progressive changes in the nutrient resources in the lake, *Asterionella* has dominated in all but one of those years. Two trends over the period have been unmistakable. One is that the *Asterionella* maximum, which, in all years, has been contained ultimately within the capacity of the silicon availability (~30 μmol Si L^{-1} (Lund, 1950) has, on average, been coming earlier each year (by an average of 30 days over 40 years). Second, during the same period, the MRP available to phytoplankton at the inception of the annual growth increased from ~0.1 to almost 1 μmol P L^{-1}. Early in the period, *Asterionella* may have seem well suited to the high Si : P conditions, although, in reality, its dominance in any individual year also invoked the size of the inocula and its ability to grow on low daily light doses. By the time of the maximum (and for a substantial period beforehand), MRP levels were below the limit of detection. As the lake has become more enriched with phosphorus, the *Asterionella* has continued to dominate the early growth stages but it has been able to maintain an accelerating growth rate, all the way to the division that finally reduces the silicon to concentrations limiting its ability to take it up (Reynolds, 1997b). Under these limiting and now relatively low Si : P conditions, should we not expect resource competition to alter the outcome of the spring growth? Reynolds' (1998a) simple model envisaged a typical standing population of *Asterionella* of 4 × 10^6 cells L^{-1} at the time that the silicon concentration is lowered to the point (8 μmol L^{-1}) where its growth rate is increasingly regulated by the rate of silicon uptake. Its next and possibly last doubling (it might take 4–7 days to complete) would require all of the remaining silicon in the water. At the same time, the subdominant *Cyclotella*, at no more than 0.8 × 10^6 cells L^{-1}, is experiencing neither phosphorus nor silicon limitation and maintains the maximum growth rate that the temperature and the light regime will allow, as predicted by the Tilman model. The difficulty is that before the altered competitive basis can be expressed at the level of community composition, the silicon is effectively exhausted. There is no advantage to better competitors for a non-existent resource: *Asterionella* still dominates, despite its (by now) competitive inadequacies.

Table 5.2 Phytoplankters tolerant or indicative of chronically oligotrophic conditions in lakes and the upper mixed layers of tropical oceans

Lakes

Cyanobacteria	Synechococcoid picoplankton
Chlorophyceae	*Chlorella minutissima, Coenocystis, Coenochloris (Eutetramorus), Sphaerocystis, Oocystis* aff. *lacustris, Willea wilhelmii, Cosmarium, Staurodesmus*
Chrysophyceae	*Chrysolykos, Dinobryon cylindricum, Mallomonas caudata, Uroglena* spp.
Bacillariophyceae	*Aulacoseira alpigena, Cyclotella comensis, C. radiosa, C. glomerata, Urosolenia eriensis*

Sources: Pearsall (1932), Findenegg (1943); Reynolds (1984b, 1998a); Hino *et al.* (1998), Huszar *et al.* (2003).

Oceans

Cyanobacteria	*Prochlorococcus, Synechococcus, Trichodesmium* spp.
Bacillariophyceae	*Rhizosolenia, Bacteristrum, Leptocylindrus*
Haptophyta	*Emiliana, Gephyrocapsa, Umbellosphaera*
Dinophyta	*Amphisolenia, Dinophysis, Histoneis, Ornithocercus* spp.

Sources: Riley (1957), Campbell *et al.* (1994), Karl (1999), Smayda and Reynolds (2001), Karl *et al.* (2002).

Chronic nutrient deficiencies

There is, however, another way in which resource-based competition shapes communities, at least in oligotrophic waters in which the availability of one or more resources waters is chronically deficient and where species having high uptake affinities for the limiting nutrient(s) are likely to thrive at the expense of inferior competitors. These species are also more likely to provide the inocula in future seasons, so the competitive outcome is magnified from generation to generation. In this way it is relatively easy to hypothesise that the selective traits most favoured in chronically oligotrophic systems – high affinity for limiting nutrients (Sommer, 1984) and small organismic size that is independent of high turbulent diffusivities for their delivery (Wolf-Gladrow and Riebesell, 1997; Section 4.2.1) – select for the relative absence of large species, for the high incidence of nanoplankters (Gorham *et al.*, 1974; Watson and Kalff, 1981) and, especially, picoplankters, both in lakes (Zuniño and Diaz, 1996; Agawin *et al* 2000, Pick 2000) and the open sea (Chisholm *et al.*, 1992; Campbell *et al.*, 1994; Karl, 1999; Karl *et al.*, 2002; Li, 2002). Thus, distinctive groups of species come to characterise severely P-deficient lakes and ultraoligotrophic oceans (see Table 5.2).

It may well be the case that the lakes can be described accurately as having high N : P

ratios. The extent to which the N : P gradient underpins, or even correlates with, compositional patterns in natural lakes has still to be fully resolved. Increasing the phosphorus loads relative to those of nitrogen (thus lowering N : P) does result in compositional shifts, often towards dominance by nitogen-fixing Cyanobacteria, as has been shown in numerous whole-lake fertilisations by Schindler (see, e.g., 1977) and his co-workers. The possible responses to simultaneous enrichment with nitrate or ammonium (raising or maintaining N : P) include frequent incidences of enhanced production of green algae (especially Chlorococcales and Volvocales) but they are frequently confounded by altered carbon dynamics and altered trophic effects attributable to the activities of grazers (Chapter 6). Seasonal changes in resourcing (especially with respect to changing availabilities with depth) may prompt compositional shifts that may be influenced less by ratios than by such mechanisms as highly specialised affinities, alternative sources of nutrients (nitrogen fixation, facultative bacterivory and mixotrophy) or vertical migration.

The idea that species that are simultaneously limited by different resources do not compete but, rather, coexist successfully has a wide following among plankton ecologists. It conforms to Hardin's (1960) *principle of competitive exclusion*, which states a long-standing ecological tenet that true competitors cannot coexist. One of its corollaries (that, in a steady state, each truly coexistent species occupies a distinctive niche Petersen 1975) has been advanced to account for the multiple species composition of natural phytoplankton species assemblages. The resource-based competition model appears to be powerfully supportive and, provided the limiting conditions persist uninterrupted and for long enough, will lead to the competitive exclusion of all but the fittest species. However, models attempting to verify this provision by simulating rather more than two species competing for more than two limiting resources seem to break down into chaos with unpredictable outcomes (Huisman and Weissing, 2001).

Discontinuous nutrient deficiences

The pattern of change in many pelagic systems is that, as a result of seasonal mixing, storm episodes or periodic inflows, the levels of all nutrients may be raised sufficiently to support the vigorous growth of various species of phytoplankton. The consequence is that the available resources are depleted, one or more to a level that represents a threshold of limitation and the ability of some or all species to grow becomes subject to stress. The initial combination of relative resource-richness and high insolation supports strong growth of many species in the surface mixed layer. Consequential resource depletion tends to proceed from the mixed layer downwards, leading eventually to the progressive uncoupling of resources from light. The water column progressively segregates into an increasingly resource-depleted upper layer and to deeper, increasingly light-deficient layers, wherein available nutrients persist pending exploitation by autotrophs.

Such sequences of events are held to differentiate among the adaptive traits and species-specific performances of phytoplankton, deeply influencing the species selection and seasonal shifts in species dominance. These progressions of composition and dominance are sufficiently striking, in some cases, to have been analogised to classical ecological successions in terrestrial plant communities (Tansley, 1939; Reynolds, 1976a; Holligan and Harbour, 1977). The most suc-

cessful of the early colonist and pioneer species of the pelagic succession have to be poised to invade or to exploit the favourable conditions that open to them. They may ascend to pre-eminence through an ability to grow faster than their rivals (to 'outperform' them but not to 'outcompete' them). Sooner or later, however, growing demands impinge upon the supply of one or more essential components, after which continued dynamic success requires slightly more specialist adaptations for resource gathering. For instance, noticeably more 'eutrophic species' might start to exclude oligotrophic ones on the basis that the reserve of carbon dioxide is depleted as a consequence of plankton growth. Another possibility is that sufficient growth is accommodated for the light availability in the surface mixed layer to become contested by superior light-harvesters. A third likelihood, to which particular attention is now given, is that algae that are able to penetrate deeper in the water column gain access to nutrients located at depth in the increasingly segregated vertical structure.

The surface mixed layer is an important entity, in limnology as in oceanography. The frequency with which it is turned over ($\sim$45 minutes or less: see Section 2.6.5) is much shorter than the generation time of the plankters embedded within the layer. It is quite proper, when considering planktic populations, to regard the surface mixed layer as a single, isotropic environment. In cold or shallow waters exposed to moderate wind stress, the surface mixed layer extends to the bottom of the water column, or to within a millimetre or two of the boundary layer therefrom. Beyond the density gradient separating mixed layer from deeper, denser water masses, many properties of the habitat can differ markedly from those of the surface mixed layer: renewal rate, temperature, insolation, gas and nutrient exchange rates, and so on. Thus, the variability in its vertical extent can also be a critical determinant of performance, alternately entraining and randomising the planktic population through a column of uniform and increasingly low insolation, then disentraining it through stagnating layers. The frequency of the alternation can be critical too: segregation may last for a few hours (as in diel ly stratifying systems), a few days (polymictic and atelomictic systems), months on end (seasonal stratification) or more or less continuously (as in meromictic systems). The longer is the separation, the greater can be the differentiation with, instead of one, unique environment, a continuous and widening spectrum of individualistic microhabitats (Reynolds, 1992c; Flöder, 1999). Simultaneous feedbacks, including the transfer of settling biomass, cadavers and the faecal pellets of algal consumers to the uninsolated layers below, may further enhance the developing structural discontinuity.

Among the species attempting still to assemble biomass and replicate their numbers, there is a progressive transfer of advantage from those adept at exploiting the mixed-layer resource base (high-sv^{-1} species, sustaining high replication rates) to those which are equipped to benefit from the separation of the resource base. The situation is reminiscent of the progressive elevation of productive terrestrial foliage from the herb layer to the woodland canopy. In the case of the pelagic, however, it is the downreach of the resource-gathering capacity that is responsible for the functional separation rather than the uplift of the light-harvesting apparatus! The investment by (appropriately adapted) terrestrial plants is in building the mechanical connection. Among the microscopic pelagic plants the adaptive investment is in migration.

Two interrelated sets of adaptations give the advantage to pelagic canopy species. One is the power of motility: if the alga is to have any prospect of covering the vertical distance separating light and nutrient resources, it must be able to determine the direction of its movement. Flagellate genera of the Cryptophyta, the Pyrrhophyta, the Chrysophyta, the Euglenophyta and the Chlorophyta would appear to have the essential preadaptations, although the buoyancy-regulating mechanism of gas-vacuolate Cyanobacteria is just as effective a means of propelling migratory movements. The second adaptation, however, is the one that turns the ability to move into the ability to perform substantial vertical migration, i.e. the size-determined capacity to disentrain from residual turbulence (Sections 2.7.1, 2.7.2). Just as small size and maximun form resistance are essential to embedding, so large size and low form resistance are advantageous to ready disentrainment from weakening turbulence and to commencing controlled, directed excursions through the water column.

The velocities achieved can nevertheless seem impressive. Smaller gas-vacuolate, bloom-forming Cyanobacteria (including *Microcystis*, *Gomphosphaeria*, *Gloeotrichia*, *Nodularia*, *Cylindrospermopsis*, raft-forming *Aphanizomenon* and the species of *Anabaena* and *Anabaenopsis* which typically aggregate into secondary tangles of filaments) whose buoyant velocites may reach 40–100 μm s^{-1} normally sink three to six times more slowly (say 0.6–3.0 m d^{-1}) (Reynolds, 1987a). The movements of large freshwater dinoflagellates like *Ceratium hirundinella* and *Peridinium gatunense* can cover 8–10 m in a single night (Talling, 1971; Pollingher, 1988). The rates quoted for many marine species exceed 100 μm s^{-1}; one or two exceed 500 μm s^{-1} (Smayda, 2002), although the distances they travel are not given. *Volvox* undertakes the longest reported circadian migration of any freshwater flagellate, traversing 17 m in either direction of the Cabora Bassa Dam, Mozambique (Sommer and Gliwicz, 1986), at an average velocity scarcely under 1.5 m h^{-1}. In absolute terms, this is modest but, at the scale of the organism, progression at the rate of 1 to 3 colony-diameters per second is impressive.

The question has to be asked whether these migrations do actually yield a harvest of nutrients, sufficient to provide a dynamic advantage over non-migrating species. According to Pollingher (1988), patterns of movement are dominated by a strong, positive phototaxis in the early part of the day (though, generally, their movements will avoid supersaturating light intensities) but they show a decining photo-responsiveness during the course of the solar day. The quality of the light and temperature gradients, the extent of nutrient limitation and the age of the population also influence patterns of movement. The extent of dinoflagellate migrations in a given lake are said to increase with decreasing epilimnetic nutrients, provided the segregated structure persists and the excursions into deeper water provide reward. It is interesting, too, how diminishing nutrient resources should determine that less of the photosynthate produced by buoyancy-regulating Cyanobacteria ends in new cytoplasm and more goes to offsetting buoyancy, forcing organisms to sink *lower* in the water column (Sas, 1989). Note that shortage of carbon, like shortage of light, means less photosynthate is produced, so organisms become lighter and float closer to the surface. This principle has been demonstrated well in the observations and experiments of Klemer (1976, 1978; Klemer *et al.*, 1982, 1985) and Spencer and King (1985). These self-regulated movements of large plankters certainly seem to open the access to deep-seated nutrient stores. It is apparent, too, that their growth may be enhanced when conditions of near-surface nutrient depletion obtain and the range of vertical migration extends to depths offering replenishment.

There are few studies that provide compelling evidence that this is always the case (Bormans *et al.*, 1999). However, Ganf and Oliver (1982) showed, through observation and careful experimental translocation, that *Anabaena* filaments picked up substantial amounts of nutrient on their buoyancy-regulated excursions in the Mount Bold Reservoir, South Australia. Raven and Richardson (1984) considered the extra nutrients derived by a migrating marine *Ceratium* to be only weakly attributable to movement per se (see Section 4.2.1) and much more to the encounters with unexploited nutrients in the (to them) accessible parts of the water column. Deep-water reserves of phosphorus were shown to be within the facultative swimming ranges of *Ceratium* in Esthwaite Water, UK (Talling, 1971) and within the 'vertical activity ranges' of *Ceratium* (and, for a time, *Microcystis*) intercepted by sediment traps in Crose Mere, UK (Reynolds, 1976b).

Thus, strong, self-regulated motility is considered to offer significant advantages, providing opportunities for the selective garnering of the diminishing resources of a structured environment. Adapted species are enabled access to parts of the water column that other algae do not reach, or do not do so sufficiently quickly, or, having done so, cannot reverse their motion to recover a position in the euphotic zone. It should not be assumed that this is the only advantage. The ability to swim strongly and in controlled

direction enables an alga to recover vertical station very quickly in the wake of disruptive mixing events, when smaller flagellates or solitary buoyany regulating filaments take hours or days to do so (Reynolds, 1984c, 1989b). Another is that in the face of weak wind- or convective-mixing, the alga can be quite effective in self-regulating its vertical position in order to balance its photosynthetic production and its resource uptake with the rate of cell growth and replication. In this way the cell saves energy in fixing photosynthate, which, if it could not be made into proteins and new cell material, would otherwise have to be voided from the cell. Organisms which do this very well, such as *Microcystis*, not only rise to dominance but remain dominant for months (and even years on end) when the appropriate conditions persist (Zohary and Robarts, 1989).

Low insolation and growth-rate regulation

Under conditions of short photoperiods and low aggregate insolation, the problem for phytoplankton is defined by the point that the alga is no longer able to intercept and harvest sufficient light energy, or to invest it the recruitment of new protoplasm and daughter biomass, at a rate that the temperature and the nutrient supply will allow. Below this level, growth rate is, indeed, light-limited. The curves inserted in Fig. 5.4 differentiate among plankters on the basis of their shape and their capacity for low-light adaptation. The point to notice is that the species that are capable of the fastest rates of growth under relatively high insolation are not necessarily the best adapted to live on small light doses. The limnetic species that do this well include the diatom *Asterionella* and solitary filamentous species of the Oscillatoriales (*Planktothrix agardhii*, *Limnothrix redekei*), in which the capability is correlated with relative morpholgical attenuation (high msv^{-1}: Fig. 5.5). There is often a high capacity for auxillary and accessory pigmentation as well. Thus, their successful contention to perform relatively well in poorly insolated, natural mixed layers owes most to their extraordinary abilities to open the angle of r on I (Fig. 5.4; α_r in Fig. 5.5) and, thus, to lower the light intensity at which growth rate can be saturated. In the open, mixed-water column, this extends the actual depth through which growth-saturating photosynthesis may be maintained and, in turn, lengthens the aggregate of probabilistic photoperiod, t_p, over that expected for unadapted species in the same water layer. From the least-squares regression fitted to the data in Fig. 5.5,

$$\alpha_r = 0.257(msv^{-1})^{0.236} \qquad (5.8)$$

it may be predicted that the stellate colony of *Asterionella* generates a slope of $\alpha_r = 0.86$ (mol photon)$^{-1}$ m^2, while for a 1-mm thread of *Planktothrix agardhii*, $\alpha_r = 1.12$. For the small spherical cell of *Chlorella*, the slope is predicted to be only 0.39 (mol photon)$^{-1}$ m^2. Analogous to the interrelationships among photosynthetic rate and the onset of light saturation of photosynthesis (Eq. 3.5, Section 3.3.1), the lowest light dose that will sustain maximal growth rate at 20 °C is indicated by the quotient, r'_{20}/α_r. Thus, on the basis of the assembled data (see Table 5.1), we may deduce that *Chlorella* growth will be saturated by a photon flux of 3.34 mol photons m^{-2} d^{-1} (equivalent to a constant $\sim$39 µmol photons m^{-2} s^{-1}), that of *Asterionella* by 2.07 mol photons m^{-2} d^{-1} (24 µmol photons m^{-2} s^{-1}), and that of the *Planktothrix* by 0.77 mol photons m^{-2} d^{-1} (9 µmol photons m^{-2} s^{-1}).

The inference is emphasised: at irradiance levels exceeding levels of 3.34 mol photons m^{-2} d^{-1}, *Chlorella* is the 'fittest' of the three, and its regression-predicted growth rate of 1.84 d^{-1} outstrips those of *Asterionella* (1.78 d^{-1}) and *Planktothrix* (0.86 d^{-1}). However, at light levels equivalent to 0.77 mol photons m^{-2} d^{-1}, *Planktothrix* can still be argued to be able to maintain its maximum growth rate (0.86 d^{-1}), and when that of *Asterionella* is cut back to 0.66 d^{-1} and *Chlorella* is severely light-limited at not more than 0.42 d^{-1}. When the low temperatures of high-latitude winters are taken into account, the impact of surface-to-volume relationships modify the relative fitnesses of these organisms. At 5 °C, the predicted resource-saturated growth rates for *Chlorella*, *Asterionella* and *Planktothrix* are, respectively, 0.375, 0.335 and 0.163 d^{-1} and the respective saturating fluxes are calculated to be 0.96, 0.39 and 0.15 mol photons m^{-2} d^{-1}. Thus, under an average irradiance (I^*) of 0.4 mol photons m^{-2} d^{-1} photon flux (equivalent to a constant 5 µmol photons m^{-2} s^{-1}), the light-limited *Asterionella* might still increase at a rate of 0.335 d^{-1}, which is twice as fast as that of the temperature-limited *Planktothrix* or the light-limited *Chlorella*. It is at once appreciable how subtle are the conditions distinguishing among species performances under low doses of light. We might also speculate that although there is an apparent discretion in favour of diatoms in cold, energy-deficient, mixed layers, a little more vigourous mixing (I^* falls) or a less severe winter temperature might favour filamentous Cyanobacteria instead. Reduced mixing and better near-surface insolation immediately favours faster growing nanoplankters such as *Chlorella*.

Trait interaction and functional differentiation in phytoplankton

The real world of phytoplankton is a blend of deficiencies of differing intensity and frequency, especially with respect to the availability and accessibility of nutrient resources and the solar energy needed to process them. Specialist adaptatations, both in terms of physiological responses at the scale of the life cycle and the traits distinguished at the evolutionary scales, may increase the relative fitness of some species along particular gradients of environmental variability but none is well suited to all conditions. For instance, we may suppose that the most competitive adaptation would be to enable the phytoplankter to self-replicate more rapidly than other species that might be present; hence, a morphology conducive to rapid surface exchanges of nutrients should be favoured, that is, one that maintains a large sv^{-1}. The opposite trend of increasing size (and reducing sv^{-1}) carries advantages of motility, storage and persistence (see also Section 6.7), where the ability to influence vertical position, to gain access to nutrient resources unavailable to other species and to avoid consumption by herbivores offer superior prospects of survival. One advantage has been 'traded' against another, at the price of lowered habitat flexibility: some environments will be better tolerated, or even preferred, by a given species than will others. Such differentiation is, of course, the basis of patterns in the spatial

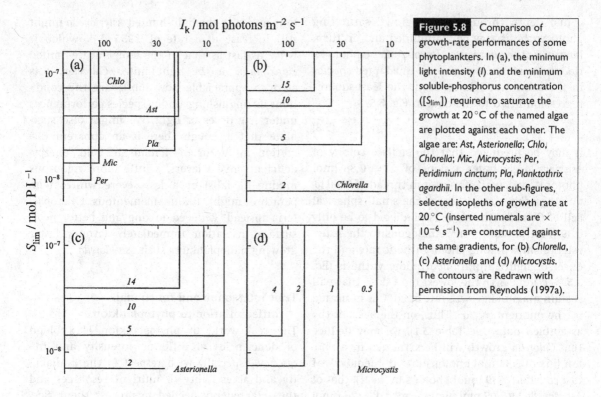

Figure 5.8 Comparison of growth-rate performances of some phytoplankters. In (a), the minimum light intensity (I) and the minimum soluble-phosphorus concentration ($[S_{lim}]$) required to saturate the growth at $20\,°C$ of the named algae are plotted against each other. The algae are: Ast, Asterionella; Chlo, Chlorella; Mic, Microcystis; Per, Peridimium cinctum; Pla, Planktothrix agardhii. In the other sub-figures, selected isopleths of growth rate at $20\,°C$ (inserted numerals are × 10^{-6} s^{-1}) are constructed against the same gradients, for (b) Chlorella, (c) Asterionella and (d) Microcystis. The contours are Redrawn with permission from Reynolds (1997a).

and temporal distribution of species, whereby some are more clearly associated with particular conditions than are others. The further adaptive option for larger algae – that of shape distortion that increases the surface area bounding a given cytovolume – provides not so much a compromise between small and large size but the enhanced ability to process resources into biomass in relatively short periods of exposure to light.

Based upon the growth rates of various species of algae against chosen dimensions in the foregoing sections, we are now able to devise comparative graphical representations of the replicative performances of algae against the two key axes of resource availability and insolation. In Fig. 5.8a, growth-rate contours of several algae are drawn in space defined by light and phosphorus saturation of growth-rate potential. The result is broadly similar to those built on mean underwater light levels and K_U values with respect to species-specific phosphorus uptake rates (Reynolds, 1987c) or of light supply and nutrient supply (Huisman and Weissing, 1995). The plots making up the rest of Fig. 5.8 show species-specific replication-rate contours

against axes representing steady-state concentrations of phosphorus and the photon flux of white light, at $20\,°C$. The high levels of resource and light required to saturate the most rapid growth of Chlorella show well against the requirements of four others species (Fig. 5.8a). The sensitivity of Chlorella performance to both light and phosphorus relative to that of Asterionella or of the poorly performing Microcystis against these two criteria is evident (Fig. 5.8b, c, d).

Growth and reproductive strategies

When growth under persistently low levels of light and nutrient are considered simultaneously, the basis for some of the very interesting patterns alluded to by Tilman et al. (1982) and by Sterner et al. (1997) may be readily appreciated. Now, for example, we may envisage circumstances under which growth rate in Asterionella is encountering (say) silicon limitation when the growth rate of Cyclotella is constrained by light, and when the growth rate of Planktothrix is too constrained by low temperature or low phosphorus to be able to take full advantage.

There are probably sufficient data to be able to simulate these interactions more rigorously. This is less interesting to pursue than it is to abstract the generalities about the differing adaptations shown by the algae considered here and the broad properties that underpin their strategies for growth and survival. The use of the word 'strategy' in the context of the evolution of life histories is open to criticism on etymological grounds, as it implies that their differention is planned or anticipated in advance (Chapleau et al., 1988). In reality, different patterns for preserving and reproducing genomes have evolved along with the organisms they regulate and, just as certainly, have been shaped by the same forces of natural selection. The patterns *are* distinctive, separating life histories that, for example, permit opportunistic exploitation of resources and photon energy (as does *Chlorella* in the example in the previous section) or, alternatively, may provide high adaptability to a low nutrient or to low energy supply (as does *Planktothrix*). The comparative efficiencies and flexibilities of investment of harvested energy and gathered resources into species-specific biomass define the *growth and reproductive strategies* of phytoplankton (Sandgren, 1988a).

So far, the discussion has identified three basic sets of strategic adaptations, involving morphologies, growth rates and associated behaviours. The first is the *Chlorella* type of exploitative or invasive strategy, in which organisms encountering favourable resource and energy fluxes can embark upon the rapid resource processing, biosynthesis and genomic replication (reproduction) that constitute growth. They necessarily have a high growth rate, r, based on an ability to collect and convert resources before other species do and, in this sense (the one followed by most plant ecologists), they are 'good competitors'. Curiously, plankton ecologists reserve this term for the 'winners' of the competition, applying it to those species that specialise in the efficient garnering, conserving and assembling the limiting resource base (K) into as much biomass as it will yield. Thus, the second set of strategic adaptations variously combines high resource affinity and/or specialist mechanisms for obtaining scarce or limit-

ing resources in short supply with their retention among a high survivorship. Unlike the obligately fast-growing, r-selected category, resource-(K-)selected species do not share the constraint of maintaining a high surface-to-volume (sv^{-1}) ratio. However, the acquisitive garnering of diminishing resources sometimes favours significant powers of migratory motility, for which a relatively large size (with attendant penalties in reduced sv^{-1}, slow growth rate and impaired light-absorption efficiency, ε_a) is essential (see Sections 2.7.1, 3.3.3). *Microcystis aeruginosa* provides a good example of this second type of strategy that identifies 'winners', or like Aesop's fabled tortoise, the 'good competitor' in the sense understood by most plankton ecologists (Kilham and Kilham, 1980).

The ability to harvest and process energy from low or diminishing irradiances or from truncated opportunities at higher irradiance is favoured by small size or by attenuation of larger sizes (in one or possibly two) planes. These traits represent a high photon affinity, which is not bound exclusively to either r- or K-selection, and to which Reynolds et al. (1983b) applied the term w-selection.

There are clear similarities and apparent analogies in these broad distinctions with the three primary ecological and evolutionary strategies identified among terrestrial plants (Grime, 1977, 1979, 2001). Grime's concept was built around the tenability of habitats according to (i) the resources available and the levels of stress on life cycles that resource shortages might impose on plant survival and (ii) the duration of these conditions, pending their disruption or obliteration by habitat disturbance. Of the four possible permutations of stress and disturbance (Table 5.3), one, the combination of continuous severe stress and high disturbance results in environments hostile to the establishment of plant communities is untenable. These are deserts! The three tenable contingencies are variously populated by plants specialised in either (a) rapid exploitation of the resources available ('competitors' in the original usage of Grime 1977, 1979), which he dubbed '*C*-strategists'; or (b) tolerance of resource stress, by efficient matching of the limited supply to managed demand, and

Table 5.3 | Basis for evolution of three primary strategies in the evolution of plants, phytoplankton and many other groups of organisms

Habitat duration	Habitat productivity	
	High	Low
Long	Competitors, invasive (**C**)	Stress-tolerant (**S**)
Short	Disturbance-tolerant ruderals (**R**)	No viable strategy

Source: Original scheme of Grime (1979; modified after Reynolds, 1988a; Grime, 2001).

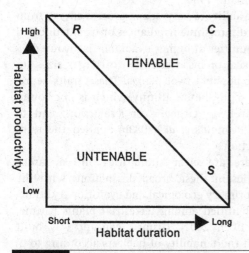

Figure 5.9 Grime's model of tenable and untenable habitats, and noting the primary (**C, S** or **R**) life-history strategies required to secure survival. Redrawn, with permission from Grime (2001).

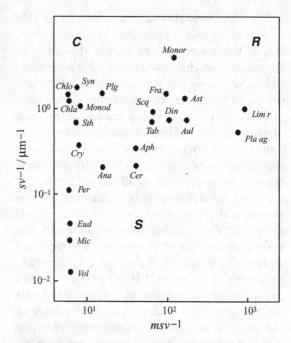

Figure 5.10 Morphological ordination of some species of freshwater phytoplanton, against axes invoking maximal linear dimension (m), surface area (s) and volume (v) of the vegetative units, with the **C**-, **S**- and **R**-strategic tendencies. The algae are: *Ana, Anabaena flos-aquae; Aphan, Aphanizomenon flos-aquae; Ast, Asterionella formosa; Aul, Aulacoseira subarctica; Cer, Ceratium hirundinella; Chla, Chlamydomonas; Chlo, Chlorella sp.; Cry, Cryptomonas ovata; Din, Dinobryon divergens; Eud, Eudorina unicocca; Fra, Fragilaria crotonensis; Lim r, Limnothrix redekei; Mic, Microcystis aeruginosa; Monod, Monodus sp.; Monor, Monoraphidium contortum; Per, Peridinium cinctum; Pla ag, Planktothrix agardhii; Plg, Plagioselmis nannoplanctica; Scq, Scenedesmus quadricauda; Sth, Stephanodiscus hantzschii; Syn, Synechococcus sp.; Tab, Tabellaria flocculosa var. asterionelloides; Vol, Volvox aureus*. Redrawn with permission from Reynolds (1997a).

to which he gave the label '*S*-strategists'; or (c) tolerance of disturbance, through making good opportunity of transient habitats and interrupted opportunities to process resources into biomass ('*R*-strategists').

The three primary strategies of Grime's **CSR** model form the apices of a triangular ordination (Fig. 5.9), which representation readily allows the accommodation of numerous intermediates and trait-combinations. Reynolds (1988a, 1995a) found only minor difficulties in analogising the *r*-, *K*- and *w*-selected groups to exemplifying,

respectively, **C, S** or **R** strategies, on the satisfying basis of agreement among the morphological properties, growth rates and life-history traits. The distribution of phytoplankton species according to their individual morphologies plotted against axes of sv^{-1} and msv^{-1} (Fig. 5.10). Just as with Grime's (1979) scheme, species are not exclusively **C** or **S** or **R** in their strategic adaptations. Many species of phytoplankton show intermediate characters. Interestingly, intermediacy in morphological and physiological characters matches well the intermediacy of their ecologies. The *C–S* gap is spanned by genera such as *Dinobryon, Dictyosphaerium, Coenochloris, Pseudosphaerocystis, Eudorina* and, arguably, *Volvox* (Reynolds, 1983b), and by *Aphanocapsa* and *Aphanothece*. The series spans diminishing sv^{-1} ratios, maximum growth rates and low-temperature tolerance but increasing ability to exploit and conserve nutrient resources. Algae in the *C–R* axis include predominantly centric diatoms of varying tolerance of turbidity and the *Scenedesmus–Pediastrum* element of enriched shallow ponds and rivers). The *R–S* possibility is represented by the slow-growing, long-surviving, acquisitive but highly acclimated species of density gradients, like *Planktothrix rubescens* and *Lyngbya limnetica*. Certain (not all) members of the genus *Cryptomonas* show a blend of the characteristics of all three primary strategies in being unicellular, having cells of moderate size ($1–4 \times 10^3$ μm^3) and of intermediate sv^{-1} ($0.3–0.5$ μm^{-1}), and being capable of intermediate replication rates ($r_{20} \sim 10 \times 10^{-6}$ s^{-1}; $r_0 \sim 0.9 \times 10^{-6}$ s^{-1}).

It is right to point out that Grime's *CSR* concept of plant stategies is not universally accepted and it has been subject of vehement and challenging debate (see Tilman, 1977, 1987, 1988; Loehle, 1988, *a.o.*). Although there is much common ground shared by the adversaries and, in truth, the differences are more of perspective and emphasis (Grace, 1991), the differences have never been entirely resolved. The application to plankton has not been so criticised and some (Huszar and Caraco, 1998; Fabbro and Duivenvorden, 2000; Gosselain and Descy, 2000; Kruk *et al.*, 2002; Padisák, 2003) but by no means all (Morabito *et al.*, 2002), have found the arguments convincing and helpful to interpretation. The applicability of a scheme devised for plant species is not a barrier: it is now quite evident that the idea has a long pedigree among other ecological schools (Ramenskii, 1938) and has been applied successfully to the 'violent', 'patient' and 'explerent' strategies of zooplankton (Romanovsky, 1985). The *CSR* model has been applied to fungi (Pugh, 1980) and periphyton (Biggs *et al.*, 1998).

An updated application to phytoplankton is set out in Box 5.1. A notable modifica-

tion recognises that motility and large size are not necessary adaptations to function in chronically very resource-depleted pelagic environments. Indeed, resource gathering in spatially continuous, rarefied environments is favoured by small size, whereas the low levels of diffuse biomass is an unattractive resource for direct grazing by mesozooplankters (see Chapter 6). The adaptive strategies for surviving the 'resource desert' of the ultraoligotrophy of the oceanic pelagic are accorded the additional stress-tolerant category *SS*.

The original ascriptions of *C*, *S* and *R* categories to phytoplankton (Reynolds, 1988a) separate quite satisfactorily on the plot of the areas projected by various species of phytoplankton and the product of maximum dimension and surface-to-volume ratio (msv^{-1}) Fig. 3.12). Near-spherical forms align close to msv^{-1} $[d \times 4\pi(d/2)^2 \div 4\pi(d/2)^3/3] = 6$ but separate broadly in to *C* and *S* species according to size, because the carbon and chlorophyll contents vary with $v = 4\pi(d/2)^3/3$ but the light interception increases as a function of the disk area, $a = \pi(d/2)^2$. The morphological attenuation of the *R* species pulls out the plot to much higher msv^{-1} values. Thus, we distinguish species that are capable of rapid growth in benign, resource-replete environments, those that are able to go on squeezing out increased biomass from diminishing light income and those who are physiologically or behaviourally adapted to function in spite of developing nutrient stress. The model appears in various guises later in the book, demonstrating the power and flexibility of the strategy–process–ecosystem interactions. It even provides the bridge to the light : nutrient hypothesis (Sterner *et al.*, 1997) in so far as the species best adapted to cope with low doses of I^* are most able to cope with high particulate content in the water and the C : P ratio of the seston available to secondary consumers.

5.4.6 Resource exhaustion and survival

It is reasonable to assume that the growth of phytoplankters distinguished by efficient, high-affinity resource-gathering capabilities may continue until they deplete their growth-limiting resource to near exhaustion. It was often and

---

Box 5.1 | Summary of behavioural, morphometric and physiological characteristics of growth and survival strategies of freshwater phytoplankton

With little adjustment, the primary strategies (otherwise, functional types) of plants devised by Grime (1979, 2001) are known to apply to other types of organism, including phytoplankton. The application of the scheme to plankton (Reynolds, 1988a) required some modest adjustment but the relevant morphological and physiological characteristics are, of course, peculiar to planktic algae. These are listed below, following Reynolds (1988a, 1995a) but include the features of a new subcategory (**SS**) to accommodate features of permanent stress-tolerant algae of ultraoligotrophic oceans.

C strategists

Grime's name	Competitors
Reynolds' (1995a) label	Invasive opportunists
Dispersal	Highly effective, cosmopolitan; mechanisms sometimes obscure
Selection	r
Cell habit	Mostly unicellular
Unit sizes	10^{-1}–10^3 μm^3
msv^{-1}	6–30
Cell projection	>10 m^2 (mol cell C)$^{-1}$
r'_{20}	$>10 \times 10^{-6}$ s^{-1}; >0.9 d^{-1}
r'_0	$>2 \times 10^{-6}$ s^{-1}; >0.18 d^{-1}
Q_{10}	<2.2

Species experience low growth thresholds for light (Section 5.5), have generally low rates of sinking (some are motile; Section 6.3) and are highly susceptible to grazing zooplankton (Section 6.4). Representative genera *Chlorella, Ankyra, Chlamydomonas, Coenocystis, Rhodomonas*.

R strategists

Grime's name	Ruderals
Reynolds' (1995a) label	Attuning or acclimating (also processing–constrained)
Dispersal	Widely distributed, mechanisms sometimes obscure
Selection	r and K (w in Reynolds et al., 1983b)
Cell habit	Some unicellular; many coenobial
Unit sizes	10^3–10^5 μm^3
msv^{-1}	15–1000
Cell projection	8–30 m^2 (mol cell C)$^{-1}$
r'_{20}	$>10 \times 10^{-6}$ s^{-1}; >0.85 d^{-1}
r'_0	0.08–2×10^{-6} s^{-1}; >0.1 d^{-1}
Q_{10}	2.0–3.5

Species force very low growth thresholds for light (Section 5.6), sinking rates low to high, most are non-motile (Section 6.2); some susceptible to grazing zooplankton

(Section 6.3). Representative genera *Asterionella, Aulacoseira, Limnothrix, Plank-tothrix.*

S strategists

Grime's name	Stress-tolerators
Reynolds' (1995a) name	Acquisitive (also resource-constrained).
Dispersal	Tendency to discontinuous distribution, mechanisms better known
Selection	Strongly K
Cell habit	Some unicellular; many coenobial
Unit sizes	10^4–10^7 μm^3
msv^{-1}	6–30
Cell projection	<2.5 m^2 (mol cell C)$^{-1}$
r'_{20}	<8 × 10^{-6} s^{-1}; <0.7 d^{-1}
r'_0	<1 × 10^{-6} s^{-1}; <0.09 d^{-1}
Q_{10}	>2.8

Species contend effectively with diminishing, or increasingly distant nutrient resources, either through exploiting alternative sources (nitrogen fixation, phosphatase production, phagotrophy on bacteria or particulate organic material). Most are motile, some are strongly so, sinking rates low but may undertake controlled migrations over large vertical distances. Also referred to as resource 'gleaners' (Anderies and Beisner, 2000). Representative genera *Microcystis, Anabaena, Gloeotrichia, Ceratium, Peridinium, Chrysosphaerella, Uroglena.*

SS strategists

Grime's name	(not applicable)
Reynolds' (1995a) name	Chronic-stress tolerant
Dispersal	Cosmopolitan
Selection	Ultimately K
Cell habit	Exclusively unicellular
Unit sizes	≤4 μm^3
msv^{-1}	6–8
Cell projection	~50–60 m^2 (mol cell C)$^{-1}$
r'_{20}	unknown, probably >20 × 10^{-6} s^{-1}; >1.8 d^{-1}
r'_0	Possibly up to 0.5 × 10^{-6} s^{-1}, 0.4 d^{-1}
Q_{10}	Not known, probably ~2

This newly separated group of species tolerant of chronic nutrient stress accommodates the prokaryotic picoplankters that dominate the rarefied environments of the tropical seas and which, increasingly, have been shown to be active in the open waters of the world's largest and most oligotrophic lakes (Reynolds *et al.*, 2001). They are non-motile but have very low sinking rates. Their small size is the key to living on very dilute nutrient sources. It would leave them very vulnerable to grazing by filter-feeders, except that they inhabit environments that fail to sustain filter-feeding zooplankton. Representative genera *Prochlorococcus* (in the sea; *Cyanobium, Cyanodictyon* are considered to be limnetic analogues).

commonly supposed by many plankton ecologists that nutrient exhaustion is followed by mass clonal mortalities. This view was perhaps encouraged by numerous observations of 'bloom collapse', of diatoms running out of silicon (e.g. Moed, 1973) or the photolysis of surface scums of Cyanobacteria (e.g. Abeliovich and Shilo, 1972). These relatively impressive eventualities apart, however, phytoplankters are rather better prepared than this to be able to avoid sudden death and disintegration. Depletion of one of the essential resources usually leads to a cellular reserve of the others and the cell may be able to use stored carbohydrate, polyphosphate or protein reserves to maintain some essential activity. However, it is quite clear that it is better for the cell to lower its metabolism and to close down those processes not directly associated with actually staying alive. Earlier chapters have emphasised the mechanisms for internal communication of nutrient-uptake activity of the membrane- transport system (Section 4.2.2), the activation of inhibitory nucleotides (such as ppGpp) in response to falling amino-acid synthesis (Section 4.3.3) and the suspension of nuclear division (Section 5.2.1). Each represents a step in the biochemical procedure by which the cell senses its environmental circumstances and organises its appropriate defences to enhance its survival prospects. These may include the inception of a 'cytological siege economy' and the structural reorganisation of the protoplast into resting cells, with or without thickened walls.

The biological forms of most kinds of resting cell are well recognised by plankton ecologists and, in many cases, so are the environmental attributes which induce them. Equally, the implicit benefit of survival of resting stages is widely accepted as a means to recruit later populations from an accumulated 'seed bank'. They need to recognise and respond to their reintroduction into favourable environments or to ameliorating conditions by embarking upon a phase of renewed vegetative growth. However, it has to be stated that, in marked contrast to the efforts that have been made to observe and understand the mechanisms generating the spatial and temporal patterns of phytoplankton occurrence, detailed information on the significance of reproductive and resting propagules has been mainly confined to studies on particular phylogenetic groups (Sandgren, 1988a). It is not inappropriate to give a brief perspective at this point.

Resting stages come in a variety of forms and are stimulated by a variety of proximate events and circumstances, and their success in 'carrying forward' biomass and genomes is also quite variable. Among the simplest resting stages are the contracted protoplasts produced in such centric diatoms as *Aulacoseira* (Lund, 1954) and *Stephanodiscus* spp. (Reynolds, 1973a). These form quite freely in cells falling into aphotic layers and may be prompted by microaerophily and low redox, which conditions may be tolerated for a year or more. The contents pull away from the wall, abandon the central vacuole and shrink to a tight ball, a micrometre or so in diameter. Individual cells or filaments containing resting stages litter the surface sediments. If seeded sediment is placed under low light in the laboratory, *Aulacoseira* will 'germinate' and produce swathes of new filaments *in situ*. Germination in nature may be only a little less spectacular but it always depends upon the resuspension of filaments and cells by entrainment from sediments accessible to turbulent shear. Thus, formation and germination of the resting stages is governed by the activity or otherwise of its photosynthetic capacity. Perhaps 5–20% of the sedimenting population may form resting stages. The percentage of these that return to the plankton is probably small but they can provide quantitatively important inocula to future populations (Reynolds, 1988a, 1996b).

The Cyanobacterium *Microcystis* has the ability to control its own vertical migrations through regulating its buoyancy and, in warm latitudes, it may move frequently (perhaps dielly) between sediment and water, very much as part of its vegetative activity (May, 1972; Ganf, 1974a; Tow, 1979). In temperate lakes, *Microcystis* is frequently observed to overwinter on the bottom sediments (Wesenberg-Lund, 1904; Gorham, 1958; Chernousova *et al.*, 1968; Reynolds and Rogers, 1976; Fallon and Brock, 1981; and many others reviewed in Reynolds, 1987a). There is a massive autumnal recruitment of vegetative colonies from the plankton to depth (Preston *et al.*, 1980), where they enter a physiological resting stage.

No physical change occurs (they do not encyst) and chlorophyll, as well as a latent capacity for normal, oxygenic photosynthesis, is retained (Fallon and Brock, 1981). Curiously, the cells also remain gas-vacuolate. Despite being (initially) loaded with glycogen, other carbohydrates, proteinaceous structured granules and polyphosphate (Reynolds et al., 1981), they would be buoyant but for the precipitation of iron hydroxide on the colony surfaces, which acts as ballast and causes the organisms to sink (Oliver et al., 1985). Once on the sediments, in very weak light and at low temperatures, they experience considerable mortalities, although some cells live on under these conditions, apparently for several years in some cases (see Livingstone and Cambray, 1978). The surviving cells function at a very low metabolism and are tolerant of sediment anoxia (and consequent re-solution of the attached iron) but there are, by now, too few of them to lift the erstwhile colonial matrix back into the water column.

Reinvasion of the water column follows a phase of in-situ cell division, in which clusters of young cells are formed, constituting a pustule-like structure that buds out of the original, 'maternal' mucilage matrix, until it is released or it escapes into the water. The process was described originally by Wesenberg-Lund (1904), but the information was largely ignored. The 'nanocytes' found by Canabeus (1929) and, later, 'rediscovered' by Pretorius et al. (1977), seem to refer to the young, budding colonies. Sirenko (pesonal communication quoted in Reynolds, 1987a) has viewed the entire sequence, claiming that the potential mother cells are identifiable in advance by their larger size and more intense chlorophyll fluorescence. The process has also been reproduced under controlled conditions in the laboratory (Cáceres and Reynolds, 1984), using material sampled from autumnal sediment. It requires the exceedence of a temperature and insolation threshold and it occurs more rapidly while anaerobic conditions persist. These conditions have to be mirrored in natural lakes of the temperate regions before Microcystis colonies begin to be recruited to the water column in the spring. Sediments have to retain colonies through the winter period, where colonies apparently need the low temperatures and low oxygen levels for their maturation. Ultimately, they also require low oxygen levels and simultaneous low insolation to persuade them to initiate the formation of the new colonies that recolonise the water column in the following year (Reynolds and Bellinger, 1992; Brunberg and Blomqvist, 2003). The completion of this cyclical process depends on interactions among light, temperature and sediment oxygen demand. Whereas upwards of 50% of the colonies constituting the previous summer maximum number of colonies may settle to the sediments, $\leq 10\%$ might contribute to the re-establishment of a summer population the following year (Preston et al., 1980; Brunberg and Blomqvist, 2003; Ishikawa et al., 2003).

It may be noted that Microcystis colonies also survive in microenvironments created by downwind accumulations of surface scums on large lakes and reservoirs, especially where warm summers, high energy inputs and high upstream nutrient loadings are simultaneously prevalent. Good examples come from the reservoirs of the Dnieper cascade (Sirenko, 1972) and the Hartbeespoort Dam in South Africa (Zohary and Robarts, 1989). The conditions in these thick, copious 'crusts' or 'hyperscums' are effectively lightless and strongly reducing (Zohary and Pais-Madeira, 1990) but, save those actually baked dry at the surface, Microcystis cells long remain viable and capable of recovering their growth.

Many species respond to the fabled 'onset of adverse conditions' by producing morphologically distinct resting propagules. Among the best known are the cysts of dinoflagellates, which are sufficiently robust to persist as a fossils of palaeontological significance (for a review, see Dale, 2001). Some 10% of the 2000 or so marine species are known to produce resting cysts. In some instances, they are known, or are believed, to be sexually produced hypnozygotes. The cell walls in many species contain a heavy and complex organic substance called dinosporin, chemically similar to sporopollenin of higher-plant pollen grains. Some species deposit calcite. In the laboratory, cyst formation may, indeed, be induced by nutrient deprivation and adverse conditions but the regular, annual formation of cysts in nature (coastal waters, eutrophic lakes) possibly occurs in response to cues that anticipate 'adverse' conditions rather than the actual onset of those adversities. The protoplasts of newly formed cysts usually contain conspicuous reserves of lipid and carbohydrate, accumulated during stationary growth (Chapman et al., 1980). The number of cysts produced by freshwater Ceratium hirundinella in autumn has been estimated from the sedimentary flux to account for ≤35% of the maximum standing crop of vegetative cells (Reynolds et al., 1983b). The success in recruiting vegetative cells from excysting propagules in the following spring is, in part, proportional to the abundance of spores retained from the previous year (Reynolds, 1978d; Heaney et al., 1981).

The excystment of vegetative cells from cysts was first described by Huber and Nipkow (1922). Much detail has been added from such landmark micrographic investigations as those of Wall and Dale (1968) and Chapman et al. (1981). A naked flagellate cell, or gymnoceratium, emerges through an exit slit and soon acquires the distinctive thecal plates of the vegetative cell. Heaney et al. (1981) noted a sharp, late-winter recruitment of new, vegetative cells of Ceratium to the plankton of Esthwaite Water, UK, after the water temperature exceeded 5 °C, and coincident with an abrupt increase in the proportion of the empty cysts recoverable from the bottom sediments of the lake.

Among the Volvocales, sexually produced zygotes of (e.g.) Eudorina (Reynolds et al., 1982a) and Volvox (Reynolds, 1983b) have the robust appearance of resting cysts and, indeed, serve as perennating propagules between population maxima. Deteriorating environmental conditions may trigger the onset of gametogenesis but formation of the eventual resting stages cannot be claimed certainly to have been consequential on resource starvation. Among the Chrysophyceae, there has evolved an opportunistic perennation strategy, involving zygotic and asexual cysts that are produced early in the growth cycle, when conditions are supposedly good (Sandgren, 1988b). This pattern of encystment apparently ensures the production of resting stages during what often turn out to be short phases of environmen-tal adequacy but which are tenanted briefly by vegetative populations.

In contrast, nostocalean Cyanobacteria produce their asexual akinetes in rapid response to the onset of physiological stress. Akinetes are the well-known 'resting stages' of such genera as Anabaena, Aphanizomenon and Gloeotrichia (Roelofs and Oglesby, 1970; Wildman et al., 1975; Rother and Fay, 1977; Cmiech et al., 1984). These, too, have typically thickened external walls, within which the protoplast remains viable for many years. Livingstone and Jaworski (1980) germinated akinetes of Anabaena from sediments confidently dated to have been laid down 64 years previously. On the other hand, rapid akinete production has been stimulated in the laboratory by the sort of carbon : nitrogen imbalance that occurs as a consequence of surface blooming, and from which conditions an effective means of escape is offered (Rother and Fay, 1979). Moreover, substantial germination can take place shortly (days rather than months or years) after akinete formation, provided the external conditions (temperature, light and, possibly, nutrients) are suitable (Rother and Fay, 1977). Reynolds (1972) observed that Anabaena akinetes were regularly resuspended by wind action in a shallow lake but failed to germinate before a temperature or insolation threshold had been surpassed. In other years, vegetative filaments surviving the winter were sufficient to explain the growth in the following season. These thresholds could be important to the distributions of individual species. The current spread of Cylindrospermopsis raciborskii from the tropics to continental lakes in the warm temperate belt may be delimited by a germination threshold temperature of 22 °C (Padisák, 1997). The akinetes of Gloeotrichia echinulata are able to take up phosphate through their walls and colonies germinating the following year can sustain substantial growth even when limnetic supplies are small (Istvánovics et al., 1993).

As suggested above, regenerative strategies are not uniform among the phytoplankton, neither is the production of spores and resting stages exclusively brought on by 'adverse conditions'. However, the existence of resting propagules of a given species are likely tolerant of more severe conditions than vegetative cells and they do increase the probability of survival through difficult times and also perhaps raise the scale of the infective inoculum when favourable conditions return.

5.5 | Growth of phytoplankton in natural environments

The rates of cell replication and population growth that are achieved in natural habitats have long been regarded as being difficult to determine. This is primarily due to the fact that what is observable is, at best, a changing density of population, expressed as species rate of increase (r_n, in Eq. 5.2) which falls short of the rate of cell replication because of unquantified dynamic losses of whole cells sustained simultaneously. The net rate of change can be negative ($-r_n$) without necessarily signifying that true growth has failed, merely that the magnitude of r_L, the rate of loss noted in Eq. (5.3), exceeds that of replication, r'. The problem of patchiness and advection (Section 2.7.2) provides the further complication of compounded sampling errors, in which even the observed rate of population change ($\pm r_n$) may prove an inadequate base. From the other direction, the true replication rate cannot be estimated from measurable photosynthetic or nutrient-uptake capacities, unless it can be assumed with confidence that the actual rate of growth is constrained by the capacity factor concerned.

There are ways around these problems and there are now several quite reliable, if somewhat cumbersome, methods for estimating growth rates *in situ*. Some of these approaches are highlighted below, through the development of an overview of dynamic trait selection in natural habitats.

5.5.1 Estimating growth from observations of natural populations

On the same basis that replication rates cannot be sustained at a faster rate than cell division can be resourced, it is clear that the observable rates of population increase cannot exceed the rates of recruitment through cell replication. The corollary of this is that attestably rapid phases of population increase, independent of recruitment by importation from horizontally adjacent patches or from germinating resting stages, are indicative of yet higher simultaneous rates of cell replication.

Growth rates from episodic events
Generically, these accumulative phases fall into two categories. One of these is the annually recurrent and broadly reproducible event, such as the spring increase of phytoplankton in temperate waters, in response to strong seasonally varying conditions of insolation (see Section 5.5.2). The second is the stochastic event, when, perhaps, a sharp change in the weather, resulting in the fortuitous stagnation of a eutrophic water column, or the relaxation from coastal upwelling, or the deepening of a nutrient-depleted mixed layer with the entrainment of nutrient-rich metalimnetic water, or some abrupt consumer failure through herbivore mortality, leads to the realisation of potential respondent growth. In this second category, the phases of increase may be brief and sensing them, accurately and with reasonable precision, requires the close-interval sampling of well-delimited populations. The study of *in-situ* increase rates of phytoplankton in Bodensee (Lake of Constance), assembled by Sommer (1981), was one that satisfied these conditions. The research based on the large (1630 m^2), limnetic enclosures in Blelham Tarn, English Lake District (variously also referred to as 'Blelham Tubes', 'Lund Tubes' (Fig. 5.11), being isolated water columns of ~12–13.5 m in depth and including the bottom sediment from the lake; for more details, see Lund and Reynolds (1982), carried out in the period 1970–84, has similarly provided many insights into phytoplankton population dynamics. Examples of specific increase rates noted from either location are included in Table 5.4.

The evident interspecific differences are partly attributable to the time period of observation, and the seasonal changes in water temperature and in the insolation attributable to seasonally shifting day length and vertical mixing. In some instances, these environmental variations are reflected in intraspecific variability in

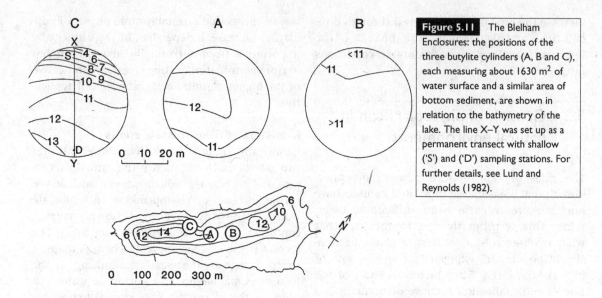

Figure 5.11 The Blelham Enclosures: the positions of the three butylite cylinders (A, B and C), each measuring about 1630 m² of water surface and a similar area of bottom sediment, are shown in relation to the bathymetry of the lake. The line X–Y was set up as a permanent transect with shallow ('S') and ('D') sampling stations. For further details, see Lund and Reynolds (1982).

increase rate. Where species are common to both locations, maximal performances are similar; species are either fast-growing or slow-growing in either location.

The observed rates of increase are also plausible in terms of the dynamic behaviours of the algae in culture. Allowing for winter temperatures and short days, only half of which might be passed in the photic zone, a vernal growth rate of 0.15 d^{-1} for *Asterionella* is perfectly explicable. For small, unicellular species such as *Ankyra* and *Plagioselmis* to be able to double the population at least once per day in summer (when they can manage it twice in grazer-free, continuously illuminated culture) also seems to be a reasonable observation. The growth-rate performances of the bloom-forming Cyanobacteria and the dinoflagellate *Ceratium* are about half those noted in culture at 20 °C (cf. Table 5.1).

Frequency of cell division

Relatively rapid growth rates, sustained over the equivalent of several cell divisions, lead assuredly to the establishment of populations making up a significant part of the biomass, if not actually coming to dominate it. It is equally probable that the same species may be relatively inactive for the quite long periods of their scarcity. Is their increase prevented by lack of light, or lack of resources, or losses to grazers, parasites or to the consignment to the depths? Obviously, more precise means of investigating the *in-situ* physiological activity of numerically scarce phytoplankters are needed to answer this question. One of the best-known and most precise techniques for estimating the species-specific growth of sub-dominant populations is to estimate the frequency of dividing cells. This works best for algae whose division is phased (i.e. it occurs at certain times of day or night) and it may need close-interval sampling (every 1–2 h) of the field population. It works especially well with algae (e.g. desmids, dinoflagellates, coccolithophorids) that have complex external architecture which has to be reproduced at each division and often requires several hours to complete. Then the numbers of cells before and after the division phase is increased by a number that should agree with, or be within, the increment deduced from the frequency of dividing cells. Pollingher and Serruya (1976) gave details of the application of this method to the seasonal increase of the dominant dinoflagellate in Lake Kinneret, now called *Peridinium gatunense*. During the period of its increase (usually February to May), the number of cells in division on any one occasion was found to be variable between 1% and 40%. They showed that the variability was closely related to wind speed. While daily average wind velocities exceeded 8 m s^{-1}, the frequency of dividing cells (FDC) was always <10%. This accelerated to 30–40% during the spring period of weak winds (and, hence, weak vertical advection) averaging <3 m s^{-1}. Successful recruitment of new cells

Table 5.4 | *Some maximal in-situ rates of increase $(r_n \; d^{-1})$ of some species of freshwater phytoplankton reported from Bodensee (Sommer, 1981) and from large limnetic enclosures in Blelham Tarn (Reynolds et al., 1982a; Reynolds, 1986b, 1998b), together with some reconstructed rates of replication (r') where available (see text)*

	Phase[a]	Bodensee (r_n)	Blelham Enclosure (r_n)	Blelham Enclosure (r')
Ankyra judayi	MS		0.86	1.09
Plagioselmis nannoplanctica	V		0.17	
	ES	0.56	0.71	
Cryptomonas ovata	V		0.15	
	ES	0.46		
	MS	0.89	0.61	0.62
Dinobryon spp.	ES, MS	0.45	0.27	
Eudorina unicocca	ES		0.43	0.48
Pandorina morum	MS	0.52		
Coenochloris fottii	ES, MS	0.64	0.43	
Peridinium cinctum	MS	0.18	0.16	
Ceratium hirundinella	MS	0.17	0.13	
Anabaena flos-aquae	MS	0.41	0.34	0.34
Aphanizomenon flos-aquae	MS	0.43		
Microcystis aeruginosa	MS		0.24	0.24
Planktotrix mougeotii	V		0.06	
	ES, MS		0.33	0.33
Asterionella formosa	V		0.15	0.24
	MS	0.36	0.34	0.50
Fragilaria crotonensis	V		0.10	
	MS	0.27	0.24	0.58
	LS		0.10	
Aulacoseira granulata	MS		0.43	
Closterium aciculare	LS		0.18	
Staurastrum pingue	LS	0.13	0.28	

[a] 'Phase' refers to part of the year: V, vernal, or spring bloom period, temperatures 5 ± 2 °C; ES, early-stratification phase (temperature 11 ± 3 °C); MS, mid-stratified (temperature 17 ± 3 °C); LS, late-stratified period (temperature generally 12 ± 2 °C).

divided off at this rate yields maximum rates of population increase equivalent to 0.26–0.34 d^{-1}. The typical net rate of population increase of *Peridinium* in Kinneret over a sequence of 20 consecutive years was found to be 0.22 ± 0.03 d^{-1} (Berman *et al.*, 1992).

Heller (1977) and Frempong (1984) estimated FDC of *Ceratium* in Esthwaite Water to be between 2% and 10%, occasionally 15%, sufficient to explain *in-situ* seasonal increase rates of 0.09–0.14 d^{-1}. Alvarez Cobelas *et al.* (1988) estimated growth rates from afternoon FDC peaks in the population of *Staurastrum longiradiatum* in a eutrophic

reservoir near Madrid to be between 0.13 and 0.16 d^{-1}. More recently, Tsujimura (2003) has estimated *in-situ* growth rates from FDC among cell suspensions of *Microcystis aeruginosa* and *M. wesenbergii* from Biwa-Ko (prepared by ultrasonication of field-sampled colonies). In both species, the frequency of diving cells varied between 10% and 15% in offshore stations and between 15% and 40% at inshore stations, with the average duration of cytokinesis varying from 25 h to 3–6 h. Growth rates of 0.34 d^{-1} thus appear sustainable in the near-shore harbour areas of Biwa-Ko, whence they are liable to become more widely distributed in the circulation of the lake (Ishikawa *et al.*, 2002).

Frequency of nuclear division

For phytoplankton species that are less amenable to the tracking of cell division, the principle may be extended to the monitoring the frequency of karyokinesis (Braunwarth and Sommer, 1985). The success of this method relies on the good fixation of field samples followed by careful staining with the DNA-specific fluorochrome, 4,6-diamidino-2-phenylindole (DAPI) (Coleman, 1980; Porter and Feig, 1980). This precise and sensitive method has been applied to a natural *Cryptomonas* population (Ojala and Jones, 1993), the results being broadly predictable on the basis of growth rates under culture conditions. Like any methods based upon the events in the cell cycle, its prospects for measuring replication accurately are high (cf. Chang and Carpenter, 1994).

There is also keen interest in sensing the DNA replication itself. Since the groundbreaking study of Dortch *et al.* (1983), microbial ecologists have been debating the validity of DNA to cell carbon as an index of the rate of DNA replication. As an indicator of the capacity for protein synthesis, the RNA : DNA ratio is already in use as a barometer of the cell growth cycle in marine flagellates (Carpenter and Chang, 1988; Chang and Carpenter, 1990) and bacteria (Kemp *et al.*, 1993; Kerkhof and Ward, 1993).

Growth from the depletion of resource

In contrast to monitoring growth-cycle indicators, methods for reconstructing growth rates from resource consumption are unsophisticated in approach and notoriously imprecise. However, methods invoking uptake of resources deployed in specific structures, such as silicon for diatom frustules (Reynolds, 1986a) or sulphur for protein synthesis (Cuhel and Lean, 1987a, b), offer more promise. Reynolds and Wiseman (1982) were able to combine the advantages of the spatial constraints of enclosure, offered by the Blelham tube, with frequent serial sampling of the plankton and careful accounting of the amounts of replenishing sodium silicate, in order to measure the true growth rate of a diatom population. Several datasets were collected and presented (partly also in Reynolds *et al.*, 1982a; Reynolds,

1996b); just the case of *Asterionella formosa* in Blelham Enclosure B in 1978 is highlighted here.

The inflatable collar of the enclosure was lifted on 2 March of that year, isolating part of the lake population comprised almost wholly of *Asterionella*, then itself already actively increasing, at a concentration of $\sim$630 cells mL^{-1}). Over the next 19 days, the population increased exponentially at an average rate of 0.147 d^{-1}, then more slowly to its eventual maximum (a total of 24,780 cells mL^{-1}, though by then including 1950 cells mL^{-1} judged to be dead or moribund) on 4 April. The decline in the standing population was slow at first but accelerated enormously as warmer and sunnier weather mediated the thermal stratification of the Tarn and the enclosure. Nutrients were added to the enclosure each week, by dispersal into solution across the enclosure surface, in measured doses respectively designed to restore the levels of available resource to 300 μg N, 20 μg P, 100 μg TFe and 1000 μg SiO_2 (466.7 μg Si) L^{-1}. None of these fell to growth-limiting thresholds. However, there was no artificial relief for either high pH or probable carbon limitation. The consumption of silicon was calculated as the sum of the observed decline in the initial concentration on 2 March, aggregated with the Si added, and averaged out across the whole volume. The conversion to *Asterionella* cells between additions was calculated using contemporaneous routine measurements of the Si content of cells sampled from the growing population (consistently within the range 51.8–61.1 pg Si $cell^{-1}$). As there was no other significant diatom 'sink' at the time, the consumption was assumed to be equal to its deposition in new *Asterionella* frustules. The rate of growth, $r'_{(Si)}$, was approximated from the numbers of new cells that the observed silicon depletion could have sponsored. Estimates were comparable with the observed rates of increase (r_n) and with the rates of growth reconstructed by correcting for the simultaneous loss processes (discussed fully in Chapter 6). Simultaneous sinking losses were 'monitored' in two ways: using the flux of settling cells into sediment traps placed near the bottom of the enclosure; and using a technique of coring and subsampling the semi-liquid superficial deposit (see Reynolds, 1979a). The possible losses to grazers were estimated from contemporaneous measurements of

Table 5.5 | *Comparison of the observed rates of increase in a natural population of* Asterionella *in a large limnetic enclosure with the growth rate as estimated by silicon uptake. The difference is equated with the rate of loss of live cells* (r_L) *and which may, itelf be compared with simultaneous observations of settling rates* (r_s) *and estimates of the rates of loss to grazers* (r_g) *and to death* (r_d)

	2 Mar–21 Mar	21 Mar–4 Apr	4 Apr–25 Apr	25 Apr–15 May
obs r_n	0.147	0.065	−0.032	−0.242
calc r'_{Si}	0.154	0.072	0.012	0
calc r_L	0.007	0.007	0.044	0.242
obs r_s	0.009–0.027	0.016–0.021	0.032–0.126	0.150–0.189
est r_g	0.001–0.009	0–0.011	0.001–0.035	0.002–0.040
est r_d	<0.001	<0.002	<0.031	<0.064
calc r'	0.157–0.184	0.081–0.099	0.001–0.160	0–0.051

Table 5.6 | *Comparison of the observed biomass (as cells per unit area) of* Asterionella *in Enblosure B, Spring 1978, with reconstructed production (from silicon uptake), the total eventually sedimented, the intercepted flux and the estimated loss to grazers at the time*

	Cells (10^{-9} m^{-2})
Maximum standing crop	272.6 (±19.9)
Production from Si uptake	335.0 (±27.7)
Estimated consumption by grazers	10.35 (±9.10)
Sedimentary flux	341.2 (±12.3)
Recruitment to sediments	319.0 (±58.0)

Source: Data of Reynolds and Wiseman (1982) and Reynolds et al. (1982a), as presented in Reynolds (1996b).

filter-feeding by zooplankton (chiefly of *Daphnia galeata*, though activity was minimal until late April: Thompson et al., 1982).

The various outcomes are tabulated and compared in Table 5.5. At first, almost all the silicon investment is realised in observable additions to the extant population. Moreover the slowing rate of increase is explained almost wholly by a declining rate of replication, for whatever reason this might be. However, the rates of loss mount and rates of recruitment through cell division slow, until the former exceed the latter and the population goes into decline. In this particular

population, it is possible to calculate a budget for the entire phase of production. These are entered in Table 5.6, but are calculated for the full water column (mean depth, 11.0 m). The entries are subject to wide margins of error but the agreement is encouraging: the lower ends of the confidence intervals are mutually compatible. From the aggregate loss of silicon from solution (equivalent to 18.788 g Si m^{-2}), the number of *Asterionella* cells that could have been produced was estimated to have been between 307.3 and 362.7 × 10^9 cells m^{-2}. Had they all been in suspension simultaneously, the concentration would have been between 27.94–32.97 × 10^3 mL^{-1} (cf. observed maximum, 24.78 × 10^3 mL^{-1}; 272.6 × 10^9 cells m^{-2}). The numbers that were recruited to the sediment, the settling flux that was intercepted and the range of fatalities to grazing are each inserted in Table 5.6.

5.5.2 The spring increase in temperate lakes: the case of Windermere

It is appropriate now to consider an example of the role of growth in the development of phytoplankton maximum, taking the case of a familiar and annually recurrent event, observed over a sequence of several consecutive years and responding to a substantial degree of interannual environmental variability. The example of the *Asterionella*-dominated spring increase in Windermere is chosen for the length and detail of the observational record and because the year-to-year variations in the dynamics of growth

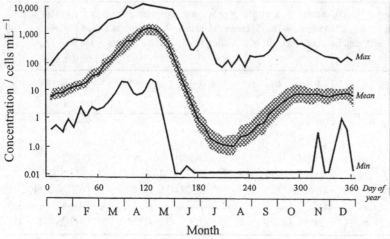

Figure 5.12 The seasonal change in the standing population of live *Asterionella* cells in the upper 7 m of the South Basin of Windermere (heavy line), averaged over the period 1946 to 1990 inclusive, together with the envelope of 95% confidence (shown by hatching) and the maximal and minimal values recorded (the lighter lines). Modified from an original figure in Maberly *et al.* (1994) and redrawn with permission from Reynolds (1997a).

and population increase (and subsequent decline) have been intensively analysed. Windermere is a glacial ribbon lake in the English Lake District, covering 14.76 km². It is not far short from being two lakes, a shallow morainic infill separating the lake into two distinct but contiguous basins. The North Basin holds nearly two-thirds of the total volume (314.5 × 10⁶ m³) and has a mean depth of 25.1 m (maximum is 64 m); the mean depth of the smaller South Basin is 16.8 m (maximum 42 m) (data of Ramsbottom, 1976). The southward water flow through the lake is generated mainly from the catchment in the central uplands of the Lake District. The mean annual discharge from Windermere (437 × 10⁶ m³ a⁻¹) corresponds to a theoretical mean retention time of 0.72 a. The upper catchment is based upon hard and unyielding volcanic rocks, the lower foothills comprising younger and slightly softer Silurian slates. Both are poor in bases. Thin soils, mostly cleared of the natural woodland covering and replaced with rather poor, leached grazing land, contribute little in the way of bases or nutrients to the lake. Nutrient loads have increased substantially over the last 50 years or so, through the increased use of inorganic fertilisers on the land. The sheep population that the added fertilty supports has roughly doubled over the same period. However, it is the increase in the human population, together with the introduction (in 1965) of secondary sewage treatment, which has most affected the phytoplankton-carrying capacity of Windermere (for full details, reconstructed loadings and the effects of the 'restoration' mea-

sures, commenced in 1991, see Reynolds and Irish, 2000). By the late 1980s, autumn–winter concentrations of MRP increased from a pre-1965 average of 2–2.5 µg P L⁻¹ (0.06–0.08 µM) to ~8 µg P L⁻¹ in the North Basin and to about 30 µg P L⁻¹ in the South Basin. In both basins, autumn–winter DIN levels more or less doubled over the same period, from ~350 to ~700 µg N L⁻¹ (25–50 µM), but SRSi levels have remained steady (at 0.9–1.1 mg Si L⁻¹; 32–39 µM; 1.9–2.3 mg SiO₂ L⁻¹).

This information permits us to establish the physical–chemical characters of the habitat of the spring bloom. It is a relatively clear (ε_{wmin} ~ 0.31 m⁻¹) – and soft(alkalinity <0.26 meq L⁻¹) – water, barely mesotrophic lake, experiencing a mostly cool, temperate, oceanic climate, but incurring, between 1965 and 1991, substantial anthropogenic enrichment. For almost 200 years, *Asterionella* has been a conspicuous member of the plankton in both basins and, during at least the last 60 of those, the almost unchallenged dominant species of the spring-bloom period.

Maberly *et al.* (1994) carried out a thorough statistical analysis of the (then) complete run of data collected from (mostly) weekly samplings of the South Basin, initiated by J. W. G. Lund in 1946 (see Lund, 1949) and maintained, with only detailed methodological variation, over the 45 years to 1990. A simplified version of their main figure is reproduced here, as Fig. 5.12, to illustrate the reproducibility of the main features of the development. The heavy line represents the 45-year mean of the standing population (on a logarithmic scale) as a function of the day of the year. The narrowness of the 95% confidence interval (shown by hatching) attests to the strong interannual comparability of the growth, even though the boundaries of the extreme records over the 45 years (delimited by the lines either side of the mean plot) cover 2 or more orders of magnitude of variation. Maberly et al. (1994) diagnosed several cardinal characteristics of the growth curve, including the size of the extant population at the beginning of the year (mean 9.7 ($\times/\div$ 3.85) cells mL^{-1}; range 0.6–330 mL^{-1}); the maximum (mean 3940 ($\times/\div$ 2.28) cells mL^{-1} range 330–11500 mL^{-1}), the date of its achivement (day 124 $\pm$ 16.7 (17 April–21 May)), the start (day 52 $\pm$ 24 (28 January–17 March)) and end of the period of rapid exponential increase (day 106 $\pm$ 17 (30 March–3 May)); the mean rate of increase achieved (0.0925 ($\pm$ 0.0357) d^{-1}), as well as the date of the commencement of the steep post-maximum exponential decrease (day 142 $\pm$ 16 (6 May–9 June)).

The source of the bloom is essentially the standing stock in the water at the turn of the year (Lund, 1949; Reynolds and Irish, 2000). Inter-annual variations in this survivor stock (mean 6.6 ($\times/\div$ 4.83) cells mL^{-1}) are influenced by the size of the late summer maximum of the previous year and the extent of its net dilution by autumnal wash-out. There is a modest increase in the standing crop detectable throughout the first few weeks of the year, so there is, in no sense, any part of the year when growth is not possible. This is close to, or just prior to, the time of greatest nutrient availability in Windermere, so the restriction on biomass increase has long been supposed to be physical. The lowest water temperatures are generally encountered in late January (days 14–28, weeks 3 or 4, of the year) but usually remain <7 °C until week 12 (day 84 and several weeks after the inception of the main exponential increase phase). On the other hand, Talling's (1957c; see Section 3.3.3) extrapolations of column-integrated photosynthetic production in the lake especially in a fully mixed water column, strongly indicate that light is the main growth-regulating factor. The early growth of Asterionella in Windermere has been observed to be relatively weak in the stormiest winters predominated by south-westerly winds and stronger in anticyclonic winters. However,

general trends over the full period are towards smaller overwintering inocula but faster rates of exponential rise in the spring. Of particular interest is the fact that, although the size of the maximum crop seems not to have increased over the 45 years, the date of its attainment has tended to be reached earlier in the year, as a consequence of the trend towards sustained faster growth rates.

The maxima of Asterionella in the North Basin of Windermere have been similar in magnitude to those observed in the South Basin in the corresponding years but, typically, have always been reached one to two weeks later. The greater mean depth of the North Basin appeared to be the most likely reason for this relative delay. However, the maxima here are also now reached significantly earlier ($\sim$10 d) in the year. The proximal explanation is, again, that a faster average rate of increase is maintained but why this should have changed when light is alleged to be the rate-regulating factor is not immediately apparent.

Circumstantial evidence and some simple modelling reveal a complex factor interaction at work. Starting either from the premise of sustainable growth rates (Reynolds, 1990) or photosynthetic behaviour (Neale et al., 1991b), it may be shown that the long-term average of the underwater light integral, I^*, sets much lower constraints on Asterionella growth than does water temperature, SRSi or SRP concentration. The time track of I^* in Fig. 5.13 is reconstructed from long-term (42-year) records of daily integrals of irradiance (I_0) and wind run, and is hypothesised to express the integral daylight period experienced by entrained algae. Comparison of the modelled growth rates, r, as determined by each of the constraining factors in turn, shows that, initially, the least are those set by I^*. They move from about 0.04 to 0.08 d^{-1} during the first 50 days (7 weeks) of the year, before accelerating up to 0.21 d^{-1} by week 15 (days 98–105). Depending upon the size of the starting inoculum, this is sufficient to take the number of formed cells into the range 10^3 to 10^4 cells mL^{-1} and incipient Si exhaustion. The supportive capacity of the initially available silicon is 14–19 $\times 10^3$ cells mL^{-1}, if it is assumed that SRSi is drawn down to extinction and shared at 60–65 pg Si cell^{-1} produced. Only for the purpose of modelling is it assumed that there is no concomitant loss of cells from suspension, or that

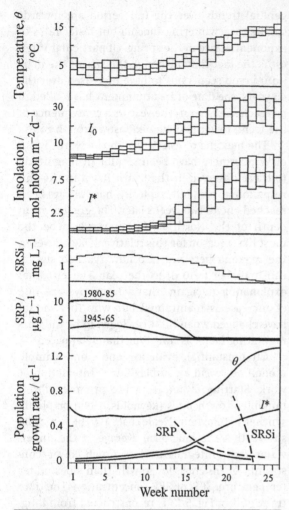

Figure 5.13 Features of the environment of the North Basin of Windermere, during the first half of the year, influencing the spring increase of *Asterionella*. The top panel shows the water temperature (θ °C), averaged for each week of each of 43 consecutive years (1946 to 1988 inclusive, with the 95% confidence interval). The second panel shows similar averages for I_0, the insolation at the water surface, and for I^*, the average light level in the mixed layer. Typical (not statistically averaged) changes in the content of SRSi (soluble reactive silicon) are shown in the third panel, while the fourth reflects long-term increases in SRP (soluble reactive phosphorus) levels by providing typical trend lines for two separate periods (1945–65 and 1980–85). The lower panel shows plots of modelled growth rate capacities of θ, I^*, SRSi and SRP. The sustainable growth rate is that of the limiting component, which, clearly, is I^* until week 16 or 17, when diminishing SRP or SRSi levels become critical. Revised from Reynolds (1990) and redrawn with permission from Reynolds (1997a).

there is no demand for silicon from any other agency. In fact, observable populations of $9\text{–}10 \times 10^3$ *Asterionella* cells mL^{-1} will not have formed without using most of the available SRSi, at least down to the critical half-saturation level of 23 μM (see Section 5.4.4).

The key deduction is that, prior to 1965, the initially available phosphorus (say 2.5 μg P L^{-1}) would have had to have been shared among a maximal population of 10×10^3 cells mL^{-1}, each having a mean residual quota of ≤ 2.5 pg $cell^{-1}$. By halving the quota again, the next cell division will submit the growth rate to P-limitation (see Section 5.4.4). It follows that the diatom-supportive capacity of Windermere is, indeed, set by the available silicon. However, the rate of its assimilation and conversion into *Asterionella* biomass is closely regulated first by the light income availability but, as soon as this begins to be relieved by lengthening days and weaker vertical mixing, phosphorus availability starts to squeeze the attainable growth rate instead. The final twist in this changing factor interaction comes with the recent phosphorus enrichment of the lake. As first noted by Talling and Heaney (1988), enrichment with phosphorus relieves the growth-rate constraint, which now continues to the limits permitted by I^*, until the ceiling of silicon exhaustion is reached, now rather earlier in the year. The sketches in Fig. 5.14 summarise the shifting date of maximal attainment. The silicon limit remains inviolable. Interactions among interannualy varying constraints have influenced the extent of capacity attainment. Over a number of years of enrichment, it became the case that the *Asterionella* maximum failed to exhaust all the phosphorus in solution, instead leaving it available to be exploited by other phytoplankton. As will be argued later (Section 8.3.2), this is a defining stage in the eutrophication of temperate lakes.

There is much more to this story, and to the next one about the rapid post-maximum collapse of the spring bloom. Both concern the magnitude of the loss terms making up r_L in Eq. (5.3), discussion of which is developed in Chapter 6. All the time that the population is increasing, the replication rate is sustaining losses of formed cells to mortality – to such physical agents as outwash and settling beyond the resurrecting limits

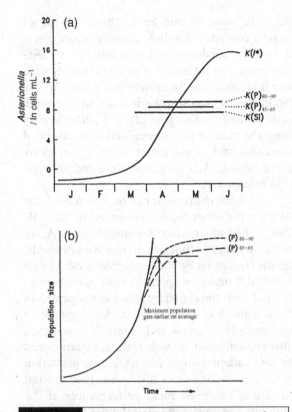

of entrainment and to the biological demands of grazing and parasitic consumers. The spring bloom in Windermere, as elsewhere, is sustained to within the limits that light and available nutrients can support, net of ongoing sinks, recognising that the latter may be more or less critical in commuting the size of the maximum crop below that of the chemical capacity.

5.5.3 Selection by performance

The example of *Asterionella* in Windermere may, or may not, have direct analogues to other diatom blooms in other systems or even to the behaviour of other kinds of phytoplankton. The main illustrative point is that species have to make the best of the environments in which they find themselves and, often, they must press their specific traits and adaptations to perform adequately under environmental conditions verging on the hostile. In this way, we might interpret the ability of *Asterionella* to dominate the vernal plankton of Windermere as being dependent upon certain attributes. The first is so obvious that it is easily overlooked: *it is there!* The success of a species in a habitat is a statement that the habitat is able to fulfil its fundamental survival requirements and, *of the species that have arrived there*, this species will be, relatively, the most efficient in exploiting the opportunity offered. Absence of a species does not inform a deduction that the habitat is not suitable; it may just not have had the opportunity to grow there. No species dominates a habitat just because theory argues it to be the most suitable. However, to be able to to outperform other species at a given point in space and in time must suggest a favourable combination of inoculum and a relatively superior exploitative efficiency under the conditions obtaining.

The fossil record shows that *Asterionella* dominance is a relatively new phenomenon in Windermere, having occurred only since the nineteenth century (Pennington, 1943, 1947). Previously, *Cyclotella* species had dominated a more oligotrophic period in the lake's history (Haworth, 1976). *Asterionella* is able to grow faster than other diatoms under the poor vernal underwater light conditions and faster than *Planktothrix* at low temperatures. It also manages to carry over substantial winter populations from which spring growths can expand. *Asterionella* does not have matters exclusively to itself – *Aulacoseira* species (*A. subarctica*, *A. islandica*) overwinter well, though they grow less rapidly than *Asterionella*; flagellates such as *Plagioselmis* grow relatively well in winter anticyclones (with frosts, sunshine and weak wind-driven vertical mixing) (Reynolds and Irish, 2000).

Small interannual variations in these environmental features may not make decisive intrasystem differences in outcomes but they may assist us to understand the differences in timing, the magnitude of crops and the species dominance of populations elsewhere. We have shown, through the comparison of growth responses and their sensitivities to environmental deficiencies, how the dynamic performances differ among species in experiments. Can we now discern differences of performance in nature that will confirm – or help us to recognise – the traits that select for some species and against others at a given location? If so, how much does this tell us about the ways in which natural communities are put together and shape trophic relationships?

The answers to these questions are clouded by the usual problems of accurate measurement of population dynamics in the field (see Section 5.5.1). Work with captive wild populations of phytoplankton in the Blelham enclosures, growing within a defined space, subject to well-characterised and, in part, artificially controlled conditions, subject to separately quantified loss rates of cell loss and, above all, sampled at high frequencies (3–4 days), does provide some insights. From data collected from numerous growth phases, observed in three enclosures over 6 years, Reynolds (1986b) assembled a series of in-situ replication rates for each of a number of common species. Summaries are shown in Fig. 5.15. Each datum point is calculated from a minimum of three serial measurements on an increasing population and is corrected for the contemporaneous estimates of loss rates to sinking and grazing (details to be highlighted in Chapter 6). These points were then plotted against a common scale of analogous insolation, this being the product of the day length (Γ, from sunrise to sunset, in hours) and the ambient ratio of Secchi-disk depth to mixed depth (z_s/z_m, with the proviso that solutions >1 are treated as 1). Finally, the points are grouped according to the approximate contemporaneous water temperatures.

The plot does reveal an encouraging level of intraspecific consistency of performance and significant interspecific differentiation. Taking the observations on Fragilaria, for instance, replication rates in the field, between 13 and 17 °C, reveal a common dependence upon the aggregate-by-analogy photoperiod, with a slope that appears steeper than (two) observations applying to temperatures between 9 and 11 °C, yet less steep that the indicated photoperiod depen-

dence between 18 and 20 °C. There also seems to be a common threshold light exposure ($\sim$4 h d^{-1} on the analogue scale) applying at all temperatures. The information sits comfortably with our understanding of growth sensitives in the laboratory. It may also be noted that the replication rates in the field do not differ widely from the maximum resource- and light-saturated rates observed in culture at 20 °C, if the appropriate adjustments for temperature and photoperiod are applied.

Analogous deductions can be drawn from the data for the other algae represented in Fig. 5.15. The ability of another R-strategist alga, Asterionella, to adapt to function on low average insolation is confirmed by observed growth rates of up to 0.15 d^{-1} on a low aggregate daily photoperiods ($\leq$4 h d^{-1} on the contrived scale) at temperatures from 5 to 15 °C. Collected data for Cryptomonas spp. (mostly C. ovata) and Planktothrix mougeotii also confirm that growth rates are maintained by photoadaptation to low aggregate insolation (with thresholds of 1–3 h d^{-1}) but they are not so fast as the most rapid performances of the diatoms. The most rapid growth rates observed in the enclosures have been attained by C-strategists such as Ankyra, which, on several occasions, has been observed to self-replicate at $>$0.8 d^{-1} (doubling its mass in less than a day!). These short-lived episodes have been possible in warm, clear, usually water that is restratifying and supplied with nutrients well in excess of growth-rate limiting concentrations. Lack of carbon, self-shading or increased vertical mixing contribute to a slowing growth rate in these instances: note the apparent threshold at $\sim$8 h d^{-1} on the contrived scale of daily photoperiod. The plots for the C–S species Eudorina unicocca seem to point to an even greater photoperiod response, little influenced by the (somewhat narrow range of) water temperatures available. The two strongly S-strategist Cyanobacteria (Anabaena and Microcystis) are generally slow-growing ($<$0.36 d^{-1}); as a function of photoperiod, there is an intermediate threshold of 4–6 h d^{-1} on the artificial scale.

Many other observations on the growth performances of phytoplankton have emerged from the studies using the Blelham enclosures. Some relate to the dynamics of loss and the way these

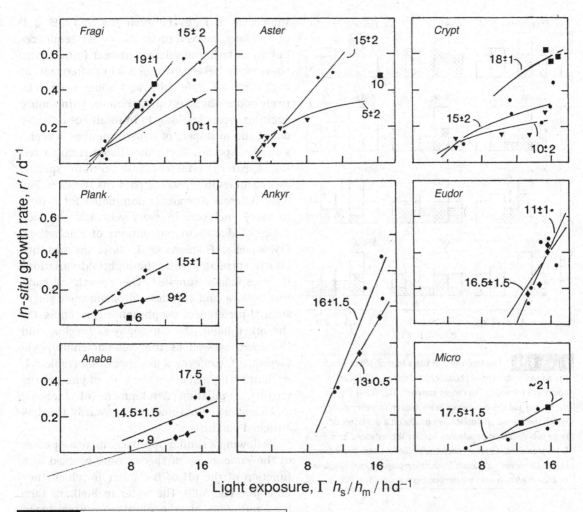

Figure 5.15 Approximations of the daily specific growth rates ($r' = r_n + r_G + r_S$) reconstructed from detailed observations on the dynamics of populations of named phytoplankters in Blelham enclosures, plotted against the contemporaneous products of the length of the daylight period (Γ, in h) and the ratio of the Secchi-disk depth to the mixed depth (h_s/h_m), being an analogue of I^* (in instances where $h_s/h_m > 1$, h_s/h_m is put equal to 1). Curves are fitted to data blocked according to contemporaneous temperatures, as stated. The algae are: *Anaba*, *Anabaena flos-aquae*; *Ankyr*, *Ankyra judayi*; *Aster*, *Asterionella formosa*; *Crypt*, *Cryptomonas ovata*; *Eudor*, *Eudorina unicocca*; *Fragi*, *Fragilaria crotonensis*; *Micro*, *Microcystis aeruginosa*; *Plank*, *Planktothrix mougeotii*. Modified and redrawn with permission from Reynolds (1986b).

interact with differential growth rates in influencing community assembly and succession, to which reference will be made in subsequent chapters. Of particular interest to the question of selection by growth performances is the collective overview of species-specific development in relation to chemical factors. The enclosures have been subject to differing levels of fertilisation, and variation in the frequency and the scale of nutrient supplied. Against the naturally soft-water, relatively P-deficient water of Blelham Tarn, it is not surprising that manipulations of the phosphorus and the carbon content of the enclosed water should have yielded the most satisfying outcomes. Reynolds (1986b) contrasted the yield of phytoplankton, in terms of biomass and species composition, through six enclosure–seasons

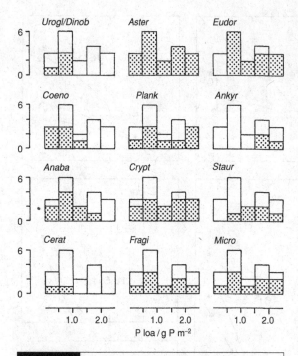

Figure 5.16 The frequency of years (out of 18) in which the annual areal load of phosphorus added to a Blelham enclosure fell within the ranges stated (0–0.5, 0.5–1.0 g P m⁻², etc.) and the number of years in each category during which the named phytoplankters produced a dominant or large sub-dominant population. Species abbreviations as in Fig. 5.15, plus: *Cerat, Ceratium hirundinella; Coeno, Coenochloris fottii; Dinob, Dinobryon spp.; Staur, Staurastrum pingue; Urogl, Uroglena cf. lindii.* Redrawn with permission from Reynolds (1986b).

subject to P loads between 0.3 and 2.5 g P m^{-2}. Many species occurred in all sequences but in differing absolute and real proportions. Some were relatively much more frequent at high rates of P fertility and some were relatively more abundant at low rates. Using more enclosure-years (making 18 in total) but in lesser detail and in respect of a small number of representative species, Reynolds (1986b) summarised the apparent preferences of certain specific ascendencies to reveal the patterns shown in Fig. 5.16. Whereas *Asterionella* dominated for a time in every enclosure in every year and the incidence of dominant populations of *Planktothrix, Cryptomonas, Fragilaria* and *Microcystis* was not clearly correlated with the eightfold variations in phosphorus supplied, the growth of *Eudorina, Ankyra* and the desmid *Staurastrum pingue* showed preferences for phosphorus richness. On the other hand, the chrysophytes *Uroglena* and *Dinobryon*, as well as the colonial chlorophyte *Coenochloris*, conformed to supposition (Table 5.2) by flourishing well down the scale of phosphorus fertility. Population development of *Anabaena* and *Ceratium* also shows bias towards the less enriched conditions.

Following a similar approach, increase phases of the same group of species were blocked as a function of the pH of the water in which they grew (see Fig. 5.17). The water in Blelham Tarn has a rather low bicarbonate alkalinity (<0.4 meq L^{-1}) and, in the enclosures, it is isolated from

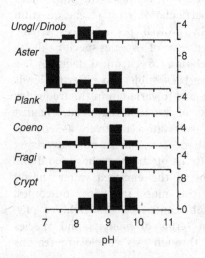

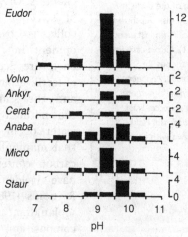

Figure 5.17 The frequencies with which the increase phases of dominant phytoplankton species in the Blelham enclosures were halted in the pH ranges indicated (pH 7.0–7.5, 7.5–8.0, etc). Species abbreviations as in Figs. 5.15 and 5.16, plus: *Coeno, Coenochloris fottii.* Redrawn with permission from Reynolds (1986b).

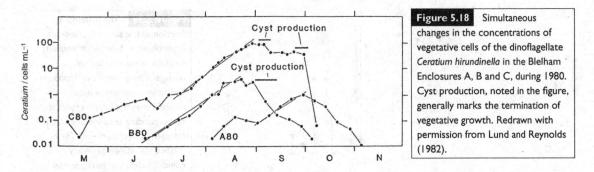

Figure 5.18 Simultaneous changes in the concentrations of vegetative cells of the dinoflagellate *Ceratium hirundinella* in the Blelham Enclosures A, B and C, during 1980. Cyst production, noted in the figure, generally marks the termination of vegetative growth. Redrawn with permission from Lund and Reynolds (1982).

terrestrial replenishment (Fig. 5.11). Thus, the inorganic carbon supply to phytoplankton in these experiments depended upon internal recycling, augmented by whatever dissolved from the air at the water surface. Thus, rising pH is a useful surrogate of carbon deficiency in the enclosures (experiments in 1978 attempted to relieve the deficiency with additions of bicarbonate; they did increase carbon but did not reduce pH). The species-specific distributions of histograms in Fig. 5.17 peak in range 9.0–9.5, simply because that is also the range reached most frequently during the maxima encouraged by fertiliser additions; it is not necessarily indicative of any algal preference for this range. However, it is evident that *Fragilaria, Cryptomonas, Eudorina, Ankyra, Ceratium* and, especially, *Anabaena, Microcystis* and *Staurastrum pingue* were all able all to function at pH levels up to 1 point higher. The observation matches those of Talling (1976) concerning differential insensitivity to carbon dioxide shortage; all these species are known or suspected for the efficiency of their carbon-concentration mechanisms (Section 3.4.3). The apparent failure of the chrysophytes *Uroglena* and *Dinobryon* to maintain growth at pH levels ≤9 aroused the suspicion that these algae might be obligate users of carbon dioxide (Reynolds, 1986b), as indeed, has since been verified in the laboratory (Saxby-Rouen *et al.*, 1998).

These response patterns require careful interpretation and their subjectivity to the experimental design must be taken into account. The species responding to the conditions contrived are, almost exclusively, the ones that are already well established in the lake and/or the experimental enclosures. The observed performances are not necessarily those of the nature's 'best-fit'

organism so much as those of them that were there were able to contend effectively with the conditions imposed. Yet, in many ways, these relative performances do, most probably, distinguish sufficiently among the traits and adaptabilities of a number of common types of plankter for their basic ecological preferences and sensitivities to be recognised and identified in further, more focussed testing.

We see that enrichment with nutrient seems to be beneficial to most species, raising the ceiling of attainable biomass and, in many instances, releasing the growth rate from the restriction of nutrients. This may not be universally so, for some of the inherently slow-growing, self-regulating species, like *Ceratium hirundinella*, have adaptations for supporting their growth requirements under nutrient-segregated conditions and growth rate is not necessarily enhanced by nutrient abundance. The growth-rate performances achieved by *Ceratium* in each of the three enclosures during 1980 were ultimately comparable (see Fig. 5.18: 0.092 d^{-1} in Enclosure A, 0.098 d^{-1} in Enclosure B and 0.105 d^{-1} in Enclosure C), even though the non-physical growth conditions were quite disparate. That the yields were quite different is influenced by the length of time that cell division was maintained *in situ*. This was partly influenced by resource availability and, as is now known, by resource renewal in the graded Enclosure C (Fig. 5.11; Reynolds, 1996b). The major influence, however, is the source of excysting inocula. The uniformly deep sediments of Enclosures A and B supported many fewer surviving cysts than those of Enclosure C, whereas recruitment of 'germinating' gymnoceratia (see Section 5.4.6) was also relatively stronger in Enclosure C. It is interesting, indeed, that Lund (1978) had

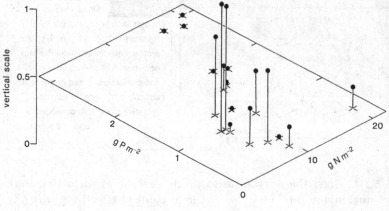

Figure 5.19 The maximum fraction of the summer standing phytoplankton biomass in Blelham enclosures contributed by nitrogen-fixing species of *Anabaena* and *Aphanizomenon* in each enclosure–year, plotted against the corresponding areal loading rates of nitrogen and phosphorus. Their relative abundance generally coincides with low nitrogen and moderate phosphorus availability but not on a low N : P ratio per se. Redrawn with permission from Reynolds (1986b).

regarded the enclosures as being somehow hostile to *Ceratium*, for the alga had never become so numerous as in the lake outside. As Fig. 5.18 demonstrates, there is nothing about these enclosures that interferes with growth. Isolation of the water from that of the lake, except during winter opening, evidently made it difficult for *Ceratium* to invade in numbers. The point about Enclosure C is that the presence of shallow sediments assisted the success of perennation and reinfection of the water column in the spring. This pertinent biological observation was, obviously, quite peripheral to the design and purpose of the experiment.

The behaviour of *Anabaena*, a supposed indicator of carbon-deficient eutrophic waters, and its apparent preference for less-enriched conditions may also be explained from more detailed analysis. Suspecting a performance influenced by the ratio of available nitrogen to available phosphorus, the maximum fractional abundance of *Anabaena* spp. (plus the smaller amounts of *Aphanizomenon*) in any given enclosure year was plotted against the coordinates of the nitrogen and phosphorus supplied (Fig. 5.19). *Anabaena* spp. featured in most annual sequences observed in the enclosures but only occasionally did it produce dominant populations. Moreover, when levels of DIN had fallen much below 100 µg N l^{-1} (~7 µM), populations developed substantial heterocyst frequencies (maximum, 7% of all cells). In terms of N and P loads, however, *Anabaena* and *Aphanizomenon* abundances are clustered within one corner of the N ver-

sus P field of Fig. 5.19, corresponding to loadings of 6–10 g N m^{-2} (0.45–0.71 mol m^{-2}) and 0.19–1.2 g P m^{-2} (6–38 mmol m^{-2}), but with no clear dependence on N : P (actual ratios 12–118). At higher loads of N and P (but with ratios still in the order of 10–20), the species were hardly represented at all. Thus seasonal dominance by these species has failed to come about under conditions in which neither nitrogen nor phosphorus was likely to have been limiting plankton growth. What can be said is that, at the low nitrogen concentrations limiting non-N-fixing species, the yields of *Anabaena* and *Aphanizomenon* are broadly proportional to phosphorus loadings up to ~1 g (or ~30 mmol) P m^{-2}. At higher levels of N and P, there will always be faster-growing species – such as *Ankyra*, *Chlorella*, *Plagioselmis*, *Cryptomonas*, even *Eudorina* – poised to outperform them.

Finally, additional information is available from the enclosure work to amplify the specific growth performances under persistent and relatively low phosphorus concentrations. *Coenochloris fottii* featured prominently in a number of the sequences (it was then referred to *Sphaerocystis schroeteri*). It was relatively more common in the summer periods when phosphorus was strongly regulating biomass and growth (see especially Reynolds *et al.*, 1985) but absolutely larger populations and sustained growth rates were observed at greater nutrient availabilities (Reynolds *et al.*, 1983b, 1984). With the imposition of artificial mixing, the alga was again found to be tolerant but only for so long as the water was clear and the ratio of Secchi-disk depth to mixed depth (z_s/z_m) was near or greater than 1. Noting rather smaller numbers of *Oocystis* aff. *lacustris*, *Coenococcus*, *Crucigeniella* and the tetrasporalean *Pseudosphaerocystis lacustris* (formerly *Gemellicystis neglecta*) showed similar responses to imposed variations in nutrients and insolation, Reynolds (1988a) grouped all these non-motile, mucilaginous colonial Chlorophyceae in a single morphological–functional group of non-motile, light-sensitive, mucilage-bound species. It is their behavioural traits with respect to threshold light levels that tend to exclude them from turbid or deep-mixed water columns and to give them a common association with oligotrophic lakes. For the common chrysophyte species of *Dinobryon*, *Synura* and *Uroglena*, the apparently similar association with nutrient poverty in the enclosures is not due to any intolerance of high nutrients but to an unrelieved dependence upon the supply of carbon dioxide, which, in the soft-water confines of the Blelham enclosures, is readily outstripped by demand (Saxby-Rouen *et al.*, 1998).

5.5.4 Temporal changes in performance selection

There is clearly a good match between how the various species studied in the controlled field conditions of the Blelham enclosures and the sorts of trait characteristics and strategic adaptations identifiable among the properties revealed in earlier sections (especially pp. 31–34). In particular, a distinction is to be made between the manner in which species develop their populations in response to what are perceived by them to be favourable conditions. On the one hand, there are species that specialise in rapid, invasive growth, building up stocks at the expense of freely available resources and high photon fluxes, and which, by analogy to the terminology of Grime (1979, 2001), we have defined *r*-selected *C*-strategists (see Box 5.1). Besides the examples of *Ankyra* and *Chlorella*, *Asterionella* and some of the other freshwater diatoms that are tolerant of intermittent and poor average insolation (*R*-strategist traits) are also strongly *r*-selected over other *R*-type species such as *Planktothrix agardhii*. The slower growth but often

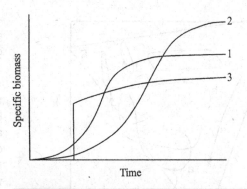

Figure 5.20 Differing demographic behaviours in exploiting favourable growth opportunities: species grow either rapidly and invasively (1) or more slowly to build a conserved, acquisitive population (2). Slow accumulation of biomass may be offset by the recruitment of pre-formed propagules from a perennial seed bank (3). Modified after Reynolds (1997a).

higher-biomass-achieving *K*-selected trait of many larger algae, represented by Curve 2 in Fig. 5.20, which, initially, lags behind the performance of more *r*-selected species (Curve 1) is characteristic of the performances of *S*- and *CS*-strategists. However, it is also clear that some of the self-regulating *S*-strategists, such as *Ceratium* and *Microcystis*, are obliged to grow so relatively slowly that eventual abundance in the plankton is influenced by the recruitment of sufficient perennating propagules at the initiation of the next period of growth. This very strong *K*-selected feature is represented by Curve 3 in Fig. 5.20.

We may venture further than this by superimposing the triangular ordination of species traits (e.g. of Fig. 5.9) and overlaying this on a notional plot to describe the interaction of mixing and nutrient availability in the near-surface waters (see Fig. 5.21). The loops and arrows are inserted to show how temporal seasonal variations in the coordinates of nutrient availability and mixing might select for particular performances and relevant traits and adaptations. These are notional and unquantified at this stage of the development and the representation is qualitative but they illustrate some general points that need to be made. The two large loops (marked a and b) reflect the transitions that might be observed over a year in a seasonally stratified (not necessarily temperate) lake. At overturn, there is a

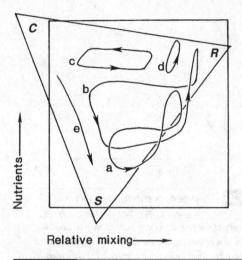

Relative mixing———▶

Figure 5.21 Notional representation of Grime's **CSR** triangle on axes representing relative nutrient abundance and water-column mixing, showing the adaptive traits most likely to be selected by changing environmental conditions. The loops represent time tracks of selective pressures acting through the year in (a) oligotrophic lakes, (b) eutrophic lakes, (c) and (d) in smaller, enriched systems; (e) is the anticipated course of autogenic succession. Redrawn from Reynolds (1988a).

rapid rightward shift on the mixing axis with an upward drift as dissolved nutrients are redispersed from depth. The best-performing species here and throughout the bloom period are likely also to show the traits and growth responses of *R*-strategist species: the limits of their morphological and behavioural adaptations are more suited to coping with low average insolation. The onset of themal stratification is represented by a lurch to the left, where conditions of low relative mixing and high relative nutrients obtain and which, initially, are open to exploitation by fast-growing *C*-strategist species. Their activity depletes the resources and may lead, eventually, to the partitioning of availability and to the dependence upon increasingly effective *S*-strategist adaptations to access them. Note that the more severe and ongoing is the insolation or resource deficiency, the closer are the trajectories to the apices of the *CSR* triangle, where the relevant adaptations become ever more important. As a consequence, the few that have them are alone able to perform successfully at all. The potential diversity of surviving species is finally 'squeezed out'

at the extremes. The converse is that variability is good for maintaining high diversity as more specific performances are accommodated. The constrained cycles of (say) an enriched water column subject to variable stratification or of persistently mixed, resource-cycling systems may also be represented in this scheme (respectively, c and d). Other factors notwithstanding, the effects of population growth should follow the direction of the arrow (marked e) as nutrients are withdrawn from the water and the increasing biomass reduces light penetration and the relative mixed depth is increased.

The traces provide adequate summaries of changes in seasonal dominance in given lakes and, to an extent, they may reflect longer-term changes in phytoplankton in response to nutrient enrichment or restoration measures. In the naturally eutrophic, nutrient- and base-rich kataglacial lakes of northern Europe and North America (see, for instance, Nauwerck, 1963; Lin, 1972; Kling, 1975; Reynolds, 1980a), the annual cycle of phytoplankton dominance features (i) vernal diatoms (which may include any or all of *Asterionella formosa*, *Fragilaria crotonensis*, *Stephanodiscus rotula*, *Aulacoseira ambigua* from the *R* apex of the triangle), followed by (ii) a burst of readily grazeable, plainly *C*-strategist nanoplankton (e.g. chlorellids, *Ankyra*, *Chlamydomonas*, *Plagioselmis*), and/or (iii) by populations of colonial Volvocales (e.g. *Eudorina*, *Pandorina* – best classed as *CS* strategists) and increasingly more *S*-strategist *Anabaena* spp., *Microcystis* or *Ceratium*). The cycle is completed by (iv) assemblages of diatoms (*Fragilaria*, *Aulacoseira granulata*) and desmids (*Closterium aciculare* and several species of *Staurastrum*).

In the nutrient- and base-deficient lakes of the English Lake District (Pearsall, 1932), the oligotrophic subalpine lakes of Carinthia (Findenegg, 1943) and the more oligotrophic lakes of New York and Connecticut studied by Huszar and Caraco (1998), as well, in all probability, similar lakes throughout the temperate regions (Reynolds, 1984a, b), the vernal plankton is typically dominated by *Cyclotella–Urosolenia* diatom associations (*R* or *CR* strategists). These may be replaced typically by such chrysophytes as *Dinobryon*, *Mallomonas* or *Synura* (*RS* strategists) and/or by colonial Chlorophyceae (*CS* strategists), then by *S*-strategist dinoflagellates (*Peridinium umbonatum, P. willei* or *Ceratium* spp.) and, finally, by *R* or *SR* desmids such as *Cosmarium* and *Staurodesmus*.

Between the oligotrophic and eutrophic systems are ranged the lakes of intermediate status (mesotrophic lakes), as well as several deep alpine lakes of central Europe (Sommer, 1986; Salmaso, 2000; Morabito *et al.*, 2002) Here, the vernal phase is characterised by *R*-strategist diatoms featuring perhaps *Aulacoseira islandica* or *A. subarctica*, as well, perhaps, as *Asterionella, Fragilaria* or *Cyclotella radiosa*. There is a late-spring phase of *C* strategists (e.g. *Plagioselmis, Chrysochromulina*), followed either by a phase of colonial Chlorophyceae–Chrysophyceae or, especially in deeper lakes, *S*-strategist *Ceratium* or *Peridinium*, or by the Cyanobacteria *Gomphosphaeria* or *Woronichinia*, or again perhaps by potentially nitrogen-fixing *Anabaena solitaria, A. lemmermannii* or *Aphanizomenon gracile*. Late-summer mixing may favour *R*-strategist diatoms (notably including *Tabellaria flocculosa* or *T. fenestrata*), desmids or the filamentous non-diatoms (such as *Mougeotia, Binuclearia, Geminella*). The outstanding algae of the deep mesotrophic systems, however, are the *RS*-strategist *Planktothrix rubescens/mougeotii* group which both tolerate winter mixing and exploit deep stratified layers in summer.

The cycles in Fig. 5.21 are not tracked at an even rate, neither are they precisely identical each year. Progress may proceed by a series of lurches, whereas interannual variability can divert the sequence to differing extents. However, the growth and potential dominance of phytoplankton adheres closely to the model tracking (Reynolds and Reynolds, 1985). The cycle may be completed in less than a year: the description of Berman *et al.* (1992) of the periodicity of phytoplankton of Lake Kinneret follows a mesotrophic path before stalling in summer deep on the left-hand (nutrient) axis of Fig. 5.21. It does not really move on until wind-driven mixing or autumn rains relieve the severe nutrient (nitrogen and phosphorus) deficiency. The cycle may also be recapitulated: Lewis' (1978a) detailed description of the seasonal changes in the plankton of Lake Lanao, Philippines, could reasonably represented by track (b) in Fig. 5.21 but it would be completed within 4–6 months before being short-cut back to an earlier stage. Even in temperate lakes and reservoirs subject to extreme fluctuations in mixed depth on scales of 5–50 days, alternations between phases of increase and dominance by *R* species (*Stephanodiscus, Synedra*) and *C–S* groupings (cryptomonads, *Chlamydomonas, Oocystis, Aphanizomenon*) are clearly distinguishable (Haffner *et al.*, 1980; Ferguson and Harper, 1982). Again, in nutrient-rich lakes where the alternations result in the lake being either predominantly mixed or stratified, so the dominating species would be (respectively) *R* species (as in Embalse Rapel, Chile: Cabrera *et al.*, 1977) or *C* species (as in Montezuma Well, Arizona: Boucher *et al.*, 1984). These possibilities comply with the track marked (c) in Fig. 5.21. Examples of enriched shallow or exposed lakes that are more or less continuously mixed seem to be dominated by the *K*-selected *R* strategists (*Planktotrix agardhii, Limnothrix redekei, Pseudanabaena* spp.: Gibson *et al.*, 1971; Berger, 1984, 1989; Reynolds, 1994b) are represented in Fig. 5.21 by track (d).

It is not yet possible to apply the same approach to temporal changes in the marine phytoplankton with a similar level of investigative evidence, as the resolution of temporal changes is less clear. On the other hand, it is a testable hypothesis that similar performance-led drivers, influenced by similar morphological adaptations to analogous liqiud environments, govern the spatial and temporal differences in the growth of phytoplankton in the sea. There is good supporting evidence that this may be the case. Smayda (2000, 2002) has shown that the wide diversity among the dinoflagellates may be rationalised against an ecological pattern that invokes morphology. Whereas the smaller, non-armoured adinophytes (such as *Prorocentrum*) and gymnodinioids (*Gymnodinium, Gyrodinium, Heterocapsa* spp.) that are characteristic of shallow, enriched coastal waters have unmistakeably *C*-like morphologies and growth-rate potential, the larger, armoured and highly motile ceratians have clear *S* tendencies. In the open ocean, Smayda (2002) distinguishes among dinoflagellates associated preferentially with fronts and upwellings (*Alexandrium, Karenia*) and those of post-upwelling relaxation waters (*Gymnodinium* catenatum, *Lingulodinium polyedrum*). *S*-strategist dinoflagellates are also prominent in the oligotrophic, stratified tropical oceanic flora, where self-regulation, high motility and photoadaptive capabilities distinguish such dinoflagellates as *Amphisolenia* and *Ornithocercus*. The buoyancy-regulating adinophytes of the genus *Pyrocystis* are most remarkable in showing parallel adaptations to limnetic *Planktothrix rubescens* and in similarly constituting a mid-water shade flora, deep in the light gradient of the tropical ocean.

It is interesting to speculate on the range of adaptations among the planktic diatoms of the sea. Most are non-motile and (presumably) reliant upon vertical mixing for residence in the near-surface waters. *Thalassiosira nordenskioldii*, *Chaetoceros compressus* and *Skeletonema costatum* are all chain-forming diatoms featuring in the spring blooms of North Atlantic shelf waters. The attenuate forms of species of *Rhizosolenia*, *Cerataulina*, *Nitzschia* and *Asterionella japonica* are conspicuous in the neritic areas. All these diatoms can be accommodated in the understanding of *R*-strategist ecologies. However, there are also centric diatoms such as *Cyclotella capsia* that occur predominantly in shallow eutrophic coastal waters and estuarine areas, along with, arguably, *C*-strategist green algae (*Dunaliella*, *Nannochloris*), cryptomonads, the dinoflagellate *Prorocentrum*, the euglenoid *Eutreptia* and the haptophytes *Chrysochromulina* and *Isochrysis*. In another direction of adaptive radiation, the very large, self-regulating *Ethmodiscus rex*, belonging to the tropical shade flora, shows the typical properties of an *S*-strategist species.

Many coccolithophorids are small and are regarded, somewhat 'uncritically' (Raymont, 1980), as nanoplankton. *Emiliana huxleyi*, *Gephyrocapsa oceanica* and *Cyclococcolithus fragilis* are, indeed, small *C*-like species of open water, which they inhabit with nanoplanktic flagellates, including the prasinophyte *Micromonas*, the chlorophytes *Carteria* and *Nannochloris*, the cryptophytes *Hemiselmis* and *Rhodomonas* and the haptophytes *Pavlova* and *Isochrysis*. The nitrogen-fixing, vertically migrating Cyanobacterium *Trichodesmium* displays strong *S*-strategist characteristics. The picoplanktic Cyanobacteria *Synechococcus* and, especially, *Prochlorococcus*, reign supreme over vast areas of ultraoligotrophic ocean, as archetypes of the newly proposed *SS* strategy.

Seasonal changes in the plankton flora of the English Channel, described by Holligan and Harbour (1977), show a clear tendency for vernal diatom–dominated (supposedly *R*-strategist) assemblages to be displaced by more mixed diatom-dinoflagellate (*CR*?) associations (*Rhizosolenia* spp.; *Gyrodinium*, *Heterocapsa*, *Prorocentrum*) in early summer and by green flagellates (*Carteria*, *Dunaliella*, *Nannochloris*) or (*S*?) ceratians (*Ceratium fusus*, *C. tripos*) in mid to late summer. In enriched near-shore areas, the haptophyte *Phaeocystis*, in its colonial life-history stage, may dominate the early summer succession, in a manner strongly reminiscent of the abundance of volvocalean *CS* strategists in eutrophic lakes.

A satisfying aspect of the performances of phytoplankton in both the sea and fresh waters is the superior influence of morphological and (presumably) physiological criteria over phylogenetic affinities. This is a powerful statement attesting to evolutionary adaptations for relatively specialised lifestyles and for the radiative potential latent within all major phyletic divisions of the photosynthetic microorganisms.

5.5.5 Modelling growth rates in field

There has for long been a requirement for robust, predictive models of phytoplankton. Nowadays, the element of stochasticity of environmental events is appreciated as a near-insuperable bar to accurate predictions of sufficient precision. However, considerable use can be made of the regressions fitted to growth performances in the laboratory and the philosophy of strategic adaptations to drive predictive solutions to which of certain kinds of alga will grow in particular water bodies and under which conditions.

This section is not intended to provide a guide or a review of different modelling approaches. These are available elsewhere (Jørgensen, 1995, 1999). The purpose here is to refer to some of the approaches addressed specifically to modelling growth and performance of phytoplankton and to promote the use of models that invoke them. It is worth first repeating the obvious, however, that different models attempt to do different things. These may be in-and-out 'black-box functions', such as Eq. (4.15), where an input (in this case, biologically available phosphorus) generates a yield (in this case, phytoplankton chlorophyll) on the basis of pragmatic observation, without any attention to the explicative processes. These internal linkages may be investigated, imitated and submitted to empirical model explanations, for instance, those which link the generation of phytoplankton biomass to photosynthetic behaviour in the underwater light field. The various explanative equations of (say) Smith (1936), Talling (1957c), Pahl-Wostl and Imboden (1990) predict, with accuracy, precision and increasing detail, the photosynthetic carbon yield as a function of light and respiration. They are, nevertheless, restricted in their effectiveness to cases where anabolic processes are simultaneously constrained by some other factor (carbon or nutrient supply). What is then needed is the further sets of precisely quantifiable algorithms that will describe these further processes (many of which are available) and their incorporation into a supermodel to simulate the interactions among all the components. In a third type of model, the broad function of the system ('the box') is predicted from a knowledge of the fundamental mechanisms and limitations (such as genomic information and energy efficiency), as elegantly employed in Jørgensen's (1997, 2002) own structural–dynamic models of ecosystem organisation and thermodynamics.

Each of these approaches, even when applied directly to phytoplankton ecology, has its inherent weaknesses and these have long been recognised (Levins, 1966). The first type simulates an indirect relationship with accuracy and some precision but lacks general applicability. The second has a small number of variables and yields accurate and often precise information but only under very conditioned circumstances. The third achieves accuracy and applicability through generality, at the cost of all precision.

Modelling philosophy and (certainly) computing power has moved on. Several attempts to compound specific process models of the second type into more comprehensive growth simulations have been rather unsuccessful, except where one or other of the components contin-uously overrides the others. This was the case in the models of growth of filamentous Cyanobacteria in an enriched, monomictic lake (Jiménez Montealegre et al., 1995) or deep in the light gradient of the Zürichsee (Bright and Walsby, 2000). A most promising modern development has come through the exploration of linkages (stimulus, responses) and the probabilistic analysis of effects (likely reaction) through artificial neural networks (ANNs) (see Recknagel et al., 1997). Like the nerve connectivities they resemble, these models can be 'trained' against real data in order to predict outcomes with modified variables. Recknagel's (1997) own application to interpret the variability in the phytoplankton periodicity in Kasumigaura-Ko in Japan provides an excellent indication of the power of this approach. Its further development is at an early stage but the use of 'supervised' and 'unsupervised' learning algorithms to interpret field data, through 'self-organising maps' of close interrelationships (Park et al., 2003), promises to overcome some of the difficulties experienced with other compound models. Prediction of 'top-line' outcomes based on 'bottom-line' capacities is generally difficult without knowledge of intermediate processes. The fundamental truth is that algal growth rate is not a continuous function of nutrient supply or uptake, or of the ability to fix carbon in the light. Below the 'threshold' values discussed in Section 5.4.5, growth cannot for long exceed the weakest capacity. On the other hand, capacity in excess of the threshold saturates processing: it does not make organisms grow faster.

So, how can the growth rates of natural populations in the field be modelled? The approach advocated by Reynolds and Irish (1997) was to suppose that the photoautotrophic plankter does not grow anywhere better than it does in the contrived culture conditions in the laboratory. Given that the best growth performances of given species occur under ideal culture conditions, that they are consistent and that between-species differences in growth rate are systematic (Reynolds, 1984b), an 'upper base line' for simulating natural population growth could be proposed (Reynolds, 1989b). Three equations were invoked to predict attainable growth rate in the field. In the best traditions of Eppley's (1972) model of phytoplankton growth, two of the equations set the growth potential to water temperature. The first equation (5.5) predicts replication rate at a standard 20 °C as a function of algal morphology. The second equation (5.6) provides the information to adjust specific growth rate to other temperatures. These predictions are applied to an inoculum (or, in reiterations) to the incremented standing crop to simulate day-on-day accumulation. This has to be linked, through a loop in the model logic, to an inventory of resource supply, which checks that a given daily increment is sustainable and, if so, to permit the growth step to be completed. A further feedback loop deducts the consumption from the pool of available resources.

Insofar as the depth-integration of light intensity and the duration of daylight impose, almost everywhere on the surface of the planet, a sub-ideal environment with respect to continuously light-saturated cultures, the sensitivity of species-specific growth sensitivity to insolation is written into the third of the model equations. The original model supposed that the insolation-limited specific growth rate is in proportion to the fraction of the day that the alga spends in the light:

$$r_{(\theta, I)}d^{-1} = r_\theta \Sigma t_p / 24 \qquad (5.9)$$

where the daily sum of photoperiods, Σt_p, comes from:

$$\Sigma t_p = \Gamma h_p / h_m \qquad (5.10)$$

where h_m is the mixed depth and h_p is the height of the light compensated water column, which, following Talling (1957c; see also Section 3.5.3) is solved as:

$$h_p = \ln(I'_0 / 0.5 I_k)\varepsilon^{-1} \qquad (5.11)$$

where I'_0 is the daily mean irradiance immediately beneath the water surface (in μmol photon m^{-2} s^{-1}) and ε is the coefficient of exponential light attenuation with depth. The onset of light limitation of growth, I_k is here related specifically to the alga via Eq. (5.12):

$$I_k = r_\theta \alpha_r^{-1} \qquad (5.12)$$

where $\alpha_r = 0.257 \, (msv^{-1})^{0.236}$ (Eq. 5.8).

Thus, the solution to Eq. (5.9) incorporated in to the growth-rate model is:

$$r_{(\theta, I)}d^{-1} = [r_\theta \Gamma (24 \, h_m)^{-1}] \cdot \ln[2I'_0 \cdot$$
$$0.257(msv^{-1})^{0.236} \cdot r_\theta^{-1}]\varepsilon^{-1} \qquad (5.13)$$

Equation (5.13) is thus the third of the three equations written into the model that eventually became known as PROTECH. It remained under development and testing for several years, but this central core has remained intact. An important adjustment in respect of dark respiration was incorporated into the model that was eventually published (Reynolds et al., 2001). This followed the important steps of sensitivity testing, authentication and validation (Elliott et al., 1999a, 2000a). It has been used to make realistic reconstructions of phytoplankton cycles of abundance and composition (by functional type) in lakes and reservoirs (Elliott et al., 2000a; Lewis et al., 2002). It has been applied to simulate succession in undisturbed environments (Elliott et al., 2000b) and to investigate the minimum size of inoculum for the growth rate still to enable an alga to attain dominance (Elliott et al., 2001b). Vertical mixing can be used as a variable to disturb community assembly (Elliott et al., 2001a) and to evaluate selective impacts of intearctions between variations in mixing depth and in surface irradiance (Elliott et al., 2002). PROTECH models exist with differing physical drivers (PROTECH-C, PROTECH-D), that work in coastal waters (PROTECH-M) and which are dedicated to (specific) rivers (RIVERPROTECH). Versions have been prepared for numerous UK and European lakes and reservoirs, with accumulating success. At time of the writing, summary papers are still in press.

5.6 | Summary

Cell replication and population growth are considered as a unified process and accorded exponential logarithmic units with the dimensions of time. The observed rates of population change in nature ($\pm r_n$) are net of a series of in-situ rates of loss (r_L, treated in Chapter 6). However, the true rates of replication (r') must always be within (and are sometimes well within) the least rate that is physiologically sustainable on the basis of the resources supplied. Cell replication is regulated internally and cannot occur without the prior mitotic division of the nucleus. Nuclear division is prevented if the cell does not have the resources to complete the division. If it does, replication can proceed at a species-specific maximum rate.

In general, division rates are strongly regulated by temperature. Species-specific division rates at 20 °C (r'_{20}) are correlated with the surface-to-volume (sv^{-1}) ratios of the algal units, as are the slopes of the temperature sensitivity of growth. The light sensitivity of growth is subject to physiological photoadaptation but, ultimately, the shape of the alga influences its effectiveness as a light interceptor. Species which offer a high surface-to-volume ratio through distortion from the spherical form (such that the maximum linear dimension, m, is rather greater in one plane than in one or two of the others) and show high values of the product msv^{-1} are indicative of probable tolerance of low average insolation. Given that the potential daily photon harvest becomes severely constrained by vertical mixing through variously turbid water to depths beyond that reached by growth-saturating levels of downwelling irradiance, enhanced light interception becomes vital to maintaining growth. Below a threshold of about 1–1.5 mol photons m^{-2} d^{-1}, growth-rate performances are maintained relatively better in plankters offering the combination of relevant morphological preadaptation and physiologically mediated photoadaptative pigmentation.

Growth rates of phytoplankton species are variously sensitive to nutrient availability, though not at concentrations exceeding 10^{-6} M DIN or 10^{-7} M MRP. Above these 'critical' levels, growth rates are neither nitrogen nor phosphorus limited. Neither, it is argued, are they dependent upon the ratio of either of these resources to the other. However, differing species-specific nutrient-uptake affinities influence the potential growth performances when nutrient availabilities fall to, or are chronically below, the 'critical' levels. Species with high affinity for phosphorus are able to maintain a faster rate of growth than

species having weaker affinities and may outperform them by building up larger inocula than potential competitors. Chronic or eventual deficiencies in nitrogen may have an analogous selective effect, although, subject to the fulfilment of other criteria (available phosphorus, high insolation and a supply of relevant trace metals; see Chapter 4), dinitrogen fixers may experience selective benefit.

In the frequent cases in which nutrient depletion is experienced first in the near-surface waters of the upper mixed layer and proceeds downwards, high affinity may be less helpful than high mobility. The beneficial ability to undertake controlled downward migration with respect to the relevant 'nutricline' or to perform diel or periodic forays through an increasingly structured and resource-segregated medium, is conferred through the combination of self-regulated motility and large organismic size. The greater is the isolation of the illuminated, nutrient-depleted, upper column from the dark but relatively resource-rich lower column, the greater is the value of such performance-maintaining adaptations.

In the face of silicon shortages, diatoms are unable to perform at all, whereas the performances of non-diatoms is normally considered insensitive to fluctuations in supply. The amounts of silicon required vary interspecifically, as do the affinities for silicic-acid uptake. Because diatom nuclei divide before the silicic acid needed to form the daughter frustules is taken up, it is possible for otherwise resource-replete and growing populations to experience big mortalities as a consequence of sudden encounter with Si limitation. With other nutrients, planktic algae are able to 'close down' vegetative growth and to adopt a physiological (or perhaps morphological) resting condition, which improves the survival prospects for the genome.

The environments of phytoplankton may be classified, following Grime (1979), upon their ability to sustain autotrophic growth, in terms of the production that the resources will sustain and the duration of the opportunity. Most algae will grow under favourable, resource-replete conditions during long photoperiods. The species that perform best rely on an early presence, rapidity in the conversion of resources to biomass and a high frequency of cell division and recruitment of subsequent generations. By analogy with Grime's (1979) functional classification of plants, these algae are considered to be C strategists; they are typically small ($<10^3$ μm^3), usually unicellular, have high sv^{-1} ratios (>0.3 μm^{-1}) and sustain rapid growth rates ($r'_{20} > 0.9$ d^{-1}). Algae whose growth performance is relatively tolerant of and adaptable to progressively shorter photoperiods and aggregate light doses are comparable to Grime's ruderal (R) strategists: their sizes are varied (10^3–10^5 μm^3) but all offer favourable msv^{-1} ratios (range 15–1000). Algae whose growth performances are maintained in the face of diminishing nutrient availability are equipped to combat resource stress. Their conservative, self-regulating S strategies are served by the properties of large size (10^4–10^7 μm^3) and motility but at the price of low sv^{-1} (<0.3 μm^{-1}), low msv^{-1} (<30) and slow rates of growth ($r'_{20} <0.7$ d^{-1}).

Neither large size nor motility offer any advantage to survival in chronically resource-stressed environments of the ultraoligotrophic oceans and the largest lakes. Moreover, the extreme resource rarification also offers a respite from direct consumption. Not only are there resource-gathering constraints against large size but the smallest picoplanktic sizes of photoautotrophs have these provinces of the aquatic environment very nearly to themselves. It is proposed here that they be henceforth referred to as 'SS strategists'.

Many algae have adaptations and biologies that represent intermediate blends of C-, S- or R-strategist adaptations and lifestyles. The spatial and temporal distributions of particular types and species of phytoplankton, and the opportunities of replication that lead to population development, are shown to be closely correlated with the extent of their C, S or R attributes. Though best demonstrated among the freshwater phytoplankton, the functional–strategic approach of Grime appears to hold just as well for the ecologies of the marine plankton. In both the sea and fresh waters, morphological and (presumably) physiological criteria are better predictors of ecology than are phylogenetic affinities. In a concluding section, this finding is reversed to show how functional properties of phytoplankton and their respnses to environmental drivers can be used to predict the structure of ascendent phytoplankton communities on the basis of their likely strategic growth responses and not the stochasticity of the processes befalling individual species.

Chapter 6

Mortality and loss processes in phytoplankton

6.1 | Introduction

This chapter considers the sinks and, more particularly, the dynamic rates of loss of formed cells from phytoplankton populations. Several processes are involved – hydromechanical transport, passive settlement and destruction by herbivores and parasites – which, separately or in concert, may greatly influence the structuring of communities and the outcome of competitive interactions among phytoplankton. Moreover, these same processes may contribute powerfully to the biogeochemical importance of pelagic communities, through their role in translocating bioproducts from one point of the planet's surface to another.

Before expanding upon these processes, however, the opportunity is taken to emphasise that the losses considered in this chapter are those that affect the dynamics of populations. The (sometimes very large) loss of photosynthate produced in excess of the cell's ability to incorporate in biomass is not considered here. The topic is covered in a different context in Chapter 3 (see especially Section 3.5.4). The emphasis is necessary as the term 'loss rates' was applied collectively to the dynamics of almost all measurable photosynthetic production that did not find its way into increased producer biomass (Jassby and Goldman, 1974a). It had been supposed by many workers at the time that the realised shortfall was attributable to grazing and sedimentation of biomass. However, with the demonstration that, very often, production in some systems

was almost wholly and precisely compensated by simultaneous bulk loss rates (Forsberg, 1985), when the rates of grazing or sedimentation might only rarely explain the disappearance of the equivalent of the day's new product, it became clear that some further separation of the 'losses' was necessary, together with some refinement of the terminology. Here, adjustments to the photosynthate content that the cell is unable to deploy in new growth or to store intracellularly and which must be dispersed through accelerated respiration, or photorespiration, or secretion as glycollate or other extracellular product, are considered to be 'physiological'. The adjustments are as much to protect the intracellular homeostasis as to supply any other component of the pelagic system. On the other hand, successfully replicated cells in growing populations are continuously but variably subject to physical or biological processes that deplete the pelagic concentrations in which they are produced. Detracting from the numbers of new cells added to the population, these losses are 'demographic' and, as such, are the proper focus of the present chapter. Its objective is to establish the quantitative basis for estimating the drain on the potential rate of recruitment, provided by cell replication, that is represented by the counteracting processes which, effectively, dilute the recruitment of phytoplankton biomass. In the sense of Eq. (5.3), the task is to quantify the magnitude and variability in the rates of dilution of finished cells (r_L).

As already suggested, the principal loss processes are hydomechanical dispersion (wash-out from lakes, downstream transport in rivers, patch dilution at sea), sedimentation and consumption by grazers. Attention is also accorded to mortalities through parasitism (a specialised consumption) and physiological death and wastage. Although highly disparate in their causes, each process has the effect of diluting the locally randomised survivors. Hence, each is describable by an exponent, summable with other loss and growth terms. Just as Eq. (5.1) explains the rate of population change, $\delta N/\delta t$, by reference to the first-order multiplier, $e^{r_n t}$, and where, from Eq. (5.3), it may be asserted that $r_n = r' - r_L$, it may now be proposed that:

$$r_L = r_W + r_S + r_G + r \ldots \qquad (6.1)$$

where r_W, r_S, r_G ... are the respective exponents for the instantaneous rates of biomass loss due to wash-out, sedimentation, grazing, etc. It is accepted that these terms, either individually or in aggregate, may raise $r_L > r'$, in which case, r_n is negative and symptomatic of a declining population.

The following sections will consider the magnitudes and variabilities of the loss terms.

6.2 | Wash-out and dilution

The hydraulic displacement and dispersion of phytoplankton is best approached by considering the case of algae in small lakes or tidal pools in which the volume is vulnerable to episodes of rapid flushing. Inflow is exchanged with the instantaneous lake volume and embedded planktic cells are removed in the outflowing volume that is displaced. In this instance, the algae thus removed from the water body are considered 'lost'. It may well be that the individuals thus 'lost' will survive to establish populations elsewhere. Indeed, this is an essential process of species dispersal. The balance of the original population that remains is, of course, now smaller and, occupying the similar volume of lake, on average, less concentrated. The predicted net rate of change in the depleting population may be offset or possibly compensated by the simultaneous rate of cell replication but, for the moment, we shall consider just the effects of biomass loss on the residual population. The essential question is whether the inflow simply replaces the original volume by direct displacement or by a flushing action, in which the inflow volume mixes extensively with the standing volume, displacing an equivalent volume of well-mixed water. In the latter case, the original volume and the algal population that it entrained will have been depleted less completely and will now be, on average, less dilute.

6.2.1 Expressing dilution

The mathematics of dilution are well established. Dilution of the standing volume and its suspended phytoplankton is described by an exponential-decay function. Until Uhlmann's (1971) consideration of the topic, there had been few attempts to express the dilution of phytoplankton by displacement. He was quick to see and to exploit the opportunity to put losses in the same terms as recruitment (cf. Eq. 5.3) and simply sum the instantaneous exponents. As depletion rates to settlement and some forms of grazing succumb to analogous exponential functions, it is perhaps helpful to rehearse the logic that is invoked.

In the present case of dilution by wash-out, we suppose that a population of uniform, nonliving, isopycnic particles (N_0) is fully entrained and evenly dispersed through the body of a brimfull impoundment of volume, V. The introduction of a volume of particle-free water, q_s, in unit time t, displaces an equal volume into the outflow from the impoundment. Thus, the theoretical retention time of the impoundment, t_r, is given by V/q_s. The outflow volume will carry some of the suspended particles. From the initial population (N_0), particles will be removed in the proportion $-q_s/V$. After a given short time step, t, the population remaining, N_t, is calculable from:

$$N_t = N_0(1 - q_s t/V) \qquad (6.2)$$

During a second time period, of identical length to the first, an equal proportion of the original might be removed but only if the remainder original population has not been intermixed with and diluted by the inflowing water. If there is mixing sufficient to render the residue uniform at t, then, at t_2, we should have:

$$N_2 = N_t(1 - q_s t/V)$$
$$= N_0(1 - q_s t/V)(1 - q_s t/V).$$

Thus, after i such periods of length t,

$$N_i = N_0(1 - q_s t/V)^i \qquad (6.3)$$

and after one lake retention time (t_w)

$$N_{t_w} = N_0(1 - q_s t/V)^{tw/t} \qquad (6.4)$$

This is, of course, an exponential series, which quickly tends to

$$N_{t_w} = N_0\, e^{-1} \qquad (6.5)$$

where e is the natural logarithmic base. Equation (6.5) has a direct mathematical solution ($N_{tw} = 0.37 N_0$) which predicts that, at the end of one theoretical retention time, the volume diluted by flushing retains 0.37 of the original population. Had the same volume simply been displaced by the inflow, the retained population would have fallen to zero. These possibilities represent the boundaries of probable dilution, lying between complete mixing with, and complete displacement by, the inflow volume.

Supposing the tendency is strongly towards the flushing of algae by inflow, Eq. (6.5) may be used to estimate the residual population at any given point in time, t, so long as the same rates of fluid exchange apply:

$$N_t = N_0\, e^{-t/tw}$$
$$= N_0\, e^{-q_s t/V} \qquad (6.6)$$

To now derive a term for the rate of change in the standing population that is attributable to outwash (r_w in Eq. 6.1) is quite straightforward, provided that the time dimension of the inflow rate (s^{-1}, d^{-1}) is compatible with the other terms.

$$r_w = q_s/V \qquad (6.7)$$

6.2.2 Dilution in the population ecology of phytoplankton

At first sight, the magnitude of hydraulic displacement rates (q_s) relative to the scale of standing volume in large lakes and seas seems sufficiently trivial for outwash to be discounted as a critical factor in the population ecology of phytoplankton. At the mesoscales of patch formation, where recruitment by growth is pitched against dilution through dispersion, it is possible to consider V as the patch size and q_s as the rate of its horizontal diffusion as being critical to the maintenance of patchiness (the models of Kierstead and Slobodkin, 1953; Joseph and Sendner, 1958; see also Section 2.7.2). In lakes much smaller than 10 km^2 in area, wherein the maintenance of large-scale developmental patches is largely untenable (that is, critical patch size usually exceeds the horizontal extent of the basin and any small-scale patchiness is very rapidly averaged) the proportion of q_s to V assumes increasing importance. While $t_w > 100$ days, small differences in hydraulic throughput may remain empirically inconsequential in relation to potential growth rate: r_w is <0.01 d^{-1}. Its doubling to 0.02 d^{-1} is still small in relation to the rates of growth that are possible. However, when the latter are themselves severely limited by environmental conditions, dilution effects can become highly significant. In the instance of *Planktothrix agardhii* considered in Section 5.4.5, performing at its best to grow under the mixed conditions of a temperate lake in winter, its replication rate of 0.16 d^{-1} will be insufficient to counter outwash losses when $t_w \leq 7$ days ($r_w \geq -0.16$ d^{-1}). If it is growing less well, the sensitivity to flushing clearly increases. Temperate lakes regularly experiencing retention times less than about 30 days seem not to support *Planktothrix* populations. In the persistently spring-flushed Montezuma's Well, Arizona, studied by Boucher *et al.* (1984) ($t_w < 9$ d), the distinctive phytoplankton comprises *only* fast-growing nanoplanktic species. In the English Lake District, some of the lakes have volumes that are small in relation to their largely impermeable, thin-soiled mountainous catchments, and which episodically shed heavy rainfall run-off (1.5–2.5 m annually). In Grasmere ($t_w \sim 24$ d annual average but, instantaneously, ranging from 5 d to ∞), Reynolds and Lund (1988) showed that the phytoplankton had almost to recolonise the water column after wet weather, while it required a long dry summer for *Anabaena* to become established. Wet winters also keep phytoplankton numbers low but, in the relatively dry winter of 1973, the autumn maximum of *Asterionella* persisted to merge with the spring maximum.

6.2.3 Phytoplankton population dynamics in rivers

The most highly flushed environments are rivers. Larger ones often do support an indigenous phytoplankton, usually in at least third- or fourth-order affluents, and sometimes in very great abundance (perhaps an order of magnitude greater concentration than in many lakes; I have an unpublished record of over 600 µg chl$a$ L^{-1} measured in the River Guadiana at Mourão, Portugal, under conditions of late summer flow). The ability of open-ended systems, subject to persistent unidirectional flow, to support plankton is paradoxical. It is generally supposed to be a function of the 'age' of the habitat (length of the river and the time of travel of water from source to mouth), for there is no way back for organisms embedded in the unidirectional flow. On the other hand, the wax and wane of specific populations in given rivers seem fully reproducible; they are scarcely stochastic events. Moreover, some detailed comparisons of the mean time of travel through plankton-bearing reaches of the River Severn, UK, with the downstream population increment would imply rates of growth exceeding those of the best laboratory cultures (Reynolds and Glaister, 1993). Downstream increases in the phytoplankton of the Rhine, as reported by de Ruyter van Steveninck *et al.* (1992), would require specific growth performances paralleling anything that could be imitated in the laboratory. It was also puzzling how the upstream inocula might be maintained and not be themselves washed out of a plankton-free river (Reynolds, 1988b).

These problems have been raised on many previous occasions and they have been subject to some important investigations and critical analyses (Eddy, 1931; Chandler, 1937; Welch, 1952; Whitton, 1975). However, it was not until relatively more recently that the accepted tenets advanced by the classical studies of (such as) Zacharias (1898), Kofoid (1903) and Butcher (1924) could be verified or challenged. New, quantitative, dynamic approaches to the study of the physics of river flow and on the adaptations and population dynamics of river plankton (potamoplankton) developed quickly during the mid-1980s. These were reviewed and synthesised by Reynolds and Descy (1996) in an attempt to assemble a plausible theory that would explain the paradoxes about potamoplankton. The principal deduction is that rivers are actually not very good at discharging water. Not only is their velocity structure highly varied, laterally, vertically and longitudinally, but significant volumes (between 6% and 40%) may not be moving at all. A part of this non-flowing water is, depending on the size of the stream, explained in terms of boundary friction with banks and bed but a large part is immobilised in the so-called 'fluvial dead-zones' (Wallis *et al.*, 1989; Carling, 1992). These structures are sufficiently tangible to be sensed remotely, either by their differentiated temperature or chlorophyll content (Reynolds *et al.*, 1991; Reynolds, 1995b). They are delimited by shear boundaries across which fluid is exchanged with the main flow. The species composition of such plankton they may contain is hardly different from that of the main flow but the concentration may be significntly greater. It has also been shown that the enhancement factor is a function of the fluid exchange rate and the algal growth rate: the longer cells are retained, the greater is the concentration that can be achieved by growing species before they are exchanged with the flow (Reynolds *et al.*, 1991). Each dead-zone has its own V and q_s characteristics and its own dynamics. The analogy to a little pond, 'buried in the river' (Reynolds, 1994b), is not entirely a trite one. Reynolds and Glaister (1993) proposed a model to show how the serial effects of consecutive fluvial dead-zones contribute to the downstream recruitment of phytoplankton. The recruitment is, nevertheless, sensitive to changes in discharge, fluid exchange and turbidity. It proposes a persistent advantage to fast-growing r-selected opportunist (*C*-strategist) phytoplankton species or to process-constrained (*CR*-strategist) ruderals, as later confirmed by the categorisation of potamoplankters of Gosselain and Descy (2002).

The model does not cover the many other fates that may befall river plankton or influence its net dynamics, especially consumption by filter-feeding zooplankton and (significantly) zoobenthos, including large bivalves (Thorp and Casper, 2002; Descy et al., 2003). Neither does it cover explicitly the issue of the perennation of algal inocula. However, in reviewing the available literary evidence, Reynolds and Descy (1996) argued for the importance of the effective meroplankty to centric diatoms whose life cycles conspicuously include benthic resting stages (*Stephanodiscus* and *Aulacoseira*, both common in potamoplankton generally, have proven survival ability in this respect). They also cited the remarkable studies of Stoyneva (1994) who demonstrated the role of macrophytes as shelters and substrata for many potamoplanktic chlorophyte species. The presence of such plants, in headwaters and in lateral dead-zones, provides a constant source of algae that can alternate between periphytic and planktic habitats. This is little different from the proposition advanced by Butcher (1932) over 60 years before. His prognoses about the sources of potamoplankton remain the best explanation to the inoculum paradox and the one aspect still awaiting quantitative verification.

6.3 | Sedimentation

6.3.1 Loss by sinking

Most phytoplankters are normally heavier than the water in which they are dispersed and, therefore, tend to sink through the adjacent medium. The settling velocity (w_s) of a small planktic alga that satisfies the condition of laminar flow, without frictional drag ($Re < 0.1$), may be predicted by the modified Stokes equation (2.16; see Sections 2.4.1, 2.4.2):

$$w_s = g(d_s)^2(\rho_c - \rho_w)(18\eta\phi_r)^{-1} \quad \text{m s}^{-1} \quad (2.16)$$

where d_s is the diameter of a sphere of identical volume to the alga, ($\rho_c - \rho_w$) is the difference between its average density and that of the water, and ϕ_r is the coefficient of its form resistance owing to its non-sphericity; η is the viscosity of the water, and g is the gravitational force attraction. In completely still water, particles may be expected to settle completely through a column of water of height h_w within a period of time (t')

not exceeding the quotient h_w/w_s (Section 2.6.2):

$$t' = h_w/w_s \quad \text{s} \quad (6.8)$$

The point that has been made at length and on many previous occasions is that planktic algae do not inhabit a static medium but one that is subject to significant physical movement. Forcing of its motion by buoyancy, tide, wind, Coriolis effect is resisted by the viscosity of the water. This resistance is largely responsible for the characteristically turbulent motion that predominates in surface waters of the sea, in lakes and rivers (see Section 2.3). Moreover, the turbulent velocities so overwhelm the intrinsic velocities of phytoplankton sinking that the organisms are effectively entrained and randomised through turbulent layers. However, it may be emphasised again that turbulent entrainment does not overcome the tendency of heavy particles to sink relative to the adjacent medium and, in boundary layers and at depths not penetrated by turbulence, particles are readily disentrained and more nearly conform to the behaviour expressed in Eq. (2.16).

Following Humphries and Imberger (1982), the criterion for effective entrainment (Ψ) is set by Eq. (2.19) (i.e when $w_s < 15[(w')^2]^{1/2}$) and is illustrated in Fig. 2.16. The depth of the mixed layer over which it applies (h_m) may be calculated from the Monin–Obukhov and Wedderburn formulations. It may often be recognisable from the vertical gradient of density ($\delta\rho w \geq 0.02$ kg m^{-3} m^{-1}) (Section 2.6.5) or, casually, from inspection of the vertical distribution of isotherms.

The estimation of sinking loss rates from a fully mixed water column (H) or a mixed layer (h_m) applies logic analogous to the dilution by wash-out of a fully dispersed population of particles subject to leakage across its lower boundary. Moreover, as sinking loss is the reciprocal of prolonged suspension, it is relatively simple to adapt Eqs. (2.20–2.25) to the sequence of steps traced in Eqs. (6.3–6.6) and to be able to assert that population remaining in the column, h_m, at the end of a period, t, of sustained and even sinking losses is predicted by:

$$N_t = N_0 e^{-t/t'}$$
$$= N_0 e^{-w_s t/h_m} \quad (6.9)$$

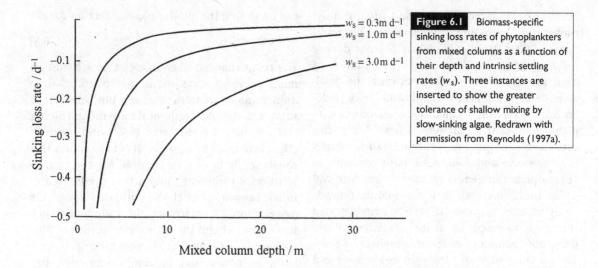

Figure 6.1 Biomass-specific sinking loss rates of phytoplankters from mixed columns as a function of their depth and intrinsic settling rates (w_s). Three instances are inserted to show the greater tolerance of shallow mixing by slow-sinking algae. Redrawn with permission from Reynolds (1997a).

Whence, the rate of change in the standing population that is attributable to sedimentation (r_s in Eq. 6.1) is:

$$r_s = w_s / h_m \qquad (6.10)$$

The equation expresses the sinking loss rate sustained by a population dispersed in a mixed layer.

It may be deduced that the continued residence of non-motile particles in the pelagic is dependent not only on having maximum entrainability (low w_s) but also on the settling velocity being small in relation to the mixed depth, h_m. As discussed in Section 2.6, the depth of mixing is an extremely variable quantity. Disentrainment is not a disadvantage for a swimming organism, especially not a large one, but non-motile organisms are highly vulnerable to variations in mixed depth (see Fig. 6.1). The growth rate of an alga with an intrinsic sinking rate of 3.5 μm s^{-1} (or ~0.3 m d^{-1}) may be able to exceed the leakage of sinking cells across the base of a 10-m mixed layer ($r_s \sim -0.03$ d^{-1}) but, in just 2 m ($r_s \sim -0.15$ d^{-1}), the sinking loss rate may become unsustainable. Species with greater settling rates experience proportionately more severe loss rates from any given mixed layer. Thus, they require yet deeper mixed layers to sustain positive net growth. On the other hand, greater mixing depths quickly begin to impose constraints of inadequate photoperiod-aggregation (see Sections 5.4.1, 5.5.3), and the difficulties of sustaining net intracellular carbon accumulation and its deployment in growth are increased accordingly (Section 3.3.3; see also Sverdrup, 1953; Smetacek and Passow, 1990; Huisman et al., 1999). It must be recognised that, in qualitative terms, larger non-motile plankters experience mixing that is 'too shallow' for growth to overcome their sinking velocity (because h_m is relatively small in relation to w_s in Eq. 6.10) or is 'too deep' (because h_m is relatively much larger than h_p in Eq. 5.10; see also Section 3.5.3 and Fig. 3.18) (O'Brien et al., 2003; see also Huisman et al., 2002).

6.3.2 Mixed depth and the population dynamics of diatoms

Included among the larger non-motile plankters are some of the larger freshwater desmids and, especially, the diatoms of the seas and of inland waters. The additional ballast that is represented by the complement of skeletal polymerised silica merely compounds the density difference component, $(\rho_c - \rho_w)$, in Eq. (2.16). Accepting that most species of phytoplankton are required to be either small or motile or to minimise excess density if they are to counter the inevitability of mixed-layer sinking losses, it is striking how poorly the diatoms represent all three attributes. Yet more perplexing is the fact that the planktic diatoms of freshwaters are relatively more silicified that their marine cousins (effectively raising ρ_c; Section 1.5.2 and Fig. 1.9), while the density of many non-saline inland waters (ρ_w) is less than that of the sea. Thus, the density difference of some freshwater diatoms may exceed $100-200$ kg m^{-3} (cf. Table 2.3). How are we to explain how this group of organisms, so relatively young in evolutionary terms, became so conspicuously successful as a component of the phytoplankton of both marine and fresh waters, when it has not only failed to comply to Stokes' rules but has actually gone against them by placing protoplasts inside a non-living box of polymerised silica? There is no simple or direct answer to this question, although, as has been recognised, sinking does have positive benefits, provided that subsequent generations experience frequent opportunities to be reintroduced into the upper water column (Section 2.5). In general, however, many of the ecological advantages of a siliceous exoskeleton were experienced first among non-planktic diatoms. As the diatoms radiated into the plankton, morphologies had to adapt rapidly: siliceous structures mutated into devices for enhancing form resistance and entrainability within turbulent eddies (Section 2.6). As was demonstrated in the case of Asterionella in the experiments of Jaworski et al. (1988) (see also Section 2.5.3), the configuration of the structures is overriding. Despite order-of-magnitude variations in colony volume and dry mass, as well as an approximate twofold variation in cell density, the sinking behaviour of Asterionella remains under the predominating influence of colony morphology. The corollary must be that the advantage of increased form resistance, and its benefits to entrainability, is greater than the disadvantage of increased sinking speed incumbent upon coenobial formation. The counter constraint, however, is that these diatoms are continuously dependent upon turbulence to disperse and to randomise them within the structure of the surface-mixed layer. As predicted by Eq. (6.10), positive population recruitment is always likely to be sensitive to the absolute mixed-layer depth (Reynolds, 1983a; Reynolds et al., 1983b; Huisman and Sommeijer, 2002).

The impact of this interplay between settlement and population dynamics of diatoms on their distribution in space and time is elegantly expressed in the study of Lund et al. (1963) of the seasonal variations in the vertical distribution of Asterionella formosa in the North Basin of Windermere (Fig. 2.21). The build-up in numbers during the month of April and, especially, towards the maximum in May reflect the general decline in vertical diffusivity. In the end, a near-surface concentration maximum is reached, followed by a rapid decline. In this instance, recruitment through growth was impaired by nutrient deficiencies (Lund et al. cited critical silicon levels but phosphorus is now seen likely to have been the more decisive; see Section 5.5.2). However, it is quite clear from the isopleths that the decline in concentration in the upper 10 m or so is extremely rapid. It is compensated, to an extent, by a temporary accumulation in the region of the developing pycnocline. This behaviour is entirely consistent with elimination through sedimentation from the mixed layer and slow settlement through the weak diffusivity of the metalimnion, revealed in the case of non-living Lycopodium spores (Fig. 2.20). Particles continue to settle through the subsurface layers for many weeks after the population maximum and, indeed, after the surface layer has become effectively devoid of cells.

Heavy sinking losses are not exclusive to nutrient-limited diatom populations. The sensitivity of the population dynamics of diatoms to the onset and stability of thermal stratification in Crose Mere, a small, enriched lake in the English north-west Midlands, rather than to nutrient limitation, was shown by Reynolds (1973a). Diatoms such as Asterionella, Stephanodiscus and Fragilaria were lost from suspension soon after the lake stratified in late spring, even though the concentrations of dissolved silicon and phosphorus remained at growth-saturating levels. Lund (1966) had already argued for the positive role of turbulent mixing in the temporal periodicity of Aulacoseira populations. He showed, in a field-enclosure experiment in Blelham Tarn (incidentally, the one that inspired the construction of the renowned tubular enclosures in the same lake) that the periodicity could be altered readily by superimposing episodes of mechanical mixing by aeration (Lund, 1971).

A little later, the eventual Blelham enclosures (Fig. 5.11) were the site of numerous quantitative

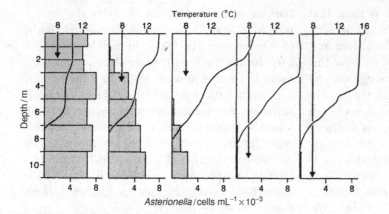

Figure 6.2 Instances in the loss from suspension of *Asterionella* cells in Blelham Enclosure A during 1980, in response to intensifying thermal stratification (shown by the temperature plots). Algal concentration is sampled in 1-m integrating sampler (Irish, 1980) and counted as an average for a 1- or 2-m depth band. The vertical arrows represent the depth of Secchi-disk extinction on each occasion. Redrawn from Reynolds (1984a).

studies of the fate of phytoplankton populations. One early illustration, cited in Reynolds (1984a), shows the depletion by settlement of a thitherto-active *Asterionella* population, following the onset of warm, sunny weather and the induction of a stable, near-surface stratification, and despite the availability of inorganic nutrients added to the enclosure each week (Fig. 6.2). In a further season-long comparison of loss processes in these enclosures (Reynolds *et al.*, 1982a), during which changes in extant numbers, vertical distribution, growth (as a function of silicon uptake) and sedimentary accumulation rates into sediment traps and at the enclosure bottom were all independently monitored, a steady 'leakage' of *Asterionella* cells was demonstrated over the entire cycle of net growth and attrition (see Table 5.5 and Reynolds and Wiseman, 1982). In the Blelham Enclosure B, *Asterionella* increased at a rate of 0.147 d^{-1} during its main phase of growth, net of sinking losses calculated to have been ~0.007 d^{-1}. The sedimenting cells intercepted by the traps were calculated to have been sinking at an average rate of ($w_s =$) 0.08 m d^{-1} (just under 1 μm s^{-1}) through a water column ($h_m =$) 11.7 m. As the population reached its maximum, the net rate of increase slowed (to 0.065 d^{-1}) but the sinking loss rate remained steady (−0.007 d^{-1}). However, shortening of the mixing depth led to an accelerated rate of sinking loss (to −0.044 d^{-1}; from a mixed depth of now only 7.5 m, a faster sinking rate of $w_s = 0.33$ m d^{-1} is also implied). More remarkably, as the epilimnion shrank to 4 m, the loss rate then rose to −0.242 d^{-1}, sustained by a sinking rate of 1.02

m d^{-1} (equivalent to 11.8 μm s^{-1}). The data are plotted in Fig. 6.3.

The accelerated sinking loss was contributed, in part, by an accelerated sinking rate. This was not unexpected. Reynolds and Wiseman (1982) had noted the altered physiological condition of the cells at the time, both in the plankton and in the sediment traps, drawing attention to the contracted plastids and 'oily' appearance of the contents. They attributed the changes to the sudden increase in insolation of cells caught in a stagnating and clarifying (cf. Fig. 6.2) epilimnion at the same time that temperature and light intensity were increasing. They suggested that the changes were symptomatic of photoinhibition. When Neale *et al.* (1991b) made similar observations on diatom populations in other lakes, they made the similar deduction. A positive feedback implied by the sequence of more insolation → more stratification → more inhibition → faster sinking rates → faster sinking loss rates has a satisfying ring of truth. However, a modified interpretation would see the accelerated sinking rate as a withdrawal of the vital mechanism of reducing sinking rate (see Section 2.5.4) for the very positive purpose of escaping the high levels of near-surface insolation.

The bulk 'production-and-loss budgets' compiled by Reynolds *et al.* (1982a) for phytoplankton populations in the Blelham Enclosures and exemplified in Table 5.6 offer a clear account of the the fate of the total production. In the example given, 81% (confidence interval, 70–95%) of the *Asterionella formosa* produced in Enclosure B in the spring of 1978 constituted the observed

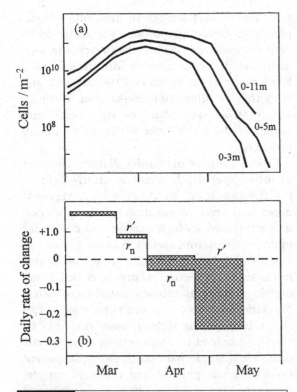

Figure 6.3 Net increase and attrition of an *Asterionella* population cells in Blelham Enclosure B during spring 1978. (a) Changes in the instantaneous areal cell concentrations in the water to 3, 5 and 11 m; (b) changes in the silicon-specific replication rate (r'_{Si}) and the net rate of population change (r_n); the hatched areas correspond to the rate of population loss, almost wholly to sinking. Redrawn with permission from Reynolds (1997a).

maximum. Around 95% (confidence interval, 72–123%) of the production was recruited intact to the sediment. The proportion of the cells produced that was estimated to have been lost to herbivores was probably <6% (see Sections 6.4.2, 6.7). The use of the adjective 'intact' is taken to include cells that may well have been dead by the time they reached the sediment surface. Judged from weekly recoveries from sediment traps placed about ~1 m above the sediment (and to which preservative was added), Reynolds and Wiseman (1982) observed that the proportion of live cells was always greater than 89% throughout the course of the population rise and decline. The proportion of live cells in the superficial sediment (supposedly dominated by the most recently recruited material) was 92% at the beginning of April. By the end of the month, it had fallen to 67%, to <2% by the end of July and to zero by the first week in September.

As part of the same investigation, Reynolds and Wiseman (1982) compared the rates of production, sedimentary fluxes and sediment recruitment of several other species forming major populations in the same enclosures. Of the estimated summer production of another diatom, *Fragilaria crotonensis*, at least 49% (statistically, possibly all) of the production was recruited to the sediments. In contrast, sedimentation could explain the fate of no more than 4% of the observed population maxima of *Ankyra*, *Chromulina* or *Cryptomonas*. Intermediate between the extremes of heavy diatoms and nanoplanktic unicells, sediment and trap recoveries of *Eudorina* accounted for 55 ± 15% of the maximum standing crops. For *Microcystis*, the sedimentary behaviour was strongly seasonal, increasing from 8% to 100% through the autumn.

From the measurements of the production and eventual fate of phytoplankton in confined, flat-bottomed Blelham enclosures, at least, the assertion that most of the larger diatoms are destined to be lost to sedimentation is strongly supportable. Scaling up to larger and deeper systems, subject to significant horizontal diffusive transport, the deduction requires some caution. In a 2-year study of sedimentary fluxes in the South Basin of Windermere (maximum depth 42 m), Reynolds *et al.* (1982b) found good, order-of-magnitude agreement between the annual fluxes into deep sediment traps and the maximal standing crops of five species of planktic diatom (*Asterionella formosa*, *Aulacoseira subarctica*, *Cyclotella praetermissa*, *Fragilaria crotonensis*, *Tabellaria flocculosa* var. *asterionelloides*) and two of desmid (*Cosmarium abbreviatum*, *Staurastrum* cf. *cingulum*). Interestingly, the magnitude of the fluxes (in numbers of cells m^{-2} d^{-1}) varied with the size of extant poulations but measurable fluxes to depth persisted through most of the year. This is presumed to reflect the relative proportion of the particle settling rates to the vertical distance to be traversed; this also fits with the observations of Lund *et al.* (1963) for the North Basin and the distribution of population isopleths plotted in

Fig. 2.21. Incidentally, the proportion of live *Asterionella* cells trapped fell from around 95% at the time of the May population maximum to just 3% in August and September. In the 100-or-so days that it takes some diatoms to settle through 40 m, many must perish, leaving only the empty frustules to continue downwards.

In contrast to the diatoms, the sedimentary fluxes of three colonial chlorophyte species (*Coenochloris fotti, Pseudosphaerocystis lacustris, Radiococcus planctonicus*), three Cyanobacteria (*Anabaena flos-aquae, Woronichinia naegeliana, Pseudanabaena limnetica*) and the dinoflagellate *Ceratium hirundinella* were 1–3 orders of magnitude smaller than the potential of the maximum standing crop. All these species either sink very slowly or they have sufficient motility to avoid being sedimented for long periods. Cryptomonads and nanoplankters were virtually unrecorded in any trap catches; they are presumed to have been subject to loss processes other than settlement.

These various findings supported the earlier deductions of Knoechel and Kalff (1978), who had applied a dynamic model to compare the effects of measured rates of growth, increase and settlement in order to calculate sinking loss rates of planktic populations in Lac Hertel, Canada. Their calculations showed that the rates of sinking loss were sufficient to explain most of the discrepancy between growth and the contemporaneous rate of population change, be it up or down. They were also able to provide quantified support for the idea that, whatever fate may befall them (nutrient, especially silicon, exhaustion, grazing, parasitism), planktic diatoms remain crucially sensitive to the intensity and extent of vertical mixing. Other workers who espoused this explanation for the seasonal fluctuations in diatom development and abundance in limnoplankton include Lewis (1978a, 1986), Viner and Kemp (1983), Ashton (1985) and Sommer (1988a).

There is now also ample evidence to support the qualitative contention of Knoechel and Kalff (1975) that sedimentation is a key trigger to the seasonal replacement of dominant diatoms by other algae. It is also plain that sedimentation is the principal loss to which limnetic diatoms are subject. In most other phytoplankton, greater proportions are either eaten or decompose long before they reach the sediment. The deduction concurs with the studies of losses made by Crumpton and Wetzel (1982) and with that of Hillbricht-Ilkowska *et al.* (1979) in Jezioro Mikołajske, Poland, on the seaonal variations in the main sinks of limnetic primary products.

The sensitivity of marine diatom dynamics to mixed-layer depth is not so clearly defined. On the one hand, net population increase is dependent upon an enhancement in insolation above thresholds which may be lower than for many other marine species (Smetacek and Passow, 1990) but the diminution of the mixed layer in the sea to the 1–3 m that may be critical to net diatom increase is inconclusively documented. Nevertheless, oceanic diatom populations experience considerable sinking losses that may be sustained only at or above certain levels of productivity. It is inferred that these are dependent upon adequate physical and chemical support (Legendre and LeFèvre, 1989; Legendre and Rassoulzadegan, 1996; see also Karl *et al.*, 2002).

As to the question posed by Huisman *et al.* (2002) about the long-term persistence of sinking phytoplankton, we have shown that there are obvious short-term benefits in being able to escape surface stagnation and resultant damaging levels of insolation in the near-surface waters (Reynolds *et al.*, 1986). Provided there is an opportunity for surviving propagules to be re-established within the photosynthetic range, the sooner may the longer-term benefit of population re-establishment be realised. Particle aggregation and, especially, the formation of 'marine snow' (Alldredge and Silver, 1988) may contribute effectively to accelerated sinking and to the escape from high-insolated surface layers. Aggregation may also serve to provide microenvironments that slow down the rate of respirational consumption and resist frustular dissolution of silicon (Passow *et al.*, 2003). The mechanisms of accelerated sinking may also add to the longevity of clone survival and facilitate the improved prospect of population re-establishment when more suitable growth conditions are encountered.

6.3.3 Accumulation and resuspension of deposited material

As has already been discussed, settling is not exclusively a loss process in the population dynamics of phytoplankton: the recruitment of resting propagules to the bottom deposits is recognised to constitute a 'seed bank' from which later extant populations of phytoplankters may arise (see Section 5.4.6). For this to be an effective means of stock perennation and mid- to long-term persistence in a given system, however, there has to be a finite probability of settled material both surviving on the sediments and, thence, of re-entering the plankton. The species-specific regenerative strategies of phytoplankton – roughly their ability to survive at the bottom of the water column and the means of 'escape' to the overlying water column – are extremely varied, ranging from the conspicuous production of morphological and/or physiological resting stages, with an independent capacity for germination, regrowth and reinfection of the water column (as in the case of *Microcystis* or *Ceratium*), through a range of resting cysts and stages whose re-establishment in the water depends upon still-suspended or resuspended propagules encountering tolerable environmental conditions (as is true for akinetes of nostocalean Cyanobacteria, certain species of volvocalean and chrysophyte resting cysts and the distinctive resting stages of centric diatom), to those that seem to make virtually no such provision at all (see Section 5.4.6).

In most instances, the settlement of vegetative crops should be regarded as terminal. Vegetative cells sinking onto deep, uninsolated sediments have little prospect but to slowly respire away their labile carbohydrates, pending depth. Resting cysts may remain viable for many years (64 a is a well-authenticated claim of viability of *Anabaena* akinetes: Livingstone and Jaworski, 1980) but without the mechanical resuspension of the resting spores in insolated, nutrient-replete water, the reinfective potential remains unrealised. Once settled to the bottom of a water column, the most likely prospect is progressive burial by the subsequent sedimentary recruitment of further particulate material, including fine, catchment-derived silts and particulate organic matter, the exuviae and excreta of aquatic animals and a rain of sedimenting phytoplankters, especially of non-motile diatoms.

Several studies have attempted to focus on the nature of the freshly sedimented material in lakes and its immediate fates. For a time, the newest recruited material remains substantially uncompacted and floccular, like a fluff, on the immediate surface. It comprises live or moribund vegetative cells, often bacterised or beset with saprophytic fungal hyphae, and resembles on a smaller scale, the structure of 'marine snow' (Alldredge and Silver, 1988; see Section 6.3.2). As its substance diminishes, however, it does become slowly compressed by later-arriving material. At the base of the semifluid layer, the same materials are progressively lost to the permanent sediment (Guinasso and Schink, 1975): compacting, losing water, perhaps leaching biominerals, the first stages of sediment diagenesis and formation are engaged.

Accordingly, the manner in which strictly ordered, laminated sediments might flow from the sequenced deposition of specific phytoplankton populations seems obvious. However, direct sampling of the semifluid layer from intact cores of the sediment water interface (Reynolds and Wiseman, 1982, used a syringe inserted into predrilled plastic tubes fitted to a Jenkin surface-mud sampler, as described in Ohnstad and Jones, 1982) reveals that sedimenting material undergoes a kind of sorting process. Once recruitment to the semifluid layer from the water column is effectively complete, its presence in the semifluid layer is found to decay exponentially. Moreover, the rates of dilution from the semifluid layer are not uniform but vary interspecifically, according to size and shape (Haworth, 1976; Reynolds, 1996b): long cells of *Asterionella*, filaments of *Aulacoseira* and chains of *Fragilaria* are diluted less rapidly from the semifluid layer than centric unicells of *Cyclotella* or *Stephanodiscus*.

Relative persistence in the surface layer improves the prospect of live specimens being restored to suspension in the water column, supposing that the physical penetration of adequate resuspending energy obtains. In general, friction in the region of the solid sediments creates a velocity gradient and a boundary layer of reduced water velocities, in which freshly settled plankton can accumulate (see Section 2.7.1). Resuspension is thus dependent upon the application of sufficient turbulent shear force to compress the boundary layer to the dimensions of the settled particles or even beyond the resistance of the unconsolidated sediment to penetration by a shear force, by then competent to entrain and resuspend it (Nixon, 1988). Quantitative observations confirm the intuition that shallow sediments are rather more liable to resuspension than sediments beneath a substantial column of water, although the actual depth limits vary with sediment type and the energy of forcing (Hilton, 1985). In many small lakes, sediments at a depth greater than 5 m beneath the water surface are protected from wave action and from most wind-generated shear. In the short to mid term, resuspension may require physical forcing of seismic proportions, or depend upon disturbance by burrowing invertebrates or foraging behaviour of fish or diving animal (Davis, 1974; Petr, 1977). In contrast, shallow sediments (substantially <5 m) may be rather more routinely exposed to resuspension of sediment and, incidentally, the redispersion of sediment interstitial water that may be relatively enriched, with respect to the open water, with nutrients released in decomposition (see also Section 8.3.4). In the Blelham enclosures (see Fig. 5.11), very little resuspension of live phytoplankton, resting spores or even empty diatom frustules was observed from the universal deep sediments of Enclosures A or B but it was observed on numerous occasions in the graded Enclosure C (Reynolds, 1996b). Moreover, disturbance or removal of the semifluid sediment from the shallow-water station, C_S (Fig. 5.11, depth ~4.5 m), occurred at such times, whereas, the deeper station, C_D (depth ~12.5 m) was exempt from this. In the wake of such resuspension events, material was perceived to resettle uniformly at both stations. Over a series of resuspensions, a net transport of once-settled material from shallow areas to deep sites was deduced.

So far as the accumulation of sedimenting phytoplankton is concerned, near-permanent deposition follows analogous patterns to non-living particulate matter. At depths typically greater than 5 m, sedimenting material accumulates and builds up in layers, undergoing diagenesis under substantially anoxic conditions. Neither live vegetative cells nor most resting spores enjoy much prospect of return to suspension and regeneration. In contrast, similar materials settling onto shallow sediments are liable to resuspension. The viable fractions (vegetative cells, resting spores) may well fulfil their infective potential and contribute directly to the establishment of extant, vegetative populations. This has been many times observed in the case of *Aulacoseira* populations (Lund, 1954, 1966, 1971) and is inferred on other occasions involving other species (Carrick *et al.*, 1993; Reynolds *et al.*, 1993a). For the non-viable detritus, including empty diatom frustules, redeposition is the most likely consequence but with a finite proportion settling into deeper water. This is precisely the mechanism of the process of 'sediment focusing' (Lehman, 1975) whereby fine particulate material is moved progressively away from lake margins and towards greater basin depths (Hutchinson, 1941; Likens and Davis, 1975; Hilton, 1985)

6.4 | Consumption by herbivores

Sharing an apparently refugeless, open-water habitat with a variety of phagotrophic animals, phytoplankton is generally vulnerable to severe physical biomass losses and, at best, to the dynamic drain on the potential recruitment of biomass. In fact, there are many types of consumer, each with differing food preferences and habitat demands, making for an extremely wide range of possible outcomes. The subject of food and feeding is, indeed, a broad one, and rather beyond the remit of the present chapter, the focus of which will remain trained on the dynamic consequences for the producers. However, even this modest ambition must take some account of the biologies of the consumers and how their impacts fluctuate in time and space.

What follows here is necessarily selective, emphasising those aspects of zooplanktic biology which relate to phytoplankton dynamics and to the shaping of pelagic ecosystems. Numerous books and monographs describing the biology and ecology of particular zooplankton groups are available. Of the more general accounts, none had equalled those of Hutchinson (1967) or Raymont (1983), until the recent publication of Gliwicz's (2003a) superb overview, to which the reader is happily referred. The emphasis here is on crustacean herbivory, with only acknowledgement of the part played by small herbivorous fish in cropping phytoplankton in (usually) small tropical lakes (see Fernando, 1980; Dumont, 1992). The present account also recognises that planktic primary products are consumed not only as the particulate foods of herbivores but also as the dissolved substrates of aquatic microorganisms.

6.4.1 The diversity of pelagic phagotrophs and their foods

Zooplankton comprises small animals suspended in the water. Some (nanozooplankton and microzooplankton, all <200 µm) are truly planktic in the sense of being fully embedded in the eddy spectrum (Section 2.3.4) but even most larger forms of mesoplankton (0.2–2 mm) are too small and too weak to escape entrainment by open-water currents. A characteristic of all zooplankters is that they are partly or wholly phagotrophic – much or all of their organic carbon and energy requirements are satisfied by feeding on live or detrital organic particles. Another feature of zooplankton, which, broadly, is shared with phytoplankton, is the cosmopolitan distribution of many genera and even some species. This may seem obvious in contiguous seas but it applies no less to the plankton of inland waters. The main difference between the zooplankton of seas and lakes is the much poorer phylogenetic representation in the latter. Whereas certain life-history stages of certain species from almost every animal phylum occur in the marine plankton, limnoplankton mainly comprises protists, rotifers and crustaceans. Major groups and representative planktic genera are summarised in Table 6.1.

In briefly surveying the diversity of zooplankton within the context of its quantitative impacts on the phytoplankton and the flow of carbon through the pelagic towards its larger metazoan beneficiaries, it is useful to adopt a functional approach in preference to a taxonomic one. There are important differences in the composition, ecological function and key selective mechanisms between the nano-/micro-planktic and meso-/macro-planktic components and, indeed, among the principal types of mesozooplanktic association.

Protistan microzooplankton

In terms of numbers, the most abundant and most common zooplankters, both in lakes and in the sea, belong to the category of nano-/microzooplankton. This includes all planktic heterotrophs in the size range 2–200 µm, with the exception of the bacteria, actinomycetes and moulds (Sorokin, 1999). Rather than be pedantic about the nano–micro separation, it is convenient to follow Sorokin's (1999) usage of the word 'microzooplankton' to apply to all heterotrophic protistans and metazoans smaller than 200 µm. This then includes representatives of protistan groups already listed as phytoplankton in Table 1.1 (especially Chrysophyta and Dinophyta). As pointed out in Section 1.3, many of these are photoautotrophs with a phagotrophic capability (that is, that they are mixotrophic) but the mesoplanktic, colourless marine dinoflagellates, such as *Noctiluca* and *Oxyrrhis*, are obligate phagotrophic consumers, feeding on nanoplankters, including the haptophyte *Prymnesium* (Tillmann, 2003). Freshwater mixotrophs in the nanoplanktic size range include *Chromulina*, *Chrysococcus* and *Ochromonas*. *Chrysochromulina* spp. and *Prymnesium* spp. fulfil this role in the sea (Riemann *et al.*, 1995). They ingest particles in the picoplanktic size range, including bacteria and algae. The nanoheterotrophs also include numerous small flagellated protists, classified in Table 6.1 as Zoomastigophora. These have well-developed cytostomes and they are able to ingest substantial food particles, up to about 5–8 µm in size. Free-living bodonids and protomonadids are represented in the nanoplankton of lakes and seas, where they can be effective consumers of nanophytoplankton. The group includes the choanoflagellates and bicosoecids that are usually attached to the surfaces of larger plankters such as diatoms. At this scale, deep within the viscous range of the eddy spectrum

Table 6.1 *Zooplankton in marine and freshwater habitats*

Phylum: **Zoomastigophora**

Four orders of free-living and epiphytic or epizoic flagellates.
Includes: *Bodo, Monas, Peranema* (marine and freshwater);
Bicosoeca, Salpingooeca and *Monosiga*

Phylum: **Dinophyta**

Several families of marine dinoflagellate are mixotrophic or primarily heterotrophic.
Includes: *Dinophysis, Noctiluca, Oxyrrhis, Protoperidinium*

Phylum: **Rhizopoda (Sarcodina)**

Four main divisions.

Order: AMOEBINA

Naked, lobose protists.
Plantic genera include: *Asterocaelum, Pelomyxa* (freshwater)

Order: FORAMINIFERA

Amoeboid protists with non-calcareous shells.
Includes: *Globigerina* (marine); *Arcella, Difflugia* (freshwater)

Order: RADIOLARIA

Marine planktic sarcodine protists having central capsule and usually a skeleton of siliceous spicules.
Includes: *Acanthometra*

Order: HELIOZOA

Mostly freshwater sarcodines with axopodia and, typically, vacuolated cytoplasm and a siliceous skeleton.
Includes: *Actinophrys*

Phylum: **Ciliophora**

Class: CILIATA

Non-amoeboid protists that possess cilia during part of their life cycle: several planktic orders, including:

Order: HOLOTRICHA

Uniformly ciliated.
Includes: *Colpoda, Prorodon, Pleuronema* and freshwater *Nassula*

Order: SPIROTRICHA

Ciliates posessing gullet and undulating membrane.
Includes many common genera of marine and freshwaters: *Euplotes, Halteria, Metopus, Strobilidium, Strombidium, Stentor, Tintinnidium*

Order: PERITRICHA

Ciliates, usually attached to surfaces. Cilia reduced over body and confined to oral region.
Includes: *Epistylis, Vorticella, Carchesium*

Class: SUCTORIA

Ciliophorans lose cilia in adult stage. Possess one or more suctorial tentacles.
Includes: *Acineta*

Phylum: **Porifera**

Amphiblastula larvae temporally in marine plankton.

Table 6.1 | (cont.)

Phylum: **Coelenterata**

Subphylum: Cnidaria

Coelenterates with stinging nematocysts. Several orders have genera that live, or appear, in (mostly marine) plankton

Class: HYDROZOA

Order: LEPTOMEDUSAE

Hydrozoan coelenterates with horny perisarc. Medusa stage in plankton.
Includes: *Obelia, Plumularia*

Order: ANTHOMEDUSAE

Hydrozoan coelenterates with horny perisarc that does not cover polyp base. Medusa stage in plankton.
Includes: *Hydractinia*

Order: HYDRIDA

Hydrozoan coelenterates without a medusa.
Includes: *Hydra*, young hydroids of which disperse through freshwater plankton

Order: TRACHYLINA

Hydrozoan coelenterates with a relatively large medusa and minute hydroid stage.
Includes: freshwater *Limnocnida, Craspedacusta*

Order: SIPHONOPHORA

Large, free-moving colonial hydrozoans.
Includes: *Velella, Physalia*

Class: SCYPHOZOA

Cnidaria that exist mostly as medusae. Several orders.
Includes: *Aurelia, Cyanea, Pelagia*

Subphylum: Ctenophora

Swimming coelenterates lacking nematocysts

Class: TENTACULATA

Ctenophores with tentacles ('sea gooseberries').
Includes: *Pleurobrachia*

Class: NUDA

Ctenophores lacking tentacles.
Includes: *Beroë*

Phylum: **Platyhelminthes**

Acoelomate metazoans (flatworms, many parasitic) with a few free-living representatives in the marine plankton.

Class: TURBELLARIA

Includes: *Convoluta, Microstomum*

Phylum: **Nemertea**

Flattened unsegmented worms with a ciliated ectoderm and eversible proboscis. Several orders, one with planktic genera.

Order: HOPLONEMERTINI

Pilidium larvae of several genera are dispersed as marine plankton. Some genera remain bathypelagic, beyond the continental shelf, in their adult stages.
Includes: *Pelagonemertes*

Table 6.1 | *(cont.)*

Phylum: **Nematoda**

Unsegmented round worms. Some shelf-water species have been reported but these may not be truly planktic.

Phylum: **Rotatoria**

Acoelomate metazoans with planktic genera widespread in the sea and in lakes. Most planktic forms belong to one order, mainly in fresh waters.

Order: MONOGONONTA

Sub-order: Flosculariacea

Free-swimming, soft-bodied.

Includes: *Conochilus*, *Filinia*.

Sub-order: Ploima

Free-swimming, usually with firm lorica but some illoricate.

Includes: *Asplanchna*, *Brachionus*, *Kellicottia*, *Keratella*, *Notholca*, *Synchaeta*, *Trichocerca*

Phylum: **Gastrotricha**

Minute unsegmented acoelomate metazoans. May be encountered in plankton of small freshwater bodies.

Includes:*Chaetonotus*

Phylum: **Annelida**

Coelomate segmented worms. One class has planktic representatives.

Class: POLYCHAETA

Several planktic genera, one of which is very well adapted to a pelagic existence. Trochospere larvae of some polychaetes are also temporarily resident in the plankton.

Includes: *Tomopteris*

Phylum: **Crustacea**

Large group of segmented, jointed-limbed arthropods, with many planktic representatives.

Class: BRANCHIOPODA

Free-living small crustaceans with at least four pairs of trunk limbs, flattened, lobed and fringed with hairs (phyllopods).

Two orders have planktic genera.

Order: ANOSTRACA

Branchiopodans lacking a carapace, phyllopods numerous.

Includes *Chirocephalus*, *Artemia*

Order: DIPLOSTRACA

Branchiopodans with a compressed carapace enclosing fewer than 27 pairs of phyllopods. Two sub-orders, one of which (Cladocera) includes several genera important in the plankton of sea and in lakes:

Evadne, *Podon* (marine); *Sida*, *Diaphanosoma*, *Holopedium*, *Bosmina*, *Daphnia*, *Ceriodaphnia*, *Moina*, *Simocephalus*, *Chydorus*, *Bythotrephes*, *Leptodora* (freshwater)

Class: OSTRACODA

Free-living small crustaceans with a bivalve shell, few trunk limbs, none being phyllopods.

Includes: *Gigantocypris* (marine), *Cypris* (freshwater)

Table 6.1 | *(cont.)*

Class: COPEPODA

Free or parasitic crustaceans lacking carapace or any abdominal limbs. Of some seven orders, two provide most free-living planktic forms. A third is well represented in the benthos and species are encountered in the pelagic.

Order: CYCLOPOIDEA

Copepods with short antennules of <17 segments.

Includes: *Oithona* (marine); *Mesocyclops, Tropocyclops* (freshwater)

Order: CALANOIDEA

Copepods with long antennules of >20 segments. Major group of mesozooplankters in the sea and in many lakes.

Includes: *Calanus, Temora, Centropages* (marine); *Eudiaptomus, Eurytemora, Boeckella* (freshwater)

Order: HARPACTICOIDEA

Benthic copepods with similar thoracic and abdominal regions and very short first antennae.

Includes: *Canthocamptus* (freshwater)

Class: BRANCHIURA

Crustacea, suctorial mouth, capacace-like lateral expansion of the head. Temporary parasites of fish.

Includes: *Argulus* (freshwater and estuaries)

Class: CIRRIPEDIA

Crustacea, sedentary and plated as adults; wholly marine; several orders, one of which (Order THORACICA: barnacles) whose cypris larvae are dispersed in the marine plankton.

Includes: *Balanus*

Class: MALACOSTRACA

Mostly larger crustacea, usually with distinct thoracic and abdominal regions. Thorax generally covered by a firm carapace; most abdominal segments bear appendages. Five major orders, two of which have typically planktic genera or ones whose juvenile stages are passed in the pelagic.

Order: LEPTOSTRACA

Primitive malacostracans with large thoracic carapaces. One genus is present in the bathypelagic beyond the continental shelf.

Includes: *Nebaliopsis* (marine)

Order: HOPLOCARIDA

Benthic malacostracans with shallow carapace fused to three thoracic somites. Larvae are temporary entrants to the plankton of warm seas.

Includes: *Squilla*

Order: PERACARIDA

Malacostracans in which the carapace is fused with no more than four thoracic segments. Several suborders include:

Sub-order: Mysidacea

Peracaridans with well-formed carapace. Mostly marine, 'opossum shrimps' are typically planktivorous in the deep layers of shallow seas. Some relict or invasive species in fresh waters.

Includes: *Leptomysis, Gastrosaccus* (marine); *Mysis* (freshwater)

Table 6.1 (cont.)

Sub-order: Cumacea
Mostly benthic in marine sublittoral sand and mud, whence animals are entrained into pelagic samples.
Includes: *Diastylis*

Sub-order: Isopoda
Peracaridans with a carapace covering three or four thoracic segments only. Most are not planktic. An exception is:
Eurydice (marine)

Sub-order: Amphipoda
Peracarida with no carapace. Body laterally compressed. Most of the shrimp-like animals are littoral or benthic but some euplanktic genera.
Includes: *Apherusa* (marine); *Macrohectopus* (freshwater genus of Lake Baykal)

Order: EUCARIDA
Malacostracans in which the capace is fused to all thoracic segments. Two main sub-orders:

Sub-order: Euphausiacea
Eucarida in which the maxillary exopodite is small and the thoracic limbs do not form maxillipeds. Large meso- and macroplankters, including krill, important as a food for whales.
Includes: *Euphausia, Nyctiphanes* (marine)

Sub-order: Decapoda
Eucarida in which the maxillary exopodite is large (the scaphognathite) and the first three pairs of thoracic limbs are modified as maxillipeds and the next five as 'legs'. These are the lobsters, prawns and crabs. Zoea, megalopa and phyllosoma larvae are temporary entrants to the marine plankton.
Includes: *Carcinus, Palinurus*

Phylum: **Arthropoda**
Class: HEXARTRA (INSECTA)
Larvae of two orders show distinct adaptations to planktic existence.
Order: MEGALOPTERA
First (especially) and second instars of alder flies are dispersed in the limnoplankton.
Includes: *Sialis*
Order: DIPTERA
Larvae of the Culicidae are typically associated with the littoral of lakes and many feed in very small, still bodies of water. Larvae of the subfamily Chaoborinae (or Corethrinae) are adapted for a larval life in the open, deeper waters of small lakes and lagoons.
Includes: *Chaoborus, Pontomyia*

Phylum: **Mollusca**
Unsegmented coelomates with a head, a ventral foot and a dorsal visceral hump, developed to varying extents. Of the five major classes, only two have typically planktic representatives.
Class: GASTROPODA
Slugs and snails. Molluscs with distinct head and tentacles.

Table 6.1 | *(cont.)*

Order: PROSOBRANCHIATA

Gastropods in which adults show torsion. Some marine genera dispersed by pelagic trochophore larvae.

Includes: *Patella*

Order: OPISTHOBRANCHIATA

Gastropod line showing secondary 'detorsion' and shell reduction. Several marine genera of pteropods or 'sea butterflies'.

Includes: *Limacina, Clio*

Class: LAMELLIBRANCHIATA

Bilaterally symmetrical molluscs more or less enclosed in a hinged bivalve shell. Certain genera dispersed by pelagic trochophore or veliger larvae.

Includes: *Ensis, Ostrea* [trochophore] (marine); *Dreissenia* [veliger] (freshwater).

Class: CEPHALOPODA

Bilaterally symmetrical molluscs, with well-developed head, and foot modified into a crown of tentacles. Some marine pelagic species release paralarvae into the plankton.

Includes: *Loligo*

Phylum: **Chaetognatha**

Slender, differentiated coelomates with distinctive head, eyes and chitinous jaws. Chaetognathes (arrow worms) are exclusive to the marine plankton, where they are carnivorous.

Includes: *Sagitta*

Phylum: **Ectoprocta**

Unsegmented sedentary coelomates, usually colonial. One division is exclusive to marine habitats and reproductive propagules are dispersed as trochophore-like cyphonautes larvae (freshwater ectoprocts produce dispersive statoblasts).

Order: PHYLACTLAEMATA

Marine ectoprocts producing planktic cyphonautes larvae.

Includes: *Flustra*

Phylum: **Phoronidea**

Small group of unsegmented tubicolous coelomates with affinities to the ectoprocts. The free-swimming actinotrocha larvae are encountered in the marine plankton.

Includes: *Phoronis*

Phylum: **Echinodermata**

Large group of coelomates in which adults show radial symmetry. Modern genera are exclusively marine, usually littoral or benthic but many are dispersed by planktic larvae. One genus known to have planktic adults.

Class: ASTEROIDEA

Starfish and seastars; pentagonal free-living benthic and littoral echinoderms. Many genera dispersed as pelagic bipinnaria larvae.

Includes: *Asterias*

Table 6.1 (cont.)

Class: OPHIUROIDEA

Brittle stars; discoid free-living benthic and littoral echinoderms with five radial arms. Many genera dispersed as pelagic pluteus larvae.
Includes: *Ophiura*

Class: ECHINOIDEA

Sea urchins; globular or discoid armless echinoderms. Many genera dispersed as pelagic pluteus larvae.
Includes: *Echinus*

Class: HOLOTHUROIDEA

Sea cucumbers; sausage-like, armless echinoderms. Many genera dispersed as pelagic auricularia larvae. One adult is bathypelagic in the South Atlantic Ocean.
Includes: (larval) *Holothuria*, adult *Pelagothuria*

Phylum: **Chordata**

Coelomate animals with notochord and gill slits and possessing a dorsal hollow central nervous system. One 'protochordate' subphylum and the vertebrate subphylum have planktic representatives.

Subphylum: UROCHORDA (TUNICATA).

Unsegmented, boneless chordates, notochord restricted to larval tail.

Class: ASCIDIACEA

Sedentary tunicates. Motile, appendicularia larvae ('ascidian tadpole') has well-developed notochord. They enter and are briefly resident in the the marine plankton but do not feed.
Includes: *Ciona, Clavelina*

Class: THALIACEA

Salps; pelagic tunicates of warm seas. Circumferential muscle bands are used to pump water through body, providing food and propelling animals forward. Tadpoles develop into zooids that are eventually set free as young salps.
Includes: *Doliolum, Salpa, Pyrosoma*

Class: LARVACEA

Neotenic pelagic tunicates in which the appendicularian tadpole becomes the sexual form. These are planktic but live within a secreted 'house', which itself resembles a salp, and which is equipped with filter. The larvacean's movements produces the filtration current.
Includes: *Oikopleura*

Subphylum: VERTEBRATA

Chordates that develop an articulated, bony or cartilaginous backbone. Of the eight or so extant classes, only one is considered to have typically planktic representatives.

Class: ACTINOPTERYGII

Bony fishes. Many types of pelagic, marine demersal and limnetic littoral feeding fish grow either from pelagic eggs and larvae or from larvae hatched from eggs on the sea bottom. Initially, at least, the larvae are truly microplanktic.

Compiled from several sources: Hardy (1956), Borradaile *et al.* (1961), Donner (1966), Reynolds (2001b).

and where feeding relies mainly on encounters of food organisms by consumers, the availability of a substratum to which to attach does not necessarily improve feeding efficiency. The heterotrophic nanoflagellate genera *Bodo*, *Monas*, *Bicosoeca*, *Monosiga* are represented by species in both the marine and the fresh water plankton.

Other microplanktic protists include sarcodines and ciliophorans, which may range in size between 10 and 200 μm. The radiolarians and foraminiferans are mainly marine (though the latter are represented in fresh waters). Most are phagotrophic on detritus and picoplankters but radiolarians also harbour algal symbionts. Amoebae are not major players in the plankton, generally contributing <5% of the nanoheterotroph biomass (Sorokin, 1999) but dense populations are now recognised in the region of hydrothermal vents.

Ciliophorans are often numerous and well represented in the microplankton of lakes and seas by some common and highly cosmopolitan species (Finlay and Clarke, 1999; Finlay, 2002). These include species of the naked spirotrichs (such as *Coleps*, *Strombidium*, *Strobilidium*) and holotrichs (*Prorodon*, *Pleuronema*), as well as the loricate *Tintinnidium*. In freshwater, vorticellids are epiphytic on large algae and use their oral cilia to move particles into the cytostome. Microzooplankters feed principally on nanoplanktic autotrophs and heterotrophs, thus fulfilling a key linkage in the transfer of carbon through the microbial food web (Sherr and Sherr, 1988) (see also Section 3.5.4). Ciliates may become the dominant animal in micraerophilous or anoxic sea water (such as in the Black Sea, at depths >30 m) or in the metalimnia of eutrophic lakes (Fenchel and Finlay, 1994). Other, mainly benthic or deepwater ciliates are frequently encountered in plankton samples from shallow water columns.

Among fresh waters, planktic ciliates that feed on larger or more specialised foods sometimes rise to become, usually for short periods, dominant consumers of algae or cyanobacteria or flagellates (Dryden and Wright, 1987). Some cases have been reported (Reynolds, 1975; Heaney et al., 1990) in which *Nassula* effectively removed the biomass of floating *Anabaena* from the surface layer of a lake, in one case the animals

ingesting so many gas vesicles that they became irreversibly buoyant. The remarkable ability of *Nassula* to ingest long filaments of green algae (*Binuclearia*) and Cyanobacteria (*Planktothrix*) by sucking them in, spaghetti-like, and coiling them intracellularly (Finlay, 2001, and personal communication) provides a striking case of the feeding adaptations of this animal.

Multicellular microzooplankton

Metazoans contribute to the composition of both the marine and freshwater microzooplankton but, despite some common features, the differences in the principal organisms represented are substantial. In the sea, they are dominated by larval crustaceans (especially the nauplii of resident copepods), rotifers and larvaceans as well as the larvae of other groups, such as molluscs and echinoderms (see Table 6.1). There is a wide range of food preferences (nanoplanktic autotrophs, heterotrophs and algae) as well as detrital particles. Cilia around the oral region, aided by muscular contractions around the feeding apparatus, or (in the case of rotifers and nauplii), mandibular mouthparts may be used to handle captured particles into the digestive tract but the main feeding strategy is still largely encounter. Larger microplankters may have some limited influence over orientation and exploration of the local environment but, mostly, they are still too small to overcome the problem of viscosity. The larvacean, *Oikopleura*, perhaps comes closest to generating and filtering a significant feeding current through its secreted 'house' (see Table 6.1).

The multicellular microzooplankton of lakes is often dominated by rotifers and copepod nauplii. Feeding among the freshwater rotifers has been studied extensively; the comprehensive reviews of Donner (1966) and Pourriot (1977) continue to provide helpful guides. Production of the most common pelagic rotifers in lakes (*Keratella*, *Brachionus*, *Synchaeta*, *Polyarthra*) responds well to abundant populations of nanophytoplankton but there is an evident selectivity which is influenced by the size of the food organisms (Gliwicz, 1969). For instance, the relatively robust *Keratella quadrata* experiences an upper size limit of ingestion of 15–18 μm but, for the smaller *K. cochlearis*, it is only 1–3 μm. Thus, it is hardly surprising that, in the experiments conducted by Ferguson *et al.* (1982), both species coexisted and were, simultaneously, able to increase while in the presence of a phytoplankton with abundant populations of *Chlorella*, *Rhodomonas* and *Cryptomonas*. However, only *K. quadrata* flourished among a substantially unialgal *Cryptomonas* nanoplankton.

Freshwater mesozooplankton

The mesozooplanktic element (0.2–2 mm) embraces what are, to many people, the most familiar components of the zooplankton. These animals are big enough not just for their movements to escape the constraints of viscosity but for them to be able to exploit turbulence and current generation to optimise contact with potential foods (Rothschild and Osborn, 1988). The groups of animals that are principally involved are really quite diverse and, beyond some shared adaptations, they are fitted to mutually distinct life modes and fulfil different ecological roles. Any residual notion that zooplankton just eat phytoplankton must be rejected as being crassly oversimplistic! The common adaptations include the tendency to transparency (to minimise their visibility to planktivorous fish, crustacea and other potential predators) and the ability to propel themselves through the water. Individual adaptations concern their means of movement and, especially, their means of gathering their required food intake.

In most fresh waters, the mesozooplankton comprises crustaceans from two main classes (see Table 6.1): copepods (from two orders in particular, the cyclopoids and the calanoids) and branchiopods (of the sub-order Cladocera). Adult (final-instar) planktic cyclopoids are faintly pear-shaped and streamlined with the paddle-like thoracic legs held under the body. They have short biramous antennules and several caudal rami of varying lengths. Cyclopoids swim gently by reciprocation of the antennules and antennae but they can 'pounce' as well, through the simultaneous use of the thoracic limbs. They are raptorial feeders (they seize their food) on a range of nano- and micro-planktic particles including algae, rotifers and detrital particles. There are numerous species in several genera; many are confined to ponds, some are primarily benthic or littoral in distribution. *Cyclops vicinus* and *Mesocyclops leuckarti* are species that are common and widely distributed in the plankton of larger lakes.

Adult calanoids are more cylindrical, have long antennules which are extended laterally to 'hang' on a rotational current generated by the beating of the paddle-like thoracic limbs. The animal moves abruptly to a new position by a single beat of the antennules (Strickler, 1977). Calanoids are able feed on small (~4-μm) algae and bacterised particles which are filtered by bristle-like setae on the maxillae from the same currents generated by the swimming appendages (Vanderploeg and Paffenhöfer, 1985). The filtration rates may reach 10–20 mL per individual adult per day (Richman *et al.*, 1980; Thompson *et al.*, 1982), although most reported averages are an order of magnitude smaller. In addition, animals feed on larger algae and ciliates, up to 30 μm in size, which are actively captured and manipulated with the maxillae and maxillipeds (and probably other appendages too). Prior to making its strike, the animal will have detected and deliberately oriented itself towards its quarry. The incidence of successful encounters is high (Strickler and Twombley, 1975). Whether or not calanoids should still be regarded as being primarily herbivorous, they have a demonstrable potential to control the numbers of ciliates in the plankton (Burns and Gilbert, 1993; Hartmann *et al.*, 1993). The two feeding modes afford to calanoids an enhanced dietary flexibility, while the measure of electivity allows them to survive lower concentrations of food.

The cladocera are specialised filter-feeders. The body is much modified from the basic crustacean form: a short thorax and abdomen carries a compressed, shell-like carapace that forms a chamber in which four to six pairs of flattened, setae-bearing trunk limbs ('phyllopods') beat rhythmically. Their motion draws water through the variable carapace gape and the setae filter particles from the inhalant current (Fryer, 1987; Lampert, 1987). The animals swim by beating the large, biramous antennae. Several cladoceran families are represented in the fresh-water plankton. The Sididae, which have six pairs of phyllopods and strong, branched antennae, are found mostly in vegetated margins and ponds. The

Holopedidae are represented by a single genus, which has a reduced carapace and is instead embedded in a mass of jelly. Bosminids are planktonic in ponds and small lakes and have a wide geographical distribution. The macrothricid and chydorid cladocerans have relatively small antennae, and are mainly feeders on hard surfaces. *Chydorus sphaericus* is extremely common in the bottom water and among weeds of small lakes and it is frequently encountered in the plankton of the water bodies in which they are present. In two other families, the polyphemids and the leptodorids, the carapace is small and covers little more than the brood pouch; planktic species of *Bythotrephes* and *Leptodora* are macroplanktic (2–20 mm) predators and do not comply with the generalisation about cladoceran filter-feeders.

There is one further family, the Daphniidae, whose species can be extremely prominent in limnoplankton and which plays a major role in regulating the structure and function of lacustrine ecosystems. Daphniids have five pairs of phyllopods within the carapace and, like the sidids, have powerful swimming antennae. They are efficient and more-or-less obligate filter-feeders. They can remove all manner of foods on the filtering setae, within defined size ranges. The upper limit is set by the width of the carapace gape (which is species-specific and varies with the size of the animal). The lower limit is governed by the spacing of the phyllopod setae (Gliwicz, 1980; Ganf and Shiel, 1985). As will be further explored below, individuals are able to filter such relatively large volumes of water that, under favourable conditions, it is likely that a significant population of maturing daphniids may be able to filter the entire volume of a lake in a day or less. The implications for their food organisms, and for the other organisms using the same food resource, and for those other microplanktic feeders that can be themselves be ingested by the daphniids are formidable and complex.

There is considerable further differentiation of the habits, predilections and dynamics within the Daphniidae and within the type genus, *Daphnia*. *Simocephalus*, *Ceriodaphnia* and some of the larger *Daphnia* species (*D. lumholtzi*, *D. magna*) are more common in ponds or at the weedy margins of lakes than in the open water of lakes.

Moina species are most closely associated with small water bodies with a tendency to offer suitable habitat conditions on a temporary basis but explosive growth phases afford a dynamic advantage when favourable conditions obtain (Romanovsky, 1985). In the open pelagic of permanent larger lakes, dominance among the daphniids is contested by such species as *D. cucullata*, *D. galeata*, *D. hyalina* and *D. pulicaria* (Gliwicz, 2003a).

Marine mesozooplankton

The principal groups of mesoplanktic herbivores in the sea are the calanoids, the cladocerans and the thaliacean tunicates (the salps and their allies) (Sommer and Stibor, 2002). In fact, the calanoids are the most familiar and such temperate shelf-water species as *Calanus finmarchius*, *Acartia longiremis*, *Temora longicornis* and *Centropages hamatus* have long been regarded as the main food organisms of commercially important surface-feeding fish, like herring (*Clupea harengus*) and mackerel (*Scomber scombrus*). Accordingly, they have been studied in some detail (e.g. Hardy, 1956; Cushing, 1996). A long-enduring understanding of a three-link food chain (phytoplankton – zooplankton – fish) places the calanoids at the fulcrum between the harvest of fish the primary producing phytoplankton. The relative proportions of the respective annual production by these three components also fitted well with the contemporaneous appreciation of the pyramidal Eltonian relationship, with an approximate 10% transfer of the energy acquisition at each trophic level being transferred to the one above (Elton, 1927; Cohen *et al.*, 1990). This paradigm was seriously challenged by the realisation of how large a proportion of pelagic photosynthate is transferred, as dissolved organic carbon, through microbes and their microplanktic consumers to ciliates (Williams, 1970; Pomeroy, 1974; Porter *et al.*, 1979; Sherr and Sherr, 1988) (see also Section 3.5.4).

On this basis, ciliate-consuming calanoids are already the fifth stage in the food chain. However, the close coupling of the components and the functional integrity of microbial food webs (e.g. Šimek *et al.*, 1999) are known to achieve a high ecological efficiency of energy transfer (10–35%:

Gaedke and Straile, 1994a). Even so, the real problem very often relates to the low supportive capacity of the nutrient resources available and the low biomasses that can ever be maintained. Selectively browsing calanoids are simply the most efficient harvesters of the carbon flux. Moreover, a relatively low fecundity and modest rate of investment in egg production enables them to satisfy their minimum food requirements at low POC concentrations, in the range of 5–80 µg C L^{-1} (Hart, 1996) (as algae, this is roughly equivalent to a chlorophyll content in the range 0.1–1.6 µg chla L^{-1}). Except when food resources are truly limiting, the dynamics of calanoid growth are most likely to be governed by temperature (Huntley and Lopez, 1992). It is even possible that the distinctively oceanic calanoids (such as *Acartia clausi*, *Centropages typicus*) function at yet lower resource availabilities than the shelf-water species.

In the more enriched coastal waters, receiving nutrient inputs from rivers or as a conequence of deep-oceanic upwellings, several resource-driven effects are evident. There are absolutely more of the biomass-constraining nutrients (N, P, Fe), at once raising the potential to support more primary producers and to maintain a higher algal biomass. The additional nutrients may also alleviate the dependence of large algae upon turbulent diffusivity to fulfil their nutrient demands (cf. Riebesell and Wolf-Gladrow, 2002) (see also Section 4.2.1). There may well be a lowering effect on the Si–N and Si–P relationships, which some consider relevant to changes in the species composition of the phytoplankton, although these really depend on the absolute nutrient levels (Sommer and Stibor, 2002).

The combination of these effects results in potentially greater concentrations of high-quality, primary foods that will support direct filter-feeding. It is not just the fact that cladocerans, such as *Evadne*, *Podon* and *Penilia*, can filter more water and, thus, harvest more food than calanoids (Sommer and Stibor, 2002). They also have faster rates of metabolism and growth, with proportionately more of their greater energy intake (60–95%) being invested in parthenogenetic reproduction (Lynch *et al.*, 1986; Stibor, 1992). At the physiological level, this requires

a relatively rich tissue content of ribosomal RNA, bringing with it a consequent high cell-phosphorus content. This is, incidentally, the major factor influencing the relatively low C:P stoichiometries (averaging about 80:1) that are typical among these (Gismervik, 1997) and other planktic cladocera (Elser *et al.*, 2000).

The third main group of marine meso-planktic animals comprises the pelagic, free-living tunicates, especially the salps, pyrosomans and doliolids that represent the Thaliacea (see Table 6.1). These near-transparent, gelatinous animals have a low body mass, comprising little more than an open barrel-shaped tube, a filtering gill and circumferential muscle bands whose systematic contraction and relaxation refresh the current of water across the filter screen. All pelagic tunicates are filter-feeders, straining perhaps the entire nanoplanktic–microplanktic size range of particles (Sommer and Stibor, 2002). In additon, they are, collectively, ubiquitous components of the pelagic fauna, from coasts to the deep sea; however, they are perhaps best known for their presence in the plankton of warm, ultraoligotrophic oceans. The low body mass requires absolutely modest resources for maintenance, while they expend little energy in keeping their near-isopycnic structures from sinking. The architecture and physiology of these animals is substantially geared to function at very low concentrations of assimilable POC.

Planktivorous macroplankton, megaplankton and nekton

Although this section has so far, taken a 'bottom–up' view of the structure of the zooplankton, in emphasising the nature of the resource base and the evolutionary adaptations of the main pelagic groups to be able to exploit it, it is only half the story. For, as many authorities tirelessly point out (e.g. Gliwicz, 1975, 2003a; Banse, 1994), active net production of each of the components of the plankton of seas and lakes is regulated by its consumers. Thus, the abundance of ciliates in the open plankton might be expected to increase on an abundant resource of nanoflagellates but the capacity to do so may be severely constrained by the numbers of calanoids or other predators (see, e.g., Thouvenot *et al.*,

2003). Equally, the growth of microphytoplankton in a lake might be constrained by the feeding of herbivorous *Daphnia* but the vulnerability of *Daphnia* to consumption by planktivorous fish might not only reduce their impact upon the phytoplankton but the latter would be allowed to increase to something like its ungrazed potential. Such dynamic 'cascading' interactions (Carpenter *et al.*, 1985) may now seem to be self-evident phenomena but they were not formally described before the publication of now-classic quantitative studies of Hrbáček *et al.* (1961). Since then, trophic cascades and their manipulation have been studied in great detail (see Carpenter and Kitchell, 1993) and exploited as the basis of system management by what has become known as biomanipulation (Shapiro *et al.*, 1975; see also Section 8.3.6). For the moment, however, I seek only to make the point that the effects of mesoplanktic herbivory and microphagy on the dynamics and standing crops of the primary producers are subject to cascading impacts of planktivory. Again, the point is elegantly made in Gliwicz's (2003b) observation that the structural composition and size distribution of the zooplankton is very different between systems with and without the presence of zooplanktivorous fish. The point is especially pertinent, as zooplankters have nowhere to hide in the open water: survival depends on not being seen or eaten by planktivorous cosumers. Apart from the protection that comes from fortuitously low predator densities, much depends upon either being less visible (which exacts a premium on zooplankton that might grow larger) or less accessible to visual predators by descending to the lightless depths. The need to return to the surface waters to feed (at night!) invokes an unavoidably high energetic cost.

Besides adult fish such as herring and mackerel (see above), marine consumers of mesozooplankton include invertebrates of the next two size divisions – macroplankton (2–20 mm) and megaplankton (>20 mm). Among the most significant of these are the chaetognaths (arrow worms), ctenophores (sea combs and sea gooseberries) and the polychaete *Tomopteris*. There is also a variety of crustaceans at this scale – the mysids, the pelagic amphipods and the impor-

tant euphausids. The latter are entirely pelagic throughout their lives; they live in all the oceans but the key place of *Euphausia frigida* and *E. tricantha* in the southern oceans is renowned for their being the main food of the great baleen whales. Quantitatively less important are the carnivorous crustacean larvae of decapods (crabs, lobsters) and of the mantis shrimps, *Squilla* spp. celebrated by Hardy (1956) as being the 'most beautiful of larvae'. Among fresh waters, non-vertebrate planktivores other than mysids occur mainly in inshore waters, and include larval megalopterans, hemipterans (e.g. *Notonecta*) and dipterans.

Among the larger pelagic animals (fish, squid), size and muscular strength take them out of the plankton and into the swimming nekton. For these to be truly pelagic planktivores, the ability to sample large volumes of water and strain from this food items of only 1–2 mm is essential and demanding of a very efficient means of food filtration. In the herring and some other closely related clupeoid species, for example, this capacity is provided by gill rakers – comprising numerous, long and slender close-set bristles borne on the gill arches. The basking shark, *Cetorhinus*, is a relatively very large elasmobranch but its gill arches are analogously set with many close-set, flattened strips that function analogously to the rakers of bony fish (Greenwood, 1963). The direct sustenance of a creature attaining a length of 10–12 m and a body mass of several tonnes through feeding on animals individually ten thousand times smaller and 10^{-9} of its body mass is a truly impressive example of emergy gain through trophic linkage. In another, fresh-water case, the productive basis of the fisheries of the meromictic rift valley Lake Tanganyika, founded mainly on two species of planktivorous clupeid (*Stolothrissa tanganicae*, *Limnothrissa miodon*) and their endemic centropomid predator (of the genus *Lates*: Lowe-McConnell, 1996, 2003), has been shown to be quantitatively dependent upon the pelagic food web of the lake (Sarvala *et al.*, 2002). Yet more remarkable is the fact that the annual stimulus for production in this oligotrophic lake relies heavily upon the recycling of nutrients during the period of increased wind action and monimolimnetic deepening and the enhanced production of picoplanktic cyanobacteria. Bacterial consumption of primary DOC yields between 25 and 130 g C m^{-2} a^{-1} to the pelagic food web, which, in turn sustains the producton of up to 23 g copepod C m^{-2} a^{-1}. Much of this (16 g C m^{-2} a^{-1}) is eaten by planktivores in generating the net annual production of 1.1 g C m^{-2} a^{-1} *Stolothrissa* plus *Limnothrissa* and 0.3 g C m^{-2} a^{-1} *Lates*.

Adults of many families of fish will eat zooplankton when it is sufficiently abundant to be an attractive and satisfying resource. The majority of adult pelagic fish (and of many that inhabit shallower margins) are carnivorous on other fish and/or on other large prey. However, many of these species produce large numbers of small eggs that give rise to juvenile stages, which are initially mesoplanktic. They feed on microplankton (Sarvala *et al.*, 2003), often in direct competition with and exposed to predation by adult planktivores (e.g. O'Gorman *et al.* 1987). According to the systematic simulation modelling of Letcher *et al.* (1996), metabolic growth capacity, rather than foraging ability or resistance to starvation, is the leading bottom–up component in larval survival. Predator size is a powerful influence on survival but has only a weak effect on the variability in the composition of the available prey. Letcher *et al.* (1996) also deduced that whether young fish died through starvation or predation usually depended most on the availability of their smallest prey organisms.

Reference should also be made to pelagic squid (*Loligo* spp.) whose juvenile hatchlings (or paralarvae) are released into the open water. They are free-living and self-propelling, using rhythmic contractions of the mantle to force a series of alternating bursts of water flow and recovery. Being barely 2 mm in length, paralarvae are, unmistakeably, initially (albeit briefly) mesoplanktic (Barón, 2003).

On the basis of this brief survey, it is clear that the precise structure of the pelagic web of consumers is highly variable and subject to dynamic forces. So far as the impacts upon the phytoplankton is concerned, outcomes hinge on the numbers and sizes of the herbivores present and the sustainability of the feeding modes available.

6.4.2 Impacts of filter-feeding on phytoplankton

Moving on from qualitative description of the structural components of the phagotrophic plankton, consideration is now given to the quantitative impacts of their feeding on the producer mass. It is conceptually easier to deal first with the impacts of filter-feeders. Although this method of food gathering is far from universal, it can be the most striking and complete in its impact. Moreover, its effects are relatively easy to model. These are good enough reasons to explain the additional fact that a large literature on filter-feeding has accumulated.

Supposing that, to the potential planktic consumer, the relatively most abundant food resource is the nanoseston – algae, large bacteria, detrital particles measuring 2–20 μm across – and that particles are generally well dispersed within the medium, then the development of some means to sieve and to concentrate such particles is likely to be favoured by evolution (Gliwicz, 2003a). The coupling of a filter and the means of generating a water current across is a characteristic of the feeding apparatus of many zooplankters, including ciliates and rotifers. However, it is at the mesoplanktic scale, of crustaceans and tunicates, that filter-feeding has a significant impact on the availabilty of food in the entire medium, or at least beyond the immediate environment of the individual animal. This difference is most due to viscosity. Where the smallest turbulent eddy is in the order of 1 mm or so, the typical microplankter (<200 μm in length) experiences a wholly viscous environment in which to move itself and, more importantly, to influence the encounter with particles ≤20 μm. An analogy that comes to mind is of a human trying to collect bananas while both are immersed in a swimming pool filled with a liquid having the consistency of molasses or setting concrete. Scaling upwards, the same human would be rather more successful in picking out kidney beans from the same liquid but now dispersed in a vessel the size of a bath tub. It is only by being big enough and strong enough to overcome frictional drag and to generate turbulent currents that it is possible to increase the rate of particle contact (Rothschild and Osborn, 1988).

Quantitatively, the best studied of the three characteristic mesozooplanktic groups of filter-feeders is the Cladocera and, especially, those of the fresh-water genus *Daphnia*. Collectively, these also illustrate well the factors that most influence the dynamic impacts on the microplanktic food organisms: how many feeders there are, what and how much food they remove, and what the consequences might be, both for themselves (in terms of biomass increase) and for the food (in terms of how much more cropping it can withstand). Each of these components is pursued exhaustively in the available literature. Much of this amplifies the findings of some of the earliest of the modern investigations. Indeed, many of these were of such enduring quality that they provide an ideal base for this section.

Filtration rates

Certainly for the larger individual filter-feeders, the basic quantity of interest is the volume of water that animals are able to strain per unit time. The usual means of its determination is to measure the rates of depletion from known concentrations of radio-labelled ingestible foods. To obtain sensible results in a short period of time, it is necessary to make the measurements on suspensions of known numbers of animals, which, in turn, necessitates that the per capita volumes of water processed are nearly always mean values. Moreover, as the individual volumes filtered vary conspicuously with the size of the animal, it is necessary always to pre-sort the animals beforehand. Possible methodological shortcomings attach to the effects of handling on the animals' performances. Another potential difficulty that must be addressed is that the removal of radio-labelled particles is incomplete or is temporary (i.e. food is selected or rejected, leading to an underestimate of the volume filtered). Further, allowance has to be made for the possibilty that, once satiated, the individual may slow its filtration rate.

Experience has shown that all these are real constraints. Consensus has clarified their magnitude and also spawned a terminology. All meaningful measurements take place at known temperatures and within short, timed exposures to known concentrations of size-categorised ani-

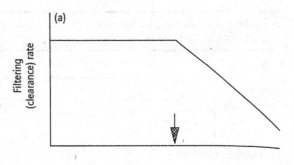

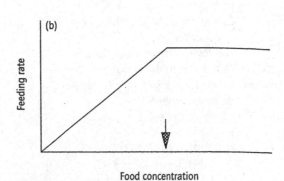

Figure 6.4 Functional responses of (a) filtering rate and (b) feeding rate of filter-feeding animals with respect to food concentration. The arrow defines the incipient limiting concentration, as defined by McMahon and Rigler (1963). Redrawn with permission from Gliwicz (2003a).

mals that are 'starved' for some hours before introduction into a suspension of radio-labelled food organisms. In this case, what is measured is strictly the 'particle clearance rate'. If all the above conditions are satisfied, the mean individual clearance rate should coincide with (but not exceed) the mean volume of water processed per unit time, which is the mean individual 'filtration rate' (F). The rate at which food particles are captured is obviously dependent upon the filtration rate but the 'feeding rate' (or 'ingestion rate') is a proportion of the food concentration in the inhalant volume. Moreover, it has been demonstrated (McMahon and Rigler, 1963) that, at high concentrations of food, clearance rates of *Daphnia* slow down as the animals are sated for less filtering effort. The relationships between filtration and feeding as a function of food concentration are sketched in Fig. 6.4. Below the 'incipient limiting food concentration', *Daphnia* is likely to

Table 6.2 | *Individual filtration rates, F_i (in mL d^{-1}), for various planktic animals as reported in, or derived from relationships in, the literature*

Species	F_i	References
Rotifers		
Generally	0.02–0.2	Pourriot (1977)
Brachionus calyciflorus	0.1–0.2	Halbach and Halbach-Keup (1974); Starkwether et al.(1979)
Freshwater calanoids		
Eudiaptomus gracilis (12 °C)	0.6–1.8	Kibby (1971)
Eudiaptomus gracilis (20 °C)	1.3–2.5	Kibby (1971)
Eudiaptomus gracilis (adults, 17 ± 3 °C)	0.5–10.7	Thompson et al. (1982)
(copepodites, 17 ± 3 °C)	0.5–6.7	Thompson et al. (1982)
Diaptomus oregonensis (adults, temperatures various)	2.4–21.6	Richman et al. (1980)
Freshwater cladocerans		
Bosmina longirostris	<3.0	Thompson et al. (1982)
Chydorus sphaericus	0.5–2.6	Thompson et al. (1982)
Daphnia galeata[a]		
(<1.0 mm, 17 ± 3 °C)	1.0–7.6	Thompson et al. (1982)
(1.0–1.3 mm, 17 ± 3 °C)	3.1–19.3	Thompson et al. (1982)
(1.3–1.6 mm, 17 ±3 °C)	3.1–30.7	Thompson et al. (1982)
(1.6–1.9 mm, 17 ± 3 °C)	14.0–60.0	Thompson et al. (1982)
Daphnia spp.		
(<1.0 mm, 15–20 °C)	<5.0	From Burns' (1969) regression
(1.0–1.3 mm, 15–20 °C)	3.6–10.4	From Burns' (1969) regression
(1.3–1.6 mm, 15–20 °C)	6.4–18.6	From Burns' (1969) regression
(1.6–1.9 mm, 15–20 °C)	10.1–30.1	From Burns' (1969) regression

[a]The organism was reported as *D. hyalina* var. *lacustris*. Under current nomenclature, the identity *D. galeata* is to be preferred (D. G. George, *personal communication*).

be hungry ('food-limited') and to filter-feed as fast as it is able.

It was another of Frank Rigler's colleagues, Carolyn Burns, who made some of the first accurate measurements of the filtration rates in *Daphnia*. Her work (Burns 1968a, b, 1969) has been well supported by subsequent determinations by others (e.g. Haney, 1973; Gliwicz, 1977; Thompson et al., 1982), to the extent that her original detailed findings are used for this quantitative development. What remains remarkable is the enormous capacity of the feeding current created by the rhythmic beating of the thoracic limbs to draw water into the carapace chamber and across the filtering setae on the third and fourth thoracic limbs. Comparison with other zooplankters is made in Table 6.2.

The superior filtration capacity of *Daphnia* (especially larger individuals) is explicit in the relationship that Burns (1969) demonstrated between the filtration rates (F_i) of four species of *Daphnia* (*D. magna*, *D. schoedleri*, *D. pulex*, *D. galeata*) and their carapace lengths (L_b). The plots in Fig. 6.5b and the entries in Table 6.2 are calculated from her regressions. The measurements of Thompson et al. (1982) on *D. galeata* are especially well predicted. At 15 °C, the hourly filtration rate is given by

$$F_i = 0.153 L_b^{2.16} \qquad (6.11)$$

and at 20 °C

$$F_i = 0.208 L_b^{2.80} \qquad (6.12)$$

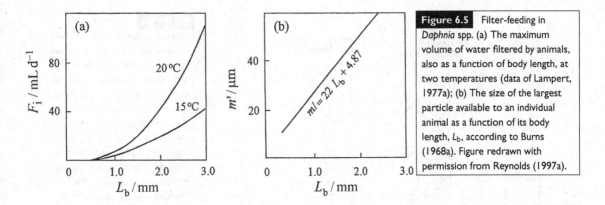

Figure 6.5 Filter-feeding in *Daphnia* spp. (a) The maximum volume of water filtered by animals, also as a function of body length, at two temperatures (data of Lampert, 1977a); (b) The size of the largest particle available to an individual animal as a function of its body length, L_b, according to Burns (1968a). Figure redrawn with permission from Reynolds (1997a).

Possibly the most pertinent deduction is the fact that as few as 20 large *Daphnia* per litre of lake water, or 200 neonates L^{-1}, is sufficient a population to process daily the entire volume in which they are suspended. More generally, the aggregate volume of water that is potentially filtered each day $(\sum F_i)$ is equivalent to:

$$\sum F_i = (N_1 \cdot F_{i_1}) + (N_2 \cdot F_{i_2}) + \cdots \cdot (N_i \cdot F_{i_i}) \quad (6.13)$$

where F_{i_i} is the filtration rate and N_i is the standing population of the *i*th species size category.

Food availability

The lower end of the size range of particles available to the filter-feeder is determined by the filter itself. In the case of the daphniids, it is set by the spacing and orientation of the setules, the short branches besetting the filtering setules fringing the third and fourth thoracic phyllopods (Gliwicz, 1980; see also Gliwicz, 2003a, and references cited therein). Comparing this character with the filterable particles recovered by rotifers, calanoids and other cladocerans (Geller and Müller, 1981; Reynolds, 1984a), there seems to be little interphyletic variation, 0.2–2.0 μm being a general value. In fact, the efficiency of retention of the smallest particles may be lowered in *Daphnia*, owing to some leakage from the filter, from the median chamber and the food groove, before they get to the animal's mouth. Fewer particles in the range 1–3 μm are retained than in the range 5–10 μm (Gliwicz, 2003a).

The upper size limit of particle that can be ingested is, according to Burns (1968a), strongly correlated to animal size (Fig. 6.5b). The relationship makes the important statement that the range of potential foods of filter-feeding cladocerans increases with the maturity of the consumer. Whereas a 1.0-mm *Daphnia* is probabilistically restricted to foods <25 μm in size, a 2.0-mm animal can take food particles up to 50 μm across. Conversely, for smaller animals the food availability may be rather more restricted and the feeding rate falls below the potential of the filtration rate. Those items that are too large are simply rejected and do not enter the filter chamber. As Gliwicz (1980) recognised, the filtering daphniid is very easily able to regulate the size of food particle that reaches the phyllopods through its control over the carapace gape. Obviously it cannot accept particles above a size-specific maximum (Gliwicz and Siedlar, 1980) but it may, however, self-impose a lower maximum, perhaps to avoid ingesting mucilaginous colonies. Thompson *et al.* (1982) observed sharply reduced filtering rates of *D. galeata* when the phytoplankton was dominated by *Microcystis* colonies of varied size. The possibility that the behaviour was in some way attributable to toxicity of the phytoplankter seemed a less likely explanation than the direct observation that the rate of phyllopod beating was impaired only if a *Microcystis* colony became ensnared in the apparatus or blocked the median chamber. At this point, the post-abdominal claw was flexed and used to scrape out the blocking mucilage. Thompson *et al.* (1982) argued that so many animals were spending so long purging the filtering apparatus that the aggregate filtration rate became severely depressed.

It is not just the size and texture but the shape of the phytoplankton that influences the proportion of food strained from the filtration

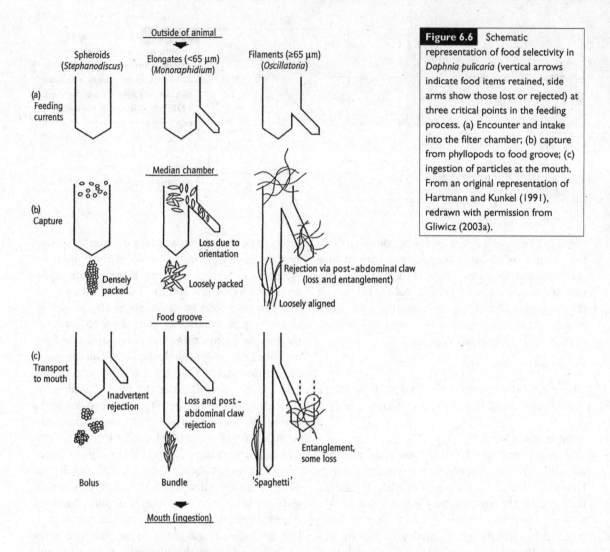

Figure 6.6 Schematic representation of food selectivity in *Daphnia pulicaria* (vertical arrows indicate food items retained, side arms show those lost or rejected) at three critical points in the feeding process. (a) Encounter and intake into the filter chamber; (b) capture from phyllopods to food groove; (c) ingestion of particles at the mouth. From an original representation of Hartmann and Kunkel (1991), redrawn with permission from Gliwicz (2003a).

current. This topic was elegantly investigated and summarised by Hartmann and Kunkel (1991). The diagram in Fig. 6.6 distinguishes the progress to ingestion of three differing algal shapes. Small sphaeroid and cuboidal unicells are easily drawn in to the median chamber and efficiently compacted by the phyllopods, and most of that food will be ingested. Slender algal cells are more difficult to orientate and compress and there is some loss from the median chamber and some rejection from the food groove. Long filaments are really quite difficult. However, as observed by Nadin-Hurley and Duncan (1976), *Daphnia* is able to arrange foods into spaghetti-like bundles (though not without significant rejection losses).

All these facets contribute to feeding rates that are below (in some instances, well below) the potential of the aggregate filtration rate. Reynolds *et al.* (1982a) proposed a food-specific correction factor (here designated as ω') in order to relate removal rates to aggregate filtration rate (ΣF). As removal of algae from the water is reciprocated by the random redispersion of the survivors in ostensibly the same volume of water, the reaction of the extant algal population corresponds to another exponential series, analogous to dilution (cf. Eq. 6.7). The analogous grazing loss rate term, r_G, inflicted upon the algal population is given by

$$r_G = \omega'(\Sigma F_i)/V \qquad (6.14)$$

Thompson *et al.* (1982) found that the removal of nanoplanktic unicells and *Cryptomonas* from the feeding current of *D. galeata* is highly effective, so that the value of ω' is close to 1 and r_G is little different from that predicted directly from the filtration rate. The value of ω' falls to $\sim$0.3 for eight-celled *Asterionella* colonies and rapidly from $\leq$0.3 to zero for *Fragilaria* colonies comprising more than 6–13 cells per colony. Large *Daphnia* are capable of feeding on small *Eudorina* colonies, short *Planktothrix* filaments and young *Microcystis* colonies while they are still quite small (m' generally <50 μm) but have great difficulty with larger colonies. Thus, the rate of removal, r_G, varies among the algae in mixed populations, even when they are simultaneously subject to the activities of the same set of filter-feeders. Combined with the age–size structure of the filter-feeding populations, precise values of r_G are scarcely easy to calculate and should carry such a wide margin of error that, for most purposes, it is acceptable to work with approximations. It should also be borne in mind also that if vertical migration takes the main filter-feeding components of the zooplankton to depths beyond the visibility of fish, and they return to the alga-dense surface waters only during darkness, then Eq. (6.14) is yet more difficult to evaluate. The maximum filtration rates of *Daphnia* are no higher by night than by day (Thompson *et al.*, 1982).

Before concluding this section, it is also important to make reference to the adaptations that algae invoke to make themselves less palatable to the filter-feeding consumers with which they do come in contact. In mathematical terms, these help them to reduce the instantaneous value of ω'. Hessen and van Donk (1993) observed a tendency of some species of alga to maintain larger coenobia (*sensu* more cells per coenobium) in the presence of *Daphnia* than without. They showed experimentally that growing clones of *Scenedesmus* maintain higher proportions of colonies comprising eight or more cells in media in which *Daphnia* had been present but since removed. Their findings have been repeated by Lampert *et al.* (1994).

Another interesting discovery is the reaction of picoplanktic cyanobacteria to the presence of grazers. On strict grounds of cell morphology, the status of a number of cyanobacterial genera has been puzzling. Work with pure cultures led Rippka *et al.* (1979) to take a strongly reductionist view of cyanobacterial diversity, proposing that a small number of named genera could accommodate the few essential distinctions among the unicellular strains (size, shape of cells and planes of cell division). Since then, new techniques and a new generation of researchers have scarcely added to the range of known picocyanobacteria. Most known inland-water forms conform to being short rods (*Synechococcus*), ellipsoids (*Cyanobium*) or spheroids dividing in one or two planes (*Cyanothece*) (Komárek, 1996; Callieri and Stockner, 2002). Moreover, the similarity of the cells of those of various species of colonial genera inhabiting mildly eutrophic waters (such as *Aphanocapsa*, *Cyanodictyon* and *Synechocystis*) has been noted on many occasions and attribution to the same or close genetic lines has been proposed. The experiments of Komárková and Šimek (2003) imitated transformations of growing strains of colonial *Aphanocapsa* and *Synechocystis* into unicellular suspensions, and back, stimulated by the presence or absence of herbivorous brine shrimp *Artemia* (or to chemicals in medium in which *Artemia* had been present).

Interestingly, this behaviour mirrors the known morphological anti-predator defences (lengthened spines, bulbous heads) that are inducible in cladocerans and rotifers exposed to water in which fish or even *Chaoborus* larvae have been present (see Gliwicz (2003a) for a detailed overview of the literature). Chemosensory perception in locating and selecting prey is probably vital in the viscous world of microzooplanktic consumers and the capability may well be widespread among the planktic protists (Weisse, 2003). It is reasonable to expect that chemoperception works in the opposite direction and that predator detection and reaction is similarly influenced. The ciliate *Euplotes* is known to react to predator-specific chemical factors produced by its amoeboid predators by producing giant, uningestible cells (Kusch, 1995). The chemical nature of the substances involved is not well known. They are collectively referred to as kairomones, though their mutual similarities owe to the ability of potential prey to sense the threat of potential predators, and not to any known chemical likenesses. The production of kairomones is distinct from the production of overtly toxic substances by potential prey, whose effects on would-be consumers may be quite general or incidental unintentional in their effects on consumers. The distinctions blur somewhat in the interesting interrelationship between the toxin-producing mixotrophic haptophyte *Prymnesium* and its phagotrophic dinoflagellate predator, *Oxyrrhis*. According to Tillmann (2003), when *Oxyrrhis* was introduced into dense cultures of P-limited *Prymnesium*, its initial feeding rate was quickly depressed (to <0.1 cell grazer^{-1} h^{-1}) below that of control animals fed on similar-sized cryptophytes (2.75 cells grazer^{-1} h^{-1}), a direct response to toxicity. Poisoned *Oxyrrhis* cells then lysed and were attacked by phagotrophic *Prymnesium*, reversing the direction of the carbon flow! The *Prymnesium*-free medium also invoked lysis in *Oxyrrhis*, though the effect reduced when the *Prymnesium* culture was more dilute or the *Oxyrrhis* were more abundant.

Many green algae, Cyanobacteria and some chrysophytes are normally invested in mucilaginous sheaths. Among many other functions that mucilage might serve (see Box 6.1), the package increases the size of the algal particle and decreases the likelihood of its entry into the filter chamber of *Daphnia*, or its retention on the filter or its successful ingestion. Thompson *et al.* (1982) found depressed values of ω' for *Daphnia* feeding on *Eudorina* colonies, except those newly released daughters being <25 μm in diameter. Even if they are ingested, mucilaginous colonies are resistant to digestion. They are not only capable of viable passage (see Canter-Lund and Lund, 1995 for examples) but they are said to use the opportunity of exposure to high nutrient concentrations to absorb and store them and to use them effectively after release from the anus (Porter, 1976). Gliwicz (2003a) considered that, despite doubts about the extent of digestive resistance, the net profit of nutrient uptake by algae surviving gut passage might well compensate. This might explain the frequent observation that the numbers of mucilaginous colonies often increase when the density of filter-feeding Cladocera is high. However, Reynolds and Rodgers

(1983), working with *Eudorina*, found that larger colonies escaped ingestion and uncropped colony growth and recruitment were such to explain the dynamics of population increase. It could be said that, by removing smaller algae, larger units are positively selected by moderate aggregate filtration rates.

Algal removal and grazer nutrition
Even a broad picture of the effects of filter-feeding on the resource cannot be completed without reference to the dynamic limits of filter-feeding activity, or to the way that this changes through time. The answers to both problems depend upon the relation of nutrition, growth and recruitment of the consumers and the quantity and quality of the food available. The food requirements of *Daphnia pulex* were exhaustively investigated by Lampert (1977a, b, c). He first derived equations to describe the maximum hourly assimilation rates of *Daphnia*, as a function of the length (L_b) and mass (w_b) of individual animals at various temperatures. At 15 °C, a 0.8-mm neonate assimilates up to 2.4 μg C d^{-1}, whereas a 2.1-mm adult is accumulating at 15.7 μg C d^{-1}. The amount of food needed to supply such requirements varied with the quality of the food. The highest assimilation efficiencies (~60%) came on such readily filterable and digestible algae as *Asterionella* and *Scenedesmus*. Thus, to satiate the metabolic capacity requires the feeding to yield upwards of 4.0 and 26.2 μg C d^{-1} (to the smaller and larger animal respectively). On the other hand, the respirational expenditure for the same individuals (respectively, 0.6 and 4.3 μg C d^{-1}) defines the minimum daily intake that will just maintain metabolism, below which they will lose weight and eventually die of starvation.

The volume of water that can be filtered defines the external food concentrations that will sustain the minimum and saturation requirements of the *Daphnia*. From the appropriate entries in Table 6.2, it is unlikely that a 0.8-mm *D. pulex* might filter more than 5 mL of lake water each day, while the 2.1-mm animal may be capable of procesing 30 mL. Then the minimum concentration of filterable food necessary to fulfil the neonate's minimum requirement is thus 0.6 μg C per 5 mL, i.e. ~0.12 μg C mL^{-1}; however, a

Box 6.1 | On the sticky question of mucilage

Sheaths or investments of mucilage are produced characteristically around the cells of many kinds of planktic algae and bacteria, often binding them in colonial structures (as in chroococcoids such as *Microcystis*, in sulphur bacteria, among representatives of several orders of chlorophyte, chrysophyte and haptophyte). Mucilage also covers filamentous Cyanobacteria (*Anabaena, Planktothrix*) and some unicellular algae, including certain diatoms. Mucilaginous threads are trailed by *Thalassiosira* filaments and, possibly, by many other kinds of planktic alga (Padisák *et al.*, 2003a). It is a feature of both marine and freshwater species. Mucilage investments vary in texture, from being robust and readily visible under the light microscope (e.g. *Microcystis wesenbergii*) to being, as in some desmids, so tenuous to require negative staining in (e.g.) Indian ink preparations to reveal the existence as a translucent halo around the organism (John *et al.*, 2002). On other occasions, the extent of a mucilaginous sheath is identifiable under the microscope by the numbers of bacteria and detrital particles that cling to the perimeter.

Despite the ubiquitous occurrence of algal mucilage and the fortuitous assistance it gives to microscopists attempting to deduce the identity of the organisms they encounter, there are surprisingly few general accounts in the literature that consider the functions and benefits that mucilage might provide or that question how the many other species of alga seem to manage quite well without it. The probability is that there is no consensus answer anyway. There are certainly several measurable benefits that the presence of mucilage imparts to the organisms that produce it. These are considered in detail at appropriate points in this book. However, it has not been resolved that any of these is a primary or an original function of mucilage production or merely opportunistic adaptive applications of some ancient trait.

Mucilages (the plural is probably reasonable) are hygroscopic lattice-like polymers of carbohydrate and substances resembling acrylic. In the literature, the product is sometimes referred to as mucus, though this term is generally applied to similar polysaccharides produced in many groups of animals (especially coelenterates, molluscs, annelids and many kinds of vertebrate). The elemental composition (C, H, O) of the secretions involves little of intrinsic value and, in this sense, may be considered as a by-product of metabolism. The observation has been made (Margalef, 1997) that there is little difference biochemically between producing mucilage and any other unused extracellular photosynthetic derivative, save that mucilage is not released in solution. The possibility that mucilage production originated as a mechanism for regulating the accumulation of photosynthate in cells that cannot be assimilated into amino acids and proteins has some resonance with statements suggesting that mucilage-bound algae are more common in nutrient-poor waters than in enriched systems and that organisms that produce variable amounts of mucilage (such as *Phaeocystis*) produce more when nutrient (especially phosphorus) concentrations are depleted (Margalef, 1997). The production of gelatinous polysaccharides has been observed among marine phytoplankton populations in the photic layer that have become 'aged by nutrient deficiency' (Fraga, 2001, citing Vollenweider *et al.*, 1995; Williams, 1995). Margalef (1997) makes the point that mucilage production is more frequent over long days and in shallow mixed layers. This, too, might be indicative of causal imbalances in the intracellular metabolism of carbon, nitrogen and phosphorus.

Mucilage certainly has a high water content and, in some cases, it must disperse rapidly into the medium (whilst being replaced by secretion from the proximal side). In spite of the high density of the polysaccharide, the mean density of the mucilage in live *Microcystis* colonies is within 0.07% of the density of the water in which they are suspended (Reynolds *et al.*, 1981) (see Section 2.5.2). This has for long nurtured the supposition that mucilage contributed to the suspension of phytoplankton by reducing average density. This it certainly can do, but it is not effective in reducing sinking rate unless the overall dimensions comply with Eq. (2.17) (Hutchinson, 1967; Walsby and Reynolds, 1980).

Other functions proposed to be fulfilled by mucilage include the following.

Streamlining

Almost in direct contradiction to the principle of reducing sinking rate, a large mucilage investment enhances floating and sinking responses to self-regulated buoyancy changes in colonial Cyanobacteria such as *Microcystis* and *Woronichinia*, making controlled migrations in natural water columns feasible (See pp. 68, 81).

Nutrient storage

The mucilage has been supposed to provide a repository for the concentration and storage of essential nutrients (e.g. Lange, 1976). No mechanism for this has been suggested; it is not clear how outward diffusion gradients and progressive dilution and dissipation of the mucilage effects could be countered.

Nutrient sequestration and processing

In nutrient-dilute environments, encounter with sufficient limiting nutrients is an empirically demonstrable problem (Wolf-Gladrow and Riebesell, 1997) (see Section 4.2.1). It is possible that a mucilaginous coat provides a cheap mechanism for increasing the size of the algal target whilst simultaneously providing a microenvironment wherefrom the rapid uptake of the nutrients across the cell wall (Section 4.2.2) maintains a yet more dilute than the exterior environment and a helpful inward gradient. No compelling demonstration of this nutrient scavenging has been offered. For cells producing phosphatases designed to work externally (Section 4.3.3), there is a need to confine the activity close to the sites of intracellular uptake, which function could arguably be fulfilled by a mucilaginous boundary layer. To be valid, however, the entry of organic solutes must be faster than the loss of phosphatase. Again, no compelling experimental evidence is available to verify this.

Metabolic self-regulation

Nutrient-deficient cells may be prevented from completing their division cycle (Vaulot, 1995) (see Section 5.2.1) but they cannot stop photosynthesis. Margalef (1997) proposed that sheaths slow down diffusion and minimise unnecessary metabolic activity. Although this idea fits with some of the field observations and also matches to the one well known to algal culturalists, that mucilage is usually lost quickly in laboratory strains, it is not clear that colonial species (e.g. of *Coenochloris*: Reynolds et al., 1983b) fail to attain maximum rates of growth under favourable supplies of nutrients and light. This they do without apparent loss or dimnution of the colonial form.

Defence against oxygen

Sirenko and her co-workers (reviewed in Sirenko, 1972) demonstrated a low-redox microenvironment is maintained within the mucilage of several species of Cyanobacteria, apparently through the production of sulphydryl radicals. They have argued that this helps to protect against oxidative processes and leads to tolerance of high external concentrations of oxygen (similar comments were also made by Gusev, 1962, and Sirenko et al., 1969).

Defence against metal poisoning

The selective permeability of mucilaginous envelopes might provide a defence against the uptake of toxic cations in the acidic environments tolerated by some desmids (Coesel, 1994). This idea has been investigated by Freire-Nordi et al. (1998), applying electron paramagnetic resonance to compare the decay rates of hydrophobically labelled tracers in normal sheathed cells of *Spondylosium* and in cells divested of their mucilage by ultrasound. Decay was slower in cells with mucilage, because of a suspected interaction with –OH groups in the polysaccharide. They concluded that such interactions could play a decisive role in uptake selectivity.

Defence against grazing

Mucilaginous sheaths reduce the palatability of algae by making them too large for microplankters to ingest, more difficult for mesoplanktic raptors to grasp and less filterable and more mechanically obstructive for cladocerans (see Sections 6.4.2, 6.4.3). There is also recent evidence that free-living picocyanobacteria are stimulated to form colonial structures in response to the presence of herbivores and to the chemicals in the water that herbivores have recently vacated (Komárková and Šimek, 2003) (see Section 6.4.2).

Defence against digestion

If mucilaginous algae fail to avoid ingestion, they may resist digestion during the period of their passage through the guts of some (but not all) consumers. The original observations of Porter (1976) have been verified by others, including Canter-Lund and Lund (1995). The simultaneous scavenging of nutrients by viable algae during passage and their deployment is a bonus function of mucilage (see Section 6.4.2).

What may be concluded? There is no single clear function of mucilage, and not all those suggested could be considered valued judgements. The buoyant properties, the reducing microenvironment, the selective permeability and the grazing deterrence are backed by good empirical, experimental verification. These may all be different, positively selected adaptations to what may have been originally homeostatic mechanisms for balancing cell stoichiometry.

concentration of 0.8 µg C mL^{-1} is necessary to saturate its maximum assimilation. The corresponding minimal and saturating concentrations for the 2.1-mm adult are quite similar, at 0.14 and 0.87 µg C mL^{-1}.

The figures make no allowance for wastage or rejection (Fig. 6.6) or for the slightly different outcomes at higher temperatures. A further aspect that was appreciated by Lampert (1977c) and developed by Lampert and Schober (1980) and Gliwicz (1990) is that, for any filter-feeder, there is an approximate upper threshold concentration of edible, filterable foods when the maximum physiological demands of growth and reproduction are satisfied. This corresponds to the inflection point shown in Fig. 6.4. Of rather greater significance, however, is the threshold food concentration that must be surpassed before an individual animal can balance respiration and maintenance and at which growth is zero. A slightly higher threshold than this is necessary to maintain the population, allowing some maturation and egg production to offset mortalities (Lampert and Schober, 1980).

To quantify the dynamic variability in the relationship between food and feeders was a major objective of the studies in the Blelham enclosures (Section 5.5.1, Fig. 5.11). In the early part of the programme to measure loss rates of primary product (Reynolds et al., 1982a), phytoplankton was allowed to develop free of the constraint of nutrient supplies and the herbivore populations developed in an environment that was substantially free of predators. The deep-water enclosures (A, B in Fig. 5.11) were devoid of any significant numbers of fish, while, fortuitously, the restriction of enclosure opening to just a few weeks in winter prevented the inward migration of chaoborid larvae from their shallow overwintering sites (Smyly, 1976). Thus, it has been assumed that the herbivore periodicities that were observed were susbstantially driven from the bottom up, being regulated mainly by temperature and variable food availability. An example is included in Fig. 6.7. Once water temperatures exceeded 7–8 °C, daphniids rapidly established themselves as the most significant consumers, responding quickly to an abundance of algae of filterable size food resources

(*Asterionella* and, especially, *Cryptomonas*, *Ankyra*). With maturing individuals producing abundant parthenogenetic eggs and more animals recruiting freely to the standing stock, the aggregate daily filtration quickly appreciated to some 600–800 mL L^{-1} d^{-1} (Thompson et al., 1982; see also top frame of Fig. 6.8). During this particular phase, the mean body length of a cohort of individuals, feeding mainly on growing *Ankyra* cells and at water temperatures close to 20 °C, increased from 0.8 to 1.7 mm in 13 days and, simultaneously, recruited a fivefold increase of neonates of the next generation. Applying the contemporaneous direct estimates of Thompson et al. (1982) to Eq. 6.14, there was, over the same period, a 12-fold increase in the collective filtration volume, ΣF_i. Reynolds (1984a) used these measurements to express the resource-replete growth in *Daphnia* filtration as an exponent: from (ln 12) / 13 days, the specific rate of increase in ΣF_i is 0.1911 d^{-1} at ~20 °C. In contrast, recruitment of *Daphnia* had been negligible in April, when water temperatures were still below 8 °C, despite an apparent abundance of food at that time (Fig. 6.8, top panel: over 2 µg C mL^{-1}). Supposing a geometric scaling of temperature dependence of growth and recruitment capacity between 8 and 20 °C, Reynolds (1984a) approximated a rate of increase in the resource-saturated maximum aggregate filtration of 0.0159 K^{-1}.

The rate of change in the aggregate filtration capacity (ΣF) is not the same thing as the population growth rate for *Daphnia*, although there are analogies and some shared information. Growth rates of *Daphnia* spp. under comparable conditions noted in Gliwicz (2003a: ~0.2 d^{-1}) are of similar magnitude. However, the seasonal fluctuations in ΣF_i provide a ready and reasonably sensitive indicator of the impact of grazing on the food resources and the consequences of food depletion on the survival of *Daphnia*. Of the examples from the Blelham enclosures plotted in Fig. 6.8, reference has already been made to events in Enclosure A during 1978 (A78). Subject to satisfaction of the temperature constraint, the episodes of increase in ΣF_i (at one point, to >1 L L^{-1} d^{-1}) were observed in the presence of a finite resource base of filterable algae (defined in the study as being individually <10^4 µm^3; mean

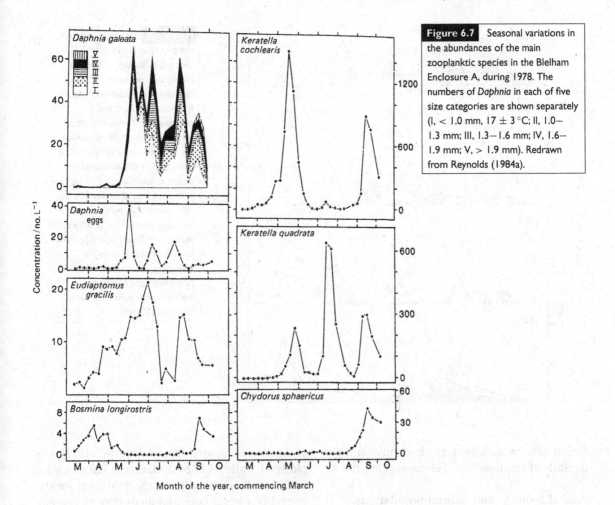

Figure 6.7 Seasonal variations in the abundances of the main zooplanktic species in the Blelham Enclosure A, during 1978. The numbers of *Daphnia* in each of five size categories are shown separately (I, < 1.0 mm, 17 ± 3 °C; II, 1.0–1.3 mm; III, 1.3–1.6 mm; IV, 1.6–1.9 mm; V, > 1.9 mm). Redrawn from Reynolds (1984a).

Month of the year, commencing March

$d = 26$ μm). These increases were not sustained, however, as the calculated aggregate filtration capacity periodically collapsed, owing, on each occasion, to severe reductions in the numbers of filter-feeders present. In almost all of these instances, the herbivore collapses followed severe depletion in the filterable algal mass. In Enclosure B in 1979 (B79), aggregate filtration rate was generally reduced and confined to a narrower time window than in A78, owing to the dominance of the algal biomass by *Planktothrix* (in spring) and *Microcystis* (in summer). Both were considered to be inaccessible food resources to the filter-feeders on the grounds of size. In B81, grazing was generally modest throughout a year when filterable foods were conspicuously low throughout the year. In A83, filterable foods were abundant in the spring but not in summer; the

response of the aggregate filtration rate could scarcely be better correlated.

Overall, there is manifestly no reciprocity between food and feeders, whereas, under the contrived conditions, the latter plainly tracked the former. Baldly, without a resource, filter-feeding is unsustainable. Filter-feeders can increase, provided the minimum threshold concentration of edible, filterable foods is exceeded. The evidence from Fig. 6.8 is that the threshold is pitched between 0.10 and 0.13 μg C mL^{-1}. Substantially greater concentrations than this will support growth of filter-feeders. Once the aggregate rate of loss of algal foods to filter-feeders (r_G) is in excess of the net rate of their recruitment (r', or r' net of all other simultaneous loss rates), rapid exhaustion of the food resource (to <0.10 μg C mL^{-1}) is inevitable. This is followed,

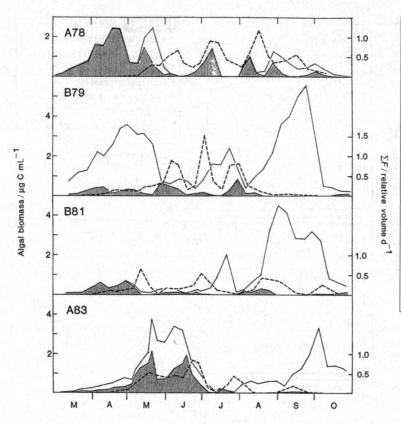

Figure 6.8 Seasonal variations in the phytoplankton biomass (as carbon) in Blelham enclosures during selected years and its approximate fractionation between filterable (hatched) and non-filterable (unshaded) size categories. The bold dashed line is the aggregate daily filtration rate generated by the zooplankton, which, at times, exceeded 1.0 (*sensu* 1 litre per litre of enclosed water volume per day). Note that the filtration volume mainly reflects the numbers of filter feeders present and that it responds to the availability of suitable foods (abundance *and* filterability). Redrawn with permission from Reynolds (1986b).

almost as inevitably, by starvation and death, particularly of the neonates (Ferguson *et al.*, 1982).

Food thresholds and natural populations

For completeness, the lower and upper threshold concentrations impinging on the ability of *D. galeata* populations to survive or to saturate their capacity for growth are expressed in terms of the populations of some common fresh-water plankters (Table 6.3). The concentrations cited should be interpreted as defining the fufillment or saturation or otherwise of the *Daphnia*'s nutritional requirements, were the food resource monospecific. For example, were *Cryptomonas ovata* the only food available, a population of 175 mL^{-1} would be the minimum that will could meet the feeder's basic maintenance needs at 15 °C. In the case of the 0.8-mm neonate, filtering up to 5 mL of water during each 24 h, it would be expected to encounter 875 algal cells in a day, at an average frequency of one every 99 s. A 2.1-mm adult, feeding in the same suspension, should be capable of ingesting 5250 algae during the day, or one every

16.5 s. At food concentrations below this threshold level, animals of both sizes will fail to ingest sufficient food to satisfy their minimum needs. In reality, several potential foods may be present but, if their combined carbon content (supposing it to be wholly filterable and assimilable) is substantially below the threshold concentration, the animals will starve.

While it is proposed that these thresholds set firm boundaries within which the mutual dynamics of phytoplankton and zooplankton interact, it is also fair to say that, in the real world, they are not often easily recognisable. There are several reasons for this and they need to be acknowledged. First, not all filter-feeding zooplankton need experience the same threshold concentrations. Second, there is the subsidiary issue that larger individuals of any filter-feeding species can ingest larger prey. This means that the lower metabolic threshold can be fulfilled by a wider size range of foods, so that the resource base of the larger filter-feeder is greater than that of the smaller one. Moreover, the algal food

Table 6.3 | *Equivalent concentrations of selected algal and bacterial foods representing the minimum and saturating threshold requirements of* Daphnia *maintenance, growth and reproduction*

Food species	Volume (μm^3)	Cell C (pg)	Populations equivalent to	
			$0.1\ \mu g\ C\ mL^{-1}$	$0.8\ \mu g\ C\ mL^{-1}$
Cryptomonas ovata cell	2710	569	175	1 406
Scenedesmus quadricauda (four-cell coenobium)	1000	225	444	3 560
Asterionella formosa cell	645	85	1 176	9 411
Plagioselmis nannoplanctica cell	72	15	6 667	53 333
Synechococcus cell	<30	<7	>14 285	>114 280
Ankyra judayi cell	24	5	20 000	160 000
Free-living bacteria	–	0.013	7.7×10^6	6.2×10^7

resource of all filter-feeders may be supplemented to a degree by a resource of suitably sized particulate organic matter (detritus) and by free-living bacteria. Finally, predation of planktic herbivores distorts the impression of community structure, function and the outcome of dynamic processes.

Gliwicz (1990) compared the body-growth rates of a number of *Daphnia* and *Ceriodaphnia* species as a function of food concentration. Some modest interspecific differences in the lower thresholds were evident (in certain cases, these could be as low as 0.03 µg C mL^{-1}), with the best survivorship being noted among the larger species of *D. magna* and *D. pulicaria*. Given that larger animals also filter more water and ingest larger particles, we may deduce that, potentially, adults of the larger species, together with mature adults of the intermediate size classes, enjoy a wider filterable resource base and may be better adapted to survive instances of periodic starvation than juveniles or adults of small species.

On balance, ecological evidence is supportive of the physiological deduction. Daphniids of all kinds seem to be relatively scarce in the plankton of lakes in which the carbon supportive capacity is generally ~0.1 mg C L^{-1}. Moreover, development in waters where the supportive capacity may only coincide with filterable POC concentrations >0.05–0.1 mg C L^{-1}. If fulfilled exclusively by algal plankton, the corresponding threshold is equivalent to a chlorophyll concentration of 1–2 µg chl*a* L^{-1}. If substituted by bacteria, the minimum threshold is some 4–7 × 10^6 mL^{-1}. Even in these instances, any further inroad into the filterable resource leaves filter-feeders severely food limited. Almost all the other components of the microbial food web, including the nanoflagellates and ciliates, as well as fine, suspended detritus, will have been eliminated by daphniids feeding in significant numbers (Porter *et al.*, 1979; Jürgens *et al.*, 1994; Sanders *et al.*, 1994; Wiackowski *et al.*, 1994).

Do these thresholds not then determine that the abundance of zooplankton in pelagic systems having a biological supportive capacity substantially greater than 0.1 mg C L^{-1} would tend to alternate between glut and self-inflicted dearth of the largest filter-feeding species present? In the case of Blelham Enclosure A in 1978 presented above, the surges and collapses in the *Daphnia* population would appear to conform to this supposition. Elsewhere, of course, the zooplankton dynamics are moderated by planktivory (and especially by fish) which tends to damp the fluctuations between famine and plenty. Furthermore, the zooplankters that are both more visible and more rewarding to the planktivore are the larger ones, and, other things being equal (see Sections, 6.4.5, 8.2.2, when interactions and entire pelagic resources are discussed), selection is against large animals and in favour of the survival of smaller animals and of smaller species that reproduce at smaller body sizes (Gliwicz, 2003a, b). As a consequence, the survival prospects of each species in the zooplankton and the resultant structure of the community are poised between the influences of food limitation and herbivore predation. Zooplankton ecologists have, for long, recognised the interaction between, on the one hand, the ability to feed, assimilate, grow and reproduce under conditions of resource stress and, on the other, the vulnerability to predation. The balance of these counteracting influences was expressed in the elegant size-efficiency hypothesis (SEH) of Brooks and Dodson (1965). Planktivory intervenes in the tendency of the large-bodied species of zooplankton to monopolise the resources to the exclusion of small-bodied species (including, it might be added, those of the microzooplankton). Selective predation of large-bodied species favours larger populations of small-bodied species that reproduce when they are still small. The relative abundance of large- and small-bodied species is thus strongly influenced by the intensity of predation by planktivores (cf. Hrbáček et al., 1961).

Although the broad thrust of the SEH remains wholly acceptable, it is important to recognise Gliwicz's (2003b) distinction between the dynamic effects of the immediacy of predation and the gradual debilitation forced by food shortages. The relationship is further complicated by the fact that, in shallow or marginal areas of water bodies, predators may easily switch their foraging efforts to the frequently more rewarding resources of the littoral and sub-littoral benthos. In this way, the intensity of planktivory on sparse zooplankton populations may be less than might be deduced from the abundance of predatory species. Thus, it may be only in the true pelagic, well away from the influences of shores and sediments, that planktivory exerts the expected continuous constraint on the growth and recruitment pattern in the zooplankton. Elsewhere, the effects of planktivory on the zooplankton may be a more casual or more opportune constraint, at least until the zooplankton offers a sufficiently abundant and attractive alternative food refuge. The corollary of this deduction is that there is a further conditional threshold for the feeding pressure exerted on the phytoplankton that is dependent upon herbivore engagement. Below it, the benthic alternatives remain the more attractive and the effects of filter-feeding zooplankton on phytoplankton are less intensively regulated than they are above it.

The elegant study of the feeding selectivity of young roach (*Rutilus rutilus*) undertaken by Townsend et al. (1986) revealed the progressive switching between benthivory and planktivory in response to the abundance of planktic crustaceans in the range 0.1–0.5 mg C L^{-1}. The implied threshold is not a fixed one, for it depends upon the availability and accessibility of benthos, as well as upon the numbers of fish competing for it. Its existence explains tacit upper limits on the concentrations of filter-feeding zooplankton (equivalent to as few as 10–20 large *Daphnia* L^{-1} but 100–200 small-bodied species: Kasprzak et al., 1999), that could be set as much by their attractivity to predators as by the sustainability of their phytoplanktic food resources. The consequences impinge on the structure of planktic food webs, the predominant energy pathways in aquatic systems and the role of the phytoplankton in sustaining either. These issues are the subject of Section 8.2.

6.4.3 Selective feeding

In fresh waters, mesoplanktic filter-feeding clearly has a lower threshold of viability, that falls in the range 0.03–0.1 µg filterable C mL^{-1} (30–100 mg C m^{-3}). This value is set by the biology of the most successful cladocerans and not by some physiological override. After all, tunicates survive in oceanic waters supporting chlorophyll concentrations habitually in the range 0.1–0.3 µg chl*a* L^{-1} (say, 5–15 mg C m^{-3}). In fairness, it must be added that the filtering rate to body mass in tunicates has to be high and their rates of growth are modest compared with those of cladocerans (Sommer and Stibor, 2002). The only viable foraging alternative in the rarefied, low-biomass worlds of the open pelagic, of both marine and fresh-waters, is to be able to locate and select prey of high nutritive value and to be successful in its capture.

The point has been made above that this ability is particularly developed among the planktic copepods. The majority of cyclopoids are exclusively raptorial, feeding on algae in their juvenile naupliar stages, then becoming more carnivorous in the later copepodite and adult stages, when they actively select rotifers and small cladocerans. Adult calanoids have the ability to filter-feed but they have the further facility to supplement the spectrum of available foods by seizing, grappling and fragmenting microplanktic algae. When both options are available, adult diaptomids supplement the yield of filter-feeding with larger items that they capture, and usually ingest more food than could have been supplied contemporarily through filter-feeding alone (Friedman, 1980). The benefit of doing so presumably increases inversely to deficiency in the concentration of nanoplanktic particles available. Moreover, upward extension of the size range of ingestible foods to 30 μm or more permits adults of *Eudiaptomus* to feed on algae as large as *Cosmarium* and *Stephanodiscus* (Gliwicz, 1977; Richman *et al.*, 1980). However, it is the selection of ciliates (e.g. Strickler and Twombley, 1975; Hartmann *et al.*, 1993) that most broadens the main diet of calanoids. This may be critical to their survival in very oligotrophic marine and fresh-water systems, wherein it also proves to be the decisive trophic linkage in the carbon metabolism of lakes and seas of low biomass-supporting capacity (Cole *et al.*, 1989; Weisse, 2003) (see also Section 8.2).

The undoubted effectiveness of calanoid foraging under these circumstances combines the animals' well-developed ability to select the sizes of food attacked and ingested with their impressive facility to be able to chemolocate (De Mott, 1986), then orientate towards and strike out at their potential prey organisms (Strickler, 1977; Alcaraz *et al.*, 1980; Friedman, 1980). The ability of calanoids to survive on modest rations is a further factor contributing to their frequent dominance of the oligotrophic mesozooplankton (Sterner, 1989). To judge from Hart's (1996) data (see P. 262), calanoids are able to draw sufficient food to satisfy their maintenance demands from POC concentrations that are an order of magnitude more dilute than those tolerated by cladocerans. Their demands for energy and reproductive investment may well be saturated at food concentrations (0.08 μg C mL^{-1}) that would hardly keep *Daphnia* from slowly starving to death.

There is thus considerable complementarity in the respective dynamics, spatial and temporal distributions of calanoids and daphniids. Though they frequently coexist in mesotrophic and mildly eutrophic lakes, the obligately filter-feeding cladocerans are unable to satisfy their metabolic needs at chronically low food concentrations. Thus, calanoid dominance is relatively common among oligotrophic lakes, usually in association with an underpinning microbial food web. If the supportive capacity of POC (be it detrital, bacterial or algal) is significantly >0.1 mg C L^{-1}, *Daphnia* should be able to thrive and, for so long as the food concentration satisfies the demand, to reproduce and recruit new individuals rapidly. The condition may persist until, potentially, the *Daphnia* population is processing such large volumes of water each day that not just the recruitment of detritus, bacteria and filterable algae but also the components of the entire microbial food web are exhausted (Lampert, 1987; Weisse, 1994). Other things being equal, such occurrences are followed by significant mortalities of cladocerans and, to a lesser extent, other mesozooplankters too (Ferguson *et al.*, 1982). These observations fit well with established features of limnetic zooplankton ecology, especially those relating to shifts in dominance. Cladocerans generally respond positively to anthropogenic eutrophication (Hillbricht-Ilkowska *et al.*, 1979), bringing greater amplitude of fluctuations in the biomass of *Daphnia* and its foods (McCauley *et al.*, 1999; Saunders *et al.*, 1999).

Not all cladoceran families are obligate filter-feeders. Some of the littoral-dwelling chydorids and macrothricids have thoracic limbs that are adapted to scrape periphyton from the surface of leaves, etc. *Chydorus sphaericus* is frequently seen among larger microphytoplankton in small, eutrophic lakes, often clutching onto colonies such as *Microcystis* and scraping epiphytes from the mucilage (Ferguson *et al.*, 1982; Ventelä *et al.*, 2002). Though the bosminids have filtering phyllopods and use them for this purpose for much of the time, they are also capable of supplementing nanoplanktic particles from the water by seizing larger individual prey items. To do this, they use the (fixed) antennules, abdominal claw and carapace gape, fragmenting the potential food to pieces of more ingestible size (De Mott, 1982; Bleiwas and Stokes, 1985). Selectivity of food is aided by chemoreception (De Mott and Kerfoot, 1982).

Neither are the advantages of food selectivity confined to the exploitation of scarce resources in oligotrophic conditions. In waters rendered turbid through the repeated resuspension of clay and fine silt particles (1–20 μm), as many or more of the food-sized particles are inert and near useless and greatly diminish the benefits of filter-feeding. Selective feeding provides a more productive food return for the foraging effort. This is a functional explanation for the common observation that the zooplankton of turbid lakes, even quite eutrophic ones, is typically dominated by calanoids or small cladocerans (Allanson and Hart, 1979; Arruda et al., 1983). High densities of large or mucilaginous algae in productive waters also constitute a nuisance to large, obligate filter-feeders such as Daphnia, especially when the algae are close to the upper limiting size. Although Daphnia species exercise their limited powers of food selection (varying the carapace gape), they do less well than bosminids that filter-feed on abundant nanoplankton and bacteria nourished by organic solutes released from the algae. Cascading reactions among the heterotrophic nanoflagellates are also noted (Ventelä et al., 2002). An analogy comes to mind of small animals grazing or browsing among trees, which are, nevertheless, of sufficient size and density to exclude larger animals. In the end, these same larger algae may share part of the role of planktivorous fish in defending the entire community against total elimination by overwhelming Daphnia filtration rates.

The demand made on the resources by selective feeders is more difficult to gauge than is that of less specialised filter-feeding. The carbon intake required by calanoids saturates at close to the minimum requirements of filter-feeding cladocerans. In consideration of a small (0.8-mm) and large (2.1-mm) Daphnia, the supposed minimum daily carbon requirements (0.6 and 4.3 μg C respectively) service projected body masses equivalent to 3.2 and 42 of μg C respectively (Box 6.2), to provide daily carbon-specific intakes of between 0.1 and 0.2 μg C (μg C)$^{-1}$ d^{-1}. The imposition of selective herbivory on the relevant size ranges of specific foods may be approximated from the sum of the individual number–mass products of consumers per unit volume. By analogy to Eq. (6.13),

$$\Sigma G = (N_1 \cdot G_1) + (N_2 \cdot G_2) + \cdots (N_i \cdot G_i) \quad (6.15)$$

where G_i is the ingestion rate and N_i is the standing population of the ith species-size category of selective grazer. Finally, a rate of grazing on the food species affected, to be set against other exponents of change for the food species affected.

$$r_G = (\Sigma G)/V \quad (6.16)$$

By way of an example, Ferguson et al. (1982) observed populations of Eudiaptomus gracilis in planktivore-free enclosures of up to 20 adults L^{-1}. Supposing an average length of 1.5 mm and an individual mass equivalent to 9.3 μg C, food intake may be approximated to be <2 μg C animal^{-1} d^{-1}, with the population demand peaking at <40 μg C L^{-1} d^{-1}. In terms of the principal food ingested at that time, Asterionella (85 pg C cell^{-1}), the maximum demand could have been equivalent to a loss of not more than 470 cells mL^{-1} d^{-1}. For smaller populations of smaller animals at lower temperatures, the food demand would normally have been rather lower than this maximum (perhaps 10–20 μg C L^{-1} d^{-1} is a more likely optimum). Numbers may be sustainable on the relatively smaller daily rations of just 1–2 μg C L^{-1} d^{-1}. Relative to algal standing crops, the removal of 470 cells mL^{-1} d^{-1} is scarcely sustainable by anything under 500 cells mL^{-1} doubling each day but, to a standing crop of 10,000 cells mL^{-1}, it represents an exponential loss rate of $r_G < 0.05$ d^{-1}.

6.4.4 Losses to grazers

The combined effects of selection and filter-feeding impose differential rates of removal upon the species composition of the phytoplankton. Some (mostly smaller species, generally <50 μm in maximum dimension, but often mainly nanoplanktic species) are removed primarily by cladoceran filter-feeding but also by other crustaceans. Others, generally the smaller microphytoplankters (20–100 μm in maximum dimension),

Box 6.2 | Carbon equivalents of planktic components

In interrelating observations on the biomass and production of plankton with the distribution and flow of organic carbon through pelagic ecosystems, it is helpful to have a ready base for interconversion. Many such conversions are available in the literature and some have been invoked in this book. The selection of conversions below is the one used throughout this book. The veracity of the relationships is not always well known and, for any particular species, something better can generally be found. On the other hand, for generalisations and order-of-magnitude flux estimates, the following statements will be found to be helpful.

Free-living bacteria

Conversions linking carbon content and biovolume were explored by Lee and Fuhrman (1987); more general averages (0.01–0.02 pg C cell^{-1}) are given in Sorokin (1999). In calculations here, I follow the determination adopted by Ferguson et al. (1982) for freshwater bacteria in a eutrophic reservoir: 1×10^6 cells / 13 ng , or carbon content of bacteria, 0.013 pg C (cell)$^{-1}$.

Non-diatomaceous phytoplankton

As developed in Section 1.5 the carbon content of most phytoplankters (save for diatoms) conforms well to the regression $C = 0.225\ v^{0.99}$ (Fig. 1.8) where v is the cell volume. Thus, the carbon content of planktic algae is ~0.225 pg C μm^{-3}. (Individual cells range from >1 pg to 10 ng C.)

Diatoms

The large vacuole and high ash content of the cell wall make it difficult to rely on the above relation. A more reasonable estimate is to take C = 50% of the ash-free dry mass. The ash content can be approximated on the basis of cell volume according to the regression in Fig. 1.9. Table 1.5 suggests an average Si of 0.1 pg Si (or 0.21 pg ash) per μm^3 of cell volume. In this case, 0.15pg C μm^{-3} is also a helpful approximation.

Nanoflagellates

Assumed to conform to the relationship for phytoplankton. Thus, carbon content of planktic algae is ~0.225 pg C μm^{-3} (individual cells in range 1–900 pg C).

Ciliates

Pending better information, individual cells probably range from 1 to 100 ng. Herein, 'small planktic ciliates' are considered <30 μm, with a carbon content of ~3 pg C; 'large ciliates' (60 μm) may easily comprise 50 ng C.

Rotifers

Examples from Bottrell et al. (1976), assuming carbon is 50% dry weight. All means subject to margin of at least ±30%.

Polyarthra spp.	35 ng C (individual)$^{-1}$
Keratella cochlearis	50 ng C (individual)$^{-1}$
Kellicottia longispina	100 ng C (individual)$^{-1}$
Brachionus calyciflorus	110 ng C (individual)$^{-1}$
Filinia longiseta	150 ng C (individual)$^{-1}$
Keratella quadrata	150 ng C (individual)$^{-1}$
Asplanchna priodonta	250 ng C (individual)$^{-1}$

Cladocera: *Daphnia*

Bottrell et al. (1976) collected, presented and pooled a wide range of regressions relating body mass to body length in *Daphnia*, of the form $\ln (W_{animal}) = \ln a + b \ln (L_{animal})$. With L_b in mm, W (in μg) is generally well predicted by the slope $b = 2.67$ and the intercept $\ln a = 2.45$. Subtraction of $\ln 2$ (0.693) gives the prediction

$$\ln[C] = 2.45 - 0.693 + 2.67 \ln(L_b).$$

Some examples:

$L_b =$	C (individual)$^{-1} =$
0.8 mm	3.2 μg C
0.95 mm	5.0 μg C
1.3 mm	11.7 μg C
1.6 mm	20.3 μg C
2.1 mm	42 μg C
2.24 mm	50 μg C
4.0 mm	234 μg C

Cladocera: *Bosmina*

Bottrell et al. (1976) pooled several regressions relating body mass to body length in *Bosmina*, of the form $\ln (W_{animal}) = \ln a + b \ln (L_b)$. With L_b in mm, W (in μg) is generally well predicted by the slope $b = 3.04$ and the intercept $\ln a = 3.09$. Subtraction of $\ln 2$ (0.693) gives the prediction

$$\ln[C] = 3.09 - 0.693 + 3.04 \ln(L_b).$$

Some examples:

$L =$	C (individual)$^{-1} =$
0.3 mm	0.3 μg C
0.7 mm	3.7 μg C
1.0 mm	11.0 μg C

Copepoda

Bottrell et al. (1976) pooled several regressions relating body mass of copepods to body length of the form $\ln (W_{animal}) = \ln a + b \ln (L_b)$. With L_b in mm, W (in μg) is generally well predicted by the slope $b = 2.40$ and the intercept $\ln a = 1.95$. Subtraction of $\ln 2$ (0.693) gives the prediction

$$\ln[C] = 1.95 - 0.693 + 2.4 \ln(L_b).$$

Some examples:

$L_b =$	$C\ (individual)^{-1} =$
0.1 mm	0.014 µg C
0.3 mm	0.2 µg C
1.0 mm	3.7 µg C
1.5 mm	9.3 µg C
2.0 mm	18.5 µg C
2.5 mm	31.7 µg C

Satapoomin (1999) published relationships of mass and carbon content for several marine species of copepods. For body lengths of 0.5 to 1.4 mm, *Centropages* carbon contents of 0.5–16.6 µg C (individual)$^{-1}$ and *Temora* 1–33 µg C (individual)$^{-1}$ are found.

Euphausiids

Lindley *et al.* (1999) presented regressions relating body mass of some euphausians to various allometric length measures, in the form log $(W_{animal}) = a + b\ \log\ (L_c)$, where L_c is the body length from the rostrum to the telson tip. Putting $a = 0.508$ and $b = 2.723$, and the typical carbon content of 40% dry mass, [C] is predicted as:

$$\ln[C] = 0.508 - 0.4 + 2.723\log(L_c).$$

Some examples:

$L_c =$	$C\ (individual)^{-1} =$
2 mm	8.5 µg C
6.3 mm	193 µg C
20 mm	13 620 µg C

are taken mainly by calanoids. Larger microplankton may be immune from either, although even these may be liable to specialist predators, which may include rotifers and protists.

In the size range, 20–50 µm, phytoplankton may be liable to attack by both main kinds of crustacean. The dynamic rates of their removal from their respective populations should be formally restated by combining Eqs. (6.14) and (6.16):

$$r_G = [\omega'(\Sigma F_i) + (\Sigma G)]/V \qquad (6.17)$$

In relative terms, selective feeding is the more prevalent loss process where filter-feeders fail to operate. However, as has been suggested, the dynamic effect of filter-feeding, *where it is sustainable*, is potentially much greater than selective feeding in absolute terms. Only cladoceran filter-feeding is normally capable of clearing the water of small algae and can almost wholly account for the demise of extant populations.

Because it still seems to generate surprise among some limnology students, it needs to be emphasised again that the simultaneous presence of grazers and grazed species is in no way incompatible. In fact, with the one being dependent upon the other, no other possibility is tenable. Referring back to Eqs. (5.3) and (6.1), grazing is tolerated by the grazed population so long as $r' > r_L$ and r_n remains positive. True, the rate of loss, r_L, at least for some species, may be very largely attributable to the rate of removal by grazers, r_G. Moreover, rising removal rates, especially from recruiting populations of cladocerans, can easily exceed declining rates of algal recruitment. These are the circumstances of the rapid elimination of the food species. Certainly, under these conditions, loss to grazers becomes

Table 6.4 | *Derived rate constants for two consecutive development phases of the nanoplanktic alga* Ankyra judayi *in Blelham Enclosure A during the summer of 1978*

Period	Observed r_n	$r_S{}^a$	$r_G{}^b$	Back-calculated r'
12 Jun–3 Jul	0.497	0–0.143	0.193–0.622	0.690–1.262
3 Jul–10 Jul	0.122	c	0.644–1.416	>0.766
10 Jul–24 Jul	−1.043	0–0.082	0.595–1.341	0–0.380
14 Aug–28 Aug	0.493	0–0.344	0.418–2.022	0.911–2.859
28 Aug–4 Sep	−0.150	0.046–0.068	0.253–0.556	0.149–0.474
4 Sep–18 Sep	−0.643	0	0.244–0.714	0–0.071

[a] Estimated from sediment traps set at depth.

[b] Calculated from aggregate filtration rate only, assuming no rejection ($\omega' = 1$).

[c] No traps set.

the main fate of the phytoplankton population in question.

In mesotrophic and in mildly eutrophic lakes, grazing losses are evidently heavy among species vulnerable to filter-feeding. Reynolds *et al.* (1982b) commented on the poverty of nanoplankton and small microplankton that were recoverable in deep sediment traps in Windermere. In the much shorter water columns of the experimental limnetic enclosures in Blelham Tarn, too, abundant growths of *Cryptomanas*, *Plagioselmis*, *Chromulina* and, especially, *Ankyra* that, at various times, were stimulated (Reynolds, 1986b) contributed little to the sedimentary flux. Within the confines of the enclosures, some reasonable approximations of the rates of grazing removal were possible. The entries in Table 6.4 track the changes in the rates of increase and decrease through two consecutive peaks of *Ankyra* in Enclosure A in 1978 (during which weekly fertilising averted the likelihood of nutrient limitation and the main grazers, *Daphnia galeata*, were almost unconstrained by predators). The estimates necessarily carry large error margins: these are properly shown, although it may be confidently stated that the lower ends of the ranges shown are more realistic in each case. The true rate of replication (r') of *Ankyra* under the general conditions of light and temperature obtaining during these times could scarcely have much exceeded 1.0–1.1 d^{-1} in the earlier phase whereas a rate between 0.9 and 1.0 would have applied during the second.

The upper sinking loss rates predict a bulk sedimentation rather greater than was actually measured during the same experiments (Reynolds and Wiseman, 1982). Thus, the additions to the one precise statistic, the observed net rate of increase, r_n, should err on the low side.

The exponents are used to back-calculate the bulk additions or subtractions in each phase. Thus, supposing

$$N_t = N_0 \exp(r' - r_S - r_G)t \quad (6.18)$$

then the increment or decrement is given by:

$$N_t - N_0 = N_0\{[\exp(r' - r_S - r_G)t] - 1\} \quad (6.19)$$

whence the number of cells produced (P) is calculated as

$$P = [r'/(r' - r_S - r_G)] N_0\{[\exp(r' - r_S - r_G)t] - 1\} \quad (6.20)$$

By analogy, the number of cells sedimented (S) and grazed (G) are calculated:

$$S = [r_S/(r' - r_S - r_G)] N_0\{[\exp(r' - r_S - r_G)t] - 1\} \quad (6.21)$$

$$G = [r_G/(r' - r_S - r_G)] N_0\{[\exp(r' - r_S - r_G)t] - 1\} \quad (6.22)$$

In either of the depicted instances, *Ankyra* cells were first detected in near-surface water at concentrations in the order of 20–40 cells mL^{-1}. In July, this built to a maximum of over 300,000 cells mL^{-1}, still mainly confined to the

Table 6.5 | *Calculated production (P) of Ankyra cells and losses attributable to sedimentation (S) and grazing (G) in each of the periods noted in Table 6.4*

Period	P (cells m^{-2} × 10^{-9})	S (cells m^{-2} × 10^{-9})	G (cells m^{-2} × 10^{-9})
12 Jun–3 Jul	905–1655	0–188	253–816
3 Jul–10 Jul	5520–11084	c	4641–10204
10 Jul–24 Jul	0–556	0–120	871–1962
Σ[a]	6425–13295	0–308	5765–12982
RW82[b]		<14	
14 Aug–28 Aug	1466–4601	0–554	673–3255
28 Aug–4 Sep	513–1631	158–234	871–1914
4 Sep–18 Sep	0–31	0	106–309
Σ[a]	1979–6263	158–788	1650–5478
RW82		4–5	

[a] Σ is the total over the wax and wane period.

[b] RW82 is the direct measurement of Reynolds and Wiseman (1982) over the whole population.

epilimnion (1531 × 10^9 m^{-2}). The contemporaneous development of a *Daphnia* population (see Fig. 6.7) was shown by gut-content analysis to be largely sustained on *Ankyra* (Ferguson *et al.*, 1982). Solution of Eq. (6.22) indicates that not fewer than 4891 × 10^9 *Ankyra* cells m^{-2} would have been eaten by then. Thus, not fewer than 6422 × 10^9 cells m^{-2} would have been produced (for comparison, the potential production of cells over the same period and at a sustained rate of 0.69 d^{-1} would have led to the production of 49 × 10^{15} cells m^{-2}!) By the end of the population, which was also rapidly cleared by grazers, losses to *Daphnia* of >5765 × 10^9 cells m^{-2} accounted (Table 6.5) for 90% of the inferred production.

In the second phase of *Ankyra* growth, which built to a maximum of 671 × 10^9 cells m^{-2}, the deduced production was not less than 1979 × 10^9 cells m^{-2}, of which > 1650 × 10^9 m^{-2} would have been cropped by grazers, whereas the directly measured sedimentary export was <5 ×10^9 m^{-2} (<0.3% of the total production).

6.4.5 Phytoplankton–zooplankton interactions

In order to draw a general perspective on the effects of consumers on the phytoplankton and on the export of primary production to the aquatic food webs, it is helpful to formulate some overview of nature of phytoplankton–zooplankton interactions. It is generally true that neither functions independently of the presence of the other, even if there is no direct linkage between the components. Excepting large colonial phytoplankters (like *Microcystis* or *Volvox*) whose consumers are more likely to be epiphytic than planktic, or those microzooplankters whose diet may be exclusively bacterial, most phytoplankton is liable to become the food of some zooplankton at some time. Removal of primary-producer biomass as the food of herbivores inevitably means that phytoplankton biomass is smaller than it would have been had there been no grazing. The converse might be that there would be a smaller consumer biomass if the consumers were denied access to primary foods and that these were not supplemented by an energetically equivalent organic carbon source.

If that much 'states the obvious', it has to be said that there are few other popular assumptions about the interactions between phytoplankton and zooplankton that are passed unchallenged. For instance, there is no clear reciprocity between the abundance of phytoplankton and of zooplankton (see the examples in Fig. 6.8), although there may be an element of the 'tracking' by zooplankton numbers of fluctuations in edible biomass. There is little evidence, either, that grazing by zooplankton necessarily controls the dynamics of the phytoplankton in any

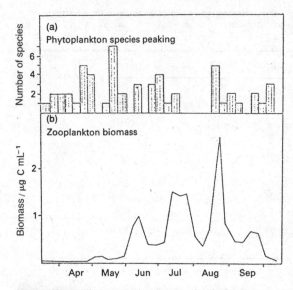

Figure 6.9 Season-long comparison between (a) the numbers of phytoplankton species attaining maximal population within a given week of the year and (b) the contemporaneous zooplankton biomass. There is no correlation, in spite of the fact that the phytoplankton development was never nutrient-limited and the zooplankton was almost entirely unpredated. Data from Ferguson et al. (1982) for Blelham Enclosure A 1978, and redrawn with permission from Reynolds (1987b).

predictable way. To judge from the example in Fig. 6.9, there is no correlation between the numbers of phytoplankton species achieving their respective population maxima and the biomass of unpredated herbivores.

Competitive interactions

Part of the complexity of these otherwise obvious relationships is due to a wide range of interspecific interactions between key consumers and key primary producers; another part is the consequence of forcing by the consumption at higher trophic levels. There are also some effects transmitted in the opposite direction, whereby phytoplankters benefit from the presence of zooplankters. Consideration reveals some interesting discernable patterns in the linkages among the main food organisms and the main groups of feeders. Gathering sufficient food to support survival and reproduction is a general problem for most animals, while those that are most efficient in converting extra energy into extra biomass are

also fittest to exploit opportunities. In this sense, either the best competitors (*sensu* Keddy, 2001) or the most exploitative animals will be favoured to succeed.

This generalisation is well exemplified by the two major groups of herbivores – calanoids and cladocerans – in lake plankton. The daphniid cladocerans, especially, are efficient in gathering and concentrating uniformly small food particles from dense suspensions and, also, in then turning carbon resource into new daphniid biomass. Subject to the satisfaction of an overiding condition of food adequacy, daphniids are exploitative and expansive, and tend to dominate overwhelmingly all other zooplankters, if not by eating them, then by eating out their resources. Their sensitivities relate to these same adaptations: having the potential to grow to large numbers and to relatively large individual body sizes makes them disproportionately vulnerable to planktivorous predators. Their restriction to filterable foods leaves them vulnerable to diminished and to chronically low concentrations of algae and, perhaps surprisingly, also to an abundance of algae of varied or of larger sizes, that interfere with systematic filtration of the resources. In stark contrast, the calanoid copepods fill in many of the gaps left by cladocerans through superior economy on a small resource base and through greater discrimination in feeding on large food particles. Analogous principles probably apply in the open sea, save that conditions of resource deficiency are more general and more persistent. Cladoceran filter-feeding is seen to bring smaller returns than are earned by calanoid selectivity.

The interactions between the abundance and species structure of the phytoplankton and the zooplankton are similarly contrasted according to whether the phytoplankton is cropped mainly by calanoids or cladocerans. Using small enclosures (mesocosms) placed in a mesotrophic lake, Sommer *et al.* (2001) observed the effects on the composition of the summer phytoplankton of experimental adjustments to the relative abundances of *Daphnia* and *Eudiaptomus*. The suppression of small phytoplankton by dominant *Daphnia* and the suppression of larger species by *Eudiaptomus* occurred much as predicted, although the authors noted that neither was able to graze down the phytoplankton to the low concentrations frequently observed in cladoceran-dominated lakes in spring (Lampert, 1978; Sommer et al., 1986) (see also Fig. 6.8). Yoshida et al. (2001) also undertook mesocosm experiments on an oligotrophic lake plankton, testing the effect of fertility and macrozooplankton on community structure. Raising the fertility alone led to a higher biomass of algae in the nano- and microplanktic size ranges and to more numerous heterotrophic nanoflagellates. The response to higher populations of *Eodiaptomus japonicus* conformed to a cascading impact, with fewer ciliates and bacteria but more heterotrophic nanoflagellates. With *Daphnia galeata*, algae, nanoflagellates and microzooplankton were all suppressed.

With enrichment raised to the level of eutrophy, when nutrient availability can support high levels of producer biomass, cladoceran feeding may eventually select against smaller algae and in favour of the larger forms that are often difficult for large *Daphnia* to ingest. Although these may not present such a problem to diaptomids, their relatively weak growth potential may delay any substantial control on the dynamics of the larger algae. Under these conditions, there may be limited scope for microbial pathways to re-establish and for them to be exploited by abundant microzooplankton (including rotifers, such as *Keratella* and *Polyarthra*) or by small cladocerans, either feeding selectively (*Bosmina*, *Chydorus* spp.). or by filtration in the nanoplanktic size range (perhaps involving small daphniids, such as *D. cucullata* or *Ceriodaphnia* spp.). These organisms are frequently prominent in the plankton of small eutrophic lakes (Gliwicz, 2003a) where they represent a third kind of interactive structure in the plankton.

Feedbacks

The nature of phytoplankton–zooplankton relationships is not without benefits passing in the reverse direction. Perhaps the most important of these is the return of limiting nutrients to the medium as a consequence of the elimination of digestive wastes. This is not always a close loop, in so far as copepod faecal pellets presumably fall some way before critical nutrients are cycled

through bacteria and consumers. This might also apply to undigested foods rejected (but not necessarily undamaged) by so-called 'sloppy' filter-feeding. Cladocerans produce a rather diffuse and amorphous waste which seems more recyclable in the immediate vicinity of its release. It seems likely that, even then, bacterial mineralisation plays a large part in regenerating nutrients that, once again, become available to phytoplankton. Probably of greater importance are the quantities of nitrogenous or phosphatic metabolites that may be excreted across the body surfaces of zooplankters. These are soluble and also readily bioavailable to algae (Peters, 1987). The restoration of nutrients to primary producers is one of the ways in which pelagic ecosystems contribute to their own self-regulation and maintain their own resource base (e.g. Pahl-Wostl, 2003).

The excretory wastes will always be dominated by the metabolites in excess. It has been pointed out (Hessen and Lyche, 1991; Elser et al., 2003) that animals will retain the amino acids that are deficient in their food and metabolise those that are in excess. Thus, *Daphnia* that meet their carbon intake requirements from planktic foods that are (say) P but not N deficient, will retain relatively more of the P content than of the N content, so that the $N:P$ ratio of their excretory products is likely to be yet higher than that of the food intake. Thus, the nutrients that are recycled do not necessarily assist in the correction of deficiencies in the growth environment of the food organisms but rather accentuate them (see also Sterner, 1993) (see also Section 5.4.5).

Bottom–up and top–down processes in oligotrophic systems

During the late 1980s, there developed vehement debate about whether phytoplankton biomass was controlled mainly by nutrients or by its grazing consumers. Both sides recognised that algal biomass was not some continuous function of the nutrients available and that herbivory was capable of reducing phytoplankton biomass to very low levels. The contentious issues were the twin possibilities that consumers might continuously regulate producer biomass at artificially low levels or that low producer biomass is always attributable to grazing. Most now seem to accept that steady-state relationships are rare and that, within the bounds of normal variability, many outcomes are possible, under the influence of several factors that are not confined to resource supply and consumption. Although the debate has receded, the vocabulary is permanently enriched (or, possibly, just stuck with) the adjectives 'bottom–up' and 'top–down'. Now these refer less to ongoing controls and much more to the processes themselves – the extent to which the biological structure and function are shaped by the resources or by the impact of consumers. Moreover, the terms are no longer applied only to phytoplankton but are now used freely in the context of an implicit hierarchy of trophic levels.

So far as the original debate about controls is concerned, it is helpful to regard the issue in terms of supportive capacity. A low resource base (whether determined by low phosphorus concentrations in a lake, or very low iron concentrations in the sea, or low concentrations of combined inorganic nitrogen in either, providing the additional resources needed to fix atmospheric nitrogen are also rare) is absolutely inescapable. Thus, on the basis of stoichiometry, a low concentration of biologically available phosphorus (BAP) of (say) 0.3 μg L^{-1} (10^{-8} M) cannot reasonably be expected to support more than a relatively small biomass (in this case, perhaps 12–15 μg biomass C L^{-1}). Invested exclusively in phytoplankton, an equivalent maximum concentration of chlorophyll a can be predicted (in this example, 0.2–0.3 μg chl$a$ L^{-1}), as can the adequacy to meet the minimum requirements of fresh-water filter-feeders (here, it fails). Photosynthetically fixed carbon that cannot be turned into biomass may be wholly or part respired, with the balance being excreted as DOC. This is, of course, useable substrate to bacteria, which, given their higher affinity for the phosphorus, are sufficiently competitive to be able to coexist with phytoplankton, yet within the same capacity limitation on their aggregate biomass. Both bacteria and small algae are liable to become the food of nanoplanktic phagotrophs and, thus, to be cropped down, but the same constraint on the total biomass persists. In the three-component system, consumer abundance might alternate with food abundance but only within the biomass capacity. As successive consumers are accommodated, the potential oscillations become more complex but, at each step, the same supportive capacity is just shared among more species representing more trophic levels. The assembled linkages making up this food chain might well process and transport large masses of carbon (in the example, as much as 4–5 mg C L^{-1} a^{-1}) but without ever raising the aggregate biomass of the participating components of the microbial food web above the supportive capacity. As it cannot be accumulated in the standing crop, the carbon thus processed is returned directly to the pool of dissolved carbon dioxide pool and/or exported in the sedimentary flux of faecal pellets and cadavers, to be returned less directly through local or global circulations (Legendre and Rivkin, 2002a).

These continuous severe constraints demand that the trophic components, from the producers and heterotrophic consumers right through to juvenile fish, are powerfully selected by their functional strengths and adaptations to deal with the rarefied resources (Weisse and MacIsaac, 2000). The overall control is within anyone's understanding of 'bottom–up' regulation. Yet little other than trophic interaction controls the relative masses of the components at any given moment: small oscillations in the effectiveness of calanoid feeding cascade through heavier predation on ciliates, better survivorship among the nanoflagellates and harder cropping of the bacterial mass. Just as easily, fish feeding on calanoids might trigger an effect upon ciliate numbers, depress the numbers of nanoflagellates, with cascading top–down effects on the balance of algal and bacterial masses (e.g. Riemann and Christoffersen, 1993).

Bottom–up and top–down processes in enriched systems

Much the same model applies in oligotrophic lakes and seas, where top–down mechanisms regulate the interspecific and interfunctional composition of the food web, even when the resource-limited carrying capacity sets a powerful bottom–up constraint on behaviour. Where the bottom–up resource constraint is less severe, rather more latitude is available to fluctuations among the components, with more opportunities to alternative species and a greater variety of possible responses. Raising the supportive capacity to (say) 150 µg biomass C L^{-1} (against 3 µg BAP L^{-1} or 10^{-7} M) opens the field to a greater average biomass of more algal species, to alternative means of harvesting and consuming them and, so, to a wider variety of consumers. Now the cascading effects influence not just how much carbon is resident in which functional level but may play a strong part in selecting the survivors.

Increasing the carrying capacity by another order of magnitude (say, to 1.5 mg biomass C L^{-1}, against ~30 µg BAP L^{-1}) would take the supportable system to well within the bounds of direct herbivory by filter-feeders and to the range of the best-known incidences of top–down control of phytoplankton. The point is now well made about the efficiency of direct consumption and assimilation of a sufficient number of *Daphnia* feeding on phytoplankters up to 30–50 µm is capable of grazing down the food resource (including most of the heterotrophic bacteria, flagellates and ciliates too) to extremely low levels indeed, leaving the water very clear (Hrbáček et al., 1961; Lampert et al., 1986; Sommer et al., 1986) (see also Fig. 6.8). Reynolds et al. (1982a) claimed this situation is unsustainable, at least in the pelagic, unless alternative sources of food can be exploited, (say) in the littoral or sublittoral benthos. In the Blelham enclosures, no such alternative was available; the consequence was the massive mortality of *Daphnia*, especially among the most juvenile cohort (Ferguson et al., 1982) and a collapse in top–down pressures. Algae were able to increase again, at the expense of nutrient resources, before supporting the resurgence of the next *Daphnia* episode.

What happens in many small, enriched lakes is that the tendency to stage these wide fluctuations in the biomass of grazed and grazer is damped by planktivory. With many species of juvenile fish and the adults of some feeding obligately or opportunistically on zooplankton, the intensity of planktivory may fluctuate seasonally, in response to temperature, to the recruitment of young fish and to continued food availability (e.g. Mills and Forney, 1987). There are, nevertheless, some top–down effects evident as the larger zooplankters are selectively removed and the smaller algae are more susceptible than large ones to grazer control. Seasonal changes in the dominance of the phytoplankton, in the direction of small to large algae, may be partly attributed to this mechanism (Sommer et al., 1986).

It is not always easy to separate the impact of bottom–up resource control from top–down predator control on the abundance of *Daphnia*. We can point to the impact of low or zero fish populations, not just in limnetic enclosures, but in natural lakes following some mass mortality of fish. Following the mortality of a major planktivore in Lake Mendota (cisco, *Coregonus artedii*), Vanni et al. (1990) observed a big increase in the *Daphnia pulicaria* over the previously more abundant *D. galeata*, enhancing grazing pressure on the phytoplankton and bringing an extended phase of clear water.

In this context, events in the mesotrophic North Basin of Windermere illustrate interannual fluctuations in the predominance of predation- and resource-driven forcing. George and Harris (1985) noted that striking interannual differences in the numbers of *Daphnia* in the plankton during June and July were inversely correlated with the temperature of the water in June. The correlation with the biomass increase in the young-of-the-year (YOY) recruits of perch (*Perca fluviatilis*) is weak, even though the latter is broadly correlated with summer water temperatutes (Le Cren, 1987). The later analyses of Mills and Hurley (1990) confirmed only a small dependence on annual perch recruitment. The stronger influence of the physical conditions on both the zooplankter and planktivore prompted deeper studies that related the biological fluctuations to subtle year-to-year variations in the annual weather patterns. These are themselves driven by significant year-to-year fluctuations in the position of the northern edge of the North Atlantic Drift Current (the 'Gulf Stream') (George and Taylor, 1995) and, especially, to interannual oscillations in the average atmospheric pressure difference between southern Iceland and the Azores (the now well-known North Atlantic Oscillation or NAO) (Hurrell, 1995). Throughout Europe, limnological behaviours are now being shown to be correlatives of the NAO (Straile, 2000; George, 2002).

So far as the linking mechanism in the dynamics of *Daphnia* in Windermere are concerned, Reynolds (1991) showed that the years of good recruitment coincided with the persistence of *Asterionella* in the plankton, as a direct consequence of the cool, windy weather associated in this region with the dominance of Atlantic airstream. In warmer years, the earlier onset of thermal stratification robs the putative *Daphnia* population of the one food source that is capable of sustaining its growth and recruitment in this lake. Indeed, it was shown that the size of the *Asterionella* crop was less important than was the last date in the year that its numbers in the surface water exceeded the threshold of adequacy to sustain the requirements of *Daphnia* (some 1300 cells mL^{-1}, or close to 0.1 µg C mL^{-1}) (cf Table 6.3). Thereafter, other filterable algae were generally too few in number to make up the shortfall.

Clearly, peak *Daphnia* numbers in any individual year will have been influenced by size of the overwintering stock, its early rate of recruitment and to the feeding choices of the standing fish stock. The point is that the years of successful recruitment of the *Daphnia* are strongly influenced by the coincidence of bottom–up and top–down forces blending in a fortuitous manner. The food resource and the exploitative potential of the *Daphnia* are the most reliable of the many stochastic variables that bear upon the quantitative strength of either their precise timing or whether these coincide significantly or at all. Thus, great interannual variations in the net production and recruitment to pelagic communities of particular locations in seas and lakes come to depend upon the interaction of small and more subtle interannual variations. This is, in essence, the 'match–mismatch hypothesis' advanced by Cushing (1982) to explain interannual fluctuations in the breeding success of fish of commercial importance. It is an extremely important principle of population and community ecology. Its workings can be explained retrospectively and its outcomes can be anticipated on a probabilistic basis but precise outturns cannot be predicted. The sketches in Fig. 6.10 summarise a series of possibilities of bottom–up tracking responses to a routinely pulsed resource opportunity but of variable magnitude and duration. However, if the duration of the opportunity is sufficiently extended, the control switches to the consumer, becoming strongly 'top–down'.

Intervention in food-web interactions

As a footnote to this section, the nature of the existing or recent interreactivity of the food-web components may be exposed not just by catastrophic interventions (fish kills, spills of toxic substances, etc.) but also through the impacts of successful invaders. The historic overexploitation of native piscivorous lake trout (*Salvelinus namaycush*) in Lake Michigan and its susceptibility to attack by the invasive sea lamprey (*Petromyzon marinus*) led to a catastrophic decline in the top carnivore niche during the 1950s and to an unopposed niche for the invasive planktivore the alewife (*Alosa pseudoharengus*) during the 1960s (Christie, 1974). A rigorous programme of lamprey eradication and salmonid restocking, helped by several severe winters, brought a substantial reduction in the alewife populations (Scavia and Fahnenstiel, 1988). There were significant repercussions in the lower trophic levels, notably an increase in large-bodied *Daphnia pulicaria*, absolutely and relative to small-bodied zooplankton (Evans and Jude, 1986). Water clarity improved as phytoplankton was cropped more heavily, aggregate abundances actually paralleling the changes in the alewife numbers (Brooks *et al.*, 1984).

Growing strength in the recruitment and near-shore activity of yellow perch (*Perca flavescens*) has since obscured one set of cascading effects and superimposed another. There have been other spontaneous changes among the lower trophic levels triggered by other alien introductions, notably the mesoplanktic carnivorous cladoceran *Bythotrephes*, which has established itself through much of the Laurentian Great Lakes (Lehman and Cáceres, 1993) and by the zebra mussel, *Dreissenia polymorpha*. Though essentially benthic, the huge aggregate filtration capacities represented by well-established populations of the bivalve have measurable effects on the planktic food web of lakes as large as Lake Erie (Beeton and Hageman, 1992; Holland, 1993).

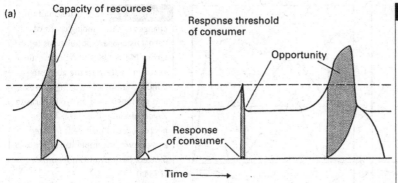

(a)

Capacity of resources

Response threshold of consumer

Opportunity

Response of consumer

Time →

Figure 6.10 (a) The (bottom–up) responses of a consumer to a regularly pulsed supply of resource but of varying magnitude relative to its requirement and, thus, of exploitative opportunity. In (b), the resource availability is absolutely greater and of longer duration; the opportunity is such to allow the consumer to control resource availability (from the top–down). Redrawn with permission from Reynolds (1991).

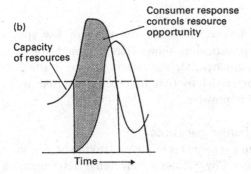

(b)

Consumer response controls resource opportunity

Capacity of resources

Time →

Food-chain length

The examples given attest to the importance of the web of interacting consumers in affecting the biomass and species composition of the producer base. In later sections, the role of the food web in regulating of ecosystem functions, resource storage and recycling, and the yield to commercial fisheries will be evaluated. The growth in appreciation of the complex mechanisms has necessitated a substantial rethink on the number of viable linkages in the web and the flows of materials and energy across them. Conventional ecological energetics developed and perpetuated the case that food-chain length is determined by the stability of the key components and by the available resource base and the usable energy influx. From a given solar area-specific energy input, investment in carbon bonds and the high cost of its transfer up the food chain, there is plainly little 'left' after just a few links. In Section 3.5.2, it was argued that the net photosynthetic production of 50–800 g C m^{-2} a^{-1} represented an energy investment of 2–32 MJ m^{-2} a^{-1}, or less than 2% of the PAR flux available. If successive consumers were to achieve a transfer efficiency of even 20%, the rate of dissipation of the potential energy (mainly as heat) through a producer → herbivore → carnivore sequence means that the investment in carnivores is probably already less than 30 g C m^{-2} a^{-1}. Extension of this argument suggests that the available food base becomes so diluted that it is impossible to support more than four or five links. Energy-based food webs are supposed to be constrained by the number of times is moved up to the next trophic level (Cohen *et al.*, 1990).

Recent studies (Spencer and Warren, 1996; Vander Zanden *et al.*, 1999; Post *et al.*, 2000a; Post, 2002a) using experimental enclosures and stable-isotope analyses, have shown the need to adopt an alternative metric for describing food-web structure. This is based on the strength of component interaction (which reflects the food choices that are actually made) and the measurement of dominant food-chain linkages. Neither resource availability nor dynamical constraints play a controlling role in these functional measures. On the other hand, it is the size of the ecosystem and the

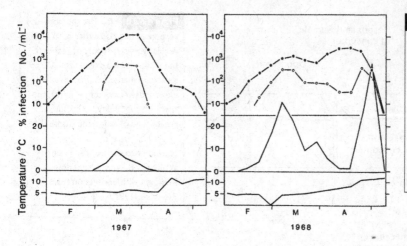

Figure 6.11 Comparison of the changes in the standing crops of, vernal *Asterionella* populations (●) in Crose Mere, UK, during consecutive years, in relation to the absolute concentrations of infecting spores and sporangia of the chytrid parasite *Zygorhizidium affluens* (○). The percentages of host cells infected and the water temperatures are also plotted. Redrawn from Reynolds (1984a).

totality of its resources that have emerged as the crucial determinant of chain length in aquatic food webs (Post *et al.*, 2000b). Whereas just two or three trophic levels can be accommodated in very small aquatic systems, from the phytotelmata of plant foliage to small pothole lakes (10^{-3} to 10^2 m^3), the linkages increase and become successively more varied and adaptable in small lakes (10^5 to 10^8 m^3), large lakes (10^{12} to 10^{14} m^3) and oceans (10^{16} to 10^{17} m^3) (Post, 2002b).

6.5 | Susceptibility to pathogens and parasites

The existence of pathogenic organisms and parasitic fungi and protists capable of infesting planktic algae and causing their death has been known for a long time. The range of the relationships that they strike with their planktic 'hosts' is wide. On the one side, it may be a form of herbivory, where the animal lives within the body of the plant, like the rhizopods which inhabit fresh-water *Microcystis* colonies (Reynolds *et al.*, 1981) or the rotifers such as *Cephalodella* and *Hertwigia*, which (respectively) eat *Uroglena* and *Volvox* from the inside out (Canter-Lund and Lund, 1995; van Donk and Voogd, 1996). At the other, there are obligately parasitic fungi (Sparrow, 1960; Canter, 1979; Canter-Lund and Lund, 1995) and pathogenic bacteria and viruses (Bratbak and Heldal, 1995; Weisse, 2003). The role played by these agents in reducing host biomass is, in general,

poorly known, although the work of few specialist investigators shows that it should not be underestimated, for the effects of fungal attack and bacterial lysis that have been reported are indeed impressive.

6.5.1 Fungal parasites

Most fungal parasites of phytoplankton belong to the order Chytridiales or are biflagellate members of the Phycomycetes (see the review of Canter, 1979). Moreover, infections can ascend to epidemic proportions, becoming the proximal cause of death of large proportion of the extant hosts (Canter-Lund and Lund, 1995). However, it is often the case that separate instances of infection but ones that involve the same species of hosts and parasites nevertheless result in quite different mortality outcomes (see Fig. 6.11). Some of the most exhaustive investigations of the epidemiology and ecology of fungal infections of phytoplankton have been carried out in the English Lakes, where chytrid infections of dominant diatoms had been observed from the outset of the research on phytoplankton dynamics. This early work established the identities of some of the parasites and the general correlation of the dynamics of their infectiveness with the environmental conditions governing other aspects of the dynamics of their hosts. They also showed that, in the case of *Asterionella* in Windermere, parasitism was not merely a constraint on the dynamics of the spring bloom but a significant control on size and species composition of the smaller autumn bloom (Canter and Lund, 1948, 1951, 1953).

The most conspicuous fungal parasites on *Asterionella* belong to the genera *Rhizophydium* and *Zygorhizidium*, which also infect many other groups of fresh-water phytoplankters. It is generally difficult to distinguish among individual species except as a consequence of their host specificity. In the presence of multi-species planktic assemblages, the incidence of infective chytrids will usually be confined to just one of them: 'fungal parasites are excellent taxonomists' (J. W. G. Lund, personal communication). Individual genera are separated on the basis of sporangial structure, dehiscence, spore characters and mycelial development. *Zygorhizidium* species constitute epidemics on mucilage-bound chlorophytes such as *Coenochloris*, *Chlamydocapsa* and *Pseudosphaerocystis*. *Rhizophydium* species also occur on *Eudorina*, on desmids and on the cysts of *Ceratium*. Cyanobacteria such as *Anabaena* are variously attacked by chytrids of the genera *Blastocladiella*, *Chytridium* and *Rhizosiphon*. Other species of *Chytridium* are found on *Microcystis*, on certain desmids (including *Staurodesmus*) and diatoms (including *Tabellaria*). *Pseudopileum* infects chrysophytes such as *Mallomonas*. *Podochytrium* is represented by species that parasitise diatoms. In all these instances, infective spores are dispersed in the medium, are attracted to and attach themselves the cells of host organisms, where they establish absorptive hyphae into the host cells, enlarge and develop new sporangia. The host cells are almost always killed. Epidemics will destroy a large number of host cells and impinge on the dynamics and perennation of the hosts.

Elucidation of the host–parasite relationships that impinge upon the ecology of phytoplankton could progress only so far on the basis of observation. Real advances came once it had become possible to isolate the fungi and to maintain them in dual-clonal cultures (Canter and Jaworski, 1978). In a series of remarkable experiments and observations, Canter and Jaworski (1979, 1980, 1981, 1982, 1986) did much to explain the life cycles of *Rhizophydium planktonicum* emend., *Zygorhizidium affluens* and *Z. planktonicum*, host range, host–parasite compatibility, zoospore behaviour in relation to light and darkness and the variability governing the onset of fungal epidemics. In

particular, hosts need to be numerous and spores have to be infective for the parasite to take hold. Of special interest is the fact that, under conditions of low light or darkness, infective zoospores are not attracted to potential host cells of *Asterionella* as they are in good light. They rarely attach to the host cells and are quite ineffective in the dark.

The effects of light on the infectivity and development of the parasite contribute to the complexity of the circumstances of epidemic initiation. The condition of both hosts and parasites formed a key part of Bruning's (1991a, b, c) systematic investigation of the response of infectivity to light and temperature. Under low levels of light, with algal growth constrained, fungal zoospores of *Rhizophydium* are only weakly infective. On the other hand, the greatest production of zoospores (as spores per sporangium) occurs at ~2 °C. Production is light saturated at ~100 μmol photons m^{-2} s^{-1} over a wide range of temperatures. Regression equations were devised to determine the development times of sporangia and the survival time of infective spores. Thence, the limiting frequency of hosts (*Asterionella*) could be calculated. By plotting the ('threshold') concentrations required to facilitate (a) a positive infection development rate of growing hosts and (b) spore suvival, Bruning (1991c) discerned the environmental conditions most likely to lead to epidemic infections. The plot of epidemic thresholds of host densities of <200 cells mL^{-1} (Fig. 6.12a) shows a plateau at temperatures above 7 °C and at light intensities down to ~15 μmol photons m^{-2} s^{-1}, which values are mildly limiting to host growth. This area extends into the lower left corner of low light and temperature but it is otherwise bounded by steep gradients of increasing thresholds, where low light or low temperature preclude epidemic development (areas shown in black). The minimum threshold, i.e the conditions most amenable to epidemic, occurs at ~11.0 °C and ~19 μmol photons m^{-2} s^{-1} (which conditions might support a host growth rate of ~0.66 d^{-1}). The contours in Fig. 6.12b show the host density necessary to parasite survival. The lowest threshold values, indicating optimum conditions for the persistence of the parasite within the host population, are encountered under

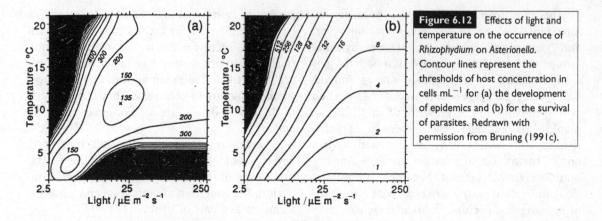

Figure 6.12 Effects of light and temperature on the occurrence of *Rhizophydium* on *Asterionella*. Contour lines represent the thresholds of host concentration in cells mL^{-1} for (a) the development of epidemics and (b) for the survival of parasites. Redrawn with permission from Bruning (1991c).

conditions of saturating light and low temperatures. Respite from parasites under low light or in darkness may assist host survival.

Other parasites and hosts possibly strike analogous relationships but the details must be expected to differ. Looking again at the plots in Fig. 6.11, the observations on *Zygorhizidium* are compatible with the Bruning model for *Rhizophydium* infectivity but the better *Asterionella* survivorship in 1967 as compared to 1968 clearly benefitted from the later intervention of the parasite into the host growth cycle.

It is also appropriate to recall that host–parasite relationships may be confounded by factors other than the 'normal' behaviours of the proximal players. One of these is host hypersensitivity (cf. Canter and Jaworski, 1979) to infective spores, where algal cells die so soon after infection that sporangia fail to develop and the progress of the epidemic becomes stalled. Another is the remarkable phenomenon of hyperparasitism: *Zygorhizidium affluens*, for example is itself frequently parasitised by another chytrid, *Rozella* sp. (Canter-Lund and Lund, 1995).

6.5.2 Protozoan and other parasites

The dividing line between a consuming phagotroph and true parasite is a fine one. Whereas the typical feeding mode in amoebae involves the pseudopodial engulfing of food organisms and intracellular digestion from a food vacuole, the amoeboid *Asterocaelum* comprises of little more than a few long, fine pseudopodia. According to the description of Canter-Lund and Lund (1995), these are wrapped around the prey organism (often a centric diatom). They fall empty and transparent and the whole becomes surrounded by mucilage, as the animal contracts to an ornately spined cyst. Digestion of the food proceeds until one or more small amoebae emerge from the cyst, to recommence the life cycle. Towards the end of a period of abundant hosts, the contents of some maturing digestion cysts, instead of releasing amoebae, are transformed into a rounded resting spore. The animal can survive in this condition for many months until hosts once again become abundant.

Vampyrella is another interesting amoeboid rhizipod that uses its slender pseudopodia to attach to the cells of algae (usually filamentous chlorophytes such as *Geminella*) but not to wrap around its prey. Instead, *Vampyrella* dissolves a little area of host cell wall and extracts ('sucks out') the contents of the host cell into its own body. It may suck out another host, perhaps simultaneously, before turning itself into a digestion cyst (though, in this genus, it is not spined). The cysts eventually break to release new infective amoebae or a resting spore. Species that attack other algae also seem to excise similar holes in the host cell wall. It also appears that individual vampyrellid species are usually consistent in their choice of hosts (Canter-Lund and Lund, 1995).

Canter-Lund and Lund (1995) also gave details of an aberrant, plasmodial organism that attaches to *Volvox* colonies and sets up a mycelial-like network, consuming individual host cells as it grows inexorably through the colony. These will extend beyond the host confines, presumably, to facilitate transfer to another. Canter-Lund and Lund (1995) believe the alga to be one of the slime moulds, or Myxomycetes.

Most of the non-fungal parasites (and many of the fungi too) escape without mention in many ecological studies of phytoplankton. This may reflect a true rarity but it seems unlikely that the profusion of parasites is confined to the sites where their presence has been diligently sought. Thus, it is difficult to form an objective view of the role of losses to parasitic attack in the population dynamics of phytoplankton generally. In terms of carbon pathways and sinks at the ecosystem scale, the probability is that parasites interfere little, as infected hosts are consumed as are the uninfected algae, while the survivorship of host propagules may be such to prevent local elimination from participation in subsequent bursts of growth and recruitment. At the level of individual species, however, parasites can exact a high cost in terms of dominance. From being the regular summer dominant of the phytoplankton in Esthwaite Water from the 1950s to the early 1980s (Harris et al., 1979; Heaney and Talling, 1980b), *Ceratium* was all but eliminated by a sequence of parasitic epidemics (Heaney et al., 1988). An infection of the overwintering cysts by a *Rhizophydium* species, may have been the 'last straw' as it reduced the algal inoculum to extremely low numbers. It was not until the mid-1990s that *Ceratium* began to become once again abundant in the lake (although, as at 2003, still in subdominant numbers).

6.5.3 Pathogenic bacteria and viruses

The existence of viral pathogens ('bacteriophages': Spencer, 1955; Adams, 1966; 'phycoviruses': Safferman and Morris, 1963; 'cyanophages': Luftig and Haselkorn, 1967; Padan and Shilo, 1973) infecting bacteria (including Cyanobacteria) and eukarytoic algae has long been recognised. In some cases, they had been isolated from field material and have been found to be capable of lysing laboratory strains of host organisms. Until recently, however, the occurrence of the organisms had been regarded as rare and their isolation in the laboratory had scarcely been attempted (Weisse, 2003).

Then, as reports of minimal abundances of virus particles in various aquatic environments, ranging between 10^4 and 10^8 mL^{-1} (Torella and Morita, 1979; Bergh et al., 1989; Bratbak and Heldal, 1995) began to accumulate, more interest in the ecological importance of viruses has been registered.

Modern methods of ennumeration invoke plaque assays, epifluorescence microscopy and flow cytometry, using DNA-specific stains. Viruses are more numerous in lakes than in the sea and their abundance in lakes is said to broadly correlated with chlorophyll content and the numbers of free-living bacteria; the likely causal connection is trophic state. According to Maranger and Bird (1995), viral abundance among the Québec lakes they studied was most closely correlated to bacterial production. This they considered to be indicative of the greater dependence of fresh-water microbiota on allochthonous organic materials.

Viruses are responsible for phage-induced mortality of bacterial and cyanobacterial hosts but the literature considered by Weisse (2003) shows enormous variability (from 0.1% to 100%). He points out that differing assumptions and conversions distort a clear picture. Regulation of the microbial community is yet harder to demonstrate. Virus–host interactions are host-specific and dynamic. Hosts may be virus-resistant; viruses may be dormant for long periods between infective opportunities. Viruses may even fulfil a positive role in microbial food webs by releasing and recycling nutrients and dissolved organic matter from lysed cells, so stimulating new bacterial production (Thingstad et al., 1993) (see also Fig. 6.13). Much remains to be elucidated about the net role of viruses in the dynamics of phytoplankton.

Algal-lysing bacteria have also been isolated from a variety of fresh-water habitats – including sewage tanks – where they attack a wide range of algae. A few affect planktic genera but there is no known host specificity. Most of the identified species are gram-negative myxobacteria that produce a variety of hydrolytic enzymes, such as cellulases, capable of lysing other microorganisms (Shilo, 1970; Daft et al., 1975; Atlas and Bartha, 1993).

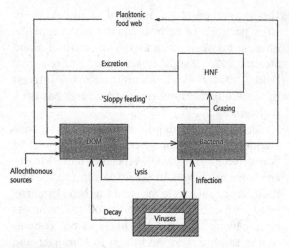

Figure 6.13 Diagram of the relationships among microbial components of the plankton (including viruses) and the pool of dissolved organic matter (DOM). HNF, heterotrophic nanoflagellates. Redrawn with permission from Weisse (2003).

6.6 | Death and decomposition

It has been implicit throughout the present chapter that the mortality of primary producers is explained principally in terms of biomass losses to other agents or to other compartments of the ecosystem. Spontaneous death – the failure of the organism to maintain its basic metabolic functions – as a consquence, for instance, of resource exhaustion or light deprivation, is rarely considered as an issue. It has to be admitted that there is some difficulty in discerning satisfactorily the boundaries between loss of vigour and physiological close-down, or between dormancy and structural breakdown. From my own experiences of tracking the dynamics of populations through temporally sequential cycles, I am bound to a view that unless an algal cell wall is entirely devoid of contents, which are not manifestly those of some invasive parasite or saprophyte, then it is not safe to assume that it is dead and incapable of physiological revival. Equally, when the other 'traditional' causes of mortality are fully quantified and attributed, the proportion that is otherwise unaccounted and, so, ascribed to cell death, is usually small.

A qualification to this statement is necessary in the context of cells sinking 'irretrievably' into deep water. Treated as 'losses to sinking', settling cells surely constitute a dynamic sink but it is not settlement that is lethal. Unless the settling cells are able to enter a physiological resting state (as do the vegetative cells of several species of centric diatom and the many types of dormant propagules, considered in Section 5.4.6), cells leaving the photic layers are likely just to respire themselves away. In deep water columns, especially during late autumnal mixing at high latitudes, much of the residual phytoplankton biomass in suspension becomes similarly unsustainable on the light energy available. Cells die and become liable to decomposition. This may be adjudged to be non-beneficial to the survival of given species but the saprophytic oxidation of cell material and the re-solution of its mineralised components are key aspects of pelagic resource recycling. Whether material recycling occurs substantially in the water column or mainly after the bulk of the algal biomass has once settled to the bottom is critical to the rate of resource renewal. Settled material may resuspended before it is finally decomposed and thus play some part in the restoration of reusable resources to organisms still in the trophogenic water layers. If not, carbon and nutrient cycling may be governed by redox transformation and solute diffusion.

It is relevant to refer here to the rates of phytoplankton disintegration and decomposition. There are reasonably attested allometric exponents of basal metabolism (Section 5.4.1) to describe the cell-specific consumption of cell carbon in self-maintenance through dark periods, generally falling within the order of magnitude 0.01 to 0.1 d^{-1}. The stoichiometric demand of oxidising 1–10% of the cell carbon each day approximates to 0.03–0.3 mg oxygen (mg cell C)$^{-1}$ d^{-1}. For night-time algal respiration to consume the air-equilibrated oxygen content of a water column (say 8–9 mg O_2 L^{-1}) requires the presence of algae at a density equivalent to ~30 mg cell C L^{-1}, or roughly 600 μg chlorophyll L^{-1}. This may give some comfort to the many managers of lakes, reservoirs and fishponds who become perplexed about the consequences of oxygen depletion by phytoplankton crops: their fears are probably exaggerated. However, in water bodies already experiencing a significant biochemical oxygen demand (BOD), the additional burden of a sudden physiological collapse and decomposition of phytoplankton might, indeed, push the oxygen dynamics to levels beyond the survival of fish and other animals.

The principal agents of decomposition are saprophytic bacteria and other microbial heterotrophs. The main groups of decomposer organisms, the organic carbon compounds that they oxidise (various carbohydrate polymers, proteins, fatty acids and lipids) and the relevant enzymes that are produced are the subject of a useful review by Perry (2002). The rates of algal decomposition have been reported in at least two dozen studies that have been published since the careful experiments of Jewell and McCarty (1971). Mostly these conform closely to the original deductions and statements. The labile materials that are readily respired and which account for about one-third of the carbon content of the healthy cell will probably have been oxidised prior to death. Much of the carbon remaining in the cadavers takes up to a year to oxidise but a fraction, comprising mostly structural polymers, including cellulose-like compounds, decomposes at only a few percent per year. First-order kinetics describe these processes quite comfortably: depending upon temperature, the exponents for oxidative carbon decay generally vary between 0.01 and 0.06 d^{-1}.

As a footnote to this section, it is proper to point to some recent findings concerning programmed cell death. Among microorganisms, perpetuation of the genomic line may be assisted by the sacrifice of quantities of vegetative cells. Examples include the differential material investment in the survivorship of overwintering cells of *Microcystis* at the expense of many that die during the benthic life-history stage (Sirenko, quoted by Reynolds *et al.*, 1981) and the reported unequal allocation of biomass between daughter cells of dividing *Anabaena*, to impose a self-control on population growth rate (Mitchison and Wilcox, 1972). Carr (1995) believed that the genetically programmed process of cell destruction, or *apoptosis*, is probably employed widely among microbes in order to achieve some measure of homeostatic function in the face of chaotic enviromental variability.

6.7 | Aggregated impacts of loss processes on phytoplankton composition

The foregoing sections demonstrate that the rates of biomass loss sustained by natural populations of phytoplankton are comparable with and, at times, exceed the scale of the anabolic processes that lead to the increase in biomass. Loss processes are important in determining whether and when given species will increase or not. Moreover, different species-specific partitioning of losses must play a major selective role in influencing day-to-day variations in the assembly and species dominance of the whole community. To be clear, loss processes affect only extant populations, the presence of which is subject to the two overriding conditions that no alga will be conspicuous in the phytoplankton unless (i) it has had both the opportunity and the capacity to exploit the resources available and (ii) that the extent to which its specific light, temperature and resource requirements for recruitment by growth are met is at least as favourable as, if not better than, that of any of its rivals. Within these twin constraints, seasonal variations in the magnitude and partitioning of species-specific losses contribute to the net dynamics of waxing and waning populations, which are the driving mechanisms of phytoplankton periodicity.

Put in its most basic qualitative terms, intensive direct filter-feeding may well suppress the net increase rates of nanoplankters and of small microplankters and favour the recruitment of larger algae instead. Similarly, contraction in the depth of the entrainment by fully developed turbulence immediately selects in favour of small or motile species at the expense of larger non-motile species, settling downwards following the loss of suspension. In many instances, the differential sensitivity to losses has an intrinsic seasonality of its own. In one of the earliest attempts to budget the fate of primary products in a lake, Gliwicz and Hillbricht-Ilkowska (1975) deduced that most

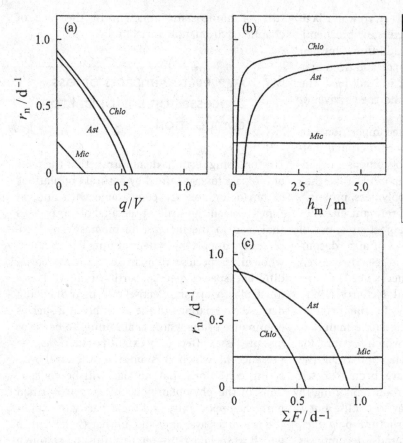

Figure 6.14 Comparative sensitivities of net increase rates of three phytoplankters (Ast, Asterionella; Chlo, Chlorella; Mic, Microcystis), each growing at their maximal rates under 12-h : 12-h light–dark alternation, to (a) dilution (q being the volume displaced each day from a water body of volume V), (b) sinking losses (h_m being the thickness of the mixed layer) and (c) filter-feeding (ΣF being aggregate volume filtered each day). Redrawn from Reynolds (1997a).

of the spring production in small, eutrophic, temperate Jezioro Mikołajske was eliminated by sedimentation. The summer production largely sustained the zooplankton while, in the autumn, most of the primary product was submitted to decomposition. The observations dovetail with the perspective of alternative survival strategies of phytoplankton (see p. 210 and Box 5.1). Thus, the vernal plankton is selected broadly by the (R-) trait of exploitation of turbulent mixing and its tolerance of low light and is vulnerable to density stratification. Early-summer (C) plankton is efficient in quickly turning resources to biomass but is usually vulnerable to herbivorous consumers. Late-summer plankton has more of the conservative (S-) characteristics of resisting sinking and grazing to the limits of its (albeit diminishing) sustainability.

Self-evidently, populations increase, stagnate or decrease in consequence of the relative quantities of the positive and the negative exponents applied. Over substantial periods (≥ 1 year was

suggested in Section 6.1), the processes balance out, so that biomass carry-over and its mean standing crop are estimable outcomes of the system that supports them. At the level of species and species-specific dynamics, however, loss processes discriminate among them and play a quantifiable part in selecting among the species aspiring to increase their populations. The large interspecific differences in the sensitivity of phytoplankton to potential loss mechanisms are exemplified in Fig. 6.14. This has been devised to illustrate the discriminatory simultaneous effects of separate alterations in flushing rate (q_s/V), mixed-layer depth (h_m) and the aggregate filtration rate (ΣF_i) of the zooplankton community upon three contrasted quite different phytoplankton species: Chlorella (representative of C strategists), Asterionella (R) and Microcystis (S). Each has been accorded a positive rate of replication, being that generated for a 12-h day of saturating light intensities and nutrient supplies at 20 °C. Despite having the slowest potential rate of growth, which prevents it from increasing at all in the face of moderate flushing, *Microcystis* nevertheless outperforms *Asterionella*, under conditions of near-surface stratification, and *Chlorella* when subject to intense grazing pressure.

Reference back to Eq. (5.1) reminds us that rates apply to standing inocula of varying sizes (N_0) and affect realised populations (N_t). A large, dominant population of *Asterionella* can still be dominant the next day, even when it has been subject to greater net loss rates than all other species present on the preceding day, simply by virtue of the large 'inoculum' carried over. However, if uninterrupted, this process of attrition in the face of net gain by another is certain to lead, eventually, to its replacement. This principle is developed later (see Section 7.3.1). For the present, it is sufficient to distinguish the dynamic processes from their effects. Most plankton biologists will accept the differences in the vulnerabilities of different kinds of phytoplankton to commonly imposed constraints. Some, however, seem to have difficulties in accepting that it *is* possible for small algae and grazers to coexist, and for *Daphnia* still to starve when filterable algae and bacteria are present in water samples submitted to microscopy.

This may also be the right place to emphasise that it is the properties of the actual organisms in relation to the contemporaneous environmental conditions that are decisive. These may, at times, outweigh behavioural differentiations anticipated by taxonomic identity alone. Thus, although it is perfectly justifiable to suggest that mature *Microcystis* colonies are quite unmanageable and reputedly unsuitable as food for most crustacean zooplankters, a substantial pre-existing population of filter-feeders is quite able to suppress the onset of *Microcystis* dominance. This is attributable to the successful removal by grazers of small colonies in their recruiting (infective) stages faster than they can grow (say, $r_G \geq 0.3$ d^{-1}). Neither is there any paradox that diatoms should *increase* in number when a large, deep lake or the sea begins to stratify and inocula are entrained over subcritical depths (Huisman et al., 1999), yet populations *decline* abruptly when a small one stagnates (Reynolds, 1973a; Reynolds et al., 1984). In both cases, it is the *absolute* depth

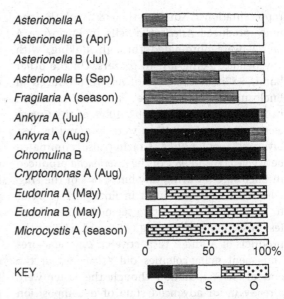

Figure 6.15 Graphical representation of the loss budgets. Each horizontal bar represents the gross cell production of the named populations in the Blelham enclosures during 1978. Appropriate shading (as shown on key) represents the minimum proportions that were grazed (G), sedimented (S) or were otherwise lost from suspension (O); intermediate shading represents the mutual overlapping of confidence intervals about the determinations. Modified and redrawn with permission from Reynolds et al. (1982a).

of the surface mixed layer that is critical to suspension.

The quantitative studies of phytoplankton losses in the Blelham enclosures have been mentioned on several occasions in the development of this chapter. Among the outcomes were the interspecific comparisons of season-long loss partitioning of phytoplankton losses. These are cited to provide an appropriate closing illustration of the differential sensitivities of phytoplankton to the various processes. The horizontal bars in Fig. 6.15 each represent the total production of the species named (calculated as *P*, according to Eq. (6.20)) in the named location and during the specified phase of growth or attrition. The minimum fractions of each product that were lost to grazing, settlement and 'other' sinks (lysis of *Microcystis* in Enclosure A, gametogenesis of *Eudorina*) are shown. Despite the width of the confidence intervals, the contrasted fates of specific production are immediately apparent from the representations. Not less than 80% of the vernally produced *Asterionella* cells in Enclosure A and B were eliminated through sinking, with any balance being lost to grazers. On the other hand, >87% of the cells of *Ankyra* produced in Enclosure A, >95% of the *Chromulina* produced in Enclosure B and probably 100% of the *Cryptomonas* produced in either were removed by herbivorous zooplankters. Grazing also eliminated between 71% and 98% of the *Asterionella* produced in Enclosure B during July but <78% of the *Fragilaria* that were produced in Enclosure A. Prior to gametogenesis, *Eudorina* seemed to have been less susceptible to grazing (<5% and <9% of the products of A and B respectively). Zygotic spores and spent male colonies did settle out at the end of the population though the latter were already in an advanced state of decomposition and disintegration. The *Microcystis* produced in Enclosure A remained loss-resistant until a significant surface scum formed overnight (on 20/21 September) and many cells were destroyed by lysis. Autumnal sedimentation as viable colonies accounted for the fraction (~44% of the total production) that survived bloom formation. Of the *Microcystis* produced in Enclosure B, a large proportion left the enclosure following a mechanical collapse of the buoyancy collar holding up the enclosure wall (7 September 1978). The residue, plus a modest recruitment through new growth, wholly sedimented as viable colonies. For further details of these populations, see Reynolds *et al.* (1982a).

6.8 | Summary

The fate of gross planktic primary production is divided between consumption and excretion at the level of physiological homeostasis (basal respiration, maintenance, voiding of unassimilable primary photosynthate, considered in Chapter 3). The materials accumulated and invested in the growth and replication of new cells experience further losses to mortality and removal of fully formed individuals (through the wash-out, sinking, consumption by animals, disruption by parasites and pathogens and physiological death of whole cells). These processes are the subject of Chapter 6.

Wash-out, especially in continuously and intermittently flushed systems, sedimentation of disentrained cells and grazing by filter-feeders can each account for major losses of phytoplankton biomass. The dynamic effects of each can be described by exponential-decay functions, analogous to dilution, and be expressed in the same, summable units as cell replication, which is a boon to simulation models of phytoplankton population dynamics. The processes discriminate among species and, hence, contribute to the selection of species in given environments.

In theory, hydraulic flushing affects all entrained species to the same absolute extent but faster-growing species are better adapted to maintain themselves against dilution. Slow-growing, larger algae tend not to be represented in the phytoplankton of rivers (potamoplankton) unless the rivers are long and have high retentivity.

Species which, for reasons of size or density difference, are most readily disentrained from turbulent layers and are simultaneously unable to swim or regulate their densities, are vulnerable to sinking losses in water columns experiencing weakened mixing and reduced mixing depth. Sinking losses of planktic diatoms may considerably exceed the rates of recruitment by replication. The losses are not necessarily permanent and it is suggested that, in certain circumstances (e.g. dielly stratifying systems at low latitudes), rapid sinking can be positively beneficial if later mixing is sufficiently vigorous and regular to return diatoms to the water surface.

Several modes of herbivory characterise the phagotrophic exploitation of the plankton. Bacteria and picophytoplankton normally occur in concentrations that offer poor energetic return for foraging effort. They are, however, harvestable by the protists (nanoflagellates and microciliates) that constitute a microplanktic–microbial food web, which is itself cropped by mesoplanktic (typically calanoid) consumers that feed efficiently on selected items of food. Nanoplanktic and microalgae, if they can be sustained in sufficient concentrations (above a threshold of ~0.1 mg C L^{-1}), represent an alternative food resource that is available to direct consumption by filter-feeders.

Filter-feeding is an efficient and effective way of foraging, provided the foods are of filterable size and sufficiently concentrated above threshold levels. While these conditions are satisfied, cladoceran filter-feeders can flourish, reproduce and recruit, potentially quite rapidly, new consumers to the assemblage. Two or more generations may be recruited over a period of a month or so, each time raising several-fold the rate of grazing loss experienced by ingestible species of phytoplankton. However, the aggregate filtration results in increasingly unsustainable loss rates to the recruiting algae and which, once surpassing the rate of cell replication, can thus bring about the collapse of the food source. All filterable particles, including the components of the microbial loop, can be very quickly cleared from the water, leading to its high clarity. Other things being equal, starvation and mortality of the herbivores results in a respite for the producers, which may increase in mass before the next cycle of consumptive tracking.

The presence of facultatively or obligately planktivorous fish may damp down considerably the potential fluctuations in feeding pressure on the phytoplankton. Resource competition generally favours the larger, more efficient herbivorous filter-feeders. Fish predation selects against large, visible filter-feeders but, presumably, only when the zooplankters also occur in sufficient numbers to attract consumer attention away from other, more rewarding food sources. Phytoplankton–zooplankton relationships vary with trophic state and with the presence and predilections of fish.

Algal populations are susceptible to epidemics of fungal parasites. Far from being stochastic events, the susceptibility of hosts to infection and the conditons favouring epidemics are now broadly predictable.

Descriptions of the destructive impacts of virus attacks on algae, cyanobacteria and bacteria have been appearing in the literature for over 40 years but it is barely more than a decade since their general abundance in waters (10^4 and 10^8 mL^{-1}) has been appreciated. The ecological role played by so many potential pathogenic organisms is still not wholly clear.

Quantifiable effects of loss processes are assembled and compared at the end of the chapter, showing how potential changes in algal dominance respond to seasonally varying rates of population depletion experienced by different species.

Chapter 7

Community assembly in the plankton: pattern, process and dynamics

7.1 | Introduction

In the pelagic, as in the great terrestrial ecosystems, space is occupied by numbers of organisms of various species forming distinct populations fulfilling differing roles. Of course, these *assemblages* of species reflect autecological aspects of preference and tolerance but they also show many synecological features of the mutual specific interactions and interdependences that characterise *communities*. The numbers of organisms, the relative abundances of the species, their biological traits and the functional roles that they fulfil all contribute to the observable community *structure*. In the plankton and in other biomes, the challenge to explain how these structures are put together, how they are then regulated and how they alter through time, falls within the understanding of community ecology.

This chapter considers the structure of phytoplankton assemblages among a broad range of pelagic systems, in the sea and among inland waters, seeking to identify general patterns and common behaviour. In the second main section (7.3), the processes that govern the assembly of communities and shape their structures are traced in detail. Because some of the terminology has been used uncritically in the literature, sometimes erroneously and often confusingly, their usage in the current work is explained in a separate text (Box 7.1).

7.2 | Patterns of species composition and temporal change in phytoplankton assemblages

7.2.1 Species composition in the sea

Despite the mutual contiguity of the marine water masses on the planet, the composition and abundance of the phytoplankton show broad and significant variations, both in space and through time. To provide some sort of review and then to offer a classification of the patterns are formidable objectives. In setting out to fulfil the present requirement, I am aware, on the one hand, of the large body of mainly miscellaneous information culled from the detailed records of particular sea cruises and, on the other, of the several monumental syntheses that have heroically attempted to sort, to classify and to draw generalities from the data (such as Raymont, 1980; Smayda, 1980). It is, of course, attractive to turn to the latter publications and to précis and paraphrase the accounts. It may seem that this is the option that has been followed but the present account is closer to the intentions of Margalef (1967) in being ordered more about the main habitat constraints. This permits its emphasis to be directed to the structural elements of the organisation of marine phytoplankton. In no sense does it provide either a catalogue of the species to be found in

Box 7.1 | Working definitions of some terms used in community and ecosystem ecology

Niche The preferred use of this term refers to exploitable resources requiring particular organismic abilities or adaptations for their utilisation. In an evolutionary context, these abilities may be acquired by some species (who may become specialist in retrieving this part of the resource pool) and, perhaps to the extent that they do so better than others, whom they may eventually exclude through competition. With sufficient division of the resource pool (niche differentiation), however, many species are able to coexist in the same assemblage (see Tokeshi, 1997). More generalist species that may broadly share the same spectrum of resources (wide niche) constitute a guild or functional group. They may also coexist, until the resource diminishes to constraining proportions and becomes the subject of competition and potential exclusion of all but the superior competitor (Hardin, 1960).

Power Power is work per unit time. The usual unit is the watt ($W = J\ s^{-1}$). In the context of power delivery of primary production to the pelagic food web, it is more convenient to express power in units of $kJ\ a^{-1}$. Salomonsen's (1992) calculations of the volume-specific primary production in the plankton is typically within the range 100–1000 of $kJ\ m^{-3}\ a^{-1}$.

Emergy Emergy is the total amount of energy invested directly or available indirectly to a system but not all of which can deliver ecologically significant power (Odum, 1986). Odum (1988) went on to compare the transformity of the energy sources to the power yield as a ratio. This ranges from up to 1 (for PAR), through wind energy (623) and hydropower (23 500) to proteinacecus foods (1 to 4 million).

Exergy Exergy is defined as the maximum flux of short-wave energy that a system can hold in the form of entropy-free chemical bonds, pending its eventual dissipation as heat (Mejer and Jørgensen, 1979). Exergy represents the information stored in the structure of the system and its ability to increase (Salomonsen, 1992). It may bolster the system against structural change when its maintenance becomes more energetically costly than its return in energy harvesting (see p. 376).

Ascendency In the context of accumulating ecosystems, this is the quantifiable synthesis of growth, development and organisation of an ecosystem. System growth is defined as the increase in total system throughput and development as the increase in information of the network of flow structures (Ulanowicz, 1986). Thus, ascendency is a measure of the structure of an ecosystem based upon the degree of organisation (information) and functioning (system activity) (Kutsch et al., 2001).

Succession. Here, usage is restricted to autogenic assembly processes that result in the substitution of species, usually in a recognisable series. The process is partly the consequence of changes to the environmental conditions wrought by the activities of earlier-establishing organisms, to the extent that they become more amenable to the later establishing of individuals of other species and less so the earlier species already established (see p. 359). Successions may culminate in a dynamic steady state, known as *climax*, where one species dominates overwhelmingly. The competitively best-fit species survives at the expense of its rivals but there is little predictable about the sequence of species participating in the succession itself. Succession is now regarded as no more than a probabilistic sequence of species replacements in a weakly variable environment.

Stability Stability is the tendency of the species composition of a community not to vary over a given significant period of time (Pielou, 1974). Observable stability carries different connotations according to context; it is better to refer to individual classes of stability by their own names (Lampert and Sommer, 1993): constancy, resistance and resilience.

Constancy Only minor fluctuations in the number of species, individuals and aggregate biomass occur. The absence of change does not distinguish whether the environment is itself is also relatively constant, or whether the structure accommodates and survives forcing by external variations. Constancy is also used in the context of cyclical consistency and the return to precisely similar community structures (see p. 381).

Resistance Resistance is the term for the situation in which despite external forcing, internal structure is preserved. This may be relative (weak forcing, resistant structure) or the forcing does not alter the current environmental constraint (see p. 376).

Resilience (elasticity) Resilience describes the situation where environmental forcing causes a deviation in structure but the system returns ('reverts': Reynolds, 1980a) to its original condition. Resilience leads to constancy over a long period (Lampert and Sommer, 1993).

Disturbance Disturbance is the situation in which the environmental forcing results in a significant shift from the original structure which is not recovered in the short term. Disturbance is a community response to the forcing. It should not be applied to the forcing (see pp. 372–9).

particular parts of the sea or an inventory of past surveys.

From the time of the earliest comparative observations (e.g. those of Gran, 1912), systematic differences in the abundance and composition of phytoplankton were recognised. At first, it was supposed that these were due to differential temperature and salinity preferences of the algae. Gradually, however, floristic assemblages became associated with the extent and longevities of water masses in the major oceanic basins and their circulations (cf. Fig 2.3), subject to the modifying effects of adjacent shelf areas and coasts, and of such local aberrations as upwellings, coastal currents and frontal activity. Some examples of the floristic distinctions among these zones are noted below, with comments on their environmental characters, constraints, fertility and variability.

Phytoplankton of the North Pacific Subtropical Gyre

The deep-water oceanic provinces of the Pacific, Atlantic and Indian Oceans cover over half the surface of the planet. Remote from the continental land masses, they are host to about one-quarter of global net primary production. Within these great basins, surface water movements constitute clockwise (in the northern hemisphere) or anticlockwise (southern hemisphere) geostrophic flows, or gyres (see Fig. 2.3). The North Pacific Subtropical Gyre, occupying an elliptical area of some 20 million km^2 located between latitudes 15° N and 35° N and between longitudes 135° E and 135° W, is the largest of these circulation features. It is also the Earth's largest contiguous biome (Karl *et al.*, 2002). The circulation substantially maintains a coherent mass of water and separates it from adjacent habitats.

Even the permanent pycnocline, located at a depth between 200 m and 1 km, isolates the surface waters from the deep, nutrient-rich layers beyond. As with the other major oceanic gyres, the severe nutrient deficiencies and low supportive capacities of the surface waters of the North Pacific have long been appreciated (TN < 3 µM, TP < 0.3 µM, SRSi < 20 µM: Sverdrup et al., 1942). On the other hand, the water has a high clarity (ε_{min} ∼0.1 m^{-1}: Tyler and Smith, 1970, quoted by Kirk, 1994). Its low planktic biomass and weak areal production have also been accepted (Doty, 1961; Beers et al., 1982; Hayward et al., 1983). The supposed constancy of these conditions nurtured an idea that the system had achieved the steady state of a successional climax (Venrick, 1995). There is certainly little doubt that the North Pacific Subtropical Gyre is a feature of great antiquity whose properties and boundaries have persisted since the Pliocene period (10 Ma B.P: Karl et al., 2002).

Since the establishment in 1988 of a monthly monitoring programme of the biomass and productivity of the phytoplankton in North Pacific Subtropical Gyre, at a station near Hawaii called ALOHA, a much clearer picture of its community dynamics has become available (Karl, 1999; Karl et al., 2002). This confirms the low depth-integrated phytoplankton biomass (averaging 22.5 mg chla m^{-2}; concentrations typically increase from < 0.2 mg chla m^{-3} near the surface to a typical deep chlorophyll maximum at between 80 and 120 m depth of up to ∼0.8 mg chla m^{-3}) and low primary production rates (484 mg C m^{-2} d^{-1}. S.D. ± 129). However, these data reveal more variability than might be expected of a fully equilibrated steady state.

The floristic composition also proved to be surprising. Whereas the earlier published accounts focused on the dinoflagellates and diatoms, much the most abundant phototrophs and the largest fraction of the standing biomass are prokaryotes, especially *Prochlorococcus* (some 50%) and smaller amounts of *Synechococcus*. Lesser numbers of eukaryote haptophyte (*Umbellosphaera*) and chrysophyte (*Pelagomonas*) nanoplankters (Letelier et al., 1993; Campbell et al., 1994) are present. The diatoms (species of *Rhizosolenia*, *Hemiaulus*) and dinoflagellates (*Prorocentrum*, *Pyrocystis*, *Ceratium*

tripos) normally make up only a small part of the plankton.

The variability revealed by the ALOHA observations occurs on several timescales. Day-to-day differences in primary production relate to differences in PAR income, especially that available to the deep chlorophyll maximum. There are broad seasonal variations in primary production rates (slightly higher in the northern-hemisphere summer) and biomass (slightly higher in the winter). However, these oscillations are also subject to larger fluctuations on supra-annual or aperiodic scales (chlorophyll range over the full study period, 13–36 mg chla m^{-2}). Longer periods of weak winds and enhanced stability, in 1989, were followed by blooms of the nitrogen-fixing Cyanobacterium *Trichodesmium*. During March–April 1997, nutrient upwelling, caused by a strong wind divergence, was followed by a significant bloom of *Rhizosolenia styliformis*. Like *Hemiaulus* (Heinbokel, 1986) this diatom is also able to supplement its supply of fixed nitrogen which, in this instance, is provided by an endosynbiotic Cyanobacterium, *Richiella*.

These fluctuations are caused proximally by events that either increase the nutrient resource or decrease the mixed depth. In turn, these follow the supra-annual 'El Niño' oscillations in wind forcing and gyre circulation brought about by high water temperatures in the western Pacific. As these events subside, vertical mixing in the gyre weakens, there is a decreased nutrient flux from depth and the upper waters become more oligotrophic. Dominance reverts to picocyanobacteria and nanophytoplankton and, potentially, to nitrogen-fixing bacteria. Annual carbon fixation settles back to ∼160 g C m^{-2} a^{-1} which is consumed mainly in turning over a microbial food web.

This pattern comprises what are manifestly the organisms best suited to the rarefied resources. Significant sedimentary losses are obviated, except during episodes of diatom abundance (Karl, 1999). The point to emphasise is that although the biomass remains generally very resource-constrained, it is not in an unmovable steady state and the events to which it is susceptible influence the rate and direction of change in community composition and function.

Phytoplankton of other low-latitude gyres

Few other modern ocean surveys match the comprehensiveness of the ALOHA series but more limited datasets suggest a common organisation. In the South Pacific Gyre, DiTullio et al. (2003) found picoplanktic Cyanobacteria to be major components of the assemblage, Prochlorococcus dominating over the scant biomass of Synechococcus and other eukaryotic plankters that included Phaeocystis and prasinophytes as significant constituents. Prochlorococcus is abundant in the low-biomass region of the central Indian Ocean (between 5° N and 30° S and 55° and 100° E; < 0.1 mg chl$a$ m^{-3}) where Trichodesmium is also common. Coccolithophorid nanoplankton, diatoms and dinoflagellates contribute to modest seasonal biomass increases during May and June, in the wake of the monsoon period (various entries in Raymont, 1980). In the North Atlantic Tropical Gyre, picoplanktic communities supporting fully developed microbial food webs have been demonstrated by Finenko et al. (2003). The dominant primary producers in the Sargasso Sea (part of the adjacent subtropical gyre) are Prochlorococcus, again being most abundant in a deep (~75 m) maximum (Moore et al., 1998). Previous investigations of the seasonality of microphytoplankton seasonality of the Sargasso Sea (Riley, 1957; Hulburt et al., 1960; Smayda, 1980) recognised that the limited abundance of spring diatoms (Rhizosolenia, Chaetoceros spp.) gives way to a long period of relative abundance of nanoplanktic Umbellosphaera and other coccolithophorids, and to Trichodesmium and ceratians such as Ornithocercus during summer.

Phytoplankton of high latitudes

Towards the edges of the great oceanic provinces, altered environmental conditions (mostly related to the proximity to land masses: depth, light and, especially, elevated nutrient availability) support alternative assemblages. In addition, the tendency towards the polar regions is to weaker and more seasonal thermal stratification. However, there are further contrasts between the circulatory patterns of the northern and southern hemispheres that relate to the distribution of land masses. The physical structure of the northern high latitudes is separable between the Arctic Polar Basin and the boreal reaches of the

Pacific and Atlantic Oceans, whereas Antarctica is wholly surrounded by an almost continuous Southern Ocean which interfaces with the South Atlantic, the Indian and the southern Pacific.

In the Arctic Basin, seasonal dynamics are supposed to be dominated by changes in day length and radiation intensity. There are few seasonal studies from which to gauge successional changes (Smayda, 1980). In the winter, underwater light availability is severely constraining in both ice-covered and ice-free areas and phytoplankton biomass is low. While there is little evidence of much seasonal growth under permanent ice, lengthening days bring significant increases in standing crop in ice-free areas (Smayda, 1980). Generally, diatoms are overwhelmingly dominant, with biomass being shared among a relatively small number of main species (Achnanthes taeniatum, Chaetoceros diadema, Corethron criophilum, Skeletonema costatum, Coscinodiscus subbulliens, Rhizosolenia styliformis and Rhizosolenia hebetata). Phaeocystis sp. is one of the few non-diatoms that can be plentiful in summer; three species of Ceratium (C. longipes, C. fusus and C. arcticum), three of Peridinium (P. depressum, P. oratum and P. pallidum) and the prasinophyte Halosphaera are also noted.

In intermediate zones, where the ice cover may be broken but its melting proceeds through much of the short summer, density differences contribute to a substantially reduced mixing depth. Within such polynia, there may be a characteristically strong seasonal development of phytoplankton. Green flagellates may appear under the thinning ice pack, to be followed by such diatoms as Achnanthes and Fragilariopsis species through the period of ice-break. These may persist until new ice is formed towards the end of summer (perhaps as little as 2–4 months later) but are often joined by other diatoms (Chaetoceros, Thalassionira spp.), Phaeocystis and Ceratium species, together forming a distinctive species assemblage (Smayda 1980).

Phytoplankton of deep boreal waters

The Atlantic to the north of the North Atlantic Drift Current, together with its arms to the Labrador and Norwegian Seas (roughly, the open ocean to the north of a line from 35° N, 75° W and 45° N, 10° W) and the Pacific Ocean to the north of the Kuroshio Current (roughly, from 35° N, 145° E to 55° N, 145° W) constitute the two main boreal oceanic provinces. The Continuous Plankton Recorder (CPR), devised originally by Sir Alister Hardy (1939) to investigate mesoscale patchiness, was used extensively and over a number of years to reveal microphytoplanktic structure in the boreal North Atlantic Ocean (Colebrook *et al.*, 1961; Robinson, 1961, 1965). These identified the ceratians *C. carriense*, *C. azoricum*, *C. hexacanthum* and *C. arcticum* and the diatom *Rhizosolenia alata* as being primarily oceanic species, quoted in a sequence from warmer southerly water to the colder northern masses. Many other species found were adjudged to have originated in adjacent shelf waters (e.g. *Ceratium lineatum*, *Thalassiothrix longissima*, *Nitzschia* spp.), yet others to have a more general occurrence (*Thalassionema nitzschioides*, *Rhizosolenia styliformis*, *R. hebetata*). Phytoplankton abundance increases with increasing day length, to a typical maximum of ∼2 mg chla m^{-3} in May or June. Among many species present, diatoms (*Thalassiosira*, *Thalassionema* and *Rhizosolenia*) generally dominate. In summer, *Ceratium fusus*, *C. furca* and *C. tripos* are relatively common.

Smayda (1980) reviewed Holmes's (1956) reconstruction of algal periodicity in the Labrador Sea. Many species are common to the CPR listings for the open Atlantic, with species of *Thalassiosira* and *Thalassionema* prominent, along with several species derived from adjacent shelfs and coasts (notably species of *Chaetoceros*, *Coscinodiscus*, *Fragilariopsis*) or from the Arctic (*Ceratium arcticum* and *Peridinium depressum*). Diatoms (especially *Fragilariopsis nana*, *Fragilaria oceanica*, *Rhizosolenia hebetata*, *Pseudonitzschia delicatissima*) dominated the single late spring/early summer maximum. Dinoflagellates, including *Peridinium depressum*, were relatively common in summer. Coccolithophorids, especially *Emiliana huxleyi*, are also numerous in the North Atlantic Ocean during summer, in some years producing significant blooms (Balch *et al.*, 1996).

In the Norwegian Sea, the sequence of dominance moves from diatoms (initially *Chaetoceros* spp., *Fragilariopsis nana* in the late spring), with *Chaetoceros convolutus*, *Corethron hystrix*, *Thalassiothrix*, *Rhizosolenia hebetata*, *R. styliformis*, *Pseudonitzschia delicatissima* and *Thalassionema*

nitzschioides becoming increasingly prominent towards the late summer. Dinoflagellates are also conspicuous in the summer months, especially *Exuviaella* spp., several *Peridinium* spp., ceratian spp. (*Ceratium longipes*, *C. fusus*, *C. furca* and *C. tripos*). Gymnodinioids are common throughout the May to November vegetation period (synthesis of Smayda, 1980, based on the observations of Halldal, 1953).

In the North Pacific, a mainly sub-Arctic coldwater community occupies the boreal waters separated from the Central Pacific by the warm Kuroshio Current (Marumo, in Raymont, 1980). The biomass supported is contrasted with the meagre populations of the gyres to the south. Typically, diatoms (species of *Chaetoceros*, *Corethron* and *Fragilariopsis*, *Rhizosolenia hebetata*, *Thalassionema nitzschioides*, *Thalassiosira nordenskioeldii*) dominated the spring–summer maxima. Summer dinoflagellates included *Ceratium fusus*.

Phytoplankton of the Southern Ocean

A discrete 'Southern Ocean' was distinguished from its contiguous oceans (Atlantic, Indian and Pacific) after its effective isolation by a series of circumpolar fronts (see below) was first fully appreciated (Deacon, 1982). The key subtropical frontal system stretches around the entire Antarctic continent, generally between the latitudes 40 and 50° S, uninterrupted but for the penetration of the Patagonian region of South America. The area enclosed represents almost 20% of the planetary ocean surface and it is now known to play an important role in regulating the planetary climate (Boyd, 2002). Variations in the synthesis of the DMSP precursor (see Section 4.6.2), the turnover of macronutrients and the passage of photosynthetically fixed carbon to Antarctic heterotrophs are all mediated by the phytoplankton. Over geological time, the carbon exchanges of the Southern Ocean have impinged significantly on planetary climate.

Within the ocean, frontal systems separate approximately concentric inner water masses, each having distinct physical and chemical identifiers. The seasonal advance and retreat of sea ice cover adds to the physical complexity of the Southern Ocean, defining permanently open-ocean and seasonally ice-covered zones (POOZ, SIZ). However, a general property of all the substructures is the degree of environmental control of phytoplankton that they exert. In short, chlorophyll concentrations (generally <0.3 mg chl$a$ m^{-3}) are rather lower than the perennial capacity of the macronutrients to support (up to 15 µM DIN, up to 2 µM BAP). The condition, now referred to as HNLC (for high nitrogen, low chlorophyll: Chisholm and Morel, 1991), is attributable to several constraints (the interaction of daily irradiance, vertical attenuation and mixed-layer depth; silica exhaustion; grazer control) but experiments (such as IRONEX, SOIREE: see Section 4.5.2) have shown iron deficiency to be often the critical factor (for a full review, see Boyd, 2002). With ambient TFe levels <10^{-9} M, the short 'window of blooming opportunity' allowed by the austral summer sustains only modest phytoplankton growth before further recruitment becomes iron-limited. Even though the severity and timing of the limitation may vary among zones and the growth of diatoms depletes the stock of dissolved silicon (data of Boyd *et al.*, 1999), the ultimate controlling role of iron is quite general across the Southern Ocean.

The dominance of Antarctic waters by diatoms was detected in the earliest plankton surveys (Gran, 1931; Hart, 1934). The major species include *Chaetoceros neglectus, C. atlanticus, Corethron criophilum* and species of *Fragilariopsis, Nitzschia* and *Rhizosolenia* (which include *R. alata* and *R. hebetata*). In the seasonally ice-covered zones, *Chaetoceros neglectus* and *Nitzschia closterium* are abundant close to the receding ice edges, before giving place to the summer assemblage. The haptophyte *Phaeocystis antarctica* also blooms in some parts of the Southern Ocean. It benefits from modest iron enrichment (as do other species) but its growth does not run into silicon limitation and it is also more efficient in adapting to variable light and carbon deployment (Arrigo *et al.*, 1999). The abundance of the prymnesiophyte flagellate *Pyramimonas* is also sensitive to light and iron levels (M. van Leeuwe, quoted by Boyd, 2002). However, picoplanktic prokaryotes become relatively more scarce with increasing latitude (Detmer and Bathmann, 1997).

Oceanic fronts

The margins of each of the major circulatory provinces so far considered in this section can interface with the land masses, or with the continental shelves of which they are part, or with the margins of another oceanic province. Each interface creates distinctive physical environments for phytoplankton growth and selection. Fronts at the mutual interface of two oceanic masses persist because differences in their temperature and salinity (and, hence, density) resist intermixing. The most significant fronts separate the permanently stratified tropical oceans from the cold, well-mixed waters of the polar seas, and which, roughly, are located at around 40° N and 40° S. The polar convergences are well defined, especially in the Pacific, by sharp gradients of surface temperature that correspond to the area in which the denser cold water is sliding under the lighter warm surface flow. A few degrees to the equatorial side of the polar front is a further, rather less abrupt subtropical convergence, formed between the warmer central portions of the gyre interface with the peripheral geostrophic flow.

In both instances, the interfaces abound with instabilities and a degree of intermixing. Phytoplankton present in either water mass are confronted with altered environmental conditions, offering improved insolation to the polar assemblage and, perhaps, more nutrient to the species in the subtropical water. In a transect across the South Pacific Ocean, following longitude 170° W from the Antarctic continent to the Equator encountered its highest chlorophyll concentrations (>0.5 mg chl$a$ m^{-3}) in the vicinity of the polar and subtropical fronts (DiTullio *et al.*, 2003). In the latter instance, there was a distinct subsurface maximum at a depth of about 40 m. Between the fronts, the dominant phytoplankters were coccolithophorids, chrysophytes and *Pelagomonas*; prasinophytes, cryptophytes and *Phaeocystis* also contributed to the standing crop. Diatoms were scarce on this occasion (the silicon concentrations were <1 µM) but other authors have found the Polar Front to be the main location of Antarctic diatom blooms (Smetacek *et al.*, 1997).

Upwellings

The major upwellings occur along continental seaboards, where the circulation current and the prevailing wind have the same direction.

The Coriolis forces acting on the flow tends to drag surface water away from the coast (Ekman transport), thus entraining deeper (usually from below the pycnocline) to the ocean surface, where it 'upwells' a short distance from the shore. This water is generally cold but relatively nutrient-rich and, entering the photic zone, becomes supportive of high biological production and prized fisheries. For example, the Canaries Current – the south-flowing arm of the North Atlantic Subtropical Gyre – coincides with the North-East Trade Wind along the coastline of Mauritania and Senegal. The matching structure in the North Pacific, the California Current, is also responsible for San Francisco's legendary fogs. Yet more striking are the Benguela and Peru Currents, respectively moving up the west coasts of Africa and South America, enhancing biological production in their wakes (to >400 g C m^{-2} a^{-1}; see Behrenfield et al., 2002) and especially where they divert from the coast lines (again respectively) near Gabon and the Galápagos Islands.

The intensity of upwelling is not continuous, fluctuating with seasonal variations in wind strength and provenance. In the Indian Ocean, upwellings in the Arabian and Andaman Seas are strongly seasonal, being related to the monsoonal episodes. The El Niño oscillations alternate suppression and enhancement of the strength of upwelling of the Peru Current. When coastal winds weaken and the sea surface calms, the mixing of the water also becomes less intense and warm, nutrient-poor water persists at the surface. This phenomenon is known as 'post-upwelling relaxation'.

Both the upwellings and the relaxations create distinctive environmental conditions for phytoplankton development. Guillen et al. (1971) refer to the periods of high primary production in the coastal Peru Current (0.3–1.0 g C m^{-2} d^{-1}), nurtured by macronutrient levels shown by others to be up to 2 μM P and 20 μM N and supporting phytoplankton biomass equivalent to the order of 2 mg chla m^{-3}. Diatoms (especially Rhizosolenia delicatula, Thalassiosira subtilis, Skeletonema costatum and Chaetoceros debilis) dominated the flora, species of other groups of organisms (dinoflagellates, coccolithophorids) making up only a small percentage of the biomass. Similar conditions obtain in the Benguela upwelling, where Chaetoceros species are similarly dominant and where primary production rates of 0.5–2.5 g C m^{-2} d^{-1} have been reported (Steemann Nielsen and Jensen, 1957). Off California, the most common phytoplankton during intense upwelling included Rhizosolenia stolterforthii, Skeletonema costatum and Leptocylindricus danicus (Eppley, 1970). The recent time series of Romero et al. (2002) on the upwelling fluxes in the region of Cap Blanc, north-west Africa, distinguished periods when characteristically oceanic diatoms (Nitzschia bicapitata, Thalassionema nitzschioides, Fragilariopsis doliolus) were dominant from those when near-shore species were ascendant (including Cyclotella litoralis, Coscinodiscus and Actinocyclus spp.).

During relaxations, production falls, diatom dominance fades as cells are lost by sinking and decay, and smaller or motile algae assume relative importance. Coccolithophorids (Umbellosphaera) and small dinoflagellates (Lingulodinium polyedrum, Gymnodinium catenatum, both notable as harmful bloom species) mark the transition to calmer, nutrient-depleting conditions (Smayda, 2002). With falling fertility at the surface, chlorophyll maxima develop at depth and in which other coccolithophorids (e.g., Florisphaera spp.: Raymont, 1980) and dinoflagellates (notably Dinophysis spp.: Reguera et al., 1995) may become relatively abundant (Raymont, 1980).

Shelf phytoplankton

What are recognised as the terrestrial land masses are blocks of continental crust that, in geological time, are moved apart or are coalesced by the tectonics of the plates of oceanic crust. Whereas the active formative ridges and the deep subduction trenches power the changing positions of the continental masses, the gaps between them are water-filled. In fact they are the repository of 97% of the planetary total. At the present time and, subject to fluctuation of ±50 m due to the changes in the volume stored as polar ice, for most of the last 250 million years, the low-lying continental perimeters have been inundated by the sea, to a depth of ≤200 m. Collectively, these are the continental shelves. Grading at first gently away from the coastline, the shelf extends to the true continental edge, the abrupt, steep slope to the ocean floor, >2 km (and locally up to 6 km) beneath the water surface.

The width of the continental shelves varies with location, literally from a few kilometres (as along the coast of central Chile, of north-east Brasil and the south-eastern seaboard of Australia) to the great expanses that characterise the Sea of Okhotsk, the East China Sea, the Arafura Sea, the Gulf of Maine, the Grand Banks of Newfoundland and the seaboards of north-west Europe (the Baltic Sea, the North Sea and the English Channel). Though mutually very different from each other in their oceanic interfaces, latitude, temperature and fluvial influence, they share common attributes of close mechanical coupling to the adjacent littoral and/or benthic habitats through wind and tidal mixing. Even so, the dilution of underwater light by mixing in shallow areas is generally less extreme than in the oceanic areas beyond. Coastal shelf waters are, indeed, distinctive from the ocean in supporting alternative planktic comunities under distinctly differing environmental conditions (Smayda, 1980).

Smayda (1980) developed his overview of phytoplankton succession in open shelf water based on well-studied examples from the Gulf of Maine and the North Sea. The phytoplankton of the Gulf of Maine, located between 41° and 44° N, is characterised by the development, commencing in March or April, of a spring diatom bloom, usually dominated by *Thalassiosira nordenskioeldii* with *Porosira glacialis* and *Chaetoceros diadema*. As water temperatures warm, other species of *Chaetoceros* (notably *C. debilis*) become relatively more abundant, before yet others (including *C. compressum*) take over dominance in early summer. Diversity is further and variably increased by the relative abundance of dinoflagellates (ceratians *Ceratium longipes*, *C. tripos*, *C. fusus* and peridinians *Peridinium faroense*, *Heterocapsa triquetra*, *Scrippsiella trochoidea*) and coccolithophorids (especially *Emiliana huxleyi*). During the autumn, diatom dominance is restored by *Rhizosolenia* species (*R. alata*, *R. styliformis*, *R. hebetata*, a.o.) and *Coscinodiscus* species.

In the shelf waters around the British Isles, the influence of penetrating Atlantic water is variable but it fails to suppress the indigenous development of a May diatom bloom (typical maximum 2–3 mg chl$a$ m^{-3}), generally featuring *Thalassionema nitzschioides*, *Rhizosolenia hebetata*, *R. delicatula*, *Chaetoceros* and *Thalassiosira* species. In the summer, dinoflagellates are relatively much more abundant, especially *Ceratium fusus*, *C. furca* and *C. tripos*; *Heterocapsa triquetra*, *Karenia mikimotoi* and *Dinophysis acuminata* also feature. *Rhizosolenia alata* and *R. styliformis* are most common in autumn (various sources, compiled by Raymont, 1980).

The Sea of Okhotsk is generally partly or wholly ice covered in winter and the onset of the spring bloom (of diatoms *Chaetoceros* spp., *Rhizosolenia hebetata*, *Thallasionema nitzschioides*, *Thalassiosira nordenskioeldii* and *Leptocylindrus danicus*) is quite abrupt. The production and phosphorus dynamics of these episodes have been the subject of recent investigations by Sorokin (2002) and Sorokin and Sorokin (2002).

Coastal and near-shore waters

Near-shore shelf waters may be further distinguished by the potential to support greater levels of biomass, production and diversity, and with more variability of abundance and dominance. Shallower water, experiencing more rapid interchange of resources with the bottom sediment, together with the inflow of 'new' resources from the land, provide more growth opportunities and more support to accumulating biomass. The phytoplankton that can be supported may be regarded as a more productive sub-set of the adjacent shelf-water assemblage but some species may benefit or express advantage more than others and there may be additional species that are rare or absent further off shore.

The relative logistic ease of sampling coastal waters also makes for the assembly of more detailed time series, although, as Smayda (1980) reminded us, the sequence of species abundances are not necessarily successional. Rather, these are temporal sequences of effects that may be under physical (wind, weather, offshore current) control or, at best, the results of successional events generated elsewhere. Smayda's (1980) own data from Naragansett Bay, Rhode Island, USA, reveal an initial spring flowering of the dominant shelf water diatom, *Thalassiosira nordenskioeldii*, though this is under way at least a month earlier than it is offshore. Later common diatoms included *Skeletonema costatum*, *Asterionella japonica* and the large-celled *Cerataulina pelagica*. In the English Channel, there is a variable influence of Atlantic water supporting dominant *Thalassionema* or *Thalassiosira* populations but it is generally masked by indigenous developments of diatoms and dinoflagellates (*Chaetoceros compressus*, *Rhizosolenia delicatula*, *Heterocapsa triquetrum*, *Karenia mikimotoi*, *Prorocentrum balticum*) in early summer and of ceratians (*Ceratium fusus*, *C. tripos*) in mid to late summer (Holligan and Harbour, 1977). Typical maximum chlorophyll concentrations were in the range 3–4 mg chl$a$ m^{-3}. At the regular sampling station used by Harbour and Holligan (depth ~70 m), areal biomass generally varied between ~20 mg chl$a$ m^{-2} in winter and ~150 mg m^{-2} during the April bloom period. However, a feature of these enriched near-shore areas is the appearance in calmer weather of the haptophyte *Phaeocystis* and occasionally abundant growths of such nanoplankters as *Carteria*, *Dunaliella* or *Nannochloris*.

One of the most significant and seminal studies of near-shore marine phytoplankton was maintained by Ramón Margalef over a number of years on the Ria de Vigo in north-west Spain. He incorporated his findings into a development of a general explanation of the mechanisms of seasonal change in community structure (Margalef, 1958, 1963, 1967). Although influenced by later observations made in the Mediterranean Sea, Margalef recognised four distinct stages of the development. Early in the northern-hemisphere year, when coastal waters are still cool, well-mixed (he called them 'turbulent') and relatively charged with nutrients, small-celled diatoms (*Skeletonema costatum*, *Leptocylindrus danicus*, *Chaetoceros socialis*, *C. radicans*, *Rhizosolenia alata* and *R. delicatula*) predominate, with small flagellates (*Eutreptia*, *Platymonas*, *Rhodomonas a.o.*), occur during this 'Stage 1'. All these algae have high surface-to-volume ratios ($sv^{-1} \geq 1$), are capable of rapid growth at low temperatures and form populations of between 10^2 and 10^3 cells mL^{-1}. 'Stage 2' tends to be dominated by a more mixed community of larger-celled diatoms *Chaetoceros* species, *Lauderia annulata*, *Eucampia cornuta*) and

peridinian (especially *Scrippsiella trochoidea*), prorocentroid and ceratian (*Ceratium furca*, *C. fusus* and *C. tripos*) dinoflagellates. As seas slacken and nutrients weaken, 'Stage 2' blends into 'Stage 3', the genera *Bacteriastrum*, *Corethron* and large species of *Rhizosolenia* (e.g. *R. styliformis*) become the main diatoms, while dinoflagellates of the genera *Ceratium*, *Dinophysis*, *Gymnodinium*, and *Lingulodinium* may increase, along with coccolithophorids such as *Emiliana*. By high summer and the onset of 'Stage 4', the ria is substantially stratified and nutrients in the surface waters are severely depleted. Only the large *Rhizosolenia* species (they may include *R. calcar-avis*) and *Hemiaulus hauckii* persist in a plankton otherwise dominated by *Ceratium*, *Peridinium* and *Prorocentrum* species. Alternatively, nitrogen-fixing Cyanobacteria may appear at this time. It is noteworthy that all these species have rather low surface-to-volume ratios ($sv^{-1} \geq 0.3$) and the populations are mostly small (~10 mL^{-1}). In enriched coastal waters, prolonged stratification may instead proceed from 'Stage 3' in which the smaller 'red-tide' dinoflagellates (including *Alexandrium tamarense*) persist and continue to grow into substantial nuisance populations (see also Section 8.3.2). Autumn cooling and renewed mixing restores the plankton back to 'Stage 1'.

Margalef's descriptions provide a basis for comparison with those of Smayda (1980) referring to the inshore waters of Norway's fjord coastline. Although some 25°–30° of latitude further north and experiencing lower summer temperatures, the sequences of phytoplankton dominance have many similarities to those of Spain's rias. In Ullsfjord (71° N), an April diatom blooming of *Chaetoceros* species, *Fragilaria oceanica*, *Thalassiosira decipiens* and *T. hyalina* is followed by a July–August bloom featuring *Chaetoceros debilis*, *Pseudonitzschia delicatissima*, *Skeletonema costatum*, *Thalassiosira nordenskioeldii*, *Leptocylindrus danicus* and *Rhizosolenia alata*. Dinoflgellates are also present in summer (see below) but they are much less abundant than diatoms. In Trondheimsfjord (64° N), blooming starts in March, with *Fragilariopsis cylindrus*, *Porosira glacialis*, *Skeletonema costatum*, *Thalassiosira hyalina* and several other species, then continues through to May/June (with *Chaetoceros debilis*, *Leptocylindrus danicus*, *Pseudonitzschia* delicatissima, Skeletonema costatum and Thalassionema nitzschioides prominent). Cerataulina pelagica, Eucampia zodiacus and Rhizosolenia fragilissima are summer species. Dinoflagellates, including Exuviaella baltica, Heterocapsa triquetra and Scrippsiella trochoideum and ceratians, approximately in the sequence Ceratium longipes, C. tripos, C. fusus, develop in summer but rather more strongly than they do further north.

In the Oslofjord (59° N), Skeletonema is the most important diatom during the March bloom, with increasing representation by Thalassiosira nordenskioeldii, various Chaetoceros species (C. debilis, C. socialis, then C. compressus), Rhizosolenia alata and Pseudonitzschia delicatissima. By summer, Lauderia borealis, Cerataulina bergonii, Cyclotella caspia and Chaetoceros species are prominent, together with Phaeocystis pouchetii and a sequence of dinoflagellates (Scrippsiella, Heterocapsa and Prorocentrum micans, and ceratians Ceratium longipes, C. tripos and C. fusus). In the inner fjord, many species of nanoplanktic flagellates are recorded, among which Micromonas, Eutreptia, Cryptomonas, Rhodomonas, Ochromonas, Pseudopedinella and Chrysochromulina can be abundant. Coccolithophorids (Emiliana, Calciopappus, Anthosphaera) are also numerous in the outer fjord at this time. In recent years, Alexandrium and other red-tide species have tended to be abundant in the locality. In the autumn, diatom dominance (Skeletonema, Leptocylindrus) is re-established.

A little to the south, in the Kattegat, the spring diatoms (Skeletonema costatum, Chaetoceros compressus, Rhizosolenia delicatula, Rhizosolenia alata and Pseudonitzschia delicatissima) give place to Cerataulina and Leptocylindrus in the summer months, together with variable amounts of Phaeocystis pouchetii, Heterocapsa triquetrum, Karenia mikimotoi, Prorocentrum minimum and Ceratium species. In the spring (May–June) of 1988, Chrysochromulina polylepis, a thitherto minor component of the nanoplankton of Norwegian and Danish coastal water, produced a significant and harmful bloom. Although equivalent to 'only' (Edvardsen and Paasche, 1998) about 10–20% of the diatom chlorophyll, the concentration of 40–80 mg chla m^{-2} proved toxic to local fish populations as well as to a variety of molluscs, echinoderms ascidians and cnidarians (Dahl et al., 1989). Unusually

large crops of this alga had been observed previously and have been since but the 1988 event was startling in its magnitude. The general eutrophication of the Danish coastal waters was blamed but the proximal cause was the combination of a strong outflow from the Baltic with calm and sunny weather in the Kattegat, favouring shallow stratification and the growth of opportunist motile algae – in this case Chrysochromulina (Edvardsen and Paasche, 1998).

Extending eastwards from the Kattegat and the Øresund, the Baltic Sea is a substantially landlocked shallow sea. It is characterised by the dilution of its salt content by an excess of precipitation over evaporation, and by some large river inflow discharges. Whilst the structure of the plankton varies with the salinity gradient across the Baltic, the tendency of warm fresh water to float on the colder, salt water emphasises the vertical stability and minimises the horizontal gradient in the summer. West of the constriction between Denmark and Sweden, this effect is normally overwhelmed by tidal and wind mixing but, as mentioned above, there have been exceptions with dramatic results. Within the Baltic and its two arms, the Gulfs of Bothnia and Finland, the sequences of phytoplankton dominance differ from those on the west side of Scandinavia. The sea being usually ice-covered in winter, the first growths of the year may be small flagellates (e.g. Chlamydomonas, Cryptomonas) beneath the ice surface. After the break-up of the ice, diatoms (including Achnanthes taeniata, Skeletonema costatum and Thalassiosira baltica) generally dominate but with developing column stability, essentially freshwater species of Oocystis, Monoraphidium a.o (Edler, 1979; Wasmund, 1994; Samuelsson et al., 2002) and picocyanobacteria (Kuuppo et al., 2003) become established. However, it is the prominent cyanobacterial flora (Anabaena lemmermannii, Aphanizomenon flos-aquae and Nodularia spumigenea are the most notable) that now most characterises the Baltic Sea plankton. Blooms of toxic species currently exercise the academics and responsible authorities alike (Kuosa et al., 1997). The autumn flora is dominated by large centric diatoms, Coscinodiscus and Actinocyclus (Edler, 1979).

Strong stratification is a feature also of the Black Sea. It has suffered intensive eutrophication from its main influent rivers (Danube, Dnestr, Dnepr and Don) since the 1970s: it is relatively nutrient-rich and its deep water is severly anoxic in summer (Aubrey *et al.*, 1996). The dominant species in spring include diatoms (*Chaetoceros curvisetus, Rhizosolenia calcaravis*) but dinoflagellates (*Prorocentrum, Heterocapsa triquetra, Scrippsiella trochoidea, Ceratium tripos, Ceratium fusum a.o.*) form a major part of the biomass, and to which *Emiliana huxleyi* can sometimes make the largest contribution (Eker *et al.*, 1999; Eker-Develi and Kideys, 2003). During the summer, diatoms become rare while coccolithophorids and red-tide dinoflagellates become dominant (Velikova *et al.*, 1999).

Another weakly flushed, river-enriched coastal area of the Mediterranean is the northern Adriatic Sea. Its main phytoplankton species are diatoms (*Chaetoceros, Rhizosolenia, Cyclotella* and *Nitzschia* spp. and dinoflagellates *Prorocentrum* and *Protoperidinium*: (Carlsson and Granéli, 1999). Finally, on the Tyrrhenian Sea (western) coast of Italy, Sarno *et al.* (1993) compared the phytoplankton of the Fusaro Lagoon with the adjacent waters of the Golfo di Napoli. Here, dinoflagellates (*Prorocentrum micans*) maintained a large winter population but this was replaced by a February–March diatom bloom, dominated by *Skeletonema costatum, Chaetoceros socialis* and other species and by *Cyclotella caspia. Alexandrium* and *Dinophysis* featured in the summer plankton as did a number of prasinophytes (e.g. *Pyramimonas* spp.) and euglenoids (*Eutreptiella* sp.). Chlorophyll concentrations varied up to a maximum of ~70 mg chl$a$ m^{-3} at the surface, and to ~50 mg m^{-3} at a depth of 4 m.

7.2.2 Species assemblage patterns in the sea

Several deductions emerge from the above excursion around the world's seas and their representative phytoplankton assemblages. One of the most self-evident of these is just how rarefied is the phytoplankton over much of the ocean. For much of the tropical ocean, the concentration of primary-producer chlorophyll is generally much lower than 1 mg chl$a$ m^{-3}, and equivalent to little more than 20 mg chl$a$ m^{-2}. In the temperate ocean, maximal concentrations may be about double this, but only during summer. Areal concentrations >100 mg chl$a$ m^{-2} are substantially confined to continental shelf waters; standing crops equivalent to >200 mg chl$a$ m^{-2} are restricted to enriched near-shore habitats and coastal lagoons.

Why is the biomass so relatively low in most places? Supposing the primary producers of the plankton to be everywhere sharing their environments with heterotrophic bacteria and phagotrophic zooplankton, top–down depletion is a less likely generic explanation than is severe bottom–up regulation by a poverty of nutrient resources and by an inadequacy of photosynthetic energy, consequential upon deep mixing and the erratic dilution of the harvestable photon flux. Only where an adequate supply of a full spectrum of essential nutrients coincides with high light income into a shallow, clear, mixed layer is there a carrying capacity sufficient to support potentially high producer mass. However, inadequacies in either restrict the carrying capacity, by imposing limitations on the ability of algae to grow and divide.

Is one of these more important than the others? Nutrient poverty and light limitation place quite different impositions on algal growth, while differences in tolerances and adaptations to survive extremes are instrumental in species selection. The effects can be represented graphically to demonstrate the interaction of these factors, both in terms of habitats and of the attributes of the species for which they select. We may start with Margalef's (1978) 'tentative' plot to illustrate the sequence (he called it a 'succession', which usage will be discussed in Section 7.3.2) of phytoplankton dominance in relation to nutrients and stratification (he used the term 'turbulence'). A simplified version is shown in Fig. 7.1. The original diagonal of his plot tracked the four developmental stages in the Ria de Vigo, progressing from the Stage-1 diatoms of the still well-mixed, nutrient-rich conditions of the early vegetative season, through the *Chaetoceros-* and *Rhizosolenia*-dominated stages to the Stage-4 preponderance of dinoflagellates capable of exploiting the well-stratified and resource-segregated water column to compensate the nutrient exhaustion of the surface water.

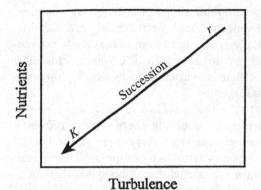

Figure 7.1 Simplified version of Margalef's diagram summarising seasonal change in phytoplankton composition in the sea as a direct function of weakening 'turbulence' and diminishing free nutrients. Redrawn with permission from Smayda and Reynolds (2001).

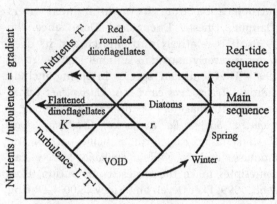

Nutrients × turbulence = productive potential

Figure 7.2 The 'mandala' (of Margalef et al., 1979), developed from Fig. 7.1, relating seasonal change to environmental selection of life forms and life-history traits. L and T are standard dimensionless units of length and time. Redrawn with permission from Smayda and Reynolds (2001).

Margalef (1978) contended that the progression moved from r-selected to K-selected species (see p. 209). The 'tentative' element was the widening of the representation to embrace more oceanic diatoms (*Thalassiosira*, at the top right of the successional diagonal), highly adapted dinoflagellates (*Ornithocercus* at the bottom left) and to insert coccolithophorids about half-way along. Thus, the entire ocean flora was potentially explicable through the relationship between mixing and fertility.

This was a remarkable and stimulating concept, relating evolutionary adaptations and survival strategies to habitat factors. It is flawed, in that to insert additional species from other locations into a single r–K continuum breaks the understanding of the succession. A more serious issue is the implication that nutrient availability and mixing are mutually correlated, whereas they are independent variables. Margalef was clearly aware of these difficulties, for his original plot (Fig. 2 in Margalef, 1978) contains a reference to 'red-tide' dinoflagellates in the upper left-hand corner ('high nutrients, low turbulence'), off the main successional diagonal. He noted their occurrence as an aberration and a sort of 'system illness'. The issue was confronted again by Margalef et al. (1979), whose 'mandala' representation (Fig. 7.2), based on a reorientation of Margalef's first plot (Fig. 7.1), accommodates a 'main sequence'

of life forms, rather than species, and a further trajectory selecting for the smaller, rounded 'red-tide' dinoflagellates.

In an early attempt to apply Margalef's (1978) conceptual model to freshwater phytoplankton, Reynolds (1980a) pointed to the frequent incidence of nutrient-rich, low-mixing conditions among shallow lakes. It was also recognised that these conditions generally promoted the rapid growth of small, exploitative organisms – in fact, precisely those with the classical attributes of r-strategists. Reynolds' (1980a) view at that time was that, against the two major axes of Margalef's model (Fig. 7.1), true $r \rightarrow K$ succession would track more or less vertically. This means that the diagonal really relates only to mixing intensity and consequential effects on nutrient redistribution of nutrients and that the distinguishing ability of the *Thalassiosira–Chaetoceros* diatoms that were considered to be exclusively r-selected by Margalef needed a new classification. Reynolds et al. (1983b) referred to them as w-selected species, based on their morphological and physiological adaptations to maintain growth on the low average light incomes incumbent upon deep column mixing, especially in high-latitude winters. Limits to the refinement of the light-harvesting adaptations required for effective operation under such conditions were

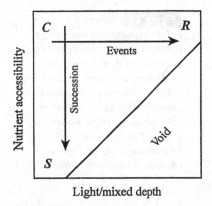

Figure 7.3 The 'intaglio' (of Smayda and Reynolds, 2001), which allows selection of species within a wide ecological space, according to their primary adaptive life-cycle strategies (**C, R, S**), except where nutrients and light are both continuously deficient (the 'void'). Redrawn with permission from Smayda and Reynolds (2001).

recognised, while the combination of low light and low nutrients was considered to be untenable as suitable habitat for phytoplankton.

There was, by now, a striking resemblance between the three viable habitat contingencies developed from Margalef's first tentative plot and Grime's (1979) (see Fig. 5.9 and Table 5.3) representation of vegetational habitats and the *C*, *S* and *R*-life-history strategies that plants needed to adopt to live in them. Reynolds (1988a) proposed the basis of fitting phytoplankton, according to their morphologies and physiological survival adaptations, to the three viable pelagic habitat combinations of mixing and resource gradients (reflecting, respectively, Grime's 'duration' and 'productivity' axes, marked in Fig. 5.9). This can now be very simply summarised in the form of a diagram (Fig. 7.3). In stable, well-insolated columns, algal uptake is expected to deplete nutrients, making the available resources more inaccessible and demanding specialist adaptations of the phytoplankton for their retrieval. This sets the direction of true autogenic succession, moving inexorably from *r*-selected *C*-strategists towards *K*-selected *S*-strategists. The income of harvestable light and its subjection to the effects on entrainment of the variable mixed depth forms the horizontal axis: mixing events or continuous deep mixing cut across the autogenic

selection, selecting increasingly for the characteristics of *R*-strategists, and which may well occur in a sequence determined by *r*–*K* selectivity. The relevant attributes of the algae favoured by the environmental conditions are shown in Box 5.1 (on p. 212). The corresponding phytoplankton performances and morphologies are also indicated in Figs. 5.8 and 5.10.

Smayda and Reynolds (2001) used the same axes to define a 'habitat template' (cf. Southwood, 1977) for marine environments. The layout, shown in Fig. 7.4, was conceptual, insofar as the axes are not precisely quantified and merely indicative of the ranges of integrated light availability and accessible nutrients. The superimposed boxes show the approximate positions of specific pelagic habitats and their phytoplankton referred to in the preceding Section (7.2.1). The broad diagonal is included as an approximate border of habitat tenability and separation from the void areas. It also serves to maintain the analogy with Grime's (1979, 2001) *C*–*S*–*R* triangular configuration (Fig. 7.3).

This representation allows us to reflect the observation that the habitats able to support substantial phytoplankton biomass are those with the least enduring or least severe constraints of energy and/or nutrient resources. Whereas nutrient availability tails off in the downward direction and harvestable light income diminishes rightwards, the high-light, high-nutrient habitats in the top left hand corner are chiefly represented by near-shore habitats and coastal lagoons, characterised by a potential for high net production and a relatively high supportive capacity for planktic biomass. These areas are also relatively rich in the range of species that they are observed to support. The striking association of such waters with outbursts of nanoplanktic flagellates, of a variety of phylogenetic affinities (prasinophytes, chlorophytes, euglenophytes, cryptophytes, chrysophytes and small haptophytes) invokes a common adaptation of small, nanoplanktic size and the potentially rapid rates of growth conferred by high unit sv^{-1} ratios. These properties are shared by the Type-I gymnodinioid and Type-II small peridinian and prorocentroid dinoflagellates (classification of Smayda and Reynolds, 2001; Smayda,

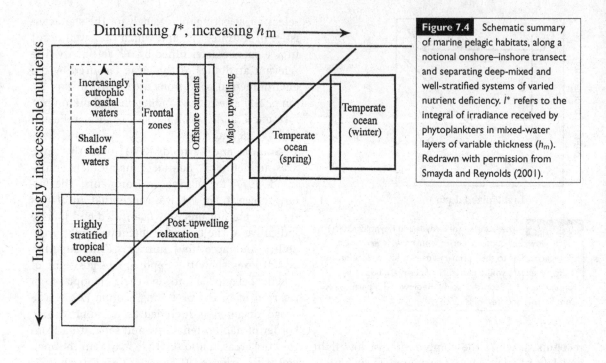

Figure 7.4 Schematic summary of marine pelagic habitats, along a notional onshore–inshore transect and separating deep-mixed and well-stratified systems of varied nutrient deficiency. I^* refers to the integral of irradiance received by phytoplankters in mixed-water layers of variable thickness (h_m). Redrawn with permission from Smayda and Reynolds (2001).

2002) that include *Gyrodinium* species, *Katodinium*, *Heterocapsa triquetra*, *Scrippsiella trochoidea* and *Prorocentrum* species and which are apparently near-cosmopolitan among coastal waters. The common Type-III ceratians (*Ceratium fusus*, *C. furca* and *C. tripos*) extend into deeper shelf waters, where they become more abundant in columns when they are at least weakly stratified, probably to within 20–40 m of the surface. Resource segregation is likely but these larger and more motile dinoflagellates are better adapted to alternate between satisfying their energy and nutrient requirements. These distributions are separately shown on the unlabelled template represented in Fig. 7.5.

The shallow mixed, moderately enriched habitats of fronts, coastal currents and upwellings are represented in the centre of Fig. 7.5. These are able to support smaller nanoflagellates (including a wealth of coccolithophorids at lower latitudes) as well as the group of distinctive small, rounded dinoflagellates that include the harmful species *Karenia mikimotoi*, *Lingulodinium polyedra* and *Pyrodinium bahamense* respectively representative of Smayda's Types IV, V and VI. The resource depletion of the upwelling relaxation zones requires the attributes of low-resource tolerance and self-regulation that characterise the Group-VII dinophysoids.

The lower left-hand apex of Fig. 7.5 covers the extreme resource-deficient environments of the stratified tropical ocean. The major constraint is to gather from the very low concentrations of essential nutrients, of which phosphorus and, especially, iron may be the most deficient (Karl, 1999). Conforming to the encounter–sufficiency relationship of Wolf-Gladrow and Riebesell (1997) (see also Section 4.2.1), the most efficient primary producers are of picoplanktic size. With a biomass capacity unsupportive of mesoplanktic phagotrophy, the arguable selective advantage in favour of a dominant *SS*-strategist picoplankton (see p. 211) is well supported by the vast extent and monotony of the *Prochlorococcus* monoculture in tropical oceans. Respite comes in the form of mixing episodes, stimulating modest growths of nanoplanktic coccolithophorids and such specialised microplankters as nitrogen-fixing *Trichodesmium*, the diatoms *Rhizosolenia styliformis*, *R. calcar-avis* and *Hemiaulus hauckii* and the dinoflagellates *Ornithocercus* and *Pyrocystis*, all of which are incumbent upon hydraulic variations in physical stability. These episodes are usually relatively short-lived, gradually reverting

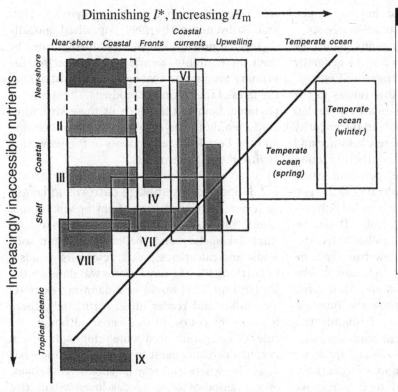

Figure 7.5 Summary of marine pelagic habitats (as shown in Figure 7.4), now populated by functional–morphological categories of phytoplankton (especially of dinoflagellates), provisionally identified by roman numerals: small, rounded gymnodinioids (I) and peridinians (II); migratory ceratians (III); frontal-, upwelling- and current-associated species (IV –VI); heterotrophic dinophysoids (VII), species of the ultraoligotrophic oceans (VIII) and tropical DCM species (IX); for further details, see text. Figure redrawn with permission from Smayda and Reynolds (2001).

to the *Prochlorococcus*-dominated ambient steady state (Karl *et al.*, 2002).

Diatoms are represented almost everywhere within the triangular space in Fig. 7.5. Although their basic requirements for light and nutrients are similar to those of all other phytoplankters, satisfaction of two diatom-specific specialist needs – a supply of skeletally progenic silicic acid (SRSi) (see Section 4.7) and frequent or continuous entrainment in a surface mixed layer >1–3 m in depth (or more if only a slow rate of growth can be sustained) (see Section 6.3.2) – is still possible over all but the extreme left-hand side of the template. Nevertheless, a reasonable first supposition is that planktic diatoms of seas and oceans should invoke the strongly *R*-strategist adaptations suited to passive entrainment in highly fluctuant, low-average-light environments.

Many of the oceanic and shelf species, indeed, comply with this anticipation. However, the great majority of these are found in coastal and near-shore waters, where they are exposed to moderately high nutrient levels and moderately high light levels. Some of these are recorded mainly

in shallow and inshore areas, including several of the large-celled, discoid species of centric diatom, such as *Cyclotella litoralis, C. caspia* and species of *Actinocyclus, Cerataulina* and *Coscinodiscus*, that are encountered also in lower-latitude upwellings. Their performances are clearly favoured by relatively high nutrient levels and may depend upon high levels of insolation. The large group of diatoms whose ranges extend into deeper, but still nutrient-rich offshore shelf areas – including species of *Thalassiosira, Chaetoceros, Leptocylindrus, Skeletonema* and the slender-celled *Rhizosolenia* species – all show the attenuated antennal morphologies of *R*-strategists (sometimes exaggerated by chain formation). Many of these same species appear in the summer plankton of the boreal oceans and polar seas. The restricted diatom flora (*Hemiaulus* spp., broad-celled *Rhizosolenia styliformis* and *R. calcar-avis*) tolerant of warm, nutrient-poor but often well-insolated waters show little tendency towards superior light-harvesting but have special adaptations to contending with chronic nutrient limitation Indeed, these diatoms show characters of more *K*-selected *S*-strategists and they may be perhaps considered as intermediate *RS*-strategists.

Several other general deductions about the composition of planktic communities generally and the functional role that they fulfil emerge from the patterns identified. One relates to the high species richness of the relatively benign environments that are not hostile to the majority of species as a consequence of severe resource and energy constraints. Extremes in either direction lead to the failure of all those species that lack the adaptations to be able to tolerate the increasingly exacting circumstances – species richness is 'squeezed out' (Reynolds, 1993b). The tolerant survivors are, by definition, well-adapted specialists and their presence in low-diversity communities constitutes a robust indicator of the particular severe conditions. More, their presence will always help to identify the function of species clusters associated with slightly less extreme circumstances (Dufrêne and Legendre, 1997). Because less exacting conditions are accessible to many more species, more outcomes are possible and, thus, they tend to lack positive species identifiers.

Secondly, species-poor, highly selected assemblages of species will dominate the behaviour of the community and control the fate of primary products. For instance, assemblages dominated by diatoms are most liable to the dynamic controls set by sinking loss rates. Biomass is mainly exported to depth, where it is processed by benthic or bathypelagic consumers through spatially large recycle mechanisms. Heavy grazing may reduce the direct sedimentation of phytoplankters but, in part, substitute a flux of zooplankton cadavers, faecal pellets and an export of particulate silica. Planktic primary products are more likely to be accumulated in the pelagic if the algae are simultaneously small and ungrazed. Export is proportionately least when most of the carbon is fixed by picophytoplankton and processed through a microbial food web (Legendre and LeFevre, 1989).

Finally, it is the matching of processes to functional groups of phytoplankton species and, in turn, to the overriding environmental circumstances biassing their selection, that leads to the elaboration and definition of macroscale spatial structures of distinct pelagic ecosystems. Platt and Sathyendranath (1999) visualised globally segregated provinces of the sea, distinguished by their susceptibility to environmental forcing, the primary production that each might sustain and the fates of their primary products. The species of phytoplankton at the hearts of these structures are, often, both the architects of the processing and the best-fitted respondents to the prevailing environmental constraints.

7.2.3 Species assemblage patterns in lakes

To review the composition of phytoplankton in a diverse range of inland waters – great lakes and small lakes, deep and shallow, saline and soft, acidic and calcareous, rivers, reservoirs, ponds – in anything like the same way as was done for the sea (Section 7.2.1) would be a daunting exercise, for author and reader alike. Fortunately, there is an easier course to be steered, although the rules of navigation need some prior explanation. Part of a personal quest to be able to define what algae live where and why (Reynolds, 1984a) has, over a period of 20 years, developed a tentative and still-evolving phytoplankton flora (Reynolds *et al.*, 2002). Cataloguing natural assemblages of phytoplankton species generally reveals interesting patterns. Not only are many species observed periodically in a given lake but their periodicity is also generally quite regular. Moreover, they often co-occur with other species whose numbers fluctuate similarly and broadly simultaneously, as if in response to the same seasonal or environmental drivers. Further, in part or in whole, the same clusters of co-occurring species are recognised in other water bodies, despite mutual hydrological isolation in many instances, and in what appear to similar kinds of water bodies but at remote distances. In between times and at many other locations, these species clusters are not represented. They may well be replaced in abundance by quite different sets of co-occurring species but which, nevertheless, form equally distinctive, recurrent clusters.

The pattern is scarcely obscure but it is difficult either to describe or to explain. Before the days of the sophisticated and readily available statistical packages, the best-known techniques were those introduced by the European school of phytosociologists to diagnose plant communities and associations (Tüxen, 1955; Braun-Blanquet, 1964). They would make a list of species in each of a series of intuitively judged small areas of uniform vegetation, called *relevés*, scoring for the relative area covered by each species. Listing the species in the same order made it easy to build up frequency tables in which regularly co-occurring species are blocked together, while those that are avoided will appear in other blocks. These blocks, or associations, can be named and classified, just as if they were individual species. The task of explaining the ecologies of the component species is arguably easier to progress if the vagaries of variable presence or relative importance of individual species is suborned to the higher level of the species cluster.

Confronting accumulating records of species counts in preserved samples collected weekly (sometimes more frequently) from each of several separate water bodies, I applied myself to the very tedious task of treating each count as a phytosociological *relevé* and to diagnosing species that co-occurred frequently, rarely or not at all. Some weighting for larger species occurring in small numbers was the only modification needed to diagnose 14 such species-clusters that were adequate to describe the entire seasonal periodicity of the phytoplankton in five contrasted lakes in north-west England and five managed annual sequences in the Blelham experimental enclosures (Fig. 5.11). The clusters were not identified beyond an alphanumeric label but the patterns and periodic sequences were conveniently rationalised in these terms (Reynolds, 1980a).

The original scheme has been much modified, mainly through the addition of more alphanumeric groups to embrace algal assemblages in other types of water and in many other global locations. Most of the new groupings have been delimited using statistical methods, which, incidentally, have been used to validate almost all the original ones. Some of these have been subdivided or realigned slightly in arriving at the 31 groups defined by Reynolds *et al.* (2002). Several independent studies have been able to apply and amplify the scheme without undue controversy and, thus, help to confirm its utility (Kruk *et al.*, 2002; Dokulil and Teubner, 2003; Leitão

et al., 2003; Naselli-Flores and Barone, 2003; Naselli-Flores *et al.*, 2003). The scheme is still evolving and two further algal groups have since been proposed (Padisák *et al.*, 2003c). A new confusion is the fact that some species are correctly classifiable in more than one cluster, according to their life histories (see p. 269).

However, the scheme is not just about recognising and giving labels to groups. The species forming the particular groups have demonstrably similar morphologies, environmental sensitivities and tolerances, and they are not necessarily confined to one phylogenetic group (Reynolds, 1984b, 1988a). They feature prominently the strategic adaptations that are required in the habitats in which they are known to be capable of good growth performances (Reynolds, 1987b, 1995a). These various aspects were summarised in tabular form in Reynolds *et al.* (2002). The coda can be used to represent seasonal changes in dominance (Naselli-Flores *et al.*, 2003), responses to eutrophication (Huszar *et al.*, 2003) and the effects of non-seasonal physical forcing (Reynolds, 1993b).

The groupings themselves are tabulated in Table 7.1 for reference. Their properties are briefly noted but these are also amplified in the context of the compositional patterns observable in the freshwater systems exemplified.

Phytoplankton of large oligotrophic and ultraoligotrophic lakes

We start with examples of the phytoplankton assemblages that are encountered in some of the world's larger lakes. According to Herdendorf (1982), 19 of the inland waters currently on planet Earth have areas greater than 10 000 km^2 and another 230 are greater than 500 km^2. Together, they contain about 90% of its inland surface water. To put these in a single category of 'large lakes' can be justified only in the present context of waters overwhelmingly dominated by open-water, pelagic habitats. Here, the grouping will exclude examples that are 'shallow'. This term itself requires careful definition; following Padisák and Reynolds (2003), shallowness is only sometimes a self-evident absolute. The statement that a lake is 'relatively shallow' is based upon the ratio between absolute

Table 7.1 | Trait-separated functional groups of phytoplankton

Group	Habitat	Typical representatives	Tolerances	Sensitivities
A	Clear, often well-mixed, base-poor lakes	*Urosolenia, Cyclotella comensis*	Nutrient deficiency	pH rise
B	Vertically mixed, mesotrophic small–medium lakes	*Aulacoseira subarctica, A. islandica*	Light deficiency	pH rise, Si depletion, stratification
C	Mixed, eutrophic small–medium lakes	*Asterionella formosa Aulacoseira ambigua, Stephanodiscus rotula*	Light, C deficiencies	Si exhaustion, stratification
D	Shallow, enriched turbid waters, including rivers	*Synedra acus, Nitzschia* spp., *Stephanodiscus hantzschii*	Flushing	Nutrient depletion
N	Mesotrophic epilimnia	*Tabellaria, Cosmarium, Staurodesmus*	Nutrient deficiency	Stratification, pH rise
P	Eutrophic epilimnia	*Fragilaria crotonensis, Aulacoseira granulata, Closterium aciculare, Staurastrum pingue*	Mild light and C deficiency	Stratification Si depletion
T	Deep, well-mixed epilimnia	*Geminella, Mougeotia, Tribonema*	Light deficiency	Nutrient deficiency
S1	Turbid mixed layers	*Planktothrix agardhii, Limnothrix redekei, Pseudanabaena*	Highly light-deficient conditions	Flushing
S2	Shallow, turbid mixed layers	*Spirulina, Arthrospira*	Light-deficient conditions	Flushing
S$_N$	Warm mixed layers	*Cylindrospermopsis, Anabaena minutissima*	Light-, nitrogen-deficient conditions	Flushing
Z	Deep, clear, mixed layers	*Synechococcus,* prokaryote picoplankton	Low nutrient	Light deficiency, grazing
X3	Shallow, clear, mixed layers	*Koliella, Chrysococcus,* eukaryote picoplankton	Low base status	Mixing, grazing
X2	Shallow, clear mixed layers in meso-eutrophic lakes	*Plagioselmis, Chrysochromulina*	Stratification	Mixing, filter-feeding
X1	Shallow mixed layers in enriched conditions	*Chlorella, Ankyra, Monoraphidium*	Stratification	Nutrient deficiency, filter-feeding
Y	Usually small, enriched lakes	*Cryptomonas, Peridinium lomnickii*	Low light	Phagotrophs!
E	Usually small, oligotrophic, base-poor lakes or heterotrophic ponds	*Dinobryon, Mallomonas (Synura)*	Low nutrients (resort to mixotrophy)	CO_2 deficiency

(cont.)

Table 7.1 (cont.)

Group	Habitat	Typical representatives	Tolerances	Sensitivities
F	Clear epilimnia	Colonial chlorophytes like *Botryococcus*, *Pseudosphaerocystis*, *Coenochloris*, *Oocystis lacustris*	Low nutrients	?CO_2 deficiency, high turbidity
G	Short, nutrient-rich water columns	*Eudorina, Volvox*	High light	Nutrient deficiency
J	Shallow, enriched lakes ponds and rivers	*Pediastrum, Coelastrum, Scenedesmus, Golenkinia*		Settling into low light
K	Short, nutrient columns	*Aphanothece, Aphanocapsa*		Deep mixing
H1	Dinitrogen-fixing nostocaleans	*Anabaena flos-aquae, Aphanizomenon*	Low nitrogen, low carbon	Mixing, poor light, low phosphorus
H2	Dinitrogen-fixing nostocaleans of larger mesotrophic lakes	*Anabaena lemmermanni, Gloeotrichia echinulata*	Low nitrogen	Mixing, poor light
U	Summer epilimnia	*Uroglena*	Low nutrients	CO_2 deficiency
L$_O$	Summer epilimnia in mesotrophic lakes	*Peridinium willei, Woronichinia*	Segregated nutrients	Prolonged or deep mixing
L$_M$	Summer epilimnia in eutrophic lakes	*Ceratium, Microcystis*	Very low C, stratification	Mixing, poor light
M	Dielly mixed layers of small eutrophic, low latitude	*Microcystis, Sphaerocavum*	High insolation	Flushing, low total light
R	Metalimnia of mesotrophic stratified lakes	*Planktothrix rubescens, P. mougeotii*	Low light, strong segregation	Instability
V	Metalimnia of eutrophic stratified lakes	*Chromatium, Chlorobium*	Very low light, strong segregation	Instability
W1	Small organic ponds	Euglenoids, *Synura, Gonium*	High BOD	Grazing
W2	Shallow mesotrophic lakes	Bottom-dwelling *Trachelomonas (e.g. T. volvocina)*	?	?
Q	Small humic lakes	*Gonyostomum*	High colour	?

Source: Updated from Reynolds *et al.* (2002).

depth and wind fetch and the ability of a lake to maintain a density-differentiated stratification for some weeks or months on end (see Fig. 2.19). The relevance of the distinction is the extent to which the plankton-bearing surface waters are affected by the internal rates of physical recycling and the frequency access of phytoplankton to resources accumulated by and discharged from the deeper sediments. For many years, the bathymetry of (and, hence, the volume of water stored in) large lakes remained less familiar than their respective areas. Several have been added to the category of 'deep lakes', even since Herdendorf's (1990) listing. These include Lago General Carrera/Buenos Aires, straddling the Chile/Argentina border, Danau Matano in Indonesia and Lake Vostok, Antarctica.

Setting aside those that are saline (Kaspiyskoye More, Aralskoye More), the natural condition of the water in most of these large, deep lakes is to be deficient in nutrient resources. They occupy large basins, fashioned either by tectonics or scraped out by glacial action, and are presently filled with water that is renewed only very slowly. Large lake volumes in relation to catchment area also make for low supportive capacities and, indeed, the phytoplankton they carry is typically dilute. Where known (or where approximated from published biovolume estimates), average seasonal maxima of chlorophyll are <4 mg chl$a$ m^{-3}, although greater concentrations may be found locally (Reynolds et al., 2000). In many instances, the paucity of phytoplankton may be determined principally by energy limitation in deep, mixed layers.

Among the most systematically studied of these large lakes is Ozero Baykal. Formed in a gap between two separating tectonic plates, Baykal is also the deepest (1741 m), stores the greatest volume (nearly 23 000 km^3) and is probably the oldest (~20 Ma) of all the world's lakes. Despite significant industrialisation and settlement of the catchment (especially around Irkutsk) and pollution of its two major inflows (Angara, Selenga Rivers), the lake remains oligotrophic in character. Retention time is estimated to be 390 a. In the offshore areas, levels of soluble phosphorus (SRP) and dissolved inorganic species of nitrogen (DIN) are <15 μg P and <100 μg N l^{-1} (Kolpakova et al., 1988; Goldman and Jassby (2001). The lake is classically dimictic. Near-surface warming of the water surface in the summer induces thermal stratification (July to September); in the winter, the lake is ice-covered from January to late May. At other times, the lake is subject to deep convective mixing that is sufficiently intense to aerate the profundal waters (Rossolimo, 1957; Votintsev, 1992). Despite substantial year-to-year variations in the production and standing biomass of phytoplankton, it is plain from each of the main overviews that increase of both is mainly

confined to the stratified periods (Kozhov, 1963; Kozhova, 1987; Kozhova and Izmest'eva, 1998; Goldman and Jassby, 2001; Popovskaya, 2001). The spring development, which takes place under ice, is subject to sharp interannual variability. In (relatively) high-production years, diatoms (especially *Aulacoseira baicalensis*, *A. islandica*, *Nitzschia acicularis*, *Synedra ulna* var. *danica* and *Stephanodiscus binderanus*), small dinoflagellates (including *Gymnodinium baicalense*) and chrysophytes (especially *Dinobryon cylindricum*) are prominent (Popovskaya, 2001). Although several of these algae are endemic, the assemblage corresponds mainly to Association-B diatoms, which have a high sv^{-1}, and whose growth is tolerant of low temperature, poor insolation and low phosphorus concentrations (Richardson et al., 2000), with some representatives of the E- and Y-groups of flagellates. In low-production years, all algae are scarce (the interannual difference in populations of dominant *Aulacoseira baicalensis* is between 100–200 cells mL^{-1} to 1 to 2 orders of magnitude fewer: (Popovskaya, 2001). The critical variable seems to be the extent of snow cover on the ice: besides letting more light through, snow-free ice allows more rapid heating of water directly beneath the ice, which then *sinks* to the point of isopycny, some 10–20 m below. This sets up a convective motion which resembles the epilimnion of a warm, ice-free lake (Rossolimo, 1957). Indeed, this structure allows rather better insolation of entrained diatoms than is possible in the ice-free column, until surface heating allows the lake to stratify directly (maximum surface temperatures may then reach 12–16 °C: Nakano et al., 2003). Then, algae tolerant of stratification, high insolation and low nitrogen and phosphorus concentrations (F-group colonial chlorophytes, including *Botryococcus*, and potentially nitrogen-fixing *Anabaeana lemmermannii* of group H2: Goldman and Jassby, 2001; Popovskaya, 2001) predominate. However, the main pelagic primary producers in summer are group-Z picoplanktic Cyanobacteria: in Baykal, populations of *Synechocystis limnetica* may exceed 10^5 cells mL^{-1}, are responsible for, perhaps, 80% of the pelagic primary production and support a well-developed microplanktic food web (Nakano et al., 2003). Like *Prochlorococcus* in the tropical sea, they are able to function in clear, well-insolated water, turning over fixed carbon to the microbial heterotrophs, whilst nevertheless maintaining a weakly grazed biomass.

Lake Superior covers a larger area than Ozero Baykal (82 100 against 31 500 km^2) but its depth (maximum 407 m, mean 149 m) and volume (12 200 km^3) are inferior (Herdendorf, 1982). Its basin is tectonic in origin but was significantly scoured by ice during the last (and possibly previous) glaciation. The present lake is little more than 15 ka in age (Gray et al., 1994). The lake area is large relative to that of the drainage basin; currently, the hydraulic retention time is around 170 a. The input of nutrients has for long been very low and, except in the vicinity of industrialised cities like Duluth and Thunder Bay, concentrations of TP (3.5–7.0 µg P L^{-1}), SRP (≤3 µg P L^{-1}) and DIN (≤300 µg N L^{-1}) are typically dilute. Annual TP loads are <0.1 g m^{-2} a^{-1} (Vollenweider et al., 1974). The phytoplankton biomass (supposed to average ~1–1.5 µg chla L^{-1}) is correspondingly modest and the water (Secchi-disk transparency 10–17 m: Gray et al., 1994). However, for much of the time, low water temperatures and weak insolation may provide more severe production constraints. Enhanced production and elevated phytoplankton biomass in the open water are noted between July and October, when surface water temperatures are sufficiently differentiated for mixing to be restricted to the upper 20–30 m, allowing weak summer stratification (Munawar and Munawar, 1986, 2000, 2001). According to these synopses, diatoms (especially *Cyclotella radiosa* and *Tabellaria fenestrata*) develop at the start of this period, followed by nanoplanktic phytoflagellates of the genera *Cryptomonas*, *Plagioselmis*, *Ochromonas* and *Chrysochromulina* and microplanktic *Uroglena*. Picoplanktic chroococcoid Cyanobacteria become numerous (Munawar and Fahnenstiel, 1982; Fahnenstiel et al., 1986) and, despite being a small part of the phytoplankton biomass, contribute significantly to the annual carbon budget (~50 C m^{-2} a^{-1}). The sequence of phytoplankton can be summarised A/B → X2/Y → U → Z.

Superior's neighbouring Great Lakes are similar in the modesty of phytoplankton they support. In Lake Huron (area 59 500 km^2, mean depth 59 m, TP 5–7 µg P L^{-1}, DIN ≤300 µg N L^{-1}, stratified July–September to within 20–30 m

of the surface), phytoplankton increases during spring to a maximum of ~7 µg chla L^{-1} in June. A second peak in August has about half this magnitude (Munawar and Munawar, 2001). In Lake Michigan (area 57 750 km^2, mean depth 85 m, TP 5–8.5 µg P L^{-1}, DIN ≤260 µg N L^{-1}, stratified July–September to within 20–30 m of the surface), a similar diacmic pattern of abundance is observed, with maxima in June and August. The earlier is larger (maximum ~11 µg chla L^{-1}) than the second. The dominant species are similar in either case: diatoms of the A-group (*Cyclotella* spp., notably *C. bodanica*, *C. radiosa*, *C. glomerata*; also *Urosolenia eriensis*) and such B-group representatives as *Aulacoseira islandica* and *Tabellaria flocculosa* are relatively abundant throughout and dominate the earlier peak. Nanoplanktic flagellates of the X2 group (*Chrysochromulina*, *Ochromonas*) and E- and U-groups of microplanktic chrysophytes (*Dinobryon* spp., *Mallomonas* spp., *Uroglena* spp.) are relatively more abundant during the second peak, when picoplanktic Chroococcoids (Z) are also at their most numerous (Munawar and Munawar, 2001). Comparison with the findings of Johnson (1975, 1994) shows that similar species assemblages make up the very sparse 'peaks' (probably <0.5 µg chla L^{-1}) of phytoplankton biomass encountered at the beginning and towards the end of the ice-free period (July–November) in Great Bear Lake (area 31 150 km^2, mean depth 72 m, SRP ≤0.1 µg P L^{-1}).

Since the late 1980s, the Laurentian Great Lakes have been experiencing the spread of a Eurasian alien, the bivalve *Dreissena polymorpha*. Its motile larvae have taken great advantage of the canal systems of Europe and the ballast tanks of ocean-going ships to spread from its native areas of the Caucasus to the Atlantic seaboard (by the mid twentieth century) and, eventually, to the St Lawrence basin. The mollusc colonises almost any firm submerged surface where, multiplied by its numbers and explosive reproductive performances, its dense aggregations can generate significant filtration capacities. *Dreissena* has been particularly successful in Lake Erie and the Saginaw Bay area of Lake Huron where its invasion has contributed to the reduction in phytoplankton and TP content of the water (arguably more so than newly imposed statutory pollution controls) and improvements in clarity and macrophyte growth (Nalepa and Fahnenstiel, 1995).

Among the other large North American lakes, the spring phytoplankton generally includes a wider range of diatoms. Various studies (Munawar and Munawar, 1981, 1986, 1996; Duthie and Hart, 1987; Pollingher, 1990) reveal that C-group *Asterionella formosa* and *Stephanodiscus binderanus* are well represented in the plankton of Great Slave Lake (area 27 200 km^2, mean depth 58 m, TP: 3–8 µg P L^{-1}, ice-free June–November, stratified to within ~30 m of the surface July/August: details from Moore, 1980), Lake Erie (area 25 660 km^2 mean depth 19 m, TP 11–45 µg P L^{-1}) and Lake Ontario (area 19 100 km^2 mean depth 86 m, TP 10–25 µg P L^{-1}, like Lake Erie, it is stratified to within ~22 m of the surface June–October). In the eastern and central basins of Lake Erie, phytoplankton biomass develops during June culminating in summer (August) maxima in the order of up to 30 µg chla L^{-1} and in which small dinoflagellates (especially *Gymnodinium helveticum*, *G. uberrimum* and *Glenodinium* spp.) and *Cryptomonas* species are common reprsentatives of a Y-type assemblage. There is also a well-developed nanoplankton in which X2-group *Plagioselmis* and *Chrysochromulina* are joined by *Chlorella*, *Monoraphidium* and *Tetraedron* species, all X1-group, overtly C-strategist species whose growth requires higher half-saturation concentrations of SRP (Section 5.4.4). Chroococcoid picoplankters (Z) are also present in the summer. In the shallower and most enriched western basin of Lake Erie, there is a substantial spring bloom dominated by *Stephanodiscus binderanus* and other B- and C-group diatoms, abundant flagellates of group Y and *Pediastrum* and *Scenedesmus*, representing the J group of eutrophic chlorococcaleans. The offshore phytoplankton of Lake Ontario behaves similarly to that of eastern Lake Erie, except that the summer plankton is more biassed towards dominance by mucilage-bound, non-motile, colonial chlorococcalean and tetrasporalean genera of green algae (group F) such as *Oocystis*, *Coenochloris* and *Pseudosphaerocystis*. Hutchinson (1967) called these 'Oligotrophic chlorococcal plankters'; it is a widespread group (see below) that survives low SRP concentrations but is apparently intolerant of poor insolation

(Reynolds *et al.*, 2002). Lake Ontario supports an abundant chroococcoid picoplankton in summer and whose dynamics and distribution were the subject of a detailed study by Caron *et al.* (1985). Diatoms occur throughout the year in Lake Ontario but, during the early part of the year, *Stephanodiscus binderanus*, especially, 'blooms', in the near-shore areas of the lake, where populations become substantially isolated by a phenomenon known as 'thermal barring'. This was described in detail by Munawar and Munawar (1975) but it is now known to be a common feature in other large, cold-water lakes, including Michigan (Stoermer, 1968), Ladozhskoye and Onezhskoye (Raspopov, 1985; Petrova, 1986) and Baykal (Shimarev *et al.*, 1993; Likhoshway *et al.*, 1996). In essence, vernal heating proceeds more rapidly in the shallow near-shore waters than in the open, offshore areas of the lake but, while temperatures remain <4 °C, the warmer water is retained within inshore circulations separated from the open lake by pronounced frontal boundaries. The slightly higher temperatures and the rather higher average insolation experienced by the algae thus retained promotes the early-season growth and recruitment of phytoplankton, dominated by group-C diatoms, group-Y flagellates and nanoplankton.

The phytoplankton of the two northern great lakes of Eurasia shows closer affinites to that of Baykal, Erie and Ontario than to that of the upper Laurentian Great Lakes. In Ladozhskoye Ozero (area 18 140 km^2, mean depth 50 m, TP: 13–40 µg P L^{-1}, ice-covered February–May, stratified July–August, but very variable stability), a spring diatom bloom of B- and C-group diatoms (including *Aulacoseira subarctica*, *A. islandica*, *Asterionella formosa*, *Diatoma* spp.), apparently nurtured in thermal-bar conditions (see above and p. 89), spreads through the lake (chla generally 1–5 µg L^{-1}). *Asterionella* continues to dominate the unstable open lake conditions, together with variable quantities of group-P *Aulacoseira granulata* and *Fragilaria crotonensis*. Often, significant quantities of filamentous Cyanobacteria (especially *Planktothrix agardhii* of group S1) and xanthophytes (*Tribonema* of group T) are selected by their tolerance of low average insolation in mixed conditions (Raspopov, 1985). Under calmer conditions, the Cyanobacteria *Aphanizomenon* and *Woronichinia* sometimes form significant blooms. I have no information to hand concerning the nano- and picoplankton but the dominant calanoid zooplankton and well-developed ciliate microplankton are suggestive of an active microbial food web. In Onezhskoye Ozero (area 9900 km^2, mean depth 28 m, TP 5–10 µg P L^{-1}, ice-covered January–May, thereafter, thermal bar formation separates the inshore from the isothermal circulation of the central water mass; stratification develops from July to September, to within 30 m from the surface), the spring assemblage is dominated by B-group *Aulacoseira* and C-group *Asterionella*. The seasonal progression passes by way of *Tabellaria* to a sequence of oligotrophic–mesotrophic assemblages featuring *Dinobryon* (E) species, *Coenochloris* (F) species, *Woronichinia* (L$_O$) and *Planktothrix agardhii* (S1) but the biomass at the summer maximum scarcely exceeds 6 µg chl*a* L^{-1} (Kauffman, 1990).

In Ozero Issyk-kul (area 6240 km^2, mean depth 279 m, TP 2–4 µg P L^{-1}, stratified April–December, to within ≥13 m of the surface: Shaboonin, 1982), relative phytoplankton abundance is diacmic. The earlier (May) peak involves group-A *Cyclotella* species and group-B *Aulacoseira* species the later (October) peak is dominated by the Cyanobacterium *Merismopedia* and the dinoflagellate *Peridinium willei* which both represent group L$_O$. Species of *Coenochloris*, *Oocystis*, *Gloeocapsa*, *Lyngbya*, as well as an 'abundant nanoplankton' also feature in this faintly saline lake (Savvaitova and Petr, 1992).

In conclusion, the phytoplankton of large, high-latitude lakes involves elements characterising either some of the ultraoligotrophic functional groups, distinguished by their high affinity for phosphorus (A, Z), or other substantially oligotrophic groups (B, X2, E, F, L$_O$, U) tolerant of low nutrient concentrations (especially of phosphorus). However, there is no clear evidence that these algae habitually fill the nutrient-determined capacity, except when average insolation allows it. Excess of nutrient capacity over average light income favours the *R*-strategies of the C, P, S and T groups of attenuated and filamentous 'antennal' algae.

Phytoplankton of smaller temperate oligotrophic lakes

The 'small-lake' category as adopted here applies to waters <10 km^2. Whilst most of the fresh water is stored in 'large' lakes (>500 km^2), most of the world's lakes (~8.4 × 10^6 in total: Meybeck, 1995) are 'small'. A further 13450 in the range 10–500 km^2 might be described as 'medium-sized'. So far as is known, a large proportion of these represents lakes in unyielding rocky basins and forested catchments and the phytoplankton carrying capacity is unambiguously constrained by the availability of nutrient resources, rather than by low temperatures and low insolation. Their smaller sizes facilitate a high frequency of sample collection and a robust picture of seasonal change in production and biomass as well as species composition. In previous work (Reynolds, 1984a), I have cited Findenegg's (1943) thorough two-year (1935–36) survey of the phytoplankton in some 15 moderately alkaline lakes of Kärnten (Carinthia), Austria, to exemplify the genre (in particular, Millstätter See: area 13.3 km^2, mean depth 86.5 m, stratified May–November to within 7–10 m of the surface, TP, not reported but SRP <1 µg P L^{-1}, DIN ≤300 µg N L^{-1}). There is a single (monacmic) summer biomass 'maximum' in summer (chl*a* ~1.5 µg P L^{-1} (Fig 7.6), when the A-group spring diatoms (*Cyclotella comensis, C. glomerata*) are supported by a mixture of other species that include self-regulating dinoflagelleates (*Peridinium willei*, some *Ceratium hirundinella*) and colonial Cyanobacteria, now known as *Woronichinia* (representing group L$_O$), and such (N-group, high-nutrient-affinity) desmids as *Staurodesmus* species Findenegg (1943) also recorded species of *Coenochloris*, *Oocystis*, *Gymnodinium*, *Rhodomonas* and other nanoplankters, as well as the non-gas-vacuolate, unicellular Cyanobacteria *Chroococcus* and *Dactylococcopsis*. Moreover, Findenegg (1943) found similar assemblages in several of the other lakes in the region: in Wörthersee, *Uroglena* and, especially, *Planktothrix rubescens*, constituted metalimnetic maxima. Findenegg's explanation for the seasonally distinctive assemblages and distributions invoked the interaction of light and temperature preferences of participating algae. The four contingencies (cold-water, low-light forms; cold-water, high-light forms; warm-water,

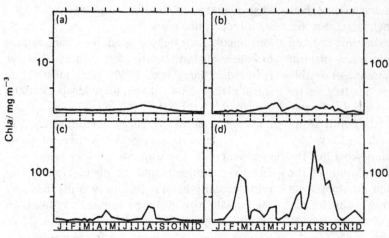

Figure 7.6 Annual cycles of phytoplankton biomass (as measured or approximated from biovolume) in some temperate lakes: (a) Millstättersee, 1935 (after Findenegg, 1943); (b) Windermere, North Basin, 1978 (Reynolds, 1984a); (c) Sjön Erken, 1957 (Nauwerck, 1963); (d) Crose Mere, 1973 (Reynolds, 1976a). Redrawn from Reynolds (1984a).

high-light forms; warm-water, low-light forms) approximated to winter, spring, summer and autumn plankton.

Other small, low-nutrient lakes show similar seasonality, involving similar species of alga. Hino *et al.* (1998) recently published the results of an investigation of a small, low-alkalinity lake in Hokkaido, Japan (Akan-Panke: area 2.8 km^2, mean depth 24 m, TP $\leq$10 μg P L^{-1}, DIN $\leq$300 μg N L^{-1}, stratified May–October to within 5 m of the surface, ice-covered November–May). This lake is classically dimictic but the overturn periods are brief; the lake is usually stratified and its resources become strongly segregated. The water has a high clarity (ε_{min} ~0.15 m^{-1}, Secchi-disk transparency 7–18 m). The spring overturn (maximum, 2–3 μg chl$a$ L^{-1}) is dominated by diatoms (A-, B- and C- group species include *Cyclotella radiosa*, *Asterionella formosa* and *Aulacoseira ambigua*). As they settle out (taking most of the bioavailable P and TP with them, the epilimnion is left with a sparse population of E-group chrysophytes (particularly *Dinobryon cylindrica*). By summer, the surface waters are substantially depleted of all but chroococcoid picoplankters and chlorococcal nanoplankters (together accounting for <1 μg chl$a$ L^{-1}). Dominant microplankters (in which U-group *Uroglena americana* and L$_O$-group *Peridinium aciculiferum* and *Merismopedia* are prominent) are based in the metalimnetic region (at a depth of 10–25 m). With increasing mixed depth towards the end of the summer, there are first more green algae

(F-group *Coenochloris*, *Oocystis*), then desmids (*Staurastrum* spp.), before *Aulacoseira ambigua* and other spring diatoms are entrained from the deepest layers.

In Finland, many small lakes are simultaneously soft-watered, oligotrophic and often strongly humic. The high latitude and short winter days greatly constrain the ability of phytoplankton to sustain net photosynthetic gain. Any growth at these times must be sustained heterotrophically; few surviving cells even contain chlorophyll (Arvola and Kankaala, 1989). However, there are some species of *Chlamydomonas*, *Chlorogonium*, *Peridinium* and *Gymnodinium* that are able to maintain high population densities just below the surface while ice cover persists (Arvola and Kankaala, 1989).

In Stechlinsee, another well-studied oligotrophic system (area: 4.3 km^2, mean depth 22.8 m, TP $\leq$16 μg P L^{-1}, DIN $\leq$95 μg N L^{-1}, stratified May–October, to within 7–10 m of the surface, ice-covered January–March: details from Gervais *et al.*, 1997), the phytoplankton biomass peaks in spring, dominated by A-group *Cyclotella* species and picoplanktic Cyanobacteria (Padisák *et al.*, 1998). Development starts in February or March and is terminated by the onset of the summer stratification and the sinking of the diatoms into the hypolimnion. Exhaustion of the epilimnetic nutrient base confines autotrophic production to the metalimnion but, in this lake, it is the picoplanktic Cyanobacteria that dominate both the biomass and the production (see also Gervais *et al.*, 1997). Originally identified as a *Synechococcus* species. the alga is now recognised as *Cyanobium* (Padisák, 2003).

In the 20 years covered by the study of Berman *et al.* (1992), the phytoplankton periodicity of Yam Kinneret (Sea of Galilee) showed great interannual similarity in the abundance, distribution and composition of the phytoplankton. This warm monomictic rift-valley lake in the upper Jordan Valley (area 168 km^2, mean depth 25.5 m, stratified March–December) experiences a typically Mediterranean climate, with rainfall confined to the winter period (Serruya, 1978). It is briefly isothermal with a minimum temperature of ~13 °C but, by April, is generally strongly stratified to within 10 m of the surface, as epilimnetic temperatures reach 30 °C. Its waters are mildly alkaline and slightly saline. The maximum bioavailability of both nitrogen and phosphorus is modest ($\leq$200 µg N L^{-1}, $\leq$5 µg P L^{-1}). In each of the years from 1970 to 1989 inclusive, the phytoplankton developed in a characteristic way. Starting with the autumnal circulation, the main components were unicellular Cyanobacteria (*Chroococcus*) and such (X2-) nanoflagellates as *Plagioselmis* and *Chrysochromulina*. With full overturn and winter flooding, coenobial and filamentous plankton develop, including the apparently warm stenothermic diatom, *Aulacoseira granulata*, of Group P. At this time, there would also be excystment of the large, self-regulating dinoflagellate now known as *Peridinium gatunense* and ascribed by Reynolds *et al.* (2002) to Group L_M. Under conditions of reduced vertical mixing, this alga typically built to a stable maximum, often lasting from March to May, which represented the greatest annual biomass. Its termination, through encystment and settlement, generally coincided with high epilimnetic temperatures (between 27 and 30 °C) and virtual exhaustion of the epilimnetic reserve of nitrogen. The summer biomass remained relatively very low, comprising a sparse nanoplankton and numerous picoplankton (Malinsky-Rushansky *et al.*, 1995). Metalimnetic layers tended to be dominated more by group-V photobacteria (*Chlorobium*) than algae.

Berman *et al.* (1992) were concerned to interpret this interannual stability against a trend of rising phosphorus loads and to fluctuating zooplankton abundance but both seemed to be incidental to the continuing predominance of *Peridinium*. The success of the alga relates to its efficient perennation and its ability to exploit its mobility to 'scavenge' (or 'glean') the water column of its nutrient reserves during the early stagnation period. Depletion of both phosphorus and nitrogen to the limits of detection attests to the efficiency of this process. The post-bloom phytoplankton is redolent of other very oligotrophic pelagic systems where the primary products of similar organisms are tightly coupled within microbial food loops. Such interannual variations in the size of the *Peridinium* crop as had been observed seem to be inverse to the fluctuations in *Aulacoseira* production: years with larger diatom maxima heralded smaller *Peridinium* maxima.

Interestingly, variations to the stable sequence have been observed in the 1990s. There have been one or two years when the excystment and spring recruitment of *Peridinium* has been poor and green algae or *Microcystis* becoming briefly abundant. The summer appearance of nitrogen-fixing *Anabaena ovalisporum*, in 1994 and in one or two subsequent years, is another departure from the stable pattern. The tempting explanation is that the increase in phosphorus loading has triggered these events but careful analysis shows a trend of decreasing winter concentrations in the lake and a lengthening of the period of thermal stratification (Hambright *et al.*, 1994). Berman and Shteinman (1998) also comment on the effect of a weakening of diffusivity in the water column working against the selective exclusivity in favour of *Peridinium* at critical points in its annual cycle.

The last example in this section is from Biwako, which, on criteria set in this chapter, is really a 'large lake' (area 674 km^2, mean depth 41 m, TP $\leq$10 µg P L^{-1}, DIN $\leq$350 µg N L^{-1}, stratified April–November to within 20 m of the surface: Nakamura and Chang, 2001). Its phytoplankton was described in detail by Nakanishi (1984) but ongoing concerns about the water quality and the appearance of *Microcystis* populations in the lake (which supplies drinking water to 15 M people in Kyoto and Osaka) have encouraged frequent monitoring of the plankton. The winter plankton of the mixed water column is sparse and dominated by *Aulacoseira solida* (Group **B**?) with increasing quantities of *Asterionella formosa* (group **C**) through spring and a summer assemblage of *Staurastrum dorsidentiferum*, *Closterium aciculare* and *Aulacoseira granulata* (group **P**). Since 1977, *Uroglena americana* has 'bloomed' each spring (maximum 6–10 μg chla L^{-1}) and, since 1983, small numbers of (group **H**) *Anabaena* and (group L_M) *Microcystis* are encountered in the water column. These are observed to form striking lee-shore scums in quiet weather (Nakamura and Chang, 2001; Ishikawa *et al.*, 2003). Rod-like picoplanktic synechococcoids (0.4–1.5 μm) become numerous in summer (reportedly, up to 10^6 cells mL^{-1}: Maeda, 1993). Close interval sampling after a typhoon suggested that microplanktic dynamics respond much more to hydrodynamic variability than do picoplankton numbers (Frenette *et al.*, 1996).

To conclude the section, the seasonal patterns in the phytoplankton of small nutrient-deficient lakes in temperate regions involve oligotrophic functional groups in the approximate sequence **B** → **X2** → **E** and/or **F** → **U** or L_O → **N**; picoplanktic group **Z** is often an important component. In the more alkaline waters, **E** and **U** are substantially missing and more eutrophic (carbon-concentrating) functional groups **C**, L_M and **P** are better represented than **B**, L_O and **N**.

A common feature of stratifying oligotrophic lakes is the tendency for phytoplankton to form deep chlorophyll layers. These are not just the consequence of sinking and formation depends upon algae being able to self-regulate their vertical position, either through their own motility or by regulation of their density. Reynolds (1992c) surmised that their maintenance is generally conditional upon low diffusivity (they have to be below the epilimnion) and adequate light penetration (they have to be within the range of net photosynthetic gain or, at worst, balance. The station also offers advantage over a position higher in the light gradient and, usually, it is access to nutrients. The likely algal components are also influenced by the trophic state and growth conditions prior to stratification. The size of the lake and relative remoteness of the metalimnion from the surface also affect the physical tenability of the structural layers. Among the 'tightest', plate-like layers are constructed by photobacteria (**V**)

(Vicente and Miracle, 1988; Guerrero and Mas-Castellà, 1995) and *Planktothrix* of the *rubescens* group (**R**) (Findenegg, 1943; Zimmermann, 1969; Bright and Walsby, 2000). *Cryptomonas* species (group **Y**) also maintain stable layers in karstic dolines (Pedrós-Alió *et al.*, 1987; Vicente and Miracle, 1988) and, at times, in larger lakes (Ichimura *et al.*, 1968). Deep chlorophyll layers involving motile chrysophytes (**E**, **U**: Pick *et al.*, 1984; Bird and Kalff, 1989), non-motile chlorococcals (**X1**: Gasol and Pedrós-Alió, 1991) and picoplankton (**Z**: Gervais *et al.*, 1997) tend to be rather more diffuse.

Phytoplankton of sub-Arctic lakes

Based on his earlier investigations in the arctic and sub-Arctic regions of Sweden and Canada (Holmgren, 1968; Schindler and Holmgren, 1971), and on a careful review of the literature, Steffan Holmgren devised a systematic and long-term experimental study of the lakes in the Kuokkel district of northen Sweden (centred on 68° 25′ N, 18° 30′ E). These lakes are small (0.01–0.03 km^2), shallow (mean depth 1–6 m) and soft-watered, are ice-covered for up to nine months per year and, at times, experience 24-h nights (December–January) and 24-h days (June–July). In the natural condition, the lakes are oligotrophic (TP <9 mg P m^{-2}, DIN <240 mg N m^{-2}) but Holmgren's enrichment experiments (+P, +N and +N+P) raised levels selectively to up to 300 mg P m^{-2} and DIN to up to 6 g N m^{-2}. Among the many interesting findings he reported (Holmgren, 1983) is a proposed ecological classification of arctic–sub-Arctic phytoplankton (Table 7.2). This recognised the ubiquity of Chrysophyceae and the differing matches with other algae, signifying between-lake and seasonal variability. Biomass varied with nutrient fertility, being well correlated to N; wind was a bigger influence on seasonality than either temperature or insolation. Greater water hardnesses supported more diatoms (*Urosolenia*, *Synedra*) and cryptophytes. Increased nitrogen levels promoted *Uroglena* and the small phagotrohic dinoflagellate *Gymnodinium*. Simultaneously elevated availabilities of higher phosphorus and nitrogen favoured *Chromulina*, *Ochromonas* and the green alga *Choricystis*. In all fertilised systems chlorococcal green alga developed.

Table 7.2 | Phytoplankton assemblages in Arctic and sub-Arctic lakes, according to the scheme of Holmgren (1983) but using the identifiers proposed by Reynolds *et al.* (2002)

	Spring	Summer	Autumn
1. Chrysophyceae lakes	**X3**,	**E, Y**	**E**
2. Chrysophyceae–diatom lakes	**X2, X3,E,Y**	**A, (B**?)	**A,Y, L$_O$**
3. Chrysophyceae–Cryptophyceae lakes	**X2, X3, E**	**B, Y**	**B,Y**
4. Chrysophyceae–Dinophyceae lakes	**Y**	**U, B, F, L$_O$**	**L$_O$**

Phytoplankton in selected regions: medium-sized glacial lakes of the European Alps

Besides distinguishing patterns from a selection of particular lake types, it is also helpful to make comparisons of phytoplankton periodicity among regional series of water bodies. Here, between-lake differences owe relatively more to edaphic distinctions among the individual lake basins (and their hydrology, hydrography and hydrochemistry) than to the general commonality of location or of formation. Regional clustering is exemplified well by the residual water bodies in upland areas where recent glaciations have scoured out the typically linear basins of ribbon lakes or finger lakes. As a generality, these lakes are usually oligotrophic or mesotrophic in character. However, it is often also true that the individual lakes of a given cluster can be ordered by productivity or mean biomass supported, or by typical composition and abundance of phytoplankton. Such arrangements are usually illustrative of the selective effects on the assembly of planktic communities within the respective regions.

Sommer (1986) undertook such a comparison of the phytoplankton of the deep lakes in the European Alps. He ably demonstrated a similarity in the year-to-year behaviours in individual lakes as well as sequential seasonal stages (spring bloom, summer stratification, summer–autumn phase of mixed-layer deepening) common to all of them. He suggested that the different phytoplankton assemblages reflected between-lake differences in trophic status and the availabilities of limiting nutrients, and which were also sensitive to between-year variability in individual lakes. Sommer reported the periodic patterns in terms of a few key dominant species and their phylogenetic groups. These patterns were further distilled by Reynolds *et al.* (2002) who ascribed Sommer's (1986) key species to their relevant trait-differentiated functional groups (see also Table 7.3). In this way, (A-group) diatoms of the *Cyclotella bodanica* – *C. glomerata* group dominated the sparse spring bloom in the oligotrophic Königsee (area 5.2 km^2, mean depth 98 m, TP 5 µg P L^{-1}); in Attersee (area 46 km^2, mean depth 84 m, TP 5 µg P L^{-1}), *Tabellaria* species were relatively more prominent. *Aulacoseira* species and *Asterionella* were abundant in the mesotrophic lakes, such as Vierwaldstättersee (area 114 km^2, mean depth 104 m, TP 20 µg P L^{-1}) and Ammersee (area 48 km^2, mean depth 38 m, TP 55 µg P L^{-1}), with increasing subdominance of (Y-group) cryptomonads but (D-group) *Stephanodiscus* in the richer waters of Bodensee (area 500 km^2, mean depth 100 m, TP (then) 100 µg P L^{-1}) and Lac Léman (area 582 km^2, mean depth 153 m, TP (then) 80 µg P L^{-1}). The onset of thermal stratification marked the end of the spring diatom bloom and, because surviving cryptomonads are always vulnerable to developing filter-feeding populations of Cladocera, a phase of high water clarity ensues. This 'vacuum' is filled by a summer assemblage whose composition is particularly sensitive to physico-chemical conditions. Continued bioavailability of phosphorus in the upper water column supports the growth of *CS*-strategist algae such as *Pandorina morum* (G-group of motile chlorophytes) and/or *Anabaena* species (H1-group of self-regulating and potentially nitrogen-fixing Cyanobacteria). Eventual phosphorus depletion is generally marked by increasingly motile (L$_M$-group *Ceratium*) and mixotrophic (E-group *Dinobryon*) algae, and the

Table 7.3 | Seasonality of dominant phytoplankton in nine lakes of the European Alps, according to Sommer (1986) but rendered in terms of the trait-differentiated functional groups of Reynolds *et al.* (2002) (see also Table 7.1)

	Spring bloom	Clear water	Summer P↑ Si↑	Summer P↓ Si↑	Summer P↓ Si↓	Late summer	Autumn
Königsee	A			N/Y			A
Attersee	N			Y			B
Walensee	N			N/Y			
Vierwaldstättersee	B/Y			N	N	H	R
Lago Maggiore	C/R	Y		P	L/R	H	L/R
Ammersee	C/R	Y	R/P/Y	P	L	H	R/P
Zürichsee	C/D/Y	Y	Y/H	P/Y			R/P
Lac Léman	C/D/Y	Y	Y/G	P	L	H	T/P
Bodensee	C/D/Y	Y	Y/G	P	L/H/E		T/P

development of deep chlorophyll maxima, dominated by (**R**-group) *Planktothrix rubescens*. On the other hand, deeper mixing in summer may support the growth of summer diatoms of the **N**-(*Tabellaria*) and **P**- (*Fragilaria*, *Aulacoseira granulata*) groups, so long as soluble reactive silicon remains available, with the appropriate desmids being more common if it does not. Indeed, nitrogen limitation may develop in these instances, when **H**-group nitrogen-fixers (*Anabaena*, *Aphanizomenon*) become active. Autumnal mixing in deep lakes enhances the effect of shortening days in imposing increasingly severe light limitation, favouring, in some instances, abundance of (group-**T**) filamentous chlorophytes and, especially, *Mougeotia*.

The classification broadly holds for the mildly alkaline (0.7–2.0 meq L^{-1}), subalpine Italian lakes (including Lago di Garda, according to Salmaso (2000), and Lago Maggiore, featured in Table 7.3). In a recent synthesis, Salmaso *et al.* (2003) analysed the structure of the phytoplankton in five of these lakes. They used different methods from Sommer (1986) and, in consequence, subdivided the vegetative season in a different way but the regional similarities are remarkable. The main features are summarised in Table 7.4 by reference to the functional groups of Reynolds *et al.* (2002). Diatoms prominent in the early part of the year in all five lakes include *Aulacoseira islandica* and *Asterionella formosa*, representing a mesotrophic

B grouping. *Tabellaria* (**N**) has a long temporal period (late autumn – spring) in Lago Maggiore (area 213 km^2, mean depth 176 m, TP ≤12 µg P L^{-1}) and *Fragilaria* and *Aulacoseira granulata* are frequent throughout the year in Lago di Garda (area 368 km^2, mean depth 133 m, TP ~20 µg P L^{-1}), Lago d'Iseo (area 62 km^2, mean depth 122 m, TP ≤68 µg P L^{-1}) and Lago di Como (area 146 km^2, mean depth 154 m, TP ≤38 µg P L^{-1}). **P**-group desmids (notably *Closterium aciculare*) and **T**-group filamentous forms (*Mougeotia* spp.) are also common in these three lakes. *Cryptomonas* species (**Y** group) are common throughout the vegetative period in all the lakes; **X2**-group nanoplankters (*Chrysochromulina*, *Ochromonas*, *Plagioselmis*) occur in summer, and **E**-group *Dinobryon*, L_M-group *Ceratium–Gomphospaeria* and stratifying **R**-group *Planktothrix rubescens* are prominent in all the lakes during summer. *Uroglena* species (group **U**) occur in Lago di Lugano (area 28 km^2, mean depth 167 m, TP ≤172 µg P L^{-1}), Lago Maggiore and Lago di Como.

Phytoplankton in selected regions: small glacial lakes of the English Lake District

Few lakes have been studied so intensively in frequency or so extensively through time as those of the Lake District of north-west England. Time series, starting in some instances in the 1930s, attesting to variable degrees of change attributable to eutrophication (Lund, 1970, 1972;

Table 7.4 Summary of phytoplankton seasonality in five deep subalpine lakes, according to Salmaso *et al.* (2003) but rendered in terms of the trait-differentiated functional-groups of Reynolds *et al.* (2002) (see also Table 7.1)

	Late winter to mid spring	Late spring	Early mid summer	Late summer – mid autumn
Lago di Garda, Lago d'Iseo	B, C, P, T, Y	Y, P, X2	Y, P, R, T, E, H, X2	Y, P, R, S
Lago di Como	B, C, N, P, T, Y	Y, P, L, U, X2	Y, P, R, T, E, U, H, X2	Y, P, R
Lago Maggiore, Lago di Lugano	B, C, N, Y	Y, P, L, U, X2	Y, P, R, E, U, H, X2	Y, P, R

Talling and Heaney, 1988; Kadiri and Reynolds, 1993), are now being analysed for sensitivity to climatic variation (George, 2002). The lakes themselves are small, glacial ribbon lakes radiating from a central dome of hard, metamorphic slates, excavating the valleys of a pre-existing radial drainage. The natural vegetation is temperate *Quercus* forest but this has been mostly cleared for pasture. The lakes themselves are universally soft-watered (alkalinity ≤0.4; most are <0.2 meq L^{-1}). Variation among their characteristics is dominated by differences in morphometry, orientation, hydraulic flushing and catchment loadings of nutrients. They were first arranged in a series of ascending 'productivity' by Pearsall (1921), which template has been used to illuminate comparisons of the solute concentrations, gross planktic photosynthesis, microbial activity and invertebrate associations of the lakes (for further details, see Sutcliffe *et al.*, 1982; Kadiri and Reynolds, 1993).

Between-lake differences in the phytoplankton have been investigated and reported by Lund (1957), Gorham *et al.* (1974) and Kadiri and Reynolds (1993). Based on extrapolations made in these works, it is reasonably easy to deduce general summaries of phytoplankton periodicity in these lakes (see Table 7.5). Among the oligotrophic lakes, the supply of MRP rarely exceeds 1 or 2 μg P L^{-1} and may remain below detection limits for months on end. Chlorophyll concentrations are also substantially <5 μg chla L^{-1}. Group-A diatoms, such as *Cyclotella comensis*, *C. radiosa* and *Urosolenia eriensis*, exhaust phosphorus rather than silicon or carbon. Nanoplankton is sparse (Gorham *et al.*, 1974; Kadiri and Reynolds, 1993). Picoplanktic Cyanobacteria survive the rarefied resource availability in the summer epilimnion, where they may form a significant, if not dominant, fraction of the biomass (Hawley and Whitton, 1991), while *Peridinium willei* and small numbers of *Ceratium* may 'glean' (*sensu* the usage on p. 327) from deeper water. Desmids of the genera *Cosmarium* and *Staurodesmus* (group **N**) are also prominent among the otherwise sparse microplankton.

Derwent Water, Coniston Water and Hawes Water are classically mesotrophic. Besides the same A-group diatoms, the spring phytoplankton of these lakes may support rather larger quantities of *Aulacoseira sub-Arctica*, *Cyclotella praeterissima* (both considered typical of Group **B**), as well as *Asterionella formosa* which also appears in Group **C** (Table 7.1). Nanoplanktic species (including *Plagioselmis*, *Chrysochromulina*, *Ochromonas*, representing Group **X2**) are also quite numerous in these lakes and may persist (and flourish) for some time after the onset of stratification. However, the dominant late-spring/early-summer plankters in these lakes are typically non-motile colonial green algae (*Coenochloris fottii*, *Dictyosphaerium pulchellum*, *Pseudosphaerocystis lacustris*, *Botryococcus braunii*, *Paulschulzia pseudovolvox*, *Radiococcus plantonicus* are among the commonly observed species) of Group **F** and/or such colonial chrysophytes as *Dinobryon divergens*, *D. sertularia*, *D. bavaricum*, *D. cylindrica*, *Mallomonas caudata* and *Synura uvella* representing group

Table 7.5 | Summary of phytoplankton seasonality in 19 stratifying lakes in the English Lake District, according to Kadiri and Reynolds (1993), rendered in terms of the trait-differentiated functional-groups of Reynolds *et al.* (2002) (see also Table 7.1)

Class	Lake	Area (km^2)	TP, c.1991 (μg P L^{-1})	Phytoplankton
Larger oligotrophic lakes	Wast Water, Ennerdale Water, Thirlmere, Crummock Water, Buttermere	0.9–3.0	3–9	**A → Z/L$_O$ → N**
Larger mesotrophic lakes	Derwent Water, Hawes Water, Coniston Water	3.9–5.4	7–11	**B(C) → X2/F/E → X3/Z/L$_O$/Y → N/R**
Well-flushed short-retention lakes	Brothers Water, Rydal Water, Grasmere, Bassenthwaite Lake	0.2–5.3	4–33	**B or C → X1/X2 → Y (E, F, H2, P)**
Eutrophied larger lakes	Windermere (North), Ullswater, Windermere (South), Esthwaite Water, Lowes Water	0.6–9.0	14–40	**C(B) →X1/X2/Y/G→ H/L$_M$/S/T → P**
Enriched small lakes	Loughrigg Tarn, Blelham Tarn	≤0.1	20–45	**C/Y → X1/X2 /E/F or H1 → L$_M$ → P/S**

E. In summer, the biomass generally falls to a phosphorus-depleted minimum during which the nanoplankton diversifies with algae such as *Chrysococcus, Monochrysis, Pseudopedinella, Bicosoeca tubiformis* and *Koliella longiseta* recruiting (from **X3**). Picoplankton may also be numerous, some of which may well be Chroococcoid (**Z**) but some is eukaryotic (including *Chlorella minutissima*, which Reynolds *et al.* (2002) placed in **X3**. These lakes have the potential to support metalimnetic maxima of (group-**R**) *Planktothrix*, although the numbers of *P. mougeotii* produced are generally small. Cyanobacteria are mainly represented by modest growths of *Anabaena solitaria* and *A. lemmermannii* (of group **H2**) and by colonies of *Woronichinia* (formerly *Gomphosphaeria*) that, together with dinoflagellates *Peridinium willei, P. inconspicuum* and *Ceratium* species, represent group **L$_O$**. *Tabellaria flocculosa* and *Cosmarium* species (including *C. abbreviatum* and *C. contractum*) make up

the phytoplankton stimulated by late-summer and autumnal mixing. Phytoplankton biomass is faintly diacmic in these lakes, the spring peak (in the order of 6–15 μg chla L^{-1}) usually being the larger. The level of MRP generally falls below detection limits for between four (May–August) and nine (March–November) months of the year. The summary notation shown in Table 7.5 for 'larger mesotrophic lakes' also fits well to the patterns among the greatest Scottish lochs (Bailey-Watts and Duncan, 1981).

Several further lakes are, to varying extents, enriched beyond the mesotrophic state. Since the mid-1960s, Windermere and Ullswater have been subject to the discharge of effluents from secondary sewage-treatment works. Esthwaite Water was already a eutrophic lake before sewage treatment affected its waters. Lowes Water has a low human population in its catchment and the reason for its enrichment is not resolved.

All these lakes can support significant numbers of (H1-group) *Anabaena flos-aquae*, *Aphanizomenon flos-aquae* and, in the case of Esthwaite Water, L_M-group *Microcystis aeruginosa* (together, the bloom-forming Cyanobacteria that have done locally much to give eutrophication its bad name). Each may also support an abundant nanoplankton in late spring, that may include species of *Ankyra*, *Chlorella*, *Chlamydomonas*, *Monoraphidium*, *Mallomonas akrokomos* (of Group **X1**) in addition to **X2** representatives, at least until grazed down by filter-feeders.

The phytoplankton response to enrichment (and to the post-1992 restoration) of Windermere has been reviewed and summarised in Reynolds and Irish (2000). Between 1965 and 1991, the winter maximum of MRP rose from about 2 to about 8 µg P L^{-1} in the larger North Basin and from 2.5 to almost 30 µg P L^{-1} in the South Basin. The greater resource has permitted the vernal diatom growth to escape the rate control imposed by phosphorus and to ascend to the silicon-determined maximum up to one month earlier (Reynolds, 1997b) (see also Section 5.5.2 and Fig. 5.13). More than that, from the early 1980s, vernal growth in the South Basin often failed to exhaust the MRP in the lake, leaving resource to support enhanced early summer production. Grazing by filter-feeders and carbon dioxide depletion in the unbuffered water biasses the outcome in favour of the larger, efficient carbon-concentrating bloom-forming Cyanobacteria. Inroads into the DIN stocks has further favoured (H1-group) *Anabaena* species, with, appropriately, high incidences of nitrogen-fixing heterocysts. However, the most successful beneficiary of the eutrophication of the South Basin of Windermere has been the solitary, filamentous, group-**S1** oscillatorian *Tychonema bourrellyi*. This particular species lacks gas vesicles and does slowly sink out to the bottom of the lake (thus exporting more oxygen-demanding reduced carbon to depth where significant anoxia triggered changes in the lake's metabolism: Heaney et al., 1996). One mark of the success of the programme of tertiary treatment of Windermere's main sewage inputs has been the near-elimination of *Tychonema* from the lake. The late-summer phytoplankton in Windermere's South Basin has returned to being dominated by P-group diatoms (especially *Fragilaria crotonensis*) and desmids (*Staurastrum pingue*)

Before 1992, peak chlorophyll concentrations in the North Basin had come close to 20 µg chl*a* L^{-1}, in the enriched South Basin, *Tychonema* populations had forced a summer maximum of up to 45 µg chl*a* L^{-1} and a distinctly diacmic annual pattern of phytoplankton abundance. Subsequently, phosphorus load reductions have restored the earlier pattern of a spring maximum followed by irregular, smaller summer peaks, all constrained within the supportive capacity of the biologically available phosphorus. The illustration in Fig. 7.6 shows the periodicity of chlorophyll concentration in North Basin during 1978.

Table 7.5 carries entries in respect of two smaller Lake District lakes that carry effects of recent changes in agricultural and domestic P loadings. In Blelham Tarn, a long-standing basic $B \rightarrow E/F \rightarrow L_M \rightarrow N$ sequence has been steadily altered by the diminution of the contributions of *Aulacoseira subarctica*, *Dinobryon*, *Ceratium* and *Tabellaria flocculosa* in favour of *Asterionella* and *Stephanodiscus minutulus* (more strongly C), several species of *Anabaena* (**H1**) species and a near year-round abundance of *Planktothrix agardhii* (**S1**).

Finally, several of the lakes have extensive catchment areas in relation to lake volume and, in this area of high annual aggregate precipitation (between 1.5 and 4 m annually), are liable to fairly frequent episodes of significant flushing. In the case of Grasmere (mean retention time 24 d, range of instantaneous rates 5–2000 d), flushing has helped to dissipate the effects of nutrient loads from sewage works commissioned in 1969. Thitherto, the lake supported a similar assemblage to that of the contemporaneous Blelham Tarn, save that the slow-growing algae of the L_M group were (and remain) poorly represented (Reynolds and Lund, 1988). The reason for this is not that the algae have difficulty in growing against the instantaneous rates of flushing but that the autumn flooding is so effective in removing pre-encystment vegetative stocks. In an effort to prevent massive algal growth in dry summer weather, the arrangements for effluent disposal were altered in 1982 so that the treated liquor was piped to the hypolimnion directly. This admirable interim solution 'locked' the summer phosphorus load in store until the autumnal breakdown of stratification and the onset of rapid hydraulic throughput, washing the phosphorus harmlessly from the lake. 'Harmless' to Grasmere, that is, for the phosphorus is moved through a second, short-retention lake (Rydal Water) to become part of the load to the relatively long-retention Windermere. 'Harmlessness' must also be judged in a temporal context: a chronic problem arising from infiltration by urban runoff during wet weather creates large volumes of dilute sewage, which is impossible to store pending any kind of treatment. It is normal practice in urban sewage works discharging to rivers to allow incompletely treated effluent into storm flows where the biological oxidation of residual organic carbon is completed naturally. In Grasmere, this residual carbon was being piped directly to, and collected in, the hypolimnion. What happened, of course, was a shift in deep-water metabolism, hypolimetic anoxia and low redox. As this is being written, a further upgrading of the sewage-treatment works is in hand. It could be argued that the real solution (though expensive and disruptive) would be to re-sewer our towns so that foul- and surface-drainage are kept completely separate, with only the former needing to be submitted to treatment.

In the meantime, the open water of Grasmere represents a highly variable environment for phytoplankton. Flushing episodes, especially during winter, are extremely effective at removing existing algal stocks from the water (benthic propagules are substantially spared). They also deplete the crustacean zooplankton and it sometimes takes many months for a significant feeding pressure to be recovered. What tends to happen after a wet winter or spring is that the water column is repopulated from meagre residual stocks of fast-growing algae (which certainly include C and P groups of diatoms, X1-group nanoplankters and, in Grasmere, *Dinobryon* spp. which develop striking populations of many-celled colonies). Typically, however, X2 and even X3 nanoplankters are prominent first. Like other rarities (chrysophytes, desmids), some of these are believed to be brought *in* by flood waters from the oligotrophic tarns and pools in Grasmere's catchment. With minimal grazing pressure, large populations of nanoplankton (over 10^5 cells mL^{-1} in some instances: Reynolds and Lund, 1988, and authors unpublished observations) develop in the physico-chemically favourable environment. These soon support correspondingly large populations of ciliates and rotifers. A quasi-stable nanoplanktic biomass of about 0.5 mg C L^{-1} and a similar biomass of microplanktic consumers make an unusual sight for a limnologist! Such associations are, however, quite transient. A generation or two of *Daphnia galeata* recruitment is eventually capable of clearing the entire nanoplanktic resource base from the water (and a lot of the ciliates too!).

One further point of interest that emerges from Table 7.5 concerns the larger cryptomonads of group Y. Because the ubiquity of such common species of *Cryptomonas* as *C. ovata*, *C. erosa* and *C. marssoni*, there is a tendency to overlook their value in comparing assemblages. They also share features of each of the three, primary *C*, *R* and *S* strategies in being colonist, in adapting well to low insolation and in their ability to constitute deep monospecific plate-like layers in the metalimnion of stratified lakes. However, their weakness is to be highly susceptible to grazing, by cladocerans, calanoids and certain rotifers (Reynolds, 1995a). In the English Lake District, their numbers are broadly proportional to trophic state: they are sparse in the oligotrophic lakes (<10 mL^{-1}); collectively, they may achieve 10–100 mL^{-1} in the mesotrophic lakes and 100–2000 mL^{-1} in the eutrophic examples. Moreover, the periods of their abundance occur progressively earlier in the year with increasing trophic state. This may be due to the interaction of growth potential and the nature and temporal phasing of phagotrophy: more nutrients nurture the development of larger populations that are detectable sooner (but are also exploited by zooplankton earlier). An alternative view is that the *CR* qualities of cryptomonad survival strategies are expressed more strongly than the *SR* traits with increasing trophic state. Among the richer Lake District lakes, a small dinoflagellate, formerly recorded as *Glenodinium* sp. (but now recognised as *Peridinium lomnickii*) is common and its dynamics coincide sufficiently closely with those of the spring *Cryptomonas* that it has been included in the spring Y-association (Table 7.1).

Phytoplankton in selected regions: glacial lakes of Araucania

Thirty-six north Patagonian lakes, situated on either side of the Andean Cordillera between the latitudes 39° and 42° S, were the subject of a careful survey carried out over 40 years ago by Thomasson (1963). Though not as detailed as some of the other reports featured in this chapter, its inclusion is urged because the striking oligotrophy of the lakes in this region is generally regulated by nitrogen availability (Soto *et al.*, 1994; Diaz and Pedrozo, 1996). Indeed, the molecular N : P ratios to be inferred from the data of Diaz *et al.* (2000) on DIN and TP levels measured in samples from several lakes on the eastern (Argentinian) side of the Andes range between 0.4 and 1.0 (far below 16, reckoned to represent parity). Yet more interesting is the fact that maximum phytoplankton biomass observed in a number of these lakes correlates well with nitrogen concentration while it is even saturated by MRP, excess of which remains measurable in lake water. From data in Diaz *et al.* (2000), it is clear that in lakes where TP exceeded 11 μg P l^{-1} (most values falling between 4 and 8 μg P l^{-1}), the levels of nitrate + ammonium nitrogen were relatively low. Almost everywhere, DIN levels were <30 μg N L^{-1}, generally, they were <14 μg N L^{-1} and, in some cases, <3 μg N L^{-1} (i.e. in the range 0.2–1 μM in which most phytoplankton are expected to experience difficulties harvesting sufficient nitrogen to support further growth; see Section 5.4.4).

What is of particular interest is that the spare phosphorus capacity seems not to be switched to the support of nitrogen-fixing Cyanobacteria. Either the phosphorus is still too low (cf. Stewart and Alexander, 1971), or there is a critical deficiency in molybdenum, vanadium or iron (Rueter and Petersen, 1987), or the energy thresholds for nitrogen fixation are unsatisfied (Paerl, 1988) (see Section 4.4.3). *Anabaena* species are recorded in several of the lakes and *Anabaena solitaria* (of the 'mesotrophic' **H2** group) occurs widely among the region's lakes. However, only in the small steppe lakes, on the eastern fringes of the region

considered, Carrilaufquen Grande (area 16 km^2, mean depth 3 m, TP 298 μg P L^{-1}) and Carrilaufquen Chica (area 5 km^2, mean depth 2 m, TP 69 μg P L^{-1}) were these algae dominant and actively fixing nitrogen (Diaz *et al.*, 2000).

Elsewhere, the lakes are steadfastly oligotrophic or slightly mesotrophic in character. It has to be recognised that the character is independent of the most remarkable feature of the region, which is the behaviour of west–east passing rain-bearing airstreams. Within a distance of 50–70 km from the Pacific seaboard, the land rises to the crest of the granitic Andean Cordillera (average altitude about 2000 m, with peaks – several of which are volcanoes – of up to 3800 m a.s.l.). Progressing a further 60–100 km eastwards, the range falls to the level of the Argentinian plateau, at an altitude of ~1000 m a.s.l. Precipitation in the mountains amounts to ~4 m annually but tapers off abruptly to barely 50 mm. The region is little affected by human settlement and a natural vegetation persists over much of the area, in distinct bands corresponding to altitude and rainfall. The highest rainforests are dominated by *Fitzroya*. These give way to *Australocedrus–Nothofagus* woodlands. Further to the east, this thins steadily, merging into *Agrostis–Cortaderia* grasslands (pampas) and, within 100 km from the Cordillera, to *Festuca–Mulinum*. This is surely one of the world's most remarkable climatic ecotones. Our interest is that most of the drainage to the lakes emanates from the mountains, flowing westwards in short rivers to the Pacific Ocean or eastwards into the Rio Negro catchment that opens to the Atlantic. The waters in the Araucanian lakes are thus almost uniformly dilute in salts, low in alkalinity and weak in nutrients. The lake waters are extremely clear (for photosynthetically active wavelengths, ε_{min} is between 0.12 to 0.2 m^{-1}), except those charged with fine material emanating from glacial melt or from the dust of recent local volcanic eruptions.

The phytoplankton assemblages represented among these lakes also show a high degree of mutual similarity. According to Thomasson's (1963) survey and the later information of Diaz *et al.* (2000), the most ubiquitous species recorded have been the diatoms *Urosolenia eriensis* (group

A) and *Aulacoseira granulata* (group **P**; Thomasson also distinguished '*Melosira hustedtii*'), dinoflagellates of the genus *Gymnodinium* and *Peridinium* (variously including *P. willei*, *P. bipes*, *P. volzii*, *P. inconspicuum*; presumed to be group **L₀**), the chrysophyte *Dinobryon divergens* (group **E**) and the desmid *Staurodesmus triangularis* (group **N**). No information is to hand on the abundance and composition of the picoplankton but a presence as part of a developed microbial food web may be inferred from Thomasson's (1963) lists of planktic ciliates and crustaceans (in which the centropagid calanoid *Boeckella gracilipes* is a prominent component). In several lakes, such as Llanquihue (area 851 km²), Ranco (408 km²) and Todos Los Santos (181 km²) on the western side (Thomasson, 1963) and Traful (area 75 km², maximum depth >100 m, TP 8 µg P L⁻¹, DIN ∼3 µg N L⁻¹) and Espejo (area 38 km², maximum depth >100 m, TP 8 µg P L⁻¹, DIN ∼11 µg N L⁻¹) to the east (Diaz *et al.*, 2000), these are also the principal species. So far as it is possible to deduce from limited sampling frequencies, maximum biomass occurs in a single summer 'peak', rarely achieving as much as 1 µg chl*a* L⁻¹. An ubiquitous nanoplanktic (**X2** group) component (*Plagioselmis*, *Chrysochromulina*) is weakly expressed in these ultraoligotrophic systems. Elsewhere, the same species may achieve larger standing crops and where which the presence of other species may be more evident. In Lago Villarrica (area 851 km²), Thomasson (1963) noted a significant (**F** group) representation of *Kirchneriella* and *Dictyosphaerium* species. The annual cycle in Lago Nauel Huapi (area 556 km², maximum depth 464 m, mean depth 157 m, TP 11 µg P L⁻¹, DIN ∼10 µg N L⁻¹) moves from a spring assemblage of *Aulacoseira–Urosolenia–Dictyosphaerium*, through *Dinobryon–Dictyosphaerium–Urosolenia* in summer to a diatom (*Aulacoseira*, *Urosolenia–Synedra ulna–Tabellaria*)-dominated autumn plankton (Thomasson, 1963). The (summer) chlorophyll maximum is still <2 µg chl*a* L⁻¹ (Diaz *et al.*, 2000). Several of the smaller lakes support an array of desmids (*Cosmarium*, *Staurastrum* spp.), as well as *Fragilaria crotonensis* (**P**) and *Mougeotia* (**T**). *Anabaena solitaria* (**H2**) also occurs quite widely; Thomasson recorded it as being dominant in autumn in Lago Correntoso (area 26 km²); he also noted the relative abundance of *Aphanocapsa* species, with *Staurastrum* species and *Aulacoseira* in Lago Lacar (area 52 km²).

It seems probable that, overall, the phytoplankton of the Araucanian lakes, with its basic A/P → E/L₀ → N or embellished A/P →X2/F/E/L₀ → N/P and A/P →E/L₀/H2 → N/T sequences, conforms to the model of pattern for ultraoligotrophic to mesotrophic lakes. In this context, it is interesting that 'oligotrophic' is just as 'oligotrophic', regardless of whether nitrogen or phosphorus is the main constraining factor. The assemblages are, in reality, typical of low-alkalinity systems, although some supposedly more 'eutrophic' species, thought to be poorly tolerant of low phosphorus conditions (not least *Aulacoseira granulata*, *Fragilaria crotonensis* and *Anabaena* spp.), are here able to function adequately.

Phytoplankton of small kataglacial lakes

Wherever glaciers are, or have been in the last 100 ka, active in eroding terrestrial landforms, there are usually to be found significant 'downstream' deposits of directly transported material, abandoned by the wasting glaciers (hence 'kataglacial'). Associated formations caused by 'freeze-thaw' cycles and solifluction in the vicinity of ice fields are known collectively as periglacial deposits. Depending upon their age, morphometry and the contemporary climatic conditions, these deposits may well enclose small lake basins. In the wake of the most recent glacial period that ended about 10.5 ka ago (known as the Devensian in northern Europe, the Weichselian in the European Alps and the Wisconsinian in North America), vast terminal-moraine systems were deposited along the southern limits of the ice sheets. These abandoned 'tidemarks' of characteristically hummocky landscapes are sometimes called moraine belts. These kataglacial moraines, peppered with generally small lakes, make up distinctive landscapes in the vicinity of Riding Mountain, Manitoba and across northern Minnesota and Wisconsin. In Europe, major moraine belts sweep through Jylland (Jutland), Holstein, Pomerania and Mazuria, and it is the morainic system striking north-eastwards across northern Russia that separates the Baltic and Black Sea watersheds in that country. The largest moraines were formed during periods of ice-front stagnation, when and where glacier recruitment and glacier melt were, for a time, in approximate equilibrium. In front of them stand massive outwash plains of fluvioglacial deposits. Behind them are the lands smoothed by the advancing ice and plastered with compacted, ground-down boulder clay, or till, which may be only lightly and locally covered with later fluvioglacial deposits. Less expansive but wholly analogous structures abound in the lowland outfalls of valley glaciers and ice sheets in other parts of the world. Those with which I am most familiar are located in the English northwest midlands, especially the Wrexham–Bar Hill moraine (Reynolds, 1979b).

The wane of the Pleistocene ice sheets was never smooth but occurred in phases of rapid retreat, alternating with phases of stagnation or even readvance (West, 1977). Lesser morainic features are widespread among areas of kataglacial deposition. The variety of structures in drift material (which term embraces all glacial deposits) is increased by proximity to solid geological features as well as in the formations themselves, such as kames (ice-deposited mounds), eskers (englacial stream beds) and pingos (periglacial frost heaves in drift outside the ice fronts). Lake basins in drifts are just as varied in the detail of their origins. Some are moraine-dammed gaps between drift hummocks and are irregular or linear in outline; others are more rounded with excentric underwater contours, corresponding to kettle-holes formed initially by the melting of detatched blocks of ice.

However, lakes in kataglacial drifts do have several generic and crucial common attributes. The unconsolidated deposits in which they are formed are generally porous, so that present-day precipitation percolates into the drift rather than runs over the topographical surface. The water collects in a zone of saturation above the basal till, which is relatively impervious. The 'surface' of the saturated zone is called the phreatic surface, or water table, and, again, represents a sort of contemporary equilbrium between its recruitment by percolation and the sluggish horizontal permeation into regional catchments. Older texts show standing surface waters wherever the topographic contours dip below the level of the local water table and, thus, merely its surface manifestation. Hydrological studies of these drift hollows confound this simplicity as annual lake levels fluctuate less than does the height of the water table. Reynolds (1979b) proposed a model in which lake basins were partially sealed by their own deposits and that ground water entered by 'inspilling' from a high water table. Lake water 'overflows' normally leak away by seepage. When the water table dropped, hydraulic exchanges were more or less confined to precipitation and evaporation from the lake surface. The fact that, on millennial timescales, basins have varied in their trophic status according to the wetness of past climates, many becoming 'terrestrialised' into fens or peat bogs (Tallis, 1973), also demands a model of restricted basin permeability.

Notwithstanding, much of the water supplied to basins in drift has percolated through subsoils and fluvioglacial deposits that are generally finely divided, offering maximum contact opportunities for the solution and leaching of salts. In this way, the chemical composition of lake water is strongly influenced by the local drifts. This does not mean that drift lakes are necessarily rich in nutrients (many are not: Stechlinsee is one that is quite nutrient-poor; see p. 326) but many certainly are. Moderately to very calcareous lakes are frequent among kataglacial series (Ca 0.4–4 meq L^{-1}, bicarbonate alkalinities $\leq$3.5 meq L^{-1}), though they may be rare locally, depending upon the provenance of the drift. Similarly, with nutrients, the dissolved silicon and orthophosphate contents of unmodified groundwater are variable from catchment to catchment but may easily reach 7–8 mg Si L^{-1} and >300 μg P L^{-1} in some instances. Other major ions may be similarly enriched in particular systems. Unless the topographical catchments have been subject to considerable anthropogenic modification, there is little scope for the enrichment of nitrogen, save through the same surface evaporative processes that affect all precipitation. This leads to another general attribute of morainic and drift lakes, that the ratios of biologically exploitable contents of nitrogen to phosphorus and to silicon are lower or much lower than in surface-fed water bodies. The capacity of the nitrogen to support algal growth may be less than that of phosphorus and algal growth rates in the field are, potentially, more likely to be limited by the external supply of nitrogen than by the phosphorus available.

The phytoplankton of a number of ground water-fed drift lakes has been described in some detail. Sjön Erken in Sweden (area 24 km^2, mean depth 9.0 m, TP 25–50 μg P L^{-1}) has been observed regularly for many years. Rodhe *et al.* (1958) and Nauwerck (1963) noted the April–May development of a spring 'bloom' (maximum ~20 μg chl*a* L^{-1}) of flagellates (including *Cryptomonas* spp., *Plagioselmis*, *Chrysochromulina* and *Dinobryon divergens*), starting under ice cover but soon giving place to dominant diatoms (including 'large' *Stephanodiscus rotula* and 'small' *Stephanodiscus hantzschii* var. *pusillus* and *Asterionella formosa*). Green algae (including *Eudorina* and *Pandorina* spp.) and Cyanobacteria appeared in the early part of summer (*Anabaena flos-aquae*, *Aphanizomenon flos-aquae* and, especially, *Gloeotrichia echinulata*) to be replaced by *Gomphosphaeria* (now *Woronichinia*) and *Ceratium hirundinella*, building to an August maximum of 30–35 μg chl*a* L^{-1} (see also Fig. 7.6). With early autumnal mixing, *Fragilaria* and *Staurastrum* species became prominent. Now, nearly 50 years later (D. Pierson, personal communication), the same floral elements (C/D/X2/Y/E → G/H → L$_O$ → P) are encountered but the dominance has changed slightly in favour of *Stephanodiscus hantzschii* var. *pusilla*, nanoplanktic flagellates, summer *Gloeotrichia* and late-summer *Asterionella* in a sequence closer to (X2/D → H2 → L$_O$ → C).

Esrum Sø (area 17.3 km^2, mean depth 12 m) is a large kettle-hole in calcareous sandy moraine in Sjælland, Denmark which has, like Erken lake, a long history as a focus for detailed studies (recently reviewed, in part, by Jónasson, 2003). Its phytoplankton, as described by Jónasson and Kristiansen (1967) develops during March under a thinning ice cover and dominated by nanoplanktic *Ankistrodesmus falcatus*, now known as *Monoraphidium contortum* and ascribed by Reynolds *et al.* (2002) to group X1. After ice break, dominance of the April spring peak passes to diatoms (predominantly B/C-group *Asterionella formosa*).

Transparency is reduced to 2–3 m. Silicon is generally heavily drawn down (from ~2.5 to <0.1 mg Si L^{-1}) while *Monoraphidium*, other green algae and the Y-group cryptophyte *Chroomonas* persist in the epilimnion (7–10 m in thickness), pending the exhaustion of DIN to <5 μg N L^{-1}. However, available phosphorus remains freely available (SRP >150 μg P L^{-1}) and, not surprisingly, nitrogen-fixing H1-group *Anabaena* species (*A. flos-aquae*, *A. planctonica*, *A. spiroides*) dominate through the summer. In autumn, *Asterionella*, *Fragilaria* and other diatoms dominate the declining biomass. Epilimnetic pH in summer rises to pH 8.8 or a little higher, with carbonate precipitation. The X1 → C/Y → H1 → C/P sequence owes much to the diminution of nitrogen and carbon levels during the vegetation season.

The drift lakes around Plön in North Germany also became classic sites in limnology, especially through the studies of August Thienemann. He distinguished their 'Baltic' biotic assemblages – of benthos as well as of plankton – from those ('Caledonian' assemblages) of European mountain lakes. The former included what we now refer to as eutrophic diatoms (C, D, P associations), cryptomonads, Cyanobacteria and dinoflagellates. Sommer's (1988c) experimental investigation of Schöhsee (area 0.79 km^2, mean depth ~9 m) demonstrated not just the seasonal progression but, through the use of a series of enrichment bioassays, also the nutrients most severely constraining contemporaneous growth capacity. Vernal diatoms (*Asterionella formosa*, *Synedra acus*, *Diatoma elongatum*, *Stephanodiscus rotula* and *S. minutula*) were, in most instances Si-limited. The supportive capacity for Cyanobacteria (*Anabaena flos-aquae*) was often P-limited, while summer dinoflagellate populations were sensitive to limiting availabilities of nitrogen (*Ceratium*) or phosphorus (*Peridinium cinctum*, *P. umbonatum*, *P. inconspicuum*). The growth of (group-F) colonial chlorophytes, (group-E) chrysophytes (*Dinobryon* spp.), cryptomonads (*Cyptomonas ovata*, *Plagioselmis nanoplanctica*) and other nanoplankters (*Ankyra judayi*, *Chrysochromulina parva*) was rarely constrained by nutrients but these would, perhaps, have experienced control through the carbon supply or through grazers or both. The summarised annual sequence

(C/D/Y → X1/X2/Y/E → H1/F → L_M) corresponds to the drawdown of silicon, then nitrogen and/or phosphorus, but always against a background of frequent carbon dioxide deficiencies.

In the small calcareous meres of the English north-west midlands, phosphorus availability is generally moderate to high, owing to a relative abundance in the drift of minerals derived from the underlying Triassic marls and evaporites. The oceanic climate to which these lakes are subject ensures that they are generally ice-free in winter (warm monomictic lakes). In Crose Mere (area 0.15 km^2, mean depth 4.8 m, alkalinity 3.2 meq L^{-1}, SRP ≤ 200 µg P L^{-1}; DIN generally ≤2 mg N L^{-1}), the phytoplankton often achieves high biomass (150–250 µg chl$a$ L^{-1} in summer) in a distinct, diacmic seasonal pattern (see Fig. 7.6). The basic periodic sequence of the phytoplankton (C/Y → G → H1 → L_M → P) begins with a February–March maximum of *Asterionella*, *Stephanodiscus rotula* and *Cryptomonas ovata* (which may, but more usually does not, exhaust the silicon). The onset of thermal stratification (in late April–early May, to within 2–6 m of the surface) leads to the rapid settlement of the diatoms (see also Fig. 6.2), while surviving nanoplankters and cryptomonads succumb to intensifying grazing rates. *Eudorina unicocca*, the next dominant, mostly escapes grazing but depletes the DIN. By late June, nitrogen-fixing *Anabaena circinalis* and/or *Aphanizomenon flos-aquae* are proponderent but *Ceratium hirundinella*, together with *Microcystis aeruginosa*, trawl into the metalimnion for the nutrients to sustain the major biomass peak of the year. Autumn mixing (or earlier summer storms) promotes renewed phases of diatom abundance (*Fragilaria crotonensis*, *Aulacoseira granulata* plus the desmid *Closterium aciculare*). For full details, see Reynolds (1973c, 1976a).

There is some interannual variability about this sequence (see especially Reynolds and Reynolds, 1985) but a core C → H1 → L_M → P is common throughout the deeper lakes of the series (shallower lakes have less *Ceratium* and, sometimes, more *Microcystis*). In Rostherne Mere (area 0.49 km^2, mean depth 13.4 m, SRP ≤ 300 µg P L^{-1}, maximum DIN generally 1.5–2 mg N L^{-1}), depth and turbidity suppress a spring bloom (unless stratification is delayed). H1 dominance may persist beyond spring (rarely resorting to heterocyst production and nitrogen fixation). *Ceratium* or *Microcystis* usually dominates the summer biomass (though not, it seems, as a function of nutrient availability but of recruitment success: Reynolds and Bellinger, 1992). P-group diatoms or even S1-group *Planktothrix agardhii* have dominated in windy summers and, in one recent case, J-group *Scenedesmus* took over a very shallow epilimnion during an unusually calm summer (Reynolds and Bellinger, 1992). J-group *Scenedesmus*, *Pediastrum* and other chlorococcalean algae are abundant in meres that have experienced recent enrichment from nitrogen-rich agricultural fertiliser run-off while some low-alkalinity meres in sandy drift support more green algae of the F group (especially *Botryococcus* in Oak Mere: Reynolds, 1979b).

Finally, it may be added that the summary C → H1 → L_M → P applies to Mazurian moraine lakes such as Mikołajskie (Kajak *et al.*, 1972) and North American prairie lakes (Lin, 1972; Kling, 1975).

Phytoplankton of large, low-latitude lakes

The phytoplankton of the great lakes of Africa differs markedly from that of the high-latitude examples considered above, but in ways that are more easily appreciated in the light of knowledge based on smaller lakes. The lakes themselves differ in character, though most of the African examples to be mentioned here are of tectonic origin, being aligned in the opening and bifurcating rift across East Africa. Its southern arm encloses Lake Malawi; a string of lakes traces the western section, including Lac Tanganyika, Lac Kivu and, at the northern end, N'zigi (formerly Lake Albert and flushed by the upper reaches of the White Nile). The eastern rift valley contains Turkana (formerly Lake Rudolf) as well as several smaller lakes. In the plateau between the eastern and western arms is a water-filled depression of a quite different character – the gigantic saucer that is Lake Victoria whose origin is not tectonic and, it is believed, quite recent.

The rift valley lakes are ancient (possibly up to 20 Ma: Coulter, 1994), deep and, because of the climate ('endless summer': Kilham and Kilham, 1990) at their respective latitudes, almost never experience full convectional overturn. Malawi, Tanganyika, Kivu and Turkana are meromictic lakes: each is vertically segregated into a perennially stagnant, lower monimolimnion and an upper mixolimnion. The monimolimnion retains most of the nutrients in the system but is severely energy-limited; the mixolimnion is frequently nutrient-deficient (Hecky et al., 1991). Even so, there is strong phytoplankton seasonality in these lakes which relates to variability in the depth of the mixolimnion through the year. Entrainment of deeper water during the period of increased mixing is extremely important to the recycling and reuse of some of the system's accumulated resources. Otherwise, new production in the pelagic relies on the supply of new resources (Hecky and Kling, 1981).

At a quoted maximum depth of 1471 m (Herdendorf, 1982), Lac Tanganyika is the second deepest lake on the planet (area 32900 km^2, mean depth 574 m). Between May and September, south-easterly winds are funnelled up the valley and drive surface water (temperature ~27 °C) to the north, where the pycnocline may be depressed to a depth of ~70 m (data of Plisnier and Coenens, 2001). When the south-east winds stop, the tilted surface of the monimolimnion continues to oscillate (or seiches) for several months until the stability is recovered (generally by February). Monimolimnetic water may be sheared off into the mixolimnion during the windy months, while its simultaneous upwelling at the southern end (and also at the northern end during the period of oscillation) augments the process. Between February and April, the only source of nutrients to mixolimnetic primary production is external. The mean mixolimnetic nutrient concentrations (0–30 m) are higher (DIN 70 µg L^{-1}, SRP 50 µg L^{-1}) in the period of wind shearing and upwelling than in the calm period (DIN 50 µg L^{-1}, SRP 20 µg L^{-1}), even though this is also the wet season of enhanced lake inflows.

Phytoplankton biomass increases gradually between May and August and progresses from south to north (Hecky and Kling, 1981), from the equivalent of <1 to ~2 µg chla L^{-1} (data of Langenberg et al., 2002). Picoplanktic Cyanobacteria contribute a large (>50%) proportion of the particulate primary production, while bacterial populations (3–6 × 10^{-6} L^{-1}) and abundant ciliates (especially Strombidium sp.) are indicative of an active microbial food web (Hecky and Fee, 1981; Hecky et al., 1991; Plisnier and Coenens, 2001). As the windy season abates, algal biomass increases at the northern end, with the development of Anabaena blooms, while pulsed upwellings at either end deliver the nutrients to sustain the main phase of diatom growth (Nitzschia spp. and Aulacoseira granulata are prominent), to maximal concentrations of ~7 µg chla L^{-1} (Langenberg et al., 2002). However, as the seiching weakens and the mixolimnetic nutrient base is dissipated in fish production and the sedimentary flux, it is green algae such as Coenochloris and Oocystis that persist longest through February to April (Hecky and Kling, 1981; Talling, 1986). From the mixing to final relaxation and the next cycle, phytoplankton composition may be summarised as P → F → Z (with the advantage to P perhaps alternating with one in favour of H with the seiche cycle).

In Lake Malawi (area 22490 km^2, mean depth 276 m), seasonal variations in the depth of the mixolimnion range from ~70 m in the cooler, windier part of the year (June to September), when its temperature falls to ≥24 °C (barely more than 1° warmer than the monimolimnion), to <30 m during the hot, calmer months between October and March (temperature >27 °C). The major nutrients occur at low concentrations, except at depth (>200 m, in the anoxic monimolimnion). Mixolimnetic concentrations are at their highest during the mixed periods (SRP ~5 µg L^{-1}, DIN ~20 µg L^{-1}: data of Irvine et al., 2001) but these are soon drawn down after the winds ease. Irvine et al. (2001) observed that planktic chlorophyll concentrations are typically <2 µg chla L^{-1} with higher levels (≤7 µg chla L^{-1}) coming during July. Diatoms are prevalent during the windier months (Aulacoseira granulata, Cyclostephanos sp.), with desmid species of Staurastrum and Closterium). Nitrogen-fixing Cyanobacteria Anabaena and Cylindrospermopsis species make up 50% of the planktic biomass in the months of stable stratification (Allison et al., 1996; Irvine et al., 2001). These authors give no specific information on either pico- or nanophytoplankton, although the latter, at least, are likely to be relatively most abundant during the early stages of the intensification of the thermal stratification. Pending confirmation of this, the phytoplankton sequences are best summarised as P → H1 (S$_N$).

The phytoplankton of Lac Kivu (area 2370 km^2, mean depth 240 m, TP ≤55 µg P L^{-1}) shows many compositional similarities to that of Lake Malawi. The stable part of the temperature gradient (and associated chemocline) begins at about 60 m, which is sufficiently shallow to support a well-developed plate of photosynthetic bacteria (Haberyan and Hecky, 1987). Eukaryote production in the mixolimnion is greatest following the periods of more active mixing, but nanoplankton and, later, the Cyanobacteria *Microcystis* and *Spirulina* species feature with *Anabaena* and *Cylindrospermopsis* in the calm period (Serruya and Pollingher, 1983). By way of summary, P → X → H1/L$_M$/S$_N$/S2) is probably a fair representation. In Lake Turkana (area 8660 km^2, mean depth 29 m), where mixing involves a much larger proportion of the lake volume, phosphorus has accumulated to high levels (TP 1.8–2.4 mg P L^{-1}, [SRP]$_{max}$ ≤786 µg P L^{-1}: Hopson, 1982). The diatom assemblage conspicuously includes 'large' species of *Surirella* and *Coscinodiscus*; F-group colonial chlorophytes appear as stability increases in January but soon give dominant place to nitrogen-fixing *Anabaena* species. *Microcystis* species also develop in some numbers (Liti *et al.*, 1991). The sequence is partly captured in the summary notation P → F → H1/L$_M$.

In contrast to the rift-valley lakes, Lake Victoria is young (possibly as little as 12 000 a) and quite shallow (mean depth 39 m, maximum depth 84 m) in relation to its area (68 800 km^2). Like the rift-valley lakes, however, it experiences seasonal alternation between a warm, wet (October–May) and a cool, dry season (June–September). The lake is wind-mixed to the bottom on frequent occasions during July and August. From the limnological information on the lake that has accumulated since systematic information started to become available (Fish, 1952; Talling, 1965; Beadle, 1974), it is quite clear that changes within and beyond the lake during the last 40 years have had a profound influence upon the biota (Hecky, 1993; Kling *et al.*, 2001). These include an eight- to tenfold increase in the phytoplankton chlorophyll, a doubling of

primary production and a doubling of the vertical coefficient of light attenuation since Talling's (1965) observations in the 1960s (Mugidde, 1993). The hypolimnion, which was previously aerated to the deepest sediment through most of the year, is now regularly anoxic. These may be responses to changes that started earlier than the 1950s but they have most certainly accelerated since. There is an increasing nutrient load originating from the activities (notably urbanisation and erosion consequential on intensifying agriculture) of a large (∼20 million) and increasing (3–4% a^{-1}) human population resident in the lake's catchment. Total phosphorus concentrations in the lake water (now 45–72 µg P L^{-1}) have roughly doubled and the area-specific sedimentation rates have gone up by a similar factor (Ramlal *et al.*, 2001). Levels of DIN were, and remain, low: the nitrogen limitation of production recognised by Talling (1965) persists. Average silicon levels have fallen considerably. Coincidentally, the ecosystem has been catastrophically altered by the introduction during this period of Nile perch (*Lates niloticus*) and Nile tilapia (*Oreochromis niloticus*) which have expanded at the expense of the lake's endemic populations of haplochromine cichlids (Ogutu-Ohwayu, 1990; Witte *et al.*, 1992).

The phytoplankton of Lake Victoria in the 1960s has been characterised by Talling (1965, 1986, 1987). Despite the great difference in basin morphometry, Lake Victoria was subject to a pattern of alternating dominance between diatoms during the mixed periods (especially by what are now referred to as *Aulacoseira granulata* and *Cyclostephanos* spp.) and nitrogen-fixers *Anabaena* and *Anabaenopsis* species in the stratified phase. Also present were *Closterium* species and a variety of chlorococcal algae associated with enriched shallow conditions (*Scenedesmus, Pediastrum, Coelastrum* spp. of group J). Chlorophyll-*a* concentration in the open lake varied between 1.2 and 5.5 µg chl*a* L^{-1}, though higher concentrations could be found in the thermocline.

By the 1990s, *Aulacoseira* had declined to the point of rarity and *Anabaena, Anabaenopsis* and *Aphanizomenon* occurred only sporadically (Kling *et al.*, 2001). *Nitzschia acicularis* had become the dominant diatom, peaking between September and November. This alga is typical of much smaller, usually enriched and turbid, bodies of water and some lowland rivers and ascribed by Reynolds *et al.* (2002) to functional group D. Though representing a larger biovolume than all the diatoms together in the 1960s, *Nitzschia* production in the Lake Victoria of the 1990s was overshadowed by that of Cyanobacteria. *Cylindrospermopsis* now dominated the nitrogen-fixer niche in the stratified period, while solitary, filamentous non-nitrogen-fixing species of *Planktolyngbya* had become abundant in the mixed period. The former P/J → H1 alternation had been usurped by a D/S1 → S_N sequence whose selection is forced presumably by the requirement for superior antennal properties.

Brief reference should be made to one or two other, well-researched tropical systems outside Africa which experience quite different patterns of seasonal forcing. Lago Titicaca (area 8559 km^2, mean depth 107 m, stratified October–July to within 40–70 m of the surface) supports a maximum biomass during the early stages of mixing (May–June, when *Aulacoseira* and other diatoms are relatively abundant: Richerson *et al.*, 1986; Dejoux and Iltis, 1992). SRP levels are ≤23 μg chl*a* L^{-1}, but DIN is low (<50 μg N L^{-1}: Vincent *et al.*, 1984). Planktic nitrogen fixation is modest, however, and nitrogen deficiency, together with light dilution, appears to most constrain production. Diurnal heating and nocturnal cooling cause wide daily fluctuations in stability. Few algae accommodate to this diel stratification but group-F colonial chlorophytes, together with self-regulating *Peridinium* species (L_O) are relatively tolerant and predominate during this part of the year.

Finally, the apparent association of diatoms and solitary filamentous cyanobacteria with mixed water columns, and (for different reasons) of rapid-growing flagellates, nitrogen-fixers and self-regulating gleaners is well demonstrated in tropical Lake Lanao, the Philippines (area 357 km^2, mean depth 60 m). Lewis' (1978a) penetrating analysis of phytoplankton periodicity in this lake provides may insights into plankton ecology elsewhere. The lake is subject year-round to frequent stormy episodes (10–15 a^{-1}; the behaviour is described as *atelomictic*) that mix the lake to depths >20 m, sometimes to >40 m and,

between November and April, to the bottom. In the intermediate quiescent episodes, microstratification develops to within a few metres of the lake surface. Although the absolute and relative abundances among them may vary among episodes, the responding populations conform to a clear and general pattern. *Aulacoseira* is ususally the dominant species among the diatoms that increase with each mixing event. As the mixing weakens, cryptomonads briefly replace the settling diatoms but fall victim to cladoceran filter-feeding. The clearing, stabilising epilimnion is populated by various colonial green algae, by *Anabaena* and then by migrating peridinoid dinoflagellates. In chemical terms, the lake is more mesotrophic than eutrophic and deficient in nitrogen rather than phosphorus. The summary sequence P → Y → F → H1 → L_M applies comfortably to the periodicity of species in many low-latitude lakes including that of another tropical lake studied by Lewis (1986), Lago Valencia, Venezuela.

Phytoplankton in shallow lakes

In terms of number, most lakes are absolutely small (as defined on p. 325) and, commonly, 'relatively shallow' (as outlined on p. 320). The essential property of a shallow lake is that much or all of the bottom sediment surface is frequently, if not continuously, contiguous with the open-water phase of the habitat (Padisák and Reynolds, 2003). The consequences may be manifest in any of several ways. Subject to particle size, depth and clarity, the bottom surfaces may support epilithic algae or rooted macrophytic plants (which, *sensu lato*, include large algae, mosses and pteridophytes, as well as angiosperms). Finer sediments are liable to entrainment by penetrating turbulence; this may well increase turbidity and light-scattering and, thus, impair light penetration. On the other hand, however, the simultaneous entrainment of interstitial water and biogenic detritus provides a mechanism for the accelerated return of resources back to the water column. From the point of view of ecosystem function, the pelagic systems of shallow lakes differ from those of deep lakes and seas in not becoming isolated from a significant store of dead, decomposing biomass.

On this basis, some 'large lakes' (>500 km², at least when they are full of water) are also 'shallow': these include the Lakes Winnipeg, Balkhash, Tchad, Eyre and Bangweolo (Reynolds *et al.*, 2000). Neither these, nor any of the greatly more numerous small lakes, support a characteristic or recognisable 'shallow water phytoplankton' – community assembly invokes other critical factors. Yet phytoplankton is important to perceptions of water quality in very many shallow lakes (Scheffer, 1998), so it is valuable to be able to extract some general principles. One of these is that, in a majority of truly 'small' (<200 m across) and absolutely 'shallow' (≤5 m) water bodies have a theoretical light-supportive capacity of ≤600 mg chl$a$ m^{-2} (say, 30 C m^{-2}), unless they are at high latitude or their waters are heavily stained with humic substances. This provides plenty of scope for the intervention of other potential limiting factors. Interestingly, Scheffer (1998) provided a plot of the average summer chlorophyll-a concentrations in each of 88 shallow lakes in the Netherlands (mean of mean depths 2.1 m) against its average summer TP concentration. Up to 0.3 mg P L^{-1}, the points cluster beneath a slope of chl$a$ = 0.9 TP. At TP concentrations >0.3 mg P L^{-1}, chlorophyll levels show a typical P-saturation response. An analogous dataset for the same lakes shows a similar behaviour with respect to TN, with the points clustering beneath a slope of chl$a$ = 0.09 (TN – 0.7), up to TN ~4 mg N l^{-1}. However, a very large number of records apply in systems with markedly lower TP and TN availabilities but, nevertheless, show marked saturation of the chlorophyll actually supported.

In many of these small lakes, there is a complex interaction with rooted, submerged plants, which, in these shallow waters, are able to compete effectively in energy- and resource-harvesting. The complexities arise through the behavioural interactions among benthic macroinvertebrates, littoral cladoceran zooplankton and their respective planktivorous and benthivorous consumers (including fish) (see Section 8.3.6). Suffice it to say here that aquatic primary production and its heterotrophic consumers can strike quite different and alternative steady states in shallow lakes (Scheffer, 1989; Scheffer *et al.*, 1993). One is a vegetated state with clear water; the other is a potentialy turbid, phytoplankton-dominated system. Moreover, many cases together indicate that small lakes can switch abruptly between these two stable states (Blindow *et al.*, 1993: Scheffer, 1998), under the influence of switches in the resource–consumer interaction.

Restricting the present discussion to the phytoplankton composition, the structural influences in shallow waters fall into several categories. Phytoplankton may be present in small concentrations if bioavailable resources in the water are modest (advantage to rooted macrophytes exploiting additional or alternative resources) or, if not, if macrophytic vegetation harbours the food web that can exert ongoing controls on the development of phytoplankton. Phytoplankton can dominate in the open water above macrophytes rooted 1–1.5 m below the water surface. In temperate waters, they can respond earlier and faster than macrophytes to lengthening days. If circumstances prevail where there is no adequate consumer control, phytoplankton have the potential to shade out rooted benthos or, at least, to contain its distribution to the shallowest margins.

Some of the representative shallow-lake phytoplankton assemblages may be noted. Unicellular nanoplanktic forms are collectively common, benefiting from *C*-type invasive, fast-growth-rate strategies. These often include nanoplanktic chlorococcals (*Chlorella, Monoraphidium* spp. of group X1), nanoplanktic flagellates (*Chlamydomonas, Plagioselmis, Chrysochromulina* of group X2) and group-Y cryptomonads and small peridinioids. A polyphyletic group-W1 association of euglenoids (*Phacus, Lepocinclis* as well as the type genus, *Euglena*), chrysophycean (especially *Synura* spp.) and small volvocalean colonies (especially *Gonium*) may be represented. On the basis of a floristic comparison of small, shallow lakes in Hungary, Padisák *et al.* (2003c) noted the frequent co-occurrence of *Phacotus* spp. with this group and its dominance in calcareous waters; they proposed a new group identity in Y$_{PH}$. Larger green colonies are represented by *Volvox, Eudorina* and *Pandorina* of group G. Colonial chrysophytes (especially *Dinobryon* spp.) are also abundant in small water bodies, even quite calcareous ones, if there is an good supply of carbon dioxide (e.g. from benthic decomposition of plant matter, including fallen leaves). Planktic diatoms are not generally abundant in the small, shallow ponds save for the small *Stephanodiscus* species, and perhaps the smaller *Synedra* species (group **D**). *Gyrosigma*, *Surirella* and *Melosira varians* that are tychoplanktic, as well as *Aulacoseira* species that are meroplanktic (the distinction is actually very fine here), may be prominent in shallow lakes that are exposed to wind and wave action (including large lakes such as Balkhash and Tchad). Elsewhere, **B**, **C**, **N** and **P** groups of diatoms and desmids are common, provided their suspension requirements are fulfilled by the absolute water-column depth (see Sections 2.6.2, 6.3.2) in small lakes. The diatoms remain sensitive to silicon exhaustion and they may be replaced by other algae, including non-vacuolate, small-celled blue-green in their colonial phases (*Aphanocapsa*, *Aphanothece* of group **K**) or by non-motile colonies of group-**F** green algae, such as *Botryococcus*. In shallow lake Śniardwy, Poland (area 110 km², mean depth 5.9 m), *Aulacoseira* species alternate with group-**J** *Scenedesmus* and *Pediastrum* species (Kajak *et al.*, 1972). During the 1970s, a similar assemblage was prominent in Budworth Mere, England (Reynolds, 1979b).

The effects of nitrogen depletion in small, shallow lakes are no different from other nutrient- and energy-rich habitats, where, accordingly, nitrogen-fixing (**H1**) *Anabaena* and *Anabaenopsis* species and, in warm lakes, (S_N) *Cylindrospermopsis* may be promoted. Both are important components in Balaton, Europe's largest shallow lake (area 593 km², mean depth 3.3 m: Padisák and Reynolds, 1998). The plankton of humic-stained Fennoscandian lakes is frequently distinguished by the presence of chrysophytes, cryptomonads and the raphidophyte *Gonyostomum semen*. Physiological studies on this alga by Korneva (2001) seemed to Reynolds *et al.* (2002) to justify its separation into group **Q**. The plankton of acidic small shallow lakes is biased towards chlorophyte (including desmid) and chrysophyte genera. An interesting observation on the plankton of Oak Mere (area 0.2 km², mean depth 2.1 m) was the replacement of a plankton dominated successively by *Ankistrodesmus*, *Lagerheimia*, *Closterium* and *Chlorella* by one in which *Asterionella*, *Pediastrum*, *Anabaena* and even *Microcystis* formed successive populations. The trigger was the introduction of base-rich water from a trial bore that raised the ambient pH of the lake from 4.5 to 6.5 (Swale, 1968; Reynolds and Allen, 1968).

Abundant populations (perhaps equivalent to 100–600 μg chla L^{-1}) of phytoplankton occurring in small, shallow and continuously nutrient-rich, hypertrophic lakes and in which, through trophic imbalances, heavy grazing by zooplankton is avoided, frequently comprise species of *Scenedesmus*, *Coelastrum* and *Pediastrum* (group **J**), often with **X1**-group nanoplankters such as *Chlorella*, *Ankyra* and/or *Monoraphidium*. The assemblage is well represented in the phytoplankton of habitats where high nutrient loading is equated with high hydraulic loads and rapid flushing rates that discount against slow-growing algae and most species of mesozooplankton. These include hypertrophic rivers, some flood-plain lakes flushed by river flow, natural ponds enriched with sewage and many artificial ponds constructed to bring about its oxidation (Uhlmann, 1971; Boucher *et al.*, 1984; Moss and Balls, 1989; Stoyneva, 1994). The assemblage is sometimes more prominent in the plankton of more substantial lakes, including Arresø, Denmark (Olrik, 1981), and during the more stable of the alternating phases in the enriched Hamilton Harbour, Lake Ontario (Haffner *et al.*, 1980).

At low latitudes, such enriched shallow systems may succumb to monospecific (or, at least, monogeneric) steady-state blooms of *Microcystis* (group **M**), where the alga's ability to regulate its buoyancy helps it to avoid excessive light near the surface during diurnal stability and to recover position after nocturnal mixing. The original demonstration of dominance through this mechanism, by Ganf (1974b), in the tropical Lake George, Uganda (area 250 km², mean depth 2.5 m), has been repeated in subsequent studies of other low-latitude, hypertrophic shallow lakes (Harding, 1997; Yunes *et al.*, 1998). My own, unpublished observations of the large but shallow lake Tai Hu (2425 km², mean depth 2.1 m) suggest that *Microcystis* similarly dominates the plankton from quite early (April) in the year.

Finally, many shallow, hypertrophic lakes in the temperate regions (especially those exposed to frequent or continuous wind mixing), experience year-round dominance by *Planktothrix agardhii* and/or *Limnothrix redekei* or other slender, solitary filamentous Cyanobacteria, such as *Pseudanabaena* species. Phytoplankters of group S1 need never normally experience nutrient limitation, and they are difficult for cladocerans to filter and for copepods to manipulate (although certain ciliates seem to have perfected a means of ingesting them). They may dominate to the extent that they exclude almost all other autotrophic phytoplankton, to the limit of the energy-determined carrying capacity set by their own highly efficient and persistent light-harvesting antennae (Reynolds, 1994b). Several examples from the literature were cited there in support, including the studies of Berger (1984, 1989; on the polder lakes Drontermeer, Wolderwijd and the pre-manipulated Veluwemeer), of Whitton and Peat (1969, on St James's Park, London) and of Gibson *et al.* (1971, on Lough Neagh, Northern Ireland). In Kasumigaura (area 220 km^2, mean depth 4 m), *Planktothrix agardhii* replaced *Microcystis aeruginosa* as one low-diversity plankton gave way to another (Takamura *et al.*, 1992) and, apparently (Recknagel, 1997), as a consequence of an ongoing expansion of the resource capacity demanding the better-performing species under increasing light limitation.

7.2.4 Species assemblage patterns in lakes

Certain consistent patterns emerge from the descriptive accounts making up Section 7.2.3, most of which begin with a recognition that phytoplankton production and the biomass that it is possible to support are often constrained by the environmental conditions obtaining. A corollary to this statement is that high rates of production, sustaining and maintaining large planktic crops, depend upon the alleviation of the typical constraining factors. Standing-crop concentrations in excess of 40 μg chl*a* L^{-1} are mainly encountered in lowland and moraine lakes. All of these offer a substantial base of bioavailable nutrients, especially phosphorus and nitrogen. Moreover, autotrophs benefit from being entrained within a layer that is absolutely shallow or is made shallow through density stratification. Such benign conditions are presumed to satisfy the minimum requirements of most phytoplankton. However, the most widely prominent trait-differentiated planktic groups achieving large, dominant and persistent populations under these conditions are epitomised by the groups L$_M$, M and R and, perhaps to a lesser extent, H1 and J (see also Naselli-Flores *et al.*, 2003).

Decreasing day length and/or deeper mixing result in diminished opportunities for light-harvesting, to the extent that the capacity for photosynthetic fixation of inorganic carbon becomes the constraint on the ability of phytoplankton to function and survive (as discussed in Section 5.4.3). It was argued there that the more sensitive species are (literally) outcompeted by those that are more robustly adaptable and predisposed to light interception and harvest. The survey in the present chapter would confirm that late-summer mixing in temperate, mesotrophic and eutrophic lakes repeatedly selects for P-group diatoms and desmids, and in some instances, for T-group *Tribonema* and *Mougeotia* and, especially, S1-group *Planktothrix agardhii*. In highly enriched, shallow waters, dominance of self-shaded assemblages by the latter may be near-perennial. Conversely, the biomass recruited during the 'spring outburst' of phytoplankton in the mixed columns of mesotrophic and mildly eutrophic temperate lakes is generally dominated by diatoms (B, C groups) and Y-group cryptomonads, gymnodinians or small peridinians.

In many instances, improving light conditions in lakes press the constraints towards the nutrient resources. Shortage of silicon is an obvious selective disadvantage to all diatom groups but it is not clear from the present review that silicon deficiency becomes a limiting factor among the most oligotrophic lakes (that is, other nutrients intervene first). The instances in which nitrogen, theoretically or by interpretation, seems to limit the supportive capacity are more common than the weight of literature might suggest, especially at low latitudes and in well-leached catchments. Nitrogen deficiency may be cited as a selective factor operating in favour of the common occurrence of H1, H2 and, in warm-water locations, S_N groups of nitrogen-fixers. However, it must be emphasised again that their prevalence is not confined to habitats that are low in nitrate or ammonium. Moreover, the argued dependence of the ability to fix nitrogen on high energy inputs and an adequate reserve of phosphorus and trace metals (see Section 4.4.3) is supported by the survey (see, especially, p. 335). In stratified lakes, conditions of low DIN and low SRP may be more amenable to motile 'gleaners', including those with known phagotrophic capabilities. This explanation fits the frequent dominance of E, U and, especially, L_O groups of algae. In water columns that are chronically deficient in nitrogen and phosphorus and, thus, in the ability to develop even a detrital store of nutrients, the most tolerant survivors appear to be the Z-group picocyanobacteria.

Phosphorus bioavailability, however, remains a major constraint upon the supportive capacity in a large number of stratifying and shallow lakes and, hence, a key factor in defining the trophic state. It has been recognised in earlier chapters that plankton algae are generally very effective in garnering their own phosphorus requirements from SRP concentrations as low as 10^{-7} M ($\sim$3 µg P L^{-1}) (see p. 158). While greater concentrations than this remain accessible, the supportive capacity of BAP has not been reached. Superior affinity for phosphorus in solution is unlikely to confer a competitive advantage to phytoplankton except in the range 10^{-9}–10^{-7} M. In those instances where BAP is always or nearly always $<10^{-7}$ M ($\sim$3 µg P L^{-1}), species with high uptake affinities experience both the immediate competitive advantage of winning resources and the longer-term advantage from being able to maintain larger inocula from one growth opportunity to the next (see p. 203). The trait is shared among species identified in Table 5.2 and which figure in groups A, E, F, N, U, X3 and Z. They are represented in lakes throughout the descriptive series but they provide the prominent components of the planktic assemblages in the low-P oligotrophic lakes, large and small. They are also well represented among those stratifying mesotrophic waters during periods in which epilimnetic BAP has been previously be depleted to levels of $\geq 10^{-8}$ M ($\sim$0.3 µg P L^{-1}). However,

among smaller lakes that stratify to within 10 m of the water surface and where accessible SRP concentrations persist in the metalimnion, the most successful species are the relatively motile gleaner species (of especially groups L_O and L_M). Given water of sufficient clarity these may well be able to operate successfully simply by remaining at depth. However, the deeper are the available resources, the more important is the light-harvesting ability and the less important is rapid motility. Ultimately, before energy and nutrient resources are finally uncoupled, the selective advantage may fall to superior, chromatically adapted light-harvesting species with a simultaneous ability to maintain and adjust vertical station – and include cryptomonads (Y) and phycoerythin-rich Cyanobacteria, especially *Planktothrix rubescens* (R).

The surveyed cases support the contention that particular adaptive traits distinguished among the phytoplankton and the species in which they are most strongly represented are better suited to particular sets of environmental conditions. The more severe are the latter, the greater the selective power that works in favour of the most tolerant species. This is the complementary deduction to that of Dufrêne and Legendre (1997) regarding the reliability of indicator species and what they may convey in terms of constraints acting upon community function. There is an obvious mutualism between, on the one hand, evolutionary strategies and adaptations of species that permit them to tolerate particular environmental conditions and, on the other, the conspicuous occurrence of these same species in locations where the critical conditions obtain. In Table 7.6, the various trait-separated functional groups are listed in terms of their reactivity to selective variables distinguishing among habitats or, for a given habitat, among seasonally varying conditions.

Of course, alternative approaches are available to establishing the link between organismic adaptations and indicative habitat preferences. For instance, it is commonly revealed through analysis of the size distributions of the totality of organisms forming the assemblage, independently of its taxonomic composition, based instead upon their functional contributions

Table 7.6 | Sensitivities to habitat properties of functional groups of phytoplankton

Group	h_m <3	l^* <1.5	θ <8	[P] <10^{-7}	[N] <10^{-6}	[Si] <10^{-5}	[CO_2] <10^{-5}	f <0.4
A	−	?	+	+	+	+	−	−
B	−	+	+	+	−	−	−	−
C	−	+	+	−	−	−	?	−
D	+	+	+	−	−	−	+	−
N	−	−	−	+	−	+/−	−	?
P	−	−	−	−	−	+/−	+	+
T	−	?	−	+/−	−	+	?	+
S1	+	+	+	−	−	+	+	+
S2	+	+	−	−	−	+	+	+
S_N	+	+	−	−	+	+	+	+
Z	+	−	+	+	+	+	?	−
X3	+	−	+	+	−	+	−	−
X2	+	−	+	?	−	+	?	−
X1	+	−	+	−	−	+	+	−
Y	+	+	+	−	−	+	?	−
E	+	+	+	+	−	+	−	−
F	+	−	+	+	−	+	−	−
G	+	−	+	−	−	+	+	+
J	+	?	+	−	−	+	?	−
K	+	?	−	−	−	+	+	?
H1, H2	+	−	−	+	+	+	+	+
U	+	−	?	+	−	+	−	+
L_o	+	−	−	+	−	+	−	+
L_M	+	−	−	−	−	+	+	+
M	+	−	−	−	−	+	+	+
R	+	+	+	−	−	+	?	+
V	+	+	+	−	−	+	?	−
W1	+	+	+	−	−	+	?	−
W2	?							
Q	+	+	+	?	?	+	−	−

Notes: Entries in table are to denote tolerance (+) or no positive benefit (−) of the environmental condition set; '+/−' is used to denote that some species in the association are tolerant; '?' denotes that tolerance suspected but not proven. Some representative genera or species only are listed. Variables signified are: depth of surface mixed layer (h_m, in m from surface); mean daily irradiance levels experienced (l^*, in mol photons m^{-2} d^{-1}); water temperature (θ, in °C); the concentration of soluble reactive phosphorus ([P], in mol L^{-1}); the concentration of dissolved inorganic nitrogen ([N], in mol L^{-1}); the concentration of soluble reactive silicon ([Si], in mol L^{-1}); the concentration of dissolved carbon dioxide ([CO_2], in mol L^{-1}); and the proportion of the water processed each day by rotiferan and crustacean zooplankton (f)

Source: Updated from Reynolds (2000a).

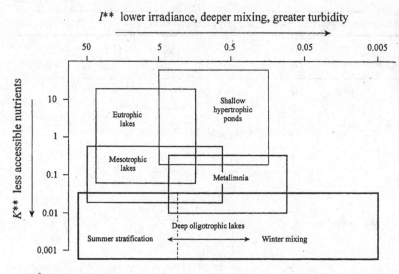

I^{**} lower irradiance, deeper mixing, greater turbidity

Figure 7.7 Schematic summary of freshwater pelagic habitats, defined in terms of I^{**} (an integral of light income and its dilution through the mixed layer) and K^{**} (an integral of nutrient accessibility), and where $I^{**} = (I'_0 \cdot I_m)^{1/2} h_m^{-1}$ and $K^{**} = [K]/(1 + \delta[K])$; for further details see text. Redrawn with permission from Reynolds (1999a).

(Bailey-Watts, 1978; Gaedke and Straile, 1994b). Size-spectral analyses are particularly sensitive to the structure of the pelagic food web and its responses to interannual enviromental variability and to fundamental alterations to the nutrient base (Gaedke, 1998). However, care is needed in making ecological interpretations from algal morphometry without allowing for its alternative indications. For instance, the many separate studies on the abundance and production of picophytoplankton support the conclusion that these organisms fulfil the major contribution to the carbon dynamics of oligotrophic pelagic systems (Stockner and Antia, 1986; Chisholm et al., 1988; Agawin et al., 2000; Pick, 2000). In contrast, the greatest concentrations of large algae like *Ceratium* and *Microcystis* are supposedly excluded from nutrient-poor systems (Wolf-Gladrow and Riebesell, 1997). Thus, a size spectrum biassed towards a predominance of smaller or larger forms should reflect a lesser or greater productive capacity. The counter to this simplicity is that, indeed, organisms at the diminutive end of the size spectrum gain full advantage of their high surface-to-volume ratio to acquire and convert resources into biomass faster than larger organisms with less favourable surface-to-volume ratio lowers (Raven, 1998) (see also Section 5.3.1). Other things being equal, then, the greater is the productive capacity, the more the size spectrum should be biassed towards smaller rather than larger forms. Recent research confirms that

this applies to photoautotrophic picoplankton as much as to nanoplankton (Carrick and Schelske, 1997). High concentrations of picophytoplankton dominated the algal assemblage of a small Antarctic pond, under the conditions of low ambient temperatures and nutrient enrichment through its use as a roost by elephant seals (Izaguirre et al., 2001).

The truth is that smaller algae have physiological and dynamic advantages over large ones. It is the intervention of other factors (most especially, the influence of grazing and the segregation of the energy and resource bases) that alters the structural balance in favour of larger or motile algae. This too impinges upon the spectral analysis. In reality, the procedure detects functional aspects of the assemblage and the extent of its successional maturity with respect to the driving variables and not necessarily what the limiting variables might be.

To summarise the compositional patterns of freshwater phytoplankton assemblages, a habitat template, analogous to the one for marine environments (Fig. 7.4), is proposed in Fig. 7.7. Habitats are characterised on axes of nutrient resources and energy distribution, the interaction of which is suggested to drive the primary selective criteria in lakes (variations in acidity/alkalinity are not addressed specifically). Unlike Fig. 7.4, however, the axes are quantified, following Reynolds (1999a), in units that were designed to embrace variabilities in both

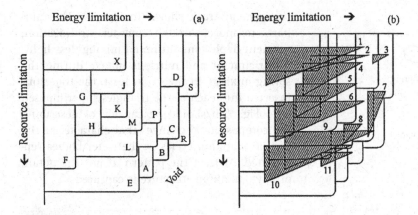

Figure 7.8 (a) Habitat template for trait-separated categories of freshwater phytoplankton (cf. Table 7.1), based on limitation gradients of decreasing energy harvest and accessible resources, as originally envisaged by Reynolds (1987b). The superimposition of one shape on another implies that the algal category is more likely to succeed than those that it obscures. (b) As (a) but unlabelled; the shaded triangles embrace approximately the floristic representation in the named systems: 1, eutrophic pools in Shropshire; 2, Montezuma's Well; 3, polder lakes such as Veluwemeer; 4, Norfolk Broads; 5, Hamilton Harbour, Lake Ontario; 6, Crose Mere; 7, Lough Neagh; 8, Volta Grande Reservoir; 9, Lagoa Carioca; 10, Windermere, pre-1965; 11, Millstättersee. Redrawn with permission from Reynolds (1997a).

deep and shallow lakes. The scale of the horizontal axis, I^{**}, integrates the income of photosynthetic energy not just through time but its dilution through the mixed layer. Whereas I^*, from Eq. (3.17), is calculated from the average vertical absorptance of photosynthetically active solar radiation over the mixed depth (h_m) [as $I^* = (I'_0 \cdot I_m)^{1/2}$, where $I_m = I'_0 \cdot h_m \exp(-\varepsilon_\lambda)$ is the residual irradiance flux reaching the base of the mixed layer], I^{**} is the average harvestable photon concentration.

Thus,

$$I^{**} = (I'_0 \cdot I_m)^{1/2} h_m^{-1} \tag{7.1}$$

If daily integrals of the photon flux are used, we can distinguish not just high from low light incomes but compare dilution of the income as a consequence of mixed layer depth and turbidity. For example, a photon flux of 60 mol photons m^{-2} d^{-1} on a shallow water body ($h_m = H = 1$ m) with a low coefficient of vertical light extinction (say $\varepsilon_\lambda = 0.2$ m^{-1}), I^{**} solves at 54.3 mol

photons m^{-3} d^{-1}. Using actual measurements (Ganf, 1974b) of average daily irradiance at equatorial Lake George (average daily irradiance $\sim$2000 J cm^{-2} d^{-1}, or 20 MJ m^{-2} d^{-1}, PAR $\sim$9.4 MJ m^{-2} d^{-1} = 43 mol photon m^{-2} d^{-1}) and vertical extinction coefficients through a 2.5-m water column of up to 7.7 m^{-1}, I^{**} falls as low as 1.45 mmol photons m^{-3} d^{-1} Similarly, at the end of the summer in the Araucanian lake Nauel Huapi (p. 336), even a low attenuation coefficient ($\varepsilon_\lambda \leq$ 0.2 m^{-1}) does not prevent convectional mixing to 60 m from diluting the light availability to a similar level ($I^{**} \sim$1.2 mmol photons m^{-3} d^{-1}).

The vertical scale (K^{**}) is based upon mixed-layer nutrient concentration, divided by a factor based on the concentration gradient ($1 + \delta[K]$) through the whole trophogenic layer. This index distinguishes chronic deficiency of the critical nutrient ([K] low, $\delta[K]$ low, so $K^{**} = [K]/(1 + \delta[K])$ is also low) from the effects of near-surface depletion (as $\delta[K]$ increases, so K^{**} decreases. In terms of phosphorus, for example, the review lakes cover a range of measured availabilities from 0.01 to 25 µM (0.3 to 750 µg P L^{-1}). However, the scale responds to the seasonal depletion of resources in the surface mixed layer to the point of critical deficiency.

Earlier versions of this template (especially in Reynolds 1987b, c; see also overview in Reynolds, 2003a) have been used to accommodate the distributions of most of the trait-separated functional groups. The shapes drawn in Fig. 7.8a are proposed to represent the respective environmental limits of each of the algal groups, with group-S species extending furthest rightwards along the I^{**} scale (shallow hypertrophic habitatats) and

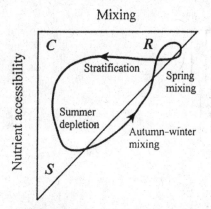

Mixing

Nutrient accessibility

C

R

Stratification

Spring mixing

Summer depletion

Autumn–winter mixing

S

Figure 7.9 Idealised year-long timetrack of the selectivity trajectory imposed by seasonal variability in a temperate system, in relation to the zones favoured by **C**, **S** and **R** primary strategists. Redrawn with permission from Smayda and Reynolds (2001).

group-E dinoflagellates furthest downwards in the K^{**} axis (depleting epilimnia of oligotrophic lakes). Less tolerant groups terminate closer to the origin in the upper left-hand corner representing resource-rich, energy-rich habitats. Superimposition of shapes, one upon another, implies superior group performances, with the fastest nutrient- and light-saturated growth rates of **X1** species dominating the top left-hand corner.

This representation is illustrative but it helps to put into perspective the floristic descriptions of many kinds of lakes on to a single two-dimensional figure. The representation in Fig 7.8b is identical to that in Fig. 7.8a, but for the stripping out of all the alphanumeric insertions. The numbered triangles that replace them are really quite effective in circumscribing the planktic flora of named lakes or lake series. These include shallow, enriched pools and well-flushed pools (1, 2), the turbid, *Plankothrix*-dominated polder lakes (3), through to oligotrophic lakes such as Millstättersee (11).

Collectively, the shapes used in Fig. 7.8 comprise the familiar triangular layout, corresponding to the disposition of the primary algal strategies (with **C**, **R** and **S** at the apices). Seasonal changes in any individual system will, to a greater or lesser extent, be tracked through the triangle, in terms of variations in I^{**} and imposed changes in resource accessibility, along the general track shown in Fig. 7.9. This purports to move through summer stratification, nutrient depletion, autumnal mixing (less light, re-enrichment and nutrient uptake during the spring mixing). In theory, at least, the trajectory inserted into each of the triangles superimposed onto Fig. 7.8b should be capable of describing the compositional changes that characterise the annual plankton cycles in the water bodies represented. At least the pattern, if not the detail, of phytoplankton structure is captured.

7.3 | Assembly processes in the phytoplankton

The purpose of the present section is to explore the active mechanisms that influence the variations in the structure of the phytoplankton in natural communities, according to circumstances and dominant functional constraints. The objective is to determine the extent of predictability of natural communities, with the processes quantified wherever possible. The key topics considered to be relevant to community assembly in the plankton – species richness and diversity, trait selection, succession and stability, structural disturbance and organisational resilience – are no different from those believed to impinge upon the community ecology of other systems. However, the dynamics of pelagic systems operate at such absolutely small timescales (when compared with those of terrestrial systems) that the outcomes are not only observable phenomena, as opposed to speculative extrapolations, but they are amenable to meaningful, controlled experimental manipulation. This is the part of the book in which the community ecology of the phytoplankton can be championed as the model for community ecological processes everywhere. Some of the terms used require clarification of usage; some working definitions are given in Box 7.1.

7.3.1 Assembly of nascent communities
The first challenge of this discussion is to establish whether the observable assemblages of different species of phytoplankton (and, for that

matter, cohabitant microbes, zooplankton and nekton) are merely fortuitous and random combinations of species that happen to be present, by chance and in sufficient concentrations to be encountered readily, or whether these are selectively biassed, invoking either mutually interspecific dependence or mutual tolerance. If the latter, then the central question becomes 'How are they put together?'

Minimal communities

The word 'community' presupposes some sort of functional differentiation, with populations of several species each doing different things but whose activities, in total, achieve some linked reactivity that defines their combined outputs – the community function. Historically, ecologists have experienced difficulties in defining the nature of communities. At one extreme, Clements (1916) anticipated a supra-organismic regulation of tightly bound organismic functions. At the other, Gleason (1917, 1927) countered that the species were present more or less by chance. Thus, species composition is fundamentally non-predictable beyond the influence of species-specific habitat preferences and the consequence of selective interspecific interactions, including predation and competition. Current thinking is rather closer to the second view than the first but the interactions are recognised to be crucial and far from straightforward. A helpful modern model is that of Ripl and Wolter (2003), which envisages community function as a series of successive organisational hierarchies, beginning with the interdependence of molecular transformations and working through to the physiological control of individual organisms and to the beneficial coordination of the separate activities of the various component organisms. They cited several illustrative examples, including the nitrogen-fixing symbionts of water plants (including diatoms) and the use of algal exudates by bacteria. They proposed the DEU (for dissipative ecological unit) as the minimal interspecific assembly within which entities contribute to an organised, functioning structure with a measurable and increasing thermodynamic efficiency. Thus, the DEU comprises five distinct components, capable of mutual coupling 'in such a

way that material cycling is internalised and, to a large extent, closed' (Ripl and Wolter, 2003, p. 300). The components are:

(1) Primary producers, which manufacture organic material and supply the energy needed for all heterotrophic structures.
(2) Water, as a moderating feedback control of production.
(3) A reservoir of organic detritus.
(4) Decomposers – bacteria and fungi – drawing energy from the oxidation of detrital stores, which process also recycles nutrients and minerals to primary producers; the store is exploited efficiently and with a low level of losses.
(5) The food web – that network of consumers comprising higher and lower animals through whose activities much of the original energetic input is finally dissipated.

It need hardly be added that the remarks of Ripl and Wolter (2003) were inspired by a view of terrestrial ecosystems. Yet, but for an alternative emphasis on the relevance of water and a tendency of open waters to store organic detritus remotely (i.e. in the sediments), the model adequately describes a self-sufficient pelagic ecosystem. Moreover, in either case, although all the biotic components are essential to the sustainable, integrated function of the whole structure, primary producers play a critical role in interceding in the dissipative flux, in synthesising the organic base and, hence, in initiating the assembly of the ecological unit. This will be taken as a sufficient justification for concentrating upon assembly processes among aquatic primary producers (and, especially, the phytoplankton) and for deferring consideration of the assembly of the heterotrophic and phagotrophic elements of the pelagic community to Section 7.3.2.

Of special interest are the answers to the questions, how many and which species of primary producer will be present? The supposition made throughout this book is that phytoplankton species will grow wherever and whenever they can, provided that (i) the supportive capacity to satisfy their minimal requirements for their growth is in place and (ii), coincidentally, viable propagules are already present and able to take advantage of the amenable conditions. Earlier chapters have probed extensively the interspecific differences in environmental requirements and tolerances but the uncertainties relating to propagule dispersion have, so far, been avoided.

Both issues are relevant to the role of species richness and diversity in assembling communities. It has to be admitted, however, that the large numbers of widely distributed species with generally rather small differences in their basic requirements do not offer clear prospects for separating critical processes. Thus, it helpful to consider the governing principles as they are expressed in the establishment phases of new, open and pristine habitats. This approach follows the powerful 'island biogeographic' concept, developed by MacArthur and Wilson (1967) to describe the processes leading to the establishment of a biotic ecosystem on a newly formed island, arising Surtsey-like from the ocean. The model translates well to the context of phytoplankton development in a small, temperate or sub-polar, inland water body, which, in this case, may be considered as an aquatic island in a terrestrial sea. Seasonality confers an element of vernal 'newness' of such a habitat. If objection be made to the bias attributable to the presence of overwintering of propagules, then the case of flood-plain lakes (varzeas) left isolated by falling rivers (García de Emiliani, 1993) or of mining subsidence pools (e.g. Dumont, 1999) could be adopted. In the open sea, the seasonal onset of thermal stratification in a water column 'sterilised' of phytoplankton during months of deep mixing also conveys the idea of open, available habitat, depauperate in indigenous planktic algae. Even highly flushed, river-fed small lakes (like Grasmere, English Lake District), may become so depleted of phytoplankton during wet periods that, effectively, they become amenable, open habitat for the next phytoplankters to arrive there (Reynolds and Lund, 1988).

The pioneer element of nascent communities

In the truly novel planktic habitat, the first colonists have to arrive, de novo, from some other external location. In other, quasi-open habitats, fluvial transport of propagules from an upstream location (including the river itself) is likely to deliver an inoculum of phytoplankton with a composition reflecting the upstream habitats that supplied it. Species that produce benthic resting spores and propagules and which were previously resident in the plankton of the flushed lake and which succeeded in recruiting a large number of overwintering propagules to the sediment prior to displacement of the water mass by throughflow will also enjoy a biassed opportunity to recruit inocula to the expanding development opportunity. In spite of this, the number of species colonising and, certainly, the relative numbers of individual organisms of each contributed is mainly a matter of chance (Talling, 1951). The larger the system and the greater its spatio-temporal connectivities, the greater is the likelihood of annually reproducible patterns.

In large and small systems alike, however, the one statement that seems true is that the species that initially become abundant either 'arrive' (with or without the benefit of a resting inoculum) in strength, or grow rapidly, or they do both. These traits are analogous to those of the pioneering, invasive, island-colonising species of MacArthur and Wilson (1967), (r-)selected by their investment in short life histories and prolific production of small, easily transmissible seeds. The corresponding phytoplankters are manifestly the small-celled, quick-growing C-strategist species (Section 5.4.5) (Box 5.1). Freshwater examples come from the X1, X2 and Y functional groups (Section 7.2.3; Table 7.1). Indeed, some of the species involved (of such genera as *Chlorella*, *Chlamydomonas*, *Chlorococcum*, *Coccomyxa*, *Diacanthos*, *Golenkinia*, *Micractinium*, *Monoraphidium*, *Treubaria*, *Westella*) are also variously encountered in small or temporary aquatic habitats, such as rain puddles of a few days' age, bird baths, rinsing water left in open containers and bottles, and in the phytotelmatic habitats of water retained in the foliage of epiphytic plants such as bromeliads. The only consistently plausible means of dispersal for propagules of these algae is through the air. The extent of spore production and the tolerance of dessication is not well known, while the viability of cells in aerosols needs further investigation. The point that needs to be emphasised is that small, potentially planktic, C-strategist algae are not just highly mobile between mutually isolated, potentially viable habitats but that, collectively, they are sufficiently abundant to have an evidently high probability of early colonisation of new planktic habitats.

The observation is relevant to one of those enduring puzzles of phytoplankton ecology, still lacking formal solution, which is just how apparently conspecific phytoplankters move among hydrologically isolated lakes and achieve near-global, cosmopolitan distributions. Many contributory investigations, for instance, those of Maguire (1963), Atkinson (1980) and Kristiansen (1996), reveal something of the variety and variability of dispersal routes for planktic organisms, especially aquatic insects and water birds. Rather less than the much-speculated 'duck's foot', guts and faeces and avian feathers retain propagules and (often) live vegetative algal cells, providing important avenues for the potential transfer of planktic inocula from one water body to another. This may apply over relatively modest distances, before desiccation or digestion destroys the viability of the potential inocula. Recent investigations by Okamura and her co-workers (see especially Bilton et al., 2001; Freeland et al., 2001) have related the spread of genetically separable clones of sessile bryozoans (small aquatic filter-feeders that produce distinctive propagules called statoblasts) to the migratory paths of wildfowl. The work makes it clear that avian movements do greatly influence the genomic composition and species richness of island freshwater communities.

Even marine phytoplankton is readily transportable via other agents. The recent diminution in the area and standing volume of the Aralskoye More, as a result of disastrous mismanagement of its hydrology over the past few decades (see e.g. Aladin et al., 1993) has been accompanied by a rise in its salinity. In the 1990s, by the time that the salinity had reached 28–30 mg L^{-1}, productivity had collapsed and all freshwater and brackish species of phytoplankton had been eliminated. According to Koroleva (1993), the depleted planktic flora included marine species of Exuviella, Prorocentrum, Actinocyclus, Chaetoceros and Cyclotella.

Each was previously unrecorded in the lake and, supposedly, recruited from the nearest sea, some 1500 km distant!

The implication of this ready colonisation by invasive species is that, for some species at least, the dispersal channels are effective and well exploited. Not all species are equally mobile: colonial Cyanobacteria such as Microcystis are relatively slow to establish in newly enriched lakes, even though the habitat might seems amenable to its growth (Sas, 1989). A number of species have very limited distributions and seem endemic to particular regions or even localities (Tyler, 1996). However, water bodies throughout most of the world, not just new ones, must be constantly assailed by the propagules of invasive species. This process helps to maintain their apparent cosmopolitan and pandemic distributions. These species turn up everywhere and, if the opportunity of habitat suitability and resource availability is there, populations may establish wherever they land.

These deductions apply to many bacteria and other microorganisms and, to a lesser extent, to mesozooplankters (Dumont, 1999). They apply at an approximately comparable level among the microzooplanktic and nanozooplanktic ciliates, studied extensively by Finlay (2002). Besides coinciding with the size ranges and body masses of dispersive phytoplankton, ciliates also occur locally in abundance and enjoy world-wide ubiquity, and, as Finlay's experiments amply demonstrate, common ecotypic species quickly establish in contrived new environments open to their colonisation from the atmosphere. Ubiquity is dependent upon mobility of propagules, which is, in turn, a function of propagule size. Then their relative transmissibilities are subject to the influence of distance and physical barriers. Mountain ranges and oceans may constrain the distributions of aquatic invertebrates and insects. Fish experience similar difficulties, even on a catchment-to-catchment basis. Aquatic mammals spread to places that they can walk or swim to. Finlay's (2002) hypothetical separation of protist species that are ubiquists from those that have biogeographies on the basis of their sizes (critical range, $\sim$1 mm) appears to hold for all pelagic organisms.

Species richness of nascent communities

Planktic colonisation of new habitat or the reopening of existing habitat through the seasonal relaxation of physical exclusions may be expressed in quantitative terms as the filling of previously unoccupied niches, or the opportunities available for the exploitation of the aggregate pool of resources. The more that the niches become occupied, the slower is the process of filling. MacArthur and Wilson (1967) formulated a differential equation (7.2) to describe colonisation, thus:

$$1/n(t)\mathrm{d}n/\mathrm{d}t = \hat{u}(\tilde{n} - n) \qquad (7.2)$$

where $n(t)$ is the number of exploitable niches in the location at a time t, $\tilde{n}$ is the number reached at $t = \infty$ and $\hat{u}$ is the rate at which they are filled. Practical solution of the equation is not possible without independent quantification of the number of potential niches ($\tilde{n}$) and experimental observations on the arrival of occupant species ($\hat{u}$). Occupancy is asymptotic to $\tilde{n}$ but its value is difficult to predict from observation, especially in the early stages of colonisation, when the resource availability (S) is likely to be large in relation to the exploitative demand (D). Thus, what may be, effectively, a single broad niche may be simultaneously and non-competitively exploited by several species, which together act as a multi-species guild, or functional group, of species (Tokeshi, 1997). However, it is readily predictable that the number of distingushable and viable niches is likely to be a function of the size of the habitat (a large lake offers more colonist opportunities than, say, a pool in a bromeliad). This is back-extrapolable from another prediction of island biogeography, pointed out by MacArthur and Wilson (1967) and supported empirically in several subsequent investigations, that there will be a positive correlation between the species richness of an island and its size. The number of species (n_{sp}) to be found in a defined area, A, conforms to the relationship

$$n_{sp} = k' A^z \qquad (7.3)$$

where k' and z are characteristics of particular categories of organisms and invasiveness. The existence of at least an approximately equivalent number of niches is a reasonable assumption. Fitted to species lists of macroscopic flora and fauna of existing islands, Eq. (7.3) has been found to account for up to 70% of the variability in species number. The slope, z, is reciprocal to the immigration rates and, in these cases, usually solves at between 0.2 and 0.35. Similar values have been obtained when the solution has been applied to the occurrences of phytophagous insects on their food-plant 'islands', though values are conspicuously influenced by the abundance (thus mutual proximity and accessibility) of hosts (Janzen, 1968; Lawton and Schröder, 1977).

Smaller values of z (0.1–0.2) represent enhanced opportunities for immigration. Interphyletic differences in immigration rates to given locations are also apparent. For instance, birds have lower z values than land snails. Finlay et al. (1998) have worked out that the slope z for freshwater planktic ciliates is yet smaller again ($z = 0.043$) as a consequence of their rapid rates of dispersal.

If a low value of z in Eq. (7.3) reciprocates a high initial dispersive value of $\hat{u}$ in Eq. (7.2), the potential richness of planktic species may approach the maximum ($\tilde{n}$) that the habitat-size constraint will allow. Certainly, in those instances where investigations of species structure has been sufficiently competent and exhaustive, the species richness of individual lake systems is demonstrably large. Dumont and Segers (1996) estimated the likely maximum species richness of cladocerans to be ∼50 and of rotifers to be ≤270. Representation by a significant proportion of the extrapolated global total of ∼3000 species of free-living ciliates is probable (Finlay et al., 1998). Compared to a supposed potential of 4000 to 5000 taxa, 435 named species of freshwater phytoplankton in tropical Lake Lanao were noted by Lewis (1978). Consecutive daily samples of the phytoplankton of shallow Lake Balaton accumulated a list of 417 separate taxa in 211 days (only ∼300 were considered to be truly planktic: Padisák, 1992). The accumulated species list of phytoplankton encountered in Windermere (Reynolds and Irish, 2000) includes 146 named taxa. In samples of phytoplankton from the upper 100 m of the central North Pacific over a 12-year period, Venrick (1990) recorded 245 species.

7.3.2 Properties of accumulating communities

Following the establishment (or re-establishment) of an assemblage of colonist primary producers, many subsequent events are set in train. Further species of primary producers may arrive and respond positively to the favourable growth conditions. Total biomass increases and individuals begin to impinge on each other's environments, inevitably setting up interspecific interactions. Moreover, the accumulation of producer biomass, detritus and organic waste soon constitutes food and substrate for the heterotrophs (phagotrophic herbivores, microbial decomposers) and the opening of primary product as a resource to consumer components of the DEU (Ripl and Wolter, 2002); the functional community is born!

Several aspects of this development command our attention. Intercepting an increasing portion of the solar flux, the accumulating system involves the deployment of greater levels of power and resources. Sooner or later, constraints in the supply of one or the other feed back, invoking adaptive reactions in the pattern of community assembly. In turn, there are consequences for species composition, richness, dominance and diversity, and key species thus favoured characterise the overall community function. These developments are sometimes referred to collectively as community 'self-organisation' (or *autopoesis*: see, for instance, Pahl-Wostl, 1995). They are, indeed, contingent upon behaviours and mutual interactions at the basic levels of community assembly (the predilections and adaptabilities of the individual participating species) and the aggregate of which is the structure and future of the emerging community. The process of increasing system throughput, information and complexity of network flows is also known as ascendency (Ulanowicz, 1986) (see also Box 7.1).

Ascendency: power and exergy

These concepts are not difficult to grasp if the appropriate analogies are used. Biological systems are organised about the managed flow of organic carbon, which is powered ultimately by short-wavelength, photosynthetically active solar energy. The role of the phototrophic primary producers is to intercept, harvest and invest energy in the production of high-energy organic carbon compounds. This process is continuously subject to the laws of thermodynamics. The second of these determines that the direction of flow is from the concentrated to the diffuse and that short-wavelength energy is dispersed irrecoverably at longer wavelengths (i.e. as heat). Short-wave electromagnetic solar radiation falling on a lifeless planet causes molecular excitation and dispersion (a rise in the surface temperature) but most is back-reflected to space at longer wavelengths. This dissipation of heat is irreversible. It follows that the effect of a pulse of energy is to reduce order and increase randomness, or, in thermodynamic terms, to raise the entropy of the recipient structures.

Against the inevitability of entropic dissipation, the ability of growing communities to assemble structure, order and complexity seems, at first sight, to be in contravention of the laws of the universe (Prigogine, 1955). In truth, ecosystems are an integral part of the thermodynamic system, doing little more than to reroute and regulate the velocity of the dissipative flux. A reasonable analogy is that of a waterwheel located on a stream (Reynolds, 2001a). Some of the kinetic energy of flow is drawn to drive the wheel (the energy-harvesting primary producers), whose motion may be exploited as a source of alternative mechanical or electrical power (the ecosystem). However, the overall direction of the flow is unaltered.

Moreover, the proportion of the available power that is realised in assembled biomass is really quite small. Chlorophyll harvests no more than ~6% of the radiative flux to which it is exposed (at sub-saturating levels); barely 1.5–2% of the available energy is captured in carbon fixation (Section 3.5.3). Much the largest proportion ($\leq$95%) of the solar flux is consumed in non-biological (but often biologically important) processes, such as heating the water and driving evaporation and cooling. The investment of the photosynthetically active radiation actually absorbed into plant biomass, typically $\geq$0.08 mol

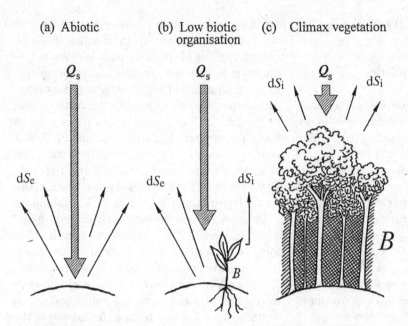

(a) Abiotic   (b) Low biotic organisation   (c) Climax vegetation

Figure 7.10 Ecosystems and the dissipative energy flux. In an abiotic world (a) all the incoming short-wave radiation, Q_s, is reflected or reradiated at longer wavelengths (i.e. as heat). This entropic disssipation is noted as dS and, in (a), is external to living organisms (hence dS_e). The intervention of photoautotrophic organisms intercepts and reroutes a fraction of the solar flux to drive the assembly of biogenic material. Much of this eventually shed in respiration and decomposition (as dS_i), in conformity with thermodynamic law, but only after the captured energy has been invested in the assembly and accumulation of biomass. This flux is equivalent to system exergy. A positive exergy flux enables ecosystems to build, perhaps to the point where all the entire solar flux is dissipated as dS_i. B, biomass. Redrawn with permission from Reynolds (1997a).

C (mol photon)$^{-1}$, or 0.2 MJ (g C)$^{-1}$, again at sub-saturating fluxes (Section 3.2.3), implies that the power requirement to assemble producer biomass is around 20 MJ (g C)$^{-1}$. There is a further energetic cost to assembling and maintaining the structure, which is met from the controlled metabolic re-oxidation (R) of part of the accumulated carbon. When producer biomass becomes part of the biomass of its consumer, some 40–90% of the power investment may be dissipated as heat. This is repeated at each successive trophic level. Alternatively, the cadavers and waste products may be exploited by decomposers for the last molecules of organic carbon and the last vestiges of the power investment. Whether we take

the community, the DEU or the ecosystem, these structures dissipate energy, just as surely as the abiotic environment, and, ultimately, they do so to the point of balanced exchanges.

The apparent affront to the second thermodynamic law is that the inward flux may not be dissipated at once but, instead, be retained within molecular bonds. While these persist, they block entropy. In effect, the biological system is siphoning off a part of the energy flux into a loop of reduced carbon, whence its eventual release is regulated and delayed. This entropy-free potential has been variously referred to as negentropy (for negative entropy) or *exergy* (Mejer and Jørgensen, 1979). Equally, it has been conceptualised in different ways, in terms of information stored (Salomonsen, 1992) or the size of the accumulated gene pool (Jørgensen *et al.*, 1995; Jørgensen, 1999).

Here, the preference is to adhere to expression in units of energetic fluxes (cf. Nielsen, 1992). Then, one of the preconditions for community ascendency is that the rate of energy harvesting and deployment in organic carbon synthesis shall exceed the aggregate of the requirement of internal maintenance and its dissipative losses. The cartoon in Fig. 7.10 illustrates the increasing biomass of an ascendent community through time, from bare ground to high forest. From wholly dissipating the flux of short-wave radiation (Q_s) as heat (units, W m^{-2} or J m^{-2} d^{-1}), the intervention of biological systems is to reroute part of the dissipative flux (dS) into the building (P) and maintenance (R) of biomass (B). That part of the abiotic flux that is now passed through organisms is now distinguished as (dS$_i$), leaving a reduced portion (dS$_e$) of the original abiotic flux external to the organisms. For B to accumulate, the sum (dS$_i$) + (dS$_e$) has to be smaller than the original flux, dS. As biomass is accumulated, R increases absolutely. Although more energy is siphoned off, absolutely more (and, eventually a bigger proportion) is dissipated as heat (dS$_i$). The achievable steady state is when the biotic component harvests the entire energy flux (dS$_e$ → 0) and consumes it exclusively in its own maintenance (R = dS$_i$ → dS). No further increase is then possible: the community has achieved its highest state (Fig. 7.10c).

Data are not available to quantify all the relevant fluxes. However, the principles can be illustrated by adopting the light-harvesting properties and phototrophic growth of the alga *Chlorella* (from Fig. 3.18) to scale the theoretical ability (and diminishing efficiency) of the producers to harvest and allocate the maximum energetic flux. The revised plot, in Fig. 7.11, shows how low, pioneer levels of biomass harvest more energy than they expend and that the resultant positive exergy flux can be allocated to increasing the energy-intercepting biomass. The size of the waterwheel paddles has been increased! Maintaining more biomass carries a higher cost (it is shown as a linear function of biomass in Fig. 7.11) but, as more light-harvesting centres that are placed in the light field begin to shade each other out, the harvesting return per unit investment declines asymptotically. The limiting condition comes when the cost of maintenance can no longer be balanced by the harvested income and the exergy flux falls to zero.

Long before that condition is reached, ascendant accumulation is led by the recruitment and expansion of species that contribute most strongly to the aggregate biomass. These are not necessarily the fastest growing (cf. Fig. 5.19) but, in the early stages of accumulation, the greatest rates of expansion are likely to come from the fastest-growing of the available pool and which

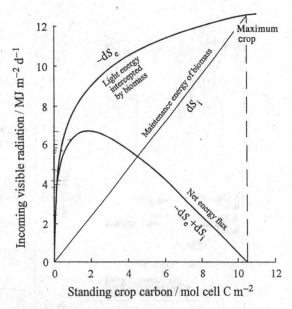

Figure 7.11 Graphical representation, based on Fig. 3.18, of the relationships sketched in Fig. 7.10. The vertical axis represents the incoming short-wave radiation. The curve represents the proportion that is 'siphoned off' into primary production, thereby (temporarily) reducing the external dissipative flux (hence, −dS$_e$). The efficiency reduces with increasing mutual interference of the producer light-harvesting but whose the maintenance costs (dS$_i$) are supposed to be proportional to the biomass. The energy difference between the curves (−dS$_e$ + dS$_i$) is available to support growth, reproduction and the ascendency of the system. The difference between the harvested energy and the internally dissipated flux of a system is equivalent to its *exergy flux*. For more information see text. Redrawn with permission from Reynolds (1997a).

are relatively most capable of overcoming the effects of small inoculum sizes. Once again, r-selected C-strategist species can be seen to hold a fitness advantage in aspiring to dominance.

Ascendency: environmental constraints

Some changes in community metabolism and resource partitioning through a phase of biomass accumulation in the Blelham enclosures (Section 5.5.1; Fig. 5.11) are illustrated in Fig. 7.12. The large panels summarise the variations in the main contributing species to the phytoplanktic biomass in the enclosure (Fig. 7.12b, as biomass carbon) and their rates of population change (Fig. 7.12c: shading separates the reconstructed

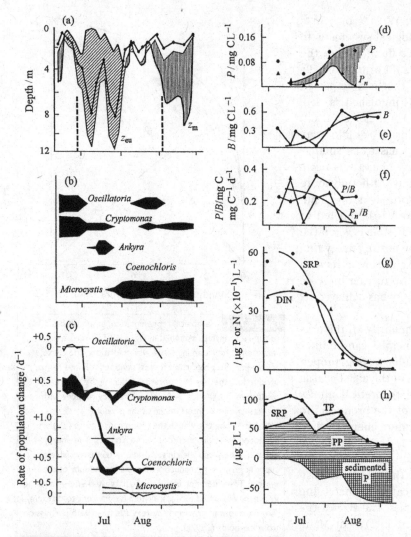

Figure 7.12 Production and dynamics of phytoplankton recruitment in a Blelham Enclosure (A, 1983), during a period (a) of weak mixing (*vide* z_m) and following high transparency (z_{eu}) and artificial fertilisation with nitrogen and phosphorus. Changes in the biomass of prominent species of phytoplankton are shown in (b) on a common scale of carbon concentration. The rates of net population change (plus or, below the horizontal line, minus) are shown in (c), relative to reconstructed growth rates, after correcting net rates for grazing losses. Changes in the net daily photosynthetic rates, measured directly (P) and as the rate of cell recruitment by growth, before grazing and sinking (P_n), each specific to the upper 5 m of water, are shown in (d) (curves fitted by eye). Changes in the standing biomass (B) are shown, and similarly summarised in (e). Productivity trends (P/B, and especially, P_n/B), are shown in (f). Changes in the concentrations of SRP and DIN (= nitrate + ammonium nitrogen) (g) reflect uptake by the phytoplankton. Changes in the stock of phosphorus (TP) in the water column and its apportionment between soluble (SRP) and particulate (PP) components are shown in (h) together with the aggregate of phosphorus collected in sediment traps. Original data of Reynolds *et al.* (1985), as reworked and presented in Reynolds (1988c) and redrawn with permission from Reynolds (1997a).

growth rate from net rate of positive or negative change), in relation to day-to-day variations in the depth of the mixed layer (z_m) and the depth of the conventionally defined photic zone (to 1% penetration of the immediate subsurface visible irradiance, z_{eu}). At the end of June 1983, the onset of a phase of very stable thermal stratification coincided with the rapid depletion in the dominant populations of *Planktothrix* (sinking and photooxidation) and *Cryptomonas* (to grazers) and an attendant increase in transparency. The enclosure was then well fertilised with solutions of nitrate, phosphate and silicate. The responses of a now-diminished phytoplankton to this relatively sharp increase in carrying capacity was followed over the next two months, after which cooler and windier conditions obtained. The changes to the community during July and August are presumed to have been substantially autogenic (self-imposed), working within the limits of the constraints of the carrying capacity of the light and nutrient resources.

At the metabolic level, gross volume-specific primary production (P, in mg C L^{-1} d^{-1}) as indicated by conventional radiocarbon fixation rates (data of Reynolds et al., 1985) and phytoplankton biomass (B, as mg C L^{-1}) increased over the two-month period (Fig. 7.12d,e). However, production net of respiratory losses (P_n) and productivity (sensu production normalised to biomass, P/B and, especially, P_n/B) decreased (Fig. 7.12 d, f). The increase in biomass took place at the expense of the nutrients in solution (Fig. 7.12g). Removal of dissolved phosphorus, ostensibly into phytoplankton, was effectively complete by early August, while the general decline in the total phosphorus fraction is explained by the sedimentary flux of planktic cadavers (mostly of the animals that had grazed on the phytoplankton).

At the compositional level, the dominance of the phytoplankton changed twice – the leading contender for the opening resource spectrum was Ankyra, whose increase in biomass was initially supported by rapid net specific rates of increase of ~0.86 d^{-1} (reconstructed cell replication rates, r' ~1.1 d^{-1}). These were not sustained and the population was reduced by zooplankton feeding, especially by filter-feeding Daphnia. The second dominant was the non-motile colonial chlorophyte Coenochloris, whose net rate of increase ($\leq$0.43 d^{-1}) was less sensitive to grazing but nevertheless weakened as z_{eu} diminished in relation to z_m. Microcystis increased steadily through July and early August ($\leq$0.24 d^{-1}), achieving dominance in place of Coenochloris. Its biomass persisted, even though there was almost no net recruitment from phytoplankton growth.

The example illustrates not just the increasing burden of accumulating biomass, as loss rates balance or exceed the rates of its recruitment, but shows well the increasing interference of the major constraining forces on energy-harvesting and resource-gathering with the process of recruiting new biomass. These effects carry other interesting hallmarks of autogenic change in developing communities. The change in composition, first in favour of strongly r-selected C-strategist species, such as Ankyra, towards slower-growing, strongly K-selected, biomass-conserving S-strategist species such as Microcystis, is reflected in the changing size spectrum of the population (Fig. 7.13a). During early ascendency (take the window for 15 July), the major part of the biomass resides in the smallest organisms: virtually all are individually smaller than 10^4 μm^3, and over 50% of them are each less than 10^2 μm^3). Towards the end of the accumulation phase (9 August and persisting until October), the major fraction is made up of units each >10^5 μm^3. Moreover, the distribution of biomass among individual species also declines, with an increasing proportion of the primary productive potential of the community residing in the assembled biomass of a decreasing number of species (Fig. 7.13b). The tendency is towards 1, as the best-adapted species in the pool progressively outperforms, to their eventual exclusion, all the others competing for the same, diminished resource. This is an issue not of species richness but of species diversity, to which the discussion will return (see p. 364).

Succession

Distinctive patterns in the developing metabolism and community organisation of major biological systems, manifest as an ordered sequence (or succession) of substitutions of species, have long been recognised by plant botanists and geographers (Margalef, 1997). Early-twentieth-century plant ecologists were espoused to the idea of an internally controlled process, culminating in a final, climactic stage of maximal persistent biomass. Moreover, some well-characterised successions (for instance, from bare soil to high forest: or the hydroseral stages of vegetation development at the edge of a lake, passing from swamp to marsh or fen through to Quercus forest: Tansley, 1939) have sharpened

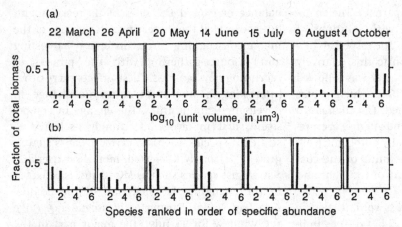

Figure 7.13 (a) The changing size distribution of phytoplankton in Blelham Enclosure A, traced through the allocation of the total live biomass among units (cells, coenobia or colonies) in the size categories shown. Note the predominance of small cells (mostly of *Ankyra*) on 15 July, during the early part of the sequence depicted in Fig. 7.12, and the predominance of larger algae later in the year. (b) The changing apportionment of biomass among species, moving from being invested in the first two to four co-dominants to being almost wholly invested in the first. Original data of Reynolds (1988c) and redrawn with permission from Reynolds (1997a).

a resolve to find deterministic explanations. This quest has had to reject some well-intentioned but erroneously rigid statements of successional governance. At the other extreme, understanding of succession has not been helped by the use of this term by students of the pelagic to refer to all temporal changes in the species composition and the abundance and relative dominance of the plankton (Smayda, 1980). The fact that succession 'conforms to no plan' and is mostly concerned with the relative probabilities of certain possible emergent outcomes (Reynolds, 1997a), might overcome the need for a precise explanation of its mechanics.

However, some aspects of the concept of succession are too valuable to reject. Change in community structure, in the plankton as in systems involving higher plants and animals, has a number of quite separate drivers. These may owe to *allogenic*, non-biological, physical forcing mechanisms (earthquake, fires, storms and floods) that destroy structure or modify the environment in favour of tolerant species. On the other hand,

changes wrought by the activities of individual organisms, to the extent that they alter the environment so that it becomes more amenable to the establishment of individuals of other species, are, in broad terms, essential and predictable aspects of succession (Tansley, 1939). 'Succession' should be reserved to refer to these wholly autogenic responses of the community (see also Reynolds, 1984b).

Successional changes, contingent upon autogenic drivers, are the principal manifestation of ascendency in developing ecosystems (Odum, 1969). We may accept 'autogenic succession' and its attributes, as circumscribed by Odum (Table 7.7), as being symptomatic of ascendency. Several of the stated attributes plainly apply to the events depicted in Fig. 7.12. The 'causes' of succession are the reactivity and interactions of the pool of the species to the environmental opportunities presented. It is these that may eventually lead to structures that, subject to the turnover of resources, achieve a steady state of energetic balance, to which the notion of successional climax is fully applicable.

Filtration and community assembly

Where strict successional selection fails to explain the mechanisms underpinning reproducible sequences, then we need to invoke another model. The opposite view from positive selection invokes the premise (see p. 351) that most species can grow anywhere they arrive at, provided the habitat is suitable and can sustain their growth and survival needs. Then, the habitat that is, or becomes, unsuitable will

Table 7.7 | Attributes of early and maturing stages of autogenic succession

Ecosystem attribute	Early stage	Maturing stage
Community energetics		
1. Gross production/community respiration (P/R)	(P/R) > 1	(P/R) → 1
2. Gross production/biomass (P/B)	high	low
3. Biomass supported/unit energy income (B/Q)	low	high
4. Net production yield (P_n)	increasing	low
5. Food chains	linear	web-like
Community structure		
6. Total organic matter	small	arge
7. Inorganic nutrients	extrabiotic	intrabiotic
8. Species diversity	rising	high, possibly falling
9. Species equitability	decreasing	low
10. Biochemical diversity	low	high
11. Structural diversity	poorly organised	well organised
Life histories		
12. Niche specialisation	broad	narrow
13. Organismic size	generally small	generally large
14. Life cycles	short, simple	long, complex
Nutrient cycling		
15. Mineral cycling	open	substantially closed
16. Exchanges among organisms and environment	rapid	slow
17. Role of detritus in nutrient	unimportant	important
Selection		
18. Growth selection	r	K
19. Production	for quantity	for quality
Community homeostasis		
20. Internal symbiosis	undeveloped	developed
21. Nutrient conservation	poor	good
22. Response to external forcing	resilient	resistant
23. Entropy	high	low
24. Information	low	high

Source: Based on Odum (1969).

simply select against the persistence of intolerant species. In this way, environments act as a sort of filter, separating off the ill-adapted species from those whose traits and adaptations allow them to pass to survival. For instance, the potential of aquatic habitats to support aerobes is constrained by a limited, finite supply of oxygen. To be able to live in water, organisms are required to be functionally or physically adapted to ensure that diffusion gradients are adequate to the needs that their size and activity dictate.

Then the organism can survive to the limits of its own adaptive capabilities. These may be superior or inferior to those of the next species and these may well determine which of them can continue to function under persistently low oxygen concentrations. Pelagic environments pose many and more subtle conditions and constraints to their exploitation by aquatic organisms and the putative systems to which their presence contributes. These may well operate simultaneously, as filters of varying coarseness, to favour or disfavour the

Table 7.8 | The rules of community assembly in the phytoplankton, proposed by the International Association of Phytoplankton Taxonomy and Ecology (see Reynolds *et al.*, 2000b) with minor modifications

(1) Provided suitable inocula are available, planktic algae will grow wherever and whenever they can and to their best potential under the conditions obtaining.

(2) Then, of those present, the species initially likely to become dominant are those likely to sustain the fastest net rates of biomass increase and/or to be recruited from the largest inocula ('seed banks').

(3) The largest autochthonous inocula (seed banks) are furnished by species that have been abundant at the location in the recent past.

(4) Environments may select, preferentially and with varying levels of intensity, for certain specific organismic attributes or traits.

(5) Species with preferred attributes are likely to build bigger populations than those that lack them and, where appropriate, to found larger inocula to carry forward.

(6) Plankton assemblages become biassed in their species composition by the conditions typically obtaining in the host water body.

(7) The species most frequently present in specific environments share common suitable attributes.

(8) The outcome of assembly processes may be subject to the food web and to other interspecific interactions.

(9) Of those species present (and quite independently of the initial conditions), the ultimate dominant is likely to be the one most advantaged by its adaptive traits.

(10) Assembly is always subject to the overriding effects of environmental variability and the resetting of the assembly processes.

relative abundances of each of the given species. The most decisive of these criteria becomes the finest of the filters operating, selecting for the fittest of the species available (Reynolds, 2001a).

In this way, the fortuitousness of the species composition of the various functional components of a community and the stochastic nature of its early dominance is squared against the increasing predictability of the outcome of autogenic succession. The mechanisms have yet to be fully elaborated. My use of the word 'filtration' (Reynolds 2001a) was inspired by the reference of Lampert and Sommer (1993) to 'environmental sieving'. However, the crucial role of specific organismic traits in, as it were, determining the filterability of species that may then participate prominently in community function is provided by the seminal work of Weiher and Keddy (1995; see also Weiher *et al.*, 1998). Their formulation of the guiding principles and rules governing community assembly has attracted the concurrence of others (see for, instance, the papers of Belyea and Lancaster, 1999; Brown, 1999; Straškraba *et al.*, 1999). Here again, the value of the short timescales of community processes in the pelagic have permitted the substantial verification of the role of trait selection in the development of pelagic communities (Rojo and Alvares Cobelas, 2000). A definitive set of rules guiding the assembly of communities in the plankton has been proposed by the International Association of Phytoplankton Taxonomy and Ecology (see Reynolds *et al.*, 2000b) (see also Table 7.8). These repeat some statements offered earlier in this chapter.

It has already been established (Chapters 5 and Sections 7.2.2 and 7.2.4) that the principal correlatives of community composition are classifiable as resource constraints and energy constraints (e.g. Fig. 7.8). It is now a simple exercise to suggest the assembly mechanisms underpinning the distribution of trait-distinguished functional groups against these constraints. The

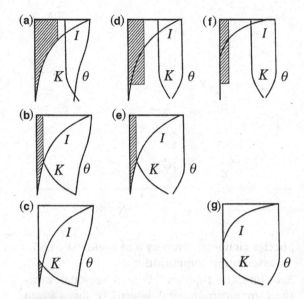

Figure 7.14 Idealised environments, defined in terms of downwelling solar irradiance (*I*) and the capacity-limiting bioavailable resource (*K*) in relation to mixed depth (represented by the temperature gradient (θ)). The potential biomass and its distribution is shown by shading. In the top left diagram (a), resources and, at the top of the water column, growth is saturated at the maximum potential of the producers present. Working downwards, many water columns correspond to (b), wherein resources are so scarce or depleted that they constrain the productive potential, possibly as far as to the metalimnion (c). Working rightwards, deeper mixing (d) relative to diminishing light penetration (f) sets the major functional constraint. Most systems probably fluctuate around a condition close to (e) but the inference of deep mixing and nutrient deficiency (g) as being untenable to primary producers is clear. Based on an original figure of Reynolds (1987b) and redrawn with permission from Reynolds (1997a).

representation of pelagic habitats shown in Fig. 7.14 is now well travelled and a little dated but it serves to illustrate how temporal habitat changes in light and nutrients intensify species filtration of species traits. The individual subfigures (a–g) making up Fig. 7.14 represent vertical profiles of light availability (*I*) and nutrient concentration (*K*), in relation to temperature (θ, included in this instance only as a correlative of structure and vertical mixing). In the first instance (Fig. 7.14a, in the top left-hand corner), nutrients and, at least near the surface, light, are shown to be capable of saturating the immediate growth requirements

(shown by hatching) of most planktic species present. Thus, they furnish an ascendent opportunity, unconstrained save by their own exploitative capabilities. Interspecific competition for the opportunity is weak but, in accord with Rules 1–3 (Table 7.8), the species that become most prominent will be the ones generating the highest exergy and investing in the fastest relative rates of growth. Working downwards to Fig. 7.14b, reduction of the base of the potentially critical nutrient to growth-limiting levels (Section 5.4.4) demands traits of high uptake affinity for the limiting nutrient, or the ability to exploit other less-accessible resources, which may be shared by relatively few of the contesting species. Once nutrient is exhausted from the upper column, the only exploitable resources are located deep in the light gradient (Fig 7.14c), requiring yet greater specialist organismic traits to be able to use them and, therefore, fewer species able to compete for them (Rules 5, 6 and 9).

Working rightwards in Fig. 7.14, the light-harvesting opportunities are diminished by weakening stratification and increased mixing depth. Light-harvesting opportunities are diminished in consequence, although they are extended and rendered more uniform through the entire mixed layer (Fig. 7.14d). Ascendent biomass also impinges increasingly upon light penetration: increased depth of mixing and raised turbidity levels impose greater adaptive efficiency of the light-harvesting apparatus and enhanced ability still to cream off the remaining exergy generation into biomass accumulation (Fig. 7.14f).

It has to be said at once that these are idealised states, none of which obtains constantly. Real systems show variability in both directions and may hover around conditions represented by Fig. 7.14e However, the inference of simultaneous segregation of the nutrient base from the energy-harvesting opportunity is supports only a desertified habitat (Fig. 7.14g), corresponding to temperate oceans in winter (Fig. 1.8).

The arrangement in Fig. 7.14 also picks up the layout of the *C–S–R* triangle, serving to amplify both the conditions and (for fresh waters) the notional ranges of some of the trait-distinguished functional groups (cf. Figs. 7.8, 7.9). For systems that are chronically resource-deficient (K^{**} always

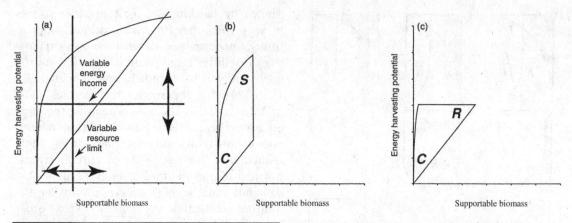

Figure 7.15 The impact on ecosystem energetics of chronic resource deficiency and low energy income, based on their representation in Fig. 7.11. The basic layout is shown as being constrained by axes delimiting low or variable resources or energy income, which, respectively reduce the exergy to the smaller shapes in (b) and (c). The preferential trend towards *S* or *R* dominance is predicted. Redrawn with permission from Reynolds (2002b).

low) or, equally, in which harvestable light (I^{**}) is habitually diluted by mixing or by turbidity, the selective bias is strongly pressed against *C* strategies and towards those of either **S** or **R** species. Moreover, the bias of inocula serves as a positive feedback to influence the floristic composition of the respective habitats.

The link to the assembly of communities in these habitats is also simply represented through reference to Fig. 7.12. If we take the segment enclosed by the curves representing potential harvest over cost as the potential for ascendent investment and subject it to persistent but variable constraints in either energy or resource, two new shapes may be derived income (Fig. 7.15a). Chronic resource limitation of supportable biomass lessens the likelihood of frequent limitation of growth by harvestable light (Fig. 7.15b). High exergy can be maintained by species that cope with resource limitation (and are highly sensitive to altered loadings of critical nutrient). For systems that are energy-limited, the biomass is unconstrained by nutrient whereas it can rise to the level of the limit of the exergy harvest by the most efficient users of the available light (Fig. 7.15c). The respective selective biasses in these extremes favour *S*- and *R*-strategist species.

Species richness, diversity and evenness in assembling communities

Recalling the representation of seasonal variations in environmental selectivity for a given individual water body (Fig. 7.9), the *C–S–R* triangle can be used to match the successional trajectories of favoured species (or at least their respective trait-selected groups) to the extent and direction of environmental variation. Some well-documented pathways are plotted in this mode in Fig. 7.16, relating compositional changes to the drift in nutrient and light availability consequential upon community ascendency. This is not prescriptive but purely an example of the kinds of self-imposed 'decision points' that may steer the assembling community towards particular outcomes. Starting at a point X, representing initially non-limiting nutrients and light, the assembling community is more likely than not to be founded upon the primary production of relatively fast-growing, *r*-selected nanoplanktic species of one or other **X** algal associations. Early dominance of these organisms is not bound to collapse: two possibilities for perpetuation are included in Fig. 7.16, where dilution by nutrient-rich throughflow ('FLUSH') and avoidance of consumers ('NO GRAZE') permit the endurance of the *status quo*, provided nutrients and light remain freely available. Modest, delayed or selective grazing in water columns sufficiently deep to be subject to self-shading may operate in favour of self-regulating cryptomonads or (eventually) of efficient energy harvesters such as attenuate diatoms and filamentous Cyanobacteria. The greater becomes the light constraint, the more

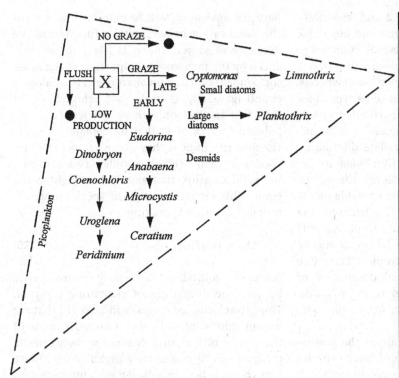

Figure 7.16 Some pathways in the development of phytoplankton composition, with some key 'decision nodes'. Starting at **X**, the predicted initial establishment of r-selected, invasive **C**-strategists may be perpetuated by flushing or may otherwise persist if there is no grazing. Early imposition of grazing may force the dominance of sequences of larger and, eventually, more versatile phytoplankters (*Eudorina* → *Ceratium*); otherwise, diminution in light availability pushes dominance towards energy-efficient **R**-species. If resources are not replaced, or are chronically deficient, sequences move vertically towards specialist resource-gleaners and, possibly, to the dominance of picophytoplankton. The enclosure within a triangle ties these temporal patterns to the habitat template. Redrawn with permission from Reynolds (1997a).

intense is the competition for the available photons, the tighter is the filter and the less is the diversity of still-functioning species. The model reconstructions of increasing light constraints on the comparative growth rates of *Chlorella*, *Asterionella* and *Planktothrix* also amply predict this outcome.

Alternatively, the early onset of heavy grazing is likely to work against self-shading and directly in favour of larger algae. Comparative growth rates, inoculum size and the relative intensity of developing constraints in the resource supply (phosphorus, nitrogen and carbon flux) each influence the compositional trends among these algae (whether as a consequence of high affinity for phosphorus, the metabolic flexibility to fix nitrogen or to resort to mixotrophy, mobility to glean or scavenge the resources available or the ability to enhance carbon uptake). The greater becomes the particular resource constraint, the more important are the specific means of its retrieval. Again, the filter tightens and the number of (increasingly *S*-strategist) species still able to function diminishes. In this way, the developing interactions increasingly exert the selective pressure in favour of the late successional participants and against the weaker competitors for the critical resource. The system moves towards a steady state, because its rate of building and the cost of its maintenance are balanced and, for so long as the constraints remain, no better competitor is available to contribute new exergy. The community has, in fact, achieved its potential, analogous to its successional climax (Reynolds, 1993b, 1997a; Naselli-Flores *et al.*, 2003). The outcome is not altogether random: the overwhelming dominance of the best competitor for the most severe constraint may be anticipated, in part, from a knowledge of the properties of the habitat (deep or shallow, hard- or soft-watered, phosphorus-rich or phosphorus-poor). However, the outcome is fashioned entirely by the assembly process. Community development pushes the selective pathway towards one or other of the selective apices (*S* or *R* in Fig. 7.9), where the growth of only a few specialist-adapted species remains possible. It is as if the richness of species potentially able to contribute to the late stages of community assembly is increasingly 'squeezed out' as the apices are approached. The patterns analysed in Section 7.2 and, especially, among some of the mesotrophic and eutrophic lakes (Section 7.2.4) provided several examples of quasi-steady states of extended dominance by one (or a very small number of) persistent species. These are, moreover, considered to be typical for the particular conditions in particular groups of lakes. These include the sequences culminating in long phases of dinoflagellate dominance of nutrient-depleted epilimnia (*Peridinium* in Kinneret, *Ceratium* in Rostherne Mere); the formation of deep chlorophyll maxima in stable metalimnia (as *Planktothrix rubescens* in Zürichsee); the association of dominant populations of *Anabaena* spp. or *Cylindrospermopsis* with hydrographically stable periods in N-deficient tropical lakes (see p. 344); and the near-perennial dominance of highly enriched, shallow polder lakes by *Planktothrix agardhii* and/or *Limnothrix redekei* (pp. 345, 349).

There may be little doubt about the dominance but competitive exclusion of less-fit species by those more adaptable is rarely so complete that small numbers of excluded species or (especially) their propagules do not persist within the system for many years. Here, they constitute a sort of system 'memory', whence they may, following some future environmental modification, have the opportunity to reassert their abundance (Padisák, 1992). Species richness is thus a less sensitive barometer of the structural organisation of communities through succession than is the relative distribution of the biomass among the species present.

Ecologists express the diversity and equitability (or evenness) of species assemblages using terms borrowed from Shannon's (1948) theory of information. Pielou (1975) proposed that the most useful estimate of biotic diversity (H'') comes from the specialised form of the function (from Shannon and Weaver, 1949):

$$H'' = -\Sigma b_i / B \log_2(b_i / B) \tag{7.4}$$

where B is the total biomass, and b_i is the biomass of the ith of the s species present. H'' increases with the number (richness) of species. Thus, we should expect the biotic diversity to be a continuous function of richness, such that $H''_{max} = \log_2 s$. The majority of the those species will, as we

have acknowledged, will be extremely rare. Thus the value of s becomes partly a function of the diligence of search and it is likely to be influenced by the taxonomic competence of the assessor. Students of phytoplankton elect to adopt a cut-off (generally, though quite arbitrarily, that they will include only those species each contributing >0.5% of B) that immediately weights the diversity indices they quote against the rare species in their original samples. This being so, the valid measure that should be sought is the equitability (or evenness) of the species representation (E_s), which approximates to:

$$E_s = H'' / H''_{max} \tag{7.5}$$

Scores for equitability are rarely quoted, partly because the difficulties of estimating s persist. Thus, the calculated diversity indices (H'') that are usually quoted provide only a proxy estimate of the course of temporal changes or between-site comparisons of community organisation. There have been other, less well-used approaches to capturing mathematically the drift in species equitability. For instance, Margalef's (1958) index of diversity, H_s, simply relates the number of species (s) to the total number of individuals (N). Applied to a fixed size of sample (volume of water), the diligence-of-search criterion is cut off in an arguably less arbitrary way. However, it has a transparent sensitivity to increased biomass and the number of species that are common. The units are bits (of information) per individual:

$$H_s = (s - 1) / \log_e N \tag{7.6}$$

Margalef's index is used to construct the quantified analogy to the changes in species evenness through the summer accumulation phase in a Blelham enclosure (Section 5.5.1) (Fig. 5.11) and the progressive onset of *Microcystis* dominance through exclusion, shown in Fig. 7.17. The observed changes in contributory specific biomass (shown in the upper part of the figure) and the rates of community change (σ_s), calculated over 3- to 4-day intervals are included for comparison. The latter is calculated, according to Equation (7.7), based on the original derivation of

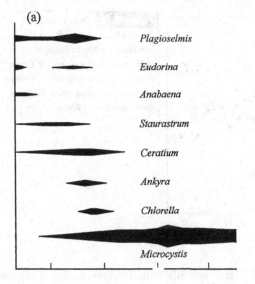

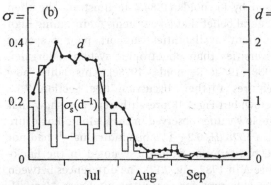

a 'succession rate index' of Jassby and Goldman (1974b):

$$\sigma_s = \Sigma\{[b_i(t_1)/B(t_1)] - [b_i(t_2)/B(t_2)]\}/(t_2 - t_1) \quad (7.7)$$

where B is the total biomass and b_i is the biomass of the ith species on each of two occasions (t_1 and t_2), so that σ_s is averaged over the period that separates them (units, d^{-1}). The plot in Fig. 7.17 confirms the increase in the number of participating species in the early phases (H_s: 1–2 bits per individual) and that diversity is kept up while species composition varies considerably (σ_s: 0.1–0.3 d^{-1}). Once *Microcystis* becomes abundant and

in a steady state of overwhelming dominance, reflected by a very low rate of community change ($\sigma_s < 0.03$ d^{-1}), any residual evenness is lost.

In spite of the implicit shortcomings of the Shannon–Weaver index, it continues to be much the more commonly used. Moreover, experience indicates that the application of the cut-off reveals very well the change in individual species contributions to biomass through the course of succession. For instance, it is a well-attested principle that, in a fully operational ecosystem, 'most' (≤95%) of the biomass at each functional level (primary producer, herbivore, decomposer, etc.) is accounted in the first six to eight species (Hildrew and Townsend, 1987). In my own career experience of counting upwards of 6000 individual samples of phytoplankton from lakes, reservoirs and rivers, I have never encountered one in which more than 20 species were simultaneously represented at >1 cell mL^{-1}, or by more than 60 species exceeding 10 L^{-1}. This is despite the fact that, in some cases, the full list of recorded species at the various given locations has exceeded 150. In the many cases where cell counts were transformed to biomass, there has not been one instance in which the eight most abundant species made up <95% of the total standing mass of the phytoplankton crop. Indeed, the 95% threshold is frequently surpassed by just one to three of the most abundant species. So it is that, even within a variabilty range of the one to eight species that will typically comprise ≥95% of the mass of phytoplankton, the truncated Shannon–Weaver index is perfectly sensitive to variations in the compositional diversity. Examples of its use are cited in the following section.

7.3.3 Species diversity and disequilibrium in natural communities

High diversity is a feature of many biological systems and it is generally considered to be important for their healthy functioning (Reynolds and Elliott, 2002). Whether this is always or only sometimes true, high species richness, equitability and genetic representation are believed to be under threat and, therefore, worthy of conservation (Lawton, 1997). It follows that the 'protection of biodiversity' must be a good thing, even if there has been an incomplete understanding

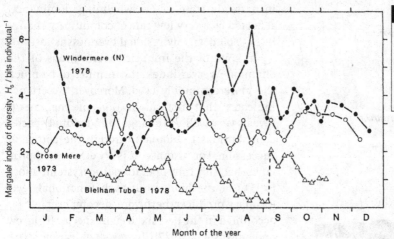

Figure 7.18 Time courses of Margalef's diversity index in the mesotrophic North Basin of Windermere, the eutrophic Crose Mere and the well-fertilised Blelham Enclosure B during 1978. Redrawn from Reynolds (1984a).

of just what this means, much less any clear strategy for bringing it about. Biodiversity (originally BioDiversity, coined as a shorthand for Biological Diversity: Wilson and Petr, 1986) has no precise definition and it has no units of measurement. As a synonym for 'species richness', it can assume fairly precise terms but it is either reliant on a sophisticated and continuously accumulating knowledge base of how many species there are. At the same time, this understanding requires an area-based focus. For instance, there is (presumably) a finite number of species on the planet, each with an instantaneously finite number of individuals; elimination of these is to extinguish the species. On a regional or local basis, the extinction of species may be reversed by subsequent invasions of new individuals of the same species from remote populations. It seems important to establish the mechanisms maintaining high local species diversity, again through reference to (apparently) cosmopolitan species whose short-generation organisms render the timescales of exclusion and dispersal conveniently measurable.

Phytoplankton assemblages in natural lakes and coastal waters often comprise several main species simultaneously and these contribute to relatively higher measures of ecological diversity, as measured by the Margalef or truncated Shannon–Weaver indices, than is suggested by the plot of H_s in Fig. 7.17 for an experimental enclosure. The plot comparing diversity indices from Windermere, Crose Mere and a well-fertilised Blelham Enclosure (Fig. 7.18) was used by Reynolds (1984a) to illustrate the then general belief that lake systems maintaining high nutrient availabilities support poorer species diversities than oligotrophic systems (Margalef, 1958, 1964; Reynolds, 1978c). This point now requires further discussion (see Section 7.3.3 below) but Fig. 7.18 presently serves to show that the low values observed in the Blelham Enclosure B in 1978 (H_s 0.24–1.75 bits individual^{-1}) had not been dissimilar from those noted in the Enclosure A in 1982 (Fig. 7.17). The differences between the measures for the eutrophic Crose Mere (H_s 2.00–4.15) and the mesotrophic North Basin of Windermere (H_s 1.78–6.44 bits individual^{-1}) are generally small but for the wide departures in the summer months. These may be compared also with Margalef's (1958) own assessments of diversity in samples taken from the Ria de Vigo, through the summer progression from diatom to dinoflagellate dominance (H_s 0.8–5.4) (see p. 311 and Fig. 7.19). Recent studies of phytoplankton diversity among a broad range of European lakes quote seasonal fluctuations in Shannon diversity (truncated H'' values) in the range 1–3.5 in stratifying lakes (Leitão et al., 2003; Salmaso, 2003), or slightly higher (2–5) in shallow lakes (Padisák, 1993; Mischke and Nixdorf, 2003).

Hutchinson (1961) famously drew attention to the paradox that high diversity among phytoplankton species competing for essentially the same limiting resources in an apparently homogeneous environment is counter-intuitive and

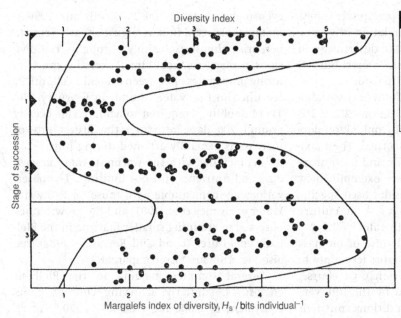

Figure 7.19 Variations in the index of diversity in a number of samples of phytoplankton collected from surface waters of the Ria de Vigo in 1955 plotted against the successional stages diagnosed by Margalef (1958). Redrawn with permission from Fogg (1975).

against Hardin's (1960) principle of competitive exclusion (see also Box 7.1). Since that time, there have been numerous attempts, often with compelling experimental evidence, to argue for a satisfactory explanation. These broadly divide among three categories. Following Paine (1966), species coexistence is promoted primarily through food-web interactions. Following Tilman (1977) and others, coexistence reflects simultaneous, niche differentiation of the surviving species. Disciples of Connell (1978) attribute a rich biota to temporal and simultaneous spatial variability of the environment. These explanations are not mutually exclusive and, indeed, they may be summative. However, the importance of the topic requires us to examine the mechanics of these concepts in more detail.

Disequilibrium

Hutchinson's (1961) own explanation of the paradox lay in the error of the first assumption of environmental homogeneity and steady-state conditions. In all the examples considered in this chapter, species composition is shown to change conspicuously through time. In every instance, this is shown to be driven by a balance of dynamic responses of planktic species that have 'arrived' from elsewhere, that are able to grow and increase specific biomass under tolerable environmental conditions but which show a net loss of biomass under conditions as they become intolerable. Just the simultaneous occurrence of species whose numbers are increasing while those of another are in net decline is sufficient to maintain species richness and local species diversity throughout the period that the two sets of responses can be detectable. For this reason, low diversity (H'', H_s) must always be accompanied by low rates of compositional change (σ_s). Equally, the maintenance of high diversity must be viewed within the context of time constraints set by changing rates of population recruitment and attrition.

The time-frame of ecologically relevant responses to an infinitely variable environment is supposed to be bounded (at the lower end) by the generation time of the shortest-lived species and (at the upper) by the time taken to reach a competitively excluded climactic state (Reynolds, 1988c, 1993b). Smaller-scale variability may demand reactivity at the biochemical and physiological levels. Provided that the cell can maintain its cycle of growth and mitotic division, the variability is smoothed to ecological constancy. Environmental changes outside the period of ascendency to climactic steady state that may destroy biomass and precipitate a new potential climax – such as inundation of vegetation by a marine incursion or the desertification of forest to scrub – have intrinsic interest but tell us little about successional maturation.

Against the many scales of physical variability in aquatic environments (Sections 2.1, 2.2.2, 2.3.2, 2.3.3), the timescales of phytoplankton population ecology are readily discerned. The maximum growth rates of the weedy and more invasive C-strategist phytoplankters – exemplified by *Chlorella* and *Ankyra* – have standardised specific replication rates (r'_{20}) of >1.8 d^{-1} in culture (Table 5.1). Verified field growth rates of 1.1 d^{-1} (Table 5.4) bring the potential time of biomass doubling to matter of several hours (certainly to <1 d under ideal field conditions). Of course, progress may well be impeded by the intervention of poor average light conditions, nutrient exhaustion or grazer responses and be truncated far short of a climactic steady state. Physical factors are also important but rapid flushing rates may benefit fast-growing organisms over slower species and their consumers. However, an adequately discretionary rate of dilution leaves the *net* rate of population increase only just positive and thus the attainment of the potential provided by growth is much protracted. The physical conditions suspend the advance of the succession, maintaining it in a sort of *plagioclimax* (Tansley, 1939). Elsewhere, selection through the foraging preferences of the zooplankters present for particular foods or particular particle sizes, constraints in the availability and accessibility of resources and light all affect the relative success of the tolerant species in moving towards dominance. A phase of apparently enhanced coexistence and (thus) higher measurable diversity is passed before the strongest competitor eventually emerges as the dominant species of a low-diversity climax.

We have considered examples of *Ceratium*, *Microcystis* and *Planktothrix* moving to this stage of community development. In Crose Mere (Fig. 7.6), *Ceratium* has several times been observed (Reynolds 1973c, 1976a) to build up to a stable (low-σ_s), low-diversity ($H_s \sim 2.0$) maximum of between 150 and 250 μg chl$a$ L^{-1} (~ 900 mg chl$a$ m^{-2}, ~ 45 mg C m^{-2}), comprising concentrations

eventually averaging 700–1000 cells mL^{-1}. These maxima, observed in late August or early September, arise almost exclusively, through serial divisions, from a base inoculum of excysting overwintering propagules of between 0.1 and 1 cells mL^{-1}, recruited to the water column in February. The 11–14 doublings required to achieve this occupy around 200 days, although the fastest rate of increase is generally attained during July (~ 0.15 d^{-1}, coinciding with the warmest water, longest days and nutrients still not limiting). The observations are comparable with those for Esthwaite Water of Heaney *et al.* (1981) and the growth rates observed in attaining smaller maxima in the Blelham enclosures (Lund and Reynolds, 1982) (see also Fig. 5.18) are also comparable.

Several summer maxima in the Blelham enclosures, similarly achieving concentrations equivalent to >600 mg chl$a$ m^{-2} (>30 C m^{-2}), have been overwhelmingly dominated by *Microcystis* populations of 300 000–400 000 cells mL^{-1}. Again the initial recruitment has come from small colonies that entered the plankton between April and June (see Section 5.4.6). Sustained exponential growth during the summer (r_n between 0.15 and 0.24 d^{-1}, slowing towards the end) would raise an invading the standing population from the equivalent of 100–200 cells L^{-1} to its maximum through some eight or nine generations in about 5–8 weeks of growth. Prior to that, three or four divisions of the overwintering cells contributing the invasive colonies whilst still on the sediment perhaps needs to be included (Preston *et al.*, 1980; Reynolds *et al.*, 1981).

It was these considerations that allowed Reynolds (1993b) to propose that the period required for a phytoplankton succession to move from initiation to a competitively excluded climax is equivalent to some 12–16 generations. Given favourable conditions and the resource capacity to sustain it, the entire process might occupy 35–60 days.

However, it is also plain that, in the plankton as on land, such low-diversity climactic, steady states are exceptional occurrences. That diversity is normally kept high is seen to be largely a consequence of processes that prevent, or delay, perhaps indefinitely, the progress to the potential steady state. Communities remain, in fact, in various states of ecological disequilibrium, having a persistent tendency to move towards a given (though possibly changing) climactic outcome but in which progress is seriously impeded or stalled. In many instances, diversity and species richness are quite unrelated to trophic state and productive capacity (Dodson et al., 2000).

Internal (biotic) mechanisms of coexistence

The activities of heterotrophs (principally, but not exclusively, phagotrophs) within the same system interact with the population dynamics of phytoplankton in a variety of ways. Easiest to understand is the concept of zooplankton feeding indiscriminately on phytoplankton. By removing individual phytoplankters from the pool, even light grazing surely delays successional progress. It is theoretically possible that individual food organisms might be removed at the same rate at which they are recruited, which would ensure zero instantaneous rates of advance. However, such balances are rarely struck (well-nourished zooplankters increase in size and recruit further generations of individuals). The more common outcome is that zooplanktic animals (at least, the filter-feeders; see Sections 6.4.2, 6.4.3) draw down their foods to concentrations at which their own population growth is impaired. The intervention of a third trophic component with a relatively slow-changing impact intensity (such as planktivorous fish or invertebrate species) is usually necessary to impart a simultaneous top–down regulation on zooplankton consumption (Reynolds, 1994c). The interactions among the consumers of the microbial food web (nanoflagellates, ciliates) with the producers (picoalgae, bacteria) are presumed to place mutual constraints on the dynamics of each of them in a way that helps to ensure the simultaneous survival of each of the components (Riemann and Christoffersen, 1993; Weisse, 2003).

Zooplankton grazing on phytoplankton is generally more selective in its effect. This selectivity has several dynamic components that need to be distinguished. Just the difference in the replication rates of two phytoplankton species may determine the relative success of one over the other whilst both are simultaneously grazed by the same consumer species. Alga A ($r' \sim 0.2$ d^{-1}) increases while alga B ($r' \sim 0.1$ d^{-1}) decreases when both are subjected to a filter-feeding consumer Z ($\Sigma F = 0.15$ d^{-1}). Food selection increases the benefit to the ungrazed alga C ($r' = 0.1$ d^{-1}). The possibilities increase greatly if depth distributions of food and feeder are considered; alga B may now increase faster than A if B swims to the surface while A fails to avoid the downward migrations of Z. Then the effects are again multiplied if consumers X and Y feed on larger or smaller foods than Z or show particular preferences for a given alga. Finally, the growth dynamics of algae and feeders are themselves fluctuating. It is not difficult to recognise the complexities of phytoplankton interactions with phagotrophs or how their fluctuating fortunes contribute to the maintenance of diversity to the extent that Paine (1966) indicated.

Phagotrophic grazers are not the only organisms that influence the structure of the planktic assemblage from within (i.e. autogenically). Parasites and pathogens are generally highly specific in the algae they affect and it is frequently the most common (or recently most common) that attract epidemics (see Section 6.5). The progressive intervention of chytrid fungi was responsible for breaking the years of summer dominance of Ceratium in Esthwaite Water, and for accordingly raising the subsequent diversity of the phytoplankton (Heaney et al., 1988) (see also Section 6.5.2).

Resource competition as a mechanism of coexistence

The theory of resource-based competition developed by Tilman and co-workers (Tilman, 1977; Tilman et al., 1982) (see also p. 197) proposed an elegant explanation of species coexistence and assemblage diversity, invoking internal responses (specific resource uptake and growth rates) to external drivers (resource availability). The experiments of Tilman and Kilham (1976) showed how it was possible for two species of diatom to grow simultaneously (and, hence, to coexist) for as long as one of them remained limited by the supply of phosphorus and the other by the supply of silicon. They reasoned that while they were not in direct competition for limiting resource, their coexistence would endure. Moreover, if two species may coexist while each is independently resource-limited, the further corollary is deduced that there may be as many coexisting, non-competing species as there are simultaneously limiting factors (cf. Petersen, 1975). The highest species diversities, accordingly, might be expected in environments of general resource deficiency and, hence, the greatest likelihood of simultaneous rate limitation.

There is no shortage of experimental information to support the differentiation of interspecific growth rates along resource-ratio gradients. Rhee and Gotham (1980) demonstrated analogous performance differences among various common species of phytoplankton with respect to the initial relative availability of nitrogen to phosphorus, while Bulgakov and Levich (1999) cited experimental evidence in support of the widely held supposition that a low N : P ratio favours the dominance of Cyanobacteria. Sommer (1988c, 1989) has argued for competitive interactions among algae for diminishing levels of Si, N and P favour the seasonality of assemblages in kataglacial lakes (see p. 338). The important role of carbon and its photosynthetic fixation is brought into the appreciation through the ratios of light to nutrient and C : P availabilites (Sterner et al., 1997) (see p. 199).

At first sight, simultaneous multiple limitation is an attractive basis for explaining coexistence. On the other hand, it is losing favour except in qualified circumstances. This is partly because simulation models of the dynamics of multiple species limitations have not borne out the prediction of stable coexistence but, rather, have pointed to instability and chaotic outcomes when more than two species compete for more than two limiting resources (Huisman and Weissing, 2001). However, some of the experimental conditons and assumptions do not necessarily obtain in natural environments. As has been pointed out previously, resources either saturate the sustainable growth rate of phytoplankton or they fall to levels that interfere with the ability of just about all species to maintain a finite rate of growth (Section 5.5.4). Persistent low levels of nutrients favour species with relatively high uptake affinities over species having weaker affinities and to which the same resources are unavailable. Species with higher affinity for limiting resources are thus able to maintain larger inocula. On the other hand, the effective exhaustion of a limiting nutrient, by species of all affinities, means that the most likely outcome of algal activity is that the uptake capacity for the limiting resource is insufficient to support further population increases. Ultimately growth, then cell division (r') is brought to a halt, and population change (r_n) falls to zero or is negative. Competition may certainly be keen but the rates of succession and exclusion are reduced to very low values. It follows that species richness and diversity endure because the mechanisms that work against them are themselves severely constrained. The low-diversity steady state is simply postponed.

It must be noted that this logic does not invalidate resource competition as a direct mechanism for sustaining coexistence, provided that the turnover of the limiting resources, between organisms and the labile pool, is demonstrably dynamic. Possibly the clearest evidence for this is the widespread co-dominance of ultraoligotrophic plankton assemblages, both in the open ocean and in lakes, by nutrient-deficient photoautotrophic picoplankton and carbon-deficient bacterioplankton. The carbon (especially) is rapidly exchanged through a labile DOC/DIC pool whereas carbon and nutrients are turned over by consumers of the microbial web. There is a thus a near-continuous, differentiating competition among the components and it maintains an ambient, diverse steady state. In these terms, the independent introduction of the erstwhile limiting nutrient is bound to help the photoautotrophs rather more than the chemoautotrophic bacteria, which will experience continuing severe or intensifying carbon deficiencies.

Disturbance as a mechanism of coexistence

The third category of processes that promote interspecific coexistence is rather generally referred to as *disturbance*. Disturbances are imposed by external (allogenic) forcing factors and are recognised by their effects in variously delaying, arresting or diverting successional sequences from achieving their low-diversity, steady-state climaxes. Commonly cited types of terrestrial disturbance include geophysical events (landslides, lava inundations) and abnormal weather conditions (hurricanes, floods, droughts: Sousa, 1984). In the pelagic, the best-known disturbance events are associated with hydraulic or hydrographic disruptions of the water column (flushing, strong mixing) or with nutrient pulsing (Sommer et al., 1993; Sommer, 1995). For example, significant wind-mixing events during summer in a small, stratifying, eutrophic lake (Crose Mere) were several times observed to cut across the usual $G \to H \to L_M$ sequence of dominant algae by restoring growth conditions suitable for diatoms (Reynolds, 1973a, 1976a). The species that grew had been represented in the vernal bloom (Association C; see Table 7.1) but others were confined to such summer growths (notably Aulacoseira granulata, since ascribed to P). However, in the post-disturbance relaxation, a tendency for the G and H elements to recapitulate their post-stratification sequences was noted. In the highly disturbed early summer of 1972, diatoms, G-chlorophytes and H-Cyanobacteria were maintained as co-dominant species (Reynolds and Reynolds, 1985). Later on, artificial mixings applied to the Blelham Enclosures imitated these effects experimentally (Reynolds et al., 1984). The imposed delay to the successional maturation, whilst retaining extant mid-successional populations and periodically resurrecting pioneer species, helps to bolster the diversity of the phytoplankton represented.

In this way, the impact of disturbance on diversity is related to the intensity and, especially, the frequency of the forcing events (Polishchuk, 1999). However, the scales of impact and timing have to be judged against the generation times and succession rates of the main species involved (Reynolds, 1993b). Without the context of successional processes, understanding of the complex mechanisms underpinning the diversity–disturbance relationship has been rather slow in development and not without controversy (Wilkinson, 1999). Even now, the mechanisms are widely misunderstood, despite the fact that most of the relevant theory was in place in the mid-1970s (Grime, 1973; Fox and Connell, 1979). The work cited most frequently is that of Connell (1978) whose strictures on species diversity in rain forests and coral reefs led to him to propose his memorable intermediate disturbance

hypothesis (IDH). In essence, it anticipates that:

(1) In the absence of any disturbance, competitive exclusion will eventually reduce the number of surviving species to minimal levels.
(2) At high frequencies of intense disturbance, only pioneer species are ever likely to re-establish themselves in the wake of each disturbance.
(3) If such disturbances are of intermediate frequency or intensity, there are more and longer opportunities for community ascendency and a greater variety of species will establish in the wake of each disturbance but these will rarely be allowed to mature to competitively excluded steady states.

The obvious consequence is that a peak of diversity should be found at intermediate frequencies and intensities of disturbance.

This is an elegant theory: its development is logical and its statements are intuitively correct. It has long remained a hypothesis for it is difficult, in most ecosystems, to collect appropriate data to support or confound it. Planktic systems provide an exception to this statement, where successions from the establishment of pioneer communities through to their total eclipse by steady-state, competitively excluded climaxes can be completed potentially in under 60 days (see p. 370). It may be deduced, incidentally, that the equivalent of 12–16 generations required for potentially climactic tree species (with generation times of 200–800 years) to achieve anything closely resembling a competitively excluded climax will occupy 2.5–100 ka. This represents is a substantial fraction of an interglacial period. Given the reconstructed climatic variability over even the last ten thousand years (10 ka) since the last great glaciation, the history of the changing dominant vegetation of the northern land masses (e.g., Pennington, 1969) may be legitimately comparable to the periodicity of phytoplankton in a small temperate lake during a single year (Reynolds, 1990, 1993b).

As has also been pointed out on many previous occasions, phytoplankton successions are not merely observable but are amenable to practical experimentation. IDH has been invoked in the explanation of periodic sequences of phytoplankton in natural lakes and reservoirs (among others, Haffner *et al.*, 1980; Reynolds, 1980a; Trimbee and Harris, 1983; Viner and Kemp, 1983; Ashton, 1985; Gaedke and Sommer, 1986; Olrik, 1994).

Reynolds' (1980a) usage of the terms 'shift' and 'reversion' to refer, respectively, to the disruption of the succession by cooling/mixing events associated with passing weather systems, and to its subsequent recapitulation, now seems too mechanistic (too Clementsian? cf. p. 351); abandonment of these terms was recommended (Reynolds, 1997a). Nevertheless, the experimental manipulations of succession and disturbance imposed by Reynolds *et al.* (1983b, 1984) attest to the importance of steady physical conditions in allowing autogenic processes to structure communities and to dominate assembly processes. Equally, they show how variations in the physical conditions may terminate one autogenic phase and initiate the next.

The physiological responses of the phytoplankton to an altered physical environment are immediate but take several hours or more to feed through to altered population recruitment rates. Accelerated rates of community change (σ_s) and higher instantaneous diversity indices (H'', H_s) are evident within days. From the cited Blelham enclosure studies, Reynolds (1988c) deduced that, depending upon the relative dominance of the disfavoured species, σ_s reaches a maximum after two to four generations of the favoured species have been recruited to the plankton. The diversity index (H_s) follows closely. In real time, this might be equivalent to 4–16 days after the stimulus is applied. This deduction agreed substantially with the findings of Trimbee and Harris (1983) and Gaedke and Sommer (1986). There is now general agreement that diversity is highest early in the disturbance response, around the second to third generation, or within 5–15 days of the imposition of the stimulus (Reynolds *et al.*, 1993b).

IDH and diversity

Despite the satisfyingly simple elegance of the concept of disturbance-induced community restructuring and of disturbance-impeded successional progression (the searching critique of Juhász-Nagy, 1993 is most apposite), the literature carries examples of instances when experiences differed from expectation. For instance, the fact that an episode of summer storms and floods failed to break the dominance of *Aphanizomenon* in the plankton of a small Danish lake, while earlier, relatively trivial, weather events had triggered upheavals of species composition was quite clearly contrary to the anticipation of Jacobsen and Simonsen (1993). Other authors have noted that the resistance of communities to disruption by external forcing is acquired progressively with increased successional complexity (Eloranta, 1993; Moustaka-Gouni, 1993; see also Barbiero *et al.*, 1999). Sommer (1993) has shown that in neighbouring lakes of similar chemistry but differing morphometry, disturbance events were more extensive and effective in increasing diversity in the deeper example. The responses of the plankton differed among three small lakes of contrasted trophic state and compared by Holzmann (1993), being weakest in the most oligotrophic of them. In reviewing each of these contributions, Reynolds *et al.* (1993b) concluded that there is no consistent or predictable relationship between diversity and external physical forcing.

For terrestrial ecologists, the development of a consistent theory of disturbance has been just as troublesome (e.g. Wilkinson, 1999). There is little difficulty in understanding the intervention of catastrophes, from fires and storms to volcanic eruptions and lava flows, in being able to arrest, not to say to obliterate, the autopoetic development of terrestrial vegetation and the reopening of the land surface to colonising propagules. Neither is there a problem in recognising that stochastic, smaller-scale forcings create just the generally fluctuating environment that prevents the extinction of opportunist strategists by persistent resource gleaners (see, for instance, Pickett *et al.*, 1989). Indeed, the simultaneous existence of sources of invasive species actually necessitates a continuity of disturbances and a continuum of patches in differing stages of successional maturation among which invasive species may migrate (Reynolds, 2001a). It is also self-evident that, were this not the case, opportunism (*r*-selection) would cease to have any viability as an adaptive survival strategy. This view of patch dynamics is explicit in Connell's (1978) hypothesis and is implicit in Hutchinson's (1961) suggested explanation of the plankton paradox, when he referred to 'contemporaneous disequilibrium'. A further corollary is that, as all species have evolved to contend with disturbed environments, they are, in a sense, matched to recurrence of perturbating events (Paine *et al.*, 1998). Thus, we may note again that dispersal constraints are as important to community assembly as is the inevitability of self-organisation and the stochasticity of external disturbances.

The problems experienced by terrestrial ecologists relate mainly to inconsistencies of pattern, especially relating to the intensity and frequency of disturbances considered to be effective in influencing diversity, and which of these might be the more critical (Romme *et al.*, 1998; Turner *et al.*, 1998). There are difficulties in the scaling of models based on field observations and differing inputs can raise or suppress the response of diversity to disturbance (Huston, 1999; Mackey and Currie, 2000; Hastwell and Huston, (2001). In another modelling approach, Kondoh (2001) purported to show the important interaction between system productivity and distubance in influencing the competitive outcome of multi-species dynamics on species richness. In essence, a relationship between diversity and disturbance is confirmed as being unimodal but the productivity level that maximises diversity itself increases with increasing disturbance.

In a way, this seems to match the deductions in respect to lake productivity (Holzmann, 1993; Reynolds *et al.*, 1993b): that productive or nutrient-rich systems are more susceptible to disturbance than nutrient-poor ones. This, indeed, would make a satisfying conclusion. However, it offers no real explanation for its mechanisms in nature and it also perpetuates one of the most persistent shortcomings in applications of the IDH. Taking the second point first, every observable disturbance is a response reaction, an outcome of internal dynamics of the species inocula present. It is to be differentiated from the forcing applied, which may, or may not, evoke a response at the level of population recruitment. Only then is a disturbance precipitated. This is where the intensity of forcing is critical – not

only must it be a signal that can feed through to species-specific growth rates but the potential respondents have to be moved from their contemporaneous limitation. Put simply, if the phytoplankton is experiencing growth-rate limitation through an effective lack of (say) iron or phosphorus, it is difficult to imagine how exposure to 1 to 2 days of strong winds and a simultaneous deepening of the surface mixed layer from <3 to ~7 m will overcome that. The forcing does not disturb the *status quo ante*. Yet this was the extent of forcing applied to Crose Mere needed to disturb the summer succession, to stimulate a significant bloom of diatoms and to raise the planktic diversity (Reynolds, 1976a) (and see p. 373).

The experience of this observation from the field should always prompt the question: is the potential of the forcing sufficient to overcome the existing structure? At the present time, this is not easy to answer from any quantified standpoint, without some common units. The potential of comparing external mechanical forcing with the current cushion of exergy flux (in $MJ\ m^{-2}\ d^{-1}$) has been explored theoretically in Reynolds (1997b). An accumulating phytoplankton was challenged by wind-mixing events of known kinetic energy. From the starting assumptions about temperature (set at 20 °C), the possible income of solar energy ($\leq 26.7\ MJ\ m^{-2}\ d^{-1}$; say, 12.6 mJ $m^{-2}\ d^{-1}$ harvestable energy), and the mechanical energy required to disperse this uniformly through 1 m (65 J $m^{-2}\ d^{-1}$; equivalent to a wind of 3.4 m s^{-1}), wind forcing was increased selectively with a view to balancing the incoming flux. By itself, mixing energy offers little challenge to the harvestable energy flux: even with a tenfold increase in wind speed (34 m s^{-1}), the kinetic energy flux is still only 6.5 kJ $m^{-2}\ d^{-1}$. However, the simultaneous deepening of the mixed layer dilutes greatly the harvestable energy available to entrained phytoplankters (see p. 138). Even a doubling of wind speed is sufficient to increase the depth of mixing of the modelled water by the cube, that is, to 8 m. Using various interpolations of existing phytoplankton biomass (1–100 mg chl*a* m^{-2}), its assumed light absorption, $\varepsilon_a = 0.01\ m^2$ (mg chl*a*)$^{-1}$, and of the background extinction of the water ($\varepsilon_w = 0.2\ m^{-1}$), the levels of energy harvestable by phytoplankton

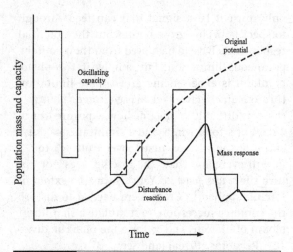

Figure 7.20 Energy exchanges and exergy accumulation under a varying capacity. The sigmoid line traces the potential accumulation (cf. Fig. 5.20, 7.12e) but this experiences setbacks every time that the sustainable biomass exceeds the capacity of the oscillating energy income. On each occasion, there has to be an adjustment (losses of structure, mass and exergy) which is manifest as system disturbance. So long as energy or resources continue to fluctuate, ecosystems go on alternating between expansion into available capacity and readjusting to contraction Redrawn with permission from Reynolds (1997a).

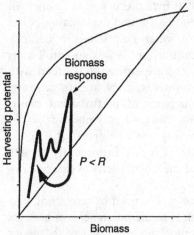

Figure 7.21 Representation, against the energy-accumulation model (Fig. 7.11) of the responses of biomass to capacity oscillations, including readjustment (a disturbance) when the accumulated mass is left unsustainable. Redrawn with permission from Reynolds (2002b).

could be reduced to <6 MJ m^{-2} d^{-1}. When the exclusive effect of increased mixing due to the doubling of mechanical stress is compounded by changes day length, cloud cover and lowered radiation intensity, the harvestable energy is quite liable to reduction to low values (often <2 MJ m^{-2} d^{-1}).

Thus, mechanical disturbance arguably carries entropic penalties of a magnitude sufficient to disturb community development in the plankton. Without more empirical data, however, it remains a conjecture. Nevertheless, the conjecture yields a further, simple conceptual model of the link between forcing and the manifestation of diversity-raising disturbances. In Fig. 7.20, a hypothetical plot tracks through time the biomass potential of an ascendent phytoplankton (and, hence, of its finite energetic requirements of maintenance) and the simultaneous supportive capacity of the harvestable energy flux. With daily fluctuations in the incoming solar flux and of daily fluctuations in its dilution

by wind mixing, variability is such that it is indeed possible that the energy that is actually harvestable may, on occasions fall below the minimum maintenance requirement of the accumulated phytoplankton biomass. Against the plot in Fig. 7.11, this is equivalent to saying there is insufficient harvestable energy to maintain a positive exergy flux. In short, the system is unsustainable. So long as the external condition persists, the requirement to restructure (lose biomass, place biomass in a lower energy resting state; switch to more light-efficient species) becomes likely to invoke a response that might be perceived as a disturbance. The energetic representation is used as the basis of Fig. 7.21, in which the variability in the potentially sustainable biomass (against variable energy income and mechanical dilution) is tracked as a function of the actual biomass. We can readily appreciate that forcing variability absorbed within the harvesting capacity and leaving a positive exergy flux is fully sustainable and does not constitute a disturbance. Should the forcing result in an energy income that is insufficient to balance maintenance, there is no exergy buffer left. Eventual reaction is inevitable, although the scale of disturbance may yet depend upon the scale of exergy shortfall (intensity), its

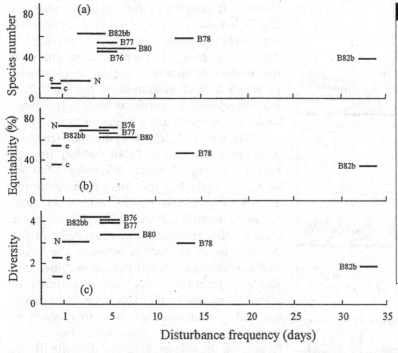

Figure 7.22 Arithmetical means of (a) species number, (b) equitability and (c) Shannon diversity in periodic phytoplankton showing disturbance responses as a function of their frequencies. Most of the data refer to events in Balaton (B) in the summers of 1976–78, 1980 and 1982, before (B82bb) and during (B82b) the enduring *Cylindrospermopsis* bloom that year. Other points are for (N) Neusiedlersee, an experimental pond (e) and its control (c). Note that all three plots peak at disturbance frequencies of 3–7 days. Original data of Padisák (1993) and redrawn with permission from Reynolds (1997a).

duration and the physiological resistance of the accumulated biomass to withstand the energetic disequilibrium. The track in Fig. 7.21 shows the necessary scaling down of biomass costs and the reversion to an earlier ascendent stage. Incidentally, it also implies a low incidence of disturbance at low biomass (as does Fig. 7.21), including those instances of severe nutrient constraints (cf. Fig. 7.15b).

With a clearer picture of what constitutes a disturbance, we may now return to the central issue of diversity being a function of the frequency of disturbance. From Connell himself to many subsequent authors elaborating on the IDH (e.g. Wilkinson, 1999), the relationship between diversity and frequency of disturbance is represented as a smooth parabola, generally referred to as 'the humpback curve'. It successfully conveys the idea of low diversity occurring at zero and at very high frequencies of disturbance and of higher diversities at intermediate frequencies (p. 373). However, there is no basis for assuming a continuous relationship with frequency, even if scaled in terms of respondent generations. Neither do most terrestrial sources of data provide much information to form any real

judgement. The evidence from Padisák's (1993) collected sequences of phytoplankton in various Hungarian lakes (see Fig. 7.22) shows species richness, equitability and Shannon diversity each to rise steeply from daily disturbances up to separation intervals of 3–7 days. Thereafter, each index decayed gradually back to low average levels at infrequent disturbances (separated by >30 days). The other valid way of expressing frequency, as a number of disturbances per year, was used by Elliott *et al.* (2001a) as a basis for applying simulated forcing events on periodic sequences modelled using PROTECH (see Section 5.5.5). External forcing to a controlled physical environment was devised (actually an instantaneous, complete mixing of a hitherto stratified 15-m water column, mixed to only 5 m) while all other variables (day length, incident radiation, temperature and saturating nutrient availability) were held constant. Grazing of algae was excluded. Eight species of phytoplankton having contrasted life forms and survival adaptations were seeded on day 1 and their dynamic changes were then simulated over a period of 1 year. Model runs were subjected to applied forcings, with varying frequency and duration, and the population

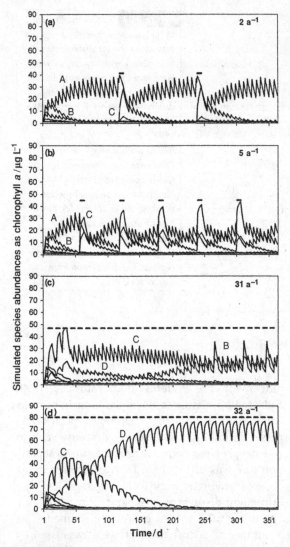

Figure 7.23 PROTECH-simulated changes in the populations of phytoplankton species to a standard physical forcings (complete and instantaneous vertical mixing of an otherwise stratified 15-m water column for 7 days) applied at various frequencies. Periods of imposed forcing are shown by the short horizontal bars. The 'species' correspond to *Chlorella* (A), *Plagioselmis* (B), *Asterionella* (C) and *Planktothrix* (D). Note the precipitous change to *Planktothrix* dominance when 32 rather than 31 disturbances are applied. Redrawn with permission from Elliott *et al.* (2001a).

responses were also monitored for 1 year. Under the ambient, unforced conditions, the community became quickly overwhelmed by the fastest growing life form (species A in Fig. 7.23, actually having the relevant properties of a *C*-strategist

Chlorella): it achieved immediate dominance and the virtual exclusion of all rivals (i.e. low diversity) within 50 days. Inserting two or three 7-day forcings during the year (Fig. 7.23a) produced the equivalent number of clear disturbances, in which a brief maximum of species B (having the relevant properties of *R*-strategist *Asterionella*) was supported. On reversion to the ambient state, species A ('*Chlorella*') duly reasserted its dominance and species B ('*Asterionella*') dropped back. Increasing the forcing frequency to five per year (Fig. 7.23b) has the effect of maintaining a more varied plankton, with '*Asterionella*' alternating in dominance with '*Chlorella*' and other species being represented more strongly. At frequencies of 25–30 forcings per year, the forced and alternating quiescent periods become quite similar in length (Fig. 7.23c) and are sufficient to allow '*Asterionella*' to dominate '*Chlorella*' throughout while other species remain in contention. Over a broad range of disturbance frequencies (6–30 a^{-1}), the average Shannon diversity (H'') of the simulated eight-species assemblage is about 1.8 of a possible 3.0 bits mm^{-3} (evenness $\sim$0.6).

Increasing the frequency yet further, a sharp alteration in behaviour was found to occur between 31 and 32 a^{-1} (i.e. quiescent periods are each <5 d (Fig. 7.23d). The system changed to persistent dominance by a third species, C (having the relevant properties of *R*-strategist but more *K*-selected *Planktothrix agardhii*), eventually to the exclusion of all the other species. Its pre-eminence was enhanced at still higher frequencies up to the point of continuous mixing to 15 m.

Changing the forcing to 5- or to 3-day periods produced analogous sequences of results, although correspondingly higher frequencies were needed to avoid the replacement of '*Planktothrix*' during the 'intermediate quiescence'. The latter term had been introduced previously by Chorus and Schlag (1993) in the context of interruptions (disturbances) to the success of *Planktothrix* in rich, turbid lakes that are normally subject to persistent mixing.

In all the series, surpassing the critical scale of forcing frequency secured an abrupt decline in Shannon diversity (to $H'' < 0.8$). Plotting diversity as a function of forcing frequency (Fig. 7.24a)

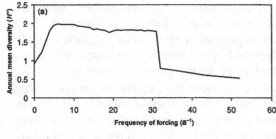

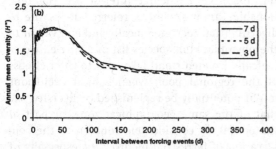

Figure 7.24 Annual mean Shannon diversity of PROTECH-simulated phytoplankton assemblages, in terms of (a) frequency of 7-day forcing events, and of (b) the time intervals separating forcing events of 3-, 5- and 7-day duration. Redrawn with permission from Reynolds and Elliott (2002).

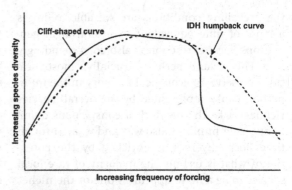

Figure 7.25 Species diversity plotted as a function of the frequency of externally forced disturbances, as the smooth 'humpback' usually supposed in many previous considerations of the IDH, and as the 'cliff-shaped curve' revealed by PROTECH modelling. Redrawn with permission from Reynolds and Elliott (2002).

shows the expected low diversity at both very low and very high frequencies of alternation. Plotted in terms of the interval between forcing events (Fig. 7.24b), diversity first increases steeply as quiescent periods lengthen, achieves a maximum in the range 10–50 days and then falls away with ongoing physical stability. The modelling plainly confirms the prejudices about intermediate disturbance and its critical times. What is also of general interest is the deduction of Elliott *et al.* (2001a) that the relationship of diversity to forcing frequency is not the smooth humpback curve that is popularly represented but one with a more cliff-like cut-off at critically high disturbance frequencies (Fig. 7.25).

Explaining the plankton paradox: where does the diversity come from?

Over 40 years of research since Hutchinson (1961) posed the paradox of the phytoplankton have succeeded in revealing many more verifications of the diversity of natural phytoplankton assemblages, whereas its explanations have emerged only gradually. For there is not one mechanism

contributing to high diversity but several. As Hutchinson himself understood well, the paradox lies wholly in the initial assumptions. Except under very well mixed conditions in a small (<10 km²) and relatively shallow lake basin, there is no single, homogeneous, isotropic environment. Even there, dynamic water movements are so variable that there is almost no opportunity for any kind of physical steady state to establish. There is a virtual simultaneity of adjacent patches, each experiencing differing conditions, yet which will themselves soon alter again, perhaps critically and perhaps in less time than it takes an alga to complete a generation. Only in the case of the most severe and ongoing deprivation of a particular nutrient, or of the harvestable energy input sufficient to satisfy the maintenance requirements of most producer biomass is the effect of this physical environmental variability overridden. Moreover, because algae are differentiated by their adaptations to function and to continue to grow in differing combinations of various environmental constraints, they show differing dynamic responses and response times to environmental variability, sufficient indeed for habitats to maintain several species in simultaneous and alternating contention. And, because local species exclusion, whether as a direct result of unsuitable prevailing conditions or through one species being outperformed by others when the prevailing conditions are suitable, requires a lapse of time equivalent to several algal generations, local extinction is also correspondingly rare. Finally (and perhaps crucially), most locations preserve an ecological memory in the information banks represented by the overall species richness. Like the handful of conspicuous dominants, their numbers also wax and wane through time; local diversity is enriched by the potential of what is usually the majority of rare local species to be able to appear in one of the niches that events conspire to make available.

It is also evident that, in the event of relatively more extreme environmental forcing events, resulting in larger-scale, catastrophic reductions in biomass (e.g. flushing, storm disruption), the recovery of a functional phytoplankton (albeit 'disturbed' as defined herein) is fully possible. In addition to a certain *resistance* of late successional stages to destruction by mixing events (that is, species-specific biomass is robust in withstanding – and recovering from – conditions that, for a time, are arguably unable to satisfy its maintenance needs), the species composition is also demonstrably *resilient* to forcing events. By this, we mean that the respondent community is based upon new growth of a similar selection of the same species that were present before the destructive forcing, so that the post-disturbance species composition may come to resemble the pre-disturbance one, albeit in new proportions. This is not surprising in itself, being attributable in many instances to the recruitment of inocula biased by the survivor vegetative biomass or by the germination of conspecific local propagules (the 'seed bank') formed in previous extant phases (Reynolds, 2002c). However, the further source of specific biomass that is important to re-establishing pelagic assemblages is the arrival of species-specific inocula from adjacent and more distant patches, exploiting effectively the channels of dispersion identified above (see p. 353). Post-disturbance phytoplankton assemblages retain an ongoing capacity to yield just those compositional surprises that make the study of phytoplankton ecology so fascinating. This is also the main mechanism of the apparently cosmopolitan and pandemic distributions of most phytoplankton.

Though strongly apparent in the phytoplankton, the dynamic nature of ecological diversity with respect to habitats or global populations had not always be appreciated or emphasised by ecologists generally. More recently, terrestrial ecologists have embraced the importance of the larger-scale processes, within the separate subdiscipline 'macroecology' (Brown and Maurer, 1989). Of particular interest are the relationships among what is referred to as the α-diversity (that recruited locally and equivalent to the phytoplankton species list of a single, hydrologically isolated small lake) and to the richness of the regional pool, whence local exclusions might potentially be replenished (γ-diversity; say that of the same geographical zone, or part of the ocean). Of considerable concern to the conservation of terrestrial biodiversity, especially of macrovegetation and vertebrate fauna, is the risk to both local and regional diversity of habitat fragmentation into smaller units (e.g. Lawton, 2000). Just as macroecological systems are separable on the basis of their habitat (β-) diversity, they too tend either to Type-I behaviour, in which local species richness is fluid and proportional to regional richness or to Type-II local richness being constrained above a certain proportion of the regional richness.

Being mutually separated, most freshwater systems have the character of isolated islands and, so, are expected to conform to the constraints of island biogeography (MacArthur and Wilson, 1967). Moreover, the obvious connectivities within predominantly linear structures are essentially gravitational. Thus, the broadest recruitment pathways for freshwater biota are provided by downstream transport, whereas mobility between catchments relates to organismic size and efficiency of dispersal. As is we have seen (Eq. 7.3), microalgae, like bacteria and microzooplankton, move very freely between catchments and are highly cosmopolitan among suitable habitats over most of the world. Given that detailed lists of species of phytoplankton recorded in individual well-studied sites may comprise between 100 and 450 species (see p. 354) and supposing that the regional species richness lies within the order of magnitude 500–5000, then habitat suitability probably distinguishes the 100 or so commonly encountered species in the local assemblage. Of the latter, fewer than eight species account for most of the local biomass at any one time. These also account for most of the measured local (α-) diversity.

The implied generality of these relationships is perhaps indicative of the intuitive Type-I phytoplankton diversity, in which local species richness represents a sub-set of the regional pool that has somehow satisfied the twin constraints of having 'passed the filters' (Keddy, 1992) of dispersal and habitat suitability. Beyond that, the actual mix of species and the relative proportions in which they appear seems stochastic and unpredictable (Rojo et al., 2000). On the other hand, planktic assemblages frequently support a high information content (H'' 3–5 bits mm^{-3}). This might equally imply that species interactions contributed only weakly to a Type-I diversity.

To be set against this are the many cases of low-diversity, near-unialgal structures wherein similar species or life forms are consistently and predictably selected in geographically quite remote locations: we think of the common dominance of Microcystis or Planktothrix or Ceratium. The organisation of species assemblages around persistent gradients of nutrients, light and redox is similar from the equator to the poles (Reynolds, 1988c). Regional variations in the participating species are evident but the functional specialisms of the dominants are wholly analogous. The uniformly low α-diversity that is achieved is independent of the regional richness and, thus, is quite unmistakably indicative of Type-II behaviour.

Poised between Type-I and Type-II mechanisms, the overall deduction about biodiversity in the plankton is that it is maintained through the combination of variable forces – environmental oscillations, disturbances and recovery from catastrophic setbacks – backed by a powerful dispersive mobility of organisms among spatio-temporal patches. The same mechanisms resist the extinction of the majority of rarer subdominant species. The conclusion concurs with the theoretical predictions of Huston (1979) and the models of Levin (2000), both based on the behaviour detected among terrestrial ecosystems. Once again, the small timescales of plankton

ecology permit the application of observations to the resolution of outstanding theories, predictions and models derived in ecological investigations of other systems.

7.3.4 Stability and fidelity in phytoplankton communities

After the preceding discussion of variability and disequilibrium in the plankton and the great distance away from steady state at which most planktic communities are poised, any exploration on the topic of their stability may seem vaguely fatuous. On the other hand, many examples of patterns in species composition and temporal change in phytoplankton assemblages in Section 7.2, both in marine and freshwater systems, were shown to apply widely among similar types of water body. Many examples of broadly repeatable annual sequences of species abundance were cited. They evidently carry a relatively low coefficient of interannual variation (CV), calculable as the ratio of variance to mean biomass. Moreover, taken at the widest geographical scales, individual species are typically either common or generally rather rare.

Given a predictable level of constancy about the relative global abundances of phytoplankton species, this high fidelity of cyclical behaviour is symptomatic of the kind of long-term stability that was once supposed to characterise all developed ecosystems. It used to be believed, for example, that the perceived stability of forest ecosystems was a consequence of their complexity (Elton, 1958; Hutchinson, 1959, quoted by Southwood, 1996) and food-web interactions (Noy-Meir, 1975). These assumptions about system stability are now seriously questioned (May, 1973; Lawton, 2000). On theoretical grounds, May (1973) showed that, far from increasing stability, complex interactive linkages increase the opportunities of chaotic behaviour and, thus, should decrease stability. Appropriate modelling of phytoplankton species composition concurs with this argument (Huisman and Weissing, 2001; Huisman et al., 2001). Yet it is perfectly evident and generally accepted that complex ecosystems conform to the ideal of low-CV stability, oscillating about one or another characteristic stable condition. May (1977) memorably analogised the behaviour to a ball in a depression on an uneven surface, gently rocked by external forcing. The movements of the ball are focused on a stable location position; however, it remains liable to a more violent, chaotic event that may dislodge the system, perhaps irreversibly, into an adjacent hollow, representing an alternative stable state.

As May (1973) had deduced, this means that ecological stability depends upon the stability of the physical habitat and that the determining interactive linkages are not at all random but are a selected set. Viewed from the standpoint of phytoplankton ecology, the first of these conditions is explicit (see p. 317). The notion that, where physical constraints allow, increasing species richness helps to stabilise aggregate community responses is also well established. Cottingham et al. (2001) recently reviewed several studies that measured species richness and variability. They concluded that the evidence for tight coupling of these properties is not as unequivocal as they had supposed previously. Methodological difficulties of sampling and scaling in the original studies may have contributed to their findings. However, they were able to verify that, while fluctuations in the populations of individual species vary independently of species richness, or may actually increase with greater species richness, the total biomass frequently aggregates around a stable level. In other words, community variability decreases with increased species richness, despite possibly increased variability in the contributions of individual species. This concurs with analyses of zooplankton data subject to planktivory. Ives et al. (1999, 2002) proposed that, subject to the condition of low environmental variability, increasing the number of species in a community decreases the coefficient of variation of the summed species concentrations (aggregate biomass).

With reference to the interannual variability of the supposedly most stable phytoplankton structures (those of the North Pacific Subtropical Gyre; see p. 304), we are now able to appreciate the disproportion of the long-term species richness. Just 21 of the 245 species recorded over 12 years by Venrick (1990) have together contributed 90% of the aggregate biomass, while the long-term dominance of any given stage of the annual cycle has remained the prerogative of just one or two species. Any of the others of the same functional group is presumably able to fulfil the role of the dominant, should the incumbent species underperform for any reason. Significant interannual differences in species composition are prompted by periodic habitat variations, such as temperature structure and wind-mixing, forced by El Niño activity (Karl, 1999, 2002).

The observation is relevant to the question, raised at the outset of Section 7.3.3, about the importance of high diversity to the functioning of complex ecosystems. The early debate lay between two extremes. The 'rivet theory' (Ehrlich and Ehrlich, 1981) supposed that because every species plays a part in the ecosystem, like the rivets in an aircraft, each one that is lost impairs performance, eventually to a point where it can no longer fly. At the other end of the spectrum, it was recognised that, provided each of its essential DEU functions is fulfilled (see p. 351), it is perfectly possible for species-poor systems to function adequately. Thus, most species tend to be functionally redundant within their habitats (B. H. Walker, 1992; Lawton and Brown, 1993). Nowadays, a more consensual understanding pervades, that the contributions of individual species to given systems are generally unequal. Some, the so-called 'keystone species' (Paine, 1980), will fulfil the role of major repositories of organic carbon or they may play a disproportionately bigger role in energy transfer. Some species may play a structural role in the sense that they modify (or 'engineer': Jones et al., 1994) the environment to the benefit of other exploiters (forest trees and the microhabitats and trophic niches they furnish come immediately to mind). This still leaves a large number of species that are functionally superfluous or, at best, mere ecosystem passengers.

The results of some ingenious field experiments, involving terrestrial herb communities, help us to resolve a general view of the functional role of species richness and local species biodiversity. Wilsey and Potvin (2000) reduced the numbers of dominant plants from old-field communities, though without reducing the overall species richness. They found that aggregate biomass increased in proportion to evenness, independently of which species had previously been dominant. In the experiments of Wardle et al. (1999), plants of the most aggressive of the grass species present, *Lolium perenne*, were removed altogether from a perennial meadow in New Zealand. There followed an increase in biomass of the species remaining, while richness was increased by germlings of invading species. For a time, at least, a broadly similar function was maintained in the grassland. When they removed all the plants, of course, there were immediate repercussions in other ecosystem components, most notably among the nematode consumers and their predators. Wardle et al. (1999) deduced that the deliberate exclusion of the dominating (high-exergy) species promotes the next fittest of the species available within the same functional group to assume the same functional role in the modified ecosystem.

I am not aware of a comparable experiment involving phytoplankton, save as an alternative to macrophytes and periphyton, as considered in Chapter 8. However, the PROTECH model (see Section 5.5.5) is eminently suited to the simulation of this kind of manipulation. Elliott et al. (1999a, 2001b) compared the rate of development, the maximum biomass attained and the diversity of phytoplankton communities serially stripped of the best-performing species. In each instance, attainment of capacity was noticeably delayed but evenness among the remaining species was increased.

The essential contribution of otherwise apparently functional redundant species to community and ecosystem function seems to lie in their potential to assume primacy when the performance of the existing dominants is impaired for any reason, either internal (due to interaction with other species) or external (imposed changes in filtration by environmental variables). A resilient ecosystem is characterised by a network of energy-flow linkages, whose individual connections may strengthen or weaken in response to fluctuations elsewhere. Interventions affecting the performance of key species and, in consequence, those to which they are trophically linked, are compensated to a great extent by the enhanced performances of hitherto subdominant species and the strengthening of their

network linkages (for further discussion, see Loreau et al., 2001).

7.3.5 Structure and dynamics of phytoplankton community assembly

In order to weave together the multiplicity of threads gathered in this exploration of the processes governing the assembly of phytoplankton communities, some conclusions now need to be aligned.

(1) Development of phytoplankton can take place, subject to the assembly rules proposed in Table 7.8, within the constraints of the environmental carrying capacity. The latter may be set by the nutrient resources available or the daily harvestable light income, as regulated under water through the interaction of mixed depth and the vertical coefficient of light extinction. While capacity remains unfilled, there can be recruitment of biomass through growth (Rule 1, Table 7.8). At such times, there need be no competition for the light and resources available. The opportunity is exploitable by all species present and whose minimal requirements are satisfied. Of these, the species with the highest exergy and fastest sustainable net rates of increase growth should benefit most (Rule 2, Table 7.8). Spare capacity is thus beneficial to net productivity (P_n/B) and to increased species richness. Equally, if biomass is equal to, or exceeds, the current capacity, recruitment is weak. Survivorship is influenced in favour of species having greater affinity for the capacity-limiting factor and/or superior adaptations to withstand the adverse conditions. Competition for the limiting factor works against species richness but competitive exclusion is slow if the development of even the superior contenders is constrained.

(2) The locally available species (richness) depends, in part, upon the local 'seed bank' of propagules (Rule 3, Table 7.8). However, the high rates of transmissibility among phytoplankton (low z value in Eq. 7.3) make for abundant immigration opportunities and high rates of invasion or reinvasion from remote sites. Relative importance of

local or invasive recruitment varies among phytoplankton, mainly with relative size and $r–K$ selectivity.

(3) Species richness may vary through the dynamics of local recruitment and local extinction. There is evidence of long system 'memories', with vestigial populations re-enacting the seasonal cycles of growth and attrition evident in previous days of abundance. Although some prominent species return to dominance year after year for long periods, there may be a slow change in the species representation and richness (Venrick, 1990). This turnover may be slow on interannual scales but it may vary over centuries (a fact exploited by palaeolimnologists reconstructing habitat changes at the millennial and greater scales: Smol, 1992; Smol et al., 2001; O'Sullivan, 2003). Species richness and/or dominance is liable to abrupt alteration in the wake of critical environmental change (eutrophication, acidification).

(4) Planktic systems are continuously liable to invasion by species that are not already present or abundant. Such invaders do not become prominent unless their adaptations are better suited to the local environment than are those of existing species (a recent example is the spread of Cylindrospermopsis raciborskii, as catalogued by Padisák (1997)) and achieve a higher exergy, with faster net recruitment rates, than existing dominants.

(5) While habitat variability remains slight, communities develop towards a steady state, generally dominated by one of a small selection of suitably adapted species (Rules 4–7, Table 7.8). This autogenic development has long been known (as *succession*); the properties of maturing successions are also well known (Table 7.7). The long-standing desire to explain successional sequences of species composition must defer to the evidence that there is no continuous sequence of species, rather a probability of certain outcomes relating to the fitness and adaptations of the species pool available (Rules 8 and 9, Table 7.8). Succession is a cycle of probabilities of replacement of dominant species. Diversity (though not necessarily overall species richness) declines through succession. Given sufficient time and with constant probabilities, the possible climactic endpoints of succession are few in number and predictable for the region, its climate, catchment and limnological characteristics.

(6) Species diversity in accumulating communities is likely to be relatively higher than in maturing successions. Species diversity is also greater where there is structural diversity of the habitat (such as in a stably stratified lake). Species richness in accumulating communities is not generally related to productivity or trophic state. According to Dodson et al. (2000), species richness shows a flattish but unimodal relation to area-specific primary production, peaking in the range 30–300 C m^{-2} a^{-1} (which range covers moderately oligotrophic to moderately eutrophic systems).

(7) Autogenic maturation of communities is exposed to disruption by allogenic physical or chemical forcing. This the communities may survive (resistance) or recover from (resilience) or the community is restructured through a forced replacement of biomass and dominant species by others more suited to the new conditions. The response is disturbance. Community disturbances imposed at a frequency of two to four algal generation times maintain the highest mix of available species and thus high levels of diversity. By virtue of the near prevention of competitive exclusion, disturbances of this frequency constitute a major driver of continuing high diversity and richness levels (p. 377). However, the mobility of species and their ability to re-establish from other locations may be critical to this process. The mechanisms resist the extinction of the majority of rarer subdominant species.

(8) Annually recurrent cycles of seasonal algal dominance suggest a high level of interannual constancy in pelagic habitats. Similar environmental conditions, roughly recapitulated each year, renew similar filtration effects upon a species pool, which in consequence becomes biassed in favour of species that have grown well in recent seasons. The development of 'commonness' among regional species pools also influences the

similarity of seasonal pattern among regionally similar types of aquatic systems, considered in Sections 7.2.1 and 7.2.3.

7.4 | Summary

Patterns in the abundance and composition of natural phytoplankton assemblages in the sea and in lakes are sought in this chapter. In the sea, characteristic floristic associations with the extent, longevity and supportive capacity of the water in the major oceanic circulations are demonstrated. Environmental distinctions from adjacent shelf and coastal areas and from localised upwellings, coastal currents and fronts are shown frequently to offer greater carrying capacities, higher levels of biomass and different planktic associations.

The tropical gyres of the Pacific, Indian and Atlantic Oceans are profoundly oligotrophic, nutrient deficient and permanently stratified beyond a depth of ~200 m. Phytoplanktic biomass is severely constrained, often <20 mg chl$a$ m^{-2}. For long periods, the dominant primary producers are picoplanktic Cyanobacteria, the fixed carbon being cycled mostly through the microbial food web. Changes in near-surface stratification, wrought by weather events or longer-term climatic fluctuations, stimulate episodes of recruitment of nitrogen-fixing cyanobacteria (*Trichodesmium*) and diatoms (*Hemiaulus*) or deep-migrating dinoflagellates.

In contrast, primary production in the high latitudes is constrained by strong seasonal fluctuations in light income, temperature and physical mixing. A small number of diatom species are overwhelmingly dominant, but dinoflagellates and the haptophyte *Phaeocystis* are sometimes abundant. In boreal and mid-latitudinal waters, diatom dominance alternates seasonally with nanoplankter and dinoflagellate abundance. In shelf and coastal waters, supportive capacity is generally much higher than in the open ocean but the more abundant phytoplankton shows strong seasonal periodicity.

Margalef's (1958, 1963, 1967, 1978) deep insight has contributed greatly to a broad understanding of the separate roles of nutrients and the physical environment in regulating both the abundance and composition of the phytoplankton. These provide the axes of the generalised habitat templates, developed by Smayda and Reynolds (2001), to which the main assemblages of marine phytoplankton are consistently aligned.

Species assemblage patterns in lakes respond to analogous drivers, with some strong coherences among major floristic components with regional climates, lake morphometries and catchment-derived nutrient loads. Among the world's largest lakes, pelagic environments resemble the open ocean in the deficiency of nutrients and the importance of mixing, and they show a similar community organisation based upon picophytoplankton and microbial processing of fixed carbon. However, at all latitudes, density stratification (including under ice) is a prerequisite of significant communal growth and assembly. Among moderate to small lakes, phytoplankton abundance and composition is demonstrably constrained by nutrient supplies and by the underwater light climate. Both are instrumental in setting limits on the carrying capacity and in influencing the adaptive requirements of the favoured phytoplankton species. Seasonal periodicities, broadly moving from (w-selected) diatom abundance in mixed columns to colonist (r-selected) nanoplankters and then to increasingly specialist (K-selected) microplankters, reveal consistent patterns. However, differences in maximum biomass and in dominant functional types usually reflect differences in resource richness. Phosphorus-deficient oligotrophic lakes support low biomass comprising distinctive diatoms (*Cyclotella bodanica* group, *Urosolenia*), desmids (*Staurodesmus*, *Cosmarium*), chrysophytes (*Dinobryon*) and dinoflagellates (*Peridinium*). Picophytoplankton may make up a substantial proportion of the biomass. In eutrophic (P-rich) lakes, other diatoms, nostocalean Cyanobacteria and self-regulating *Ceratium* species may have eventually to be able to contend with shortages in the rates of supply of carbon and nitrogen, while high turbidity may force light constraints favouring filamentous cyanobacteria, diatoms and xanthophytes.

These behaviours help to explain the fit of groups of species (trait-selected functional types) to analogous templates defined by axes scaled in limiting resources and harvestable underwater light. These templates are suggested to represent the filterability of species by pelagic habitat conditions: where light and nutrients are not constraining, many species are potentially able to grow, given that suitable inocula are present. Successful species will have evolved rapid-growth, exploitative life-history (*C*) strategies that are preferentially (*r*-) selected under these conditions. Moving down the gradients of resource and of availability is analogised to ever constricting filtration of (*R*- or *S*-) specialisms that will be increasingly *K*-selected by the ever more exacting constraint.

Analysing the mechanisms behind the observable patterns, the near ubiquity of common phytoplanton species is deduced to rely on efficient and highly effective dispersal mechanisms. These are comparable to those of bacteria, some protists and the micropropagules of sedentary invertebrates. The ability of phytoplankters to reach suitable habitat soon after it becomes available means that their ecological behaviour is readily describable in the terms of island biogeography (MacArthur and Wilson, 1967).

The assembly of biomass by colonist species (ascendency) is subject to the flux of energy and carbon, and the size of the base of resources that is available to sustain a supportable biomass. The standing mass and abundance of phytoplankton is also influenced by the fate of primary product – to metabolism, or to the loss of intact cells by sedimentation or consumption by phagotrophic or parasitic heterotrophy. There remains a strong element of fortuity about phytoplankton composition, what has arrived, how well it can function there (or how well it is allowed to do so). However, accumulating populations interact, with dynamic consequences. Assembly becomes increasingly contingent upon the sum of accumulating behaviours, conforming to patterns ('rules') summarised in Table 7.8. Of those present, the species initially likely to become dominant are those likely to sustain the fastest net rates of biomass increase. These will ultimately depend upon their specific evolutionary attributes and adaptive traits and upon their appropriateness under the environmental conditions obtaining. Conversely, it is deduced that the species compostion that is typical in a particular water body, or in a particular type of water body, is biassed by the conditions typically obtaining. Moreover, other species that are frequently present in the same specific environments, as a result of simultaneous recruitment, share common traits and are thus allied to the same trait-selected functional groups.

Phytoplankton communities can continue to develop so long as nutrients and harvestable light are available to sustain it. Once either capacity is exceeded, however, recruitment rates weaken. Adaptive traits representing greater affinity for the limiting factor, or greater flexibility in accessing or overcoming the deficiency assume premier importance. Species with the appropriate traits to withstand the deficiency survive while populations of more generalist species stagnate or regress. Competition for the limiting factor works against species richness. Given sufficient time, competitive exclusion leaves only the superior contender.

This mechanism underpins the long-recognised process of succession. The equally long-standing desire to explain successional sequences of species composition are found to be unhelpful. Autogenic maturation of communities is exposed to disruption by allogenic physical or chemical forcing. Community structure may resist or recover in the wake of such episodes, or the community is restructured through a forced replacement of biomass and dominant species by others more suited to the new conditions. The response is disturbance. Such disturbances imposed at a frequency of two to four algal generation times contribute to the endurance of high levels of diversity.

The diversity of phytoplankton, for so long considered paradoxical, is found to be maintained by a combination of variable forces – environmental oscillations (habitat instability), more severe disturbances and recovery from catastrophic forcing – backed by the powerful dispersive mobility of organisms.

Chapter 8

Phytoplankton ecology and aquatic ecosystems: mechanisms and management

8.1 | Introduction

The purpose of this chapter is to assess the role of phytoplankton in the pelagic ecosystems and other aquatic habitats. The earliest suppositions to the effect that phytoplankton is the 'grass' of aquatic food chains and that the production of the ultimate beneficiaries (fish, birds and mammals) is linked to primary productivity are reviewed in the context of carbon dynamics and energy flow. The outcome has a bearing upon the long-standing problem of phytoplankton overabundance and related quality issues in enriched systems, its alleged role in detracting from ecosystem health and the approaches to its control.

The chapter begins with an overview of the energetics and flow of primary product through pelagic ecosystems, especially seeking a reappraisal of the relationship between biomass and production.

8.2 | Material transfers and energy flow in pelagic systems

One of the essential components of ecological systems is the network of consumers that exploit the investment of primary producers in reduced organic carbon compounds. Some of these are reinvested in consumer biomass but much of the food intake is oxidised for the controlled release of the stored energy in support of activities (for-

aging, flight, reproduction) that contribute to the survival and genomic preservation of the consumer species in question. In thermodynamic terms, the food web serves to dissipate as heat that part of the solar energy flux that was photosynthetically incorporated into chemical bonds (see p. 355).

8.2.1 Fate of primary product in the open pelagic

Based upon this simple premise, the performance of pelagic systems can be quantified in units of energy dissipated (or of organic carbon reduced and re-oxidised) per unit area per unit time, by a biomass also quantified in terms of its organic carbon content (or its energetic equivalent) and partitioned according to function (primary producer, herbivore, carnivore, decomposer). For the open sea and the open water of large, deep lakes, the scales of the input components and the rates at which they are (or can be) processed have been established in the preceding chapters. In Section 3.5.1, the attainable net productive yield of pelagic photosynthesis, across a wide range of fertilities, was suggested to be typically in the range 500–600 mg C m^{-2} d^{-1} (the thickness of the photic layer compensates for differences in concentration of photosynthetic organisms). Extrapolated annual aggregates (in the order of 100–200 g C m^{-2} a^{-1}) agree well with the generalised findings of oceanographers for the open ocean, as well as those deduced from satellite-based remote sensing (Section 3.5.2). On the other hand, they clearly underestimate the cumulative production in many small to medium lakes, as well as in what were referred to as 'oceanic hotspots'. Even so, from an investment of PAR of some 40 kJ $(g\ C_{org})^{-1}$, the synthesis of even this amount of photosynthate (4–8 MJ PAR $m^{-2}\ a^{-1}$) is acknowledged to be just a tiny fraction of the annual harvestable PAR flux (less than 1% of the solar energy flux) (Section 3.5.3). So far as its role in supplying the conventional energy requirements of the dependent food chain is concerned, biological turnover represents only a small proportion of the dissipative flux.

At the same time, it is evident from the modest year-to-year changes in the photosynthetic biomass in the open sea that there is little net accumulation of primary-producer mass or carbon. Taking chlorophyll levels of 20–40 mg chl$a$ m^{-2} to be typical for the trophogenic zone of the tropical ocean, phytoplankton standing crops remain steadfastly constrained, probably in the range 1–5 g cell C m^{-2} (0.04–0.20 MJ m^{-2}). Phytoplankton concentrations in oligotrophic, high-latitude lakes may be lower still (in some cases, $\leq$0.5 g C m^{-2}) (Section 3.5.1). These obdurately low levels of producer biomass (B), like the production (P) it yields, are deemed indicative of the 'unproductive' nature of the nutrient-poor systems. Yet, as pointed out (Section 3.5.1), a P/B yield of not less than 100 g C from not more than 5 g cell carbon represents an extremely high biomass-specific productivity!

Low areal biomass and low cumulated production are traditionally attributed to resource poverty. Without a simultaneously available fund of other elements, including nitrogen, phosphorus, iron and other traces especially, new biomass and new cells cannot be built and recruited. In the ocean and, to a large extent, in deep, oligotrophic lakes, very little of the photosynthetically fixed carbon is deployed in new algal biomass: as much as 97% may be vented from the active cells (Reynolds *et al.*, 1985). Some of this is respired to carbon dioxide but a proportion is released as DOC (see Sections 3.2.3, 3.5.4). This is processed by bacteria, whose own respiration may account for 20–50% of the carbon thus assimilated (Legendre and Rivkin, 2000b; Bidanda and Cotner, 2002). In smaller oligotrophic lakes, a significant proportion of the original primary product may be exported to the sediments (but generally $\leq$100 g C $m^{-2}\ a^{-1}$) (Jónasson, 1978), to be processed there by benthic food webs. By subtraction, the potential yield of primary product available to pelagic microzooplankton or to herbivorous mesozooplankton can scarcely exceed 80% of the annual primary production (say, 30–150 g C $m^{-2}\ a^{-1}$) (Callieri *et al.*, 2002). Even these consumer pools turn over much of the carbon that they harvest in a matter of days; the proportion that is available to larger metazoans (10–25%; now rather less than 40 g C $m^{-2}\ a^{-1}$, or <1.6 MJ $m^{-2}\ a^{-1}$) (Legendre and Michaud, 1999) is yet smaller. The balance is, of course, respired to carbon dioxide.

8.2.2 Food and recruitment of consumers in relation to primary production

It is nevertheless worth emphasising again that the flow of energy and materials to higher trophic levels in the pelagic is a function of carbon turnover rate and not of biomass per se. However, it is also relevant to remind ourselves that the general uniformity of the pelagic outputs (90 $\pm$ 60 g C $m^{-2}\ a^{-1}$) is of a similar magnitude to the rate of invasion of the water column by atmospheric carbon dioxide (see Section 3.4.1). The fastest rates envisaged (up to 310 mg C $m^{-2}\ d^{-1}$, or $\leq$110 g C $m^{-2}\ a^{-1}$) depend upon winds that blow more strongly and more continuously than is generally the case. They also require a steep diffusion gradient from air to water, caused by severe depletion of CO_2 in solution. With due allowance for turnover of the carbon by the cycling between photosynthesis and respiration, it is feasible that the capacity of CO_2-limited pelagic primary production to sustain new secondary production can be up to about 40 g C $m^{-2}\ a^{-1}$ but scarcely much more.

The exploitation of even this meagre supply of particulate organic carbon by mesoplanktic consumers depends upon successful foraging. This, in turn, depends upon adequate encounter rates of suitable foods by consumers, which is often a function of food concentration. The range of concentrations of suitable food organisms providing the encounter rates representing the minimum and the saturating levels for diaptomid copepods is suggested to be 5–80 mg C m^{-3} (Section 6.4.3) (say, $\geq$5 g C m^{-2} in a 100-m layer). This is sustainable on a productive turnover of 100 g C m^{-2} a^{-1}, in the sense that it could yield up to 40 g C m^{-2} a^{-1} and 1.6 MJ m^{-2} a^{-1} to consumers. Supposing this was invested entirely in the production of diaptomids, a potential turnover of some 10×10^6 1-mm animals m^{-2} a^{-1} (or 2×10^6 2-mm animals m^{-2} a^{-1}) may be projected from the data in Box 6.2 (p. 281). Depending upon the depth of the mixed layer and the generation times of the copepods, the sustainable recruitment is in the order of 100–1000 animals l^{-1} a^{-1}.

What kind of resource is this to planktivorous fish? The foods, feeding habits and bioenergetic requirements of several commercially important species are relatively well studied. The models of Kitchell et al. (1974, 1977) and Post (1990) have been widely applied in fish management. The work of Elliott (1975a, b) showed particularly well how the maximum rate of growth of brown trout (*Salmo trutta*) varies with temperature and with the quantity of food (*Gammarus*) consumed, within the range between total satiation and the minimum needed to balance basal metabolism. These have also been the subject of sophisticated modelling (Elliott and Hurley, 1998, 1999) that simulates the earlier observations of performance of trout feeding upon invertebrates (Elliott and Hurley, 2000b) and reflects differences when other food sources are offered (Elliott and Hurley, 2000a). Information based upon only one species of fish is not assumed to apply to all others but the data serve to establish some important deductions about pelagic foraging. The maximum growth rate (and, hence, the maximum satiating food intake) of all sizes of trout investigated increased roughly fivefold between 5 °C and its maximum at around 18 °C, before tapering off quickly at temperatures >20 °C. Growth-saturating food intake was absolutely greatest in the largest fish examined (250 g), being equivalent to ~4.5 g *Gammarus* d^{-1}, or ~75 kJ d^{-1} or ~1.9 g C d^{-1}. Weight-for-weight, however, small fish eat more, 11-g fish consuming up to 400 mg *Gammarus* d^{-1}, or about 7 kJ d^{-1}. In either case, the efficiency of conversion of ingested carbon to trout biomass is 30–35% (Elliott and Hurley,

2000b). The amount of food required to cover the maintenance of body mass is about one-third the satiating ration at 18 °C, rising to about 50% at temperatures below 10 °C.

From these data, we may compare the approximate ranges between the minimum and maximum food requirements of trout at about 18 °C (2–7 kJ d^{-1} for an 11-g fish, 25–75 kJ d^{-1} for a 250-g fish), and at temperatures below 10 °C (0.4–1.5 kJ d^{-1} for an 11-g fish, 5–15 kJ d^{-1} for a 250-g fish). Were these energy requirements to be derived exclusively from copepods, a slightly lower intake is required (*Diaptomus* yields ~23.9 kJ g^{-1} dry weight) (Cummins and Wuychek, 1971). However, to consume a minimum of 0.08 g d^{-1} of copepod (for an 11-g fish at 18 °C), or up to 3.1 g d^{-1} (for a 250-g fish), requires very efficient foraging. Referring to the data in Box 8, between 2500 to 100000 2-mm calanoids would need to be captured and eaten *each day* or between 13000 and 500000 smaller (1-mm) animals. Given potential copepod recruitment rates equivalent to 10×10^6 1-mm animals m^{-2} a^{-1} (say 27.4×10^3 m^{-2} d^{-1}), the planktivorous 250-g trout would have to harvest the entire daily production under 18 m^2 of surface (i.e. up to 1800 m^3 of water) in order to fulfil its maximum growth rate. Just to maintain its body mass, it would have to forage the entire copepod production from under 6 m^2.

Though plainly approximations, they serve to show two things about the scale of demand of planktivory. First, that truly pelagic systems, exchanging carbon only with the atmosphere, are bound to be oligotrophic and capable of sustaining only very low densities of fish (averaging, perhaps, in the order of 1–10 g fresh weight m^{-2}). Second, that the ability of large pelagic consumers to harvest planktic production really does require specialist foraging adaptations. Indeed, pelagic feeding herring-like fish (clupeoids), including shads, freshwater salmonids, coregonids (whitefish and ciscos) and tropical stolothrissids, are all characteristically strong swimmers, capable of covering vast distances in the near-surface layer. All have large gill rakers that confer enhanced capabilities for straining small particles from several cubic metres of water passed over the gills each day. To adult fish of almost all other species, pelagic planktivory is normally a nutritionally unrewarding means of foraging.

8.2.3 Carbon subsidies in lakes and the sea

In coastal and shelf waters, in estuaries and embayments and in small to medium-sized lakes (according to the dimensional ranges proposed in Section 7.2.3: inland waters of less than 500 km^2 in area), cumulative net primary production often exceeds the 100–200 g C m^{-2} a^{-1}, sometimes considerably so. Values as high as 800–900 g C m^{-2} a^{-1} have been reported (see Section 3.5.1). Primary production at these levels is possible in one or both of two ways. One arises through a much-accelerated metabolic turnover of carbon synthesis and respiration; the other is the result of the system receiving additional carbon from external sources. Such subsidies come in various forms but their provenance is largely from terrestrial sources. They are tangible in the sense that they support finite sedimentary fluxes, known to amount to 100–300 g C m^{-2} a^{-1} in particular eutrophic lakes (Jónasson, 1978). Inflowing rivers are generally well equilibrated so far as their dissolved carbon dioxide content is concerned (typically 0.5–1.0 mg CO_2 L^{-1}, or ∼0.15–0.3 g C m^{-3}) (see Section 3.4.1). Delivered to the sea or to the inland water in solution, it is also immediately available for photosynthetic uptake. Maberly's (1996) investigation of the carbon speciation in Esthwaite Water, a small (1.0 km^2), eutrophic, soft-water lake in the English Lake District (Section 3.4.1), showed that the lake received much more of its inorganic carbon in the inflow streams than directly from the atmosphere. This is despite phytoplankton-driven summer episodes of high pH and CO_2 depletion when the atmosphere became the principal source. For most of the year, pH is near neutral and the concentration of free CO_2 in the lake is close to 0.12 mol m^{-3} (1.4 g C m^{-3}, or up to seven times the atmosphere-equilibrated concentration). Far from relying on CO_2 invasion from the atmosphere, the net flux is in the opposite direction, the lake venting CO_2. This relative importance of the inflows in supplying inorganic carbon is likely to be general among lakes wherever the inflows displace the lake volume in less than a year (see also Section 3.4.1 and cited work of Cole et al. (1994) and Richey et al. (2002).

To be a net source of carbon dioxide (outgassing) also indicates that the lake is oxidising more organic carbon than was fixed in situ. The short-retention lake must be a repository of organic carbon from the catchment, delivered in the inflow. Particulate organic matter (POM, of various sizes), derived from terrestrial ecosystems, especially including anthropogenic interventions (agriculture, industries, settlements) is transported in abundance by inflowing streams and rivers. Yielding slowly to microbial decomposition in the receiving water (or, more likely, on is bottom sediments), this material releases its component carbon dioxide in solution. Quantification of these contributions is still in relative infancy. External POM, derived mainly from the terrestrial plant production of open moorland, was shown to be an important food source in the oligotrophic, base-poor water of Loch Ness, Scotland, UK. Analysing seasonal variations in the stable-isotope composition, Grey et al. (2001) determined that the crustacean zooplankton, dominated by Eudiaptomus gracilis, derives some 40% of body carbon through ingestion of allochthonous particles; only the late-summer population of filter-feeding Daphnia seems to derive most of its nutrition through autochthonous pelagic producers. External POC sources are likely to be relatively less important in large lakes. Ozero Baykal is functionally oligotrophic, where at least 90% of the annual carbon and oxygen exchanges take place in the open pelagic. However, external POC sources should not be discounted. Even in a lake as large as Michigan, stream and groundwater inputs contribute as much as 20% of the usable carbon supply to the ecosystem (Bidanda and Cotner, 2002). Interestingly, the largest component of terrestrial organic carbon contributed to lakes and seas is often dissolved humic matter (DHM) (see Section 3.5.4) and, at first sight, the most likely subsidy to bacterial metabolism and the microbial food web. Although some DHM is amenable to bacterial degradation, the metabolic yield, in terms of both energy and liberated carbon dioxide, is generally slight (Tranvik, 1998), and production in oceanic bacteria remains more constrained by the turnover of carbon than nutrients (Kirchman, 1990).

Frequent recycling of inorganic carbon should be relatively neutral in the carbon profit-and-loss accounting. When the same atoms of carbon are first incorporated into carbohydrates and then released in algal, bacterial or microplanktic respiration, the net yield is tiny in comparison to the amount of carbon fixed. When this happens several times in a year, the aggregate of fixation (say 100 g C m^{-2}) may still have contributed little in the way of either producer output or export (perhaps <20 g C m^{-2}). The balance ($\leq$80 g C m^{-2}) would have been fixed, respired and refixed several times over. In this case, the 20 g C may have reached the bodies of large pelagic animals or have been consigned to the basal sediments.

Much of the carbon fixed in biomass at the bottom end of the trophic web is actually still quite labile, with about two-thirds of it being restored to inorganic components in less than a year (Jewell and McCarty, 1971) (see Section 6.6). It is interesting that, in the open sea, much of that process takes place in the upper 200 m, before the material has passed into the more permanent depths, whence its subsequent return may be very delayed indeed. In many small to medium-sized lakes, however, planktic cadavers may reach the bottom of the lake within 200 m of the surface. Here, they are substantially protected from turbulent re-entrainment and they stay more or less where they settle. Though persistently low temperatures and a lack of oxidant in the interstitial water (and perhaps the bottom part of the adjacent water column) may considerably slow the rate of decomposition, benthic detritivory, bacterial decomposers and bacterivorous protozoa nevertheless continue to mineralise sedimentary organic material and release inorganic carbon. The return of carbon and other inorganic elements to the trophogenic layers is extremely slow when compared to the open sea. However, in shallow lakes, where the bottom sediments are in direct contact with the trophogenic layer and are liable to frequent entrainment by penetrating turbulent eddies (Padisák and Reynolds, 2003), recycling of sedimentary materials back to the water column is greatly facilitated. The fact that shallowness also assists the concentration of areal production and of the founding resources, grants to shallow systems the opportunity to recycle raw materials efficiently and nearly continuously.

Shallow lakes (and the shallow margins of deeper ones) also offer potential habitat to macrophytic vegetation, which, once established, alters the turnover, internal stores and the direction and dynamics of carbon pathways. Rooted littoral plants fix carbon into carbohydrates but, unlike phytoplankton, they generally have the ability to retain and store carbohydrate polymers within their tissue, rather than venting it back to the water. Through their rooting systems, they have potential access to nutrients necessary for protein synthesis that are not available to phytoplankton. Relatively slower cycles of growth, maturation, death and decomposition also contribute to the long retention of resources in biogenic products. Macrophytes may also provide substratum for epiphytic algae as well as habitat and nutritional refuge for a wide variety of benthic invertebrates – herbivores, detritivores and their predators. Littoral and sublittoral benthos also offer far more attractive foraging opportunities to most fish, mainly because potential prey organisms are both larger and distributed more nearly in two dimensions rather than three. The reward of 4 g of *Gammarus* or *Asellus* is considerably more attainable to a 250-g trout than gathering the same mass of zooplankton. Offshore sediments, supplied in part by autochthonous organic carbon from the phytoplankton, also nourish detritivore-based chains of invertebrate consumers. These too mediate an alternative producer–fish trophic link, passing by way of benthic macroinvertebrates, and avoiding the tenuous bridge of a diffusely dispersed zooplankton.

Empirical comparisons of the biomass and production of successive trophic levels in lakes of various depths and trophic states have been used to determine the strength of phytoplankton–zooplankton linkages. Many factors are involved but energy-transfer efficiencies are found to be generally greater in deep lakes than in shallow lakes, where the alternative pathways are more likely to be available (Lacroix *et al.*,

1999). Where higher consumers are involved, modern techniques invoking the determination of the ratios of stable isotopes in the body mass and the foods of predators reveal much about the principal pathways exploited. In a recent overview, Vadeboncoeur et al. (2002) were able to confirm the relative importance of zoobenthos in the diets of a wide variety of adult fish from lakes in the north temperate zone. In their bioenergetic analysis of the fish production in the large tropical Lac Tanganyika, Sarvala et al. (2002) showed that the non-clupeid species were substantially supported other than by planktivory. Copepods were the principal food only of the dominant Stolothrissa tanganicae; alternative foods also featured in the diet of another clupeid, Limnothrissa miodon.

Although benthic pathways probably provide the dominant link in the production of most species of fish in many small and medium-sized waters, zooplankton may still be readily exploited where and when it is abundant. High concentrations of microzooplankton are necessarily sustained by correspondingly high availabilities of appropriate particulate organic carbon, which include fine detritus, bacteria, microzooplankters and, especially, planktic algae. Algal abundance requires not just the subsidised carbon fluxes but the additional biomass supportive capacity provided by an ample nutrient supply. Concentrations of ingestible algae (nanoplankters and small microplankters, generally $<25 \mu m$) (see p. 267) substantially greater than 100 mg $C m^{-3}$ are adequate to support growing populations of filter-feeding daphniids. These can grow quickly and recruit further generational cohorts, in a matter of days rather than months. Without predation, daphniids can go on to achieve aggregate filtration rates that may well exhaust the food supply altogether. This level of filtration capacity may be developed by 20–30 large ($\sim$2-mm) or 200 small ($\sim$1-mm) Daphnia L^{-1} (Kasprzak et al., 1999) (see also p. 267). Multiplied by the respective carbon contents (Box 6.2), the equivalent biomass of filter-feeders is about 1000 μg C L^{-1} (1 g C m^{-3} or $\sim$40 kJ m^{-3} of potential energy to a consumer). Now weighed against the requirements of Elliott's (1975a) trout, the 250-g fish could expect to harvest its daily require-

ments from 2 m^3 and the 11-g animal could be fully sated by clearing 200 L d^{-1}.

Under these circumstances, feeding on zooplankton once again becomes a reasonably rewarding foraging option. The progressive switching between benthivory and planktivory in the feeding of young roach (Rutilus rutilus) in response to the abundance of planktic crustaceans, as revealed by Townsend et al. (1986), has been highlighted earlier (see p. 278). The threshold of $\sim$1 g m^{-3} zooplankton carbon is not a fixed one but the behaviour helps to explain the coexistence of moderate concentrations of crustacean zooplankton and potential fish predators.

It is worth emphasising again that, quite apart from the habits and preferences of the adults, the quasi-planktic 0+ hatchlings and juveniles of a wide range of non-pelagic fish species feed almost exclusively in the plankton (Mills and Forney, 1983; Cryer et al., 1986). Seasonal peaks of phytoplankton production and zooplankton recruitment, especially the vernal pulse in temperate lakes and marine shelf waters, provide the major feeding opportunity to young-of-the-year fish. Generally, they are recruited in large numbers to the pelagic, often coinciding with the maximum recruitment phases of zooplankton. Survivorship is generally poor but individual growth is potentially rapid. In temperate lakes, the rates of recruitment and mortality experienced by the young-of-the-year are closely coupled to the seasonality of planktivore activity and the optimal exploitation of the zooplankton resource (Mills and Forney, 1983; Scheffer et al., 2000).

At the same time, planktivory inevitably depresses the numbers and impacts upon the size distribution and recruitment rates of the zooplanktic prey. On the basis of quantities considered in this section, sustainable planktivory might yield between 10 and 100 g C m^{-2} a^{-1} to consumers, depending upon the quality of zooplankton nutrition. Planktivory also has an upper limit of sustainability (in the sense of not exhausting the food supply): this may be encountered in carbon-rich ponds at fish densities equivalent to 30 g m^{-2} fresh weight (Gliwicz and Preis, 1977). Among planktic systems generally, direct planktivory is likely to sustain mean biomass in the order 1–10 g C m^{-2}. Larger consumer masses are indicative of subsidies to the carbon produced in the pelagic.

8.2.4 The relationship between energy flow and structure in the plankton

What emerges is that phytoplankton provides almost the unique capacity to supply carbon to aquatic food webs in large, truly pelagic systems of oceans and large deep lakes. They are unable to support a large biomass (as a consequence of resource poverty) and they support no greater net areal biomass production than the rate of turnover of dissolved inorganic carbon will allow. Terrestrial subsidy to the carbon may relieve somewhat the constraints upon carbon deployment: tangible production and biomass recruitment yields can be higher but are no longer reliant on phytoplankton production. The smaller and the shallower is the water body, the greater is the probability that the ecosystem energetics will be powered by the adjacent hydrological catchment, the weaker is their dependence upon phytoplankton production and the more integrated are littoral and benthic pathways into the food webs.

The predominant species and their organisational structures in the pelagic are thus closely coupled to the carbon dynamics of the entire system: oligotrophic and eutrophic systems are distinguished less by nutrients than by the carbon fluxes and the types of organisms most suited to their mediation. A provisional guide to the apparent thresholds separating these structural provinces was proposed by Reynolds (2001a) and is reproduced in Fig. 8.1. The various trophic indicators are shown against a logarithmically-scaled spectrum of carbon *availability* (scaled in mmol C L^{-1}; 1 mmol L^{-1} is equivalent to 12 g C m^{-3}) to register critical biological boundaries. Thus, the first row represents the availability of total CO_2: concentrations higher than the air-equilibrium level of CO_2 gas in water are due to the reserve of dissolved CO_2 maintained by internal recycling and external terrestrial subsidy, as well as to the store of dissolved bicarbonate ions present. Instances where CO_2 concentration falls below the air-equilibrium

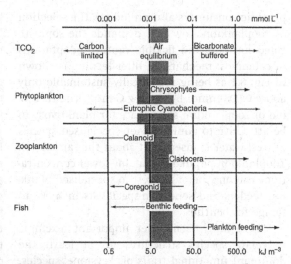

Figure 8.1 Carbon- and energy-flow constraints on the structuring of emergent pelagic communities. Accepting that the amount and distribution of native carbon sources vary over several orders of magnitude, phytoplankton composition varies with carbon dynamics, while the concentration of food particles determines the type and productivity of the zooplankton and, in turn, the resource and its relative attractiveness to fish. Shaded areas represent the transition but is generally close to a carbon availability of ~0.01 mmol L^{-1} in each instance. Redrawn with permission from Reynolds (2001a).

level of CO_2 solution are exclusively the consequence of withdrawal by photosynthetic organisms at a rate exceeding replenishment and indicate the imminence of limitation of photosynthetic rate by the carbon dioxide flux. This situation may be a rare occurrence in resource-deficient, oligotrophic, 'soft-water' systems that are chronically unable to support high levels of biomass. It would be yet more rare in 'hard-water' systems typified by a bicarbonate-enriched augmented total carbon capacity. However, in either case, a more ample supply of nutrient resources not only sustains higher levels of biomass but places greater demands on the capacity to supply carbon dioxide. Thus, enrichment ultimately selects against phytoplankters with a low affinity for carbon dioxide (like freshwater chrysophytes) and in favour of species (such as the group **H1**, **L$_M$** and **M** Cyanobacteria) (see Tables 7.1 and 7.2).

That the carbon capacity underpins the nature and abundance of the POC availability to planktic consumers also prejudices the selection of zooplankton. Low POC demands the sophisticated location and flexible capture adaptations of calanoids; mechanised filter feeding is shown in Fig. 8.1 as being realistically sustainable only above 0.008 mmol L^{-1} (0.10 g C m^{-3}). The threshold of zooplankton abundance for planktivory to be attractive to adults of non-specialised species of freshwater is inserted at about the same level. Zooplankton populations at any lower concentrations can only be of interest to specialised plankton feeders and send non-specialists inshore, to forage for benthos.

There have been other important attempts to relate species structures or, at least, the dominant functional traits of 'keystone' species, to given habitats. Setting zoogeographical constraints to one side, habitat characteristics provide the most relevant filter of regional (or γ-) diversity (see Section 7.3.3). This has been demonstrated strikingly in the species structure of local fish assemblages in different parts of contiguous rivers and to be strongly predictable from their adaptive traits (Lamouroux et al., 1999). At the level of primary producers, the direction and fate of carbon is integral to the opportunities for its use and the adaptive traits that are advantageous. The conceptual explanations originally proposed by Legendre and LeFevre (1989) and based upon hydrodynamical singularities have been refined progressively to a model of regulation by food-web 'control nodes' (Legendre and Rivkin, 2002a). In essence, the size and the trophic structure of the marine plankton direct the flow of carbon through particular, optimised channels (cycling in the web, export to the benthos or bathypelagic, etc.). The 'deciding' species structures are those whose functional traits best fulfil the opportunities of the habitat.

The idea that the habitat is the best predictor of the most-favoured species traits, survival strategies and, thus, community structures has, of course, been around for some years (Grime, 1977; Southwood, 1977; Keddy, 1992). Put in the simplest terms, different communities comprise species having certain suites of common characters and their prevalence is related to particular features of the habitat in which the communities occur. Significantly, Southwood (1977) proposed a 'habitat template' to accommodate the favoured life-history traits and ecological strategies of terrestrial communities. Reynolds (2003a) reviewed a series of attempts to develop an analogous template for freshwater communities in a way that partitioned quantitatively the distinguishing thresholds (or controlling nodes).

The outcome is presented as Fig. 8.2. Against logarithmically scaled axes representing the resource constraints upon supportable biomass and the processing constraints upon the rates of its assembly, the space is divided up according to the most likely limitations regulating the system. The resource axis is the easier to explain, with the upper limit of biomass depending upon the lowest relative bioavailability of nutrient (corresponding stoichiometrically scaled axes for bioavailable N and P are inserted). Low resource availability is thus the key to habitats in which the elaboration of biomass usually confronts 'nutrient-limiting' conditions. The processing axis is constructed on the basis of photosynthetic assimilation rates and their various dependences upon the fluxes of phota and inorganic carbon. Towards its right-hand side, assembly rates of primary producers and their dependent heterotrophs and phagotrophs are light-dependent and critically constrained by modest harvestable PAR fluxes ('photon flux-limiting' conditions). Moving leftwards, photosynthetic assimilation rates become light-saturated but the carbon-flux capacity is increasingly strained to the limits of the invasion rate of carbon dioxide from the atmosphere and the various catchment sources of carbon. With increasing external sources of organic carbon, rates of processing moves from being less dependent on the carbon supply and more so upon the supply of oxidant needed to make it available.

These axes describe the space available to ecosystem components according to their primary adaptations and so define a template for community structures (Fig. 8.3). The axes readily accommodate the typical distributions of the trait-separated functional groups of freshwater phytoplankton (Table 7.1) among waters of given habitat characteristics (see for instance Fig. 7.8) favouring particular morphological and physiological adaptations (Section 5.4.5). In this way, the

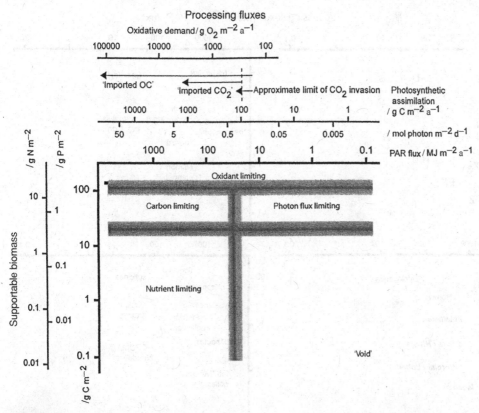

Figure 8.2 Log/log representation of aquatic habitats in terms of resource-supportable biomass (which, it is assumed can be regulated by the stoichiometrically least available nutrient; nitrogen and phosphorus axes are included as examples) and processing fluxes, which may be set by the PAR income or the rate of carbon delivery or by the flux of oxidative flux required to process the organic carbon delivered. The area thus defined can be subdivided, as shown, according to the characteristic of the habitat thus defined (nutrient-limiting, carbon-limiting, etc.). Figure combines features of figures presented in Reynolds (2002b, 2003a).

species characteristically dominating strongly nutrient-constrained systems (representatives of the Z, X3, A, E, F associations), those dominating turbid or light-deficient environments (C, D, P and especially R and S species) and those coping with carbon-flux challenges in high-nutrient, high-light environments (X1, H, L) are separated (Fig. 8.3a). The template similarly distinguishes (Fig. 8.3b) among planktic phagotrophs: the daphniid, moinid and diaptomid traits signify different interplays in the availability of resources

and the speed of processing opportunities. Potential plaktivores may also be plotted, habitual pelagic-feeding fish are distinguished from typical benthivores in sediment-retaining habitats (the psammophilic cyprinids) and those from less silted habitats (lithophilic fish). The template is also tentatively advanced to plot the distributions of macrophytes (Fig. 8.3c) and benthos (Fig. 8.3d).

8.3 Anthropogenic change in pelagic environments

The habitat template of pelagic systems, invoking resource and processing constraints, provides a useful bridge to understanding temporal variation in stability and its consequences. We have already considered how seasonal changes in habitat conditions alter the selection among stocks of primarily *C*-, *S*-, or *R*-strategist phytoplankters (Section 5.5.4; see Fig. 5.21), essentially as the coordinates on the axes defining relative mixing and nutrient are altered during the year. In the

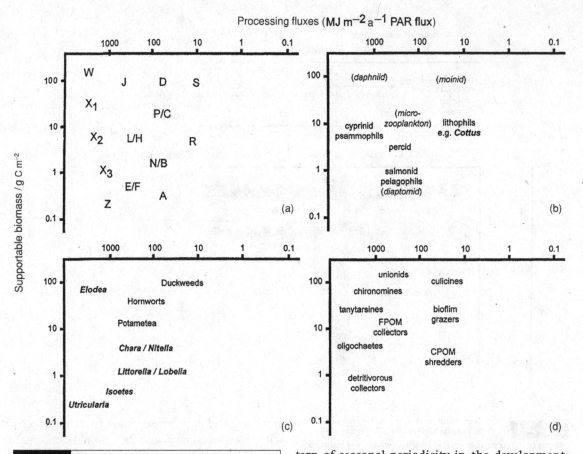

Figure 8.3 Tentative use of the resource/processing habitat template to summarise the distributions of familiar aquatic assemblages. (a) Phytoplankton shown by trait-segregated functional groups (summarised in Table 7.2); (b) representative functional groups of zooplankton and fish; (c) freshwater macrophytes; (d) freshwater benthos. CPOM, coarse particulate organic matter; FPOM, fine particulate organic matter. The figure follows ideas proposed in Reynolds (2003b).

model thus developed, variations in the nutrient availability were substantially the consequence of biogenic activity but the physical conditions were set, at least in part, by the external physical environment. We have also extended this concept in the context of predicted variations of unexpected magnitude or of variations at unexpected temporal intervals, sufficient to effect significant structural disturbances (Section 7.3.3). Over a number of years, interannual adherence to a basic pat-

tern of seasonal periodicity in the development and community assembly in the phytoplankton describes a series of loops tracking approximately similar courses through the habitat template. The sketch in Fig. 8.4 shows a notional series of such loops that, basically, follow the selection pathway 'a' in Fig. 5.21. There are modest interannual variations that may affect the numbers, relative abundance and occasional dominance changes of the phytoplankton in successive years. These loops could be said to be governed by the various behavioural contols, acting as Lorenzian 'attractors' (see, for example, Gleick, 1988). These defy randomness and are variously susceptible to circumstantial weighting of the dynamics of the individual species and, thus, the communal outcome. Significant departures from this track, forced by extreme combinations of resource and processing opportunities ('strange attractors'), then conform to conventionally defined *chaotic* responses (Gleick, 1988, *a. o.*).

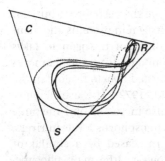

Figure 8.4 Sketch to show several annual sequences of the phytoplankton in a eutrophic lake, tracked as in Fig. 5.21, to show its Lorenzian attraction and the occasional chaotic, strange-attracted departure.

Many of the interannual differences in seasonal dynamics of phytoplankton, from oligotrophic oceans (Karl *et al.*, 2002) to eutrophic lakes (Reynolds, 2002a), can be related to chaotic displacement of the environmental attractors. It results in significant (though generally rather temporary) variation in the driving constraints, sufficient to evoke a series of population responses. These summate to community responses, which, even if not precisely explicable, may still be tracked in the habitat template. However, there may well be interannual differences that are more insidious and more systematic, that occur in response to relatively slow progressive environmental in morphometry, climate or in the anthropogenic influences to which they are subject. Prior to modern concerns about accelerated climate change as a consequence of human intervention in the global carbon cycle (see Section 3.5.2), the first two named sources were principally the concern of palaeoecologists, investigating the effects of postglacial sedimentary infill from catchments subject to changing erosion rates and vegetation cover. Environmental changes and system responses over a few decades and within the span of observation of individual laboratories and, in some cases, of individual scientists may be tracked from extant data collections. The most familiar of these changes have been the effects of eutrophication through anthropogenically enhanced nutrient supplies and increasing acidification of precipitation.

8.3.1 Eutrophication and enhanced phosphorus loading

Defining the issues

The problem of eutrophication (most especially of lakes) has been known for many years and, as a topic, has been discussed and reviewed in numerous previous publications. Here, the reader is simply referred to other works (Ryding and Rast (1989), is excellent) charting the history and the fascinating early attempts to explain the quite alarming increases in the plankton biomass that were observed in a number of prime European and North American lakes during the middle part of the twentieth century. The background to the present section is that it mainly concerns the raising of supportive capacity of otherwise clear-water lakes as a direct result of increased loadings of hitherto limiting nutrients, arising from anthropogenic activities in the hydrological catchment. The problem is not confined to lakes as analogous enrichment of coastal seas has enhanced their fertility several-fold several with respect to the pre-agricultural period (Howarth *et al.*, 1995). In both lakes and shelf waters, eutrophication has led to alterations in the species structure of the plankton and to what are generally regarded as damaging changes to aquatic ecosystems.

Cultural eutrophication was defined by the Organisation for Economic Co-Operation and Development (1982) as the nutrient enrichment of waters resulting in the stimulation of an array of symptomatic changes, among which are the increased production of algae and macrophytes, that are injurious to water quality, are undesirable and interfere with other water uses. This definition distinguished the process from natural eutrophication that many people associated with lake ageing, recognising that the undesirable changes were, in effect, merely some sort of acceleration. There are such things as naturally eutrophic lakes (especially among kataglacial outwash lakes) (see p. 336) but most of the palaeological evidence points to the opposite, that as natural catchments mature and support closed forest ecosystems, the nutrients released decrease rather than increase. Sediment deposition diminishes the water depth and eventually concentrates the nutrient load in a shallower depth of water, contributing positively to the apparent fertility of the water column. In this context, eutrophication (or its reverse, 'oligotrophication') should be seen as no more than a contemporaneous reaction to contemporaneous external nutrient loadings, which will exercise a regulatory role so long as the nutrient is critical to the supportive capacity.

It is also well known that the nutrient that is generally considered to act as the limiting control is BAP (see Section 4.3.1). Actually, dissolved sources of assimilable nitrogen are just as likely to be in short supply, relative to demand. However, owing to the advantageous trait of certain cyanobacteria to be able to fix atmospheric nitrogen (Section 4.4.3), nitrogen-fixing species may still be able to grow, often to limits imposed by the available phosphorus. This leaves phosphorus still the principal regulatory factor, even in nitrogen-deficient lakes (Schindler, 1977). Deficiencies of micronutrients (Mo, Va, Fe) may prevent much nitrogen fixation (Rueter and Petersen, 1987) (Section 4.4.3). Availability of iron constrains the plankton-supportive capacity of large parts of the ocean (Section 4.5.2) but, with the exception of those instances where plankton is simultaneously nitrogen-deficient and fixation-constrained, phytoplankton in lakes is more likely to be limited by phosphorus than any other nutrient.

Anthropogenic enhancement of phosphorus comes from several sources. The clearing of forest, the promotion of agriculture and the use of inorganic fertiliser have been trade marks of a green revolution of food production on an industrial scale. The ascendency of a human population of over 6 000 000 000 would have been quite unsustainable without this. Agriculture rejuvenates ecosystem- and biomass-specific productivity but also opens them to resource exchange (Ripl and Wolter, 2003). Often, deficiencies of nitrogen have to be overcome in order to realise productive potential and, world-wide, applications of inorganic nitrogenous fertilisers represent a major component of agricultural practice. Their high water solubility makes then vulnerable to leaching in drainage water, if applications

are not optimised to give the best opportunity for the assimilation of nitrogen in proteins and plant growth. Even so, loadings of nitrogen to lakes and coastal waters from agricultural catchments have probably been boosted two- to fivefold in the last 50 years (cf. Lund, 1972). The use of phosphatic fertilisers has also increased over the same period, during which phosphorus levels in rivers, lakes and seas have increased by a similar or still greater factor. However, the main phosphorus pathway is probably quite different. Plants 'compete' with soil chemistry for the phosphorus applied, much becoming immobilised in the soil by clay minerals, iron and aluminium hydroxides and through co-precipitation with carbonates (Cooke, 1976). Drainage from well-managed pastures and cropfields generally tends to have a relatively low soluble-P content (there are plenty that are otherwise; see, for example, Haygarth (1997), although it may carry a substantial PP load as eroded soil particles but which remains 'scarcely bioavailable' (cf. Table 4.1). On the other hand, much phosphorus remains in the soil or is incorporated into the biomass that is actually cropped by consumers.

As a metabolite of consuming animals (including humans), excess phosphorus is eliminated in solution (i.e. excreted as MRP). In the natural way of things, this is most likely to be returned to soils and to further chemical immobilisation. Changing cultural standards, driven by burgeoning human numbers, urbanisation and public health concerns, insist that most of that 'waste product' is intercepted and subjected to secondary biological treatment (in which much of the organic content is re-oxidised and re-mineralised). On average, adult humans excrete some 1.6 g P d^{-1} (0.58 kg P a^{-1}) (Morse et al., 1993); with modern standards of secondary sewage treatment, between 70% and 100% (Källqvist and Berge, 1990) of this is returned to the aquatic environment in a readily bioavailable form.

Historically augmented by soluble phosphates emanating from the hydrolysis of detergents based on sodium tripolyphosphate (STPP; see Clesceri and Lee, 1965) and boosted by other household wastes, domestic sewage has undoubtedly enhanced P-loads to rivers and pelagic systems in general. Reynolds and Davies (2001) estimated human contributions to have contributed up to 1.6 kg P individual^{-1} a^{-1} in the recent past. Certain manufacturing industries also generate significant quantities of dissolved phosphorus. However, in both its coincidence with the implementation of widespread treatment and the statistical allocation of contributions to the P budgets of contrasted river basins (e.g. Caraco, 1995; Brunner and Lampert, 1997), effluents from secondary sewage treatment are most strongly implicated as the leading proximal source of the phosphorus complicit in eutrophication (Reynolds and Davies, 2001). Lest this is taken as an accusation, it must be recalled that modern agriculture and the socio-economic structures that it sustains are founded upon the enhanced distribution, chemical mobility and biotic dissipation of (e.g. Moroccan) phosphate and (e.g. Chilean) nitrate. Whether these elements are removed to water directly or through a complex anthropogenic food chain is really academic. What matters is that these pathways are understood, that their impacts are quantified, and that they inform rational strategies for improving their future management.

Aquatic impacts of phosphorus loads to lakes
The Organisation for Economic Co-Operation and Development (OECD) sponsored a series of co-ordinated studies of the relationship between phytoplankton biomass (analogised to chlorophyll a concentration) and phosphorus availability in lakes (TP was preferred to MRP, which, if limitation means anything, is rather scarce). The first report (Vollenweider, 1968) uncovered the essence of a mathematical relationship between in-lake concentrations of chlorophyll a and TP (although, in fairness, Sakamoto (1966) and Sawyer (1966) had each previously recognised this). The final report (Vollenweider and Kerekes, 1980) provided definitive statistical fits of mean annual chlorophyll a concentration against the delivery of TP from the catchment (the 'loading'), corrected for the effects of depth and hydraulic replacement. In between, a flurry of other equations was published that (variously)

related mean summer chlorophyll a concentrations to winter–spring concentrations of TP (Dillon and Rigler, 1974; Oglesby and Schaffner, 1975) or maximum concentrations to vernal BAP (Lund, 1978; Reynolds, 1978c) or, eventually, the mean annual chlorophyll concentration to the annual P-loading factor (Vollenweider, 1975, 1976; Rast and Lee, 1978).

All these formulations yielded similar slopes (Reynolds, 1984a): the greater the phosphorus availability the greater is the phytoplankton biomass likely to be supported. However, it is the Vollenweider–OECD approach which has been most frequently applied as it incorporates the further useful empirical step of relating in-lake availability to the nutrient loading from the catchment. The full formulation models the relationship between average chlorophyll biomass and the amount of phosphorus supplied to the lake, adjusted to availability through corrections for water depth and hydraulic retention. In its final form (Vollenweider and Kerekes, 1980), the regression of the annual mean concentration of chlorophyll a ($[chla]_a$, in mg m^{-3}) is a direct function of an averaged, 'steady-state' index of phosphorus availability (Λ_P, also in mg m^{-3}):

$$\log[chla]_a = 0.91 \log[\Lambda_P] - 0.435 \qquad (8.1)$$

The derivation of Λ_P is interesting and is extremely valuable when the management of P loads becomes an issue. It begins with the dynamical relationship between the total phosphorus concentration in the lake water ($[P]$, in mg m^{-3}) and the rate of supply, less the amounts lost in the outflow and to the sediments:

$$d[P]/dt = (q_i[P]_i)/V - w_P[P] - \sigma_P[P] \qquad (8.2)$$

where q_i is the inflow rate and $[P]_i$ is the phosphorus concentration of the ith inflow stream and V is the volume of the lake; σ_P is the phosphorus sedimentation rate and w_P is the rate of loss of phosphorus in the outflow. At steady state,

$$[\Lambda_P] = \{L(P)/z\}/\{w_P + \sigma_P\} \qquad (8.3)$$

where $L(P)$ is the aggregate areal rate of phosphorus loading (in mg m^{-2} of lake area per year) and z represents mean depth. The most difficult term to estimate without detailed measurement is the phosphorus sedimentation rate; Vollenweider (1976) proposed an empirical solution in which:

$$\sigma_P \approx \sqrt{(z/q_s)}/t_q \qquad (8.4)$$

where t_q is the hydraulic residence time in years, calculated as lake volume (V) divided by the annual sum of the inflow rates ($\Sigma\ q_i$), and where q_s is the hydraulic loading rate (in m a^{-1}), approximately equivalent to z/t_q. Rewriting Eq. (8.4),

$$\sigma_P \approx \sqrt{[z/(z/t_q)]}/t_q = \sqrt{(1/t_q)}/t_q \qquad (8.5)$$

Supposing that $w_P = 1/t_q$, Eq. (8.4) may now be rewritten

$$[\Lambda_P] = \{L(P)/z\}/\{1/t_q + \sqrt{(1/t_q)}/t_q\} \qquad (8.6)$$

Multiplying out by t_q gives

$$t_q[\Lambda_P] = \{L(P)/z\}/\{1 + \sqrt{[z/(z/t_q)]}\}$$

and cancelling

$$[\Lambda_P] = \{L(P)/q_s\}/\{1 + \sqrt{(t_q)}\}$$

or

$$[\Lambda_P] = \{L(P)/q_s\}/\{1 + \sqrt{(z/q_s)}\} \qquad (8.7)$$

The log–log relationship of chlorophyll concentration to Vollenweider's index of phosphorus availability (Λ_P) is plotted in Fig. 8.5. The fitted regression is thus the definitive 'Vollenweider–OECD model'; restated,

$$\log[\text{chl}a]_a = 0.91 \log[\{L(P)/q_s\}/ \\ \{1 + \sqrt{(z/q_s)}\}] - 0.435 \qquad (8.8)$$

Great store has been placed on this formulation. Some authors have apparently been perplexed by sites which do not concur with the regression (such outliers are evident in Fig. 8.5) and have sought amendments to the equation in respect of other influences, including the effects of grazers (e.g. Prairie et al., 1989). Were serious criticisms to be made, they might be directed at the original and inadvertent bias towards deep temperate lakes. Had more shallow, heavily loaded, P-cycling or well-flushed systems been considered, a less satisfying result would have been obtained. As it

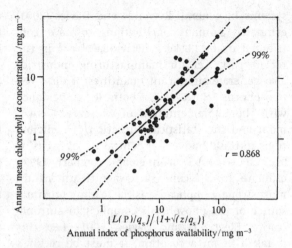

Figure 8.5 The Vollenweider–OECD relationship between average chlorophyll a concentrations in lakes and the phosphorus availability inferred from loading characteristics, with 99% confidence limits. The equation (8.8, in the text) of the fitted line relates chlorophyll to mean in-lake phosphorus concentration (log [chl$a$]$_a$ = 0.91 log Λ_P – 0.435, where Λ_P is derived as Λ_P = {$L(P)$ / q_s} / {1 +(z / q_s)} (see text and Eq. 8.7). Redrawn with permission from Reynolds (1992a).

is, the log–log format obscures a wide variation in real average chlorophyll content for a given derivation of [Λ_P]. For instance, a literal extrapolation of Fig. 8.5 is that the average chlorophyll supported in a lake having an availability index of 100 mg P m^{-3} may be predicted, with 95% confidence, to be between 10 and 53 mg chl$a$ m^{-3}. On the other hand, the Vollenweider–OECD model is a powerful, empirical statement of the long-range performance of large and medium-sized lakes subject to varied differing phosphorus loadings.

Reducing the relationship back yet more towards a semi-quantitative word model, the Organisation for Economic Co-Operation and Development (1982) proposed approximate phosphorus availabilities to coincide with the trophic-state, metabolic-dependent categories of lakes in use since the days of Naumann (1919). The slightly modified scheme of Reynolds (2003c), incorporating the continuing prevalence of phosphorus limitation is shown in Table 8.1.

Such deductions help to inform practical approaches and strategies for the manipulation

Table 8.1 | Proposed criteria for classifying lakes on the basis of trophic state and metabolic constraints, based upon OECD (1982) and Reynolds *et al.* (1998)

Category	[TP] or [Λ_P] (mg P m^{-3})	Months where MRP <3 mg P m^{-3}	Average [chl$a$]^a (mg m^{-3})
Ultraoligotrophic	<3	Always	<2
Oligotrophic	3–10	9–12 per year	0.7–4.5
Mesotrophic	10–35	4–9 per year	2–24
Eutrophic	35–100	<4 months per year	3–53
Hypertrophic	>100	Never	>10

^{a} For predicting maximum chlorophyll yield from the bioavailable phosphorus, see Section 4.3.4 and Eqs. (4.14) and (4.15).

of phosphorus loads. These are reviewed later (see Section 8.3.3). Mention may also be made of lake-specific derivations of the metabolic consequences of varied external phosphorus loads. Supposing that the portion of the bioavailable phosphorus load that is actually retained generates a stoichiometrically equivalent mass of organic carbon and that this is oxidised by a stoichiometrically equivalent quantity of hypolimnetic oxygen, Reynolds and Irish (2000) successfully simulated the differing impacts of the progressive eutrophication and its subsequent reversal on the hypolimnetic oxygen content of the North and South Basins of Windemere.

8.3.2 Blue-green algae and red tides

In both lakes and shelf waters, progressive eutrophication has resulted not just in the maintenance of a higher average phytoplankton biomass but also to shifts in species composition that apparently favour Cyanobacteria (in lakes) and small dinoflagellates (in certain seas). Neither change is straightforward. In either case, the change brings consequential effects upon environmental quality and both, by coincidence, carry implications for human health. Beyond that the direct causes of the problems are quite distinct and are referred to separately.

Cyanobacterial 'blooms'

Long before it was confirmed that certain species of Cyanobacteria (or blue-green algae) were capable of producing several extremely toxic chemicals, they had already achieved notoriety for their tendency to form surface scums. Accumulated by downwind drift, these paint-like shore-line swathes of stranded algae are considered to be, at best, unsightly. As they die and decompose, they release pigment into the water and become evil-smelling. Thus, 'leprous and fetid' (Sinker, 1962), they discourage contact with, enjoyment of and fishing from, the waters so affected. In point of fact, rather few species of Cyanobacteria do this: those belonging to the 'bloom-forming', gas-vacuolate genera that either occur in coenobial or colonial structures (such as *Microcystis*, *Woronichinia* and *Gloeotrichia*; trait-separated functional groups L, M) or in filaments that aggregate in clumps or flakes (as in many *Anabaena*, *Anabaenopsis* and *Aphanizomenon* species belonging to the H groups). Other gas-vacuolate, filamentous forms that remain solitary (*Planktothrix*, *Limnothrix*, some *Anabaena* and *Cylindrospermopsis* of the S associations) can be abundant but rarely scum. Generally, small-celled species, including picoplanktic Cyanobacteria, do not have gas vacuoles and, thus, have no means of constituting surface scums at all.

While such blooms are certainly not new (see Griffiths (1939) for a review), even entering local folklore at locations where they have been long recognised (Reynolds and Walsby, 1975), reports of scums (or blue-green algal blooms) became much more numerous since the mid twentieth Century and appeared, often spectacularly, in lakes where they had previously been virtually unknown. At first, their appearance was attributed to sudden, unexplained bursts of rapid growth and replication (Mackenthun et al., 1968). In fact, the algae grow only relatively slowly. They accumulate at the surface only during quiet weather (minimal vertical mixing) and only as a consequence of their failure to regulate the buoyancy imparted by the gas vacuoles. From being well dispersed through the water column, buoyant algae are disentrained to the surface when the wind drops. Becoming thus 'telescoped' (Reynolds, 1971) from depth to the surface and further concentrated along lee shores by light winds, the scums give a greatly exaggerated impression of abundance. Nevertheless, the increasing incidence of blooms prejudicial to water quality correlates well with increasingly enriched conditions, suggesting a powerful causal link between increased abundance of bloom-forming Cyanobacteria with increased phosphorus availability (Gorham et al., 1974). Indeed, such blue-green algal blooms have perhaps done most to give eutrophication its bad name.

Interestingly, bloom-forming Cyanobacteria need no more phosphorus to support growth or to saturate their growth-rate requirements than other common eukaryotic algae (Reynolds, 1984a) (see also Section 4.3.2 and Table 4.1). Neither does phosphorus availability guarantee blue-green algal abundance: they are relatively intolerant of acidity and low insolation (Table 7.2). They are extremely tolerant of high pH and may dominate where algal production generally strains the carbon supply (Shapiro, 1990). They commonly do well in mildly P-enriched calcareous lakes (Reynolds and Petersen, 2000).

Also relevant is the fact that the growth of the 'large' algal units that characterise the bloom-forming blue-green algae is especially sensitive to lower temperatures (see Section 5.3.2 and Fig. 5.3). At least so far as the well-studied temperate lakes of North America and Europe are concerned, it became apparent that the offending bloom-forming species are excluded in the early year by low temperatures. These are, of course, no bar to the successful early growth of diatoms and nanoplankters. By the time that the physical conditions are adequate for their growth, many temperate Cyanobacteria would be confronted by post-vernal phosphorus exhaustion. The history of eutrophication in Windermere (as told by Reynolds and Irish, 2000) illustrates very well the interannual stability of the silicon-constrained diatom crops despite year-on-year increases on annual P loadings. The record reveals the onset of late-spring P surpluses and the advent of significant summer cyanobacterial blooms. It shows the demise of both in the wake of reduced phosphorus loadings after 1992. Sas' (1989) review of successful lake restoration schemes in Europe identified a reduction in cyanobacterial crops in each of several lakes, only after BAP levels had been reduced during the spring period to levels too low to support significant blue-green recruitment during the early summer.

Cyanobacterial toxicity

Problems with cyanobacterial scums have been compounded by the recent 'discovery' (really it was a confirmation of a long-held suspicion) of their severe toxicity to humans. Again, the issue does not arise as a direct consequence of eutrophication but, on the other hand, the occurrence of toxic organisms in health-threatening concentrations is dependent upon an enriched resource base. Cyanobacteria are not the only group of phytoplankters to produce toxic metabolites: besides the 'red-tide' dinoflagellates (see p. 407 below), several kinds of marine and brackish haptophytes (including Prymnesium and Chrysochromulina) produce substances toxic to fish and other vertebrates. However, it is the self-harvesting in scums that so magnifies the potential harm that cyanobacterial toxins might pose to humans.

Many instances of sickness and death of livestock and, occasionally, humans after consuming water containing Cyanobacteria had been reported over a number of years, especially from warm, low-latitude regions. It had been supposed by many hydrobiologists working remotely that these were symptomatic of putrescence and bacterial pathogens. The possibility that the algae themselves could be harmful appears not to have been taken very seriously, prior to the pioneering investigation by a small but active group of scientists led by P. R. Gorham (see e.g. Gorham et al., 1964). The work was continued by W. W. Carmichael and his colleagues (see Carmichael et al., 1985). Many of the present perspectives are due to the efforts of G. A. Codd and his colleagues at Dundee (for a recent overview, see Codd, 1995). The eventual interest of the World Health Organisation promoted the useful handbook by Bartram and Chorus (1999).

It is now well appreciated that Cyanobacteria are capable of producing at least three classes of toxins. The acutely hepatotoxic microcystins attack the digestive tract of consumers, causing acute pneumonia-like symptoms and sickness in humans. More than 60 structural variants have been detected in cells or cell-free extracts from a range of cyanobacterial species, not just of Microcystis (K. Sivonen and G. Jones, in Bartram and Chorus, 1999). It is believed that they act by blocking phosphorylation; weight for weight, the toxicity of microcystins is comparable to that of curare and cobra venom. The neurotoxic anatoxins are also acute poisons. Chemically, they resemble the dinoflagellate saxitoxins, and are produced principally by the nostocalean genera. The third group of toxic compounds are lipopolysaccharides: these are the least well characterised but possibly the most insidious of the three, being associated with sub-lethal skin irritations as a consequence of contact with affected water or with cumulative chronic effects of frequent exposure.

The benefit to the organisms of producing such poisonous chemicals is still an unexplained paradox. It is scarcely a necessary defence against herbivorous crustacean or ciliate consumers and, besides, to kill potential grazers is as protective to other species of the phytoplankton as to the Cyanobacterium and most of these, it will be recalled, grow rather faster. There would be more point to producing toxins against the other algae. There is some evidence for the suppression of growth of common algae, either in media from which the cyanobacteria have been removed or in fresh medium, spiked with extract from spent Microcystis cultures (Reynolds et al., 1981).

What is interesting about this is that the toxin production seems most prolific at the time of population climax, after the organism has become dominant and its rivals have already been competitively excluded. It is possible that the toxicity is fortuitous and the compounds are the by-product of some unknown homeostatic step in the ageing of the cyanobacterial cells. Toxicity of Microcystis was long suspected to be a symptom of stress in natural populations and which was reversible by introducing colonies to nutrient-replete media under normal illumination and temperature (Carmichael, 1986). More recently, Orr and Jones (1998) provided convincing evidence that the production of microcystin in nitrogen-limited Microcystis cultures is proportional to the cell-replication rates. Field evidence is inconsistent, at best suggesting that toxin production is sporadic. Jähnichen et al. (2001), experimenting with Microcystis harvested from Bautzen Reservoir, Germany, showed that microcystin was synthesised only during the phase of exponential increase and only after the external pH exceeded 8.4, free CO_2 was virtually exhausted and photosynthesis was drawing on bicarbonate. The suggestion that microcystin production has some connection with the extreme affinity of cyanobacteria for carbon uptake (see Section 3.4.2) is an attractive proposition. It is also far from incompatible with the earlier findings.

Toxicity per unit of Cyanobacterial mass varies with the species of Cyanobacteria present, the potential of the resources to support their growth and the availability of DIC. However, the extent to which physical processes may have further concentrated the cyanobacterial mass magnifies the risk of human exposure to a toxic dose. As has been stated, the microcystins are themselves extremely toxic: the lethal intraperitoneal lethal dose to mice of the common microcystin-LR is 1.25 µg/25 g (Rinehart et al., 1994), or 50 µg kg^{-1}. It is not assumed that all mammals are equally sensitive but, supposing it were similar, the generally lethal oral dose would be some 10- to 100-fold greater, i.e. in the order 0.5 to 5 mg kg^{-1}. Swallowing 30–300 mg of microcystin would probably be sufficient to kill a 60-kg adult. However, the mass of microcystin produced by individual cyanobacterial cells is measurable in femtogram quantities. Lyck and Christoffersen

(2003) have recently measured ~85 ± 44 fg microcystin per *Microcystis* cell in field populations, somewhat less than the 278 ± 115 fg cell^{-1} that Christoffersen (1996) had measured in laboratory cultures. Against the dry mass of the cell (average 32 pg; Table 1.3), microcystin accounts for perhaps 1–1.5% of the typical dry mass (note, a similar amount to its chlorophyll complement). Supposing the highest measured toxin content, then the smallest number of *Microcystis* cells that would have to be ingested in order to deliver a toxic dose is close to 100 thousand million (10^{11} cells) and, in all probability, ten times that amount. When compared to the dispersed cell populations attained by natural *Microcystis* populations noted in nature (my record is ~360 000 cells mL^{-1} for a near-monospecific population in one of the Blelham enclosures and equivalent to ~120 mg chl*a* m^{-3}), it is clear that a toxic dose is equivalent to scarcely less than 28 L of lake water.

One would be entitled to conclude that the risk of drowning in the water far exceeds that of poisoning by its suspended contents, but for the phenomenon of surface-scum formation. Through the abrupt flotation of buoyant colonies to the water surface, the concentration of *Microcystis* colonies hitherto dispersed through a depth of ~5 m is quickly compacted ('telescoped': Reynolds, 1971) into a layer no thicker than 5 mm (i.e. a 1000-fold concentration). The scum is further concentrated by subsequent drift to a lee shore, where the population from (say) a 1-km fetch could be reasonably aggregated into a shoreline scum of 1 m or so in width. Now, the microcystins are perhaps concentrated by a factor in the order of 10^6, capable of supplying the toxic dose within some 28 μL of lake water!

The salutory deduction is that the shoreline scums are rather more threatening than the dispersed populations. It is interesting that our species has coexisted with bloom-forming Cyanobacteria for some millions of years, mostly oblivious to the latent hazard. On the other hand, the scums are so uninviting that they invoke the life-saving instinct of disgust: people are generally dissuaded from consuming or contacting this water by revulsion (although *Spirulina* is

reportedly harvested for food by natives of Chad: Pirie, 1969). I am not aware of a scientific study that confirms my impression that most animals avoid drinking water tainted by Cyanobacteria but there is certainly evidence that domestic livestock may be reluctant to drink until overcome by thirst (Reynolds, 1980b). It is noticeable that fish and *Daphnia* also avoid water sullied by scum. All this makes the behaviour of domestic dogs – which seem uniquely attracted to wallow and roll in shoreline scums – quite difficult to explain. The attraction proves fatal when the animals start to lick their coats.

Pets apart, there is a need to be concerned about sub-lethal or chronic exposure to the toxic Cyanobacteria, not least through drinking water purified from reservoir storages in which planktic Cyanobacteria may be abundant. Treatment processes that remove planktic cells by filtration, flocculation or dissolved-air scavenging are effective in removing intracellular toxin but engineers need to be aware of treatments that induce cell lysis and secretion of toxin into the water. Cyanobacterial toxins are effectively removed from the water passed through granular activated carbon. Reservoir managers are also often able to select from several draw-off options, in order to avoid the intake of Cyanobacteria. Current guidelines from the World Health Organisation suggest 1 μg microcystin L^{-1} as the upper safe limit (see Bartram and Chorus, 1999). If reservoir populations approached the toxicity level observed by Christoffersen (1996), it has to be admitted that many reservoirs supply water supporting Cyanobacteria rather more numerous than the equivalent of 1 μg chl*a* L^{-1}.

For recreational waters, the hazard posed to swimmers, sailors and anglers alike remains the ingestion of scum. In addition to providing periodic warnings, site managers usually seek a compromise between banning public access to water that they know may contain extant blue-green algae and allowing activities to continue until there is a significant risk of toxic algae aggregating along the shore. This equation is not just about how much alga is dispersed in the water but whether, and by how much, it is likely to self-concentrate in areas of public access. Using a reverse calculation of what was needed to generate a toxic dose in 28 L of lake water, Reynolds (1998c) started with what seemed like a reasonable estimate of a volume swallowed, how many toxic cells that that might contain and how much that factor of concentration could be sustained by flotation and leeward drifting. The outcome for buoyant *Microcystis* suggested that a lake population equivalent to 5 μg chl*a* L^{-1} is a level warning of significant risk and is a trigger for careful monitoring for downwind scums in calm weather. Most *Anabaena*, *Aphanizomenon* and *Woronichinia* species. float up only half as rapidly, permitting a doubling of the concentration triggering the warning level. *Planktothrix* and *Limnothrix* float so relatively slowly that concentrations equivalent to 100 μg chl*a* L^{-1} may be tolerated before warning levels are triggered. The dispersed cell concentrations of named Cyanobacteria equivalent to these warning thresholds are set out in Table 8.2.

Control of cyanobacteria

Interest persists in being able to eliminate and/or exclude Cyanobacteria from managed water bodies or, at least, to keep their numbers at background levels. Unfortunately, there is no simple or universal means to attack Cyanobacteria per se which is not likely to be destructive of all other biota, desirable or otherwise. Deep-rooted suppositions about the nutrient requirements of 'the Cyanobacteria' and about their susceptibility to grazing rather ignore the very wide diversification evident among the group of morphology, physiology and life history. Even when authors have focused on just the bloom-forming genera, they have tended to seek mechanistic explanations emphasising the importance of certain correlative factors, such as nutrient ratios and biological interactions, in governing population dynamics (see e.g. Levich, 1996; Bulgakov and Levich, 1999; Elser, 1999; Smith and Bennett, 1999). Other analyses suggest that cyanobacterial dominance is a more fortuitous outcome of interacting factors that include perennation, weather effects, column mixing and carbon affinity (Reynolds, 1987a, 1989a, 1999b; Dokulil and Teubner, 2000; Downing *et al.*, 2001). The main generalisations that are possible seem to be that the filamentous group-S Oscillatoriales frequently dominate shallow, turbid lakes (Scheffer *et al.*, 1997) and that nitrogen-fixing Nostocales may dominate eutrophic lakes in defiance of low levels of combined nitrogen, provided certain other conditions are satisfied (Section 4.4.3). Cyanobacterial abundance is, like that of most other forms, correlated to TP and TN (e.g. Downing *et al.*, 2001), not least because their biomass will account for much of the TP and TN that is present. In temperate lakes, growth of bloom-forming Cyanobacteria is slow in winter and they may fail to grow at all if the competitors (especially vernal diatoms) clear the water of dissolved phosphorus before the light and temperature thresholds are past. Species that overwinter on the benthos experience critical difficulties of seasonal recruitment (Reynolds *et al.*, 1981; Reynolds and Bellinger, 1992; Braginskiy and Sirenko, 2000; Brunberg and Blomqvist, 2003). If filter-feeders are abundant before the new colonies are released into the water column, then grazing can be highly effective in stemming recruitment to the plankton (Ferguson *et al.*, 1982; Reynolds, 1998c).

By deduction, effective controls against cyanobacterial dominance are few. Low concentrations of available phosphorus are beneficial in temperate lakes (in the sense of not supporting bloom-forming Cyanobacteria) but not so in tropical lakes, especially if they stratify and have long retention times (see Section 7.2.3). Of general importance, however, is the poor tolerance by bloom-forming genera, especially *Anabaena*, *Aphanizomenon* and *Microcystis*, of entrainment in mixed layers that take the algae well beyond the conventional photic depth and thus dilute and fragment their exposure to the light field. This sensitivity was detected in the early literature reviews (Reynolds and Walsby, 1975), was verified in the mixing experiments of Reynolds *et al.* (1983b, 1984) and was a demonstrable consequence of the application of destratification techniques in London's Thames Valley Reservoirs (Ridley, 1970; Steel, 1975). Since then artificial mixing techniques have been used widely in the protection of water quality in reservoirs (Steel and Duncan, 1999; Kirke, 2000); the reduction in cyanobacterial mass has been generally welcomed, even where this was not the primary

Table 8.2 Approximate biomass equivalents of potentially toxic Cyanobacteria attaining a ('warning') level that could deliver a lethal oral dose to a human adult; the derivations assume that intracellular toxin content to be the highest reported at time of writing, that the organisms are at their most buoyant and that their horizontal aggregation is subject to surface winds of 3.5 m s^{-1}. Developed from Reynolds (1998c).

Species[a]	Cell volume (μm^3)[b]	'Average' cell C (pg)	'Average' cell chla (pg)	Cells (or mm) mL^{-1} to 1 μg chla L^{-1}	Suggested warning level (μg chla L^{-1}	Number of cells (or mm) mL^{-1} equivalent to warning level
Microcystis aeruginosa	30 (**65**) 100	15	0.3	2 000–4 000	5	10 000–20 000
Woronichinia naegeliana	5(**40**)75	10	0.2	4 000–6 000	10	40 000–60 000
Aphanizomenon flos-aquae	5(**12**)20	3	0.06	15 000–18 000	10	150 000–180 000
Anabaena circinalis/ flos-aquae/ spiroides	70 (**100**) 130	22	0.45	1 500–3 000	10	15 000–30 000
Anabaena lemmer-mannii	33(**47**)113	11	0.22	3 000–6 000	10	30 000–60 000
Anabaena solitaria	270(**400**)520	90	1.8	400–700	50	20 000–40 000
Planktothrix mougeotii* (1 mm)	28 000 (**46 600**) 71 000	10 500	210	4–5.5	100	400–550
Planktothrix agardhii* (1 mm)	12 000 (**24 000**) 28000	5 500	110	8–11	100	800–1 100
Limnothrix redekei* (1 mm)	1800 (**3140**) 7500	700	14	60–80	100	6 000–8 000
Pseudanabaena limnetica* (1 mm)	780 (**1220**) 1800	275	5.5	160–200	100	16 000–20 000

[a] Asterisks indicate where instead of cell volume, the volume of a 1-mm filament is given.
[b] Bold figures indicate a typical middle-of-the-range value.
Source: Developed from Reynolds (1998c).

objective. The mechanical aspects of artificial mixing are reviewed in a later section (8.3.5).

Mention may be made here of the use of barley straw as a defence against Cyanobacteria. From a chance observation in which a length of stream was cleared of most of its detached algae downstream of an abandoned straw bale, the idea quickly spread that straw added to ponds and small reservoirs would provide effective protection from cyanobacterial growth. The implicit confidence in the technique seemed misplaced, as there was no information upon the mechanism and the instances of successful treatments represented a very small proportion of the attempts. By degrees, however, it became clear that the effective agent was a phenolic substance, presumably produced by the barley plant (as in other plants, including bark-bearing woody plants) as a live defence against fungi and microorganisms. It is released from the straw as it ages and rots and the substance will inhibit growth in the laboratory of Cyanobacteria and other algae at quite low concentrations (Newman and Barrett, 1993). Suitably aged, broken up and dispersed, barley straw has been shown to be reliably effective in preventing the growth of nuisance blue-green algae (Everall and Lees, 1996; Barrett *et al.*, 1999). Although there is inevitably some imprecision in what makes an effective dose and algae show some variation in susceptibility to barley toxins, the use of straw should be recognised as a legitimate and effective defence against Cyanobacteria, capable of inducing species-specific algal mortalities and altering dominance in mixed assemblages (Brownlee *et al.*, 2003). The comment is still inspired, however, that the use of a toxic agent to protect against another toxic agent is faintly ironic, although, as already admitted, there may be other reasons for seeking to eradicate Cyanobacteria from ponds and reservoirs.

Red tides

Several phyletic groups of marine phytoplankton have toxic representatives. Besides the dinoflagellates mentioned at the start of Section 8.3.2, they include a handful of haptophyte genera (mainly especially the Prymnesiales) that produce substances causing osmoregulatory failure and death in fish. Both *Prymnesium parvum* and *Chrysochromulina polylepis* have been implicated in fish kills around the North Sea coast, the alga having first built unusually large local populations in each instance (see, for instance, Edvardsen and Paasche, 1998). In addition, non-toxic species have been implicated in fish kills as a consequence of local algal abundance, near-simultaneous death, followed by rapid decompo-

sition and local oxygen depletion. 'Brown tides' of the picoplanktic *Aureococcus* and 'green tides' of *Phaeocystis* in its microplanktic colonial stages may cause considerable distress among fishermen. It has become fashionable to lump all these organisms together as 'harmful algae'; certainly, they are the subjects of several recent reviews and compendia (see e.g. Anderson *et al.*, 1998; Reguera *et al.*, 1998).

However, the particular issue of the red-tide dinoflagellates remains an intriguing one. A dozen or so genera are known to produce low-molecular-weight saxitoxins which are acutely neurotoxic to birds, fish and humans. Being endotoxins (that is, they are not released into the water by live cells), they are apparently far more injurious to larger organisms than to planktic assemblages. One well-known pathway leading to human poisoning is through the consumption of filter-feeding shellfish taken from areas recently affected by red-tide organisms.

The development of significant concentrations of relevant dinoflagellates is, in part, attributable to high local nutrient inputs. Over the last three decades or so, red-tide events and their constituent organisms have been becoming more frequent in the enriched coastal and shelf waters adjacent to major urban centres of the world (north-west Europe's Atlantic seaboard, the St Lawrence, the Gulfs of Maine and of Mexico, off Baja California, the north-eastern seaboard of Argentina, around Japan and Korea, and Australia and New Zealand). However, abundance is compounded locally by the ability of the algae to self-regulate in slack water and to become aggregated by weak water movements. Thus, the development of red tides is substantially consequent upon the containment of populations by near-surface stratification of enriched waters, or in shallow water columns (Smayda, 2002).

It seems that the natural habitats and probable sources of many of the more troublesome of the red-tide species are associated with the exploitation of rapid nutrient renewal in frontal zones and post-upwelling relaxations, where reduced mixing and relative resource abundance normally coincide (see Figs. 7.3, 7.4 and Section 7.2.2). *Alexandrium tamarense* and *Karenia mikimotoi*

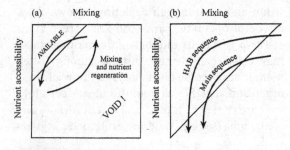

Figure 8.6 (a) Selection trajectory in the sea, as proposed by Smayda and Reynolds (2001), accepting that more of the potential resource/processing interaction is inhospitable to phytoplankton. The trajectory now closely resembles Margalef's 'main sequence' (see Fig. 7.2). (b) By analogy, the 'red-tide sequence' (here styled HAB sequence) runs parallel to the main sequence but more deeply into environments made more tenable by greater nutrients and reduced depth of mixing, representing the involvement of bloom-forming taxa in the active community. Redrawn with permission from Smayda and Reynolds (2001).

are typical members of the frontal flora in middle latitudes, while *Lingulodinium polyedrum* and *Gymnodinium catenatum* are notable toxic species of upwelling relaxations. The highly toxic *Pyrodinium bahamense* and *Karenia brevis* are adapted to entrainment and dispersal in offshore currents.

The links among size and morphology (life form), physiology and ecology that have been noted at frequent intervals in this book were explored famously by Margalef (1978), and not least in the context of an early exposition of the causes of red tides (Margalef *et al.*, 1979). In their mandala (Fig. 7.2), the 'red-tide sequence' was represented as a sort of system 'sickness', striking parallel to the successional 'main sequence', in an area of greater nutrient availability. In the context of the *C–S–R* triangle, preferred by Smayda and Reynolds (2001), increased nutrient enrichment of coastal waters permits a higher arc than the main sequence to be traced (Fig. 8.6). This is more favourable to the Type-IV, Type-V or Type-VI species associations than to the gleaning, *K*-selected *S*-strategists normally culminating Margalef's (1978) main sequence. However, Fig. 8.6 gives little new insight into why these species should be toxic or what benefit obtains. This explanation is still sought.

8.3.3 Controlling eutrophication by phosphorus load reduction

The worth of the Vollenweider–OECD regression lies in the elegant empirical statement it makes about the generalised behaviour of the lakes included in the original dataset upon which it is based. It makes no statement about any individual lake, beyond the qualitative deduction that if it becomes more enriched, it may well support more phytoplankton chlorophyll on average. Still less does it confirm that reducing current phosphorus loads will result in the support of more modest algal populations. Yet many managers have been deceived by the power of the regression to employ it as a driver and measure of schemes intended to reverse eutrophication and restore the lake to something supposed to be closer to 'pristine' conditions. This latter, rather nebulously used, term usually refers to a state that preceded the agricultural–industrial anthropogenic impacts and not to the conditions at the lake's birth. Even so, the Vollenweider–OECD relationship is not a management tool, nor was it ever intended as such. It may be seen as a guide to the determination of nutrient loads and to the prospects for the benefits to be gained from their reduction. It is 'not a slope, up or down which a given lake will progress during a period of artificial enrichment or deliberate restoration' (Reynolds, 1992a, p. 5).

On the other hand, some restoration schemes, involving the alleviation of anthropogenic phosphorus loads, either by diversion or tertiary treatment of the sewage inflows, have been spectacularly successful. One of the best-known and most-cited examples concerns the abrupt reductions in the average phytoplankton biomass in Lake Washington, in the north-west USA, following the diversion of all sewage discharges to the lake. Once the diversions began to take effect, from the mid-1960s, the phytoplankton biomass, for long dominated year-round by *Planktothrix agardhii*, was quickly reduced to below its 1933 level (the first year for which quantitaive records had been kept) (Edmondson, 1970, 1972). By 1976, the alga had effectively disappeared, leaving microplankters dominating the reduced phytoplankton biomass, and average Secchi-disk transparency cleared from <2 m to

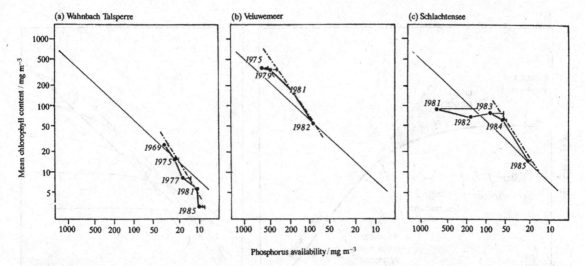

Figure 8.7 Stages in the recovery of three European lakes after reduction in external phosphorus loading. Heavy lines link the changing observed annual average chlorophyll *a* concentrations against the phosphorus availability and are superimposed upon the slope of the Vollenweider–OECD regression. The bars to the right of each point represent the difference between TP (total phosphorus) and PP (the particulate fraction). Note that the difference, approximately equivalent to unused dissolved P, is virtually exhausted before there is a response in the mean chlorophyll concentration. Redrawn with permission from Reynolds (1992a).

>4 m. Further consequential changes included the rise of *Daphnia* species and their replacement of *Diaphanosoma* and *Leptodiatomus* as dominant zooplankters, with further benefits to water clarity (Edmondson and Litt, 1982). Other associated changes in the structure of the planktic food web have been consistent with the lake recovering an essentially mesotrophic condition (Edmondson and Lehman, 1981).

The Lake Washington case has been an enduring example of what can be achieved when the eutrophication is still relatively mild (Lorenzen, 1974). Schemes elsewhere have not necessarily been quite so successful, either taking long periods to take effect or, in some cases, having still to show improvements (Marsden, 1989). The need to understand these apparently quite different sensitivities of lakes to altered phosphorus loads provided the inspiration for the analysis carried out by Instituut voor Milieu- en Systeem-Analyse (IMSA) in Amsterdam. The study, published under the editorship of H. Sas (1989), showed substantial differences in their responses to reduced external P loads, most especially between lakes judged to be 'deep' and those considered as 'shallow'. There were also systematic differences in the onset of the biological response to the diminished delivery of phosphorus to the lake (Sas' Subsystem 1) which was consistently mediated through the bioavailability to algal production (Subsystem 2). Three cases are illustrated in Fig. 8.7.

The first concerns the Wahnbach Talsperre, a mainstem reservoir on the River Sieg that supplies water to Bonn and Köln in Germany. During the early 1970s, unacceptably large, year-round crops of *Planktothrix rubescens* were proving expensive to treat for potability. The morphology of the valley in the inflow region lent itself to the construction of a small pre-reservoir and a treatment plant that would remove biogenic debris, particulate matter and dissolved phosphorus from the inflow, so that the water in the reservoir was rendered much less supportive of phytoplankton growth. This approach to tackling the problem of diffuse phosphorus loads was novel, costly and, so far, scarcely imitated, but the savings in water treatment and disinfection for distribution have repaid the investment. However, there was some nervousness in the early days following commissioning, for although the dominance of *Planktothrix* was eroded, diatom blooms became more prominent in the phytoplankton and these were still disappointingly expensive to treat (Clasen, 1979). Eventually though, as Subsystem 2 in-lake

(a)

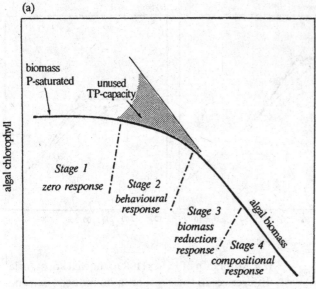

(b)

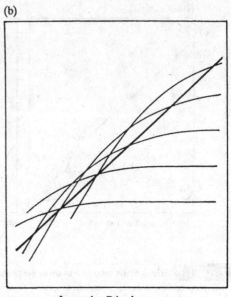

Figure 8.8 (a) Revised explanation of lake responses to phosphorus-load reductions, including changes in phytoplankton composition. Reading from right to left approximates the effects of eutrophication. (b) Hypothetical P–chl relationships for individual lakes. The fitting of an averaged regression probably underestimates the slope of chl *a* on P in those lakes where it is genuinely a limiting constraint on biomass-carrying capacity Redrawn with permission from Reynolds (1992a).

phosphorus availability slowly adjusted to the sharp depletion in subsystem 1 phosphorus delivery, so annual average chlorophyll concentrations also declined. By 1985, the reservoir had achieved an oligotrophic status (Fig. 8.7a). Of particular interest is that the line fitted to the year-on-year averages is rather steeper than the Vollenweider OECD regression. It is in fact almost linear, at 1:1, reflecting the fact that phosphorus had become the capacity-controlling resource and, so, truly limiting to phytoplankton crops supported.

In Veluwemeer (Fig. 8.7b), one of the polder lakes bordering the reclaimed Ijssel Meer, problems of excessive *Planktothrix agardhii* growth were addressed by interception of the main point sources of phosphorus and by progressive flushing with a more dilute water source (Hosper, 1984). In-lake TP concentrations, then about ten times those in the Wahnbach Talsperre, were

halved between 1975 and 1979 but with little effect either upon the average phytoplankton biomass or its species composition. A steep decline in the phytoplankton mass occurred in the early 1980s, once the available phosphorus had been lowered to a point below the demand of previous maximum crops, MRP had become a crop-limiting resource. Again, the slope of response was close to 1:1 and steeper than Vollenweider mean. In Schlachtensee (Fig. 8.7c), the lag between the sharp reduction in phosphorus loading between 1981 and 1984 (when bioavailability was also reduced eight- to tenfold) and any noticeable impact upon average phytoplankton crops was plainly related to the prior diminution of the large cushion of unused MRP. Once again, the eventual, near-linear biomass response depended upon the prior imposition of a demonstrable capacity limitation by the bioavailable phosphorus. Without that, biomass is insensitive to the actual amounts of phosphorus supplied to its growth medium.

The behaviours represented in Fig. 8.7 are generic. The general case, proposed in Sas (1989) and reproduced in Fig. 8.8a, has received wide acceptance. Reducing phosphorus loads have little effect on biomass while unused PP capacity saturates the requirements of the biomass suffering some other constraint. Movement from stage 1 to a P-led biomass reduction (stage 3) and a switch, perhaps, away from relative abundance of Cyanobacteria to (e.g.) more benign chlorophyte–chrysophyte associations (stage 4) depends upon the reduction in P availability to a point when it imposes the main constraint on biomass carrying capacity. The critical quantities are unique to each individual lake, however, so the coordinates of the response curve are lake-specific. It is not difficult to recognise that the Vollenweider–OECD regression is actually fitted to what are ultimately the biomass response curves to P loads for the individual lakes (notionally shown in Fig. 8.8b). Then, the more lakes that are featured in which the phytoplankton biomass is not severely or continuously P-limited, then the flatter would be the fitted slope. The Vollenweider–OECD coefficient of 0.91 reflects a higher representation of P-deficient oligotrophic lakes in the original dataset than (say) among the (mainly) North American lakes analysed by Rast and Lee (1978) to yield a coefficient of 0.76.

In the context of controlling Cyanobacteria, the analysis of site restorations considered by Sas (1989) unexpectedly revealed another common behavioral feature. Before any noticeable biomass or compositional response to reduced in-lake P availability had occurred, several of the lakes with large or dominant component of bloom-forming Cyanobacteria showed a considerable increase in clarity (actually, Secchi-disk transparency). This observation was explained by a deeper average vertical distribution of the Cyanobacteria and other self-regulating species. The expectation that average buoyancy in gas-vacuolate Cyanobacteria should respond to lowered ambient nutrient concentrations (see p. 206), thereby permitting the algae to scour for resources more deeply in the water column, is strongly upheld by these observations. Sas (1989) included this behavioural response in the generic curve (Fig. 8.8a) as a discrete stage 2.

It remains to explore the between-site differences in the Subsystem 1 – Subsystem 2 linkage and the difficulty, or otherwise, of depriving phytoplankton of sufficient bioavailable phosphorus for them to continue to grow and replicate. The potential unexploited BAP concentration has a direct, steady-state relationship with the aggregate of the MRP load and which is broadly predictable (Eq. 8.3, Section 8.3.1 above). Its transfer to biomass alters the MRP concentration in the water but – theoretically – without alteration to the TP content of the water and without affecting its overall bioavailability (though it is temporarily restricted to those organisms that took it up). The subsequent fate of the particulate P (and the organismic P in particular) does very much affect its future bioavailability and to whom. Whether the first beneficiary is eaten and its phosphorus incorporated in the biomass of the consumer, or its cadavers are bacterised and the phosphorus is recycled to the water, the effect of the intervention of the food web is twofold. Either phosphorus ends up in the biomass of large organisms (macrophytes, fish, birds, mammals), whence it is possibly exported from the lake, or it contributes to a flux of biogenic deposits. Typically, these products range from plant necromass (wood, rhizomes, leaves, consigned to slow, microaerophilous decay) to the rain of fine planktogenic detritus that settles down towards the bottom of the lake. On geochemical scales, such sedimentary phosphorus may eventually be liberated but, in ecological terms, the possibilities for its release and reuse are strikingly habitat-dependent. Biogenic organic materials reaching the bottom of the water column contain still-reduced carbon and a variety of other co-associated elements (including phosphorus). These become the resource of benthic food webs in which detritivorous and decomposer organisms assemble and maintain their respective biomass and perhaps sustain the assembly of benthic consumers (including fish). Collectively, the potential effect of the benthic food web is progressively to oxidise the carbon and to liberate the other elements which steadily become in stoichiometric excess.

The influence of habitat on this process operates first through depth. As has been pointed out earlier (Section 8.2.3), much of the pelagic sedimentary output in the sea is oxidised and its more labile associated nutrients (including nitrates and phosphate) are liberated within 200 m of the surface. Nutrient cycling is not necessarily impaired in lakes substantially <200 m in depth, although the largely diffusive return pathways, even for the most soluble nitrogenous components, operate more slowly than in the open turbulence of the water column. The supply of oxidant can also become problematic, insofar as annual net organic carbon fluxes of >100 mg C m^{-2} carry an oxidative demand that may well challenge the capacity of the 'deep sediments' (in this conext, anywhere between about 5 and 200 m from the surface) to satisfy, especially in periods of density stratification. Indeed, beneath an oxidised microzone, the sediments and their interstitial liquor is anaerobic and strongly reducing. Carbon oxidation and nutrient release are correspondingly slow, ecological function being truly 'oxidant limited' (also as discussed in Section 8.2.3 and represented in Fig. 8.2). With still greater contributions of organic carbon from the production of eutrophic lakes, the oxygen deficit may extend well into the hypolimnetic water above the bottom sediment, where the rate of its biological reuse and organic decomposition is equally depressed.

Against the background of differential oxidative turbulent entrainment rates, the recycling of phosphorus becomes similarly varied (see e.g. Reynolds and Davies, 2001). Orthophosphate ions released from decomposing biomass into the interstitial water of oxidised superficial sediments may travel but a short distance before being exchanged preferentially for hydroxyls on the surface of iron flocs and clay minerals in the sediment (Reynolds and Davies, 2001). Here, they are effectively immobilised, pending the onset of strongly reducing conditions (in which the ferric ion Fe^{3+} ion is reduced, biotically and abiotically, to the soluble ferrous Fe^{2+} ion; see Section 4.3.1) or increased alkalinity, in which excess hydroxyls now displace immobilised phosphate ions. Though potentially bioavailable in this form (Golterman et al., 1969), phosphate thus liberated into the interstitial of anaerobic sediments (or equally, into the anaerobic hypolimnetic water if present) is likely to be once again scavenged and immobilised by iron reprecipitating on exposure to oxygen. This is important, because the act of re-solution of iron-bound phosphate is not 'recycled' to the biota if it is just re-precipitated in the non-bioavailable form. If, in the interim, the mass of reduced iron is diminished, for instance

through its combination with excess sulphide ions (which are liberated by sulphate reduction under slightly lower redox levels than required for oxidation of the ferrous ion) and precipitated as FeS, then the potential for chemical scavenging of phosphate by ferric iron during the next oxidation is proportionately weaker. Accordingly, more phosphate will be available to biotic recycling.

What emerges is that in a majority of lakes of less than 200 m in depth, the bottom sediments normally function as a phosphorus sink. Phosphorus atoms take a single, albeit devious, trip from hydrological catchment to permanent lake sediment. Tomorrow's production effectively depends upon tomorrow's (or even today's) delivery of external phosphorus. Were this not so, neither the Vollenweider model nor the restoration methods that rest upon its central philosophy can work. It is the mechanism by which truncation of the external phosphorus load to the water column results directly lowered phytoplankton-supportive capacity. The new steady-state can be struck, usually within two to three retention times (Ryding and Rast, 1989; Reynolds and Irish, 2000). The limiting condition to this statement is the saturation of the phosphorus-binding capacity of the sediment iron, either by excessive orthophosphate recruitment or by the onset of high alkalinities (caused, for instance, by episodes of high productivity, carbon withdrawal and base generation). With eutrophication comes the prospect of accelerated recruitment to sediments, possible saturation of the P-binding capacity and its complete suppression by anoxic reducing conditions. Even then, the deeper sediments do not give up orthophosphate to the water column very readily and little is necessarily returned to the water column for biotic reuse and recycling.

8.3.4 The internal load problem

As noted by Sas (1989), the further contributory process which so delays the restoration of eutrophied shallow lakes is that the phosphorus-rich interstitial water is much more readily re-entrained than it ever is from deep sediments protected by adjacent boundary layers. Structural disruption of the shallow deposits, mostly through their physical penetration by wind- or wave-driven turbulence currents, may well result in the discharge of interstitial water and its solutes into the main water column. Applied to fine superficial sediments, turbulent shear, dissipating energy at rates in the range, $E = 10^{-6}$ to 10^{-5} m^2 s^{-3} (see Table 2.2 and Section 2.3.4), is capable of penetrating and resuspending material from a thickness 30–40 mm (Nixon, 1988). The process can be abetted through the activities of burrowing animals and foraging fish ('bioturbation'; see, for instance, Davis et al., 1975; Petr, 1977). The kinetics of 'phosphorus release' from the sediments to the water have been the subject of several major overviews (including Kamp-Nielsen, 1975; Baccini, 1985; Boström et al., 1988). Ultimately, however, it is the relative scarcity of significant chemical binding capacity in the water that permits the phosphorus thus released to become once again substantially bioavailable to planktic producers.

The further deduction that can be made is that, whereas deep lakes are likely to respond to (i.e. be sensitive to) managed reductions in nutrient loads, the prospect for successful restoration of eutrophied small shallow lakes is probably much poorer after they have once been enriched significantly. The greater likelihood is that historic phosphorus loads will, on ecological scales, thereafter be recycled indefinitely (van der Molen and Boers, 1994). A case in point is that of Søbygaard, a highly eutrophied shallow lake in Denmark, where a sixfold reduction in the external phosphorus load of $\sim$30 g P m^{-2} a^{-1} had not succeeded, even after 12 years, in reducing MRP levels in the lake to anywhere near a point where they might limit phytoplankton biomass capacity in the lake (Søndergaard et al., 1993; Jeppesen et al., 1998). By analogy, efforts to reduce external phosphorus loads on large, shallow Lake Okeechobee, Florida, USA, have so far failed to bring about restorative changes in the ecology of the lake, which continues to function on the support lent by internal P cycles (Havens et al., 1996).

The available data make plain that massive P recycling continues to be maintained quite independently of present external loadings. In another much-cited instance, a considerable municipal investment in P reduction had failed to ameliorate the eutrophication of the shallow lake Trummen, in Sweden, and nothing less than the dredging and removal of the veneer of P-saturated sediment was needed in order to restore an acceptable quality to the lake (Björk, 1972, 1988). This is an expensive, disruptive and problematic technique, not least because of the high water content and fluidity of the dredged or air-lifted material. In spite of its potential fertility as a soil conditioner, the material is not easily stored and de-watered except on level ground. Not surprisingly, lake restoration based upon sediment removal has been little practised.

There is, however, considerable interest in a recent methodological development involving the application of clay minerals to shallow water bodies prone to free cycling of phosphorus (Douglas et al., 1998). These imitate the P-binding properties of metal oxides and hydroxides but without causing the toxicity problems that direct dosing of (say) iron or alum salts might cause to water bodies. The binding performance of modified clay minerals based on lanthanum is especially effective, removing phosphorus from solution and firmly immobilising it in the particulate sedimentary fraction.

As the techniques for treating and restoring lakes through nutrient reduction become more varied in concept and suitability, it becomes increasingly important to managers to seek guidance in selecting the most effective approach and to predict its prospects of success. Such knowledge might also assist the prioritisation of investment among competing schemes of restoration. What is required is a simple basis for distinguishing among the sites where, to use the terminology of Carpenter et al. (1999), eutrophication is either readily reversible (response immediate and in proportion to the change in P loading), hysteretic (requiring profound reductions in P input over a protracted time period) or irreversible (recovery impossible by reducing P inputs alone). As indicated in the IMSA study (see Sas, 1989, above), the most sensitive direct measure of the sensitivity of the biomass to change is the amount of bioavailable P capacity in Subsystem 2 that remains outside the biomass. This 'cushion' then represents the hysteretic resistance to change that must be all but exhausted before biomass is made dependent upon altered loads. Sensitivity is then reduced to the latitude of TP–MRP transformation and the extent to which internal recycling of MRP is suppressed by sediment binding.

Following on from an earlier attempt (Reynolds *et al.*, 1998) to categorise the relative dependence of biomass to each of a number of environmental variables, Reynolds (2003c) used the inferred phosphorus thresholds to inform a sensitivity model. This involved scoring against each of a number of categories (trophic state, hydraulic retention, bicarbonate availability and the extent of shallow sediments likely to exchange phosphorus). The lower the score value, the greater is the sensitivity to change.

The first of these took the five trophic divisions in Table 8.1. Where small amounts of phosphorus are available, dependency is high and, accordingly, small changes in P loading will invoke significant responses; thus ultraoligotrophic lakes score 1, hypertrophic 5. Flushing lowers the responsiveness to nutrients; thus, lakes were considered to be sensitive (1) to P-load variation if the hydraulic retention time (t_q) was >30 d, not at all so (3) if t_q <3 d and slightly so (2) when 3 d < t_q <30 d. Bicarbonate alkalinity and the elevated frequency of opportunities for phosphorus exchanges at high pH also affect sensitivity and this is reflected in the grading of alkalinity. Lakes are sensitive (1) where bicarbonate alkalinity <0.4 meq L^{-1}, not at all sensitive (3) to alkalinities >2 meq L^{-1}, and slightly sensitive (2) when 0.4 < alkalinity < 2.0 meq L^{-1}. Similarly, recognising the propensity for shallow-water recycling, water bodies are considered to be sensitive (1) to changes in external load if <15% of the surface covered sediments <5 m deep but insensitive (3) if >50% of the lake is shallower than 5 m; lakes where >15% but <50% is under 5 m deep are considered to be slightly sensitive (2). Multiplication of the four factors yields a product between 1 (a deep, ultraoligotrophic lake with a chronically low phosphorus content, low alkalinity and long retention) and 135 (for a shallow hypertrophic river fed pool, subject to rapid flushing by nutrient rich calcreous flow).

Plainly, the lower the product, the more the lake is expected to respond to altered P loads. Reynolds (2003c) also proposed a further factor to cover the importance of good water quality at the particular site (internationally important amenity, drinking-water source, local wildlife reserve, etc.). It was urged that the total product should not determine prioritisation for corrective treatment but rather prompt further investigation of the importance of overcoming high factors; nevertheless, it was noted that low overall scores (≤ 6) would refer to high-quality sites whose trophic state should be ruthlessly defended, whereas scores up to 32 are indicative of sites that should respond hysteretically to altered nutrient loads. Sites with higher scores may err beyond the bounds of a likely reversal of eutrophication.

The UK Environment Agency now uses a scheme, based upon this approach, to assist the implementation of its Eutrophication Strategy.

8.3.5 Physical methods to control phytoplankton abundance

Alternative practical approaches to the control of phytoplankton biomass yield, the rate of capacity attainment and, in many instances, its species composition have invoked the artificial enhancement of the physical processing constraints. Methods for extending the period of full artificial mixing in lakes deep enough to stratify, either by artificial destratification or by preventing the onset of stratification at all. The principle that is invoked is the one illustrated in Fig. 3.19, where the greater is the depth of the layer in which algae are entrained, then the greater is the dilution of the light-determined supportive capacity. It is not just that the algae (say, at an areal concentration of 80 mg chl$a$ m^{-2}) may be diluted from a high near-surface concentration (80 mg chl$a$ m^{-3}, if confined to the top metre) to being spread uniformly through the top 80 m, at a concentration of 1 mg chl$a$ m^{-3}, but the plot says this is the maximum concentration that an 80- m mixed layer could possibly support, as calculated according to Eq. (3.25) in Section 3.5.3 and assuming the most favourable conditions of insolation (I_0') and background absorption ($\varepsilon_w + \varepsilon_p$). Mixed through only 40 m, the maximum supportable concentration is closer to 20 mg chla m^{-3} (i.e. 800 mg chla m^{-2}). Mixed through 10 m, the maximum supportable concentration is about 160 mg chla m^{-3} (or 1600 mg chla m^{-2}). Using this logic in reverse, increasing the depth of mixing ameliorates the concentration of algae in the reservoir water to be treated for potability, especially if the water is at all coloured or carries a significant non-algal turbidity. The several Thames Valley storage reservoirs that serve London have either been subjected to artificial mixing (Ridley, 1970; Steel, 1972, 1975) or they have been commissioned since that time with the capability built in at the design stage (Steel and Duncan, 1999). The benefits, in terms of filtration efficiency and savings on the cost of treatment, have been substantial and worthwhile and the nuisance Cyanobacteria are scarcely represented any longer in the planktic flora of the reservoirs in operation (Toms, 1987; Steel and Duncan, 1999).

The methods for overcoming and preventing thermal stratification and for extending the period of non-stratification have been in use for many years and are well established (Irwin et al., 1969; Dunst et al., 1974; Tolland, 1977; Simmons, 1998; Kirke, 2000). The most successful of these function as lift pumps, comprising a vertical cylinder with either a paddle or a injected bubble stream to draw cold deep water to within a short distance from the surface where the flow diffuses laterally. Another device, the Helixor®, uses an Archimedean screw to achieve a similar effect. By setting up entraining vortices, these devices move rather more water than flows through the cylinders. Moreover, they work reasonably efficiently so long as water is discharged from the pipe below the surface of the water body: lifting water above the surface level requires a rather greater pumping effort. Most use electrical power to drive the pumps (the alternative is wind power) and their operation carries significant costs, which have to be balanced against the savings to be made on treatment and disinfection.

Another popular method of weakening stratification is to use bubble plumes: punctured hose or tubing is laid out on the reservoir floor and compressed air is forced out of the perforations.

This also requires energy expenditure and, thus, an appropriate cost–benefit analysis; the benefits may include the direct effects of aeration, as well as mixing. Yet another method is to force the inflow through angled nozzles that direct the jets of water to maximise their homogenising effect (Toms, 1987).

Caution is urged, however, against an assumption that lack of stratification is necessarily indicative of being well mixed. My experience from several reservoirs where pumps or bubble plumes are in operation agrees with the view that temperature gradients are generally weak, or non-existent in the immediate environs of the device. Further away, phytoplankton may still be disentrained and the self-regulating species may perhaps engineer slightly better growth conditions for themselves. It may be sufficient just to keep stratification weak, so that the work of wind and cooling can more easily mix the water column from time to time. Modelling algal growth in reservoirs with PROTECH (Section 5.5.5; work still in process of publication at the time of this book going to press) has strongly suggested to me that many such reservoirs are often less well mixed than supposed. On the other hand, other reservoirs without any other source of artificial mixing also remain very weakly stratified, at those times when water is drawn off for treatment from near the bottom of the water column. Unless it is unsatisfactory on chemical or biological grounds, I believe it to be preferable to draw water from depth as it yields water of reasonable quality but (literally) undermines the tendency to stratify. The power-generating plant situated at the dam of Brasil's Volta Grande Reservoir draws its flow from the base of the reservoir and, in complete contrast to others in the cascade, fails to stratify (Reynolds, 1987b). A 70-m, near-isothermal water column in a tropical impoundment is an unforgettable surprise to a limnologist!

The phytoplankton in this reservoir also differed from that of others in the chain. The unstable water column of Volta Grande was dominated by the diatom Aulacoseira and the desmid Staurastrum, together constituting an assemblage classifiable as Association P. This is one of the groups of mainly R-strategist groups whose natural occurrence is in well-mixed water columns and whose ascendency can be stimulated in lakes and reservoirs by episodes of artificial or experimental mixing (Lund, 1971; Reynolds et al., 1983b, 1984; Reynolds, 1986b). Restratification may revert the flora to species either less dependent upon entrainment for suspension or more dependent upon disentrainment to have any chance of harvesting sufficient light (i.e. C-strategist nanoplankton and motile microplankton, including the larger, self-regulating S-strategist species). The possibility of alternating conditions, before any particular set of adaptations led one group of species to become overabundant, was specifically investigated in the study of Reynolds et al. (1984) and through the population dynamics of selected algal respondents (Reynolds, 1983a, 1983b, 1984c). This led Reynolds to propose an approach to managing reservoirs by mixing them intermittently, in order to prevent any group becoming too numerous. Reynolds et al. (1984) conceded that it was probably too cumbersome to apply safely, in the sense of maintaining close control over phytoplankton biomass. Generally (and always provided that the water column is sufficiently deep and/or the water sufficiently turbid), persistent mixing to control light-determined carrying capacity is the simpler and more efficient option. However, a potential drawback of doing this, which was of concern to Lund (1971, 1975; Lund and Reynolds, 1982), is that prolonged mixing would lead surely and ultimately to dominance by such organisms from functional group S1. Once established, prevalence of Planktothrix agardhii, Limnothrix redekei or Pseudanabaena limnetica is hard to overcome (Reynolds, 1989b). That this has not happened in the case of the routinely destratified Thames Valley reservoirs seems to be a function of their short retention times (usually <20 days) (Toms, 1987; Reynolds, 1993b). The combination of a hydrological displacement time of just 10–30 days and a fairly intensive mixing regime is perhaps decisive in the continued exclusion of Planktothrix.

Mention of the treatment or prevention of algal growth in drinking water reservoirs gives an appropriate opportunity to refer briefly to once-common practice but now much discredited technique of directly applying powdered copper sulphate to reservoirs as an algicide. The salt dissolves easily in the water, where a strength of 1 mg L^{-1} of the penthydrate is sufficient to kill most species of phytoplankton and – incidentally – more than enough to administer a lethal dose to crustacean zooplankters (my own unpublished observations). The effectiveness of the treatment against algae seemed to vary, apparently in relation to the degree of mucilage investment (Coesel, 1994; see also Box 6.1, p. 273). Most of the copper is eventually precipitated and sinks to the sediment. Therein resides the problem, for the copper remains sensitive to changes in acidity and redox potential. The salutary case of the gross copper poisoning of the Fairmont Lakes Reservoir, Minnesota, USA, following years of frequent dosing to control algal growths (Hanson and Stefan, 1984) both confirmed the worst fears about the potential threats of long-term copper applications and doubtless influenced the banning of its use in many countries.

8.3.6 The control of phytoplankton by biological manipulation

As an alternative to being able to control adequately the biomass-supportive capacity, the idea of regulating its allocation among desirable and undesirable biota has had a long-standing appeal in lake management. The term 'biomanipulation', coined originally by Shapiro et al. (1975), has become increasingly restricted to refer to the control of undesired organisms by the deliberate adjustment of the abundance of the next trophic level of consumers. The underpinning logic is the same one that distinguishes 'bottom-up' from top–down processes (see pp. 287–8). Thus, not surprisingly, it has been just as controversial and contested just as vehemently (see, for instance, Carpenter and Kitchell, 1992; De Melo et al., 1992). However, also like the 'bottom–up vs. top–down' debate, the issues ceased to be so contentious, once the critical roles of variability and site-specific factors affecting resource turnover became more clearly understood. There is, for instance, a much wider acceptance of the importance of the basic carbon pathways in determining the optimum functional adaptations and, so, the identity of the most efficient traits (discussion in Sections 8.2.3 and 8.2.4). Interestingly, the assumption of direct and manipulable linkages between phytoplankton, zooplankton, planktivorous fish and piscivores is now argued to provide an acceptable model only for truly pelagic systems. To apply the deductive reasoning that says (e.g.) the selective removal of piscivores distorts the food web by sparing planktivorous fish to prey more heavily on zooplankton and promote a larger phytoplankton biomass, requires effective management over very substantial areas of the pelagic. Similarly, to approach the problem of phytoplankton overabundance fuelled by extra nutrients simply by reducing planktivory or increasing piscivory in a small lake is rather to ignore the role of the terrestrial subsidies and benthic feedbacks in undermining the system reliance upon phytoplankton photosynthesis. Such retrospective considerations aid the appreciation that the more successful attempts to devise sustainable biomanipulative schemes for controlling phytoplankton and attaining an aesthetically pleasing water clarity have been most successful in small, shallow lakes and ponds and least so in larger, deeper lakes (Reynolds, 1994c). It is nevertheless necessary to qualify this statement again by saying that 'if to biomanipulate is simply to distort the food chain simply attain a more beneficial condition, then there is no theoretical objection to biomanipulating an area the size of the Atlantic Ocean' (Reynolds, 1997a). If, however, the intent is an alternative, stable and self-sustaining system but running on, broadly, the existing resource and energy inputs, then attention has to be given to the means of substituting alternative system components that process those inputs along quite different pathways. These alternatives are confined to small or shallow systems. Biomanipulation of the functional importance of phytoplankton is a practical and sustainable corrective to the symptoms of eutrophication only in such water bodies.

In the context of the present chapter, the adjective 'shallow' is used more in its functional sense than within any absolute constraint (Padisák and Reynolds, 2003). The most important property is the frequent or continuous contact of most of the bottom sediment with the trophogenic layer of the water column and its periodic exposure to the shear stress of the suface mixed layer. This property influences both the cycling of resources and the interactions between benthic and planktic consumers. However, the remarks apply broadly to lakes in which >50% of the area is shallower than 5 m (as in Section 8.3.4).

The supportive capacity of shallow lakes may be regulated by hydraulic throughput, turbidity or colour but, naturally, the supply and chemical binding of phosphorus are frequently such to regulate the accumulation of biomass. Modest depth and high perimeter to volume ratio, however, leave the small water body quite vulnerable to drainage and the import of solutes and particulates arising from land-use change, loss of vegetation, increased particulate loss from cultivated soil, as well as to incursion by any of a large number of organic pollutants. With the possible exception of chronically nutrient-deficient small lakes, many are susceptible to quite rapid ecological change – between clarity and high algal turbidity and between supporting extensive macrophytic vegetation, both submerged and emergent. These fluctuations among periodically steady states have been studied for some time (Scheffer, 1989; Blindow et al., 1993; Scheffer et al., 1993; Carpenter, 2001). Some of these transitions have now been empirically characterised and the outcomes are helpful in informing strategies for securing the successful biomanipulative management of shallow lakes. The following subsections recount these.

Nutrients, phytoplankton and grazers

There is no fundamental difference between large and small lakes in the supportive capacity of plant nutrients so the relationship between the potential yield of phytoplankton chlorophyll and the supportive capacity of bioavailable phosphorus (as suggested in Eq. 4.15, Section 4.3.4) is independent of lake size. Scheffer's (1998) analysis of the average summer chlorophyll yields of 88 shallow Dutch lakes, showing a slope close to 1 μg chl$a$ (μg TP)$^{-1}$ and saturating at about 300 μg TP L^{-1} is in accord with expectation. Such concentrations of chlorophyll a are unexceptional in nutrient-rich shallow water, where they may typically characterise near-unialgal populations of group-J or group-X1 chlorophytes or group-S1 Planktotricheta (see Section 7.2.3).

Whereas the latter may dominate perennially, the green algae may be subject to abrupt collapses, usually as a consequence of intensive grazing by filter-feeding zooplankton, or, subsequently, to recovery to large concentrations. The dynamics of these fluctuations are best known from systematic observations on commercially managed fishponds (Hrbácek et al., 1961; Hillbricht-Ilkowska and Weglenska, 1978; Korínek et al., 1987) but many have been imitated in controlled experiments (Shapiro and Wright, 1984; Moss, 1990, 1992; Carpenter et al., 2001). Unchecked by consumers, phytoplankton can grow to the limits of the supportive capacity (which may be set by light rather than by phosphorus). Direct herbivorous consumers include rotifers and crustaceans and, at low latitudes, certain species of fish (Fernando, 1980; Nilssen, 1984). Among temperate lakes, filter-feeding daphniids prove to be well capable of developing biomass (equivalent to ~1 mg C L^{-1}) (Section 8.2.3) with an aggregate filtration capacity (approaching 1 L L^{-1} d^{-1}) that will overhaul the rate of algal growth and rapidly clear the water of all fine particulates. Without access to filterable foods, the Daphnia quickly starve and there is mass mortality among the smaller instars especially (Fig. 6.4). Filtration collapses, and the next phase of algal dominance can commence.

This see-saw between alga and grazer mass is scarcely a management formula but the fluctuations may well be damped in the presence of significant densities of young cyprinid fish. The expectation is that fish consumption of zooplankton will reduce the exploitation of the phytoplankton yet still control it within acceptable levels. The belief is that the persistence of high algal populations is the consequence of overstocking with fish and the zooplankton is simply squeezed out. The truth is that even the three-component system is extremely difficult to balance. If a stable phytoplankton is desired, it needs to be cropped as fast as it self-replicates. If it is able to double once per day, then the zooplankton needs to filter about half the water volume each day. If the requisite biomass (roughly 0.5 mg C L^{-1}) is

to kept in check from its own potential growth and recruitment of filtration capacity (say 0.19 d^{-1}), fish feeding needs to remove about 0.1 mg zooplankton C L^{-1} d^{-1}, but with rather narrow margins of variation. If we take the model of Elliott and Hurley (2000a; see Section 8.2.2) and estimate a requirement of 40 mg zooplankton C per gram fresh-weight of fish per day, then the required stability would be struck when the fish biomass is equivalent to 1 g per 400 L or 2.5 g fresh weight m^{-3}. No allowance is made in these approximations for variable temperature, day-to-day differences in phytoplankton growth potential or for the divorced timescale of changes in fish biomass, through progressive growth and increasing appetites.

Again, there is little encouragement here for realising a reliable and stable managed state. In one of the few really clear cases of an engineered stability among three trophic levels, Kořínek et al. (1987) observed a persistent standing population of cryptomonads, of 2–4 µg chl*a* L^{-1} in a carp pond. The check was exercised by an apparently steady population of Daphnia pulicaria, numbering about 100 L^{-1}. Carp were present but at an unspecifed 'low level'. On restocking with fish in the summer, there was, predictably, an abrupt decrease in Daphnia, followed by a rapid increase in chlorophyll *a* concentration, to >100 µg chl*a* L^{-1}.

The role of macrophytes

This last example supports the conjectures about stability but it also reinforces the intuition that heavy fish stocking is conducive to the overabundance of phytoplankton in small lakes. On the other hand, the lurching, cyclical alternation of bottom–up production and exploitative, top–down responses is probably the norm in such highly managed systems. That similar behaviour is not so apparent in less impacted systems, which may support fish production but without being subject to algal blooms of a consistency approaching that of pea soup, requires us to look more closely at the trophic pathways in other types of small lake. Stability of water quality, at least at the time scales of season to season and, in many instances, from year to year, is usually associated with a strong presence of macrophytic angiosperms and pteridophytes.

A number of families is represented in the aquatic macrophytic flora (for an up-to-date account, see Pokorný and Kvet, 2003). All show secondary adaptations to one or other of the habits that they have adopted. These broadly separate among species that are either: rooted and fully submerged; rooted with emergent foliage; rooted with floating leaves; unrooted and floating. Besides supplying one of its key needs in abundance, water offers to submerged macrophytes the benefit of Archimedean support and the equability of temperature fluctuation that surpasses that experienced by their terrestrial relatives. Aerating the submerged tissues (especially the roots) was a major constraint to returning to water but its achievement was a major evolutionary advance. The large interconnected air passages and the aerenchymatous tissue that distinguish the structures of most aquatic species are illustrative of the remarkable power of adaptation. Rooted macrophytes vie with microphytic algae for light and nutrients but it is plain that they have inferior rates of growth and replication to autotrophic microorganisms. However, most macrophytes enjoy two crucial advantages over microalgae: through their roots, they have access to nutrients in the sediments that are not normally available to algae; and they are able to develop tissues for the internal storage of excess photosynthetic products and reserves of potentially limiting nutrients. Aquatic macrophytes are able to colonise standing water in all climatic zones. Their distribution within lakes is restricted to shallow water, to depths depending mostly upon its clarity (the typical Secchi-disk reading is often a good guide to the limits of colonisation by submerged plants; see Blindow, 1992). In clear, deep lakes, there is usually a vertical zonation of the principal species and life forms (Pokorný and Květ, 2003). In the horizontal, there may be many shores from which they are excluded, for reasons to do with the suitability of the substratum and its exposure to wind or wave action. Species also vary in the degrees of acidity or alkalinity they will tolerate, and many are susceptible to changing levels of productivity.

In this way, different types of macrophyte tend to be associated with particular habitats. A full spectrum of life forms may be encountered in clear, oligotrophic lakes. *Lobelia–Littorella* stands develop offshore, where, simultaneously, the sediments are sufficiently fine and the light penetration is good. Deeper and further out, dense stands of naiads and elodeids may develop, merging finally with the isoetids, the deepest-growing macrophytes. Emergent flowering plants (reeds and reed-like plants such as *Phragmites*, *Typha*, *Schoenoplectus* and some associated herbs) may develop on more silted shores, while stands of fully submerged plants (such as *Myriophyllum* and *Najas* and various *Potamogeton* spp.) and plants with floating leaves (nymphaeid water lilies) form in front of them in quieter bays. In more enriched lowland lakes, the reedswamp may be prolific but the submerged plants are compressed into narrow depth ranges. In small, shallow ponds, a submerged vegetation of *Myriophyllum* or *Ceratophyllum* may carpet the entire bottom. In clear, calcareous shallows, the dominant plants may be macroalgae – the stoneworts, such as *Chara* and *Nitella*.

In their respective localities, macrophytes may be regarded as 'system engineers' (Jones *et al.*, 1994), fulfilling a keystone role to many of associated species (see Section 7.3.4). Dense plant stands suppress turbulence and create a calm, cryptic habitat to a trophic network of invertebrates. It is an environment that traps fine silt and organic debris, whose deposition leads to a truncation of the water depth and a succession of swamp, marsh, and, finally, woodland species, whose centripetal invasion potentially obliterates open water (Tansley, 1939). The macrophytes or, more particularly, the fungal fragmentation and bacterial decomposition of seasonal dieback, provide some direct source of nourishment to the community of benthic invertebrates but their primary role in the energy flow of the lake lies in the provision of habitat to many other producers. Surface-growing (epiphytic) algae and fungi (together, the *Aufwuchs*) constitute a source of food to snails, microcrustaceans and larval ephemeropterans. These have a variety of potential predators, including flatworms, hirudineans and beetles. Accumulating organic debris and detritus is exploited by a variety of malacostracans and larval hemipterans and by dipterans, oligochaetes and bivalves. There is another layer of predatory carnivores including larval odonates and coleopterans. Macrophyte beds play host to a very complex web of trophic interactions. Ideally, their part in the economy of the lake increases relatively and inversely to the size and depth of the lake. Quite plainly, in small or shallow lakes, macrophyte beds are the focus of the most intense production and the location of most of its producers.

Effects of eutrophication on macrophytes

In many instances of eutrophication of shallow lakes, the first symptom of change has been the disruption and decline of macrophyte dominance. Charophyte communities are considered to be the most susceptible to replacement by more diverse assemblages of species of richer lakes, before these too become eliminated (De Nie, 1987). The complete transition, to a turbid, phytoplankton-dominated system, is regressive on grounds of loss of diversity and amenity, as well as a probable deterioration in the yield of fish. The mechanism of such transitions was supposed to be light-mediated: as nutrients in solution increased, so did the phytoplankton and, in consequence, light penetration decreased. Even the intermediate replacement of low-growing charophytes by tall-stalked *Potamogeton* (Moss, 1983) fits this explanation. However, it is only part of the story and it does not explain why species with floating leaves and the emergent reedswamp species should also fail to survive. Work by Sand-Jensen (1977) had suggested that the shading of epiphytes growing on eelgrass (*Zostera*) significantly reduces light penetration to the leaf blade. At about the same time, Phillips *et al.* (1978) reported field observations and some supporting tank experiments in which the effect of epiphytic algal growth was to smother the photosynthetic surfaces of the host plant. Although a film of epiphytic algae is probably a normal occurrence on the surfaces of water plants, Phillips *et al.* (1978) considered that the increasing availability of nutrients in the water column is as beneficial to epiphytes as to planktic algae.

This explanation is now widely accepted. It has been supported by some direct measurements of the light attenuation by epiphytes on eelgrass (Brush and Nixon, 2002). The critical photon flux transmitted to the photosynthetic apparatus to the host plant, before its own performance is impaired, will vary with its own depth and the levels of harvestable light. However, Brush and Nixon (2002) developed regressions predicting epiphyte densities of 10 mg dry mass per cm^2 reduced PAR transmission by 33–70%. This is slightly less than some of the values suggested in earlier studies. The nature and habit of the epiphytes also influences light transmissibility: the biomass achieved by crustose growths of naviculoid diatoms is less than what is attainable by erect cells of *Synedra*-like diatoms or the branching arbuscular formations of *Gomphonema* or many chlorophyte forms. Nevertheless, the expectation is that epiphyte densities of 20 mg dry mass per cm^2 would exceed their tolerance by macrophytes. Actually, such densities become unstable and slough off periodically, although the host plant would, by then, be unlikely to derive much benefit. Depending on species composition, the chlorophyll *a* complement of the epiphytes might be 0.5–2.0% of the dry mass: the critical epiphyte density of 10–20 mg dry mass per cm^2 is equivalent to 0.05–0.4 mg chl*a* cm^{-2}, or 0.5–4 g chl*a* m^{-2} of macrophyte leaf area. To compare this with the areal concentration of phytoplankton, or even of epilithic films on the solid surface of rocks and stones, allowance must be made for the leaf-area index (the area of leaf surface per unit water area) (LAI). For a LAI of ~5, potential densities of epiphytes (on this logic, up to 20 g chl*a* m^{-2}) would be far in excess of supportable phytoplankton or epiliths. Turning the calculation around, the theoretical light-limited maximum active photoautotrophic biomass that can be supported (~120 g C m^{-2}) Fig. 3.8, Section 3.5.3) and the maximum phytoplankton crops that are observed (range 30–50 g C m^{-2}, say, 0.8–1 g chl*a* m^{-2}) would leave submerged macrophytes already light-deficient.

Of course, the epiphytic community of the *Aufwuchs* is not exclusively algal and its chlorophyll content is normally dynamic owing to grazers. Macrophytes may be able to suppress epiphytic and some planktic algae through the production of allelopathic substances (reviewed in Scheffer, 1998), although, realistically, the principal influence over dominance is the competition for space, light and nutrients. This struggle is further influenced by the role of consumers – especially planktic filter-feeders and *Aufwuchs* browsers. The complexity of these interactions make it difficult to predict their many potential outcomes. Nevertheless, some experiences of the change from macrophyte to microplankton dominance of shallow lakes, and the reverse, focus on the critical switches.

The eventual loss of submerged macrophytes as a consequence of progressive eutrophication of shallow lakes has often been found to be quite abrupt, presumably as a result of near-simultaneous light inadequacy across a relatively flat bottom. It is often the case that macrophytes are either extensive across the lake, or they are very nearly absent from a water column that is turbid with phytoplankton. Thus, these small, shallow lakes seem typically to exist in one of two *alternative steady states* – either they are macrophyte-dominated with clear water; or they are phytoplankton-dominated, with frequent high turbidity and a dearth of macrophytes (Blindow *et al.*, 1993; Scheffer *et al.*, 1993).

From syntheses based on the eutrophication of a large number of shallow lakes in Europe, it appears that the change from macrophyte to phytoplankton dominance occurs anywhere within a wide range of aquatic total phosphorus concentrations (50–650 µg P L^{-1}). These are arguably capable of sustaining phytoplankton and/or epiphytic algae of chlorophyll *a* at concentrations likely to deprive submerged macrophytes of adequate light. According to Jeppesen *et al.* (1990), many upward transitions occur in a range 120–180 µg P L^{-1}, in which the phosphate sequestering power of macrophytes is saturated (Søndergaard and Moss, 1998). However, it does not follow that a downward shift in phosphorus availability will trigger a return to macrophyte dominance in the same range. Indeed, it seems that phosphorus concentrations need to be rather lower than 120 µg P L^{-1} for the switch to work in the opposite direction (Moss, 1990).

This hysteretic behaviour of forward and reverse switches has been successfully modelled by Scheffer (1990; see also 1998), using graphical analysis. The model has stimulated further investigative studies of the factors that most influence site-to-site differences in the trigger levels of phosphorus and the response of the whole system.

Macrophytes and trophic relationships

The influence of fish on the operation of the lower trophic levels is no less profound in the littoral world of shallow lakes than it is in the pelagic zone (Gliwicz, 2003b). In the absence of fish predation, filter-feeding crustacean zooplankton can become sufficiently numerous to clear the water of algae. In macrophyte-rich ponds, however, alternations with algal abundance are considerably damped, both in intensity and through time. The basis of this relative stability seems to be that a low concentration of large-bodied, filter-feeding cladocerans, involving genera such as *Simocephalus*, *Sida*, *Diaphonosoma* and the large *Daphnia magna*, is able to keep the water reasonably clear of phytoplankton. The supply of locally generated and/or trapped organic debris, dislodged epiphytes and abundant detritus and bacteria from near the bottom provides an alternative and sufficient resource of filterable organic carbon to support the stock of cladocera (Gulati *et al.*, 1990).

The presence of fish does not automatically disrupt this stability. Cropping of zooplankton by mature fish and 0+ (i.e. fry under 1 year in age) is not necessarily less intense in shallow lakes than it is in deeper lakes. Indeed, without the refuge provided by vertical migration, zooplankton is potentially subject to even heavier predation. Several studies have revealed that the behaviour of free-swimming cladocerans alters under the threat of fish predation, exploiting the alternative refuge from ready visibility offered by macrophyte beds. Horizontal migrations were first noted by Timms and Moss (1984), the zooplankton moving into macrophytes by day and moving into more open water in the hours of darkness. Moreover, subsequent studies have demonstrated that the variety of species and the size classes of the cladocerans behaving thus is directly related to the potential intensity of planktivory (Lauridsen and Buenk, 1996; Lauridsen et al., 1999). The experimental investigation of the behaviour of Daphnia magna (Lauridsen and Lodge, 1996) showed a natural tendency to avoid the Myriophyllum plants was reversed in the presence of sunfish (Lepomis cyanellus) and of chemicals in the water in which the fish had been recently present. In effect, the cladocerans benefit from the presence of macrophyte beds through reduced exposure to planktivory but remain sufficiently active to regulate the concentration of phytoplankton and other forms of fine particulate organic carbon. These are powerful contributions to the short-term ecological stability of macrophyte-dominated systems.

However, these are not continuous forces, neither are they insensitive to changing fish densities, species composition and feeding refuges, nor are they independent of macrophyte density. High densities of facultative, opportunist and obligate planktivores in open water (in small, shallow lakes, this may refer mainly to young percids and cyprinids, as well, of course, as the 0+ recruits of most species) will eventually overcome the capacity of cladocerans to clear the water. Encouraging piscivorous predation on planktivores (for instance, by stocking, with pike, Esox lucius) should assist in keeping their impact on the zooplankon within its critical threshold intensity and, thus, in upholding and enhancing the stable, macrophyte-dominated state (Grimm and Backx, 1990; Bean and Winfield, 1995). Otherwise, phytoplankton may be released from effective control and the balance in favour of macrophytes, supposing that they really do dominate the material transfers within the existing system, may quickly be turned against them. Moreover, with the foraging of the facultative planktivores focused on benthic invertebrates (see Section 8.2.2), there is an increasing risk that the weakened macrophytes are uprooted and dislodged. Sediments are increasingly liable to turbation with further feedbacks towards macrophyte exclusion by turbidity and loss of refuge for zooplankton. The whole system can quickly experience a rapid 'regime shift' (Carpenter, 2001), towards a simplified, low-diversity structure based upon copious planktic primary pro-

duction (usually by large and persistent populations of Planktothrix agardhii), robust, microbially mediated recycling of autochthonous and allochthonous organic matter, and a food web comprising little but the larvae of Chironomus plumosus or C. anthracinus and benthic-foraging carp, such as Cyprinus carpio. Most people find this unattractive and believe its ecosystem health to be poor but it is, nonetheless, a stable state and also one that is 'depressingly sustainable' (Reynolds, 2000b). It is for these reasons that it is also very difficult to manage sites away from this kind of species structure without applying some fairly drastic environmental engineering.

Biomanipulative management
The prospects for preventing and reversing the worst symptoms of eutrophication in small, shallow lakes probably depend upon first recognising the present state. Traditionally, the art of successful biomanipulation is built upon the control of phytoplankton abundance, which really means protecting the zooplankton. There are thus threshold levels of planktivory that should not be exceeded. Even then, phosphorus availability may still bias against macrophytes, whose collapse will precipitate the loss of a viable agent in maintaining water quality.

Quite clearly, if habitat quality once slips to that level, then recovery demands that the tight linking among the minimal components of the whole system is broken. This may require drastic reductions by netting of stocks of cyprinid fish (roach, Rutilus; bleak, Alburnus; bream, Abramis; carp, Cyprinus carpio), as well as small individuals of non-cyprinid species. It may also require the dredging out of the unoxidised sediments and/or routine flushing with less nutrient-rich water. The assisted re-establishment of appropriate macrophytes and their protection from grazing birds (such as coot, Fulica atra) may prepare the way for restocking with non-cyprinid fish.

Such restorations are cumbersome and expensive to apply and the precedents have achieved only varying degrees of success. Needless to say, instances of successful, self-sustaining biomanipulative management have been attained before the undesirable phytoplankton-dominated steady state has become too firmly established (Hosper and Meijer, 1993). Often, it is sufficient to intervene before the 'forward switches' have operated. Over-management of macrophytes and overstocking with fingerling fish, both perpetrations of single-minded but misinformed anglers, have predictably deleterious impacts on zooplankton stocks and the objective of limpid water.

Once the stability of the macrophyte-dominated state has been threatened, greater control over planktivore stocks must be exercised. Based on the investigations of Gliwicz and Preis (1977) on the direct impacts of planktivory on zooplankton, it would seem unwise to allow the fresh biomass to exceed 300 kg ha^{-1} (30 g m^{-2}, about 6 g C m^{-2}); to do so would impair the sustainability of secondary production yields (see Sections 6.4.2, 8.2.3). Stocking of benthic foragers (carp, bream and tench, *Tinca tinca*) should be avoided. It may be necessary to exclude bottom-feeding ducks, coots and other rails. Other 'reverse switches', including the reduction of nutrient loading, should be applied if feasible. For fuller advice and guidance on applying biomanipulative techniques, any of the excellent manuals now available should be consulted (Hosper *et al.*, 1992; Moss *et al.*, 1996).

8.3.7 Phytoplankton and acidification

Acidic waters are not unnatural but anthropogenic acidification of rainfall and land drainages has impacted upon aquatic ecosystems with effects quite as deleterious as those imposed by nutrient enrichment. However, the behaviour of phytoplankton is rather less central to the perceived environmental damage through acidification than it is in the case of eutrophication. The restoration of acidified water bodies is not a major concern of the present section; rather, the relationships between phytoplankton and environmental excesses in the hydrogen ion concentration are the main focus.

Natural water supplies to lakes and seas are not pure. Characteristically and distinctively, they contain varying amounts of numerous solutes, leached from the atmosphere and from the terrestrial catchments with which the water has been in contact. In much the same way as nutrients are loaded on receiving water bodies, the concentrations of acidic gases and radicals and the concentrations of neutralising bases determine the hydrogen ion concentration in lakes and seas. Lakes are, or become, acidic when the input of hydrogen ions exceeds the amounts of neutralising bases gleaned from the watershed through the weathering of rocks. Phytoplankton production in the lake also helps to reduce acidity through the uptake of nitrate, sulphate and carbon dioxide gas dissolved in the water.

The simple methodology for estimating the hydrogen-ion concentration (expressed as a negative logarithm, the pH) allows us to record it with some diffidence. The effects of photosynthetic withdrawal of carbon dioxide and the buffering provided by the dissociation of the weakly acidic bicarbonate ion have been discussed earlier (see, especially, Sections 3.4.2, 3.4.3). The carbon dioxide dissolved in precipitation (rainfall, snow, condensation) comes the atmosphere. The natural pH of rainwater may be between 5.0 and 5.8, depending upon contemporary temperature and pressure conditions and the extent of carbon dioxide saturation. The particular problem of 'acid rain' begins with the anthropogenic enhancement of the products of oxidation – oxides of carbon, sulphur and nitrogen – in the atmosphere. There, solution and reaction with liquid water results in the enhanced formation of strong acids (H_2SO_3, H_2SO_4, HNO_3). On contact with the ground, acidic precipitation may be neutralised or rendered more or less alkaline by the bases (carbonates of Ca and Mg, K$^+$, Na$^+$ and NH$_3$) that it dissolves. Alternatively, rainwater passing across impervious catchments dominated by granitic, dioritic and other weathering-resistant rocks acquires little base or alkaline buffering capacity. Accordingly, lakes and rivers draining catchments in which the buffering capacity is generally poor, achieving alkalinities of ≤ 0.2 meq L^{-1}, are particularly sensitive to 'acid rain'.

At first, the symptoms of acidity generation were recognised mainly in the immediate downwind localities of industrial and domestic fuel-burning: areas of northern England and western Germany were once very badly affected with damage to trees and buildings, loss of soil fertility and lowered pH of drainage water. Through much of the industrial period, these effects were secondary to (but they compounded) the more obvious consequences of the deposition of unburned carbon (soot). Analogous problems arose from ore-smelting in the area of Sudbury, Ontario. As much to deal with local air-pollution problems as for any other reason, a progressive switch was made to alternative, more completely burning fuels that are combusted at higher temperatures, whilst waste gases were vented higher into the atmosphere. At the same time, the total consumption of carbon-based fuels fuel has greatly expanded, with the result that local problems of poor air quality and noxious deposition were replaced by a global one of acidified precipitation. It is not, however, a straightforward case of a universal and systematic lowering of the pH of rain: depending upon the provenance of a particular airstream and its recent history of rainfall elution, precipitation is much more subject to what are sometimes termed 'acid events' (Brodin, 1995). Since the mid-1980s, some amelioration has come through better understanding and recent international concord has secured cuts in the emissions of strong acids to the atmosphere. The expedient of burning natural gas in preference to oil or coal also generates less sulphate per unit energy yield. Nevertheless, there can be few locations on the Earth where rainfall does not, at times, continue to deposit abnormally enhanced loads of acid.

In the present context, the concern lies in the precipitation after it has reached the ground. It is early on in its percolation into a well-formed soil and its throughflow to the surface drainage that the chemical composition of the drainage to rivers and lakes is mainly determined. It is quite apparent that the natural acidity of rain is normally quickly neutralised by the bases present and which, generally, are derived from the weathering of parent bedrocks. However, where rocks are hard or slopes are steep, soil cover is thin and bases are deficient, the acidity is not overcome and it registers in the lake water. Acidity in drainage may actually be enhanced by certain types of vegetation exchanging hydrogen ions for valuable nutrients. Thus, as already indicated, almost all the reported instances of progressive acidification rising to levels damaging

aquatic ecosystems have come from base-poor, high-rainfall upland areas, where the inherent neutralising capacity is least: northern Europe (Scotland, Fennoscandia and, especially, the Telemark area of Norway) and certain Shield areas of Canada.

The impacts of acidity on aquatic ecosystems invoke several mechanisms. Besides having to cope with an excess of hydrogen ions, low pH affects the chemistry of several elements of biological importance. One obvious consequence of the direct sensitivity of calcium carbonate solubility to pH affects the tolerances of shell-building molluscs and carapace formation in the crustaceans (Økland and Økland, 1986). Also critical is the solubility and aluminium ions which are extremely toxic to organisms not equipped to deal with them (Brown and Sadler, 1989; Herrmann et al., 1993). Metals, including iron and manganese, may be activated to harmful levels through the disruption of dissolved humic complexes (DHM) (see Section 3.5.4).

Aquatic organisms vary in their sensitivity to low pH and elevated concentrations of aluminium and manganese. Examples of species (or races) of fish and invertebrate that are tolerant of high acidity levels have been documented (by e.g. Almer et al., 1978). However, those close to the extremes of physiological tolerance may suffer catastrophic mortalities as a consequence of a relatively small downward drift in pH during a single acid event. Among the macroinvertebrates, only insects seem tolerant of high acidity (Henrikson and Oscarson, 1981). Investigations of known acid-tolerant species of phytoplankton have revealed a biochemical mechanism in Chlorella pyrenoidosa, Scenedesmus quadricauda (Chlorophyta) and Euglena mutabilis (Euglenophyta) for regulating internal pH against high external H^+ concentrations (Lane and Burris, 1981). However, the ability of E. mutabilis to deal with mobilised aluminium species (Nakatsu and Hutchinson, 1988) may be the decisive specialist defence against extreme acid enviromnents.

Some algae, including desmids (Cosmarium, Cylindrocystis) and eustigmatophytes (Chlorobotrys), are characteristically associated with acidic pools in wet Sphagnum bog. Among the freshwater phytoplankton, tolerance of natural high acidity is shared among a number of other species (see, for instance, Swale, 1968; Nixdorf *et al.*, 2001), many of which do happen to be flagellated chlorophytes or euglenophytes. Species of *Chlamydomonas*, *Sphaerella*, *Euglena* and *Lepocinclis* are frequently encountered in small acid lakes, one or other usually dominating a low-diversity population. Cryptophytes (*Plagioselmis*, *Cyathomonas*), chrysophytes (*Ochromonas*, *Chromulina*) and dinoflagellates (*Gymnodinium*, *Peridinium umbonatum*) are also recorded, as are certain non-motile chlorococcalean species of *Chlorella* and *Ankistrodesmus*. Some species of the diatom genera *Eunotia* and *Nitzschia* are also known to be acid tolerant but most planktic species of Cyanobacteria do not grow at pH levels much below 6.0 ($H^+ \geq 1$ μmol L^{-1}).

Swale (1968) recorded occasional large populations of *Lagerheimia genevensis* and a 14-month period of dominance by *Botryococcus braunii* in Oak Mere, England (contemporaneous pH 4.7–4.9). In the extremely acid lakes left by lignite mining in Lusatia, Germany (pH 2.3–2.9), Lessmann *et al.* (2000) noted the presence of *Scourfieldia* and a *Nannochloris*, and the eugleniod *E. mutabilis*. In the highly acidic caldera Lago Caviahue, Patagonia, Argentina (pH 2.5, i.e. acidity ≥ 4 mmol H^+ L^{-1}), the phytoplankton was found by Pedrozo *et al.* (2001) to be dominated by a single green alga, *Keratococcus raphidioides*. Indeed, this alga, a subdominant *Chlamydomonas* and a single bdelloid rotifer comprised the full list of recorded planktic species. Species selection in the lake may also be influenced by the low nitrogen content of the water and its relatively high concentrations of the metals Fe, Cr, Ni and Zn.

The application of restorative treatments to lakes that have become artificially acidified is generally considered if it benefits commercially important fish species or viable fisheries. The usual method is to apply lime to the water or to relevant parts of the catchment (Henrikson and Brodin, 1995; Hindar *et al.*, 1998), Lakes can be assisted to move back towards neutrality and to acquire a fully functional, healthy, pelagic ecosystem. Acidity is consumed and progressively more of the carbon dioxide enters the carbonate/bicarbonate reserve. Another interesting and controversial alternative approach that has been devised recently involves the deliberate *addition* of phosphorus to the water (Davison, 1987). It was successfully applied to secure the restoration of a series of highly acidified former sand workings in eastern England (Davison *et al.*, 1989) and, later, to the restoration of an acidified upland tarn in the English Lake District (Davison *et al.*, 1995). Davison's (1987) original stoichiometric calculations led him to propose that phosphorus is 47 times more effective in generating base than the molecular equivalent of calcium carbonate. In extreme cases, lime may still be needed to bring the system to a pH level at which phosphorus addition will promote the development of a biotic community and the consumption of acidic anions. In each of the restorations thus attempted, a single or pulsed addition of phosphate was sufficient to maintain an alternative planktic community, often with new species of algae and crustaceans, submerged macrophytes and fish, for long periods of time measurable in units of hydraulic residence time.

This may not be a universally acceptable approach to managing lakes, especially among managers who have craved the virtues of phosphorus reduction as a route to high water quality (Reynolds, 1992a). Their perplexity is easy to understand; phosphorus addition could easily contribute to a worse problem than the one to be overcome. On the other hand, overzealous pursuit of phosphorus limitation strategies has sometimes reduced the ability of base-poor lakes to resist the effects of strong-acid precipitation and the loss of their fish populations. Careful analysis of the respective loads and fates of hydrogen and orthophosphate ions could lead to more imaginative strategies for management offering longer-lasting benefits to biological water quality.

8.3.8 Anthropogenic effects on the sea

Nutrient enrichment of the oceans, at least as a by-product of human activities, has not been widely considered as an issue on the scale that it has become among inland waters. This does not mean that anthropogenic eutrophication of the sea is necessarily a lesser problem, nor that we should not question a popular notion of raising the fertility of the ocean in order to enhance its role as a sink for atmospheric carbon dioxide. However, both are probably overshadowed by problems arising from gross anthropogenic distortion of the structure and geometry of the marine food webs. Let us examine these issues in sequence.

Eutrophication of the seas

Given the vastness of the oceans, compared both to the tiny total area of inland standing waters and to the scale of human impacts on global nutrient cycles, it is probably not at all surprising that oceanic systems remain resolutely ultraoligotrophic and support such low levels of biomass. On the other hand, the modest levels of biomass and production that they do support are not controlled primarily by the fluxes of nitrogen and phosphorus but, more probably, by the supply of bioavailable iron or of inorganic carbon and, for long periods in most latitudes, by the hydrodynamics of vertical mixing and the deprivation of light. It is in the relative shallows of the continental shelves and at the coastal interfaces with the land masses, where the iron, carbon and light-dilution constraints are simultaneously assuaged, that the enriching effects of terrestrial sources of nitrogen and phosphorus are already well known. For instance, in their review of transport processes and fates of terrestrial phosphorus, Howarth et al. (1995) estimated that current inputs to the sea have trebled since the pre-agricultural period. There are likely to have been significant increases in inputs of inorganic nitrogen (chiefly as nitrate) during the twentieth century but, as concentrations in shelf waters and oceanic surface waters typically continue to be depleted to <2 μmol N L^{-1} (<30 μg N L^{-1}), increasing biomass capacity remains in the control of nitrogen availability. Interestingly, recent increases in planktic biomass in the Baltic Sea have been among the nitrogen-fixing Cyanobacteria (notably *Anabaena lemmermannii*, *Aphanizomenon flos-aquae* and *Nodularia spumigenea*) in response to progressive increases in phosphorus loads from adjacent land masses (Kuosa et al., 1997).

Away from the shelves and distant from terrestrial influences, the nutrient content of the upper oceans is plainly subject to depletion through uptake by autotrophs but the variations remain within levels representing approximate contemporary maxima (~2 μM P, 20–40 μM N and ~160 μM Si (see Sections 4.3.1, 4.4.1 and 4.7). Actual concentrations in surface waters are generally rather lower than these levels, with, typically, the residual concentrations suggesting that available nitrogen is more vulnerable to exhaustion by autotrophic growth than phosphorus. However, as acknowledged and verified by experimentation, the capacity-regulating element in the upper ocean is iron, where maximum levels may be one or two orders of magnitude less concentrated than 10^{-3} μM (Martin et al., 1994; Section 4.5.2).

Thus, the open oceans remain steadfastly oligotrophic in potential and in their performance. On the principle that, at low levels of limiting nutrient availability, any increase in supply will raise the carrying capacity by a corresponding margin, then the pelagic biomass is most sensitive to a change in iron availability. Without this, indeed, enrichment with other nutrients will not stimulate classic eutrophication effects. One corollary is that autotrophic production and the biomass-carrying capacity of the ocean (and thus its role as a carbon sink) can be stimulated by seeding the ocean with bioavailable iron (see pp. 428–31 below).

Anthropogenic impacts upon marine communities and food webs

At present, the greater anthropogenic impacts to the health and functionality of marine ecosystems come from a series of devastating assault on the structure of marine ecosystems. These include destructive alterations to coastal habitats, pollution and increasingly industrialised methods of protein harvesting from the sea. Mostly, they are relatively local in concept and in execution but, cumulatively, their impacts are alarming. In a thoughtful review, Jackson and Sala (2001) summarised the effects of widespread coastal engineering, pollution, turbidity and fishing activities in damaging coral-reef habitats, the truncation of temperate kelp forests and their replacement by 'sea urchin barrens' of crustose algae. In Europe, competing demands for the exploitation of the land surface in valley flood plains are driving the winning of sands, gravels and other building aggregates into increasing intensities of offshore coastal dredging. It is inevitable that the process removes entire benthic food webs and generates clouds of turbidity in its wake. These may be short-lived impacts while there are adjacent 'islands' whence biota can re-establish but it is not clear how well current practices allow this.

At another level, Jackson and Sala (2001) catalogued the diminution in large animals that were once 'keystone species', linking structure to energy flow (see Section 8.2.4) and maintaining the pristine coastal ecosystems of which they were once part. Loss of these animals has contributed to significant habitat changes. Declines in grazing and foraging by manatees (*Trichechus* spp.) and dugongs (*Dugong dugon*) have allowed former sea-grass meadows to develop into dense stands. The top layer of predators in coral reefs (tiger sharks (*Galeocerdo cuvievi*), monk seals (*Monachus* spp.)) has been reduced with cascading effects on coral consumers. Destruction of cod (*Gadus morhua*) stocks and sea otters (*Enhydra lutris*) has released from control the sea urchins that now interfere with kelp regrowth.

Some of these changes are the direct effects of human overexploitation of the animals in question, either for food or fur. Others are the indirect consequences of the diminution of the exploitation, either through the vacation of a foraging niche to competitors, or to the relaxation of cropping of lower trophic levels. It is perhaps 4000 years (4 ka) ago that *Homo sapiens* entered the marine food web by catching predatory fish and mammals in their natural environments (Cushing, 1996). However, it is mainly within the last century and, especially, the last 30 years that industrial scales of sea fishing, backed by such technological sophistications as sonar and satellite tracking that the hunter–gatherer approach to foraging has descended to plunder and looting. The trouble is threefold. One is the economic drive for a return (through the sale of catches) on the investment in ships and crews. The second is the lack of ownership of the resource and,

thus, the lack of an economic incentive to better manage it in a way that does not lead to its overexploitation. The third is that the industrial momentum is in advance of the ecological knowledge to set sustainable levels of exploitation on the underpinning relationships of biomass, production and energy flow through pelagic ecosystems.

It is not even clear that, were such knowledge available, informed guidance would be followed. Long traditions in the study of growth and recruitment in fish populations (Beverton and Holt, 1957; Ricker, 1958; Cushing, 1971, 1988: Cushing and Horwood, 1977) and their match or otherwise to environmental conditions (Cushing, 1982, 1990, 1995) have not been able to avert well-documented stock-recruitment collapses of the herring (*Clupea harengus*) in the North Sea or of the northern cod in the north-west Atlantic, around the Labrador Grand Banks (see Cushing, 1996). The contested imposition of quotas now looks like actions that were 'too little, too late' as the symptoms of collapse show very early. The ultimate truth of the maxim that 'fisheries that are unlimited become unprofitable' (Graham, 1943) is upheld.

In recent decades, the issues have only intensified. Fishing fleets now scour continental shelves. Trawl nets are built to be dragged along the sea floor where they sweep up all in their path for the prize of a few more demersal fish. The imposition quotas for size and species catches can only be controlled after sorting, when most of the 'illegal excess' is returned to the water dead. Indeed the catching methods are still so coarse that fishing can scarcely be directed to catch preferentially the relatively plentiful fish. Conventional ecological theory suggests that, where interventions are targetted successfully at a particular dominant species, then another with similar environmental requirements and dietary preferences and the next highest exergy potential (i.e. the next best competitor) should be poised to assume that role (Wardle *et al.*, 1999; Elliott *et al.*, 2001b) (explanation in Section 7.3.4). This is roughly what happened in the North Sea: following the severe reduction in herring, there was an 'outburst' of several gadoid species, including cod and haddock (*Megalogrammus aeglifinus*). The reasons are complicated, however, and include the possibility of reduced herring predation on cod larvae and the greater availability of *Calanus* larvae to young cod, now relatively freed from consumption by herring. One of the problems of the largely unselective trawling of those same cod stocks is that it is unlikely to leave many demersal contenders to fulfil its role. Thus, a further trophic linkage is compromised.

Following similar logic, the industrial fishing of species with other commercial value (for instance, for the purpose of manufacturing fishmeal and the extraction of fish oils and proteins) may be depleting shelf waters of their diversity and of the ability to support other maritime species (birds, seals, whales). Of yet further concern should be the 'creaming off' of the higher levels of the oceanic food chain of the oligotrophic oceans; catches of such species as tuna (*Thunnus*), barracuda (*Sphyraena*), sea-bass (*Morone*) and others are commercially very attractive but probably unsustainable at present rates.

The science and the economics of overexploitation of fisheries are not the principal concern of this book, though it does recognise that urgent action is required if the excellent source of healthy protein and oils is to remain available. It is clear that farming of commercially important fish species (as opposed hunting and gathering them) is no more sustainable while it requires the continued fishing of other species to provide feed. Nothing less than the total exclusion of all fishing fleets from large stock-recruiting areas is going to assist the survival of viable and functionally intact food webs. The consequence of doing nothing is likely to be a fairly rapid slide towards a pelagic ecosystem that effectively comprises little more than phytoplankton, mesozooplankton and macroplankton.

Anthropogenic countermeasures to the atmospheric accumulation of greenhouse gases

The irrefutable evidence for the increase in atmospheric carbon dioxide during the industrial period (caused by the oxidation of fossil fuels and of humic matter in agriculturally improved land) and its likely role in current trends in global warming has been referred to in Section 3.5.2. Burning coal, gas and oil is generating a flux of carbon dioxide from the Earth to the atmosphere equivalent to 7.1 (± 0.5) Pg a^{-1}. Despite the Kyoto Accord, this is likely to go on increasing until demand or depletion of the relatively readily accessible parts of the global reserve (estimated to be ~5000 Pg) raise the cost sufficiently to make it competitive to exploit other energy sources. Apart from its contribution to acidity, carbon dioxide is one of the atmospheric gases (with water vapour and ozone) that absorb much of the long-wave terrestrial radiation back-reflected from the Earth (chiefly in the wavebands 2–8 and >14 µm). Thus, its accumulation in the atmosphere might enhance the so-called 'greenhouse effect' and contribute to global warming. The extent of effects directly attributable to the accelerated oxidation of organic carbon is still not clear. On the other hand, such a scale of intervention into the natural planetary cycling of carbon is unlikely to avoid significant consequences on average planetary temperatures, climatic patterns and sea levels.

Many insist that the changes have already commenced and there is a growing, general (but not universal) political consensus that 'something must be done'. Simply cutting back on the consumption of fossil fuels would be an obvious step but cheap and available energy is a drug, addiction to which is economically too painful to abandon. There are problems of political equitability over the consequences of therapies and over the cost-sharing of the alternative energy sources.

Another idea, seriously advanced as a means of stabilising atmospheric carbon dioxide levels, is that we could augment the flux of carbon dioxide from atmosphere to ocean. The suggested mechanism is to raise the fertility of the sea, so that more carbon might be fixed in photosynthesis into more biomass, and with more being consigned as export to the deep ocean. By implication, the key players invoked are members of the marine phytoplankton, so the idea must command the attention of readers of this book, requiring a careful consideration of its logic and the likely consequences of its implementation.

Starting with what we know, the global stores and fluxes between atmosphere and ocean are already impressive. Using the tabulated data of Margalef (1997), the atmospheric store (presently around about 650 Pg C) is currently increasing by about 3.3 Pg a^{-1}. Thus, the net burden of anthropogenic carbon emissions to the atmosphere is currently countered by an annual removal of carbon dioxide of some 3.8 ($\pm$0.5) Pg C a^{-1}. These are not trivial amounts but, on the scale of the natural biogenic fluxes, they are comparable with the error terms of primary production (Table 3.3). Terrestrial net primary production (56 Pg C a^{-1}) maintains a biomass (including wood) of about 800 Pg C and a store of other organic necromass of about twice that (1600 Pg C). These are not static amounts but present rates of forest clearance and land drainage probably impede any current net increment to the storage capacity. By deduction, much of the missing 3.8 Pg C a^{-1} is already dissolving in the sea, assisted by increasing partial pressure. In any case, the quantity of carbon dioxide in solution in the sea (some 40 000 Pg) makes it by far the largest store of carbon in the biosphere. This does not mean that the additional input to the dissolved carbon dioxide pool is necessarily insignificant. We should bear in mind that the additional gaseous load is to the immediate surface layer, where its involvement in the carbon dioxide–bicarbonate system (see Section 3.4.1 and Fig. 3.17) and a net depressive impact upon pH will make it increasingly difficult for carbonate-deposting phytoplankters (such as coccolithophorids) and animals to form their calcareous shells.

The desired fate of the additional carbon dioxide flux is that it should be taken up by enhanced phytoplankton photosynthesis. Currently, net primary production in the sea (48 Pg C a^{-1}), impressively generated by a producer mass equivalent to $\leq$0.7 Pg C, supports an average oceanic biomass of $\sim$10 Pg C and which is, self-evidently from these figures, turned over very rapidly. A large proportion of the organic carbon fixed in phytoplankton photosynthesis is respired back to carbon dioxide without leaving the euphotic zone. Our first question might be: 'How is such a small part of the global biomass, itself already involved in 40% of the short-term movements of biospheric carbon, supposed to increase the capacity for its storage in the sea?'

One part of the answer is another question – 'If so little does so much, then how much more could a little more do?' Put another way, 'Will a bigger primary producer biomass not increase the flux of carbon dioxide to the ocean, perhaps to the point of balancing the anthropogenic excesses?' How, indeed, can we be sure that adding nutrients does not simply accelerate the cycle, with the wheel spinning faster rather than increasing in mass? This must be a serious possibility if we follow the deduction that the net yield of exportable primary product across most of the open ocean is tolerably close to the likely invasion rate of carbon dioxide across the surface of the sea (Sections 3.4.1, 8.2.2). A rising flux of carbon dioxide, driven by an increasing partial pressure in the atmosphere, might raise the sedimentary flux by the same few percentage points but, mainly dependent upon microbial processing, enhanced exports depend as much upon faster turnover through the web. Then, raising the fertility of the medium might lift the supportable biomass at successive trophic levels, including the larger consumer categories that are likely to provide the desired increases in carbon exports. Moreover, the deep oceans are capable of storing a lot more carbon dioxide in solution without the net 'outgassing' evident among small lakes (see Section 8.2.3). What is required is the minimisation of C cycling in the microbial loop of the surface waters and the numerous opportunities it offers for the venting of carbon dioxide as a primary metabolite and, instead, to direct fixed carbon into exportable, sedimentary biomass

This is easily stated but it is a formidable hurdle to overcome. Let us recall the viscous environment of oceanic primary producers (Section 2.2.1), in which viscosity is a more significant force than gravity, and vertical transport is relevant primarily in the context of exposure to underwater irradiance (Section 3.3.3). The supply of nutrients (some albeit very scarce) is mediated primarily by molecular diffusion (Section 4.2.1). Though species composition may vary, the structure of the food web and the relative proportionality of mass among its components (the producers, heterotrophs and microphagotrophs) are, within certain limits, highly conserved (a point elegantly emphasised by Smetacek, 2002). The effect of pulsed enrichment by an otherwise limiting resource should benefit the growth rates of all microbes alike. However, photoautotrophs are uniquely able to invest resource in new biomass whilst simultaneously cutting the overspill of carbon to the heterotrophs. At the same time, bacterial populations suffer the ongoing constraints of protist grazing and viral pathogens. Thus, for a variety of reasons, the structure of the fertilised microbial web is distorted in favour of pico- and nanoplanktic producers and disproportionately against the smooth flow of carbon through to an exportable flux of mesoplanktic consumers or food to fish. More and more sustainable phytoplankton is not at all an ideal carbon sink while it largely stimulates only producer biomass. Above unspecified thresholds (though probably in the range $0.01–0.1$ mg C L^{-1}), the concentration of phytoplankton carbon (together with other detrital sources of POC) may permit the short-circuiting of the microbial loop and the imposition of direct mesoplanktic herbivory on algae. Above 0.1 mg C L^{-1}, this is still more likely to be true, as the threshold concentrations of many types of planktic filter-feeder are saturated (Section 6.4.2).

Thus, the critical cue (singularity: Legendre and LeFevre, 1989; see also Legendre and Rassoulzadegan, 1996) to enhanced carbon export from surface waters is to promote sustainable pelagic structures producing either readily grazeable levels of nanoplankton (and the attendant flux of faecal pellets) or predominantly sedimentary microplankton. In the contemporary ocean, large-celled diatoms are usually the main agents of the sedimentary export flux of organic carbon (Falkowski, 2002), especially at times of accelerated production, beyond the control of the microbial food web (Legendre and Rassoulzadegan, 1996).

This behaviour is generally attributable to the relatively high sinking rates of diatoms (Smayda, 1970; see also Section 2.5). However, the fastest sinking rates of settling individual cells are probably inadequate to prevent the death and decomposition of the protoplast (and even the re-solution of the siliceous walls) before they have settled more than a few tens to hundreds of metres through the ocean (Reynolds, 1986a). Of course, the aggregation of dying cells into larger flocs (up to and including 'marine snow' (Alldredge and Silver, 1988; see also Section 2.5.4) may accelerate the rates of sinking and slow the rates of decay (Smayda, 1970). In this way, the net export of diatom and other POC in the form of 'phytodetritus' is enhanced (Legendre and Rivki, 2002a).

While oceanic production is concentrated in the picoplanktic prokaryotes and is intimately coupled to the microbial web, it is easy to understand why diatom-based export of unoxidised POC beyond the upper ocean is abnormal and event-led (Karl, 2002; Karl et al., 2002). In this context, the phytoplankton biomass of the coastal and shelf waters, which is responsible for about a quarter of the net primary production of the sea (Table 3.3), is more continuously and more conspicuously populated by diatoms, dinoflagellates and coccolithophorids. Globally, coastal and shelf systems are already likely to be playing a significant part in the export of biogenic carbon out of the surface circulation (Bienfang, 1992). Moreover, parts of these same shelves are already sites for the accumulation of inwashed terrestrial detritus and POC, and where the concern over nutrient limitation of carrying capacity has been more to prevent its increase rather than to encourage it.

To recap, the uncertainties about the likely effects of oceanic fertilisation on the atmospheric carbon dioxide content are too many and too great to advocate its deployment as a strategy. The impressive results of the IRONEX and SOIREE fertilisations of the iron-deficient Pacific (Martin et al., 1994; Bowie et al., 2001; see also Section 4.5.2) might offer a persuasive case for emulation on the basin scale. However, supplementing the iron supply in a low-latitude ocean would only raise the ceiling on producer mass to the scale of the next capacity limitation (nitrogen) and not at all where iron is not already limiting growth. Under these circumstances, the most likely consequence of iron enrichment would be, presumably, the promotion of nitrogen-fixing prokaryotes, with little benefit either to the carbon-exporting elements of the pelagic food web or to the growth of diatoms. According to Legendre and Rivkin (2002b), the frequency, quantity and composition of nutrient additions would greatly bias the structure of the planktic community. Frequent additions of nutrients offering high ratios of Si to both N and to P are necessary to promote diatom dominance. For the production of blooms to sink to depth ungrazed, any matching of nutrient additions to grazer demand must be avoided. It is clear that there is no simple strategy for enriching tropical seas that does not risk chaotic or undesirable consequences. Outside the tropics, the most likely capacity constraint upon phytoplankton is an insufficiency of light, which no amount of fertilisation will overcome. Shallow coastal waters offer the only real prospect for accepting more carbon into biomass carbon and then transporting it to depth. It is not clear that such waters are strongly nutrient deficient. However, the grave distortions to the food webs of the continental shelves, mentioned in the previous section, make the impacts of further fertilisation on the structure and function of their surviving components still more difficult to predict.

A better candidate for taking forward the principle of enhanced carbon dioxide sequestration by marine primary producers is surely furnished by kelps and other large seaweeds. These at least have the merit of accumulating (rather than simply metabolising) the carbon they absorb and, moreover, in a form that is conveniently harvestable, compostable and combustible as biofuel. Seaweed growth is naturally even more confined to unpolluted shallow shelf waters than is phytoplankton. However, if the provision of suitable artificial substrata could be devised, such as floating mats on which the seaweeds could establish, then the potential of the marine system to remove of more carbon dioxide from the atmosphere might begin to be achievable.

Phytoplankton and the future

The balance of wisdom should be against deliberate manipulation of the marine phytoplankton to mop up anthropogenic carbon dioxide. Present knowledge provides an inadequate basis for reliable and precise anticipation of the outcome. Even the extrapolations on which the current fears about climate change are based may be quite wrong, with the response of global temperatures to the rise in greenhouse gases being either overestimated or frighteningly underestimated in its effects on human civilisations. The latter might yet persuade us to adopt appropriate countermeasures. However, doing nothing but allow the planet to adjust its carbon fluxes and stores might be a prudent option, even though it probably carries enormous social and economic discomforts to humankind.

The contemporary view of the role of the oceanic biota in the global carbon cycles has been elegantly encapsulated in Falkowski's (2002) overview. The diatoms are the major exporters of organic carbon (with silica) to the sediments; the coccolithophorids are major exporters of calcite (calcium carbonate). This arrangement has persisted for over 100 Ma, really since tectonic upheavals and continental drift effectively opened up new niches for exploitation (see also Section 1.3). Throughout that period, there has been a progressive reduction in the atmospheric carbon dioxide content (from about 700 to 280 p.p.m.). This may be due largely to the rate of subduction of carbonate-rich oceanic sediments exceeding the rate of volcanic and orogenic carbon dioxide generation, throughout the Cenozoic period. The planet has cooled sufficiently for the ice caps to form, for the sea level to fall, exposing areas of continental shelf, and for the planet to become drier.

Following the establishment of polar ice, the Milankovitch cycle of climatic oscillations, caused by rhythmic orbital variations of the Earth around the Sun (with a frequency of 90–120 ka) and resultant variation in solar radiative forcing, has been marked by significant glaciations. The last four, at least, saw the expansion of the Arctic ice cover over large parts of the northern hemisphere. The interactive role of the biotic oceanic carbon cycle with climatic variations through the last 400–500 ka are now well understood, through the combination of radiocarbon dating and stable-isotope analysis of biogenic deposits with atmospheric fingerprints of contemporaneous layers in ice cores. With the onset of a glacial period, terrestrial plant production decreases and significant amounts of carbon are transferred from the land to the sea. The process is reversed during interglacials. Sea levels fall as ice volume increases, exposing coastal lowlands to oxidation and erosion and the release of nitrogen and other nutritive elements to marine ecosystems. Increased nutrient delivery to inshore waters stimulates production of large-celled phytoplankton and the export of more carbon to depth. Drier climates also allow more wind-blown dust and the aeolian fluxes of minerals, including iron, to the sea are raised. These changes in the nutrient availability drive accelerated carbon dioxide exchanges, with increased leakage back to the atmosphere. Falkowski (2002) analogised these exchanges to the ocean 'breathing' in and out on a 100-ka cycle, inhaling carbon dioxide from terrestrial systems during glacial periods and exhaling it during the interglacials. Because these exchanges feed back positively on the Milankovich-driven variations, it is reasonable to attribute to biological processes a dominant contribution in the contemporary carbon cycle.

Just as certainly, they will also impose a balanced distribution of the contemporary anthropogenic contribution to carbon dioxide 'inhalation' by the sea. This being a warm (and warming) phase, the rate of inhalation may be less rapid than during a glacial phase. On the other hand, the flooding of coastal plains rich in nutrients may accelerate carbon withdrawal into diatoms and other 'large' phytoplankters. Moreover, the rate of exploitation of the finite and increasingly inaccessible remnants of the fossil-fuel carbon is, in any case, bound to diminish, so the fluxes from atmosphere to ocean will, inevitably, reduce. Within the context of the last 100 Ma, the present carbon crisis would seem relatively trivial and unlikely to disturb the pattern of prokaryote-mediated carbon cycling in the open ocean and eukaryote-mediated exports from shelf waters. The pre-industrial quasi-steady state might be re-established within 100–200 years and, if not, before the onset of the next glacial phase.

It is prudent, nevertheless, to follow my own stricture and recognise the weak base for 'reliable and precise anticipation'. Nobody can be sure that the rate of global warming will not overtake the scale of Cenozoic fluctuations and match the temperature rises of the late Permian, which brought life on earth close to extinction (Section 1.3).

Neither prospect – being baked or suffering violent climate change, lowland inundation and massive economic disruption – is at all comforting to humankind. However, the prospects for the microbial engineers of atmospheric and oceanic composition, including the phytoplankton, are mostly good. Their ability to go on regulating the planet, while other species have come through and caused desolation, seems to be assured.

8.4 | Summary

The chapter seeks to evaluate the importance of phytoplankton to pelagic function and to the biogeochemical role of pelagic domains to the behaviour of various categories of aquatic ecosystem. It moves on to examine the responses of pelagic systems to changes, especially those wrought through deliberate and unthinking anthropogenic activities.

In the open water of the sea and of the largest lakes (those >500 km^2 in area), phytoplankton are the only photoautotrophic primary producers, upon which everything else in the water is dependent for its nutrition (energy as fixed carbon and the other elements of biomass). This long-standing view is correct but the secondary processing of primary product depends less on the grazing of phytoplankton, as once supposed, so much as on the microbial uptake of fixed carbon released into the medium by the producers. Severe poverty of nutrient resources controls the biomass at all trophic levels but carbon continues to be traded through the food-web components. The bacteria $\rightarrow$ nanoflagellate $\rightarrow$ ciliate $\rightarrow$ copepod pathway delivers efficiently a yield, perhaps nearly 10% of the original primary product. A further 10–20%, up to 100 mg C m^{-2} a^{-1}, may be exported by sedimentation. The balance of the carbon budget is simply recycled through the pelagic network, which thus constitutes a fairly closed cycle. The export is argued to be similar in magnitude to the net annual invasion of atmospheric carbon. The system is a quasi-steady state, a low, stable, average biomass being run on the available flux of carbon. The activity is analogised to a small wheel spinning rather fast (20 to 70 revolutions per year, according to the criteria used).

In the pelagic zones of smaller lakes, coastal seas, shelf waters and upwelling areas of the ocean, primary production benefits from additional supplies of inorganic carbon, including that dissolved in inflowing streams and rivers and the metabolic gases released through the oxidation of sedimentary organic carbon. The delivery to the water-body margins of terrestrial particulate organic carbon can provide a direct input to littoral and sub-littoral consumption. The interaction with the adjacent terrestrial ecosystems contributes to enhanced fixation of inorganic carbon (net yields of primary production range are frequently in the range 200–800 g C m^{-2}) as well as providing an additional income of organic carbon to limnetic heterotrophs. The shallow littoral margins of lakes are important in other ways, not least because of their ability to support macrophytic plant production. Stands of aquatic plants accumulate organic debris (some of it by filtration from the circulation), offer shelter and microhabitats to invertebrates. Linked to the pelagic through the horizontal movements of animals, especially fish, macrophyte stands influence the material and energy flow to a considerable distance into the lake. The additional carbon flux represents a supportive subsidy with respect to the truly pelagic system. The smaller is the water body then the relatively greater is the potential subsidy to the overall function of the lake. More carbon, proportionately and absolutely, may reside in the biomass of aquatic producers, phagotrophic consumers and other heterotrophs of smaller lakes and ponds, provided always that it is within the stoichiometric capacity of the limiting bioavailable nutrient(s).

Resource availability and processing constraints determine the relationship between energy flow and the structure (*sensu* the functional groupings of the main participants). The abilities of phytoplankton to gather resources (whether through diffuse demand, high uptake affinities or scavenging propensity) and of zooplankton to forage (to encounter, to hunt, to filter feed) are geared to the availability of their nutritional requirements. The corollary is that the relatively most abundant forms will be those most capable of fulfilling their requirements. This is another way of saying that species composition (at least at the level of functional types) depends, in part, on the resource base. For example, greater nutrient availability raises the autotrophic supportive capacity, enabling photoautotrophic phytoplankton to retain more of their own photosynthate in their own biomass, and with a lesser proportion garnered by heterotrophic plankton. Greater nutrient availability makes it easier for larger species of algae to flourish in the pelagic. Mesozooplanktic feeding on absolutely larger nano- and microplanktic algal fractions short-circuit the microbial web; filter-feeding zooplankton are capable of its destruction. The optimal pelagic pathway changes to microalgae → cladoceran.

How the pelagic community processes its resources and directs the flow of energy is shown to be strongly conditioned by the activities of the main species (that is, how the community functions is a consequence of what is there). Examples are presented that show that the principal components of the phytoplankton and zooplankton, as well as the foraging activities of fish, can be matched to the carbon concentration (subsidised or otherwise) and the energy transferred among trophic levels. Critical boundaries of ∼0.01 mmol C L^{-1} (∼0.12 mg C L^{-1}) distinguish classically eutrophic lakes from classically oligotrophic lakes. A similar abundance of carbon separates those lakes that support Cyanobacteria from those in which chrysophytes are abundant; those whose zooplankton is dominated by calanoids from those in which cladocerans are abundant; and those supporting coregonids from those in which benthivorous fish dominate. It is also suggested that, in small to medium-sized lakes where the carbon balance exceeds this threshold, a large, possibly dominant, part of the authochthonous organic carbon is transmitted to higher trophic levels by way of the sub-littoral sediments, benthic invertebrates and browsing benthivorous fish.

The chapter also considers the susceptibility and dimensions of changes in pelagic communities consequential upon anthropogenic changes. Eutrophication of lakes through enhanced phosphorus loading is considered in some detail, developing the underlying processes linking catchment sources ('loads') of phosphorus to in-lake concentrations and the mean algal biomass (as chlorophyll *a* concentration) in the lake, as represented in the well-known Vollenweider–OECD regression.

The relationship remains a powerful descriptor of the biomass capacity of the deep, high-latitude lakes that dominated the original dataset, but it is not a general management model. Mean phytoplankton abundance in low-latitude lakes, in shallow lakes, especially those experiencing fast rates of hydraulic exchange, and those in which phosphorus is manifestly not the capacity-regulating factor, is not well simulated by the regression. Special problems arising from the abundance of particular types of phytoplankton are considered here. Even though their dominance may have only tenuous connections with nutrient eutrophication, their abundance is potentially greater where there is significant enrichment. Conspicuous among the organisms generating nuisance on aesthetic grounds are the bloom-forming Cyanobacteria. These have the added hazard of often being highly toxic. They are difficult to eradicate from lakes but, with an understanding of their preferences and weaknesses, some kinds of lake and reservoir can be managed in ways that keep these algae in check. Cyanobacterial blooms are not confined to inland waters, having become common in recent years in the Baltic Sea. Elsewhere, harmful algal blooms in the sea comprise 'brown tides' of *Aureococcus*, 'green tides' of *Phaeocystis* and, especially, 'red tides' of toxic dinoflagellates. In all events, local abundances may depend upon concentration by water movements but the incidence of such events is increasing and are considered indicative of nutrient enrichment of the shelf waters from which they are best known.

There are few instances where eutrophication is considered beneficial; mostly it is abhorred on aesthetic grounds of appearance, the detriment usually being directly attributable to increased biomass of phytoplankton and littoral epiliths. Greening of the water, loss of clarity and sully-

ing of the shorelines are unwelcome symptoms of eutrophication. Phytoplankton abundance prejudicial to water quality, to treatment for potability and to the quality of fishing is likely to inspire and justify the costs of reversal of eutrophication ('oligotrophication'), and the restoration of past quality or rehabilitation of the water body to an acceptable ecological state. A series of successful schemes in which phosphorus-load reductions have led to reduced plankton biomass and better perceived water quality is highlighted. The biological response is usually dramatic but it is never manifest before the soluble, MRP fractions, readily available to support further phytoplankton growth, are effectively drawn down to the limits of detection. Phosphorus-reduction schemes that, so far, have been unsuccessful in invoking the desired biomass response have not achieved, or have progressed too slowly towards, the satisfaction of this criterion. These poor responses are observed frequently among shallow lakes, where the persistent recycling of phosphorus from already enriched sediments goes on sustaining phytoplankton independently of external loads.

Methods for anticipating the sensitivity of individual sites to altered external loadings (up as well as down) are available. If control over the internal loads (or adequate contol over the external loads) is impractical, other rehabilitative approaches are available. Provided the water column is deep enough to make a difference, artificial circulation of water bodies (chiefly reservoirs) to break or to prevent thermal stratification reduces the light-carrying capacity and rates of growth of phytoplankton. Subject to the same provision, the overriding imposition of deficient insolation pushes species composition in favour of functional types with acknowledged light-harvesting specialisms (diatoms, filamentous green, yellow-green and blue-green algae, of trait-separated groups P, T and, especially, S). Several examples are presented where mixing of deep reservoirs assures reasonable supplies of water for treatment and within predictable limits of phytoplankton abundance.

There is a good prospect of restoring or rehabilitating those 'small' lakes (i.e. up to 10 km^2 in area) and, especially, 'shallow' lakes (in which

>50% of the area is <5 m in depth) through the biomanipulation of the ecosystem components. Investment of carrying capacity is moved from phytoplankton to macrophyte and/or fish biomass. The necessary changes in the consumption pressures may be imposed from the top down by adjusting the abundance of fish population (lower the density of planktivores, raise the density of piscivores; both supposedly reduce the controls on zooplankton which, obligingly, graze down the phytoplankton). Managers still find themselves having to repeat treatments to enforce the imposed, non-steady state. Left to themselves, water bodies will often gravitate to a system characterised by high POC/dissolved nutrient concentrations → *Planktothrix agardhii* → chironomids → cyprinids. The task is to induce the system into a valid alternative, macrophyte → benthos → fish, steady state, in which better habitat and water quality are more nearly self-maintaining. Even here, there are approximate critical boundaries (in terms of phosphorus rather than carbon). These may not be easy to pass in the downward direction, usually requiring quite drastic interventions into the numbers of planktivorous and benthivorous fish. Below the critical boundaries, however, handbook guidance extols the importance of macrophyte dominance to attractive ecosystems. Put simply, macrophyte beds support macroinvertebrate populations adequate to interest many species of adult fish, and a sufficient refuge for cladocerans to keep the water reasonably clear of phytoplankton. The boundary ('switch') in the upward direction is mediated by the algal biomass sponsored by bioavailable phosphorus concentrations exceeding 120–180 μg P L^{-1}, whether it is present as phytoplankton or epiphytic growth. Further stability may be brought by stocking with piscivores such as pike.

Brief consideration is directed towards the selective responses of phytoplankton to anthropogenic acidification. The impacts on some waters affected by acid precipitation are catalogued. However, in this instance, algal behaviour in highly acidic waters filling some Lusatian mining hollows and a natural, sulphate-rich caldera in Argentina illustrate the remarkable tolerances of a few green-algal species down to pH values of <2.5. The problem of metal (Al, Zn, Cu) toxicity is generally critical.

Apart from some estuaries, coastal waters and part-landlocked shelf areas, the open seas have suffered less from anthropogenic nutrient enrichment than from severe food-web distortions as a consequence of exploitative industrial fisheries. Indeed, deliberate fertilisation of the oceans, with a view to increasing carbon fluxes from atmosphere to sea water, has been seriously suggested as a counter to the net accumulation of greenhouse gases from the oxidation of fossil fuels. Such action is argued to be misguided. While the carbon storage capacity of the ocean is not exhausted, the *net* carbon flux is possibly as fast as it can be. If fertilisation could secure the rapid deep transport of POC (for instance, in diatoms or faecal pellets) balances might be altered a little. Possible outcomes are considered. The consequences for human societies are uncomfortable to contemplate.

8.5 A last word

On the time and space scales of the biogeochemistry of the Earth, the current 'carbon crisis' is modest, and more serious shifts in the planetary distribution of carbon have occurred in the Mesozoic. Indeed, the groups of phytoplankton that evolved during that era – diatoms, dinoflagellates and coccolithophorids – may have done so in response to contemporaneous ocean–atmosphere interactions, have persisted for 100 million years and may still be best equipped to operate through the worst anthropogenically induced changes that can be anticipated.

It is perhaps humbling that precise extrapolations cannot be made. Through the past century and a half, there have been several reasons for studying phytoplankton, not least for the beauty (that first attracted Haeckel), the desire to catalogue and name many thousands of species and the challenge of sorting out their phylogenetic affinities and evolutionary development. These persist today. The physiology of cell growth and replication and collective impacts in the global carbon and oxygen cycles represent extremes of a scale covering a dozen orders of magnitude and nearly as many biological disciplines. The ecology of populations and communities is relevant to many aspects of human existence, from the safety of drinking water to the sustainability of fisheries. The accumulated knowledge is both broad and deep but it is far from complete. Common tenets and basic understanding of the pelagic ecosystem have undergone comprehensive revision more than once, the most recent occasion having been only in the last two decades. Completely new organisms and new organisations are being found, even now, in the deep oceans and in the vicinity of hydrothermal vents. New tools, from satellite-based remote sensing to the analysis of gene sequences from individual chromo-

somes are providing ever greater dimensions of precision to the knowledge base. Yet it is quite evident that, far from knowing all there is to know about phytoplankton, relevant and politically sensitive questions persist, such as predicting and understanding the impacts of global warming on the fundamental life-supporting systems of the planet. The answers elude us. Much needs to be done; the scientific study of phytoplankton will continue. Tomorrow's leaders need an appreciation of the breadth and limits of existing knowledge of our home planet. It is my fervent hope that this book may contribute to the base of reference and the stimulus for the future research.

Glossary

Text boxes are used to explain the usage of certain terms specific to plankton (Box 1.1) and their ecology (Box 7.3). The meanings of some other less familiar terms used in the book are noted below.

abyssal pertaining to the abyss, or the very deep parts of the ocean

aeolian of the wind, referring to the transport and deposition of dust particles from the land to lakes and oceans

amictic of lakes that are scarcely wind mixed or which are weakly and incompletely mixed for very long periods

anoxygenic photosynthesis carbon fixation that takes place in the absence of oxygen and during which no oxygen is produced; photosynthesis that does not split water to generate oxygen

atelomictic of water columns, usually in low latitudes, in which the depth of wind mixing varies conspicuously in extent but at frequencies of days to weeks rather than at diel or annual cycles

autotrophy the capacity of organisms to grow and reproduce independently of an external supply of organic carbon; the ability to generate organic carbon by reduction of inorganic sources using light or chemical energy

auxospore sexually reproduced propagule of diatoms that also enables organismic size to be recovered after a period of asexual replication

bathypelagic of the deep open water of the ocean

benthic of the solid surfaces at the physical bottom of aquatic habitats; this can be very shallow as well as far beneath the water surface

bioassay a technique for identifying the supportive capacity of water and for identifying nutritive components that are deficient therein; if the addition of a particular nutrient raises the response relative to the unmodified control then that nutrient is deemed to have been limiting the full response

biogenic of the ability of living materials to create or generate a particular substance; of substances thus generated by living organisms

cadavers the corpses and remains of dead organisms

carboxylation the essential step in photosynthesis involving the combination of carbon dioxide, water and ribulose biphosphate and yielding the initial fixation product that incorporates the high-energy phosphate bond

chemotrophy the capacity of organisms to obtain energy through chemical oxidation and using either inorganic compounds as electron donors (chemolithotrophs) or preformed organic compounds (chemoorganotrophs)

chromophore intracellular organelle containing photosynthetic pigments

coccoliths flattened, often delicately fenestrated scales, impregnated with calcium carbonate

coenobium (plural *coenobia*) a group of monospecific cells forming a single, often distinctive unit

compensation point the point in a water column at which the rate of photosynthesis of an alga just balances its respiration; or the notional point when the daily depth-integrated photosynthesis compensates the daily depth integral of respiration

dimictic of lakes and water bodies, generally at mid to high latitudes, that are fully mixed during two separate periods of the year (typically spring and autumn)

emergy the total amount of energy required to produce the usable energy of a final product. In the ecological context, a grown elephant contains biomass equivalent to a potential yield of direct energy but a lot more energy has been lost in foraging for food, as well as in its production Emergy represents the sum of these contributions

endosymbiosis a symbiosis between two organisms in which one lives entirely within the body of the other but to the mutual benefit of both partners

epilimnion the upper part of a seasonally stratified lake; the part between the water surface and the first seasonal pycnocline

eukaryote a cellular organism having a membrane-bound nucleus within which the chromosomes are carried

euphotic zone that (upper) part of the water column wherein there is sufficient light to support net photosynthetic gain (often approximated as requiring 1% of surface light intensity but, in reality, variable according to species and light history)

exergy the extent of high-quality, short-wave energy that an individual, population or community can invest in the synthesis of biomass (see also Box 7.3) *form resistance* the resistance to movement through water effected by shape distortion with respect to a sphere of similar volume and density

frustule the siliceeous case of a diatom, comprising two similar valves

haptonema distinctive additional appendage, said to be characteristic of the Haptophyta, carried anteriorially, between the flagella

heterocysts (or *heterocytes*) specialised cells of Cyanobacteria (Nostocales) dedicated to the fixation of nitrogen (see Fig. 1.4h)

heterotrophy the means of nutrition of organisms that grow and reproduce that is dependent upon externally produced organic carbon compounds as a source of carbon and energy

hypolimnion the lower part of a seasonally stratified lake; the part between the seasonal pycnocline and the lake floor

isopycny the condition of one entity having the same density as another or as the medium in which it is suspended

karyokinesis the apportionment of the mitotically reproduced genetic material of a cell; nuclear division

kataglacial of processes or deposits associated with the melting of glaciers

kinase an enzyme that mediates the transfer of phosphate groups from (e.g.) ATP to a specific substrate or target molecule

laminar flow ordered fluid motion, molecules moving in layers, one upon another

lorica hard part of the surface of certain flagellates, especially euglenophytes and chrysophytes; from the Latin word for breastplate

meromixis condition of (usually) deep tropical lakes in which mixing energy is inadequate to overcome stratification and which, therefore, remain stratified for many years on end

meroplankton organisms that are planktic for a short part of their life history, the rest of which they pass in the benthos or in the periphyton

metalimnion the intermediate part of the water column, separating epilimnion and hypolimnion and characterised by density gradients and much weaker vertical diffusivity than either of the layers it separates

monomictic of lakes and water bodies, generally at low to mid latitudes that are fully mixed during only one part of the year (typically between autumn and spring)

monimolimnion the lower part of a meromictic lake, between the perennial pycnocline and the lake floor

mixis wind- or convection-driven integration of the water masses of a lake; complete mixing is called *holomictic*; frequent complete mixing is described as *polymictic*

mixolimnion the upper part of a meromictic lake; the part between the water surface and the perennial pycnocline, that is actually or potentially mixed during the year

mixotrophy the capacity of an autotroph to supplement its carbon and/or nutrient requirement from externally produced organic carbon compounds

neritic pertaining to the shallow, inshore regions of the sea

osmotrophy the capacity of certain microorganisms to absorb selected dissolved organic compounds across the cell surface

oxidation chemical reaction involving the removal of electrons and/or hydrogen atoms with, usually, a release of energy

phagotrophy case of heterotrophy in which organic carbon is ingested in the form of another organism or part thereof, requiring digestion prior to assimilation

photoautotrophy autotrophy using light energy to generate organic carbon by reduction of inorganic sources

phycobilin accessory biliprotein pigment of Cyanobacteria, also found in Rhodophyta, some cryptophytes and glaucophytes

prokaryote cellular organism lacking membrane-bound nucleus and organelles

phytoplanktont or *phytoplankter* terms applied to an individual organism of the phytoplankton

pseudotissue tissue-like structure of a fungus but comprising hyphae in close mutual application

pycnocline a gradient (usually vertical) separating water masses of different densities. The density difference may be due to a difference in temperature or solute (e.g. salt) content

reduction chemical reaction involving the addition of electrons and/or hydrogen atoms with, usually, an investment of energy; the opposite of oxidation

redox potential the balance of electrochemical potential between oxidative and reducing reactants

reductant chemical substance that will yield electrons and hydrogen ions

solar constant the energy flux from the Sun to the Earth, estimated as that reaching a notional surface held perpendicular to the solar rays and at a point above Earth's atmosphere, before there is any reflection, absorption or consumption. It is *not* constant but an average, about 1.36 kW m^{-2}

thermocline a gradient (usually vertical) separating water masses of different temperature

thylakoid membrane supporting the light-harvesting complexes and sites of photosynthesis

tychoplankton organisms that are not planktic but which may, fortuitously, be introduced to the plankton from adjacent habitats. The distinction from meroplankton is narrow but the criterion is whether the suspended phase is essential to the life cycle or purely incidental

Units, symbols and abbreviations

UNITS

Dimension	Unit	Notation	Equivalent
mass	kilogram	kg	
	gram	g	10^{-3} kg
	gram molecule	mol	
length	metre	m	
area	square metre	m^2	
volume	cubic metre	m^3	
	litre	L	10^{-3} m^3
density	kilogram per cubic metre	kg m^{-3}	
time	second	s	
	day	d	86 400 s
	year	a	
velocity	metres per second	m s^{-1}	
acceleration	metres per second per second	m s^{-2}	
force	newton	N	kg m s^{-2}
pressure	pascal	Pa	N m^{-2} $= $ kg m^{-1} s^{-2}
viscosity	poise	P	N m^{-2} s^{-1} $= $ kg m^{-1} s^{-1}
work	joule	J	N m $= $ kg m^2 s^{-2}
power	watt	W	J s^{-1}
absolute temperature	kelvin	K	
customary temperture	degree Celsius	°C	K + 273
Arrhenius temperature	reciprocal Kelvin	A	1000/K
electrical potential	volt	V	
molecular weight	dalton	Da	
concentration	mass per volume, e.g.	kg m^{-3}	
		M	mol L^{-1}

Multiple units

Many units are managed numerically by the use of prefixes, as in Pg (for petagrams), μm (for micrometres) or nM (for nanomols L^{-1}). The SI notations are used throughout: $P_- = 10^{15}$ times the basic unit; $T_- = 10^{12}$; $G_- = 10^9$; $M_- = 10^6$; k(or K)$_- = 10^3$; $m_- = 10^{-3}$; $\mu_- = 10^{-6}$; $n_- = 10^{-9}$; $p_- = 10^{-12}$; $a_- = 10^{-15}$.

Symbols

A	reciprocal kelvins $\times$ 1000
Da	molecular weight, especially of large molecules
Eq	equivalent mass of a reactant ion or substance
J	joule, the basic unit for expressing work
K	kelvin, the basic unit for expressing absolute temperature

L litre, a customary unit of volume; $1 \text{ m}^3 = 1000 \text{ L}$

M molar, being a concentration of 1 gram molecule per litre of solvent (herein, always water)

N newton, the basic unit for expressing force

N nitrogen

P poise, the basic unit for expressing viscosity

P phosphorus

$[P]_i$ phosphorus concentration of the ith inflow stream to a receiving water

Pa pascal, the basic unit for expressing pressure

V volt, the basic unit for expressing electric potential

a year

d day

d (as a prefix) indicating a very small increment, as in:
dN population rise
dt time increment
dS_e extrabiotic dissipative flux of solar energy
dS_i biotic dissipative energy flux

e the base of natural logarithms

g gram

ln natural logarithm (i.e. power to base e)

log standard logarithm (i.e. power to base 10)

m metre

p_{CO_2} the partial pressure exerted by atmospheric gaseous carbon dioxide

s second

Variables

A a defined area of the Earth's surface

A_x the cross-sectional area of a river channel (in m^2)

B biomass

B_{max} biomass-supportive capacity of the available resources or, effectively, the maximum biomass that can be supported by the available resources

B_Q energy of penetrative convection

C_o solute concentration, as in Fick's equation (3.19).

D demand for a resource

D_x coefficient of horizontal diffusivity (see Eq. 2.37)

E rate of energy dissipation, in $\text{m}^2 \text{ s}^{-3}$

E_s equitability or eveness of species representation in a multi-species community

F_a solute flux to the vicinity of a cell (see Eq. 4.4)

F_C flux of CO_2 across the water surface, in mol $\text{m}^{-2} \text{ s}^{-1}$ (see Eq. 3.18)

F_i filtration rate (sensu volume processed per unit time) of an individual filter-feeding zooplankter

F_0 emitted chlorophyll fluorescence, when all light-harvesting centres are open, compared to the fluorescence following a subsequent saturating flash (F_m)

F_m emitted chlorophyll fluorescence, when all light-harvesting centres are closed by a saturating flash of light

F_v variable fluorescence ($= F_m - F_0$),

G material removed by consumers from a given volume and in a given time period

G_C the gas exchange coefficient, or linear migration rate m s^{-1} (see Eq. 3.18)

H the depth of water in a flow or lake

H'' Shannon diversity in bits

I irradiance, or the flux of visible light (or PAR), or photon flux density, in mol per unit area per unit time

I_0' instantaneous irradiance flux penetrating the surface of the sea or a lake, usually expressed in μmol photon $\text{m}^{-2} \text{ s}^{-1}$ or integrated over the day as mol photon $\text{m}^{-2} \text{ d}^{-1}$

I_c notation for the intercept of photosynthetic rates (P) against (I) derived in Fig. 3.3

I_k irradiance intensity that just saturates photosynthetic rate

I_m residual irradiance intensity at the base of the contemporary mixed layer

$I_{P=R}$ irradiance intensity that sustains just enough photosynthesis to meet respirational demand, and which

	defines the water-column compensation point
I_z	irradiance intensity at a given depth in the water column, z
I_{530}	irradiance intensity at the subscripted wavelength
I^*	instantaneous integral of fluctuating irradiance intensities applying in a mixed layer
I^{**}	integral of fluctuating irradiance intensities applying in a mixed layer over the day
J_b	buoyancy force in the surface mixed layer, owing to the difference in its density with that of the water immediately beneath
J_k	the kinetic energy available to overcome buoyancy of the surface mixed layer
K_i	Steady-state concentration of the ith nutrient resource
K_r	half-saturation constant of resource-limited growth rate (i.e. the concentration of resource at which the rate of cell replication, r', is half the maximum when growth rate uptake is nutrient-saturated)
K_U	half-saturation constant of resource uptake (i.e. the concentration of resource at which the rate of uptake, V_U, is half the maximum when uptake is nutrient-saturated $(V_{U_{max}})$
L	Horizontal downwind distance over a lake or other defined water body
LD	'light divisions'; Talling's (1957c) integral of light intensity in a water column capable of supporting net photosynthesis: LD = [ln $(I'_{0\,max}/\,0.5\,I_k)$ / ln 2]
LDH	Talling's (1957c) 'light division hours', expressing the daily integral irradiance received by a water body that is deemed to support phtosynthesis (see Eq. 3.20)
$L(P)$	aggregate areal rate of phosphorus loading to a lake (in mg m^{-2} of lake area per year)

N	population of particles or organisms or their biomass per unit area or volume (as specified)
N_0	initial population, N (i.e. at $t = 0$) and subject to dynamic change
N_t	population remaining after a period of time, t, has elapsed
$N_{t/m}$	specifically to settling particles, to show an alternative outcome at t, had the settlement been subject to m complete and instantaneous mixings
NP	communal photosynthetic rate, being the product of the chlorophyll-specific photosynthetic rate (P) and the population, as chlorophyll, present (N)
P	individual, usually biomass-specific, photosynthetic rate, expressed by mass of carbon fixed or oxygen generated per unit biomass carbon or chlorophyll per unit time
P	primary production
P_g	gross primary production
P_n	net primary production
P_{max}	maximum observed or extrapolated photosynthetic rate
PQ	photosynthetic quotient, as mol O_2 evolved per mol CO_2 assimilated
Pe	Péclet number, being the ratio of the momentum of a moving particle to diffusive transport in the medium
Q_s	short-wave solar radiation
Q_T	net radiative heat flux reaching the planetary surface (units, W m^{-2} or J m^{-2} d^{-1})
Q_T^*	that fraction of Q_T that penetrates beyond the top millimeter or so of the water column
Q_z	net radiative heat flux reaching a given point in the water column
R	rate of respiration (or biomass maintenance)
R_a	rate of community respiration contributed by photoautotrophs
R_{a20}	steady R_a at 20 °C
R_h	rate of respiration of heterotrophs
Re	Reynolds number, being the ratio between the driving forces of water flow and the viscous forces resisting them

Ri_b bulk Richardson number, being the ratio between the driving forces of motion in a water body and the buoyant resistance to entrainment of deep waters

S supply or concentration of a resource

Sh Sherwood number, being the ratio between the total flux of a nutrient solute arriving at the surface of a cell in motion and the wholly diffusive flux

U wind velocity

V volume of a lake or other defined body of water

V_U rate of uptake of a given nutrient

$V_{U_{max}}$ nutrient-saturated uptake rate of a given nutrient

W Wedderburn number, being the ratio between the Richardson number of a structure and its aspect ratio (see Eq. 2.34)

W_b dry mass of a zooplankter

W_c dry mass of an algal cell

a factor of increased diameter, used in Eq. (2.17)

a area term used in Fick's equation (3.19)

b number of beads in a chain as a variable in sinking rate (see Eq. 2.18)

b_i biomass of the ith of s species present in a community or sample

c speed of light

c_d coefficient of frictional drag

d diameter of a spherical cell

d_c diameter of a cylindrical cell

d_s diameter of a sphere of equal volume to an irregularly shaped particle

g gravitational acceleration (here taken as a constant 9.8081 m s^{-2})

h' Planck's constant, having the value 6.63×10^{-34} J s

h_c height of a cylidrical cell

h_m height of the mixed layer, from its base to the water surface

h_p height of the photic layer, being from the depth of $I_{P=R}$ to the water surface

h_s height of the water layer from the surface to the depth of Secchi-disk extinction (z_s)

h_w height of a theoretically static water layer

i,j identifiers in a multiple component list of variables

k exponential coefficient of heat absorption (see Eq. 2.27)

k' exponent of invasiveness of dispersing organisms (see Eq. 7.3)

k_a area-specific light interception of an alga

k_n rate of population change in a horizontal patch (see Eq. 2.37)

l_a length dimension available for the dissipation of the energy, usually the depth of the flowing water layer

l_e length dimension of the largest turbulent eddies

l_m smallest eddy size supported by the available mechanical energy before it is overwhelmed by viscous forces

m number of mixing events (Section 2.6.2)

m maximum cell dimension

m' maximum size of particle that is available to a given filter-feeder

n number of moles of a solute that will diffuse across an area

$n(t)$ number of exploitable niches in the location at a time t

$\tilde{n}$ number of occupied niches when $t = \infty$

n_{sp} number of species to be found in a defined area, A, as influenced by arrivals and extinction rates

p wetted perimeter of a vertical section through a stream channel

q cell-specific content of a given nutrient (or quota) (see Eq. 4.12)

q_0 minimum cell quota of a given nutrient, below which it is deemed to be no longer viable

q_{max} replete cell content of a given nutrient

q_i the inflow rate, or inflow volume per unit time (usually) to a lake

q_s discharge, or outflow volume per unit time (usually) from a lake

r'	exponent of the rate of cell replication, or rate of cell recruitment through growth	t_q	theoretical hydraulic retention time (V/q_s)
r'_{max}	resource-saturated rate of cell replication	t_w	measured hydraulic retention time (V/q_s)
r'_{20}	rate of cell replication at temperature 20 °C	u	vector or flow velocity in the horizontal (x) plane
r'_θ	rate of cell replication at the field temperature θ	$\bar{u}$	mean horizontal velocity of turbulent flow
$r'_{\theta,I}$	rate of cell replication at the field temperature θ and the natural light conditions	$\pm u'$	velocity variations in turbulent flow in relation to $\bar{u}$
$r'_{(Si)}$	rate of cell replication of diatoms calculated from uptake of soluble reactive silicon from solution	u_s	horizontal advective velocity around patch
r_c	critical radius of a patch able to maintain distinctive pelagic populations (see Eq. 2.37)	u^*	turbulent velocity, the friction velocity or the shear velocity, as derived in Section 2.3.2 and Eq. (2.4)
r_p	roughness of the stream bed (*sensu* the height of projections) (see Eq. 2.6)	$(u^*)^2$	turbulent intensity, as derived in Section 2.3.2
r_n	exponent of the rate of increase of a self-replicating population net of simultaneous rates of loss	$\hat{u}$	arrival rate of colonist species at a given habitat
r_L	sum of the exponents detracting from a self-replicating population	v	volume of a phytoplankton cell or colony
r_G	exponent of the rate of loss from a population of cells to consumers	w	vector or flow velocity in the vertical (z) plane
r_S	exponent of the rate of loss from a population of cells to settlement	$\pm w'$	velocity variations in turbulent flow in the vertical direction
r_W	exponent of the rate of loss from a population of cells to wash-out	w_c	sinking rate of chain
s	surface area of a cell or colony	w_s	sinking rate of algal cell or colony
s_b	gradient of the stream bed (in m m^{-1})	w_P	rate of loss of phosphorus in the outflow from a lake
s^2	statistical variance	x	concentration of organisms in a sample, used in the calculation of Lloyd's crowding index (see Sections 2.7.1 and 2.7.2)
t	a period of time or a fixed point in time		
t'	a shorter period of time within t or an intermediate fixed point	$\bar{x}$	mean of concentration, x, in a series of samples
t_c	limiting closure period of LHC intercepting light (in s photon^{-1})	x^*	Lloyd's crowding index, based upon the variance in x among individual samples. The greater the variance, the greater is the difference among individual samples. In Sections 2.7.1 and 2.7.2, the index is used to gauge the effect of wind upon small-scale patchiness
t_e	time to achieve total elimination (or a defined proportion) of particles from suspension		
t_m	probabilistic time of travel of an entrained particle through a mixed layer	z	used to identify points in the vertical direction; a specific depth
t_p	photoperiod or time spent in euphotic zone	$z_{(I_k)}$	depth beneath the water surface at which I_k is located; depth below which photosynthetic rate becomes light-dependent

$z_{(0.5I_k)}$ depth beneath the water surface at which photosynthetic rate is 50% that at light saturation

z_{eu} the depth of the conventionally defined euphotic layer

z_m the depth to which the water is mixed; the base of the mixed layer

z_s the Secchi-disk depth, being the distance beneath the surface at which a Secchi disk becomes obscured from the observer above the water surface (see Box 4.1)

Γ day length from sunrise to sunset

Δ to indicate difference, as in $\Delta\rho w$ (of density between top and bottom water in a stratified water body)

Δ_N integral of simultaneous concentration gradients ($\delta/\delta x$, $\delta/\delta y$, $\delta/\delta z$) in the x, y and z planes (nabla operator) (see Section 4.2.1)

Λ_P mean availability of phosphorus in a lake taking account of loading and hydraulic exchanges

Σ to indicate 'sum of', as in ΣF_i, the cumulative filtration rate of filter-feeders in a unit volume of medium, and Σt_p ($= \Gamma h_p / h_m$), the daily sum of photoperiods

ΣNP area-integrated photosynthetic rate, expressed as the sum of NP

$\Sigma\Sigma NP$ daily area-integrated photosynthetic rate

ΣNR area-integrated respiration rate, expressed as the sum of NR

$\Sigma\Sigma NR$ daily area-integrated respiration rate

Ψ quotient relating sinking behaviour to turbulent diffusivity, used as an index of entrainment (explained in Section 2.6.1)

α coefficient of slope of a regression of y on x, used herein mainly to describe light dependence of photosynthesis and growth ($\alpha = P/I$)

α_r growth efficiency on low light incomes, derived from the slope of the regression of r' on I

β slope of maximum replication rate on Arrhenius temperature scale

γ temperature coefficient of thermal expansion, given by $- (1/\rho_w) (d\rho_w / d\theta)$ K^{-1}

δ to indicate 'increment in'

ε exponent of vertical light attenuation in water (units, m^{-1})

ε_a exponent of that part of the vertical light attenuation due to the presence of planktic algae

ε_p exponent of that part of the vertical light attenuation due to the presence of tripton

ε_w exponent of that part of the vertical light attenuation due to absorption by the water

ε_z exponent of vertical light attenuation at a given depth

ε_{440}, ε_{530}, etc.   exponent of attenuation at the particular wavelengths

ε_{min} coefficient of attenuation in least-absorbed waveband

ε_{av} coefficient of attenuation averaged over a number of wavebands

$^w\varepsilon_{av}$ is a weighted average attenuation coefficient average (see Section 3.3.3)

ε' the energy of a single photon; ε' varies with the wavelength (see Eq. 3.2)

η absolute viscosity of a fluid, in kg m^{-1} s^{-1}

ν kinematic viscosity of a fluid, in m^2 s^{-1}

θ temperature, usually in °C

λ wavelength (e.g. of light, in nm)

ξ gas solubility coefficient (in mol m^{-3} atmosphere^{-1})

π the ratio of the circumference of a circle to its diameter

ρ density, usually in kg m^{-3}

ρ_a density of air

ρ_c density of a cell

ρ_m density of mucilage

ρ_w density of water, in kg m^{-3}

σ specific heat of water, in J kg^{-1} K^{-1}

σ_s succession rate index

σ_P phosphorus sedimentation rate

σ_X cross-sectional area of a light-harvesting centre

τ force or stress, in N

φ	quantum yield of photosynthesis, in mol C (mol photon)$^{-1}$
φ_r	coefficient of form resistance (dimensionless)
ω	coefficient of selectivity of ingestible particles from the inhalant current of a filter-feeder
$\int$	integral sign, as in $\int_\infty^0$: integrate from zero to infinity, or to a large number

Abbreviations used in the text

A	Photon acceptor for photochemical system I
ANN	artificial neural network
APAR	absorbed photosynthetically active radiation
ATP	adenosine triphosphate
BAP	biologically available phosphorus
BOD	biochemical oxygen demand
CCM	carbon-concentration mechanism
CRP	cAMP receptor proteins
CV	coefficient of interannual variation
DAPI	4,6-diamidino-2-phenylindole (see Section 5.5.1)
DCM	deep chlorophyll maximum
DCMU	3-(3,4-dichlorophenyl)-1, 1-dimethyl urea
DEU	dissipative ecological unit
DHM	dissolved humic matter
DIC	dissolved inorganic carbon
DIN	dissolved inorganic nitrogen (combined; does not include dissolved nitrogen gas)
DMS	dimethyl sulphide
DMSP	dimethyl-sulphonioproponiate
DNA	deoxyribonucleic acid
DOC	dissolved organic carbon
DOFe	dissolved organic iron (complexed with DOC)
DON	dissolved organic nitrogen
EDTA	ethylene diamine tetra-acetic acid
FDC	frequency of dividing cells
GPP	gross primary production
G3P	glycerate 3-phosphate
G3AP	glyceraldehyde 3-phosphate
HAB	harmful algal bloom

HNF	heterotrophic nanoflagellates
IDH	intermediate disturbance hypothesis
LAI	leaf area index
LHC	light-harvesting complex
MLD	maximum linear dimension
MPF	maturation-promoting factor (see Section 5.2.1)
MRP	molybdate-reactive phosphorus
NADP	nicotinamide adenine dinucleotide phosphate
NAO	North Atlantic Oscillation
NPP	net primary production
NTA	nitrilotriacetic acid
OECD	Organisation for Economic Co-Operation and Development
PAR	photosynthetically active radiation
PCR	polymerase chain reaction
POC	particulate organic carbon
POM	particulate organic matter
POOZ	permanently open-ocean zone (of high-latitude waters)
PP	particulate phosphorus
PQ	plastoquinone
PSI	photochemical system I
PSII	photochemical system II
Q_A, Q_B	quinones in PSII
RNA	ribonucleic acid
RUBISCO	ribulose 1,5-biphosphate carboxylase
RuBP	ribulose 1, 5-biphospate
SEH	size-efficiency hypothesis
SIZ	seasonally ice-covered zone (of high-latitude seas)
SRP	soluble reactive phosphorus
SRSi	soluble reactive silicon
STPP	sodium tripolyphosphate
TFe	total iron (*sensu* all chemical forms present)
TN	total nitrogen (*sensu* all chemical forms present)
TP	total phosphorus (*sensu* all chemical forms present)
a.s.l	above sea level
cAMP	cyclic adenosine monophosphate
chl*a*	chlorophyll *a*
[chl*a*]	concentration of chlorophyll *a*
[chl*a*]$_{max}$	maximum supportable concentration of chlorophyll *a*
ppGpp	guanosine 3′,5′-bipyrophosphate
tris	trishydroxymethyl-aminomethane

References

Abeliovich, A. and Shilo, M. (1972). Photo-oxidative death in blue-green algae. *Journal of Bacteriology*, **111**, 682–9.

Achterberg, E. P., Berg, C. M. G. van den, Boussemart, M. and Davison, W. (1997). Speciation and cycling of trace metals in Esthwaite Water, a productive English lake with seasonal deep-water anoxia. *Geochimica et Cosmochimica Acta*, **61**, 5233–53.

Ackleson, S. G. (2003). Light in shallow waters: a brief research review. *Limnology and Oceanography*, **48**, 323–8.

Adams, M. H. (1966). *Bacteriophages*, 3rd edn. New York: Interscience.

Agawin, N. S. R., Duarte, C. M. and Agustí, S. (2000). Nutrient and temperature control of the contribution of picoplankton to phytoplankton biomass and production. *Limnology and Oceanography*, **45**, 591–600.

Ahlgren, G. (1977). Growth of *Oscillatoria agardhii* in chemostat culture. I. Nitrogen and phosphorus requirements. *Oikos*, **29**, 209–24.

(1978). Growth of *Oscillatoria agardhii* in chemostat culture. II. Dependence of growth constants on temperature. *Mitteilungen der internationalen Vereinigung für theoretische und angewandte Limnologie*, **21**, 88–102.

(1985). Growth of *Oscillatoria agardhii* in chemostat culture. III. Simultaneous limitation of nitrogen and phosphorus. *British Phycological Journal*, **20**, 249–61.

Aladin, N. V., Plotnikov, I. S. and Filippov, A. A. (1993). Alteration of the Aral Sea ecosystem by human impact. *Hydrobiological Journal*, **29**(2), 22–31. (English translation of original paper in *Gidrobiologeskii Zhurnal*.)

Alcaraz, M., Paffenhofer, G.-A. and Strickler, J. R. (1980) Catching the algae: a first account of visual obeservations on filter-feeding calanoids. In *Evolution and Ecology of Zooplankton Communities*, ed. W. C. Kerfoot, pp. 241–8. Hanover, NH: University of New England Press.

Algéus, S. (1950). Further studies on the utilisation of aspartic acid, succinamide and asparagine. *Physiologia Plantarum*, **3**, 370–5.

Allanson, B. R. and Hart, R. C. (1979). Limnology of the P. K. le Roux Dam. *Reports of the Rhodes University Institute for Freshwater Studies*, **11**(7), 1–3.

Alldredge, A. L. and Silver, M. W. (1988). Characteristics, dynamics and significance of marine snow. *Progress in Oceanography*, **20**, 41–82.

Allen, E. D. and Spence, D. H. N. (1981). The differential ability of aquatic plants to utilise the inorganic carbon supply in fresh waters. *New Phytologist*, **87**, 269–83.

Allison, E. H., Irvine, K., Thompson, A. B. and Ngatunga, B. P. (1996). Diets and food consumption rates of pelagic fish in Lake Malawi, Africa. *Freshwater Biology*, **35**, 489–515.

Almer, B., Dickson, W., Ekström, C. and Hörnström, E. (1978). Sulfur pollution and the aquatic ecosystem. In *Sulfur in the Environment*, ed. O. Nriagu, pp. 271–86. New York: John Wiley.

Alvarez Cobelas, M., Velasco, J. L., Rubio, A. and Brook, A. J. (1988). Phased cell division in a field population of *Staurastrum longiradiatum* (Conjugatophyceae: Desmidaceae). *Archiv für Hydrobiologie*, **112**, 1–20.

Amezaga, E. de, Goldman, C. R. and Stull, E. A. (1973). Primary productivity and rate of change of biomass of various species of phytoplankyton in Castle Lake, California. *Verhandlungen der internationalen Vereinigung für theoretische und angewandte Limnologie*, **18**, 1768–75.

Anderies, J. M. and Beisner, B. E. (2000). Fluctuating environments and phytoplankton community structure: a stochastic model. *American Naturalist*, **155**, 556–9.

Anderson, D. M., Cembella, A. D. and Hallegraeff, G. M. (1998). *Physiological Ecology of Harmful Algal Blooms*. Berlin: Springer-Verlag.

Anderson, L. W. J. and Sweeney, B. M. (1978). Role of inorganic ions in controlling sedimentation rate of a marine centric diatom, *Ditylum brightwelli*. *Journal of Phycology*, **14**, 204–14.

Antia, N. J., Harrison, P. J. and Oliveira, L. (1991). The role of dissolved organic nitrogen in phytoplankton nutrition, cell biology and ecology. *Phycologia*, **30**, 1–89.

Armbrust, E. V., Bowen, J. D., Olsen, R. J. and Chisholm, S. W. (1989). Effect of light on the cell cycle of a marine *Synechococcus* strain. *Applied and Environmental Microbiology*, **55**, 425–32.

Arrigo, K. R., Robinson, D. H., Worthern, D. L. *et al.* (1999). Phytoplankton community structure and the drawdown of nutrients and CO_2 in the Southern Ocean. *Science*, **283**, 365–8.

Arruda, J. A., Marzolf, G. R. and Faulk, R. T. (1983). The role of suspended sediments in the nutrition of zooplankton in turbid reservoirs. *Ecology*, **64**, 1225–35.

Arvola, L. and Kankaala, P. (1989). Winter and spring variability in phyto- and bacterioplankton in lakes with different water quality. *Aqua Fennica*, **19**, 29–39.

Assaf, G., Gerard, R. and Gordon, A. L. (1971). Some mechanisms of oceanic mixing revealed in aerial photographs. *Journal of Geophysical Research*, **76**, 6550–72.

Ashton, P. (1985). Seasonality in southern-hemisphere phytoplankton assemblages. *Hydrobiologia*, **125**, 179–90.

Atkin, D. J. (1949). Seasonal fluctuations in planktonic distribution in a tropical impounding reservoir. In *Proceedings of the Conference on Biology and Civil Engineering, 21–23 September 1948*, pp. 235–8. London: Institution of Civil Engineers.

Atkinson, K. M. (1980). Experiments in dispersal of phytoplankton by ducks. *British Phycological Journal*, **15**, 49–58.

Atlas, R. M. and Bartha, R. (1993). *Microbial Ecology*, 3rd edn. Redwood City, CA: Benjamin-Cummings.

Aubrey, D., Moncheva, S., Demirov, E., Diaconu, V. and Dimitrov, A. (1999). Environmental changes in the western Black Sea related to anthropogenic and natural conditions. *Journal of Marine Systems*, **7**, 411–25.

Aubriot, L., Wagner, F. and Falkner, G. (2000). The phosphate uptake behaviour of phytoplankton communities in eutrophic lakes reflect alterations in the phosphate supply. *European Journal of Phycology*, **38**, 255–62.

Axler, R. P., Gersberg, R. M. and Goldman, C. R. (1980). Stimulation of nitrate uptake and photosynthesis by molybdenum in Castle Lake, California. *Canadian Journal of Fisheries and Aquatic Sciences*, **37**, 707–12.

Azam, F. and Chisholm, S. W. (1976). Silicic acid uptake and incorporation by natural marine phytoplankton populations. *Limnology and Oceanography*, **21**, 427–35.

Azam, F., Fenchel, T., Field, J. G. *et al.* (1983). The ecological role of water-column microbes in the sea. *Marine Ecology Progress Series*, **10**, 257–63.

Baccini, P. (1985). Phosphate interactions at the sediment–water surface. In *Chemical Processes in Lakes*, ed. W. Stumm, pp. 189–205. New York: John Wiley.

Badger, M. R. and Gallacher, A. (1987). Adaptation of photosynthetic CO_2 and HCO_3: accumulation by the cyanobacterium *Synechococcus* PCC6301 to growth at different inorganic carbon concentrations. *Australian Journal of Plant Physiology*, **14**, 189–210.

Badger, M. R., Kaplan, A. and Berry, J. A. (1980). Internal inorganic carbon pool of *Chlamydomonas reinhardtii*: evidence for a carbon-dioxide concentrating mechanism. *Plant Physiology*, **66**, 407–13.

Badger, M. R., Andrews, T. J., Whitney, S. M. *et al.* (1998). The diversity and co-evolution of Rubisco, plastids, pyrenoids and chloroplast-based CO_2-concentrating mechanisms in algae. *Canadian Journal of Botany*, **76**, 1052–71.

Bailey-Watts, A. E. (1978). A nine-year study of the phytoplankton of the eutrophic and non-stratifying Loch Leven (Kinross, Scotland). *Journal of Ecology*, **66**, 741–71.

Bailey-Watts, A. E. and Duncan, P. (1981). The phytoplankton. In *The Ecology of Scotland's Largest Lochs*, ed. P. S. Maitland, pp. 91–118. The Hague: W. Junk.

Balch, W. M., Kilpatrick, K. A. and Trees, C. C. (1996). The 1991 coccolithophore bloom in the central North Atlantic. I. Optical properties and factors affecting distribution. *Limnology and Oceanography*, **41**, 1669–83.

Bannister, T. T. and Weidemann, A. D. (1984). The maximum quantum yield of phytoplankton photosynthesis. *Journal of Plankton Research*, **6**, 275–94.

Banse, K. (1976). Rates of growth, respiration and photosynthesis of unicellular algae as related to cell size: a review. *Journal of Phycology*, **12**, 135–40.

(1994). Grazing and zooplankton production as key controls of phytoplankton production in the open ocean. *Oceanography*, **7**, 13–19.

Barber, J. and Anderson, J. M. (2002). Introduction [to papers of a discussion meeting on photosystem II]. *Philosophical Transactions of the Royal Society of London B*, **357**, 1325–8.

Barber, J. and Nield, J. (2002). Organization of transmembrane helices in photosystem II: comparison of plants and cyanobacteria. *Philosophical Transactions of the Royal Society of London B*, **357**, 1329–35.

Barber, R. T. and Hilting, A. K. (2002). History of the study of plankton productivity. In *Phytoplankton*

Productivity, ed. P. J. leB. Williams, D. N. Thomas and C. S. Reynolds, pp. 16–43. Oxford: Blackwell Science.

Barbiero, R. P., James, W. F. and Barko, J. W. (1999). The effects of disturbance events on phytoplankton community structure in a small temperate reservoir. *Freshwater Biology*, **42**, 503–12.

Barón, P. J. (2003). The paralarvae of two South American sympatric squid: *Loligo gahi* and *Loligo sanpaulensis. Journal of Plankton Research*, **25**, 1347–58.

Barrett, P. R. F., Littlejohn, J. W. and Curnow, J. (1999). Long-term algal control in a reservoir using barley straw. *Hydrobiologia*, **415**, 309–13.

Bartram, J. and Chorus, I. (1999). *Toxic Cyanobacteria*. London: E. and F. Spon.

Bautista, M. F. and Paerl, H. W. (1985). Diel N_2 fixation in an intertidal marine cyanobacterial mat community. *Marine Chemistry*, **16**, 369–77.

Beadle, L. C. (1974). *The Inland Waters of Tropical Africa: An Introduction to Tropical Limnology*. London: Longman.

Bean, C. W. and Winfield, I. J. (1995). Habitat use and activity patterns of roach (*Rutilus rutilus*), rudd (*Scardinius erythrophthalmus* (L.), perch (*Perca fluviatilis*) and pike (*Esox lucius*) in the laboratory: the role of predation threat and structural complexity. *Ecology of Freshwater Fish*, **4**, 37–46.

Beard, S. J., Handley, B. A., Hayes, P. K. and Walsby, A. E. (1999). The diversity of gas-vesicle genes in *Planktothrix rubescens* from Lake Zürich. *Microbiology*, **145**, 2757–68.

Beard, S. J., Davis, P. A., Iglesias-Rodriguez, D., Skulberg, O. M. and Walsby, A. E. (2000). Gas vesicle genes in *Planktothrix* spp. from Nordic lakes: strains with weak gas vesicles possess a longer variant of gvpC. *Microbiology*, **146**, 2009–18.

Beardall, J., Berman, T., Heraud, P. *et al.* (2001). A comparison of methods of detection of phosphate limitation in microalgae. *Aquatic Sciences*, **63**, 107–21.

Beers, J. R., Reid, F. M. H. and Stewart, G. L. (1982). Seasonal abundance of the microplankton population in the North Pacific cental gyre. *Deep-Sea Research*, **29**, 227–45.

Beeton, A. M. and Hageman, J. (1992). Impact of *Dreissena polymorpha* on the zooplankton community of western Lake Erie. *Verhandlungen der internationalen Vereinigung für theoretische und angewandte Limnologie*, **25**, 2349.

Behrenfeld, M. J. and Falkowski, P. G. (1997). Photosynthetic rates derived from satellite-based chlorophyll concentration. *Limnology and Oceanography*, **42**, 1–20.

Behrenfeld, M. J., Esaias, W. E. and Turpie, K. R. (2002). Assessment of primary production at the global scale. In *Phytoplankton Productivity*, ed. P. J. leB. Williams, D. N. Thomas and C. S. Reynolds, pp. 156–86. Oxford: Blackwell Science.

Béjà, O., Suzuki, M. T., Heidelberg, J. F. *et al* (2002). Unsuspected diversity among marine aerobic anoxygenic phototrophs. *Nature*, **415**, 630–3.

Belcher, J. H. (1968). Notes on the physiology of *Botryococcus braunii* Kützing. *Archiv für Mikrobiologie*, **61**, 335–46.

Bellinger, E. (1974). A note on the use of algal sizes in estimates of population standing crop. *British Phycological Journal*, **9**, 157–61.

Belyea, L. R. and Lancaster, J. (1999). Assembly rules within a contingent ecology. *Oikos*, **86**, 402–16.

Berger, C. (1984). Consistent blooming of *Oscillatoria agardhii* Gom. in shallow hypertrophic lakes. *Verhandlungen der internationalen Vereinigung für theoretische und angewandte Limnologie*, **22**, 910–16.

(1989). *In situ* primary production, biomass and light regime in the Wolderwijd, the most stable *Oscillatoria agardhii* lake in the Netherlands. *Hydrobiologia*, **185**, 233–44.

Bergh, O., Børsheim, K. Y., Bratbak, G. and Heldal, M. (1989). High abundances of viruses found in aquatic environments. *Nature*, **340**, 467–8.

Berman, T. and Bronk, D. A. (2003). Dissolved organic nitrogen: a dynamic participant in aquatic ecosystems. *Aquatic Microbial Ecology*, **31**, 279–305.

Berman, T. and Shteinman, B. (1998). Phytoplankton development and turbulent mixing in Lake Kinneret (1992–1996). *Journal of Plankton Research*, **20**, 709–26.

Berman, T., Hadas, O. and Kaplan, B. (1977). Uptake and respiration of organic compounds and heterotrophic growth in *Pediastrum duplex* (Meyen). *Freshwater Biology*, **7**, 495–502.

Berman, T., Walline, P. D., Schneller, A., Rothenberg, J. and Townsend, D. W. (1985). Secchi-disk depth record: a claim for the eastern Mediterranean. *Limnology and Oceanography*, **30**, 447–8.

Berman, T., Yacobi, Y. Z. and Pollingher, U. (1992). Lake Kinneret phytoplankton: stability and variability during twenty years (1970–1989). *Aquatic Sciences*, **54**, 104–27.

Berner, R. A. (1980). *Early Diagenesis: A Theoretical Approach*. Princeton, NJ: Princeton University Press.

Bertilsson, S. and Tranvik, L. J. (1998). Photochemically produced carboxylic acids xsas substrates for freshwater bacterioplankton. *Limnology and Oceanography*, **43**, 885–95.

(2000). Photochemical transformation of dissolved organic matter in lakes. *Limnology and Oceanography*, **45**, 753–62.

Besch, W. K., Ricard, M. and Cantin, R. (1972). Benthic diatoms as indicators of mining pollution in the Northwest Miramichi River System, New Brunswick, Canada. *Internationale Revue des gesamten Hydrobiologie*, **57**, 39–73.

Beverton, R. J. H. and Holt, S. J. (1957). On the dynamics of exploited fish populations. *Fishery Investigations*, **19**, 1–532.

Bhattacharya, D. and Medlin, L. (1998). Algal phylogeny and the origin of land plants. *Plant Physiology*, **116**, 9–15.

Bidanda, B. P. and Cotner, J. B. (2002). Love handles in aquatic ecosystems: the role of dissolved organic carbon drawdown, resuspended sediments and terrigenous inputs in the carbon balance of Lake Michigan. *Ecosystems*, **5**, 431–45.

Biddanda, B. A., Ogdahl, M. and Cotner, J. B. (2001). Dominance of bacterial metabolism in oligotrophic relative to eutrophic waters. *Limnology and Oceanography*, **46**, 730–39.

Bienfang, P. K. (1992). The role of coastal high-latitude ecosystems in global export production. In *Primary Productivity and Biogeochemical Cycles in the Sea*, ed. P. G. Falkowski and A. Woodhead, pp. 285–97. New York: Plenum Press.

Biggs, B. J. F., Stevenson, R. J. and Lowe, R. L. (1998). A habitat matrix coeneptual model for stream periphyton. *Archiv für Hydrobiologie*, **143**, 21–56.

Bilton, D. T., Freeland, J. R. and Okamura, B. (2001). Dispersal in freshwater invertebrates. *Annual Review of Ecology and Systematics*, **32**, 159–81.

Bindloss, M. E. (1974). Primary productivity of phytoplankton in Loch Leven, Kinross. *Proceedings of the Royal Society of Edinburgh B*, **74**, 157–81.

Bird, D. F. and Kalff, J. (1989). Phagotrophic sustenance of a metalimnetic phytoplankton peak. *Limnology and Oceanography*, **34**, 155–62.

Björk, S. (1972). Swedish lake restoration program gets results. *Ambio*, **1**, 153–65.

(1988). Redevelopment of lake ecosystems: a case-study approach. *Ambio*, **17**, 90–98.

Bleiwas, A. H. and Stokes, P. M. (1985). Collection of large and small food particles by *Bosmina*. *Limnology and Oceanography*, **30**, 1090–2.

Blindow, I. (1992). Decline of charophytes during eutrophication: comparison with angiosperms. *Freshwater Biology*, **28**, 9–14.

Blindow, I., Andersson, G., Hargeby, A. and Johansson, S. (1993). Long-term pattern of alternative stable states in two shallow eutrophic lakes. *Freshwater Biology*, **30**, 159–67.

Blomqvist, P., Pettersson, A. and Hyenstrand, P. (1994). Ammonium nitrogen: a key regulatory factor causing dominance of non-nitrogen-fixing Cyanobacteria in aquatic systems. *Archiv für Hydrobiologie*, **132**, 141–64.

Booker, M. and Walsby, A. E. (1979). The relative form resistance of straight and helical blue-green algal filaments. *British Phycological Journal*, **14**, 141–50.

(1981). Bloom formation and stratification by a planktonic blue-green alga in an experimental water column. *British Phycological Journal*, **16**, 411–21.

Booker, M. J., Dinsdale, M. T. and Walsby, A. E. (1976). A continuously monitored column for the study of stratification by planktonic organisms. *Limnology and Oceanography*, **14**, 915–19.

Booth, K., Bellinger, E. D. and Sigee, D. (1987). Use of X-ray microanalysis and atomic-absorption spectroscopy for monitoring metal levels in algal and bacterial plankton. *British Phycological Journal*, **22**, 300–1.

Bormans, M., Sherman, B. S. and Webster, I. T. (1999). Is buoyancy regulation in Cyanobacteria an adaptation to exploit separation of light and nutrients? *Marine and Freshwater Research*, **50**, 897–906.

Borradaile, L. A., Eastham, L. E. S., Potts, F. A. and Saunders, J. T. (1961). *The Invertebrata*, 4th edn (revised by G. A. Kerkut). Cambridge: Cambridge University Press.

Boström, B., Andersen, J. M., Fleischer, S. and Jansson, M. (1988). Exchange of phosphorus across the sediment–water interface. *Hydrobiologia*, **170**, 229–44.

Bottrell, H. H., Duncan, A., Gliwicz, Z. M. et al. (1976). A review of some problems in zooplankton production studies. *Norwegian Journal of Zoology*, **24**, 419–56.

Boucher, P., Blinn, D. W. and Johnson, D. B. (1984). Phytoplankton ecology in an unusually stable environment (Montezuma Wall, Arizona, USA). *Hydrobiologia*, **199**, 149–60.

Bowen, C. C. and Jensen, T. E. (1965). Blue-green algae: fine structure of the gas vacuoles. *Science*, **147**, 1460–3.

Bower, A. S., Le Cann, B., Rossby, T. *et al.* (2002). Directly measured mid-depth circulation in the north-eastern Atlantic Ocean. *Nature*, **419**, 603–7.

Bowie, A. R., Maldonado, M. T., Frew, R. D. *et al.* (2001). The fate of added iron during a mesoscale fertilisation experiment in the Southern Ocean. *Deep-Sea Research Part II, Topical Studies in Oceanography*, **48**, 2703–43.

Bowie, A. R., Achterberg, E. P., Sedwick, P. N., Ussher, S. and Worsfold, P. J. (2002). Real-time monitoring of picomolar concentrations of iron(II) in marine waters using automated flow injection-chemiluminescence instrumentation. *Environmental Science and Technology*, **36**, 4600–7.

Boyce, F. M. (1974). Some aspects of Great Lakes physics of importance to biological and chemical processes. *Journal of the Fisheries Research Board of Canada*, **31**, 689–730.

Boyd, P. W. (2002). Environmental factors controlling phytoplankton processes in the Southern Ocean. *Journal of Phycology*, **38**, 844–61.

Boyd, P. W., LaRoche, J., Gall, M., Frew, R. and McKay, R. M. L. (1999). Role of iron, light and silicate in controlling algal biomass in Subantarctic waters, south east of New Zealand. *Journal of Geophysical Research*, **104**, 13395–408.

Braginskiy, L. P. and Sirenko, L. A. (2000). Vegetation cycle of *Microcystis aeruginosa* Kütz. emend. Elenk. *International Journal on Algae*, **2**(3), 78–91. (English translation of original publication in Russian, *Algologiya*, **7**(2), 153–65.)

Bratbak, G. and Heldal, M. (1995). Viruses – the new players in the game: their ecological role and could they mediate genetic exchange by transduction. In *Molecular Ecology of Aquatic Microbes*, ed I. Joint, pp. 249–64. Berlin: Springer-Verlag.

Braun-Blanquet, J. (1964). *Pflanzensociologie*. Vienna: Springer-Verlag.

Braunwarth, C. and Sommer, U. (1985). Analyses of the in-situ growth rates of Cryptophyceae by use of the mitotic index. *Limnology and Oceanography*, **30**, 893–7.

Bright, D. I. and Walsby, A. E. (1999). The relationship between critical pressure and width of gas vesicles in isolates of *Planktothrix rubescens* from Lake Zürich. *Microbiology*, **145**, 2769–75.

(2000). The daily integral of growth of *Planktothrix rubescens* calculated from growth in culture and irradiance in Lake Zürich. *New Phytologist*, **146**, 301–16.

Brodin, Y.-W. (1995). Acidification of lakes and water courses in a global perspective. In *Liming of Acidified Surface Waters: A Swedish Synthesis*, ed. L. Henrikson and Y.-W. Brodin, pp. 45–62. Berlin: Springer-Verlag.

Brook, A. J., Baker, A. L. and Klemer, A. R. (1971). The use of turbidimetry in studies of the population dynamics of phytoplankton populations, with special reference to *Oscillatoria agardhii* var. *isothrix*. *Mitteilungen der internationalen Vereinigung für theoretische und angewandte Limnologie*, **19**, 244–52.

Brooks, A. S., Warren, G. J., Boraas, M. E., Scale, D. B. and Edington, D. N. (1984). Long-term phytoplankton shifts in Lake Michigan: cultural eutrophication or biotic shifts. *Verhandlungen der internationalen Vereinigung für theoretische und angewandte Limnologie*, **22**, 452–9.

Brooks, J. L. and Dodson, S. I. (1965). Predation, body size and composition of plankton. *Science*, **150**, 28–35.

Brown, D. J. A. and Sadler, K. (1989). Fish survival in acid waters. In *Acid Toxicity and Aquatic Animals*, ed. R. Morris, E. W. Taylor, D. J. A. Brown and J. A., Brown, pp. 31–44. Cambridge: Cambridge University Press.

Brown, J. H. (1999). Macroecology: progress and prospect. *Oikos*, **87**, 3–14.

Brown, J. H. and Maurer, B. A. (1989). Macroecology: the division of food and space among species on continents. *Science*, **243**, 1145–50.

Brownlee, E. F., Sellner, S. G. and Sellner, K. G. (2003). Effects of barley straw (*Hordeum vulgare*) on freshwater and brackish phytoplankton and cyanobacteria. *Journal of Applied Phycology*, **15**, 525–31.

Brunberg, A. K. and Blomqvist, P. (2003). Recruitment of *Microcystis* (Cyanophyceae) from lake sediments: the importance of littoral inocula. *Journal of Phycology*, **39**, 58–63.

Bruning, K. (1991a). Infection of the diatom *Asterionella* by a chytrid. I. Effects of light on reproduction and infectivty of the parasite. *Journal of Plankton Research*, **13**, 103–17.

(1991b). Infection of the diatom *Asterionella* by a chytrid. II. Effects of light on survival and epidemic development of the parasite. *Journal of Plankton Research*, **13**, 119–29.

(1991c). Effects of temperature and light on the population dynamics of the *Asterionella– Rhizophydium* association. *Journal of Plankton Research*, **13**, 707–19.

Brunner, P. H. and Lampert, C. (1997). The flow of nutrients in the Danube River Basin. *EAWAG News*, **43**, 15–17.

Brush, M. J. and Nixon, S. W. (2002). Direct measurement of light attenuation by epiphytes on eelgrass *Zostera marina*. *Marine Ecology – Progress Series*, **238**, 73–9.

Brush, M. J., Brawley, J. W., Nixon, S. W. and Kremer, J. N. (2002). Modelling phytoplankton production: problems with the Eppley curve and an empirical alternative. *Marine Ecology – Progress Series*, **238**, 31–45.

Bulgakov, N. G. and Levich, A. P. (1999). The nitrogen : phosphorus ratio as a factor regulating phytoplankton community structure. *Archiv für Hydrobiologie*, **146**, 3–22.

Burger-Wiersma, T., Stal, L. J. and Mur, L. R. (1989). *Prochlorothrix hollandica* gen. nov., sp. nov., a filamentous oxygenic photoautotrophic prokaryote containing chlorophyll *a* and chlorophyll *b*. *International Journal of Systematic Bacteriology*, **39**, 250–7.

Burkhardt, S., Riebesell, U. and Zondervan, I. (1999). Effects of growth rate, CO_2 concentration and cell size on the stable carbon isotope fractionation in marine phytoplankton. *Geochimica and Cosmochimica Acta*, **63**, 3729–41.

Burmaster, D. (1979). The continuous culture of phytoplankton: mathematical equivalence among three steady-state models. *American Naturalist*, **113**, 123–34.

Burns, C. W. (1968a). The relationship between body size of filter-feeding cladocera and the maximum size of particle ingested. *Limnology and Oceanography*, **23**, 675–8.

(1968b). Direct observations of mechanisms regulating feeding behaviour of *Daphnia* in lakewater. *Internationale Revue des Gesamten Hydrobiologie*, **53**, 83–100.

(1969). Relation between filtering rate, temperature and body size in four species of *Daphnia*. *Limnology and Oceanography*, **14**, 693–700.

Burns, C. W. and Gilbert, J. J. (1993). Predation on ciliates by freshwater calanoid copepods: rates of predation and relative vulnerabilities of prey. *Freshwater Biology*, **30**, 377–93.

Burris, J. (1981). Effects of oxygen and inorganic carbon concentrations on the photosynthetic quotients of marine algae. *Marine Biology*, **65**, 215–19.

Butcher, R. W. (1924). The plankton of the River Wharfe. *The Naturalist, Hull*, April–June 1924, 175–214.

(1932). Studies in the ecology of rivers. II. The microflora of rivers with special reference to the algae on the river bed. *Annals of Botany new series*, **46**, 813–61.

Bykovskiy, V. I. (1978) The effect of current velocity on the phytoplankton (a review). *Hydrobiological Journal*, **14**, 34–40.

Cabrera, S., Montecino, V., Vila, I. *et al.* (1977). Caracteristicas limnologicas del Embalse Rapel, Chile Central. *Seminario sobre medio ambiente y represas*, **1**, 40–61.

Cáceres, O. and Reynolds, C. S. (1984). Some effects of artificially enhanced anoxia on the growth of *Microcystis aeruginosa* Kütz. emend. Elenkin, with special reference to the initiation of the annual growth cycle in lakes. *Archiv für Hydrobiologie*, **99**, 379–97.

Callieri, C. and Stockner, J. G. (2002). Freshwater autotrophic picoplankton: a review. *Journal of Limnology*, **61**, 1–14.

Callieri, C., Karjalainen, S. M. and Passoni, S. (2002). Grazing by ciliates and heterotrophic nanoflagellates on picocyanobacteria in Lago Maggiore, Italy. *Journal of Plankton Research*, **24**, 785–96.

Campbell, L., Nolla, H. A. and Vaulot, D. (1994). The importance of *Prochlorococcus* to community structure in the central North Pacific Ocean. *Limnology and Oceanography*, **39**, 954–61.

Canabeus, L. (1929). Über die Heterocysten und Gasvakuolen der Blaualgen. *Pflanzenforschung*, **13**, 1–48.

Canter, H. M. (1979). Fungal and protozoan parasites and their importance in the ecology of phytoplankton. *Report of the Freshwater Biological Association*, **47**, 43–50.

Canter, H. M. and Jaworski, G. H. M. (1978). The isolation, maintenance and host range studies of a chytrid, *Rhizophydium planktonicum* Canter emend., parasitic on *Asterionella formosa* Hassall. *Annals of Botany*, **42**, 967– 79.

(1979). The occurrence of a hypersensitive reaction in the planktonic diatom *Asterionella formosa* Hassal parasitised by the chytrid *Rhizophydium planctonicum* Canter emend. in culture. *New Phytologist*, **82**, 187–206.

(1980). Some general observations on the zoospores of the chytrid *Rhizophydium planctonicum* Canter emend. *New Phytologist*, **84**, 515–31.

(1981). The effect of light and darkness upon infection of *Asterionella formosa* Hassal by the chytrid, *Rhizophydium planktonicum* Canter emend. *Annals of Botany, New Series*, **47**, 13–30.

(1982). Some observations on the alga *Fragilaria crotonensis* Kitton and its parasitism by two chytridaceous fungi. *Annals of Botany, New Series*, **47**, 13–30.

(1986). A study on the chytrid *Zygorhizidium planktonicum* Canter, a parasite of the diatoms *Asterionella* and *Synedra*. *Nova Hedwigia*, **43**, 269–98.

Canter, H. M. and Lund, J. W. G. (1948). Studies on plankton parasites. I. Fluctuations in the numbers of *Asterionella formosa* Hass. in relation to fungal epidemics. *New Phytologist*, **47**, 238–61.

(1951). Studies on plankton parasites. III. Examples of the interaction between parsitism and other factors determining the growth of diatoms. *Annals of Botany, New Series*, **15**, 359–71.

(1953). Studies on plankton parasites. II. The parasitism of diatoms with special reference to the lakes in the English Lake District. *Transactions of the British Mycological Society*, **36**, 13–37.

Canter-Lund, H. M. and Lund, J. W. G. (1995). *Freshwater Algae: Their Microscopic World Explored*. Bristol: Biopress.

Caperon, J. and Meyer, J. (1972). Nitrogen-limited growth of marine phytoplankton. II. Uptake kinetics and their role in nutrient-limited growth of phytoplankton. *Deep-Sea Research*, **19**, 619–32.

Caraco, N. F. (1995). Influence of human poulations on P transfers to aquatic systems: a regional-scale study using large rivers. In *Phosphorus in the Global Environment*, ed. H. Tiessen, pp. 235–44. Chichester: John Wiley.

Carling, P. A. (1992) In-stream hydraulics and sediment transport. In *The Rivers Handbook*, vol. 1, ed. P. Calow and G. E. Petts, pp. 101–25. Oxford: Blackwell Scientific Publications.

Carlsson, P. and Granéli, E. (1999). Effects of N : P : Si ratios and zooplankton grazing on phytoplankton communities in the northern Adriatic Sea. II. Phytoplankton species composition. *Aquatic Microbial Ecology*, **18**, 55–65.

Carmichael, W. W. (1986). Algal toxins. In *Advances in Botanical Research*, vol. 12, ed. J. Callow, pp. 467–501. London: Academic Press.

Carmichael, W. W., Jones, C. L. A., Mahoud, N. A. and Thiess, W. C. (1985). Algal toxins and water-based diseases. *Critical Reviews in Environmental Control*, **15**, 275–313.

Caron, D. A., Pick, F. R. and Lean, D. R. S. (1985). Chroococcoid Cyanobacteria in Lake Ontario: vertical and seasonal distributions during 1982. *Journal of Phycology*, **21**, 171–5.

Carpenter, E. J. and Chang, J. (1988). Species-specific phytoplankton growth rates via diel and DNA synthesis cycles. I. Concept of the method. *Marine Ecology – Progress Series*, **43**, 105–11.

Carpenter, E. J. and McCarthy, J. J. (1975). Nitrogen fixation and uptake of combined nitrogenous nutrients by *Oscillatoria* (*Trichodesmium*) *thiebautii* in the western Sargasso Sea. *Limnology and Oceanography*, **20**, 389–401.

Carpenter, E. J. and Price, C. C. (1976). Marine *Oscillatoria* (*Trichodesmium*): explanation for aerobic nitrogen fixation without heterocysts. *Science*, **191**, 1278–80.

Carpenter, E. J. and Romans, K. (1991). Major role of the Cyanobacterium, *Trichodesmium*, in nutrient cycling in the North Atlantic Ocean. *Science*, **254**, 1356–8.

Carpenter, S. R. (2001). Alternate states of ecosystems: evidence and some implications. In *Ecology: Achievement and Challenge*, ed. M. C. Press, N. J. Huntly and S. Levin, pp. 357–83. Oxford: Blackwell Science.

Carpenter, S. R. and Kitchell, J. F. (1992). Trophic cascade and biomanipulation: interface of research and management. A reply to the comment by de Melo *et al.* *Limnology and Oceanography*, **37**, 208–13.

(1993). *The Trophic Cascade in Lakes*. Cambridge: Cambridge Univerity Press.

Carpenter, S. R., Kitchell, J. F. and Hodgson, J. R. (1985). Cascading trophic interactions and lake productivity. *BioScience*, **35**, 634–9.

Carpenter, S. R., Ludwig, D. and Brock, W. A. (1999). Management of eutrophication for lakes subject to potentially irrevresible change. *Ecological Applications*, **9**, 751–71.

Carpenter, S. R., Cole, J. J., Hodgson, J. R. *et al.* (2001). Trophic cascades, nutrients and lake productivity: whole-lake experiments. *Ecological Monographs*, **71**, 163–86.

Carr, N. G. (1995). Microbial cultures and natural poulations. In *Molecular Ecology of Aquatic Microbes*, ed. I. Joint, pp. 391–402. Berlin: Springer-Verlag.

Carrick, H. J. and Schelske, C. L. (1997). Have we overlooked the importance of small phytoplankton in productive waters? *Limnology and Oceanography*, **42**, 1613–21.

Carrick, H. J. F., Aldridge, J. and Schelske, C. (1993). Wind influences phytoplankton biomass and composition in a shallow, productive lake. *Limnology and Oceanography*, **38**, 1179–92.

Cassie, R. M. (1959). Micro-distribution of plankton. *New Zealand Journal of Science*, **3**, 26–50.

Cavicchioli, R., Ostrowski, M., Fegatella, F., Goodchild, A. and Guixa-Boixereu, N. (2003). Life under nutrient limitation in oligotrophic marine environments: an eco/physiological perspective of *Sphingopyxis alaskensis* (formerly *Sphingomonas alskensis*). *Microbial Ecology*, **45**, 203–17.

Cembella, A. D., Antia, N. J. and Harrison, P. J. (1984). The utilization of inorganic and organic phosphorus compounds as nutrients by eukaryotic microalgae: a multidisciplinary perspective. Part I. *Critical Reviews in Microbiology*, **10**, 317–91.

Chandler, D. C. (1937). Fate of typical lake plankton in streams. *Ecological Monographs*, **7**, 445–79.

Chang, J. and Carpenter, E. J. (1990). Species-specific phytoplankton growth rates via diel and DNA synthesis cycles. IV. Evaluation of the magnitude of error with computer-simulated cell populations. *Marine Ecology – Progress Series*, **65**, 293–304.

(1994). Active growth of the oceanic dinoflagellate *Ceratium teres* in the Caribbean and Sargasso Seas estimated by cell cycle analysis. *Journal of Phycology*, **30**, 375–81.

Chapleau, F., Johansen, P. H. and Williams, M. M. (1988). The distinction between pattern and process in evolutionary biology: the use and abuse of the term 'strategy'. *Oikos*, **53**, 136–8.

Chapman, D. V., Dodge, J. D. and Heaney, S. I (1980). Light- and electron-microscope observations on cysts and cyst formation in *Ceratium hirundinella*. *British Phycological Journal*, **15**, 193.

Chapman, D. V., Livingstone, D. and Dodge, J. D. (1981). An electron-microscope study of the excystment and early development of the dinoflagellate *Ceratium hirundinella*. *British Phycological Journal*, **16**, 183–94.

Chernousova, V. M., Sirenko, L. A. and Arendarchuk, V. V. (1968). Localisation and physiological state of species of blue-green algae during late-autumn and spring periods. In *Tsvetenie Vody*, ed. A. V. Topachvskiy, L. P. Braginskiy, N. V. Kondra'eva, L. A. Kulskiy and L. A. Sirenko, pp. 81–91. Kiev: Naukova Dumka. (In Russian.)

Chisholm, S. W. and Morel, F. M. M. (1991). What controls phytoplankton production in nutrient-rich areas of the open sea? *Limnology and Oceanography*, **36**, i.

Chisholm, S. W., Olson, R. J., Zettler, E. R. *et al.* (1988). A novel free-living prochlorophyte abundant in the the oceanic euphotic zone. *Nature*, **334**, 340–3.

Chisholm, S. W., Frankel, S. L., Goericke, R. *et al.* (1992). *Prochlorococcus marinus* nov. gen. nov. sp.: an oxyphototrophic marine prokaryote containing divinyl chlorophyll *a* and *b*. *Archives of Microbiology*, **157**, 297–300.

Chorus, I. and Schlag, G. (1993). Importance of intermediate disturbances for species composition and diversity in two very different Berlin lakes. *Hydrobiologia*, **249**, 67–92.

Christie, W. J. (1974). Changes in the fish species composition of the Great Lakes. *Journal of the Fisheries Research Board of Canada*, **31**, 827–54.

Christoffersen, K. (1996). Ecological implications of cyanobacterial toxins in aquatic food webs. *Phycologia*, **35** (Suppl. 6), 42–50.

Chu, S. P. (1942). The influence of the mineral composition of the medium on the growth of planktonic algae. I. Methods and culture media. *Journal of Ecology*, **30**, 284–325.

(1943). The influence of the mineral composition of the medium on the growth of planktonic algae. II. The influence of the concentrations of inorganic nitrogen and phosphorus. *Journal of Ecology*, **31**, 109–48.

Clasen, J. (1979). Das Ziel der Phosphoreliminierung am Zulauf der Wahnbachtalsperre in Hinblick auf die Oligotrophierung diese Gewässers. *Zeitschrift für Wasser- und Abwasser-Forschung*, **12**, 65–77.

Clements, F. E. (1916). *Plant Succession: An Analysis of the Development of Vegetation*, Publications of the Carnegie Institute No. 242. Washington, DC: Carnegie Institute.

Clesceri, N. L. and Lee, G. F. (1965). Hydrolysis of condensed polyphospates. I. Non-sterile environment. *Air and Water Pollution*, **9**, 723–42.

Cloern, J. E. (1977). Effects of light intensity and temperature on *Cryptomonas ovata* (Cryptophyceae) and nutrient uptake. *Journal of Phycology*, **13**, 389–95.

Cloern, J. E., Grenz, C. and Vidergar-Lucas, L. (1995). An empirical model of the phytoplankton chlorophyll : carbon ratio: the conversion factor between productivity and growth rate. *Limnology and Oceanography*, **40**, 1313–21.

Cmiech, H. A., Reynolds, C. S. and Leedale, G. F. (1984). Seasonal periodicity, heterocyst differentiation and sporulation of planktonic Cyanophyceae in a shallow lake, with special reference to *Anabaena solitaria*. *British Phycological Journal*, **19**, 245–57.

Codd, G. A. (1995). Cyanobacterial toxins: occurrence, properties and biological significance. *Water Science and Technology*, **32**(4), 149–56.

Coesel, P. F. M. (1994). On the ecological significance of a mucilaginous envelope in planktic desmids. *Algological Studies*, **73**, 65–74.

Cohen, J. E., Briand, F. and Newman, C. M. (1990). *Community Food Webs: Data and Theory*. New York: Springer-Verlag.

Cole, J. J. (1982). Interactions between bacteria and algae in aquatic ecosystems. *Annual Reviews of Ecology and Systematics*, **13**, 291–314.

Cole, J. J. and Caraco, N. F. (1998). Atmospheric exchange of carbon dioxide in a low-wind oligotrophic lake. *Limnology and Oceanography*, **43**, 647–56.

Cole, J. J., Likens, G. E. and Strayer, D. L. (1982). Photosynthetically produced dissolved organic carbon: an important carbon source for planktonic bacteria. *Limnology and Oceanography*, **27**, 1080–90.

Cole, J. J., Findlay, S. and Pace, M. L. (1988). Bacterial production in fresh and saltwater ecosystems: a cross-system overview. *Marine Ecology – Progress Series*, **43**, 1–10.

Cole, J. J., Caraco, N. F., Strayer, D. L., Ochs, C. and Nolan, S. (1989). A detailed carbon budget as an ecosystem-level calibration of bacterial respiration in an oligotrophic lake during midsummer. *Limnology and Oceanography*, **34**, 286–96.

Cole, J. J., Caraco, N. F., Kling, G. W. and Kratz, T. W. (1994). Carbon dioxide supersaturation in the surface waters of lakes. *Science*, **265**, 1568–70.

Colebrook, J. M. (1960). Plankton and water movements in Windermere. *Journal of Animal Ecology*, **29**, 217–40.

Colebrook, J. M., Glover, R. S. and Robinson, G. A. (1961). Continuous plankton records: contribution towards a plankton atlas of the noth-eastern Atlantic and the North Sea: general introduction. *Bulletins of Marine Ecology*, **5**, 67–80.

Coleman, A. (1980). Enhanced detection of bacteria in natural environments by fluorochrome staining of DNA. *Limnology and Oceanography*, **25**, 948–51.

Connell, J. H. (1978). Diversity in tropical rain forests and coral reefs. *Science*, **199**, 1302–10.

Conway, K. and Trainor, F. R. (1972). *Scenedesmus* morphology and flotation. *Journal of Phycology*, **8**, 138–43.

Cooke, G. W. (1976). A review of the effects of agriculture on the chemical composition and quality of surface and underground waters. *Technical Bulletin, Ministry of Agriculture Fisheries and Food*, **32**, 5–57.

Cordoba-Molina, J. F., Hudgins, R. R. and Silverston, P. L. (1978). Settling in continuous sedimentation tanks. *Journal of the Environmental Division of the American Society of Civil Engineers*, **104**(EE6), 1263–75.

Corner, E. D. S. and Davies, A. G. (1971). Plankton as a factor in the nitrogen and phosphorus cycles in the sea. *Advances in Marine Biology*, **9**, 101–204.

Cottingham, K. L., Brown, B. L. and Lennon, J. T. (2001). Biodiversity may regulate the the temporal variability of ecological systems. *Ecology Letters*, **4**, 72–85.

Coulter, G. W. (1994). Speciation and fluctuating environments, with special reference to ancient African lakes. *Ergebnisse der Limnologie*, **44**, 127–37.

Coveney, M. F. and Wetzel, R. G. (1995). Biomass, production and specific growth rate of bacterioplankton and coupling to phytoplankton in an oligotrophic lake. *Limnology and Oceanography*, **40**, 1187–1200.

Crawford, D. W. and Lindholm, T. (1997). Some observations on vertical distribution and migration of the phototrophic ciliate *Mesodinium rubrum* (*Myrionecta rubra*) in a stratified brackish inlet. *Aquatic Microbial Ecology*, **13**, 267–74.

Crawford, R. M. and Schmid, A.-M. M. (1986). Ultrastructure of silica deposition in diatoms. In *Biomineralization in the Lower Plants and Animals*, ed. B. S. C. Leadbeater and R. Ridings, pp. 291–314. Oxford: Oxford University Press.

Croome, R. L. and Tyler, P. A. (1984). Microbial microstratification and crepuscular photosynthesis in meromictic Tasmanian lakes. *Verhandlungen der internationalen Vereinigung für theoretische und angewandte Limnologie*, **22**, 1216–23.

Crumpton, W. G. and Wetzel, R. G. (1982). Effects of differential growth and mortality in the seasonal succession of phytoplankton populations of Lawrence Lake, Michigan. *Ecology*, **63**, 1729–39.

Crusius, J. and Wanninkhof, R. (2003). Gas transfer velocities measured at low wind speed over a lake. *Limnology and Oceanography*, **48**, 1010–17.

Cryer, M., Pierson, G. and Townsend, C. R. (1986). Reciprocal interactions between roach, *Rutilus rutilus*, and zooplankton in a small lake: prey dynamics and fish growth and recruitment. *Limnology and Oceanography*, **31**, 1022–38.

Csanady, G. T. (1978). Water circulation and dispersal mechanisms. In *Lakes: Chemistry, Geology and*

Physics, ed. A. Lerman, pp. 21–64. New York: Springer-Verlag.

Cuhel, R. L. and Lean, D. R. S. (1987a). Protein synthesis by lake plankton measured using in situ carbon dioxide and sulfate assimilation. *Canadian Journal of Fisheries and Aquatic Sciences*, **44**, 2102–17.

(1987b). Influence of light intensity, light quality temperture and daylength on uptake and assimilation of carbon dioxide and sulfate by lake plankton. *Canadian Journal of Fisheries and Aquatic Sciences*, **44**, 2118–32.

Cummins, K. W. and Wuychek, J. C. (1971). Caloric equivalents for investigations in ecological energetics. *Mitteilungen der internationalen Vereinigung für theoretische und angewandte Limnologie*, **18**, 1–158.

Cushing, D. H. (1971). The dependence of recruitment on parent stock in different groups of fishes. *Journal du Conseil internationale pour l'Exploration de la Mer*, **33**, 340–62.

(1982). *Climate and Fisheries*. London: Academic Press.

(1988). The study of stock and recruitment. In *Fish Population Dynamics*, ed. J. A. Gulland, pp. 105–28. Chichester: John Wiley.

(1990). Plankton production and year-class strength in fish poulations: an update on the match/mismatch hypothesis. *Advances in Marine Biology*, **26** 249–93.

(1995). The long-term relationship between zooplankton and fish. *ICES Journal of Marine Science*, **50**, 611–26.

(1996). *Towards a Science of Recruitment of Fish Populations*. Oldendorf: ECI.

Cushing, D. E. and Horwood, J.W (1977). Development of a model of stock and recruitment. In *Fisheries Mathematics*, ed. J. H. Steele, pp. 21–35. London: Academic Press.

Daft, M. J., McCord, S. B. and Stewart, W. D. P. (1975). Ecological studies on algal-lysing bacteria in fresh waters. *Freshwater Biology*, **5**, 577–96.

Dahl, E., Lindahl, O., Paasche, E. and Throndsen, J. (1989). The *Chrysochromulina polylepis* bloom in Scandinavian waters during spring 1988. In *Novel Phytoplankton Blooms: Causes and Impacts of Recurrent Brown Tides and Other Unusual Blooms*, ed. E. M. Cosper, M. Bricelj and E. J. Carpenter, pp. 383–405. Berlin: Springer-Verlag.

Dale, B. (2001). The sedimentary record of dinoflagellate cysts: looking back into the future of phytoplankton blooms. *Scientia Marina*, **65** (Suppl. 2), 257–72.

Darwin, C. (1859). *On the Origin of Species by Means of Natural Selection*. London: John Murray.

Dau, H. (1994). Molecular mechanisms and quantitative models of variable photosystem II fluorescence. *Photochemistry and Photobiology*, **60**, 1–23.

Dauta, A. (1982). Conditions de développement du phytoplancton: étude comparative du comportement de huit espèces en culture. I. Détermination des paramètres de croissance en fonction de la lumière et de la température. *Annales de Limnologie*, **18**, 217–62.

Davey, M. C. and Walsby, A. E. (1985). The form resistance of sinking algal chains. *British Phycological Journal*, **20**, 243–8.

Davies, P. S. (1997). Fluxes of phosphorus in lakes relevant to phytoplankton ecology. Ph.D. thesis, University of Leeds.

Davis, P. A., Dent, M. M., Parker, J. M., Reynolds, C. S. and Walsby, A. E. (2003). The annual cycle of growth rate and biomass change in *Planktothrix* spp. in Blelham Tarn, English Lake District. *Freshwater Biology*, **48**, 852–67.

Davis, R. B. (1974). Tubificids alter profiles of redox potential and pH in profundal lake sediment. *Limnology and Oceanography*, **19**, 342–6.

Davis, R. B., Thurlow, D. L. and Brewster, F. E. (1975). Effects of burrowing tubificid worms on the exchange of phosphorus between lake sediment and overlying water. *Verhandlungen der internationalen Vereinigung für theoretische und angewandte Limnologie*, **19**, 382–94.

Davison, W. (1987). Internal element cycles affecting the long-term alkalinity status of lakes: implications for lake restoration. *Schweizerische Zeitschrift für Hydrologie*, **49**, 186–201.

Davison, W., Reynolds, C. S., Tipping, E. and Needham, R. F. (1989). Reclamation of acid waters using sewage sludge. *Environmental Pollution*, **57**, 251–74.

Davison, W., George, D. G. and Edwards, N. J. A. (1995). Controlled reversal of lake acidification by treatment with phosphate fertiliser. *Nature*, **377**, 504–7.

Deacon, G. E. R. (1982). Physical and biological zonation in the Southern Ocean. *Deep-Sea Research*, **29**, 1–16.

Dejoux, C. and Iltis, A. (1992). *Lake Titicaca*. Dordrecht: Kluwer.

Dekker, A. C., Malthus, T. J. and Hoogenboom, H. J. (1995). The remote sensing of inland water quality. In *Advances in Environmental Remote*

Sensing, ed. F. M. Danson and S. E. Plummer, pp. 123–42, New York: John Wiley.

Delwiche, C. (2000). Tracing the thread of plastid diversity through the tapestry of life. *American Naturalist*, **154**, 164–77.

De Melo, R., France, R. and McQueen, D. J. (1992). Biomanipulation: hit or myth? *Limnology and Oceanography*, **37**, 192–207.

De Mott, W. R. (1982). Feeding selectivities and relative ingestion rates of *Daphnia* and *Bosmina*. *Limnology and Oceanography*, **27**, 518–27.

(1986). The role of taste in food selection by freshwater zooplankton. *Oecologia*, **69**, 334–40.

De Mott, W. R. and Kerfoot, W. C. (1982). Competition among cladocerans: nature of the interaction between *Bosmina* and *Daphnia*. *Ecology*, **63**, 1949–66.

De Nie, H. W. (1987). *The Decrease in Aquatic Vegetation in Europe and its Consequences for Fish Populations*, EIFAC/CECPI Occasional Papers No. 19. Rome: Food and Agriculture Organisation.

Demmig-Adams, B. and Adams, W. W. (1992). Photoprotection and other responses of plants to high light stress. *Annual Reviews in Plant Physiology and Plant Molecular Biology*, **43**, 599–626.

Denman, K. and Gargett, A. E. (1983). Time and space scales of vertical mixing and advection of phytoplankton in the upper ocean. *Limnology and Oceanography*, **28**, 801–15.

Denman, K. L. and Platt, T. (1975). Coherences in the horizontal distributions of phytoplankton and temperature in the upper ocean. *Mémoirs de la Société royale des Sciences, Liège, 6e Série*, **7**, 19–30.

Descy, J.-P., Everbecq, E., Gosselain, V., Viroux, L. and Smitz, J. S. (2003). Modelling the impact of benthic filter-feeders on the composition and biomass of river plankton. *Freshwater Biology*, **48**, 404–17.

Detmer, A. E. and Bathmann, U. V. (1997). Distribution patterns of autotropic pico- and nano-plankton and their relative contribution to algal biomass during spring in the Atlantic sector of the Southern Ocean. *Deep-Sea Research Part II Topical Studies in Oceanography*, **44**, 299–325.

Diaz, M. M. and Pedrozo, F. L. (1996). Nutrient limitation in Andean–Patagonian lakes at latitude 40–41° S. *Archiv für Hydrobiologie*, **138**, 123–43.

Diaz, M., Pedrozo, F. L. and Baccala, N. (2000). Summer classification of southern-hemispere tropical lakes (Patagonia, Argentina). *Lakes and Reservoirs*, **5**, 213–29.

Dillon, P. J. and Rigler, F. H. (1974). The phosphorus–chlorophyll relationship in lakes. *Limnology and Oceanography*, **19**, 767–73.

Dinsdale, M. T and Walsby, A. E. (1972). The interrelations of cell turgor pressure, gas vacuolation and buoyancy in a blue-green alga. *Journal of Experimental Botany*, **23**, 561–70.

DiTullio, G. R., Geesey, M. E., Jones, D. R. *et al.* (2003). Phytoplankton assemblage structure and primary productivity along 170° W in the South Pacific Ocean. *Marine Ecology – Progress Series*, **255**, 55–80.

Dobbins, W. E. (1944). Effect of turbulence on sedimentation. *Transactions of the American Society of Civil Engineers*, **109**, 629–56.

Dodson, S. I., Arnott, S. E. and Cottingham, K. L. (2000). The relationship in lake communities between primary productivity and species richness. *Ecology*, **81**, 2662–79.

Dokulil, M. T. and Teubner, K. (2000). Cyanobacterial dominance in lakes. *Hydrobiologia*, **438**, 1–12.

(2003). Steady-state phytoplankton assemblages during thermal stratification in deep alpine lakes: do they occur? *Hydrobiologia*, **502**, 65–72.

Donato-Rondon, J. C. (2001). *Fitoplancton de los lagos andinos del norte de Sudamérica*, Colección Jorge Alvares Lleras No. 19. Bogotá: Academia Colombiana de Ciencias Exactas, Fisicas y Naturales.

Donk, E. van and Voogd, H. (1998). Control of *Volvox* blooms by *Hertwigia*, a rotifer. *Verhandlungen der internationalen Vereinigung für theoretische und angewandte Limnologie*, **26**, 1781–84.

Donner, J. (1966). *Rotifers*. (Translated by H. G. S. Wright.) London: Frederick Warne.

Doolittle, W. F., Boucher, Y., Nesbo, C. L. *et al.* (2003). How big is the iceberg of which organellar genes in nuclear genomes are but the tip? *Philosophical Transactions of the Royal Society of London B*, **358**, 39–57.

Dortch, Q., Roberts, T. L., Clayton, J. R. and Ahmed, S. I. (1983). RNA/DNA ratios and DNA concentrations as indicators of growth rate and biomass in planktonic marine organisms. *Marine Ecology – Progress Series*, **13**, 61–71.

Doty, M. S. (ed.) (1961). *Proceedings of the Conference on Primary Productivity Measurement, Marine and Freshwater*. Washington, DC: US Atomic Energy Commission.

Douglas, A. E. and Raven, J. A. (2003). Genomes at the interface between bacteria and organelles. *Philosophical Transactions of the Royal Society of London B*, **358**, 5–17.

Douglas, G. B., Coad, D. N. and Adeney, J. A. (1998). Reducing phosphorus in aquatic systems using modified clays. *Water*, **25**(3), 42–3.

Downing, J. A., Watson, S. B. and McCauley, E. (2001). Predicting cyanobacteria dominance in lakes. *Canadian Journal of Fisheries and Aquatic Sciences*, **58**, 1905–8.

Dring, M. J. and Jewson, D. H. (1979). What does ^{14}C-uptake by phytoplankton really measure? A fresh approach using a theoretical method. *British Phycological Journal*, **14**, 122–3.

(1982). What does ^{14}C-uptake by phytoplankton really measure? A theoretical modelling approach. *Proceedings of the Royal Society of London B*, **214**, 351–68.

Droop, M. R. (1973). Some thoughts on nutrient limitation in algae. *Journal of Phycology*, **9**, 264–72.

(1974). The nutrient status of algae cells in continuous culture. *Journal of the Marine Biological Association of the United Kingdom*, **54**, 825–55.

Drum, R. W. and Pankratz, H. S. (1964). Post-mitotic fine structure of *Gomphonema parvulum*. *Journal of Ultrastructural Research*, **10**, 217–23.

Dryden, R. C. and Wright, S. J. L. (1987). Predation of cyanobacteria by protozoa. *Canadian Journal of Microbiology*, **33**, 471–82.

Ducklow, H. W. (2000). Bacterial production and biomass in the Ocean. In *Microbial Ecology of the Oceans*, ed. D. L. Kirchman, pp. 47–84. New York: Wiley-Liss.

Ducklow, H. W., Kirchman, D. L. and Anderson, T. R. (2002). The magnitude of spring bacterial production in the North Atlantic Ocean. *Limnology and Oceanography*, **47**, 1684–93.

Dufrêne, M. and Legendre, P. (1997). Species assemblages and indicator species: the need for a flexible asymmetric approach. *Ecological Monographs*, **67**, 345–66.

Dugdale, R. C. (1967). Nutrient limitation in the sea: dynamics, identification and significance. *Limnology and Oceanography*, **12**, 685–95.

Dugdale, R. C., Dugdale, V. A., Neess, J. C. and Goering, J. J. (1959). Nitrogen fixation in lakes. *Science*, **130**, 859–60.

Dumont, H. (1972). The biological cycle of molybdenum in relation to primary production. *Verhandlungen der internationalen Vereinigung für theoretische und angewandte Limnologie*, **18**, 84–92.

(1992). The regulation of plant and animal species and communities in African shallow lakes and wetlands. *Revue d'Hydrobiologie Tropicale*, **25**, 303 346.

(1999). The species richness of reservoir plankton and the effect of reservoirs on plankton dispersal (with particular emphasis on rotifers and cladocerans). In *Theoretical Reservoir Ecology and its Applications*, ed. J. G. Tundisi and M. Strakraba, pp. 477–91. São Carlos: Institute of Ecology.

Dumont, H. and Segers, H. (1996). Estimating lacustrine zooplankton species richness and complementarity. *Hydrobiologia*, **341**, 125–32.

Dunst, R. C., Born, S. M., Uttormark, P. D. *et al.* (1974). *Survey of Lake Rehabilitation Techniques and Experiences*, Technical Bulletin No. 75. Madison, WI: Wisconsin Department of Natural Resources.

Duthie, H. C. (1965). Some observations on the ecology of desmids. *Journal of Ecology*, **53**, 695–703.

Duthie, H. C. and Hart, C. J. (1987). The phytoplankton of subarctic Canadian great lakes. *Ergebnisse der Limnologie*, **25**, 1–9.

Duysens, L. N. M. (1956). The flattening of the absorption spectrum of suspensions as compared to that of solutions. *Biochimica et Biophysica Acta*, **19**, 1–12.

Eberley, W. R. (1959). The metalimnetic oxygen maximum in Myers lake. *Investigations of Indiana Lakes and Streams*, **5**, 1–46.

(1964). Primary production in McLish Lake (northern Indiana), an extreme plus-heterograde lake. *Verhandlungen der internationalen Vereinigung für theoretische und angewandte Limnologie*, **15**, 394–401.

Eddy, S. (1931). The plankton of the Sangamon River in the summer of 1929. *Bulletin of the Natural History Survey of Illinois*, **19**, 469–86.

Edler, L. (1979). Phytoplankton succesion in the Baltic Sea. *Acta Botanica Fennica*, **110**, 75–8.

Edmondson, W. T. (1970). Phosphorus, nitrogen and algae in Lake Washington after diversion of sewage. *Science*, **169**, 690–1.

(1972). The present condition of Lake Washington. *Verhandlungen der internationalen Vereinigung für theoretische und angewandte Limnologie*, **19**, 606–15.

Edmondson, W. T. and Lehman, J. T. (1981). The effect of changes in the nutrient income on the condition of lake Washington. *Limnology and Oceanography*, **26**, 1–29.

Edmondson, W. T. and Litt, A. H. (1982). *Daphnia* in Lake Washington. *Limnology and Oceanography*, **27**, 272–93.

Edvardsen, B. and Paasche, E. (1998). Bloom dynamics and physiology of *Prymnesium* and *Chrysochromulina*. In *Physiological Ecology of Harmful*

Algal Blooms, ed. D. M. Anderson, A. D. Cembella and G. M. Hallegraeff, pp. 193–208. Berlin: Springer-Verlag.

Ehrlich, P. R. and Ehrlich, A. H. (1981). *Extinction: The Causes and Consequences of the Disappearance of Species*. New York: Random House.

Einsele, W. (1936). Über die Beziehungen den Eisenkrieslaufs zum Phosphatkrieslauf im eutrophen See. *Archiv für Hydrobiologie*, **33**, 361–87.

Einsele, W. and Grim, J. (1938). Über den Kieselsäurgehalt planktischer Diatomeen und dessen Bedeutung für einiger Fragen ihrer Ökologie. *Zeitschrift für Botanik*, **32**, 545–90.

Eker, E., Georgieva, L., Senichkina, L. and Kideys, A. E. (1999). Phytoplankton distribution in the western and eastern Black Sea in spring and autumn 1995. *ICES Journal of Marine Science*, **56** (Suppl.), 15–22.

Eker-Develi, E. and Kideys, A. E. (2003). Distribution of phytoplankton in the southern Black Sea in summer 1996, spring and autumn 1998. *Journal of Marine Systems*, **39**, 203–11.

Elliott, J. A., Reynolds, C. S., Irish, A. E. and Tett, P. (1999a). Exploring the potential of the PROTECH model to investigate phytoplankton community theory. *Hydrobiologia*, **414**, 37–43.

Elliott, J. A., Irish, A. E., Reynolds, C. S. and Tett, P. (1999b). Sensitivity analysis of PROTECH, a new approach in phytoplankton modelling. *Hydrobiologia*, **414**, 45–51.

(2000a). Modelling freshwater phytoplankton communities: an exercise in validation. *Ecological Modelling*, **128**, 19–26.

Elliott, J. A., Reynolds, C. S. and Irish, A. E. (2000b). The diversity and succession of phytoplankton communities in disturbance-free environments, using the model PROTECH. *Archiv für Hydrobiologie*, **149**, 241–58.

Elliott, J. A., Irish, A. E. and Reynolds, C. S. (2001a). The effects of vertical mixing on a phytoplankton community: a modelling approach to the intermediate disturbance hypothesis. *Freshwater Biology*, **46**, 1291–7.

Elliott, J. A., Reynolds, C. S. and Irish, A. E. (2001b). An investigation of dominance in phytoplankton using the PROTECH model. *Freshwater Biology*, **46**, 99–108.

Elliott, J. A., Irish, A. E. and Reynolds, C. S. (2002). Predicting the spatial dominance of phytoplankton in a light-limited and incompletely mixed eutrophic water column using the PROTECH model. *Freshwater Biology*, **47**, 433–40.

Elliott, J. M. (1975a). The growth rate of brown trout (*Salmo trutta* L.) fed on maximum rations. *Journal of Animal Ecology*, **44**, 805–21.

(1975b). The growth rate of brown trout (*Salmo trutta* L.) fed on reduced rations. *Journal of Animal Ecology*, **44**, 823–42.

Elliott, J. M. and Hurley, M. A. (1998). A new functional model for estimating the maximum amount of invertebrate food consumed per day by brown trout, *Salmo trutta*. *Freshwater Biology*, **39**, 339–49.

(1999). A new energetics model for brown trout, *Salmo trutta*. *Freshwater Biology*, **42**, 235–46.

(2000a). Daily energy intake and growth of piscivorous brown trout, *Salmo trutta*. *Freshwater Biology*, **44**, 237–45.

(2000b). Optimum energy intake and gross efficiency of energy conversion for brown trout, *Salmo trutta*, feeding on invertebrates or fish. *Freshwater Biology*, **44**, 605–15.

Eloranta, P. (1993). Diversity and succession of the phytoplankton in a small lake over a two-year period. *Hydrobiologia*, **249**, 25–32.

Elser, J. J. (1999). The pathway to noxious cyanobacteria blooms in lakes: the food web as the final turn. *Freshwater Biology*, **42**, 537–43.

Elser, J. J., Sterner, R. W., Gorokhova, E. *et al.* (2000). Biological stoichiometry from genes to ecosystems. *Ecology Letters*, **3**, 540–50.

Elser, J. J., Kyle, M., Makino, W., Yosshida, T. and Urabe, J., (2003). Ecological stoichiometry in the microbial food web: a test of the light: nutrient hypothesis. *Aquatic Microbial Ecology*, **31**, 49–65.

Elton, C. S. (1927). *Animal Ecology*. London: Macmillan.

(1958). *The Ecology of Invasions by Animals and Plants*. London: Methuen.

Elzenger, J. T., Prins, H. B. A. and Stefels, J. (2000). The role of extracellular carbonic anhydrase activity in inorganic utilization of *Phaeocystis globosa* (Prymnesiophyceae): a comparison with other marine algae using an isotopic disequilibrium technique. *Limnology and Oceanography*, **45**, 372–80.

Emsley, J. (1980). The phosphorus cycle. In *The Handbook of Environmental Chemistry*, vol. 1A, ed. O. Hutzinger, pp. 147–67. Berlin: Springer-Verlag.

Eppley, R. W. (1970). Relationships of phytoplankton species distribution to the depth distribution of nitrate. *Bulletin of the Scripps Institute of Oceanography*, **17**, 43–9.

(1972). Temperature and phytoplankton growth rate in the sea. *Fisheries Bulletin*, **70**, 1063–85.

Eppley, R. W., Holmes, R. W. and Strickland, J. D. H. (1967). Sinking rates of marine phytoplankton measured with a fluorometer. *Journal of Experimental Marine Biology and Ecology*, **1**, 191–208.

Eppley, R. W., Coatsworth, J. L. and Solorzano, L. (1969a). Studies of nitrate reductase in marine phytoplankton. *Limnology and Oceanography*, **14**, 194–205.

Eppley, R. W., Rogers, J. N. and McCarthy, J. N. (1969b). Half-saturation constants for uptake of nitrate and ammonium by marine phytoplankton. *Limnology and Oceanography*, **14**, 912–20.

Evans, G. T. and Taylor, F. J. R. (1980). Phytoplankton accumulation in Langmuir cells. *Limnology and Oceanography*, **25**, 840–5.

Evans, M. S. and Jude, D. J. (1986). Recent shifts in *Daphnia* community structure in south-eastern Lake Michigan: a comparison of the inshore and offshore regions. *Limnology and Oceanography*, **31**, 56–67.

Everall, N. C. and Lees, D. R. (1996). The use of barley straw to control general and blue-green algal growth in a Derbyshire reservoir. *Water Research*, **30**, 269–76.

Fabbro, L. D. and Duivenvorden, L. J. (2000). A two-part model linking multi-dimensional environmental gradients and seasonal succession of phytoplankton assemblages. *Hydrobiologia*, **438**, 13–24.

Fahnenstiel, G. L., Sicko-Goad, L., Scavia, D. and Stoermer, E. F. (1986). Importance of picoplankton in Lake Superior. *Canadian Journal of Fisheries and Aquatic Sciences*, **51**, 2769–83.

Falconer, R. A., George, D. G. and Hall, P. (1991). Three-dimensional numerical modelling of wind-driven circulation in a shallow homogeneous lake. *Journal of Hydrology*, **124**, 59–79.

Falkner, G., Falkner, R. and Schwab, A. (1989). Bioenergetic state characterization of transient state phosphate uptake by the cyanobacterium *Anacystis nidulans*. *Archives of Microbiology*, **152**, 353–61.

Falkowski, P. G. (1997). Evolution of the nitrogen cycle and its influence on the biological sequestration of CO_2 in the ocean. *Nature*, **387**, 272–5.

(2002). On the evolution of the carbon cycle. In *Phytoplankton Productivity*, ed. P. J. leB. Williams, D. N. Thomas and C. S. Reynolds, pp. 318–49. Oxford: Blackwell Science.

Falkowski, P. G. and Raven, J. A. (1997). *Aquatic Photosynthesis*. Oxford: Blackwell Science.

Falkowski, P. G., Barber, R. T. and Smetacek, V. (1998). Biogeochemical controls and feedbacks on ocean primary production. *Science*, **281**, 200–6.

Faller, A. J. (1971). Oceanic turbulence and the Langmuir circulation. *Annual Review of Ecology and Systematics*, **2**, 201–34.

Fallon, R. D. and Brock, T. D. (1981). Overwintering of *Microcystis* in Lake Mendota. *Freshwater Biology*, **11**, 217–26.

Fay, P., Stewart, W. D. P., Walsby, A. E. and Fogg, G. E. (1968). Is the heterocyst the site of nitrogen fixation in blue-green algae? *Nature*, **220**, 810–12.

Fenchel, T. and Finlay, B. J. (1994). *Ecology and Evolution in Anoxic Worlds*. Oxford: Oxford University Press.

Ferguson, A. J. D. and Harper, D. M. (1982). Rutland Water phytoplankton: the development of an asset or a nuisance? *Hydrobiologia*, **88**, 117–33.

Ferguson, A. J. D., Thompson, J. M. and Reynolds, C. S. (1982). Structure and dynamics of zooplankton maintained in closed systems, with special reference to the food supply. *Journal of Plankton Research*, **4**, 523–43.

Fernando, C. H. (1980). The species and size composition of tropical freshwater zooplankton with special reference to the oriental region (South East Asia). *Internationale Revue des gesamten Hydrobiologie*, **76**, 149–67.

Field, C. B., Behrenfeld, M. J., Randerson, J. T. and Falkowski, P. G. (1998). Primary production of the biosphere: integrating terrestrial and oceanic components. *Science*, **281**, 237–40.

Findenegg, I. (1943). Untersuchungen über die Ökologie und die Produktionsverhältnisse des Planktons im Kärnter Seengebeite. *Internationale Revue des gesamten Hydrobiologie*, **43**, 388–429.

(1947). Über die Lichtanspruche planktischer Süsswasseralgen. *Sitzungsberichte der Akademie der Wissenschaften in Wien*, **155**, 159–71.

Finenko, Z. Z., Piontkovski, S. A., Williams, R. and Mishonov, A. V. (2003). Variability of phytoplankton and mesozooplankton biomass in the subtropical and tropical Atlantic Ocean. *Marine Ecology – Progress Series*, **250**, 125–44.

Finkel, Z. V. (2001). Light absorption and size scaling of light-limited metabolism in marine diatoms. *Limnology and Oceanography*, **46**, 86–94.

Finlay, B. J. (2001). Protozoa. In *Encyclopedia of Biodiversiy*, vol. 4, ed. S. R. Levin, pp. 569–99. San Diego, CA: Academic Press.

(2002). Global dispersal of free-living microbial eukaryote species. *Science*, **296**, 1061–3.

Finlay, B. J. and Clarke, K. J. (1999). Ubiquitous dispersal of microbial species. *Nature*, **400**, 828.

Finlay, B. J., Esteban, G. F. and Fenchel, T. (1998). Protozoan diversity: converging estimates of the global number of free-living ciliate species. *Protist*, **149**, 29–37.

Fish, G. R. (1952). Hydrology and algology. In *Annual Report of the East African Fisheries Research Organisation (1951)*, pp. 6–11.

Fleming, R. H. (1940). Composition of plankton and units for reporting populations and production. In *Proceedings of the 6th Pacific Science Congress*, vol. 3, pp. 535–40.

Flöder, S. (1999). The influence of mixing frequency and depth on phytoplankton mobility. *International Review of Hydrobiology*, **84**, 119–28.

Flöder, S., Urabe, J. and Kawabata, Z. (2002). The influence of fluctuating light intensities on species composition and diversity of natural phytoplankton communities. *Oecologia*, **133**, 395–401.

Fogg, G. E. (1971). Extracellular products of algae in freshwater. *Ergebnisse der Limnologie*, **5**, 1–25.

(1975). *Algal Cultures and Phytoplankton Ecology*, 2nd edn. London: University of London Press.

(1977). Excretion of organic matter by phytoplankton. *Limnology and Oceanography*, **22**, 576–7.

Fogg, G. E. and Thake, B. (1987). *Algal Cultures and Phytoplankton Ecology*, 3rd edn. London: University of London Press.

Forsberg, B. (1985). The fate of planktonic primary production. *Limnology and Oceanography*, **30**, 807–19.

Fox, J. F. and Connell, J. H. (1979). Intermediate disturbance hypothesis. *Science*, **204**, 1344–5.

Foy, R. H. and Gibson, C. E. (1982). Photosynthetic characteristics of planktonic blue-green algae: changes in photosynthetic capacity and pigmentation of *Oscillatoria redekei* Van Goor under high and low light intensities. *British Phycological Journal*, **17**, 183–93.

Foy, R. H., Gibson, C. E. and Smith, R. V. (1976). The influence of daylength, light intensity and temperature on the growth rates of planktonic blue-green algae. *British Phycological Journal*, **11**, 151–63.

Fraga, F. (2001). Phytoplanktonic biomass synthesis: application to deviations from Redfield stoichiometry. *Scientia Marina*, **65** (Suppl. 2), 153–69.

Frankignoulle, M. (1988). Field measurements of air–sea CO_2 exchange. *Limnology and Oceanography*, **33**, 313–22.

Freeland, J. R., Rimmer, V. K. and Okamura, B. (2001). Genetic changes within freshwater bryozoan populations suggest temporal gene flow from statoblast banks. *Limnology and Oceanography*, **46**, 1121–9.

Freeman, C., Evans, C. D. and Monteith, D. T. (2001). Export of organic carbon from peat soils. *Nature*, **412**, 785.

Freire-Nordi, C. S., Vieira, A. A. H. and Nascimento, O. R. (1998). Selective permeability of the extracellular envelope of the microalga *Spondylosium panduriforme* (Chlorophyceae) as revealed by electron paramagnetic resonance. *Journal of Phycology*, **34**, 631–7.

Frenette, J.-J., Vincent, W. F., Legendre, L. *et al.* (1996). Biological responses to typhoon-induced mixing in two morphologically distinct basins of Lake Biwa. *Japanese Journal of Limnology*, **57**, 501–10.

Friedman, M. M. (1980). Comparative morphology and functional significance of copepod receptors and oral structures. In *Evolution and Ecology of Zooplankton Communities*, ed. W. C. Kerfoot, pp. 185–97. Hanover, NH: University of New England Press.

Frempong, E. (1984). A seasonal sequence of diel distribution patterns for the planktonic dinoflagellate *Ceratium hirundinella* in a eutropic lake. *Freshwater Biology*, **14**, 301–21.

Fromme, P., Kern, J., Loll, B. *et al.* (2002). Functional implications on the mechanism of the function of photosystem II including water oxidation based on the structure of photosystem II. *Philosophical Transactions of the Royal Society of London B*, **357**, 1337–45.

Fryer, G. (1987). Morphology and the classification of the so-called Cladocera. *Hydrobiologia*, **145**, 19–28.

Fuhrman, J. A., Griffith, J. F. and Schwalbach, M. S. (2002). Prokaryotic and viral diversity patterns in marine plankton. *Ecological Research*, **17**, 183–94.

Gaarder, T. and Gran, H. H. (1927). Investigations of the production of plankton in the Oslo Fjord. *Rapports et Procès-Verbaux des Réunions, Conseil international pour l'exploration de la Mer*, **42**, 1–48.

Gaedke, U. (1998). Functional and taxonomical properties of the phytoplankton community of large and deep Lake Constance: interannual variability and response to re-oligotrophication (1979–93). *Ergebnisse der Limnologie*, **53**, 119–41.

Gaedke, U. and Sommer, U. (1986). The influence of the frequency of periodic disturbances on the maintenance of phytoplankton diversity. *Oecologia*, **71**, 98–102.

Gaedke, U. and Straile, D. (1994a). Seasonal changes in the quantitative importance of protozoans in a large lake: an ecosystem approach using mass-balanced carbon-flow diagrams. *Marine Microbial Food Webs*, **8**, 163–88.

(1994b). Seasonal changes of trophic transfer efficiencies in a plankton food web derived from biomass size distributions and network analysis. *Ecological Modelling*, **75/76**, 435–45.

Gálvez, J. A., Niell, F. X. and Lucena, J. (1988). Description and mechanism of formation of a deep chlorophyll maximum due to *Ceratium hirundinella* (O. F. Müller) Bergh. *Archiv für Hydrobiologie*, **112**, 143–56.

Ganf, G. G. (1974a). Diurnal mixing and the vertical distribution of phytoplankton in a shallow equatorial lake (Lake George, Uganda). *Journal of Ecology*, **62**, 611–29.

(1974b). Incident solar radiation and underwater light penetration as factors controlling the chlorophyll *a* content of a shallow equatorial lake (Lake George, Uganda). *Journal of Ecology*, **62**, 593–609.

(1975). Photosynthetic production and irradiance–photosynthesis relationships of the phytoplankton from a shallow equatorial lake (L. George, Uganda). *Oecologia*, **18**, 165–83.

(1980). *Factors Controlling the Growth of Phytoplankton in Mount Bold Reservoir, South Australia*, Technical Paper No. 48. Canberra: Australian Water Research Council.

Ganf, G. G. and Oliver, R. L. (1982) Vertical separation of light and available nutrients as a factor causing replacement of green algae by blue-green algae in the plankton of a stratified lake. *Journal of Ecology*, **70**, 829–44.

Ganf, G. G. and Shiel, R. J. (1985). Feeding behaviour and limb morphology of the cladocerans with small intersetular distances. *Australian Journal of Marine and Freshwater Research*, **36**, 69–86.

Ganf, G. G., Heaney, S. I. and Corry, J. (1991). Light absorption and pigment content in natural populations and cultures of a non-gas-vacuolate cyanobacterium, *Oscillatoria bourrelleyi* (= *Tychonema bourrellyi*) in Windermere. *Journal of Plankton Research*, **13**, 1101–21.

García de Emiliani, M. O. (1993). Seasonal succession of phytoplankton in a lake of the Paraná flood plain, Argentina. *Hydrobiologia*, **264**, 101–14.

Garcia-Pichel, F. and Castenholz, R. W. (1991). Characterisation and biological implications of scytonemin, a cyanobacterial sheath pigment. *Journal of Phycology*, **27**, 395–409.

Gasol, J. M. and del Giorgio, P. A. (2000). Using flow cytometry for counting ntural planktonic bacteria and understanding the structure of planktonic bacterial communities. *Scientia Marina*, **64**, 197–224.

Gasol, J. M. and Pedrós-Alió, C. (1991). On the origin of deep chlorophyll maxima: the case of Lake Cisó. *Verhandlungen der internationalen Vereinigung für theoretische und angewandte Limnologie*, **24**, 1024–28.

Gasol, J. M., Guerrero, R. and Pedrós-Alió, C. (1992). Spatial and temporal dynamics of a metalimnetic *Cryptomonas* peak. *Journal of Plankton Research*, **14**, 1565–79.

Gattuso, J.-P., Peduzzi, S., Pizay, M.-D. and Tonolla, M. (2002). Changes in freshwater bacterial community composition during measurements of microbial and community respiration. *Journal of Plankton Research*, **24**, 1197–206.

Geider, R. J. and La Roche, J. (2000). Redfield revisited: variability of C : N : P in marine microalgae and its biochemical basis. *European Journal of Phycology*, **37**, 1–17.

Geider, R. J. and MacIntyre, H. L. (2002). Physiology and biochemistry of photosynthesis and algal carbon acquisition. In *Phytoplankton Productivity*, ed. P. J. leB. Williams, D. N. Thomas and C. S. Reynolds, pp. 44–77. Oxford: Blackwell Science.

Geider, R. J. and Osborne, B. A. (1992). *Algal Photosynthesis: The Measurement of Algal Gas Exchange*. New York: Chapman and Hall.

Geider, R. J., Delucia, E. H., Falkowski, P. G. *et al.* (2001). Primary productivity of planet Earth: biological determinants and physical constraints in terrestrial and aquatic habitats. *Global Change Biology*, **7**, 849–82.

Geller, W. and Müller, H. (1981). The filtration apparatus of Cladocera: filter mesh sizes and their implications on food selectivity. *Oecologia*, **49**, 316–21.

Gentry, D. R., Hernandez, V. J., Nguyen, L. H., Jensen, D. B. and Cashel, M. (1993). Synthesis of stationary phase sigma factor sS is positively regulated by ppGpp. *Journal of Bacteriology*, **175**, 5710–13.

George, D. G. (1981). Zooplankton patchiness. *Report of the Freshwater Biological Association*, **49**, 32–44.

(2002). Regional scale influences on the long-term dynamics of plankton. In *Phytoplankton Productivity*, ed. P. J. leB. Williams, D. N. Thomas and C. S. Reynolds, pp. 265–90. Oxford: Blackwell Science.

George, D. G. and Edwards, R. W. (1973). *Daphnia* distributions within Langmuir circulations. *Limnology and Oceanography*, **18**, 798–800.

(1976). The effect of wind on the distribution of chlorophyll *a* and crustacean zooplankton in a shallow eutrophic reservoir. *Journal of Applied Ecology*, **13**, 667–90.

George, D. G. and Harris, G. P. (1985). The effect of climate on long-term changes in the crustacean zooplankton biomass of Lake Windermere, UK. *Nature*, **316**, 536–9.

George, D. G. and Heaney, S. I. (1978). Factors influencing the spatial distribution of phytoplankton in a small, productive lake. *Journal of Ecology*, **66**, 133–55.

George, D. G. and Taylor, A. H. (1995). UK lake plankton and the Guf Stream. *Nature*, **378**, 139.

Gerloff, G. C., Fitzgerald, G. P. and Skoog, F. (1952). The mineral nutrition of *Microcystis aeruginosa*. *American Journal of Botany*, **39**, 26–32.

Gervais, F., Padisák, J. and Koschel, R. (1997). Do light quality and low nutrient concentration favour picocyanobacteria below the thermocline of the oligotrophic Lake Stechlin? *Journal of Plankton Research*, **19**, 771–81

Gibson, C. E. (1971). Nutrient limitation. *Journal of the Water Pollution Control Federation*, **43**, 2436–40.

Gibson, C. E., Wood, R. B., Dickson, E. L. and Jewson, D. M. (1971). The succession of phytoplankton in L. Neagh, 1968–1971. *Mitteilungen der internationalen Vereinigung für theoretische und angewandte Limnologie*, **19**, 146–60.

Gismervik, I. (1997). Stoichiometry of some marine planktonic cladocerans. *Journal of Plankton Rearch*, **19**, 279–85.

Gjessing, E. T. (1970). Ultrafiltration of aquatic humus. *Environmental Science and Technology*, **4**, 437–8.

Gleason, H. (1917). The structure and development of the plant association. *Bulletin of the Torrey Botanical Club*, **44**, 463–81.

(1927). Further views on the succession concept. *Ecology*, **8**, 299–326.

Gleick, J. (1988). *Chaos: Making a New Science*. London: Heinemann.

Gliwicz, Z. M. (1969). The food sources of lake zooplankton. *Ekologia Polska, Seria B*, **15**, 205–23.

(1975). Effect of zooplankton grazing on photosynthetic activity and composition of the phytoplankton. *Verhandlungen der internationalen Vereinigung für theoretische und angewandte Limnologie*, **19**, 1490–97.

(1977). Food size selection and seasonal succession of filter-feeding zooplankton in an eutrophic lake. *Ekologia Polska, Seria A*, **25**, 179–225.

(1980). Filtering rates, food size selection and feeding rates in cladocerans: another aspect of interspecific competition in filter-feeding zooplankton. In *Evolution and Ecology of Zooplankton Communities*, ed. W. C. Kerfoot, pp. 282–91. Hanover, NH: University of New England Press.

(1990). Food thresholds and body size in cladocerans. *Nature*, **343**, 638–40.

(2003a). Zooplankton. In *The Lakes Handbook*, vol. 1, ed. P. O'Sullivan and C. S. Reynolds, pp. 461–516. Oxford: Blackwell Science.

(2003b). *Between Hazards of Starvation and Risk of Predation: The Ecology of Offshore Animals*. Oldendorf: ECI.

Gliwicz, Z. M. and Hillbricht-Ilkowska, A. (1975). Ecosystem of Mikoajskie Lake: elimination of phytoplankton biomass and is subsequent fate in the lake through the year. *Polskie Archiwum Hydrobiologii*, **22**, 39–52.

Gliwicz, Z. M. and Preis, A. (1977). Can planktivorous fish keep in check planktonic crustacean populations? A test of the size–efficiency hypothesis. *Ekologia Polska, Seria A*, **25**, 567–91.

Gliwicz, Z. M. and Siedlar, E. (1980). Food-size limitation and algae interfering with food collection in *Daphnia*. *Archiv für Hydrobiologie*, **88**, 155–77.

Glover, H. E. (1977). Effects of iron deficiency on *Isochrysis galbana* (Chrysophyceae) and *Phaeodactylum cornutum* (Bacillariophyceae). *Journal of Phycology*, **13**, 208–12.

Goericke, R. (2002). Bacteriochlorophyll *a* in the ocean: is anoxygenic bacterial photosynthesis important? *Limnology and Oceanography*, **47**, 290–5.

Goldman, C. R. (1960). Molybdenum as a factor limiting primary productivity in Castle Lake, California. *Science*, **132**, 1016–17.

(1964). Primary productivity and micro-nutrient limiting factors in some North American and New Zealand lakes. *Verhandlungen der internationalen Vereinigung für theoretische und angewandte Limnologie*, **15**, 365–74.

Goldman, C. R. and Jassby, A. D. (2001). Primary productivity, phytoplankton and nutrient status in Lake Baikal. In *The Great Lakes of the World (GLOW): Food Web, Health and Integrity*, ed. M. Munawar and R. E. Hecky, pp. 111–25. Leiden: Backhuys.

Goldman, J. C. (1977). Steady-state growth of phytoplankton in continuous culture: comparison of internal and external nutrient concentrations. *Journal of Phycology*, **13**, 251–8.

(1980). Physiological processes, nutrient availability and the concept of relative growth rate in marine phytoplankton ecology. In *Primary Productivity in the Sea*, ed. P. G. Falkowski, pp. 179–94. New York: Plenum Press.

Goldman, J. C. and Dennett, M. R. (2000). Growth of marine bacteria in batch and continuous culture under carbon and nitrogen limitation. *Limnology and Oceanography*, **45**, 789–800.

Goldman, J. C., McCarthy, J. J. and Peavey, D. G. (1979). Growth rate influence on the chemical composition of phytoplankton in oceanic waters. *Nature*, **279**, 210–15.

Golterman, H. L., Bakels, C. C. and Jakobs-Möglin, J. (1969). Availability of mud phosphates for the growth of algae. *Verhandlungen der internationalen Vereinigung für theoretische und angewandte Limnologie*, **17**, 467–79.

González, N., Anadón, R. and Viesca, L. (2003). Carbon flux through the microbial community in a temperate sea during summer: role of bacterial metabolism. *Aquatic Microbial Ecology*, **33**, 117–26.

Gorham, E. (1958). Observations on the formation and breakdown of the oxidised microzone at the mud surface in lakes. *Limnology and Oceanography*, **3**, 291–8.

Gorham, E. and Boyce, F. M. (1989). Influence of lake surface area and depth upon thermal stratification and the depth of the summer thermocline. *Journal of Great Lakes Research*, **15**, 233–45.

Gorham, E., Lund, J. W. G., Sanger, J. E. and Dean, W. E. (1974). Some relationships between algal standing crop, water chemistry and sediment chemistry in the English lakes. *Limnology and Oceanography*, **19**, 601–17.

Gorham, P. R., MacLachlan, J., Hammer, U. T. and Kim, W. W. (1964). Isolation and culture of toxic strains of *Anabaena flos-aquae* (Lyngb.) Brèb. *Verhandlungen der internationalen Vereinigung für theoretische und angewandte Limnologie*, **15**, 796–804.

Gorlenko, V. M. and Kuznetsov, S. I. (1972). Über die photosynthesirenden Bakterien der Kononjer-Sees. *Archiv für Hydrobiologie*, **70**, 1–13.

Gosselain, V. and Descy, J.-P. (2000). Phytoplankton of the River Meuse (1994–1996): form and size structure analysis. *Verhandlungen der internationalen Vereinigung für theoretische und angewandte Limnologie*, **27**, 1031.

Gosselain, V., Hamilton, P. B. and Descy, J.-P. (2000). Estimating phytoplankton carbon from microscopic counts: an application for riverine systems. *Hydrobiologia*, **438**, 75–90.

Goussias, C., Boussac, A. and Rutherford, A. W. (2002). Photosystem II and photosynthetic oxidation of water: an overview. *Philosophical Transactions of the Royal Society of London B*, **357**, 1369–81.

Grace, J. B. (1991). A clarification of the debate between Grime and Tilman. *Functional Ecology*, **5**, 583–7.

Graham, M. (1943). *The Fish Gate*. London: Faber.

Gran, H. H. (1912). Pelagic plant life. In *The Depths of the Ocean*, ed. J. Murray and J. Hjort, pp. 307–87. London: Macmillan.

(1931). On the conditions for the production of plankton in the sea. *Rapports du Conseil internationale pour l'exploration de la Mer*, **75**, 37–46.

Gray, C. B., Neilsen, M., Johannsson, O. *et al.* (1994). Great lakes and the St. Lawrence lowland lakes. In *The Book of Canadian Lakes*, ed. R. J. Allan, M. Dickman, C. B. Gray and V. Cromie, pp. 14–98. Burlington: Canadian Association on Water Quality.

Greenwood, P. H. (1963). *A History of Fishes*. London: Ernest Benn.

Grey, J., Jones, R. I. and Sleep, D. (2001). Seasonal changes in the importance of the source of organic matter to the diet of zooplankton in Loch Ness, as indicated by stable isotope analysis. *Limnology and Oceanography*, **46**, 505–13.

Griffiths, B. M. (1939). Early references to water blooms in British lakes. *Proceedings of the Linnean Society of London*, **151**, 12–19.

Grime, J. P. (1973). Competitive exclusion in herbaceous vegetation. *Nature*, **242**, 344–7.

(1977). Evidence for the existence of three primary strategies in plants and its relevance to ecological and evolutionary theory. *American Naturalist*, **111**, 1169–94.

(1979). *Plant Strategies and Vegetation Processes*. Chichester: John Wiley.

(2001). *Plant Strategies, Vegetation Processes and Ecosystem Properties*. Chichester: John Wiley.

Grimm, M. P. and Backx, J. J. G. M. (1990). The restoration of shallow lakes and the role of northern pike, aquatic vegetation and nutrient concentration. *Hydrobiologia*, **200/201**, 557–66.

Gross, F. and Zeuthen, E. (1948). The buoyancy of planktonic diatoms. a problem of cell physiology. *Proceedings of the Royal Society of London B*, **135**, 382–9.

Grossman, A. R., Schaefer, M. R., Chiang, G. C. and Collier, J. L. (1993). The phycobilisome, a light-harvesting complex responsive to environmental conditions. *Microbiological Reviews*, **57**, 725–49.

Grover, J. P. and Chrzanowski, T. H. (2000). Seasonal patterns of substrate utilisation by bacterioplankton: case studies in four temperate lakes of different latitudes. *Aquatic Microbial Ecology*, **23**, 41–54.

Grünberg, H. (1968). Über das Zeta-Potential zweier synchron kultiviert *Chlorella*-Stämme. *Archiv für Hydrobiologie* (Suppl.), **33**, 331–62.

Guerrero, R. and Mas-Castellà, J. (1995). The problem of excess and/or limitation of the habitat conditions: do natural assemblages exist? In *Molecular Ecology of Aquatic Microbes* ed. I. Joint, pp. 191–203. Berlin: Springer-Verlag

Guikema, J. A. and Sherman, L. A. (1984). Influence of iron deprivation on the membrane composition of *Anacystis nidulans*. *Plant Physiology*, **74**, 90–5.

Guillen, O., Rojas de Mendiola, B. and Izaguirre de Rondan, R. (1971) Primary productivity and phytoplankton. In *Fertility of the Sea*, vol. 1, ed. J. D. Costlow, pp. 187–96. New York: Gordon and Breach.

Guinasso, N. L. and Schink, D. R. (1975). Quantitative estimates of biological mixing rates in abyssal sediments. *Journal of Geophysical Research*, **80**, 3022–43.

Gulati, R. D., Lammens, E. H. R. R., Meijer, M.-L. and Donk, E. van (1990). *Biomanipulation: Tool for Water Management*. Dordrecht: Kluwer.

Gurung, T. B. and Urabe, J. (1999). Temporal and vertical differences in factors limiting growth rate of heterotrophic bacteria in Lake Biwa. *Microbial Ecology*, **38**, 136–45.

Gurung, T. B., Urabe, J. and Nakanishi, M. (1999). Regulation of the relationship between phytoplankton *Scenedesmus acutus* and heterotrophic bacteria by the balance of light and nutrients. *Aquatic Microbial Ecology*, **17**, 27–35.

Gusev, M. V. (1962). Effect of dissolved oxygen on the development of blue green algae. *Doklady Akademia Nauk SSSR*, **147**, 947–50. (In Russian.)

Haberyan, K. A. and Hecky, R. E. (1987). The late Pleistocene and Holocene stratigraphy and palaeolimnology of Lakes Kivu and Tanganyika. *Palaeo*, **61**, 169–97.

Häder, D.-P. (1986). Effect of solar and artificial UV-radiation on motility and phototaxis in the flagellate *Euglena gracilis*. *Photochemistry and Photobiology*, **44**, 651–6.

Haeckel, E. (1904). *Kunstformen der Natur*. Leipzig: Bibliographisches Institut.

Haffner, G. D., Harris, G. P. and Jarai, M. K. (1980). Physical variability and phytoplankton communities. III. Vertical structure in phytoplankton populations. *Archiv für Hydrobiologie*, **89**, 363–81.

Halbach, U. and Halbach-Keup, G. (1974). Quantitaive Beziehungen zwischen Phytoplankton und der Populationsdynamik des Rotators *Brachionus calyciflorus* Pallas. Befunde aus Laboratoriumsexperimenten und Freiland untersuchungen. *Archiv für Hydrobiologie*, **73**, 273–309.

Halldal, P. (1953). Phytoplankton investigations from weather ship M in the Norwegian Sea (1948–49). *Hvalrdets Skrifter*, **38**, 1–91.

Hambright, K. D., Gophen, M. and Serruya, S. (1994). Influence of long-term climatic changes on the stratification of a subtropical, warm monomictic lake. *Limnology and Oceanography*, **39**, 1233–42.

Hammer, U. T., Walker, K. F. and Williams, W. D. (1973). Derivation of daily phytoplankton production estimates from short-term experiments in some shallow eutrophic Australian saline lakes. *Australian Journal of Marine and Freshwater Research*, **24**, 259–66.

Haney, J. F. (1973). An in situ examination of the grazing activities of natural zooplankton communities. *Archiv für Hydrobiologie*, **72**, 88–132.

Hanson, M. J. and Stefan, H. G. (1984). Side effects of 58 years of copper sulphate treatment of the Fairmont Lakes, Minnesota. *Water Resources Bulletin*, **20**, 889–900.

Happey-Wood, C. M. (1988). Ecology of freshwater planktonic green algae. In *Growth and Reproductive Strategies of Freshwater Phytoplankton*, ed. C. D. Sandgren, pp. 175–226. Cambridge: Cambridge University Press.

Hardin, G. (1960). The competitive exclusion principle. *Science*, **131**, 1292–7.

Harding, W. R. (1997). Phytoplankton primary production in a shallow, well-mixed, hypertrophic South African Lake. *Hydrobiologia*, **344**, 87–102.

Hardy, A. C. (1939). Ecological investigations with the Continuous Plankton Recorder: objects, plans and methods. *Hull Bulletin of Marine Ecology*, 1, 1–57.

(1956). *The Open Sea: The World of Plankton*. London: Collins.

(1964). *The Open Sea*, Part 1, *The World of Plankton*. revd edn. London: Collins.

Harris, G. P. (1978). Photosynthesis, productivity and growth. *Ergebnisse der Limnologie*, 10, 1–163.

(1986). *Phytoplankton Ecology: Structure, Function and Fluctuation*. London: Chapman and Hall.

Harris, G. P. and Lott, J. N. A. (1973). Observations of Langmuir circulations in Lake Ontario. *Limnology and Oceanography*, 18, 584–9.

Harris, G. P. and Piccinin, B. B. (1977). Photosynthesis by natural phytoplankton populations. *Archiv für Hydrobiologie*, 80, 405–57.

Harris, G. P., Heaney, S. I. and Talling, J. F. (1979). Physiological and environmental constraints in the ecology of the planktonic dinoflagellate *Ceratium hirundinella*. *Freshwater Biology*, 9, 413–28.

Hart, R. C. (1996). Naupliar and copepodite growth and survival of two freshwater calanoids at various food levels: demographic contrasts, similarities and food needs. *Limnology and Oceanography*, 41, 648–58.

Hart, T. J. (1934). On the phytoplankton of the Southwest Atlantic and the Bellingshausen Sea, 1929–1931. *Discovery Reports*, 8, 1–268.

Hartmann, H. J. and Kunkel, D. D. (1991). Mechanisms of food selection in *Daphnia*. *Hydrobiologia*, 225, 129–54.

Hartmann, H. J., Taleb, H., Aleya, L. and Lair, N. (1993). Predation on ciliates by the suspension-feeding calanoid copepod *Acanthodiaptomus denticornis*. *Canadian Journal of Fisheries and Aquatic Sciences*, 50, 1382–93.

Harvey J. G. (1976). *Atmosphere and Ocean: Our Fluid Environments*, London: Artemis Press.

Hastwell, G. T. and Huston, M. A. (2001). On disturbance and diversity: a reply to Mackey and Currie. *Oikos*, 92, 367–71.

Havens, K. E., Aumen, N. G., James, R. T. and Smith, V. H. (1996). Rapid ecological changes in a large subtropical lake undergoing cultural eutrophication. *Ambio*, 25, 150–5.

Hawlay, G. R. W. and Whitton, B. A. (1991). Seasonal changes in chlorophyll-containing picoplankton populations of ten lakes in northern England. *Internationale Revue des gesamten Hydrobiologie*, 76, 545–54.

Haworth, E. Y. (1976). The changes in composition of the diatom assemblages found in the surface sediments of Blelham Tarn in the English Lake District during 1973. *Annals of Botany*, 40, 1195–205.

(1985). The highly nervous system of the English Lakes: aquatic sensitivity to external changes, as demonstrated by diatoms. *Report of the Freshwater Biological Association*, 53, 60–79.

Hayes, P. K. and Walsby, A. E. (1986). The inverse correlation between width and strength of gas vesicles in Cyanobacteria. *British Phycological Journal*, 21, 191–7.

Haygarth, P. M. (1997). Agriculture as a source of phosphorus transfer to water: sources and pathways. *Scope Newsletter*, 21, 1–16.

Hayward, T. L., Venrick, E. L. and McGowan, J. A. (1983). Environmental heterogeneity and plankton community structure in the central North Pacific. *Journal of Marine Research*, 41, 711–29.

Healey, F. P. (1973). Characteristics of phosphorus deficiency in *Anabaena*. *Journal of Phycology*, 9, 383–94.

Heaney, S. I. (1976). Temporal and spatial distribution of the dinoflagellate *Ceratium hirundinella* O. F. Müller within a small productive lake. *Freshwater Biology*, 6, 531–42.

Heaney, S. I. and Furnass, T. I. (1980). Laboratory models of diel vertical migration in the dinoflagellate, *Ceratium hirundinella*. *Freshwater Biology*, 10, 163–70.

Heaney, S. I. and Talling, J. F. (1980a). Dynamic aspects of dinoflagellate distribution patterns in a small, productive lake. *Journal of Ecology*, 68, 75–94.

(1980b). *Ceratium hirundinella*: ecology of a complex, mobile and successful plant. *Report of the Freshwater Biological Association*, 48, 27–40.

Heaney, S. I., Chapman, D. V. and Morison, H. R. (1981). The importance of the cyst stage in the seasonal growth of the dinoflagellate, *Ceratium hirundinella*, in a small productive lake. *British Phycological Journal*, 16, 136.

Heaney, S. I., Smyly, W. J. P. and Talling, J. F. (1986). Interactions of physical, chemical and biological processes in depth and time within a productive English lake during summer stratification. *Internationale Revue des gesamten Hydrobiologie*, 71, 441–94.

Heaney, S. I., Lund, J. W. G., Canter, H. M. and Gray, K. (1988). Population dynamics of *Ceratium* spp. in three English lakes, 1945–85. *Hydrobiologia*, 161, 133–48.

Heaney, S. I., Canter, H. M. and Lund, J. W. G. (1990). The ecological significance of grazing planktonic populations of cyanobacteria by the ciliate *Nassula*. *New Phytologist*, **114**, 247–63.

Heaney, S. I., Parker, J. E., Butterwick, C. and Clarke, K. J. (1996). Interannual variability of algal populations and their influence on lake metabolism. *Freshwater Biology*, **35**, 561–77.

Hecky, R. E. (1993). The eutropication of Lake Victoria. *Verhandlungen der internationalen Vereinigung für theoretische und angewandte Limnologie*, **25**, 39–48.

Hecky, R. E. and Fee, E. J. (1981) Primary production and rates of algal growth in Lake Tanganyika. *Limnology and Oceanography*, **26**, 532–46.

Hecky, R. E. and Kling, H. J. (1981). The phytoplankton and protozooplankton of the euphotic zone of Lake Tanganyika: species composition, biomass, chlorophyll content and spatio-temporal distribution. *Limnology and Oceanography*, **26**, 548–64.

Hecky, R. E., Spigel, R. H. and Coulter, G. (1991). The nutrient regime. In *Lake Tanganyika and its Life*, ed. G. Coulter, pp. 76–89. Oxford: Oxford University Press.

Hegewald, E. (1972). Untersuchungen zum Zetapotential von Planktonalgen. *Archiv für Hydrobiologie* (Suppl.), **42**, 14–90.

Heinbokel, J. F. (1986). Occurrence of *Richelia intracellularis* (Cyanophyta) within the diatoms *Hemiaulus haukii* and *H. membranaceus* of Hawaii. *Journal of Phycology*, **22**, 399–403.

Heller, M. (1977). The phased division of the freshwater dinoflagellate *Ceratium hirundinella* and its use as a method for assessing growth in a natural population. *Freshwater Biology*, **7**, 527–33.

Henrikson, L. and Brodin, Y.-W. (1995). Liming of surface waters in Sweden: a synthesis. In *Liming of Acidified Surface Waters: A Swedish Synthesis*, ed. L. Henrikson and Y.-W. Brodin, pp. 1–44. Berlin: Springer-Verlag.

Henrikson, L. and Oscarson, H. G. (1981). Corixids (Hemiptera – Heteroptera): the new top predators of acidified lakes. *Verhandlungen der internationalen Vereinigung für theoretische und angewandte Limnologie*, **21**, 1616–20.

Hensen, V. (1887). Über die Bestimmung des Planktons oder des im Meere treibenden materiels an Pflanzen und Tieren. *Bericht des deutschen wissenschaftlichen Kommission für Meerenforschung*, **5**, 1–107.

Herdendorf, C. E. (1982). Large lakes of the world. *Journal of Great Lakes Research*, **8**, 379–412.

(1990). Distribution of the world's largest lakes. In *Large Lakes: Ecological Structure and Function*, ed. M. M. Tilzer and C. Serruya, pp. 3–38. New York: Springer-Verlag.

Herrmann, J., Degerman, E., Gerhardt, A. *et al.* (1993). Acid-stress effects on stream biology. *Ambio*, **22**, 298–307.

Hessen, D. O. and Lyche, A. (1991). Inter- and intra-specific variations in zooplankton elemental composition. *Archiv für Hydrobologie*, **121**, 343–53.

Hessen, D. O. and van Donk, E. (1993). Morphological changes in *Scenedesmus* induced by substances released from *Daphnia*. *Archiv für Hydrobiologie*, **127**, 129–40.

Hildrew, A. G. and Townsend, C. R. (1987). Organisation in freshwater benthic communities. In *Organization of Communities, Past and Present*, ed. J. H. R. Gee and P. S. Giller, pp. 347–71. Oxford: Blackwell Scientific Publications.

Hill, R. and Bendall, F. (1960). Function of two cytochrome components in chloroplasts: a working hypothesis. *Nature*, **186**, 136–7.

Hillbricht-Ilkowska, A. and Weglenska, T. (1978). Experimentally increased fish stock in the pond-type Lake Warniak. VII. Numbers, biomass and production of zooplankton. *Ekologia Polska, Seria A*, **21**, 533–51.

Hillbricht-Ilkowska, A., Spodniewska, I. and Wegleska, T. (1979). Changes in the phytoplankton-zooplankton relationship connected with the eutrophication of lakes. *Symposia Biologica Hungariae*, **19**, 59–75.

Hilton, J. (1985). A conceptual framework for predicting the occurrence of sediment focussing and sediment redistribution in small lakes. *Limnology and Oceanography*, **30**, 1131–43.

Hilton, J., Rigg, E. and Jaworski, G. H. M. (1989). Algal identification using in vivo fluorescence spectra. *Journal of Plankton Research*, **11**, 65–74.

Hindar, A., Henrikson, L., Sandøy, S. and Romundstad, A. J. (1998). Critical load concept to set restoration goals for liming acidified Norwegian waters. *Restoration Ecology*, **6**(4), 1–13.

Hino, K., Tundisi, J. G. and Reynolds, C. S. (1986). Vertical distribution of phytoplankton in a stratified lake (Lago Dom Helvécio, south eastern Brasil), with special reference to the metalimnion. *Japanese Journal of Limnology*, **47**, 239–46.

Hino, S., Mikami, H., Arisue, J. *et al.* (1998). Limnological characteristics and vertical distribution of phytoplankton in oligotrophic

Lake Akan-Panke. *Japanese Journal of Limnology*, **59**, 263–79.

Hobson, L. A., Morris, W. J. and Pirquet, K. T. (1976). Theoretical and experimental analysis of the ^{14}C technique and its use in studies of primary production. *Journal of the Fisheries Research Board of Canada*, **33**, 1715–21.

Hoge, F. and Swift, R. (1983). Airborne dual laser excitation and mapping of phytoplankton photopigments in a Gulf-Stream warm core ring. *Applied Optics*, **22**, 2272–81.

Holland, R. (1993). Changes in planktonic diatoms and water transparency in Hatchery Bay, Bass Island area, western Lake Erie, since the establishment of the zebra mussel. *Journal of Great Lakes Research*, **16**, 121–9.

Holligan, P. M. and Harbour, D. S. (1977). The vertical distribution and succession of phytoplankton in the Western English Channel in 1975 and 1976. *Journal of the Marine Biological Association of the United Kingdom*, **57**, 1075–93.

Holm, N. P. and Armstrong, D. E. (1981). Role of nutrient limitation and competition in controlling the populations of *Asterionella formosa* and *Microcystis aeruginosa* in semicontinuous culture. *Limnology and Oceanography*, **26**, 622–34.

Holmes, R. W. (1956). The annual cycle of phytoplankton in the Labrador Sea, 1950–51. *Bulletin of the Bingham Oceanographic College*, **16**, 1–74.

Holmgren, S. (1968). *Fytoplankton och primarproduktion i Nedre Laksjon i Abisko National Park under aren 1962–1965*. Uppsala: Limnologiska Institutionene.

(1983). *Phytoplankton Biomass and Algal Composition in Natural, Fertilised and Polluted Subarctic Lakes*, Acta Universitatis Upsaliensis No. 674. Uppsala: University of Uppsala.

Holzmann, R. (1993). Seasonal fluctuations in the diversity and compositional stability of phytoplankton communities in small lakes in upper Bavaria. *Hydrobiologia*, **249**, 101–9.

Hoogenhout, H. and Amesz, J. (1965). Growth rates of photosynthetic microorganisms in laboratory cultures. *Archiv für Mikrobiologie*, **50**, 10–25.

Hopson, A. J. (1982). *Lake Turkana: A Report of the Findings of the Lake Turkana Project, 1972–1975*. London: Government of Kenya and Ministry of Overseas Development.

Horne, A. J. (1979). Nitrogen fixation in Clear Lake, California. IV. Diel studies of *Aphanizomenon* and *Anabaena* blooms. *Limnology and Oceanography*, **24**, 329–41.

Horne, A. J. and Commins, M. L. (1987). Macronutrient controls on nitrogen fixation in planktonic cyanobacterial populations. *New Zealand Journal of Marine and Freshwater Research*, **21**, 413–23.

Hosper, S. H. (1984). Restoration of Lake Veluwe, The Netherlands, by reduction of phosphorus loading and flushing. *Water Science and Technology*, **17**, 757–68.

Hosper, S. H. and Meijer, M.-L. (1993). Biomanipulation: will it work in your lake? A simple test for the assessment of chances for clear water following drastic fish-stock reduction in shallow, eutrophic lakes. *Ecological Engineering*, **2**, 63–72.

Hosper, S. H., Meijer, M.-L. and Walker, P. A. (1992). *Handleidung actief biologisch Beheer: Beoordeeling van de mogelijkhedenvan Visstandbeheer bij het Herstel van meren en plassen*. Lelystad: Rijksinstituut voor integraal Zoetwaterbeheer en Afwalwaterdehanling – Organasitie ter Verbetering van de Binnenvisserij.

Howard, A. and Pelc, S. R. (1951). Nuclear incorporation of ^{32}P as demonstrated by auto-radiographs. *Experimental Cell Research*, **2**, 178–87.

Howarth, R. W. and Michaels, A. F. (2000). The measurement of primary production in aquatic ecosystems. In *Methods in Ecosystem Science*, ed. O. E. Sala, R. B. Jackson, H. A. Mooney and R. W. Howarth, pp. 72–85. New York: Springer-Verlag.

Howarth, R. W., Jensen, H. J., Marino, R. and Postma, H. (1995). Transport to and processing of P in near-shore and oceanic waters. In *Phosphorus in the Global Environment*, ed. H. Tiessen, pp. 323–45. Chichester: John Wiley.

Hrbáček, J., Dvořáková, M., Kořínek, V. and Procházková, L. (1961). Demonstration of the effect of fish stock on the species composition of the zooplankton and the intensity of metabolism of the whole plankton association. *Verhandlungen der internationalen Vereinigung für theoretische und angewandte Limnologie*, **14**, 192–5.

Huber, G. and Nipkow, F. (1922). Experimentelle Untersuchungen über die Entwicklung von *Ceratium hirundinella*. *Zeitschrift für Botanik*, **14**, 337–71.

Hudson, J., Schindler, D. W. and Taylor, W. (2000). Phosphate concentrations in lakes. *Nature*, **406**, 54–6.

Hughes, J. C. and Lund, J. W. G. (1962). The rate of growth of *Asterionella* in relation to its ecology. *Archiv für Mikrobiologie*, **42**, 117–29.

Huisman, J. and Sommeijer, B. (2002). Maximal sustainable sinking velocity of phytoplankton. *Marine Ecology – Progress Series*, **244**, 39–48.

Huisman, J. and Weissing, F. J. (1994). Light-limited growth and competition for light in well-mixed aquatic environments: an elementary model. *Ecology*, **75**, 507–20.

(1995). Competition for nutrients and light in a mixed water column: a theoretical analysis. *American Naturalist*, **146**, 536–64.

(2001). Fundamental unpredictability in multi-species competition. *American Naturalist*, **157**, 488–94.

Huisman, J., van Oostveen, P. and Weissing, F. J. (1999). Critical depth and critical turbulence: two different mechanisms for the development of phytoplankton blooms. *Limnology and Oceanography*, **44**, 1781–7.

Huisman, J., Johanson, A. M., Folmer, E. O. and Weissing, P. J. (2001). Towards a solution of the plankton paradox: the importance of physiology and life history. *Ecology Letters*, **4**, 408–11.

Huisman, J., Arrayas, M., Ebert, U. and Sommeijer, B. (2002). How do sinking phytoplankton species manage to persist? *American Naturalist*, **159**, 245–54.

Hulburt, E. M., Ryther, J. H. and Guillard, R. R. L. (1960). The phytoplankton of the Sargasso Sea off Bermuda. *Journal du Conseil permanent international de l'exploration de la Mer*, **25**, 115–27.

Humphries, S. E. and Imberger, J. (1982). *The Influence of the Internal Structure and Dynamics of Burrinjuck Reservoir on Phytoplankton Blooms*. Nedlands: Centre for Water Research.

Huntley, M. E. and Lopez, M. D. G. (1992). Temperature-dependent production of marine copepods: a global synthesis. *American Naturalist*, **140**, 201–42.

Huntsman, S. A. and Sunda, W. G. (1980). The role of trace metals in regulating phytoplankton growth. In *The Physiological Ecology of Phytoplankton*, ed. I. Morris, pp. 285–328. Oxford: Blackwell Scientific Publications.

Hurrell, J. W. (1995). Decadal trends in the North Atlantic Oscillation: regional temperature and precipitation. *Science*, **269**, 676–9.

Huston, M. (1979). A general theory of species diversity. *American Naturalist*, **113**, 81–101.

(1999). Local processes and regional patterns: appropriate scales for understanding variation in the diversity of plants and animals. *Oikos*, **86**, 393–401.

Huszar, V. L. M. and Caraco, N. (1998). The relationship between phytoplankton composition and physical-chemical variables: a comparison of taxonomic and morphological–functional approaches in six temperate lakes. *Freshwater Biology*, **40**, 1–18.

Huszar, V., Kruk, C. and Caraco, N. (2003). Steady-state assemblages of phytoplankton in four temperate lakes (NE USA). *Hydrobiologia*, **502**, 97–109.

Hutchinson, G. E. (1941). Limnological studies in Connecticut. IV. Mechanisms of intermediary metabolism in stratified lakes. *Ecological Monographs*, **11**, 21–60.

(1959). Homage to Santa Rosalia: or, why are there so many kinds of animals? *American Naturalist*, **93**, 145–59.

(1961). The paradox of the plankton. *American Naturalist*, **95**, 137–47.

(1967). *A Treatise on Limnology*, vol. 2, *Introduction to Lake Biology and the Limnoplankton*. New York: John Wiley.

Ibelings, B. W. and Maberly, S. C. (1998). Photo-inhibition and the availability of inorganic carbon restrict photosynthesis by surface blooms of Cyanobacteria. *Limnology and Oceanography*, **43**, 408–19.

Ibelings, B. W., Kroon, B. M. A. and Mur, L. (1994). Acclimation of photosystem II in a cyano-bacterium and a eukaryotic green alga to high and fluctuating photosynthetic photon flux densities, simulating light regimes induced by mixing in lakes. *New Phytologist*, **128**, 407–24.

Ichimura, S., Nagasawa, S. and Tanaka, T. (1968). On the oxygen and chlorophyll maxima found in the metalimnion of a mesotrophic lake. *Botanical Magazine, Tokyo*, **81**, 1–10.

Ihlenfeldt, M. J. A. and Gibson, J. (1975). Phosphate utilisation and alkaline phosphatase activity in *Anacystis nidulans* (*Synechococcus*). *Archives of Microbiology*, **102**, 23–8.

Imberger, J. (1985). Thermal characteristics of standing waters: an illustration of dynamic processes. *Hydrobiologia*, **125**, 7–29.

(ed.) (1998). *Physical Processes in Lakes and Oceans*, Coastal and Estuarine Studies No. 54. Washington, DC: American Geophysical Union.

Imberger, J. and Hamblin, P. F. (1982). Dynamics of lakes, reservoirs and cooling ponds. *Annual Review of Fluid Mechanics*, **14**, 153–87.

Imberger, J. and Ivey, G. N. (1991). On the nature of turbulence in a stratified fluid. II. Application to lakes. *Journal of Physical Oceanography*, **21**, 659–80.

Imberger, J. and Spigel, R. H. (1987). Circulation and mixing in Lake Rotongaio and Lake Okaro under conditions of light to moderate winds: preliminary results. *New Zealand Journal of Marine and Freshwater Research*, **21**, 515–19.

Imboden, D. M. and Wüest, A. (1995). Mixing mechanisms in lakes. In *Lakes: Chemistry, Geology and Physics*, 2nd edn, ed. A. Lerman, D. M. Imboden and J. R. Gat, pp. 83–138. New York: Springer-Verlag.

Irish, A. E. (1980). A modified 1-m Friedinger sampler: a description and some selected results. *Freshwater Biology*, **10**, 135–9.

Irish, A. E. and Clarke, R. T. (1984). Sampling designs for the estimation of phytoplankton abundance in limnetic environments. *British Phycological Journal*, **19**, 57–66.

Irvine, K., Patterson, G., Allison, E. H., Thompson, A. B. and Menz, A. (2001). The pelagic system of Lake Malawi, Africa: trophic structure and current threats. In *The Great Lakes of the World (GLOW): Food Web, Health and Integrity*, ed. M. Munawar and R. E. Hecky, pp. 3–30. Leiden: Backhuys.

Irwin, W. H., Symons, J. M. and Roebuck, G. G. (1969). Water quality in impoundments and modifications from destratification. In *Water Quality Behaviour in Reservoirs*, ed. J. M. Symons, pp. 363–88. Cincinnati, OH: US Department of Health, Education and Welfare.

Ishikawa, K., Kumagai, M., Vincent, W. F., Tsujimura, S. and Nakahara, H. (2002). Transport and accumulation of bloom-forming Cyanobacteria in a large, mid-latitude lake: the gyre–*Microcystis* hypothesis. *Limnology*, **3**, 87–96.

Ishikawa, K., Tsujimura, S., Kumagai, M. and Nakahara, H. (2003). Surface water recruitment of bloom-forming Cyanobacteria from deep water sediments of Lake Biwa, Japan. In *Sediment Quality Assessment and Management: Insight and Progress*, ed. M. Munawar, pp. 287–305. Burlington: Aquatic Ecosystem Health and Management Society.

Istvánovics, V., Pettersen, K., Rodrigo, M. A., Pierson, D. and Padisák, J. (1993). *Gloeotrichia echinulata*, a colonial Cyanobacterium with unique phosphorus uptake and life strategy. *Journal of Plankton Research*, **15**, 531–52.

Ives, A. R. and Hughes, J. B. (2002). General relationships between species diversity and stability in competitive systems. *American Naturalist*, **159**, 388–95.

Ives, A. R. Carpenter, S. R. and Dennis, B. (1999). Community interaction webs and the response of a zooplankton community to experimental manipulations of planktivory. *Ecology*, **80**, 1405–21.

Ives, A. R., Klug, J. L. and Gross, K. (2000). Stability and species richness in complex communities. *Ecology Letters*, **3**, 399–411.

Ives, K. J. (1956). Electrokinetic phenomena of planktonic algae. *Proceedings of the Society for Water Treatment and Examination*, **5**, 41–58.

Izaguirre, I., Mataloni, G., Allende, L. and Vinocur, A. (2001). Summer fluctuations of microbial communities in a eutrophic lake – Cierva Point, Antarctica. *Journal of Plankton Research*, **23**, 1095–1109.

Jackson, J. B. C. and Sala, E. (2001). Unnatural oceans. *Scientia Marina*, **65** (Suppl. 2), 273–81.

Jacobsen, B.A. and Simonsen, P. (1993). Disturbance events affecting phytoplankton biomass, composition and species diversity in a shallow, eutrophic, temperate lake. *Hydrobiologia*, **249**, 9–14.

Jähnichen, S., Petzoldt, T. and Benndorf, J. (2001). Evidence of control of microcystin dynamics in Bautzen Reservoir (Germany) by cyanobacterial population growth rates and dissolved inorganic carbon. *Archiv für Hydrobiologie*, **150**, 177–96.

James, W. F., Taylor, W. D. and Barko, J. W. (1992). Production and vertical migration of *Ceratium hirundinella* in relation to phosphorus availability in Eau Galle Reservoir, Wisconsin. *Canadian Journal of Fisheries and Aquatic Sciences*, **49**, 694–700.

Jannasch, H. W. and Mottl, M. J. (1985). Geomicrobiology of deep-sea hydrothermal vents. *Science*, **229**, 717–25.

Janzen, D. H. (1968). Host plants as islands in evolutionary and contemporary time. *American Naturalist*, **102**, 592–5.

Jassby, A. D. and Goldman, C. R. (1974a). Loss rates from a phytoplankton community. *Limnology and Oceanography*, **19**, 618–27.

(1974b). A quantitative measure of succession rate and its application to the phytoplankton of lakes. *American Naturalist*, **108**, 688–93.

Jassby, A. D. and Platt, T. (1976). Mathematical formulation of the relationship between photosynthesis and light for phytoplankton. *Limnology and Oceanography*, **21**, 540–7.

Javorničky, P. (1958). Revise nekterých metod pro zjištováni kvantity fytoplanktonu. *Sbornk vysoké Skoly chemicko-technologické v Praze*, **2**, 283–367.

Jaworski, G. H. M., Talling, J. F. and Heaney, S. I. (1981). The influence of carbon dioxide depletion on growth and sinking rate of two planktonic diatoms in culture. *British Phycological Journal*, **16**, 395–410.

Jaworski, G. H. M., Wiseman, S. W. and Reynolds, C. S. (1988). Variability in the sinking rate of the freshwater diatom, *Asterionella formosa*: the influence of colony morphology. *British Phycological Journal*, **23**, 167–76.

Jaworski, G. H. M., Talling, J. F. and Heaney, S. I. (2003). Potassium dependence and phytoplankton ecology: an experimental study. *Freshwater Biology*, **48**, 833–40.

Jeppesen, E., Jensen, J. P., Kristensen, P. *et al.* (1990). Fish manipulation as a lake restoration tool in shallow, eutrophic, temperate lakes. II. Threshold levels, long-term stability and conclusions. *Hydrobiologia*, **200/201**, 219–28.

Jeppesen, E., Søndergaard, M., Jensen, J. P. *et al.* (1998). Cascading trophic interactions from fish to bacteria and nutrients after reduced sewage loading: an 18-year study of a shallow hypertrophic lake. *Ecosystems*, **1**, 250–97.

Jewell, W. J. and McCarty, P. L. (1971). Aerobic decomposition of algae. *Environmental Science and Technology*, **5**, 1023–31.

Jewson, D. H. (1976). The interaction of components controlling net phytoplankton photosynthesis in a well-mixed lake (Lough Neagh, Northern Ireland). *Freshwater Biology*, **6**, 551–76.

Jewson, D. H. and Wood, R. B. (1975). Some effects on integral photosynthesis of artifiial circulation of phytoplankton through light gradients. *Verhandlungen der internationalen Vereinigung für theoretische und angewandte Limnologie*, **19**, 1037–44.

Jiménez Montealegre, R., Verreth, J., Steenbergen, K., Moed, J. and Machiels, M. (1995). A dynamic simulation model for the blooming of *Oscillatoria agardhii* in a monomictic lake. *Ecological Modelling*, **78**, 17–24.

John, D. M., Whitton, B. A. and Brook, A. J. (2002). *The Freshwater Algal Flora of the British Isles*. Cambridge: Cambridge University Press.

Johnson, L. (1975). Physical and chemical characteristics of Great Bear Lake. *Journal of the Fisheries Research Board of Canada*, **32**, 1971–87.

(1994). Great Bear. In *The Book of Canadian Lakes*, ed. R. J. Allan, M. Dickman, C. B. Gray and V. Cromie, pp. 549–59. Burlington: Canadian Association on Water Quality.

Joint, I. R. and Groom, S. B. (2000). Estimation of phytoplankton production from space: current status and future potential of satellite remote sensing. *Journal of Experimental Marine Biology and Ecology*, **250**, 233–55.

Jónasson, P. M. (1978). Zoobenthos of lakes. *Verhandlungen der internationalen Vereinigung für theoretische und angewandte Limnologie*, **20**, 13–37.

(2003). Hypolimnetic eutrophication of the N-limited dimictic lake Esrom. *Archiv für Hydrobiologie* Suppl., **139**, 449–512.

Jónasson, P. M. and Kristiansen, J. (1967). Primary and secondary production in Lake Esrom: growth of *Chironomus anthracinus* in relation to seasonal cycles of phytoplankton and dissolved oxygen. *Internationale Revue des gesamten Hydrobiologie*, **52**, 163–217.

Jones, C. G., Lawton, J. H. and Shachak, M. (1994). Organisms as ecosystem engineers. *Oikos*, **69**, 373–86.

Jones, R. I. (1978). Adaptation to fluctuating irradiance by natural phytoplankton communities. *Limnology and Oceanography*, **23**, 920–6.

Jones, R. I., Grey, J., Quarmby, C. and Sleep, D. (2001). Sources and fluxes of inorganic carbon in a deep, oligotrophic lake (Loch Ness, Scotland). *Global Biogeochemical Cycles*, **15**, 863–70.

Jordan, P., Fromme, P., Witt, H. T., Klukas, O., Saenger, W. and Krauss, N. (2001). Three-dimensional structure of cyanobacterial photosystem 1 at 2.5 Å resolution. *Nature*, **411**, 909–17.

Jørgensen, S.-E. (1995). State-of-the-art management models for lakes and reservoirs. *Lakes and Reservoirs, Research and Management*, **1**, 79–87.

(1997). *Integration of Ecosystem Theory: A Pattern*, 2nd edn. Dordrecht: Kluwer.

(1999). State-of-the-art of ecological modelling with emphasis on development of structural dynamic models. *Ecological Modelling*, **120**, 75–96.

(2002). Explanation of ecological rules and observation by application of ecosystem theory and ecological models. *Ecological Modelling*, **158**, 241–8.

Jørgensen, S.-E., Nielsen, S. N. and Mejer, H. F. (1995). Emergy, environ, exergy and ecological modelling. *Ecological Modelling*, **77**, 99–109.

Joseph, J. and Sendner, H. (1958). Über die horizontale Diffusion im Meere. *Deutsches hydrographische Zeitschrift*, **11**, 51–77.

Juday, C. (1934). The depth distribution of some aquatic plants. *Ecology*, **15**, 325–35.

Juhász-Nagy, P. (1992). Scaling problems almost everywhere: an introduction. *Abstracta Botanica*, **16**, 1–5.

(1993). Notes on compositional biodiversity. *Hydrobiologia*, **249**, 173–82.

Jürgens, K., Arndt, H. and Rothaupt, K. O. (1994). Zooplankton-mediated changes of microbial food-web structure. *Microbial Ecology*, **27**, 27–42.

Kadiri, M. O. and Reynolds, C. S. (1993). Long-term monitoring of the conditions of lakes: the example of the English Lake District. *Archiv für Hydrobiologie*, **129**, 157–78.

Kahn, N. and Swift, E. (1978). Positive buoyancy through ionic control in the non-motile marine dinoflagellate *Pyrocystis noctiluca* Murray ex Schuett. *Limnology and Oceanography*, **23**, 649–58.

Kajak, Z., Hillbricht-Ilkowska, A. and Piczyska, E. (1972). The production processes in several Polish lakes. In *Productivity Problems of Freshwaters (Proceedings of the IBP-UNESCO Symposium, Kazimierz Dolny, Poland, May 6–12, 1970)*, ed. Z. Kajak and A. Hillbricht-Ilkowska, pp. 129–147. Warsaw: Polstie Wydawnictwo Naukowe.

Källqvist, T. and Berge, D. (1990). Biological availability of phosphorus in agricultural runoff compared to other sources. *Verhandlungen der internationale Vereinigung für theoretische und angewandte Limnologie*, **24**, 214–17.

Kamp-Nielsen, L. (1975). A kinetic approach to the aerobic sediment–water exchange in Lake Esrom. *Ecological Modelling*, **1**, 153–60.

Kamykowski, D., Reed, R. E. and Kirkpatrick, G. I. (1992). Comparison of sinking velocity, swimming velocity, rotation and path characteristics among six marine dinoflagellate species. *Marine Biology*, **113**, 319–28.

Kappers, F. I. (1984). *On the Population Dynamics of the Cyanobacterium* Microcystis aeruginosa. Amsterdam: University of Amsterdam.

Karl, D. M. (1999). A sea of change: biogeochemical variability in the North Pacific sub-tropical gyre. *Ecosystems*, **2**, 181–214.

(2002). Nutrient dynamics in the deep blue sea. *Trends in Microbiology*, **10**, 410–18.

Karl, D. M., Wirsen, C. O. and Jannasch, H. W. (1980). Deep-sea primary production at the Galapagos hydrothermal vents. *Science*, **207**, 1345–7.

Karl, D. M., Bidigare, R. R. and Letelier, R. M. (2002). Sustained and aperiodic variability in organic matter production and phototrophic microbial community structure in the North Pacific Subtropical Gyre. In *Phytoplankton Productivity*, ed. P. J. leB. Williams, D. N. Thomas and C. S. Reynolds, pp. 222–64. Oxford: Blackwell Science.

Karp-Boss, L., Boss, E. and Jumars, P. A. (1996). Nutrient fluxes to planktonic osmotrophs in the presence of fluid motion. *Oceanography and Marine Biology*, **34**, 71–107.

Kasprzak, P., Lathrop, R. C. and Carpenter, S. R. (1999). Influence of different-sized *Daphnia* species on chlorophyl concentration and summer phytoplankton community structure in eutropic Wisconsin lakes. *Journal of Plankton Research*, **21**, 2164–74.

Kauffman, Z. S. (1990). *The Ecosystem of Onega Lake and the Trends of its Changes*. Leningrad: Nauka. (In Russian.)

Keddy, P. A. (1992). Assembly and response rules: two goals for predictive community ecology. *Journal of Vegetation Science*, **3**, 157–64.

(2001). *Competition*, 2nd edn. Dordrecht: Kluwer.

Kemp, P. F., Lee, S. and LaRoche, J. (1993). Estimating the growth rate of slowly growing marine bacteria from RNA content. *Applied and Environmental Microbiology*, **59**, 2594–601.

Kerby, N. W., Rowell, P. and Stewart, W. D. P. (1987). Cyanobacterial ammonium transport, ammonium assimilation and nitrogenase regulation. *New Zealand Journal of Marine and Freshwater Research*, **21**, 447–55.

Kerkhof, L. and Ward, B. B. (1993). Comparison of nucleic-acid hybridization and fluorometry for measurement of the relationship between RNA/DNA ratio and growth rate in a marine bacterium. *Applied and Environmental Microbiology*, **59**, 1303–9.

Ketchum, B. H. and Redfield, A. C. (1949). Some physical and chemical characteristics of algal growth in mass cultures. *Journal of Cellular and Comparative Physiology*, **13**, 373–81.

Kibby, H. V. (1971). Energetics and population dynamics of *Diaptomus gracilis*. *Ecological Monographs*, **41**, 311–27.

Kiene, R. P., Linn, L. J. and Bruton, J. A. (2000). New and important roles for DMSP in marine microbial communities. *Journal of Sea Research*, **43**, 209–24.

Kilham, P. and Kilham, S. S. (1980). The evolutionary ecology of phytoplankton. In *The Physiological Ecology of Phytoplanktons*, ed. I. Morris, pp. 571–97. Oxford: Blackwell Scientific Publications.

Kilham, P. and Kilham, S. S. (1990). Endless summer: internal loading processes dominate nutrient cycles in tropical lakes. *Freshwater Biology*, **23**, 379–89.

Kierstead, H. and Slobodkin, L. B. (1953). The size of water masses containing plankton blooms. *Journal of Marine Science*, **12**, 141–7.

Kirchman, D. L. (1990). Limitation of bacterial growth by dissolved organic matter in the subarctic Pacific. *Marine Ecology – Progress Series*, **62**, 47–54.

Kirk, J. T. O. (1975a). A theoretical analysis of the contribution of algal cells to the attenuation of light within natural waters. I. General treatment of suspensions of pigmented cells. *New Phytologist*, **75**, 11–20.

(1975b). A theoretical analysis of the contribution of algal cells to the attenuation of light within natural waters. II. Spherical cells. *New Phytologist*, **75**, 21–36.

(1976). A theoretical analysis of the contribution of algal cells to the attenuation of light within natural waters. III. Cylindrical and sphaeroidal cells. *New Phytologist*, **77**, 341–58.

(1994). *Light and Photosynthesis in Aquatic Ecosystems*, 2nd edn. Cambridge: Cambridge University Press.

(2003). The vertical attenuation of irradiance as a function of the optical properties of the water. *Limnology and Oceanography*, **48**, 9–17.

Kirke, B. K. (2000). Circulation, destratification, mixing and aeration: why and how? *Water*, **27**(4), 24–30.

Kitchell, J. F., Koonce, J. F., O'Neill, R. V. *et al.* (1974). Model of fish biomass dynamics. *Transactions of the American Fisheries Society*, **103**, 786–98.

Kitchell, J. F., Stewart, D. J. and Weininger, D. (1977). Applications of a bioenergetic model to yellow perch (*Perca flavescens*) and walleye (*Stizostedion vitreum vitreum*). *Jounal of the Fisheries Research Board of Canada*, **34**, 1922–35.

Kjelleberg, S., Albertson, N., Flärdh, K. *et al.* (1993). How do non-differentiated bacteria adapt to starvation? *Antonie van Leeuwenhoek*, **63**, 333–41.

Klaveness, D. (1988). Ecology of the Cryptomonadida: a first review. In *Growth and Reproductive Strategies of Freshwater Phytoplankton*, ed. C. D. Sandgren, pp. 105–33. Cambridge: Cambridge University Press.

Klebahn, H. (1895). Gasvakuolen, ein Bestandteil der Zellender Wasserblütebildenden Phycochronaceen. *Flora, Jena*, **80**, 241–82.

Klemer, A. R. (1976). The vertical distribution of *Oscillatoria agardhii* var. *isothrix*. *Archiv für Hydrobiologie*, **78**, 343–62.

(1978). Nitrogen limitation of growth and gas-vacuolation in *Oscillatoria rubescens*. *Verhandlungen der internationale Vereinigung für theoretische und angewandte Limnologie*, **20**, 2293–7.

Klemer, A. R., Feuillard, J. and Feuillard, M. (1982). Cyanobacterial blooms: carbon and nitrogen have opposite effects on the buoyancy of *Oscillatoria*. *Science*, **215**, 1629–31.

Klemer, A. R., Pierson D. C. and Whiteside, M. C. (1985). Blue-green algal (cyanobacterial) nutrition, buoyancy and bloom formation. *Verhandlungen der internationale Vereinigung für theoretische und angewandte Limnologie*, **22**, 2791–8.

Kling, H. (1975). *Phytoplankton Successions and Species Distributions in Prairie Ponds of the Erickson-Elphinstone District, South-West Manitoba*. Technical Report No. 512. Ottawa: Fisheries and Marine Sciences of Canada.

Kling, H. J., Mugidde, R. and Hecky, R. E. (2001). Recent changes in the phytoplankton community of Lake Victoria in response to eutrophication. In *The Great Lakes of the World (GLOW): Food Web, Health and Integrity*, ed. M. Munawar and R. E. Hecky, pp. 47–65. Leiden: Backhuys.

Knapp, C. W., deNoyelles, F., Graham, D. W. and Bergin, S. (2003). Physical and chemical conditions surrounding the diurnal vertical migration of *Cryptomonas* spp. (Cryptophyceae) in a seasonally stratified reservoir (USA). *Journal of Phycology*, **39**, 855–61.

Knoechel, R. and Kalff, J. (1975). Algal sedimentation: the cause of a diatom–bluegreen succession. *Verhandlungen der internationalen Vereinigung für theoretische und angewandte Limnologie*, **19**, 745–54.

(1978). An *in situ* study of the productivity and population dynamics of five freshwater diatom species. *Limnology and Oceanography*, **23**, 195–218.

Knox, R. S. (1977). Photosynthetic efficiency and exciton trapping. In *Primary Production Processes of Photosynthesis*, ed. J. Barber, pp. 55–97. Amsterdam: Elsevier.

Knudsen, B. M. (1953). The diatom genus *Tabellaria*. II. Taxonomy and morphology of the plankton varieties. *Annals of Botany*, **17**, 131–5.

Kofoid, C. A. (1903). The phytoplankton of the lower Illinois River and its basin. I. Quantitative investigations and general results. *Bulletin of the Illinois State Laboratory for Natural History*, **6**, 95–269.

Kolber, Z. and Falkowski, P. G. (1993). Use of active fluorescence to estimate phytoplankton photosynthesis in situ. *Limnology and Oceanography*, **38**, 1646–65.

Kolber, Z. S., Barber, R. T., Coale, H. *et al.* (1994). Iron limitation of phytoplankton photosynthesis in the equatorial Pacific Ocean. *Nature*, **371**, 145–9.

Kolber, Z. S., Plumley, F. G., Lang, A. S. *et al*. (2001). Contribution of photoheterotrophic bacteria to the carbon cycle in the ocean. *Science*, **292**, 2492–4.

Kolmogorov, A. N. (1941). The local structure of turbulence in incompressible viscous fluid for very high Reynolds numbers. *Doklady Akademii nauk SSSR*, **30**, 301. (English summary of paper in Russian.)

Kolpakova, A., Meshcheryakova, A. I. and Votintsev, K. K. (1988). Characteristics of seasonal dynamics of chlorophyll-*a*, primary production and biogenic elements in Lake Baikal. *Hydrobiological Journal*, **24**, 1–6. (English translation of original paper in *Gidrobiologeskii Zhurnal*.)

Komar, D. (1980). Settling velocities of circular cylinders at low Reynolds numbers. *Journal of Geology*, **88**, 327–36.

Komárek, J. (1996). Towards a combined approach to the taxonomy and species limitation of picoplanktic cyanoprokaryotes. *Algological Studies*, **83**, 377–401.

Komárková, J. and Šimek, K. (2003). Unicellular and colonial formations of picoplanktonic cyanobacteria under variable environmental conditions and predation pressure. *Algological Studies*, **109**, 327–40.

Kondoh, M. (2001). Unifying the relationships of species richness to productivity and disturbance. *Proceedings of the Royal Society of London B*, **268**, 269–71.

Konopka, A. E. and Brock, T. D. (1978). Effect of temperature on blue-green algae (Cyanobacteria) in Lake Mendota. *Applied and Environmental Microbiology*, **36**, 572–6.

Konopka, A., Brock, T. D. and Walsby, A. E. (1978). Buoyancy regulation by *Aphanizomenon* in Lake Mendota. *Archiv für Hydrobiologie*, **83**, 524–37.

Kořínek, V., Fott, J., Fuksa, J., Lellák, J. and Prazáková, M. (1987). Carp ponds of central Europe. In *Ecosystems of the World*, vol. 29, *Managed Aquatic Ecosystems*, ed. R. G. Michael, pp. 29–62. Amsterdam: Elsevier.

Korneva, L. G. (2001). Ecological aspects of mass development of *Gonyostomum semen* (Ehr.) Dies. (Raphidophyta). *International Journal of Algae*, **3**, 40–54. (Originally published, in Russian, in *Algologiya*, **10**, 265–77.)

Koroleva, N. N. (1993). Phytoplankton in the northern part of the Aral Sea in in September, 1991. In *Ekologichekiy krizis na Aral'skom more*, ed. O. A. Skarlato and N. V. Aladin, pp. 52–6. St Petersburg: Zoological Institute, Russian Academy of Sciences.

Kozhov, M. (1963). *Lake Baikal and its Life*. The Hague: W. Junk.

Kozhova, O. M. (1987). Phytoplankton of Lake Baikal: structural and functional characteristics. *Ergebnisse der Limnologie*, **25**, 19–37.

Kozhova, O. M. and Izmest'eva, L. R. (1998). *Lake Baikal: Evolution and Biodiversity*. Leiden: Backhuys.

Kratz, W. A. and Myers, J. (1955). Nutrition and growth of several blue-green algae. *American Journal of Botany*, **42**, 282–7.

Krienitz, L. (1990). Coccale Grünalgen der mittleren Elbe. *Limnologica*, **21**, 165–231.

Kristiansen, J. (1996). Dispersal of freshwater algae: a review. *Hydrobiologia*, **336**, 151–7.

Krivtsov, V., Tien, C. and Sigee, D (1999). X-ray microanalytical study of the protozoan *Ceratium hirundinella* from Rostherne Mere (Cheshire, UK): dynamics of intracellular elemental concentrations, correlations and implications for overall ecosystem functioning. *Netherlands Journal of Zoology*, **49**, 263–74.

Kromkamp, J. and Mur, L. (1984). Buoyant density changes in the cyanobacterium *Microcystis aeruginosa* due to changes in the cellular carbohydrate content. *FEMS Microbial Letters*, **25**, 105–9.

Kruk, C., Mazzeo, N., Lacerot, G. and Reynolds, C. S. (2002). Classification schemes for phytoplankton: a local validation of a functional approach to the analysis of species temporal replacement. *Journal of Plankton Research*, **24**, 1191–216.

Kuentzel, L. E. (1969). Bacteria, carbon dioxide and algal blooms. *Journal of the Water Pollution Control Federation*, **41**, 1737–47.

Kühlbrandt, W. (2001). Chlorophylls galore. *Nature*, **411**, 896–8.

Kühlbrandt, W. and Wang, D. N. (1991). Three-dimensional structure of plant light-harvesting complex determined by electron crystallography. *Nature*, **351**, 130–1.

Kühlbrandt, W., Wang, D. N. and Fujiyoshi, Y. (1994). Atomic model of plant light-harvesting complex by electron crystallography. *Nature*, **367**, 614–21.

Kunkel, W. B. (1948). Magnitude and character of errors in Stokes' Law estimates of particle radius. *Journal of Applied Physics*, **19**, 1066–8.

Kuosa, H., Autio, R., Kuuppo, P., Setälä, O. and Tanskanen, S. (1997). Nitrogen, silicon and zooplankton controlling the Baltic spring bloom: an experimental study. *Estuarine, Coast and Shelf Science*, **45**, 813–21.

Kusch, J. (1995). Adaptations to inducible defense in *Euplotes daidaleos* (Ciliophora) to predation risks by various predators. *Microbial Ecology*, **30**, 79–88.

Kustka, A., Sañudo-Wilhelmy, S., Carpenter, E. J., Capone, D. G. and Raven, J. A. (2003). A revised estimate of the iron-use efficiency of nitrogen fixation, with special reference to the marine Cyanobacterium *Trichodesmium* spp. (Cyanophyta). *Journal of Phycology*, **39**, 12–25.

Kutsch, W. L., Steinborn, W., Herbst, M. *et al.* (2001). Environmental indication: a field test of an ecosystem approach to quantify self-organisation. *Ecosystems*, **4**, 49–66.

Kuuppo, P., Samuelsson, K., Lignell, R. *et al.* (2003). Fate of increased production in late-summer plankton communities due to nutrient enrichment of the Baltic proper. *Aquatic Microbial Ecology*, **32**, 47–60.

Kyewalyanga, M., Platt, T. and Sathyendranath, S. (1992). Ocean primary production calculated by spectral and broad-band models. *Marine Ecology – Progress Series*, **85**, 171–85.

Lack, T. J. and Lund, J. W. G. (1974). Observations and experiments on the phytoplankton of Blelham Tarn, English Lake District. I. The experimental tubes. *Freshwater Biology*, **4**, 399–415.

Lacroix, G., Lescher-Moutoué, F. and Bertolo, A. (1999). Biomass and production of plankton in shallow lakes and deep lakes: are there general patterns? *Annales de Limnologie*, **35**, 111–22.

Lamouroux, N., Olivier, J.-M., Persat, H. *et al.* (1999). Predicting community characteristics from habitat conditions: fluvial fish and hydraulics. *Freshwater Biology*, **42**, 275–99.

Lampert, W. (1977a). Studies on the carbon balance of *Daphnia pulex* De Geer as related to environmental conditions. II. The dependence of carbon assimilation on animal size, temperature, food concentration and diet species. *Archiv für Hydrobiologie* (Suppl.), **48**, 310–35.

(1977b). Studies on the carbon balance of *Daphnia pulex* De Geer as related to environmental conditions. III. Production and production efficiency. *Archiv für Hydrobiologie* (Suppl.), **48**, 336–60.

(1977c). Studies on the carbon balance of *Daphnia pulex* De Geer as related to environmental conditions. IV. Determination of the threshold concentration as a factor controlling the abundance of zooplankton species. *Archiv für Hydrobiologie* (Suppl.), **48**, 361–8.

(1978). Climatic conditions and planktonic interactions as factors controlling the regular succession of spring algal bloom and extremely clear water in Lake Constance. *Verhandlungen der internationale Vereinigung für theoretische und angewandte Limnologie*, **20**, 969–74.

(1987). Feeding and nutrition in *Daphnia*. *Memorie dell'Istituto italiano di Idrobiologia*, **45**, 143–92.

Lampert, W. and Schober, U. (1980). The importance of threshold food concentrations. In *Evolution and Ecology of Zooplankton Communities*, ed. W. C. Kerfoot, pp. 264–7. Hanover, NH: University of New England Press.

Lampert, W. and Sommer, U. (1993). *Limnoökologie*. Stuttgart: Georg Thieme Verlag.

Lampert, W., Fleckner, W., Rai, H. and Taylor, B. E. (1986). Phytoplankton control by grazing zooplankton. *Limnology and Oceanography*, **31**, 478–90.

Lampert, W., Rothaupt, K. O. and von Elert, E. (1994). Chemical induction of colony formation in a green alga (*Scenedesmus acutus*) by grazers (*Daphnia*). *Limnology and Oceanography*, **39**, 1543–50.

Lane, A. E. and Burris, J. E. (1981). Effects of environmental pH on the internal pH of *Chlorella pyrenoidosa*, *Scenedesmus quadricauda* and *Euglena mutabilis*. *Plant Physiology*, **68**, 439–42.

Lange, W. (1976). Speculations on a possible essential function of the gelatinous sheath of blue-green algae. *Canadian Journal of Microbiology*, **22**, 1181–5.

Langenberg, V. T., Mwape, L. M., Tshibangu, K. *et al.* (2002). Comparison of the thermal stratification, light attenuation and chlorophyll-*a* dynamics between the ends of Lake Tanganyika. *Aquatic Ecosystem Health and Management*, **5**, 255–65.

Langmuir, I. (1938). Surface motion of water induced by wind. *Science*, **87**, 119–23.

Larkum, A. W. D. and Barrett, J. (1983). Light-harvesting processes in algae. In *Advances in Botanical Research*, vol. 10, ed. H. W. Woolhouse, pp. 1–219. London: Academic Press.

Larsson, U. and Hagström, A. (1979). Phytoplankton exudate release as an energy source for the growth of pelagic bacteria. *Marine Biology*, **52**, 199–206.

Lauridsen, T. L. and Buenk, I. (1996). Diel changes in the horizontal distribution of zooplankton in the littoral zone of two eutrophic lakes. *Archiv für Hydrobilogie*, **137**, 161–76.

Lauridsen, T. L. and Lodge, D. M. (1996). Avoidance by *Daphnia magna* of fish and macrophytes: chemical cues and predator-mediated use of macrophyte beds. *Limnology and Oceanography*, **41**, 794–8.

Lauridsen, T. L., Jeppesen, E. and Mitchell, S. F. (1999). Diel variation in horizontal distribution of *Daphnia* and *Ceriodaphnia* in oligotrophic and mesotrophic lakes with contrasting fish densities. *Hydrobiologia*, **409**, 241–50.

Laurion, I., Lami, A. and Sommaruga, R. (2002). Distribution of mycosporine-like amino-acids and photoprotective carotenoids among freshwater phytoplankton assemblages. *Aquatic Microbial Ecology*, **26**, 283–94.

Laws, E. A. (1991). Photosynthetic quotients, new production and net communiy production in the open ocean. *Deep-Sea Research*, **38**, 143–67.

(2003). Partitioning of microbial biomass in pelagic aquatic communities: maximum resiliency as a food-web organising construct. *Aquatic Microbial Ecology*, **32**, 1–10.

Lawton, J. H. (1997). The role of species in ecosystems: aspects of ecological complexity and biological diversity. In *Biodiversity: An Ecological Perspective*, ed. T. Abe, S. R. Levin and M. Higashi, pp. 215–28. New York: Springer-Verlag.

(2000). *Community Ecology in a Changing World*. Oldendorf: ECI.

Lawton, J. H and Brown, V. K. (1993). Redundancy in ecosystems. In *Biodiversity and Ecosystem Function*, ed. E.-D, Schulze and H. A. Mooney, pp. 255–70. Berlin: Springer-Verlag.

Lawton, J. H. and Schröder, D. (1977). Effects of plant type, size of geographical range and taxonomic isolation on number of insect species associated with British plants. *Nature*, **265**, 137–40.

Lazier, J. R. N. and Mann, K. H. (1989). Turbulence and the diffusive layers around organisms. *Deep-Sea Research*, **36**, 1721–33.

Le Cren, E. D. (1987). Perch (*Perca fluviatilis*) and pike (*Esox lucius*) in Windermere from 1940 to 1985: studies in poulation dynamics. *Canadian Journal of Fisheries and Aquatic Sciences*, **44** (Suppl. 2), 216–28.

Leadbeater, B. S. C. (1990). Ultrastructure and assembly of the scale case in *Synura* (Synurophyceae Andersen). *British Phycological Journal*, **25**, 117–32.

Lee, S. and Fuhrman, J. (1987). Relationships between biovolume and biomass of naturally derived marine bacterioplankton. *Applied Environmental Microbiology*, **53**, 1298–303.

Legendre, L. and LeFevre, J. (1989). Hydrodynamical singularities as controls of recycled versus export production in oceans. In *Productivity of the Ocean, Present and Past*, ed. W. H. Berger, V. Smetacek and G. Wefer, pp. 49–63. New York: Wiley Interscience.

Legendre, L. and Michaud, J. (1999). Chlorophyll *a* to estimate the particulate organic carbon available as food to large zooplankton in the euphotic zone of oceans. *Journal of Plankton Research*, **21**, 2067–83.

Legendre, L. and Rassoulzadegan, F. (1996). Food-web mediated export of biogenic carbon in oceans: hydrodynamic control. *Marine Ecology – Progress Series*, **145**, 179–93.

Legendre, L. and Rivkin, R. B. (2002a). Fluxes of carbon in the upper ocean: regulation by food-web control nodes. *Marine Ecology – Progress Series*, **242**, 95–109.

(2002b). Pelagic food webs: responses to environmental processes and effects on the environment. *Ecological Research*, **17**, 143–9.

Lehman, J. T. (1975). Reconstructing the rate of of accumulation of lake sediment: the effect of sediment focusing. *Quaternary Research*, **5**, 541–50.

(1976). Ecological and nutritional studies on *Dinobryon* Ehrenb.: seasonal periodicity and the phosphate toxicity problem. *Limnology and Oceanography*, **21**, 646–58.

Lehman, J. T. and Cáceres, C. E. (1993). Food-web responses to species invasion by a predatory invertebrate: *Bythotrephes* in Lake Michigan. *Limnology and Oceanography*, **38**, 879–91.

Lehman, J. T., Botkin, D. B. and Likens, G. E. (1975). The assumptions and rationales of computer model of phytoplankton population dynamics. *Limnology and Oceanography*, **20**, 343–64.

Leibovich, S. (1983). The form and dynamics of Langmuir circulations. *Annual Review of Fluid Mechanics*, **15**, 391–427.

Leitão, M., Morata, S. M., Rodriguez, S. and Vergon, J. P. (2003). The effect of perturbations on phytoplankton assemblages in a deep reservoir (Vouglans, France). *Hydrobiologia*, **502**, 73–83.

Lessmann D., Fyson, A. and Nixdorf, B. (2000). Phytoplankton of the extremely acidic mining lakes of Lusatia (Germany) with pH ≤3. *Hydrobiologia* **433**, 123–8.

Letcher, B. H., Rice, J. A., Crowder, L. B. and Rose, K. A. (1996). Variability in survival of larval fish: disentangling components with a generalised individual-based model. *Canadian Journal of Fisheries and Aquatic Sciences*, **53**, 787–801.

Letelier, R. M., Bidigare, R. R., Hebel, D. V. *et al.* (1993). Temporal variability of phytoplankton community structure based on pigment analysis. *Limnology and Oceanography*, **38**, 1420–37.

Lévêque, C., Carmouze, J. P., Dejoux, C. *et al.* (1972). Recherches sur les biomasses et la productivité du Lac Tchad. In *Productivity Problems of Freshwaters*, ed. Z. Kajak and A. Hillbricht-Ilkowska, pp. 165–81. Warsaw:

Levich, A. P. (1996). The role of nitrogen : phosphorus ratio in selecting for dominance of phytoplankton by Cyanobacteria or green algae and its application to reservoir management. *Journal of Aquatic Ecosystem Health*, **5**, 55–61.

Levich, V. G. (1962). *Physicochemical Hydrodynamics*. Englewood Cliffs, NJ: Prentice Hall.

Levin, S. A. (2000). Multiple scales and the maintenance of biodiversity. *Ecosystems*, **3**, 498–506.

Levins, R. (1966). The strategy of model building in population ecology. *American Scientist*, **54**, 421–31.

Lewin, J. C. (1962). Silicification. In *Physiology and Biochemistry of the Algae*, ed. R. A. Lewin, pp. 445–55. New York: Academic Press.

Lewin, J. C., Lewin, R. A. and Philpott, D. E. (1958). Observations on *Phaeodactylum tricornutum*. *Journal of General Microbiology*, **18**, 418–26.

Lewin, R. A. (1981). The Prochlorophytes. In *The Prokaryotes: A Handbook on Habitats, Isolation and Identification of Bacteria*, ed. M. P. Starr, H. Stolp, H. G. Trüper, A. Ballows and H. G. Schlegel, pp. 257–66. Berlin: Springer-Verlag.

Lewin, R. A. (2002). Prochlorophyta: a matter of class distinction. *Photosynthesis Research*, **73**, 59–61.

Lewis, D. M., Elliott, J. A., Lambert, M. J. and Reynolds, C. S. (2002). The simulation of an Australian reservoir using a phytoplankton community model, PROTECH. *Ecological Modelling*, **150**, 107–16.

Lewis, W. M. (1976). Surface/volume ratio: implications for phytoplankton morphology. *Science*, **192**, 885–7.

(1978). Dynamics and succession of the phytoplankton in a tropical lake: Lake Lanao, Philippines. *Journal of Ecology*, **66**, 849–80.

(1983). A revised classification of lakes based on mixing. *Canadian Journal of Fisheries and Aquatic Sciences*, **40**, 1779–87.

(1986). Phytoplankton succession in Lake Valencia, Venezuela. *Hydrobiologia*, **138**, 9–204.

Lewitus, A. J. and Kana, T. M. (1994). Responses of estuarine phytoplankton to exogenous glucose: stimulation versus inhibition of photosynthesis and respiration. *Limnology and Oceanography*, **39**, 182–9.

Li, A. S., Stoecker, D. K. and Coats, D. W. (2000). Mixotrophy in *Gyrodinium galatheanum* (Dinophyceae): grazing responses to light intensity and inorganic nutrients. *Journal of Phycology*, **36**, 33–45.

Li, C.-W. and Volcani, B. E. (1984). Aspects of silicification in wall morphogenesis of diatoms. *Philosophical Transactions of the Royal Society of London B*, **304**, 519–28.

Li, W. K. W. (2002). Macroecological patterns of phytoplankton in the northwestern North Atlantic Ocean. *Nature*, **419**, 154–7.

Li, W. K. W. and Platt, T. (1982). Distribution of carbon among photosynthetic end products in phytoplankton of the eastern Canadian Arctic. *Journal of Phycology*, **58**, 135–50.

Liebig, J. von (1840). *Organic Chemistry and its Application to Agriculture and Physiology*. London: Taylor and Walton.

Liere, L. van (1979). *On Oscillatoria agardhii Gomont: Experimetal Ecology and Physiology of a Nuisance Bloom-Forming Cyanobacterium*. Zeist: De Niuwe Schouw.

Lijklema, L. (1977). The role of iron in the exchange of phosphate between water and sediments. In *Interactions between Sediments and Freshwater*, ed. H. L. Golterman, pp. 313–17. The Hague: W. Junk.

Likens, G. E. and Davis, M. B. (1975). Postglacial history of Mirror Lake and its watershed in New Hampshire, USA: an initial report. *Verhandlungen der internationale Vereinigung für theoretische und angewandte Limnologie*, **19**, 982–93.

Likhoshway, Ye. V., Kuzmina, A. Ye., Potyemkina, T. G., Potyemkin, V. L. and Shimarev, M. N (1996). The distribution of diatoms near a thermal bar in Lake Baikal. *Great Lakes Research*, **22**, 5–14.

Lin, C. K. (1972). Phytoplankton succession in a eutrophic lake with special reference to blue-green algal blooms. *Hydrobiologia*, **39**, 321–34.

Lindley, J. A., Robins, B. B. and Williams, R. (1999). Dry weight, carbon and nitrogen content of some euphausiids from the north Atlantic Ocean and the Celtic Sea. *Journal of Plankton Research*, **21**, 2053–66.

Lindström, E. S. (2000). Bacterioplankton community composition in five lakes differing in trophic status and humic content. *Microbial Ecology*, **40**, 104–13.

Link, J. S. (2002). What does ecosystem-based fisheries management mean? *Fisheries*, **27**(4), 18–21.

Litchman, E. (2000). Growth rates of phytoplankton under fluctuating light. *Freshwater Biology*, **44**, 223–35.

(2003). Competition and coexistence of phytoplankton under fluctuating light: experiments with two Cyanobacteria. *Aquatic Microbial Ecology*, **31**, 241–8.

Liti, D., Källqvist, T. and Lien, L. (1991). Limnological aspects of Lake Turkana, Kenya. *Verhandlungen der internationale Vereinigung für theoretische und angewandte Limnologie*, **24**, 1108–11.

Livingstone, D. and Cambray, R. S. (1978). Confirmation of ^{137}Cs dated by algal stratigraphy. *Nature*, **276**, 259–60.

Livingstone, D. and Jaworski, G. H. M. (1980). The viability of akinetes of blue-green algae recovered from the sediments of Rostherne Mere. *British Phycological Journal*, **15**, 357–64.

Lloyd, M. (1967). 'Mean crowding'. *Journal of Animal Ecology*, **36**, 1–30.

Loehle, C. (1988). Problems with the triangular model for representing plant strategies. *Ecology*, **69**, 284–6.

Löffler, H. (1988). Natural hazards and health risks from lakes. *Water Resources Development*, **4**, 276–83.

Long, S. P., Humphries, S. and Falkowski, P. G. (1994). Photoinhibition of photosynthesis in nature. *Annual Reviews in Plant Physiology and Plant Molecular Biology*, **45**, 633–62.

Longhurst, A., Sathyendranath, S., Platt, T. and Caverhill, C. (1995). An estimate of global primary production in the ocean from satellite radiometer data. *Journal of Plankton Research*, **17**, 1245–71.

Loreau, M., Naeem, S., Inchausti, P. *et al.* (2001). Biodiversity and ecosystem functioning: current knowledge and future challenges. *Science*, **294**, 804–8.

Lorenzen, C. J. (1966). A method for the continuous measurement for *in-vivo* chlorophyll concentration. *Deep-Sea Research*, **13**, 223–7.

Lorenzen, M. W. (1974). Predicting the effects of nutrient diversion on lake recovery. In *Modelling the Eutrophication Process*, ed. E. J. Middlebrooks, D. H. Falkenborg and T. E. Maloney, pp. 205–21. Ann Arbor, MI: Ann Arbor Science.

Lovelock, J. E. (1979). *Gaia: A New Look at Life on Earth*. Oxford: Oxford University Press.

Lowe-McConnell, R. (1996). Fish communities in African great lakes. *Environmental Biology of Fishes*, **45**, 219–35.

(2003). Recent research in the African great lakes: fisheries, biodiversity and cichlid evolution. *Freshwater Forum*, **20**, 1–64.

Lucas, L. V., Koseff, J. R., Monismith, S. G., Cloern, J. E. and Thompson, J. K. (1999). Processes governing phytoplankton blooms in estuaries II. The role of horizontal transport. *Marine Ecology – Progress Series*, **187**, 17–30.

Lucas, W. J. and Berry, J. A. (1985). *Inorganic Carbon Uptake by Aquatic Photosynthetic Organisms*. Rockville, MD: American Society of Plant Physiologists.

Luftig, R. and Haselkorn, R. (1967). Morphology of a virus of blue-green algae and properties of its deoxyribonucleic acid. *Journal of Virology*, **1**, 344–61.

Lund, J. W. G. (1949). Studies on *Asterionella*. I. The origin and nature of the cells producing seasonal maxima. *Journal of Ecology*, **37**, 389–419.

(1950). Studies on *Asterionella formosa* Hass. II. Nutrient depletion and the spring maximum. *Journal of Ecology*, **38**, 1–35.

(1954). The seasonal cycle of the plankton diatom *Melosira italica* subsp. *subarctica* O. Mull. *Journal of Ecology*, **42**, 151–79.

(1957). Chemical analysis in ecology illustrated from Lake District tarns and lakes. II. Algal differences. *Proceedings of the Linnean Society of London*, **167**, 165–71.

(1959). Buoyancy in relation to the ecology of the freshwater phytoplankton. *British Phycological Bulletin*, **1**(7), 1–17.

(1961). The algae of the Malham Tarn District. *Field Studies*, **1**(3), 85–115.

(1965). The ecology of the freshwater phytoplankton. *Biological Reviews*, **40**, 231–93.

(1966). The importance of turbulence in the periodicity of certain freshwater species of the genus Melosira. *Botanicheskii Zhurnal SSSR*, **51**, 176–87. (In Russian.)

(1970). Primary production. *Water Treatment and Examination*, **19**, 332–58.

(1971). An artificial alteration of the seasonal cycle of the plankton diatom *Melosira italica* subsp. *subarctica* in an English Lake. *Journal of Ecology*, **59**, 421–33.

(1972). Eutrophication. *Proceedings of the Royal Society of London B*, **180**, 371–82.

(1975). The use of large experimental tubes in lakes. In *The Effects of Storage on Water Quality*, ed. R. E. Youngman, pp. 291–311. Medmenham: Water Research Centre.

(1978). Changes in the phytoplankton of an English lake, 1945–1977. *Hydrobiological Journal*, **14**, 6–21.

Lund, J. W. G. and Reynolds, C. S. (1982). The development and operation of large limnetic encosures and their contribution to phytoplankton ecology. In *Progress in Phycological Research*, vol. 1, ed. F. E. Round and D. J. Chapman, pp. 1–65. Amsterdam: Elsevier.

Lund, J. W. G., Kipling, C. and Le Cren, E. D. (1958). The inverted microscope method of estimating algal numbers and the statistical basis of estimation by counting. *Hydrobiologia*, **11**, 143–70.

Lund, J. W. G., Mackereth, F. J. H. and Mortimer, C. H. (1963). Changes in depth and time of certain chemical and physical conditions and of the standing crop of *Asterionella formosa* Hass. in the north basin of Windermere in 1947. *Philosophical Transactions of the Royal Society of London B*, **246**, 255–90.

Lund, J. W. G., Jaworski, G. H. M. and Butterwick, C. (1975). Algal bioassay of water from Blelham Tarn, English Lake District, and the growth of planktonic diatoms. *Archiv für Hydrobiologie* (Suppl.), **49**, 49–69.

Lyck, S. and Christoffersen, K. (2003). Microcystin quota, cell division and microcystin net production of precultured *Microcystis aeruginosa* CYA 228 (Chrococcales, Cyanophyceae) under field conditions. *Phycologia*, **42**, 667–74.

Lynch, M., Wieder, L. and Lampert, W. (1986). Measurement of the carbon balance in *Daphnia*. *Limnology and Oceanography*, **31**, 17–33.

Maberly, S. C. (1996). Diel, episodic and seasonal changes in pH and concentrations of inorganic carbon in a productive lake. *Freshwater Biology*, **35**, 579–98.

Maberly, S. C., Hurley, M. A., Butterwick, C. et al. (1994). The rise and fall of *Asterionella formosa* in the South Basin of Windermere: analysis of a 45-year series of data. *Freshwater Biology*, **31**, 19–34.

MacArthur, R. H. and Wilson, J. O. (1967). *The Theory of Island Biogeography*. Princeton, NJ: Princeton University Press.

Mackereth, F. J. H. (1953). Phosphorus utilisation by *Asterionella formosa* Hass. *Journal of Experimental Botany*, **4**, 296–313.

Mackey, R. L. and Currie, D. J. (2000). A re-examination of the expected effects of disturbance on diversity. *Oikos*, **88**, 483–93.

Maeda, H. (1993). A new perspective for studies of plankton community: ecology and taxonomy of picophytoplankton and flagellates. *Japanese Journal of Limnology*, **54**, 155–9.

Maguire, B. (1963). The passive dispersal of small aquatic organisms and their colonisation of isolated bodies of water. *Ecological Monographs*, **33**, 161–85.

Malin, G., Turner, S., Liss, P., Holligan, P. and Harbour, D. (1993). Dimethylsulfide and demethylsulphoniopropionate in the northeast Atlantic during the summer coccolithophore bloom. *Deep-Sea Research*, **40**, 1487–508.

Malinsky-Rushansky, N., Berman, T. and Dubinsky, Z. (1995). Seasonal dynamics of picophytoplankton in Lake Kinneret, Israel, *Freshwater Biology*, **34**, 241–51.

Maloney, T. E., Miller, W. R. and Blind, N. L. (1973). Use of algal bioassays in studying eutrophication problems. In *Advances in Water Pollution Research*, ed. S. H. Jenkins, pp. 205–15. Oxford: Pergamon Press.

Mann, K. H. and Lazier, J. R. N. (1991). *Dynamics of Marine Ecosystems*. Oxford: Blackwell Scientific Publications.

Mann, N. H. (1995). How do cells express limitation at the molecular level? In *Molecular Ecology of Aquatic Microbes*, ed I. Joint, pp. 171–90. Berlin: Springer-Verlag.

Maranger, R. and Bird, D. F. (1995). Viral abundance in aquatic systems: a comparison between marine and freshwaters. *Marine Ecology – Progress Series*, **121**, 217–26.

Margalef, R. (1957). Nuevos aspectos del problema de la suspensión en los organismos planctónicos. *Investigación pesquera*, **7**, 105–16.

(1958). Temporal succession and spatial heterogeneity in phytoplankton. In *Perspectives in Marine Biology*, ed. A. A. Buzatti-Traverso, pp. 323–49. Berkeley, CA: University of California Press.

(1963). Succession in marine populations. *Advancing Frontiers in Plant Science*, **2**, 137–88.

(1964). Correspondence between the classic types of lake and the structural and dynamic properties of their populations. *Verhandlungen der internationale Vereinigung für theoretische und angewandte Limnologie*, **15**, 169–75.

(1967). Some concepts relative to the organisation of plankton. *Annual Review of Oceanography and Marine Biology*, **5**, 257–89.

(1978). Life-forms of phytoplankton as survival alternatives in an unstable environment. *Oceanologia Acta*, **1**, 493–509.

(1997). *Our Biosphere*. Oldendorf: ECI.

Margalef, R., Estrada, M. and Blasco, D. (1979). Functional morphology of organisms involved in red tides, as adapted to decaying turbulence. In *Toxic Dinoflagellate Blooms*, ed. D. L. Taylor and H. H. Seliger, pp. 89–94. Amsterdam: Elsevier-North Holland.

Margulis, L. (1970). *Origin of Eukaryotic Cells*. New Haven, CT: Yale University Press.

(1981). *Symbiosis in Cell Evolution*. San Francisco, CA: W. H. Freeman.

Marra, J. (1978). Phytoplankton photosynthetic response to vertical movement in a mixed layer. *Marine Biology*, **46**, 203–8.

Marra, J. (2002). Approaches to the measurement of plankton production. In *Phytoplankton Productivity*, ed. P. J. leB. Williams, D. N. Thomas and C. S. Reynolds, pp. 78–108. Oxford: Blackwell Science.

Marra, J., Haas, L. W. and Heinemann, K. R. (1988). Time course of C assimilation and microbial food webs. *Journal of Experimental Marine Biology and Ecology*, **115**, 263–80.

Marsden, M. W. (1989). Lake restoration by reducing external phosphorus loading: the influence of phosphorus release. *Freshwater Biology*, **21**, 139–62.

Marti, D. E. and Imboden, D. M. (1986). Thermische Energieflüsse an der Wasseroberfläche: beispiel Sempachersee. *Schweizerische Zeitschrift für Hydrologie*, **48**, 196–229.

Martin, J. H. (1992). Iron as a limiting factor in oceanic productivity. In *Primary Productivity and Biogeochemical Cycles in the Sea*, ed. P. G. Falkowski and A. Woodhead, pp. 137–55. New York: Plenum Press.

Martin, J. H., Coale, K. H., Johnson, K. S. *et al.* (1994). Testing the iron hypothesis in ecosystems of the equatorial Pacific Ocean. *Nature*, **371**, 123–43.

Massana, R. and Jürgens, K. (2003). Composition and population dynamics of planktonic bacteria and bacterivorous flagellates in seawater chemostat cultures. *Aquatic Microbial Ecology*, **32**, 11–22.

May, R. M. (1973). *Stability and Complexity in Model Ecosystems*. Princeton, NJ: Princeton University Press.

(1977). Thresholds and breakpoints in ecosystems with a multiplicity of stable states. *Nature*, **269**, 471–7.

May, V. (1972). *Blue-Green Algal Blooms at Braidwood, New South Wales (Australia)*. Sydney: New South Wales Department of Agriculture.

McAlice, B. J. (1970). Observations on the the small-scale distribution of estuarine phytoplankton. *Marine Biology*, **7**, 100–11.

McCarthy, J. J. (1972). The uptake of urea by natural poulations of marine phytoplankton. *Limnology and Oceanography*, **17**, 738–48.

(1980). Nitrogen. In *The Physiological Ecology of Phytoplankton*, ed. I. Morris, pp. 191–233, Oxford: Blackwell Scientific Publications.

McCarthy, J. J., Taylor, W. R. and Taft, J. L. (1975). The dynamics of nitrogen and phosphorus cycling in the open waters of the Chesapeake Bay. In *Marine Chemistry in the Coastal Environment*, ed. T. M. Church, pp. 664–81. Washington, DC: American Chemical Society.

McCauley, E., Nisbet, R. M., Murdoch, W. W., de Roos, A. M. and Gurney, W. S. C. (1999). Large-amplitude cycles of *Daphnia* and its algal prey in enriched environments. *Nature*, **402**, 653–6.

McDermott, G. A., Prince, S. M., Freer, A. A. *et al.* (1995). Crystal structure of an integral membrane light-harvesting complex from phoyosynthetic bacteria. *Nature*, **374**, 517–21.

McGowan, J. A. and Walker, P. W. (1985). Dominance and diversity maintenance in an oceanic ecosystem. *Ecological Monographs*, **55**, 103–18.

Mackenthun, K. M., Keup, L. E. and Stewart, R. K. (1968). Nutrients and algae in Lake Sebasticook, Maine. *Journal of the Water Pollution Control Federation*, **40**, R72–81.

McMahon, J. W. and Rigler, F. H. (1963). Mechanisms regulating the feeding rate of *Daphnia magna* Strauss. *Canadian Journal of Zoology*, **41**, 321–32.

McNown, J. S. and Malaika, J. (1950). Effect of particle shape on settling velocity at low Reynolds numbers. *Transactions of the American Geophysical Union*, **31**, 74–82.

Meffert, M.-E. (1971). Cultivation and growth of two planktonic *Oscillatoria* species. *Mitteilungen der internationalen Vereinigung für theoretische und angewandte Limnologie*, **19**, 189–205.

Megard, R. O. (1972). Phytoplankton photosynthesis and phosphorus in lake Minnetonka, Minnesota. *Limnology and Oceanography*, **17**, 68–87.

Mejer, H. and Jørgensen, S.-E. (1979). Exergy and ecological buffer capacity. In *State-of-the-Art Ecological Modelling*, vol. 7, ed. S.-E. Jørgensen, pp. 829–46. Copenhagen: International Society for Ecological Modelling.

Meybeck. M. (1995). Global distribution of lakes. In *Physics and Chemistry of Lakes*, 2nd edn, ed. A. Lerman, D. M. Imboden and J. R. Gat, pp. 1–35. Berlin: Springer-Verlag.

Mills, C. A. and Hurley, M. A. (1990). Long-term studies on the Windermere populations of perch, *Perca fluviatilis*, pike, *Esox lucius*, and arctic charr, *Salvelinus alpinus*. *Freshwater Biology*, **23**, 119–36.

Mills, E. L. and Forney, J. L. (1983). Impact on *Daphnia pulex* of predation by young perch in Oneida Lake, New York. *Transactions of the American Fisheries Society*, **112**, 154–61.

(1987). Trophic dynamics and development of freshwater pelagic food webs. In *Complex Interactions in Lake Communities*, ed. S. R. Carpenter, pp. 11–30. New York: Springer–Verlag.

Mischke, U. and Nixdorf, B. (2003). Equilibrium-phase conditions in shallow German lakes: how Cyanoprokaryota species establish a steady-state phase in late summer. *Hydrobiologia*, **502**, 123–32.

Mitchison, G. J. and Wilcox, M. (1972). A rule governing cell division in *Anabaena*. *Nature*, **239**, 110–11.

Moed, J. R. (1973). Effect of combined action of light and silicon depletion on *Asterionella formosa* Hass. *Verhandlungen der internationale Vereinigung für theoretische und angewandte Limnologie*, **18**, 1367–74.

Molen, D. T. van der and Boers, P. C. M. (1994). Influence of internal loading on phosphorus concentration in shallow lakes before and after reduction of the external loading. *Hydrobiologia*, **275/276**, 279–389.

Moll, R. A. and Rohlf, F. J. (1981). Analysis of temporal and spatial variability in a Long Island saltmarsh. *Journal of Experimental Marine Biology and Ecology*, **51**, 133–44.

Moll, R. A. and Stoermer, E. F. (1982). A hypothesis relating trophic status and subsurface chlorophyll maxima in lakes. *Archiv für Hydrobiologie*, **94**, 425–40.

Montagnes, D. J. S., Berges, J. A., Harrison, P. J. and Taylor, F. J. R. (1994). Estimating carbon, nitrogen, protein and chlorophyll *a* from volume in marine phytoplankton. *Limnology and Oceanography*, **39**, 1044–60.

Moore, J. W. (1980). Seasonal distribution of phytoplankton in Yellowknife Bay, Great Slave Lake. *Internationale Revue des gesamten Hydrobiologie*, **65**, 283–93.

Moore, L. R., Rocap, G. and Chisholm, S. (1998). Physiology and molecular phylogeny of coexisting *Prochlorococcus* ecotypes. *Nature*, **393**, 464–7.

Morabito, G., Ruggiu, D. and Panzani, P. (2002). Recent dynamics (1995–1999) of the phytoplankton assemblages in Lago Maggiore as a basic tool for defining association patterns in the Italian deep lakes. *Journal of Limnology*, **61**, 129–45.

Morel, A. (1991). Light and marine photosynthesis: a spectral model with geochemical and climatological implications. *Progress in Oceanography*, **26**, 263–306.

Morel, A. and Bricaud, A. (1981). Theoretical results concerning light absorption in a discrete medium and application to specific absorption of phytoplankton. *Deep-Sea Research*, **28A**, 1375–93.

Morel, A. and Smith, R. C. (1974). Relation between total quanta and total energy for aquatic photosynthesis. *Limnology and Oceanography*, **19**, 591–600.

Moroney, J. V. (2001). Carbon concentrating mechanisms in aquatic photosynthetic organisms: a report on CCM 2001. *Journal of Phycology*, **37**, 928–31.

Moroney, J. V. and Chen, Z. Y. (1998). The role of the chloroplast in inorganic carbon uptake by eukaryotic algae. *Canadian Journal of Botany*, **76**, 1025–34.

Morse, G. K., Lester, J. N. and Perry, R. (1993). *The Economic and Environmental Impact of Phosphorus Removal from Waste Water in the European Community*. London: Selper Publications.

Mortimer, C. H. (1941). The exchange of dissolved substances between mud and water in lakes. I. *Journal of Ecology*, **29**, 280–329.

(1942). The exchange of dissolved substances between mud and water in lakes. II. *Journal of Ecology*, **30**, 147–201.

(1974). Lake hydrodynamics. *Mitteilungen der internationale Vereinigung für theoretische und angewandte Limnologie*, **29**, 124–97.

Moss, B. (1967). Vertical heterogeneity in the water column of Abbot's Pond. II. The influence of physical and chemical conditions on the spatial and temporal distribution on the phytoplankton and of a community of epipelic algae. *Journal of Ecology*, **57**, 397–414.

(1972). The influence of environmental factors on the distribution of freshwater algae: an experimental study. I. Introduction and the influence of calcium concentration. *Journal of Ecology*, **60**, 917–32.

(1973a). The influence of environmental factors on the distribution of freshwater algae: an experimental study. II. The role of pH and the carbon dioxide–bicarbonate system. *Journal of Ecology*, **61**, 157–77.

(1973b). The influence of environmental factors on the distribution of freshwater algae: an experimental study. III. Effects of temperature, vitamin requirements and inorganic nitrogen compounds on growth. *Journal of Ecology*, **61**, 179–92.

(1973c). The influence of environmental factors on the distribution of freshwater algae: an experimental study. IV. Growth of test species in natural lake waters and conclusions. *Journal of Ecology*, **61**, 193–211.

(1983). The Norfolk Broadland: experiments in the restoration of a complex wetland. *Biological Reviews*, **58**, 521–61.

(1990). Engineering and biological approaches to the restoration from eutropication of shallow lakes in which aquatic plant communities are important components. *Hydrobiologia*, **200/201**, 367–77.

(1992). The scope of biomanipulation for improving water quality. In *Eutrophication: Research and Application to Water Supply*, ed. D. W. Sutcliffe and J. G. Jones, pp. 73–81. Ambleside: Freshwater Biological Association.

Moss, B. and Balls, H. (1989). Phytoplankton distribution in a temperate floodplain lake and river system. II. Seasonal changes in the phytoplankton communities and their control by hydrology and nutrient availability. *Journal of Plankton Research*, **11**, 839–67.

Moss, B., Madgwick, J. and Phillips, G. L. (1996). *A Guide to the Restoration of Nutrient-Enriched Shallow Lakes*. Norwich: Environment Agency and Broads Authority.

Moustaka-Gouni, M. (1993). Phytoplankton succession and diversity in a warm, monomictic, relatively shallow lake: Lake Volvi, Macedonia, Greece. *Hydrobiologia*, **249**, 33–42.

Mugidde, R. (1993). The increase in phytoplankton primary productivity and biomass in Lake Victoria (Uganda). *Verhandlungen der internationale Vereinigung für theoretische und angewandte Limnologie*, **25**, 846–9.

Mullin, M. M., Sloane, P. R. and Eppley, R. W. (1966). Relationships between carbon content, cell volume and area in phytoplankton. *Limnology and Oceanography*, **11**, 307–11.

Munawar, M. and Fahnenstiel, G. L. (1982). *The Abundance and Significance of Ultraplankton and Microalgae at an Offshore Station in Central Lake Superior*, Canadian Technical Report No. 1153. Ottawa: Fisheries and Marine Sciences of Canada.

Munawar, M. and Munawar, I. F. (1975). Some observations on the growth of diatoms in Lake Ontario, with emphasis on *Melosira binderana* Kütz. during thermal bar conditions. *Archiv für Hydrobiologie*, **75**, 490–9.

(1981). A general comparison of the taxonomic composition and size analyses of the phytoplankton of the North American Great Lakes. *Verhandlungen der internationale Vereinigung für theoretische und angewandte Limnologie*, **21**, 1695–716.

(1986). The seasonality of phytoplankton in the North American Great Lakes: a comparative synthesis. *Hydrobiologia*, **138**, 85–115.

(1996). *Phytoplankton Dynamics in the North American Great Lakes, vol. 1, Lakes Ontario, Erie and St. Clair*. Amsterdam: SPB Science Publishers.

(2000). *Phytoplankton Dynamics in the North American Great Lakes*, vol. 2, *The Upper lakes, Huron, Superior and Michigan*. Leiden: Backhuys.

(2001). An overview of the changing flora and fauna of the North American Great Lakes. I. Phytoplankton and microbial food web. In *The Great Lakes of the World (GLOW): Food Web, Health and Integrity*, ed. M. Munawar and R. E. Hecky, pp. 219–75. Leiden: Backhuys.

Munk, W. H. and Riley, J. A. (1952). Absorption of nutrients by aquatic plants. *Journal of Marine Research*, **11**, 215–40.

Murphy, J. and Riley, J. P. (1962). A modified single-solution method for the determination of phosphate in natural waters. *Analytica Chimica Acta*, **27**, 1–36.

Murphy, T. P., Lean, D. R. S. and Nalewajko, C. (1976). Blue-green algae: their excretion of iron-selective chelators enables them to dominate other algae. *Science*, **192**, 900–2.

Murray, A. and Hunt, T. (1993). *The Cell Cycle: An Introduction*. New York: W. H. Freeman.

Murray, A. W. and Kirschner, M. W. (1991). What controls the cell cycle? *Scientific American*, **264**, 34–41.

Murrel, M. C. (2003). Bacterioplankton dynamics in a subtropical estuary: evidence for substrate limitation. *Aquatic Microbial Ecology*, **32**, 239–50.

Nadin-Hurley, C. M. and Duncan, A. (1976). A comparison of daphnid gut particles with the sestonic particles in two Thames Valley reservoirs throughout 1970 and 1971. *Freshwater Biology*, **6**, 109–23.

Nakamura, M. and Chang, W. Y. B. (2001). Lake Biwa: largest lake in Japan. In *The Great Lakes of the World (GLOW): Food Web, Health and Integrity*, ed. M. Munawar and R. E. Hecky, pp. 151–82. Leiden: Backhuys.

Nakanishi, M. (1984). Phytoplankton. In *Lake Biwa*, ed. S. Horie, pp. 154–74. Dordrecht: W. Junk.

Nakano, S.-I., Ishi, N., Mange, P. M. and Kawabata, Z. (1998). Trophic roles of heterotrophic nanoflagellates and ciliates among planktonic organisms in a hypereutrophic pond. *Aquatic Micobial Ecology*, **16**, 153–61.

Nakano, S.-I., Mitamura, O., Sugiyama, M. *et al.* (2003). Vertical planktonic structure in the central basin of Lake Baikal in Summer, 1999, with special reference to the microbial food web. *Limnology*, **4**, 155–60.

Nakatsu, C. and Hutchinson, T. (1988). Extreme metal and acid tolerance of *Euglena mutabilis* and an associated yeast from Smoking Hills, Northwest Territories, and their apparent mutualism. *Microbial Ecology*, **16**, 213–31.

Nalepa, T. F. and Fahnenstiel, G. L. (1995). *Dreissena polymorpha* in Saginaw Bay, Lake Huron ecosystem: overview and perspective. *Journal of Great Lakes Research*, **21**, 411–16.

Nalewajko, C. (1966). Dry weight, ash and volume data for some freshwater planktonic algae. *Journal of the Fisheries Research Board of Canada*, **23**, 1285–8.

Nalewajko, C. and Lean, D. R. S. (1978). Phosphorus kinetics–algal growth relationships in batch cultures. *Mitteilungen der internationale Vereinigung für theoretische und angewandte Limnologie*, **21**, 184–92.

Naselli-Flores, L. and Barone, R. (2003). Steady-state assemblages in a Mediterranean hypertrophic reservoir: the role of *Microcystis* ecomorphological variability in maintaining an apparent equilibrium. *Hydrobiologia*, **502**, 133–43.

Naselli-Flores, L., Padisák, J., Dokulil, M. T. and Chorus, I. (2003). Equilibrium/steady-state concept in phytoplankton ecology. *Hydrobiologia*, **502**, 395–403.

Nasev, D., Nasev, S. and Guiard, V. (1978). Statistiche Auswertung von Planktonuntersuchungen. II. Räumliche Verteilung des Planktons. Konfidenzintervalle für die Individuendichte (Ind./l) des Phytoplanktons. *Wissenschaftliche Zeitschrift der Wilhem-Pieck-Universitat*, **27**, 357–61.

Naumann, E. (1919). Några synpunkter angende limnoplanktons ökologgi med särskild hänsyn till fytoplankton. *Svensk botanisk Tidskrift*, **13**, 129–63.

Nauwerck, A. (1963). Die Beziehungen zwischen zooplankton und phytoplankton in See Erken. *Symbolae botanicae Upsaliensis*, **17**(5), 1–163.

Neale, P. J. (1987). Algal photoinhibition and photosynthesis in the aquatic environment. In *Photoinhibition*, ed. D. Kyle, C. J. Arntzen and B. Osmond, pp. 39–65. Amsterdam: Elsevier.

Neale, P. J., Talling, J. F., Heaney, S. I., Lund, J. W. G. and Reynolds, C. S. (1991a). Long time series from the English Lake District: irradiance-dependent phytoplankton dynamics during the spring diatom maximum. *Limnology and Oceanography*, **36**, 751–60.

Neale, P. J., Heaney, S. I. and Jaworski, G. H. M. (1991b). Responses to high irradiance contribute to the decline of the spring diatom maximum. *Limnology and Oceanography*, **36**, 761–8.

Newman, J. R. and Barrett, P. R. F. (1993). Control of *Microcystis aeruginosa* by decomposing barley straw. *Journal of Aquatic Plant Management*, **31**, 203–6.

Nielsen, S. L. and Sand-Jensen, K. (1990). Allometric scaling of maximal photosynthetic growth rate to surface-to-volume ratio. *Limnology and Oceanography*, **35**, 177–81.

Nielsen, S. N. (1992). *Application of Maximum Energy in Structural Dynamic Models*. Copenhagen: Miljøministeriet.

Nilssen, J.-P. (1984). Tropical lakes – functional ecology and future development: the need for a process-oriented approach. *Hydrobiologia*, **113**, 231–42.

Nimer, N. A., Iglesias-Rodriguez, M. D. and Merrett, M. J. (1997). Bicarbonate utilisation by marine phytoplankton species. *Journal of Phycology*, **33**, 625–31.

Nixdorf, B., Fyson, A. and Krumbeck, H. (2001). Review: plant life in extremely acidic waters. *Environmental and Experimental Botany*, **46**, 203–11.

Nixon, S. W. (1988). Physical energy inputs and comparative ecology of lake and marine ecosystems. *Limnology and Oceanography*, **33**, 1005–25.

Noy-Meir, I. (1975). Stability of grazing systems: application of predator–prey graphs. *Journal of Ecology*, **63**, 459–81.

Nyholm, N. (1977). Kinetics of phosphorus-limited growth. *Biotechnology and Bioengineering*, **19**, 467–72.

Obernosterer, I., Reitner, B. and Herndl, G. J. (1999). Contrasting effects of solar radiation on dissolved organic matter and its bioavailability to marine bacterioplankton. *Limnology and Oceanography*, **44**, 1645–54.

Obernosterer, I., Kawasaki, N. and Benner, R. (2003). P-limitation of respiration in the Sargasso Sea and uncoupling of bacteria from P regeneration in size-fractionation experiments. *Aquatic Microbial Ecology*, **32**, 229–37.

O'Brien, K. R., Ivey, G. N., Hamilton, D. P., Waite, A. M. and Visser, P. M. (2003). Simple mixing criteria for the growth of negatively buoyant phytoplankton. *Limnology and Oceanography*, **48**, 1326–37.

Odum, E. P. (1969). The strategy of ecosystem development. *Science*, **164**, 262–70.

Odum, H. T. (1986). Energy in ecosystems. In *Ecosystem Theory and Application*, ed. N. Polunin, pp. 337–69. New York: John Wiley.

(1988). Self-organization, transformity and information. *Science*, **242**, 1132–8.

(2002). Explanations of ecological relationships with energy systems concepts. *Ecological Modelling*, **158**, 201–11.

Oglesby, R. T. and Schaffner, W. R. (1975). The response of lakes to phosphorus. In *Nitrogen and Phosphorus: Food Production, Waste and the Environment*, ed. K. S. Porter, pp. 23–57. Ann Arbor, MI: Ann Arbor Science.

Ogutu-Ohwayu, R. (1990). The decline of native fishes in Lake Victoria and Kyoga (East Africa) and the impact of introduced species, especially the Nile perch, *Lates niloticus*, and the Nile tilapia, *Oreochromis niloticus*. *Environmental Biology of Fishes*, **27**, 81–96.

O'Gorman, R., Bergtedt, R. A. and Eckert, T. H. (1987). Prey fish dynamics and salmonine predator growth in Lake Ontario, 1978–84. *Canadian Journal of Fisheries and Aquatic Sciences*, **44** (Suppl. 2), 390–403.

Ohmori, M. and Hattori, A. (1974). Effect of ammonia on nitrogen fixation by the blue-green alga *Anabaena cylindrica*. *Plant and Cell Physiology*, **15**, 131–42.

Ohnstad, F. R. and Jones, J. G. (1982). *The Jenkin Surface-Mud Sampler*, Occasional Publication No. 15. Ambleside: Freshwater Biological Association.

Ojala, A. and Jones, R. I. (1993). Spring development and mitotic division pattern of a *Cryptomonas* sp. in an acidified lake. *European Journal of Phycology*, **28**, 17–24.

Okada, M. and Aiba, S. (1983). Simulation of water bloom in a eutrophic lake. III. *Water Research*, **17**, 883–93.

Okino, T. (1973). Studies on the blooming of *Microcystis aeruginosa*. *Japanese Journal of Botany*, **20**, 381–402.

Økland, J. and Økland, K. A. (1986). The effects of acid deposition on benthic animals in lakes and streams. *Experimentia*, **42**, 471–86.

Okubo, A. (1971). Oceanic diffusion diagrams. *Deep-Sea Research*, **18**, 789–802.

(1978). Horizontal dispersion and critical scales of phytoplankton patches. In *Spatial Pattern in Plankton Communities*, NATO Conference Series, IV, No. 3, ed. J. Steele, pp. 21–42. New York: Plenum Press.

(1980). *Diffusion and Ecological Problems*. New York: Springer-Verlag.

Oliver, R. L. (1994). Floating and sinking in gas-vacuolate Cyanobacteria. *Journal of Phycology*, **30**, 161–73.

Oliver, R. L. and Whittington, J. (1998). Using measurements of variable chlorophyll-*a* fluorescence to investigate the influence of water movement on the photochemistry of phytoplankton. In *Physical Processes in Lakes and Oceans*, Coastal and Estuarine Studies No. 54, ed. J. Imberger, pp. 517–34. Washington, DC: American Geophysical Union.

Oliver, R. L., Kinnear, A. J. and Ganf, G. G. (1981). Measurements of cell density of three freshwater phytoplankters by density gradient centrifugation. *Limnology and Oceanography*, **26**, 285–94.

Oliver, R. L., Thomas, R. N., Reynolds, C. S. and Walsby, A. E. (1985). The sedimentation of buoyant *Microcystis* colonies caused by precipitation with an iron-containing colloid. *Proceedings of the Royal Society of London B*, **223**, 511–28.

Olrik, K. (1981). Succession of phytoplankton in response to environmental factors in Lake Arresø, North Zealand, Denmark. *Schweizerische Zeitschrift für Hydrologie*, **43**, 6–15.

(1994). *Phytoplankton Ecology*. Copenhagen: Miljøministeriet.

Omata, T., Takahashi, Y., Yamaguchi, O. and Nishimura, T. (2002). Structure, function and regulation of the cyanobacterial high-affinity bicarbonate transporter, BCT-1. *Functional Plant Biology*, **29**, 151–9.

Organisation for Economic Co-operation and Development (1982). *Eutrophication of Waters: Monitoring, Assessment and Control*. Paris: OECD.

Orr, P. T. and Jones, G. J. (1998). Relationship between microcystin production and cell division rates in nitrogen-limited *Microcystis aeruginosa* cultures. *Limnology and Oceanography*, **43**, 1604–14.

Osmond, C. B. (1981). Photorespiration and photosynthesis: some implications for the energetics of photosynthesis. *Biochimica et Biophysica Acta*, **639**, 77–98.

O'Sullivan, P. (2003). Palaeolimnology. In *The Lakes Handbook*, vol. 1, ed. P. E. O'Sullivan and C. S. Reynolds, pp. 609–66. Oxford: Blackwell Science.

Owens, O. van H. and Esaias, W. (1976). Physiological responses of phytoplankton to major environmental factors. *Annual Review of Plant Physiology*, **27**, 461–83.

Paasche, E. (1964). A tracer study of the inorganic carbon uptake during coccolith formation and photosynthesis in the coccolithophorid *Coccolithus huxleyi*. *Physiologia Plantarum*, Suppl. III, 1–82.

Paasche, E. (1973a). Silicon and the ecology of marine plankton diatoms. I. *Thalassiosira pseudonana* growth in a chemostat with silicate as limiting nutrient. *Marine Biology*, **19**, 117–26.

(1973b). Silicon and the ecology of marine plankton diatoms. I. Silicate uptake kinetics in five diatom species. *Marine Biology*, **19**, 262–9.

(1980). Silicon. In *The Physiological Ecology of Phytoplankton*, ed. I. Morris, pp. 259–84. Oxford: Blackwell Scientific Publications.

Padan, E. and Shilo, M. (1973). Cyanophages: viruses attacking blue-green algae. *Bacteriological Reviews*, **37**, 343–70.

Padisák, J. (1992). Seasonal succession of phytoplankton in a large shallow lake (Balaton, Hungary): a dynamic approach to ecological memory, its possible role and mechanisms. *Journal of Ecology*, **80**, 217–30.

(1993). The influence of different disturbance frequencies on the species richness, diversity and equitability of phytoplankton of shallow lakes. *Hydrobiologia*, **249**, 135–56.

(1997). *Cylindrospermopsis raciborskii* (Wolszynska) Seenaayya *et* Subba Raju, an expanding, highly adaptive Cyanobacterium: worldwide distribution and a review of its ecology. *Archiv für Hydrobiologie* (Suppl.), **107**, 563–93.

(2003). Phytoplankton. In *The Lakes Handbook*, vol. 1, ed. P. E. O'Sullivan and C. S. Reynolds, pp. 251–308. Oxford: Blackwell Science.

Padisák, J. and Reynolds, C. S. (1998). Selection of photoplankton associations in Lake Balaton, Hungary, in response to eutrophication and restoration measures, with special reference to the cyanoprokaryotes. *Hydrobiologia*, **384**, 41–53.

(2003). Shallow lakes: the absolute, the relative, the functional and the pragmatic. *Hydrobiologia*, **506/509**, 1–11.

Padisák, J., Krienitz, L., Scheffler, W. *et al.* (1998). Phytoplankton succession in the oligotrophic Lake Stechlin (Germany) in 1994 and 1995. *Hydrobiologia*, **370**, 179–97.

Padisák, J., Soróczki-Pintér, É. and Rezner, Z. (2003a). Sinking properties of some phytoplankton shapes and the relation of form resistance to morphological diversity of plankton: an experimental study. *Hydrobiologia*, **500**, 243–57.

Padisák, J., Scheffler, W., Sípos, C. *et al.* (2003b). Spatial and temporal pattern of development and decline of the spring diatom populations in Lake Stechlin in 1999. *Advances in Limnology*, **58**, 135–55.

Padisák, J., Borics, J., Fehér, G. *et al.* (2003c). Dominant species, functional assemblages and frequency of equilibrium phases in late summer phytoplankton assemblages in Hungarian small, shallow lakes. *Hydrobiologia*, **502**, 169–76.

Paerl, H. W. (1988). Growth and reproductive strategies of freshwater blue-green algae (Cyanobacteria). In *Growth and Reproductive Strategies of Freshwater Phytoplankton*, ed. C. D. Sandgren, pp. 261–315. Cambridge: Cambridge University Press.

Paerl, H. W., Tucker, J. and Bland, P. T. (1983). Carotenoid enhancement and its role in maintaining blue-green algal (*Microcystis aeruginosa*) surface blooms. *Limnology and Oceanography*, **28**, 847–57.

Pahl-Wostl, C. (1995). *The Dynamic Nature of Ecosystems*. Chichester: John Wiley.

(2003). Self-regulation of limnetic ecosystems. In *The Lakes Handbook*, vol. 1, ed. P. O'Sullivan and C. S. Reynolds, 583–608. Oxford: Blackwell Science.

Pahl-Wostl, C. and Imboden, D. M. (1990). DYPHORA: a dynamic model of the rate of photosynthesis of alga. *Journal of Plankton Research*, **12**, 1207–21.

Paine, R. T. (1966). Food-web complexity and species diversity. *American Naturalist*, **100**, 65–75.

(1980). Food webs: linkage, interaction, strength and community infrastructure. *Journal of Animal Ecology*, **49**, 667–85.

Paine, R. T., Tegner, M. J. and Johnson, E. A. (1998). Compound perturbations yield ecological surprises. *Ecosystems*, **1**, 535–45.

Park, Y.-S., Céréghino, R., Compin, A. and Lek, S. (2003). Applications of artificial neural networks for patterning and predicting aquatic insect species richness in rinning waters. *Ecological Modelling*, **160**, 265–80.

Parsons, T. R. and Takahashi, M. (1973). Environmental control of phytoplankton cell size. *Limnology and Oceanography*, **18**, 511–15.

Pasciak, W. J. and Gavis, J. (1974). Transport limitation of nutrient uptake in phytoplankton. *Limnology and Oceanography*, **19**, 881–9.

(1975). Transport-limited nutrient-uptake rates in *Ditylum brightwelli*. *Limnology and Oceanography*, **20**, 604–17.

Passow, U., Engel, A. and Ploug, H. (2003). The role of aggregation for the dissolution of diatom frustules. *FEMS Microbiology Ecology*, **46**, 247–55.

Pavoni, M. (1963). Die Bedeutung des Nannoplanktons im Vergleich zum Netzplankton. *Schweizerische Zeitschrift für Hydrologie*, **25**, 219–341.

Pearsall, W. H. (1921). The development of vegetation in the English lakes, considered in relation to the general evolution of glacial lakes and rock basins. *Proceedings of the Royal Society of London B*, **92**, 259–84.

(1922). A suggestion as to factors influencing the distribution of free-floating vegetation. *Journal of Ecology*, **9**, 241–53.

(1924). Phytoplankton and environment in the English Lake District. *Revue d'Algologie*, **50**, 1–5.

(1932). Phytoplankton in the English Lakes. II. The composition of the phytoplankton in relation to dissolved substances. *Journal of Ecology*, **20**, 241–58.

Pedrós-Alió, C., Gasol, J. M. and Guerrero, R. (1987). On the ecology of a *Cryptomonas phaseolus* population forming a melimnetic bloom in Lake Cisó, Spain: annual distribution and loss factors. *Limnology and Oceanography*, **32**, 285–98.

Pedrozo, F., Kelly, L., Diaz, M. *et al.* (2001). First results on the water chemistry, algae and trophic status of an Andean acidic lake system of volcanic origin in Patagonia (Lake Caviahue). *Hydrobiologia*, **452**, 129–37.

Pennington, W. (1943). Lake sediments: the bottom deposits of the North Basin of Windermere. *New Phytologist*, **42**, 1–27.

(1947). Studies of the post-glacial history of the British vegetation. VII. Lake sediments: pollen diagrams from the bottom deposits of the North Basin of Windermere. *Philosophical Transactions of the Royal Society of London B*, **233**, 137–75.

Pennington, W. (1969). *The History of the British Vegetation*. London: English Universities Press.

Perry, J. J. (2002). Diversity of microbial heterotrophic metabolism. In *Biodiversity of Microbial Life*, ed. J. T. Stanley and A.-L. Reysenbach, pp. 155–80. New York: Wiley-Liss.

Peters, R. H. (1987). Metabolism in *Daphnia. Memorie dell'Istituto italiano di Idrobiologia*, **45**, 193–243.

Petersen, R. (1975). The paradox of the plankton: an equilibrium hypothesis. *American Naturalist*, **109**, 35–49.

Peterson, B. J. (1978). Radiocarbon uptake: its relation to net particulate carbon production. *Limnology and Oceanography*, **23**, 179–84.

Petr, T. (1977). Bioturbation and exchange of chemicals in the mud–water interface. In *Interactions between Sediments and Freshwater*, ed. H. L. Golterman, pp. 216–26. The Hague: W. Junk.

Petrova, N. A. (1986). Seasonality of *Melosira* plankton in great northern lakes. *Hydrobiologia*, **138**, 65–73.

Pfiester, L. A. and Anderson, D. M. (1987). Dinoflagellate reproduction. In *The Biology of Dinoflagellates*, ed. F. J. R. Taylor, pp. 611–44. Oxford: Blackwell Scientific Publications.

Phillips, G. L., Eminson, D. F. and Moss, B. (1978). A mechanism to account for macrophyte decline in progressively eutrophicated freshwaters. *Aquatic Botany*, **4**, 103–26.

Pick, F. R. (2000). Predicting the abundance and production of photosynthetic picoplankton in temperate lakes. *Verhandlungen der internationale Vereinigung für theoretische und angewandte Limnologie*, **27**, 1884–9.

Pick, F. R., Nalewajko, C. and Lean, D. R. S. (1984). The origin of a metalimnetic chrysophyte peak. *Limnology and Oceanography*, **29**, 125–34.

Pickett, S. T. A., Kolasa, I., Armesto, I. I. and Collins, S. L. (1989). The ecological concept of disturbance and is expression at various hierarchical levels. *Oikos*, **54**, 129–36.

Pickett-Heaps, J. D., Schmid, A.-M. M. and Edgar, L. E. (1990). The cell biology of diatom valve formation. In *Progress in Phycological Research*, vol. 7, ed. F. E. Round and D. J. Chapman, pp. 1–168. Bristol: Biopress.

Pielou, E. C. (1974). *Population and Community Ecology*. New York: Gordon and Breach.

(1975). *Ecological Diversity*. New York: John Wiley.

Pinevich, A. B., Velichko, N. B. and Bazanova, A. V. (2000). Prochlorophytes twenty years on. *Russian Journal of Plant Physiology*, **47**, 639–43.

Pirie, N. W. (1969). *Food Resources, Conventional and Novel*. Harmondsworth: Penguin Books.

Plisnier, P.-D. and Coenens, E. J. (2001). Pulsed and dampened annual limnological fluctuations in Lake Tanganyika. In *The Great Lakes of the World (GLOW): Food Web, Health and Integrity*, ed. M. Munawar and R. E. Hecky, pp. 83–96. Leiden: Backhuys.

Platt, T. and Denman, K. (1980). Patchiness in phytoplankton distribution. In *The Physiological Ecology of Phytoplankton*, ed. I. Morris, pp. 413–31. Oxford: Blackwell Scientific Publications.

Platt, T. and Sathyendranath, S. (1988). Oceanic primary production: estimation by remote sensing at local and regional scales. *Science*, **241**, 1613–20.

(1999). Spatial structure of pelagic ecosystem processes in the global ocean. *Ecosystems*, **2**, 384-94.

Pokorný, J. and Kvĕt, J. (2003). Aquatic plants and lake ecosystems. In *The Lakes Handbook*, vol. 1, ed. P. E. O'Sullivan and C. S. Reynolds, pp. 309-40. Oxford: Blackwell Science.

Polishchuk, L. V. (1999). Contribution analysis of disturbance-caused changes in phytoplankton diversity. *Ecology*, **80**, 721-5.

Pollingher, U. (1988). Freshwater armoured dinoflagellates: growth, reproductive strategies and population dynamics. In *The Growth and Reproductive Strategies of Freshwater Phytoplankton*, ed. C. D. Sandgren, pp. 134-74. Cambridge: Cambridge University Press.

(1990). Effects of latitude on phytoplankton composition and abundance in large lakes. In *Large Lakes: Ecological Structure and Function*, ed. M. M. Tilzer and C. Serruya, pp. 368-402. New York: Springer-Verlag.

Pollingher, U. and Berman, T. (1977). Quantitative and qualitative changes in the phytoplankton in Lake Kinneret, Israel, 1972-1985. *Oikos*, **29**, 418-28.

Pollingher, U. and Serruya, C. (1976). Phased division in *Peridinium cinctum* f. *westii* (Dinophyceae) and development of the Lake Kinneret (Israel) bloom. *Journal of Phycology*, **12**, 163-70.

Pomeroy, L. (1974). The ocean's food web: a changing paradigm. *BioScience*, **24**, 499-504.

Poole, H. H. and Atkins, W. R. G. (1929). Photoelectric measurements of submarine illumination throughout the year. *Journal of the Marine Biological Association of the United Kingdom*, **16**, 297-324.

Popovskaya, G. I. (2001). Ecological monitoring of phytoplankton in Lake Baikal. *Aquatic Ecosystem Health and Management*, **3**, 215-25.

Porter, J. and Jost, M. (1973). Light-shielding by gas vacuoles in *Microcystis aeruginosa*. *Journal of General Microbiology*, **75**, xxii.

(1976). Physiological effects of the presence and absence of gas vacuoles in *Microcystis aeruginosa* Kuetz. emend. Elenkin. *Archives of Microbiology*, **110**, 225-31.

Porter, K. G. (1976). Enhancement of algal growth and productivity by grazing zooplankton. *Science*, **192**, 1332-4.

Porter, K. G. and Feig, Y. S. (1980). The use of DAPI for identifying and counting aquatic microflora. *Limnology and Oceanography*, **25**, 943-8.

Porter, K. G., Pace, M. L. and Battey, J. F. (1979). Ciliate protozoans as links in freshwater planktonic food shains. *Nature*, **277**, 563-5.

Porter, K. G., Sherr, E. B., Sherr, B. F., Pace, M. and Sanders, R. W. (1985). Protozoa in planktonic food webs. *Journal of Protozoology*, **32**, 409-15.

Post, A. F., Wit, R. de and Mur, L. R. (1985). Interactions between temperature and light intensity on growth and photosynthesis of the cyanobacterium, *Oscillatoria agarrdhii*. *Journal of Plankton Research*, **7**, 487-95.

Post, D. M. (2002a). Using stable isotopes to estimate trophic position: models, methods and assumptions. *Ecology*, **83**, 703-18.

(2002b). The long and short of food chains. *Trends in Ecology and Evolution*, **17**, 269-77.

Post, D. M., Conners, M. E. and Goldberg, D. S. (2000a). Prey preference by a top predator and the stability of linked food chains. *Ecology*, **81**, 8-14.

Post, D. M., Pace, M. L. and Hairston, N. G. (2000b). Ecosystem size determines food-chain length in lakes. *Nature*, **405**, 1047-9.

Post, J. R. (1990). Metabolic allometry of larval and juvenile yellow perch (*Perca flavescens*): *in situ* estimates and bioenergetic models. *Canadian Journal of Fisheries and Aquatic Sciences*, **47**, 554-60.

Postgate, J. R. (1987). *Nitrogen Fixation*, 2nd edn. London: Edward Arnold.

Pourriot, R. (1977). Food and feeding habits of Rotifera. *Ergebnisse der Limnologie*, **8**, 243-60.

Prairie, Y. T., Duarte, C. M. and Kalff, J. (1989). Unifying nutrient : chlorophyll relationships in lakes. *Canadian Journal of Fisheries and Aquatic Sciences*, **46**, 1176-82.

Preisendorfer, R. W. (1976). *Hydrological Optics*. Washington, DC: US Department of Commerce.

Preston, T., Stewart, W. D. P. and Reynolds, C. S. (1980). Bloom-forming Cyanobacterium *Microcystis aeruginosa* overwinters on sediment surface. *Nature*, **288**, 365-7.

Pretorius, G. A., Krüger, G. H. J. and Eloff, J. N. (1977). The release of nanocytes durin the growth of *Microcystis*. *Journal of the Limnological Society of Southern Africa*, **3**, 17-20.

Price, G. D., Maeda, S., Omata, T. and Badger, M. R. (2002). Modes of active inorganic carbon uptake in the cyanobacterium *Synechococcus* sp. PCC7942. *Functional Plant Biology*, **29**, 131-49.

Prigogine, I. (1955). *Thermodynamics of Irreversible Processes*. New York: Interscience.

Pringsheim, E. G. (1946). *Pure Cultures of Algae*. Cambridge: Cambridge University Press.

Provasoli, L. and Carlucci, A. F. (1974). Vitamins and growth regulators. In *Algal Physiology and*

Biochemistry, ed. W. D. P. Stewart, pp. 741–87. Oxford: Blackwell Scientific Publications.

Provasoli, L., McLaughlin, J. J. A. and Droop, M. R. (1957). The development of artificial media for marine algae. *Archiv für Mikrobiologie*, **25**, 392–428.

Prowse, G. A. and Talling, J. F. (1958). The seasonal growth and succession of plankton algae in the White Nile. *Limnology and Oceanography*, **3**, 222–38.

Pugh, G. J. F. (1980). Strategies in fungal ecology. *Transactions of the British mycological Society*, **75**, 1–14.

Purcell, E. M. (1977). Life at low Reynolds number. *American Journal of Physics*, **45**, 3–11.

Pushkar', V. A. (1975). Diurnal distribution of phytoplankton in nursery ponds. *Hydrobiological Journal*, **11**(5), 18–22. (English Translation of original paper in *Gidrobiologeskii Zhurnal*.)

Qiu, B. and Gao, K. (2002). Effects of CO_2 enrichment on the bloom-forming cyanobacterium *Microcystis aeruginosa* (Cyanophyceae): physiological responses and relationships with the availability of dissolved inorganic carbon. *Journal of Phycology*, **38**, 721–9.

Ramamurthy, V. D. (1970). Antibacterial activity of the marine blue-green alga *Trichodesmium erythraeum* in the gastro-intestinal contents of the sea gull, *Larus brunicephalus*. *Marine Biology*, **6**, 74–6.

Ramenskii, L. G. (1938). *Introduction to the Geobotanical Study of Complex Vegetations*. Moscow: Selkozgiz.

Ramlal, P. S., Kling, G. W., Ndawula, L. M., Hecky, R. E. and Kling, H. J. (2001). Diurnal fluctuations in P_{CO2}, DIC, oxygen and nutrients at inshore sites in Lake Victoria, Uganda. In *The Great Lakes of the World (GLOW): Food Web, Health and Integrity*, ed. M. Munawar and R. E. Hecky, pp. 67–82. Leiden: Backhuys.

Ramsbottom, A. E. (1976). *Depth charts of the Cumbrian Lakes*. Scientific Publication No. 33. Ambleside: Freshwater Biological Association.

Rao, N. N. and Torriani, A. (1990). Molecular aspects of phosphate transport in *Escherichia coli*. *Molecular Microbiology*, **4**, 1083–90.

Raspopov, I. M. (1985). *Higher Aquatic Vegetation of Large Lakes in the North-Wetern Area of the USSR*. Leningrad: Nauka. (In Russian.)

Rast, W. and Lee, G. F. (1978). *Summary Analysis of the North American (US Portion) OECD Eutrophication Project: Nutrient Loading Lake Response Relationships and Trophic State Indices*, Report No. EPA-600/3-78-008. Corvallis, OR: US Environment Protection Agency.

Raven, J. A. (1982). The energetics of freshwater algae: energy requirements for biosynthesis and volume regulation. *New Phytologist*, **92**, 1–20.

(1983). The transport and function of silicon in plants. *Biological Reviews*, **58**, 179–207.

(1984). A cost–benefit analysis of photon absorption by photosynthetic cells. *New Phytologist*, **98**, 593–625.

(1991). Implications of inorganic carbon utilisation: ecology, evolution and geochemistry. *Canadian Journal of Botany*, **69**, 908–24.

(1997). Inorganic carbon acquisition by marine autotrophs. *Advances in Botanical Research*, **27**, 85–209.

(1998). Small is beautiful: the picophytoplankton. *Functional Ecology*, **12**, 503–13.

Raven, J. A. and Richardson, K. (1984). Dinophyte flagella: a cost–benefit analysis. *New Phytologist*, **98**, 259–76.

Raven, J. A., Kübler, J. E. and Beardall, J. (2000). Put out the light, then put out the light. *Journal of the Marine Biological Association of the United Kingdom*, **80**, 1–25.

Ravera, O. and Vollenweider, R. A. (1968). *Oscillatoria rubescens* D.C. as an indicator of pollution. *Schweizerische Zeitschrift für Hydrologie*, **30**, 374–80.

Rawson, D. S. (1956). Algal indictors of trophic lake types. *Limnology and Oceanography*, **1**, 18–25.

Ray, J. M., Bhaya, D., Block, M. A. and Grossman, A. R. (1991). Isolation, transcription and inactivation of the gene for an atypical alkaline phosphatase of *Synechococcus* strain PPC 7942. *Journal of Bacteriology*, **173**, 4297–309.

Raymond, P. A. and Cole, J. J. (2003). Increase in the export of alkalinity from North America's largest river. *Science*, **301**, 88–91.

Raymont, J. E. G. (1980). *Plankton Productivity in the Oceans*, vol. 1, *Phytoplankton*, 2nd edn. Oxford: Pergamon Press.

Raymont, J. E. G. (1983). *Plankton productivity in the oceans*, vol. 2, *Zooplankton*, 2nd edn. Oxford: Pergamon Press.

Recknagel, F. (1997). ANNA: artificial neural network model for predicting species abundance and succession of blue-green algae. *Hydrobiologia*, **349**, 47–57.

Recknagel, F., French, M., Harkonenen, P. and Yabunaka, K.-L. (1997). Artificial neural network approach for modelling and prediction of algal blooms. *Ecological Modelling*, **96**, 11–28.

Redfield, A. C. (1934). On the proportions of organic derivatives in sea water and their relation to the composition of plankton. In *The James Johnstone Memorial Volume*, ed. University of Liverpool, pp. 176–92. Liverpool: Liverpool University Press.

(1958). The biological control of chemical factors in the environment. *American Scientist*, **46**, 205–21.

Reguera, B., Bravo, I. and Fraga, S. (1995). Autecology and some life-history stages of *Dinophysis acuta* Ehrenberg. *Journal of Plankton Research*, **17**, 999–1015.

Reguera, B., Blanco, J., Fernández, M. L. and Wyatt, T. (1998). *Harmful Algae*. Vigo: Xunta de Galici and Intergovernmental Oceanographic Commission of UNESCO.

Reynolds, C. S. (1971). The ecology of planktonic blue-green algae in the North Shropshire meres. *Field Studies*, **3**, 409–32.

(1972). Growth, gas vacuolation and buoyancy in a natural population of a planktonic blue-green alga. *Freshwater Biology*, **2**, 87–106.

(1973a). The seasonal periodicity of planktonic diatoms in a shallow, eutrophic lake. *Freshwater Biology*, **3**, 89–110.

(1973b). Growth and buoyancy of *Microcystis aeruginosa* Kütz. emend. Elenkin in a shallow eutrophic lake. *Proceedings of the Royal Society of London B*, **184**, 29–50.

(1973c). The phytoplankton of Crose Mere, Shropshire. *British Phycological Journal*, **8**, 153–62.

(1975). Interrelations of photosynthetic behaviour and buoyancy regulation in a natural population of a blue-green alga. *Freshwater Biology*, **5**, 323–38.

(1976a). Succession and vertical distribution on phytoplankton in response to thermal stratification in a lowland lake, with special reference to nutrient availability. *Journal of Ecology*, **64**, 529–51.

(1976b). Sinking movements of phytoplankton indicated by a simple trapping method. II. Vertical activity ranges in a stratified lake. *British Phycological Journal*, **11**, 293–303.

(1978a). Notes on the phytoplankton periodicity of Rostherne Mere, Cheshire, 1967–1977. *British Phycological Journal*, **13**, 329–35.

(1978b). Stratification in natural populations of bloom-forming blue-green algae. *Verhandlungen der internationale Vereinigung für theoretische und angewandte Limnologie*, **20**, 2285–92.

(1978c). Phosphorus and the eutrophication of lakes: a personal view. In *Phosphorus in the Environment: Its Chemistry and Biochemistry*, CIBA Foundation Symposium No. 57, ed. R. Porter and D. FitzSimons, pp. 201–28. Amsterdam: Excerpta Medica.

(1978d). *The Plankton of the North-West Midland Meres*, Occasional Paper No. 2. Shrewsbury: Cardoc and Severn Valley Field Club.

(1979a). Seston sedimentation: experiments with *Lycopodium* spores in a closed system. *Freshwater Biology*, **9**, 55–76.

(1979b). The limnology of the eutrophic meres of the Shropshire–Cheshire Plain. *Field Studies*, **5**, 93–173.

(1980a). Phytoplankton assemblages and their periodicity in stratifying lake systems. *Holarctic Ecology*, **3**, 141–59.

(1980b). Cattle deaths and blue-green algae. *Journal of the Institution of Water Engineers and Scientists*, **34**, 74–6.

(1983a). A physiological interpretation of the dynamic responses of populations of a planktonic diatom to physical variability of the environment. *New Phytologist*, **95**, 41–53.

(1983b). Growth-rate responses of *Volvox aureus* Ehrenb. (Chlorophyta, Volvocales) to variability in the physical environment. *British Phycological Journal*, **18**, 433–42.

(1984a). *The Ecology of Freshwater Phytoplankton*. Cambridge: Cambridge University Press.

(1984b). Phytoplankton periodicity: the interaction of form, function and environmental variability. *Freshwater Biology*, **14**, 111–42.

(1984c). Artificial induction of surface blooms of Cyanobacteria. *Verhandlungen der internationale Vereinigung für theoretische und angewandte Limnologie*, **22**, 638–43.

(1986a). Diatoms and the geochemical cycling of silicon. In *Biomineralization in the Lower Plants and Animals*, ed. B. S. C. Leadbeater and R. Ridings, pp. 269–89. Oxford: Oxford University Press.

(1986b). Experimental manipulations of the phytoplankton periodicity in large limnetic enclosures in Blelham Tarn, English Lake District. *Hydrobiologia*, **138**, 43–64.

(1987a) Cyanobacterial water blooms. In *Advances in Botanical Research*, vol. 13, ed. J. Callow, pp. 67–143. London: Academic Press.

(1987b). Community organization in the freshwater plankton. In *Organization of Communities, Past and Present*, ed. J. H. R. Gee and P. S. Giller, pp. 297–325. Oxford: Blackwell Scientific Publications.

(1987c). The response of phytoplankton communities to changing lake environments. *Schweizerische Zeitschrift für Hydrologie*, **49**, 220–36.

(1988a). Functional morphology and the adaptive strategies of freshwater phytoplankton. In *Growth and Reproductive Strategies of Freshwater Phytoplankton*, ed. C. D. Sandgren, pp. 338–433. Cambridge: Cambridge University Press.

(1988b). Potamoplankton: paradigms, paradoxes, prognoses. In *Algae and the Aquatic Environment*, ed. F. E. Round, pp. 285–311. Bristol: Biopress.

(1988c). The concept of ecological succession applied to seasonal periodicity of freshwater phytoplankton. *Verhandlungen der internationale Vereinigung für theoretische und angewandte Limnologie*, **23**, 683–91.

(1989a). Relationships among the biological properties, distribution and regulation of production by planktonic Cyanobacteria. *Toxicity Assessment*, **4**, 229–55.

(1989b). Physical determinants of phytoplankton succession. In *Plankton Ecology: Succession in Plankton Communities*, ed. U. Sommer, pp. 9–56. Madison, WI: Brock-Springer.

(1990). Temporal scales of variability in pelagic environments and the responses of phytoplankton. *Freshwater Biology*, **23**, 25–53.

(1991). Lake communities: an approach to their management for conservation. In *The Scientific Management of Temperate Communities for Conservation*, ed. I. F. Spellerberg, F. B. Goldsmith and M. G. Morris, pp. 199–225. Oxford: Blackwell Scientific Publications.

(1992a). Eutrophication and the management of planktonic algae: what Vollenweider couldn't tell us. In *Eutrophication: Research and Application to Water Supply*, ed. D. W. Sutcliffe and J. G. Jones, pp. 4–29. Ambleside: Freshwater Biological Association.

(1992b). Algae. In *The Rivers Handbook*, vol. 1, ed. P. Calow and G. E. Petts, pp. 195–215. Oxford: Blackwell Scientific Publications.

(1992c). Dynamics, selection and composition of phytoplankton in relation to vertical structure in lakes. *Ergebnisse der Limnologie*, **35**, 13–31.

(1993a). Swings and roundabouts: engineering the environment of algal growth. In *Urban Waterside Regeneration, Problems and Prospects*, ed. K. H. White, E. G. Bellinger, A. J. Saul, M. Symes and K. Hendry, pp. 330–49. Chichester: Ellis Horwood.

(1993b). Scales of disturbance and their role in plankton ecology. *Hydrobiologia*, **249**, 157–71.

(1994a). The role of fluid motion in the dynamics of phytoplankton in lakes and rivers. In *Aquatic Ecology: Scale, Pattern and Process*, ed. P. S. Giller, A. G. Hildrew and D. G. Raffaelli, pp. 141–87. Oxford: Blackwell Scientific Publications.

(1994b). The long, the short and the stalled: on the attributes of phytoplankton selected by physical mixing in lakes and rivers. *Hydrobiologia*, **289**, 9–21.

(1994c). The ecological basis for the successful biomanipulation of aquatic communities. *Archiv für Hydrobiologie*, **130**, 1–33.

(1995a). Succssional change in the planktonic vegetation: species, structures, scales. In *Molecular Ecology of Aquatic Microbes*, ed. I. Joint, pp. 115–32. Berlin: Springer-Verlag.

(1995b). River plankton: the paradigm regained. In *The Ecological Basis of River Management*, ed. D. Harper and A. J. D. Ferguson, pp. 161–74. Chichester: John Wiley.

(1996a). Plant life of the pelagic. *Verhandlungen der internationale Vereinigung für theoretische und angewandte Limnologie*, **26**, 97–113.

(1996b). Phosphorus recycling in lakes: evidence from large enclosures for the importance of shallow sediments. *Freshwater Biology*, **35**, 623–45.

(1997a). *Vegetation Processes in the Pelagic: A Model for Ecosystem Theory*. Oldendorf: ECI.

(1997b). Successional development, energetics and diversity in planktonic communities. In *Biodiversity: An Ecological Perspective*, ed. T. Abe, S. R. Levin and M. Higashi, pp. 167–202. New York: Springer.

(1998a). What factors influence the species composition of phytoplankton in lakes of different trophic status? *Hydrobiologia*, **369/370**, 11–26.

(1998b). Linkages between atmospheric weather and the dynamics of limnetic phytoplankton. In *Management of Lakes and Reservoirs during Global Climate Change*, ed. D. G. George, J. G. Jones, P. Punčochář, C. S. Renolds and D. W. Sutcliffe, pp. 15–38. Dordrecht: Kluwer.

(1998c). The control and management of Cyanobacterial blooms. *Special Publications of the Australian Society of Limnology*, **12**, 6–22.

(1999a). With or against the grain: responses of phytoplankton to pelagic variability. In *Aquatic Life-Cycle Strategies*, ed. M. Whitfield, J. Matthews and C. Reynolds, pp. 15–43. Plymouth: Marine Biological Association.

(1999b). Non-determinism to probability, or N : P in the community ecology of phytoplankton. *Archiv für Hydrobiologie*, **146**, 23–35.

(2000a). Phytoplankton designer or how to predict compositional responses to trophic state change. *Hydrobiologia*, **424**, 123–32.

(2000b). Defining sustainability in aquatic systems: a thermodynamic approach. *Verhandlungen der internationale Vereinigung für theoretische und angewandte Limnologie*, **27**, 107–17.

(2001a). Emergence in pelagic communities. *Scientia Marina*, **65**(Suppl. 2), 5–30.

(2001b). Plankton, status and role. In *Encyclopedia of Biodiversiy*, vol. 4, ed. S. R. Levin, pp. 569–99. San Diego, CA: Academic Press.

(2002a). On the interannual variability in phytoplankton production in freshwaters. In *Phytoplankton Productivity*, ed. P. J. leB. Williams, D. N. Thomas and C. S. Reynolds, pp. 187–221. Oxford: Blackwell Science.

(2002b). Ecological pattern and ecosystem theory. *Ecological Modelling*, **158**, 181–200.

(2002c). Resilience in aquatic ecosystems: hysteresis, homeostasis and health. *Aquatic Ecosystem Health and Management*, **5**, 3–17.

(2003a). Pelagic community assembly and the habitat template. *Bocconea*, **16**, 323–39.

(2003b). Planktic community assembly in flowing water and the ecosystem health of rivers. *Ecological Modelling*, **160**, 191–203.

(2003c). The development of perceptions of aquatic eutrophication and its control. *Ecohydrology and Hydrobiology*, **3**, 149–63.

Reynolds, C. S. and Allen, S. E. (1968). Changes in the phytoplankton of Oak Mere following the introduction of base-rich water. *British Phycological Bulletin*, **3**, 451–62.

Reynolds, C. S. and Bellinger, E. D. (1992). Patterns abundance and dominance of of the phytoplankton of Rostherne Mere, England: evidence from an 18-year data set. *Aquatic Sciences*, **54**, 10–36.

Reynolds, C. S. and Butterwick, C. (1979). Algal bioassay of unfertilised and artficially fertilised lake water maintained in Lund Tubes. *Archiv für Hydrobiologie* (Suppl), **56**, 166–83.

Reynolds, C. S. and Davies, P. S. (2001). Sources and bioavailability of phosphorus fractions in freshwaters: a British perspective. *Biological Reviews of the Cambridge Philosophical Society*, **76**, 27–64.

Reynolds, C. S. and Descy, J.-P. (1996). The production, biomass and structure of phytoplankton in large rivers. *Archiv für Hydrobiologie* (Suppl.), **113**, 161–87.

Reynolds, C. S. and Elliott, J. A. (2002). Phytoplankton diversity: discontinuous assembly responses to environmental forcing. *Verhandlungen der internationale Vereinigung für theoretische und angewandte Limnologie*, **28**, 336–44.

Reynolds, C. S. and Glaister, M. R. (1993). Spatial and temporal changes in phytoplankton abundance in the upper and middle reaches of the River Severn. *Archiv für Hydrobiologie* (Suppl.), **101**, 1–22.

Reynolds, C. S. and Irish, A. E. (1997). Modelling phytoplankton dynamics in lakes and reservoirs: the problem of *in-situ* growth rates. *Hydrobiologia* **397**, 5–17.

(2000). *The Phytoplankton of Windermere (English Lake District)*. Ambleside: Freshwater Biological Association.

Reynolds, C. S. and Jaworski, G. H. M. (1978). Enumeration of natural *Microcystis* populations. *British Phycological Journal*, **13**, 269–77.

Reynolds, C. S. and Lund, J. W. G. (1988). The phytoplankton of an enriched, soft-water lake subject to intermittent hydraulic flushing (Grasmere, English Lake District). *Freshwater Biology*, **19**, 379–404.

Reynolds, C. S. and Maberly, S. C. (2002). A simple method for approximating the supportive capacities and metabolic constraints in lakes and reservoirs. *Freshwater Biology*, **47**, 1183–88.

Reynolds, C. S. and Petersen, A. C. (2000). The distribution of planktonic Cyanobacteria in Irish lakes in relation to their trophic states. *Hydrobiologia*, **424**, 91–9.

Reynolds, C. S. and Reynolds, J. B. (1985). The atypical seasonality of phytoplankton in Crose Mere, 1972: an independent test of the hypothesis that variability in the physical environment regulates community dynamics and structure. *British Phycological Journal*, **20**, 227–42.

Reynolds, C. S. and Rodgers, M. W. (1983). Cell- and colony-division in Eudorina (Chlorophyta: Volvocales) and some ecological implications. *British Phycological Journal*, **18**, 111–19.

Reynolds, C. S. and Rogers, D. A. (1976). Seasonal variations in the vertical distribution and buoyancy of *Microcystis aeruginosa* Kütz. emend. Elenkin in Rostherne Mere, England. *Hydrobiologia*, **48**, 17–23.

Reynolds, C. S. and Walsby, A. E. (1975). Water blooms. *Biological Reviews of the Cambridge Philosophical Society*, **50**, 437–81.

Reynolds, C. S. and West, J. R. (1988). Stratification in the Severn Estuary: physical aspects and biological consequences, unpublished report to the Severn Tidal Power Group. Ambleside: Freshwater Biological Association.

Reynolds, C. S. and Wiseman, S. W. (1982). Sinking losses of phytoplankton in closed limnetic systems. *Journal of Plankton Research*, **4**, 489–522.

Reynolds, C. S., Jaworski, G. H. M., Cmiech, H. A. and Leedale, G. F. (1981). On the annual cycle of the blue-green alga *Microcystis aeruginosa* Kütz. emend. Elenkin. *Philosophical Transactions of the Royal Society of London B*, **293**, 419–77.

Reynolds, C. S., Thompson, J. M., Ferguson, A. J. D. and Wiseman, S. W. (1982a). Loss processes in the population dynamics of phytoplankton maintained in closed systems. *Journal of Plankton Research*, **4**, 561–600.

Reynolds, C. S., Morison, H. R. and Butterwick, C. (1982b). The sedimentary flux of phytoplankton in the south basin of Windermere. *Limnology and Oceanography*, **27**, 1162–75.

Reynolds, C. S., Tundisi, J. G. and Hino, K. (1983a). Observations on a limnetic *Lyngbya* population in a stably stratified tropical lake (Lagoa Carioca, eastern Brasil). *Archiv für Hydrobiologie*, **97**, 7–17.

Reynolds, C. S., Wiseman, S. W., Godfrey, B. M. and Butterwick, C. (1983b). Some effects of artificial mixing on the dynamics of phytoplankton populations in large limnetic enclosures. *Journal of Plankton Research*, **5**, 203–34.

Reynolds, C. S., Wiseman, S. W. and Clarke, M. J. O. (1984). Growth- and loss-rate responses to intermittent artificial mixing and their potential application to the control of planktonic algal biomass. *Journal of Applied Ecology*, **21**, 11–39.

Reynolds, C. S., Harris, G. P. and Gouldney, D. N. (1985). Comparison of carbon-specific growth rates and rates of cellular increase in large limnetic enclosures. *Journal of Plankton Research*, **7**, 791–820.

Reynolds, C. S., Montecino, V., Graf, M.-E. and Cabrera, S. (1986). Short-term dynamics of a *Melosira* population in the plankton of an impoundment in central Chile. *Journal of Plankton Research*, **8**, 715–40.

Reynolds, C. S., Oliver, R. L. and Walsby, A. E. (1987). Cyanobacterial dominance: the role of buoyancy regulation in dynamic lake environments. *New Zealand Journal of Marine and Freshwater Research*, **21**, 379–99.

Reynolds, C. S., Carling, P. A. and Beven, K. J. (1991). Flow in river channels: new insights into hydraulic retention. *Archiv für Hydrobiologie*, **121**, 171–9.

Reynolds, C. S., Padisák, J. and Kóbor, I. (1993a). A localized bloom of *Dinobryon sociale* in Lake Balaton: some implications for the perception of patchiness and the maintenance of species richness. *Abstracta Botanica*, **17**, 251–60.

Reynolds, C. S., Padisák, J. and Sommer, U. (1993b). Intermediate disturbance in the ecology of phytoplankton and the maintenance of species diversity: a synthesis. *Hydrobiologia*, **249**, 183–8.

Reynolds, C. S., Desortová, B. and Rosendorf, P. (1998). Modelling the responses of lakes to day-to-day changes in weather. In *Management of Lakes and Reservoirs during Global Climate Change*, ed. D. G. George, J. G. Jones, P. Punčochář, C. S. Reynolds and D. W. Sutcliffe, pp. 289–95. Dordrecht: Kluwer.

Reynolds, C. S., Reynolds, S. N., Munawar, I. F. and Munawar, M. (2000a). The regulation of phytoplankton population dynamics in the world's great lakes. *Aquatic Ecosystem Health and Management*, **3**, 1–21.

Reynolds, C. S., Dokulil, M. and Padisák, J. (2000b). Understanding the assembly of phytoplankton in relation to the trophic spectrum: where are we now? *Hydrobiologia*, **424**, 147–52.

Reynolds, C. S., Irish, A. E. and Elliott, J. A. (2001). The ecological basis for simulating phytoplankton responses to environmental change (PROTECH). *Ecological Modelling*, **140**, 271–91.

Reynolds, C. S., Huszar, V. L., Kruk, C., Naselli-Flores, L. and Melo, S. (2002). Towards a functional classification of the freshwater phytoplankton. *Journal of Plankton Research*, **24**, 417–28.

Rhee, G.-Y. (1973). A continuous culture study of phosphate uptake, growth rate and polyphosphate in *Scenedesmus* sp. *Journal of Phycology*, **9**, 495–506.

(1978). Effects of N : P atomic ratios and nitrate limitation on algal growth, cell composition and nitrate uptake. *Limnology and Oceanography*, **23**, 10–25.

Rhee, G.-Y. and Gotham, I. J. (1980). Optimum N : P ratios and co-existence of planktonic algae. *Journal of Phycology*, **16**, 486–9.

Richardson, T. L., Gibson, C. E. and Heaney, S. I. (2000). Temperature, growth and seasonal succession of phytoplankton in Lake Baikal, Siberia. *Freshwater Biology*, **44**, 431–40.

Richerson, P. J., Neale, P. J., Wurtsbaugh, W., Alfaro Tapia, R. and Vincent, W. F. (1986). Patterns of temporal variation in Lake Titicaca. I. Background, physical and chemical processes. *Hydrobiologia*, **138**, 205–20.

Richey, J. E., Melack, J. M. Aufdemkampe, A. K., Ballester, V. M. and Hess, L. L. (2002). Outgassing from Amazonian rivers and wetlands as a large tropical source of atmospheric CO_2. *Nature*, **416**, 617–19.

Richman, S., Bohon, S. A. and Robbins, S. E. (1980). Grazing interactions among freshwater calanoid copepods. In *Evolution and Ecology of Zooplankton Communities*, ed. W. C. Kerfoot, pp. 219–33. Hanover, NH: University of New England press.

Ricker, W. E. (1958). Stock and recruitment. *Journal of the Fisheries Research Board of Canada*, **1**, 559–623.

Riddolls, A. (1985). Aspects of nitrogen in Lough Neagh. II. Competition between *Aphanizomenon flos-aquae*, *Oscillatoria redekei* and *Oscillatoria agardhii*. *Freshwater Biology*, **15**, 299–306.

Ridley, J. E. (1970). The biology and management of eutrophic reservoirs. *Water Treatment and Examination*, **19**, 374–99.

Riebesell, U. and Wolf-Gladrow, D. A. (2002). Supply and uptake of inorganic nutrients. In *Phytoplankton Productivity*, ed. P. J. leB. Williams, D. N. Thomas and C. S. Reynolds, pp. 109–140. Oxford: Blackwell Science.

Riebesell, U., Wolf-Gladrow, D. A. and Smetacek, V. (1993). Carbon dioxide limitation of marine phytoplankton growth rates. *Nature*, **361**, 249–51.

Riemann, B. and Christoffersen, K. (1993). Microbial trophodynamics in temperate lakes. *Marine Microbial Food Webs*, **7**, 69–100.

Riemann, B., Havskum, H., Thingstad, F. and Bernard, C. (1995). The role of mixotrophy in pelagic environments. In *Molecular Ecology of Aquatic Microbes*, ed. I. Joint, pp. 87–105. Berlin: Springer-Verlag.

Riemann, L. and Winding, A. (2001). Community dynamics of free-living and particle-associated bacterial assemblages during a freshwater phytoplankton bloom. *Microbial Ecology*, **42**, 274–85.

Riley, G. A. (1944). Carbon metabolism and photosynthetic efficiency. *American Scientist*, **32**, 132–4.

(1957). Phytoplankton of the north central Sargasso Sea. *Limnology and Oceanography*, **2**, 252–70.

Rinehart, K. L., Namikoshi, M. and Choi, B. (1994). Structure and biosynthesis of toxins from blue-green algae (Cyanobacteria). *Journal of Applied Phycology*, **6**, 159–76.

Ripl, W. and Wolter, K.-D. (2003). Ecosystem function and degradation. In *Phytoplankton Productivity*, ed.

P. J. leB. Williams, D. N. Thomas and C. S. Reynolds, pp. 291–317. Oxford: Blackwell Science.

Rippka, R., Neilson, A., Kunisawa, R. and Cohen-Bazire, C. (1971). Nitrogen fixation by unicellular blue-green algae. *Archiv für Mikrobiologie*, **76**, 341–48.

Rippka, R., Deruelles, J., Waterbury, J. B., Herdman, M. and Stanier, R. Y. (1979). Generic assignments, strain histories and properties of pure cultures of Cyanobacteria. *Journal of General Microbiology*, **111**, 1–61.

Robarts, R. D. and Howard-Williams, C. (1989). Diel changes in fluorescence capacity, photosynthesis and macromolecular synthesis by *Anabaena* in response to natural variations in solar irradiance. *Ergebnisse der Limnologie*, **32**, 35–48.

Robarts, R. D. and Zohary, T. (1987). Temperature effects on photosynthetic capacity, respiration and growth rates of bloom-forming Cyanobacteria. *New Zealand Journal of Marine and Freshwater Research*, **21**, 391–9.

Robinson, G. A. (1961). Continuous plankton records: contribution towards a plankton atlas of the north-eastern Atlantic and the North Sea. I. Phytoplankton. *Bulletins of Marine Ecology*, **5**, 81–9.

(1965). Continuous plankton records: contribution towards a plankton atlas of the north-eastern Atlantic and the North Sea. IX. Seasonal cycles of phytoplankton. *Bulletins of Marine Ecology*, **6**, 104–22.

Rodhe, W. (1948). Environmental requirements of freshwater plankton algae: experimental studies in the ecology of phytoplankton. *Symbolae Botanicae Upsaliensis*, **10**, 5–149.

Rodhe, W., Vollenweider, R. A. and Nauwerck, A. (1958). The primary production and standing crop of phytoplankton. In *Perspectives in Marine Biology*, ed. A. A. Buzzati-Traverso, pp. 299–322. Berkeley, CA: University of California Press.

Roelofs, T. D. and Oglesby, R. T. (1970). Ecological observations on the planktonic cyanophyte *Gloeotrichia echinulata*. *Limnology and Oceanography*, **15**, 244–9.

Rojo, C. and Alvares Cobelas, M. (2000). A plea for more ecology in phytoplankton ecology. *Hydrobiologia*, **424**, 141–6.

Rojo, C., Ortega-Mayagoitia, E. and Alvares Cobelas, M. (2000). Lack of pattern among phytoplankton assemblages or what does the exception to the rule mean? *Hydrobiologia*, **424**, 133–40.

Romanovsky, Yu. I. (1985). Food limitation and life-history strategies in cladoceran crustaceans. *Ergebnisse der Limnologie*, **21**, 363–72.

Romero, O. E., Lange, C. B. and Wefer, G. (2002). Interannual variability (1988–1991) of siliceous phytoplankton fluxes off northwest Africa. *Journal of Plankton Research*, **24**, 1035–46.

Romme, W. H., Everham, E. H., Frelich, L. E., Moritz, M. A. and Sparks, R. E. (1998). Are large, infrequent disturbances quantitatively different from small, frequent disturbances? *Ecosystems*, **1**, 524–34.

Rossolimo, L. L. (1957). Temparatuniy rezhim Ozero Baykal. *Trudy baykalskoi limnologeskoi Stantsii*, **16**, 1–551.

Rother, J. A. and Fay, P. (1977). Sporulation and the development of planktonic blue-green algae in three Salopian meres. *Proceedings of the Royal Society of London B*, **196**, 317–32.

(1979). Blue-green algal growth and sporulation in response to simulated surface-bloom conditions. *British Phycological Journal*, **14**, 59–68.

Rothschild, B. J. and Osborn, T. R. (1988). Small-scale turbulence and plankton contact rates. *Journal of Plankton Research*, **10**, 465–74.

Round, F. E. (1965). *The Biology of the Algae*. London: Edward Arnold.

Round, F. E., Crawford, R. M. and Mann, D. G. (1990). *The Diatoms: Biology and Morphology of the Genera*. Cambridge: Cambridge University Press.

Rüdiger, W. (1994). Phycobiliproteins and phycobilins. In *Progress in Phycological Research*, vol. 10, ed. F. E. Round and D. J. Chapman, pp. 97–135. Bristol: Biopress.

Rueter, J. G. and Petersen, R. R. (1987). Micronutrient effects on cyanobacterial growth and physiology. *New Zealand Journal of Marine and Freshwater Research*, **21**, 435–45.

Ruttner, F. (1938). Limnologische Studien an einigen Seen der Östalpen. *Archiv für Hydrobiologie*, **32**, 167–319.

(1953). *Fundamentals of Limnology*. (Translated by D. G. Frey and F. E. D. Fry.) Toronto: University of Toronto Press.

Ruyter van Steveninck, E. D. de, Admiraal, W., Breebart, L., Tubbing, G. M. J. and Zanten, B. van (1992). Plankton in the River Rhine: structural and functional changes observe during downstream transport. *Journal of Plankton Research*, **14**, 1351–68.

Ryding, S.-O. and Rast, W. (1989). *The Control of Eutrophication in Lakes and Reservoirs*. Paris: UNESCO.

Ryther, J. H. (1956). Photosynthesis in the ocean as a function of light intensity. *Limnology and Oceanography*, **1**, 61–70.

Safferman, R. S. and Morris, M. E. (1963). Algal virus: isolation. *Science*, **140**, 679–80.

Sakamoto, M. (1966). Primary production by phytoplankton community in some Japanese lakes and its dependence upon lake depth. *Archiv für Hydrobiologie*, **62**, 1–28.

Salmaso, N. (2000). Factors affecting the seasonality and distribution of Cyanobacteria and chlorophytes: a case study from the large lakes south of the Alps, with special reference to Lake Garda. *Hydrobiologia*, **438**, 43–63.

(2003). Life strategies, dominance patterns and mechanisms promoting species coexistence in phytoplankton communities along complex environmental gradients. *Hydrobiologia*, **502**, 13–36.

Salmaso, N., Morabito, G., Mosello, R. *et al.* (2003). A synoptic study of phytoplankton in the deep lakes south of the Alps (Lakes Garda, Iseo, Como, Lugano and Maggiore). *Journal of Limnology*, **62**, 207–27.

Salomonsen, J. (1992). Examination of properties of exergy, power and ascendency along a eutrophication gradient. *Ecological Modelling*, **62**, 171–81.

Samuelsson, K., Berglund, J., Haecky, P. and Andersson, A. (2002). Structural changes in an aquatic microbial food web caused by inorganic nutrient addition. *Aquatic Microbial Ecology*, **29**, 29–38.

Sand-Jensen, K. (1977). Effect of epiphytes on eelgrass photosynthesis. *Aquatic Botany*, **3**, 55–63.

Sanders, R. W., Leeper, D. A., King C. H. and Porter, K. G. (1994). Grazing by rotifers and crustacean zooplankton on nanoplanktonic protists. *Hydrobiologia*, **288**, 167–81.

Sandgren, C. D. (1988a). General introduction. In *Growth and Reproductive Strategies of Freshwater Phytoplankton*, ed. C. D. Sandgren, pp. 1–8. Cambridge: Cambridge University Press.

(1988b). The ecology of chrysophyte flagellates: their growth and perennation as freshwater phytoplankton. In *Growth and Reproductive Strategies of Freshwater Phytoplankton*, ed. C. D. Sandgren, pp. 9–104. Cambridge: Cambridge University Press.

Sargent, J. R. (1976). The structure, metabolism and function of lipids in marine organisms. In *Biochemical and Biophysical Perspectives in Marine Biology*, ed. D. C. Malins and J. R. Sargent, pp. 150–212. London: Academic Press.

Sarmiento, J. L. and Wofsy, S. C. (1999). *A US Global Carbon Cycle Plan*. Washington, DC: US Global Change Research Program.

Sarno, D., Zingone, A., Saggiomo, V. and Carrada, G. C. (1993). Phytoplankton biomass and species composition in a Mediterranean coastal lagoon. *Hydrobiologia*, **271**, 27–40.

Sarvala, J., Tarvainen, M., Salonen, K. and Mölsä, H. (2002). Pelagic food web as the basis of fisheries in Lake Tanganyika: a bioenergetic modeling analysis. *Aquatic Ecosystem Health and Management*, **5**, 283–92.

Sarvala, J., Rask, M. and Karjalainen, J. (2003). Fish community ecology. In *The Lakes Handbook*, vol. 1, ed. P. E.O 'Sullivan and C. S. Reynolds, pp. 538–82. Oxford: Blackwell Science.

Sas, H. (1989). *Lake Restoration by Reduction of Nutrient Loading: Expectations, Experiences, Extrapolations*. St' Augustin: Academia Verlag Richarz.

Satake, K. and Saijo, Y. (1974) Carbon content and metabolic activity of microorganisms in some acid lakes in Japan. *Limnology and Oceanography*, **19**, 331–8.

Satapoomin, S. (1999). Carbon content of some common tropical Andaman Sea copepods. *Journal of Plankton Research*, **21**, 2117–23.

Saubert, S. (1957). Amino-acid utilisation by *Nitzschia thermalis* and *Scenedesmus bijugatus*. *South African Journal of Science*, **53**, 335–9.

Saunders, P. A., Porter, K. G. and Taylor, B. E. (1999). Population dynamics of *Daphnia* spp. and implications for trophic interactions in a small, monomictic lake. *Journal of Plankton Research*, **21**, 1823–45.

Savvaitova, K. and Petr, T. (1992). Lake Issyk-kul, Kirghizia. *International Journal of Salt Lake Research*, **1**, 21–46.

Sawyer, C. N. (1966). Basic concepts of eutrophication. *Journal of the Water Control Federation*, **38**, 737–44.

Saxby, K. J. (1990). The physiological ecology of freshwater chrysophytes with special reference to *Synura petersenii*. Ph.D. thesis, University of Birmingham.

Saxby-Rouen, K. J., Leadbeater, B. S. C. and Reynolds, C. S. (1996). Ecophysiological studies on *Synura petersenii* (Synurophyceae). *Beiheft, Nova Hedwigia*, **114**, 111–23.

(1997). The growth response of *Synura petersenii* (Synurophyceae) to photon-flux density, temperature and pH. *Phycologia*, **36**, 233–43.

(1998). The relationship between the growth of *Synura petersenii* (Synurophyceae) and components

of the dissolved inorganic carbon system. *Phycologia*, **37**, 467–77.

Scanlan, D. J. and Wilson, W. H. (1999). Application of molecular techniques to addressing the role of P as a key effector in marine ecosystems. *Hydrobiologia*, **401**, 149–75.

Scanlan, D. J., Mann, N. H. and Carr, N. G. (1993). The response of the picoplanktonic marine cyanobacterium *Synechococcus* species WH7803 to phosphate starvation involves a protein homologous to the periplasmic phosphate-binding protein of *Escherichia coli*. *Molecular Microbiology*, **10**, 181–91.

Scavia, D. and Fahnenstiel, G. L. (1988). From picoplankton to fish: complex interactions in the Great Lakes. In *Complex Interactions in Lake Communities*, ed. S. R. Carpenter, pp. 85–97. New York: Springer-Verlag.

Scheffer, M. (1989). Alternative stable states in eutrophic shallow freshwater systems: a minimal model. *Hydrobiological Bulletin*, **23**, 73–83.

(1990). Multiplicity of stable states in freshwater system. *Hydrobiologia*, **200/201**, 475–86.

(1998). *Ecology of Shallow Lakes*. London: Chapman and Hall.

Scheffer, M., Hosper, S. H., Meijer, M.-L., Moss, B. and Jeppesen, E. (1993). Alternative equilibria in shallow lakes. *Trends in Ecology and Evolution*, **8**, 275–9.

Scheffer, M., Rinaldi, S., Gragnani, A., Mur, L. R. and van Nes, E. H. (1997). On the dominance of filamentous Cyanobacteria in shallow, turbid lakes. *Ecology*, **78**, 272–82.

Scheffer, M., Rinaldi, S. and Kuznetsov, Yu. A. (2000). Effects of fish on plankton dynamics: a theoretical analysis. *Canadian Journal of Fisheries and Aquatic Sciences* **57**, 1208–19.

Schindler, D. W. (1971). Carbon, nitrogen and the eutrophication of freshwater lakes. *Journal of Phycology*, **7**, 321–9.

(1977). Evolution of phosphorus limitation in lakes. *Science*, **196**, 260–2.

Schindler, D. W. and Holmgren, S. K. (1971). Primary production and phytoplankton in the experimental lakes area, north-westerrn Ontario, and other low-carbonate waters and a liquid scintillation method for determining ^{14}C activity in photosynthesis. *Journal of the Fisheries Research Board of Canada*, **28**, 189–201.

Schindler, D. E., Carpenter, S. R., Cottingham, K. I. *et al.* (1996). Food web structure and littoral zone coupling to trophic cascades. In *Food Webs: Integration of Patterns and Dynamics*, ed. G. A. Polis and K. O. Winemiller, pp. 96–105. New York: Chapman and Hall.

Scott, J. T., Myer, G. E., Stewart, R. and Walther, E. G. (1969). On the mechanism of Langmuir circulations and their role in epilimnion mixing. *Limnology and Oceanography*, **14**, 493–503.

Sell, A. and Overbeck, J. (1992). Exudates: phytoplankto-bacterioplankton interactions in Plusssee. *Journal of Plankton Research*, **14**, 1199–215.

Serruya, C. (1978). *Lake Kinneret*, Monographie Biologiae Series No. 32. The Hague: W. Junk.

Serruya, C. and Pollingher, U. (1983). *Lakes of the Warm Belt*. Cambridge: Cambridge University Press.

Shaboonin, G. D. (1982). Long-term multi-year characteristics of the regime of the temperature in Issyk-kul. *Trudy Akademia Nauk Kirghiziaskaya SSR*, **3**, 39–46.

Shannon, C. E. (1948). A mathematical theory of communication. *Bell Systems Technical Journal*, **27**, 623–56.

Shannon, C. E. and Weaver, W. (1949). *The Mathematical Theory of Communication*. Urbana, IL: University of Illinois Press.

Shapiro, J. (1973). Blue-green algae: Why they become dominant. *Science*, **179**, 382–4.

(1990). Current beliefs regarding dominance by blue-greens: the case for the importance of CO_2. *Verhandlungen der internationalen Vereinigung für theoretische und angewandte Limnologie*, **24**, 38–54.

Shapiro, J. and Wright, D. J. (1984). Lake restoration by biomanipulation: Round Lake, Minnesota, the first two years. *Freshwater Biology*, **14**, 371–83.

Shapiro, J., Lamarra, V. and Lynch, M. (1975). Biomanipulation: an ecosystem approach to lake restoration. In *Proceedings of a Symposium on Water Quality Management through Biological Control*, ed. P. L. Brezonik and J. L. Fox, pp. 85–89. Gainesville, FL: University of Florida Press.

Sharp, J. H. (1977). Excretion of organic matter by marine phytoplankton: do healthy cells do it? *Limnology and Oceanography*, **22**, 381–99.

Sherr, E. B. and Sherr, B. F. (1988). Role of microbesin pelagic food webs: a revised concept. *Limnology and Oceanography*, **33**, 1225–7.

Shiba, T., Simidu, U. and Taga, N. (1979). Distribution of aerobic bacteria which contain bacterichlorophyll *a*. *Applied and Environmental Microbiology*, **38**, 43–8.

Shilo, M. (1970). Lysis of blue-green algae by *Myxobacter*. *Journal of Bacteriology*, **104**, 453–61.

Shimarev, M. N., Granin, N. G. and Zhdanov, A. A. (1993). Deep ventilation of Lake Baikal water during spring thermal bars. *Limnology and Oceanography*, **38**, 1068–72.

Sicko-Goad, L. M., Schelske, C. L. and Stoermer, E. F. (1984). Estimation of intracellular carbon and silica content of diatoms from natural assemblages, using morphometric techniques. *Limnology and Oceanography*, **29**, 1170–78.

Sieburth, J.McN., Smetacek, V. and Lenz, J. (1978). Pelagic ecosystem structure: heterotrophic compartments of the plankton and their relationship to plankton size fractions. *Limnology and Oceanography*, **23**, 1256–63.

Siever, R. (1962). Silica solubility, 0–200 °C, and the diagenesis of siliceous sediments. *Journal of Geology*, **70**, 127.

Sigg, L. and Xue, H. B. (1994). Metal speciation: concepts, analysis and effects. In *Chemistry of Aquatic Systems: Local and Global Perspectives*, ed. G. Bidoglio and W. Stumm, pp. 153–81. Dordrecht: Kluwer.

Šimek, K., Kojecká, P., Nedoma, J. *et al.* (1999). Shifts in bacterial community composition associated with different microzooplankton size fractions in a eutrophic reservoir. *Limnology and Oceanography*, **44**, 1634–44.

Simmons, J. (1998). Algal control and destratification at Hanningfield Reservoir. *Water Science Technology*, **37/2**, 309–16.

Simo, R. (2001). Production of atmospheric sulphur by oceanic plankton: biogeochemical, ecological and evolutionary links. *Trends in Ecology and Evolution*, **16**, 287–94.

Simon, M. I. (1995). Signal transduction in microorganisms. In *Molecular Ecology of Aquatic Microbes*, ed. I. Joint, pp. 205–15. Berlin: Springer-Verlag.

Simpson, F. B. and Neilands, J. B. (1976). Sidero-chromes in Cyanophyceae: isolation and characterisation of schizokinen from *Anabaena* sp. *Journal of Phycology*, **12**, 44–8.

Simpson, T. L. and Volcani, B. E. (1981). Introduction. In *Silicon and Siliceous Structures in Biological Systems*, ed. T. L. Simpson and B. E. Volcani, pp. 3–12. New York: Springer-Verlag.

Sinker, C. A. (1962). The North Shropsire meres and mosses: a background for ecologists. *Field Studies*, **1**(4), 101–38.

Sirenko, L. A. (1972). *Physiological Basis of Multiplication of Blue-Green Algae in Reservoirs*. Kiev: Naukova Dumka. (In Russian.)

Sirenko, L. A., Stetsenko, N. M., Arendarchuk, V. V. and Kuz'menko, M. I. (1968). Role of oxygen conditions in the vital activity of certain blue-green algae. *Microbiology*, **37**, 199–202. (English translation of original paper in *Microbiologiya*.)

Sirenko, L. A., Chernousova, V. M., Arendarchuk, V. V. and Kozitskaya, V. N. (1969). Factors of mass development of blue-green algae. *Hydrobiological Journal*, **5**(3), 1–8. (English translation of original paper in *Gidrobiologeskii Zhurnal*.)

Skellam, J. G. (1951). Random dispersal in theoretical populations. *Biometrika*, **78**, 196–218.

Skulberg, O. (1964). Algal problems related to eutrophication of Eurpoean water supplies and a bioassay method to assess fertilsing influences of pollution in inland waters. In *Algae and Man*, ed. D. F. Jackson, pp. 269–99. New York: Plenum Press.

Smayda, T. J. (1970). The suspension and sinking of phytoplankton in the sea. *Annual Review of Oceanography and Marine Biology*, **8**, 353–414.

(1973). The growth of *Skeletonema costatum* during a winter–spring bloom in Narragansett Bay, Rhode Island. *Norwegian Journal of Botany*, **20**, 219–47.

(1974). Some experiments on the sinking characteristics of two freshwater diatoms. *Limnology and Oceanography*, **19**, 628–35.

(1980). Phytoplankton species succession. In *The Physiological Ecology of Phytoplankton*, ed. I. Morris, pp. 493–570. Oxford: Blackwell Scientific Publications.

(2000). Ecological features of harmful algal blooms in the sea. *Limnology and Oceamography*, **42**, 1137–53.

(2002). Turbulence, watermass stratification and harmful algal blooms: an alternative view and frontal zones as 'pelagic seed banks'. *Harmful Algae*, **1**, 95–112.

Smayda, T. J. and Boleyn, B. J. (1965). Experimental observations on the flotation of marine diatoms. I. *Thalassiosira* cf. *nana*, *Thalassiosira rotula* and *Nitzschia seriata*. *Limnology and Oceanography*, **10**, 499–509.

(1966a) Experimental observations on the flotation of marine diatoms. II. *Skeltonema costatum* and *Rhizoslenia setigera*. *Limnology and Oceanography*, **11**, 18–34.

(1966b). Experimental observations on the flotation of marine diatoms. III. *Bacteriastrum hyalinum* and *Chatoceros lauderi*. *Limnology and Oceanography*, **11**, 35–43.

Smayda, T. J. and Reynolds, C. S. (2001). Community assembly in marine phytoplankton: application of recent models to harmful dinoflagellate blooms. *Journal of Plankton Research*, **23**, 447–62.

Smetacek, V. (2002). The ocean's veil. *Nature*, **419**, 565.

Smetacek, V. and Passow, U. (1990). Spring-bloom initiation and Sverdrup's critical-depth model. *Limnology and Oceanography*, **35**, 228–34.

Smetacek, V., de Baar, H. J., Bathmann, U. V., Lochte, K. and Rutgers van der Loeff, M. M. (1997). Ecology and biochemistry of the Antarctic Circumpolar Current during austral spring: a summary of Southern Ocean JGOFS cruise ANT X/6 of R. V. Polarstern. *Deep-Sea Research Part II, Topical Studies in Oceanography*, **44**, 1–22.

Smetacek, V., Montresa, M. and Verity, P. (2002). Marine productivity: footprints of the past and steps into the future. In *Phytoplankton Productivity*, ed. P. J. leB. Williams, D. N. Thomas and C. S. Reynolds, pp. 350–69. Oxford: Blackwell Science.

Smith, E. L. (1936). Photosynthesis in relation to light and carbon dioxide. *Proceedings of the National Academy of Sciences of the USA*, **22**, 504–11.

Smith, I. R. (1975). *Turbulence in Lakes and Rivers*, Scientific Publication No. 29. Ambleside: Freshwater Biological Association.

(1982). A simple theory of algal deposition. *Freshwater Biology*, **12**, 445–9.

Smith, V. H. (1983). Low nitrogen to phosphorus ratios favor dominance by blue-green algae in lake phytoplankton. *Science*, **221**, 669–71.

Smith, V. H. and Bennett, S. J. (1999). Nitrogen : phosphorus supply ratios and phytoplankton community structure in lakes. *Archiv für Hydrobiologie*, **146**, 37–53.

Smol, J. P. (1992). Palaeolimnolgy: an important tool for effective ecosystem management. *Journal of Aquatic Ecosystem Health*, **1**, 49–58.

Smol, J. P., Birks, H. J. B. and Last, W. M. (2001). *Tracking Environmental Change Using Lake Sediments*, vol. 3, *Terrestrial, Algal and Siliceous Indicators*. Dordrecht: Kluwer.

Smyly, W. J. P. (1976). Some effects of enclosure on the zooplankton in a small lake. *Freshwater Biology*, **6**, 241–51.

Sommer, U. (1981). The role of *r*- and *K*-selection in the succession of phytoplankton in Lake Constance. *Acta Oecologia*, **2**, 327–42.

(1984). The paradox of the plankton: fluctuations of phosphorus availability maintain the diversity of phytoplankton in flow-through cultures. *Limnology and Oceanography*, **29**, 633–6.

(1986). The periodicity of phytoplankton in Lake Constance (Bodensee) in comparison to other deep lakes of central Europe. *Hydrobiologia*, **138**, 1–7.

(1988a). Growth and survival strategies of planktonic diatoms. In *Growth and Reproductive Strategies of Freshwater Phytoplankton*, ed. C. D. Sandgren, pp. 227–60. Cambridge: Cambridge University Press.

(1988b). Some size relationships of phytoflagellate motility. *Hydrobiologia*, **161**, 125–31.

(1988c). Does nutrient competition among phytoplankton occur *in situ*? *Verhandlungen der internationalen Vereinigung für theoretische und angewandte Limnologie*, **23**, 707–12.

(1989). The role of competition for resources in phytoplankton succession. In *Plankton Ecology: Succession in Plankton Communities*, ed. U. Sommer, pp. 57–106. Madison, WI: Brock-Springer.

(1993). Disturbance–diversity relationships in two lakes of similar nutrient chemistry but contrasted disturbance regimes. *Hydrobiologia*, **249**, 59–65.

(1994). *Planktologie*. Berlin: Springer-Verlag.

(1995). An experimental test of the intermediate disturbance hypothesis using cultures of marine phytoplankton. *Limnology and Oceanography*, **40**, 1271–7.

(1996). Plankton ecology: the past two decades of progress. *Naturwissenschaften*, **83**, 293–301.

Sommer, U. and Gliwicz, Z. M. (1986). Long-range vertical migration of *Volvox* in tropical lake, Cabora Bassa (Mozambique). *Limnology and Oceanography*, **31**, 650–3.

Sommer, U. and Stibor, H. (2002). Copepoda–Cladocera–Tunicata: the role of the three major mesozooplankton groups in pelagic food webs. *Ecological Research*, **17**, 161–74.

Sommer, U., Gliwicz, Z. M., Lampert, W. and Duncan, A. (1986). The PEG-model of seasonal succession of planktonic events in lakes. *Archiv für Hydrobiologie*, **106**, 433–71.

Sommer, U., Padisák, J., Reynolds, C. S. and Juhász-Nagy, P. (1993). Hutchinson's heritage: the diversity–disturbance relationship in phytoplankton. *Hydrobiologia*, **249**, 1–7.

Sommer, U., Sommer, F., Santer, B. *et al.* (2001). Complementary impacts of copepods and cladocerans on phytoplankton. *Ecology Letters*, **4**, 545–50.

Søndergaard, M. (2002). A biography of Einar Steeman Nielsen: the man and his science. In *Phytoplankton Productivity*, ed. P. J. leB. Williams, D. N. Thomas and C. S. Reynolds, pp. 1–15. Oxford: Blackwell Science.

Søndergaard, M. and Moss, B. (1998). Impact of submerged macrophytes on phytoplankton in shallow lake. In *The Structuring Role of Submerged Macrophytes in Lakes*, ed. E. Jeppesen, M. Søndergaard and K. Christoffersen, pp. 115–32. New York: Springer-Verlag.

Søndergaard, M., Kristensen, P. and Jeppesen, E. (1993). Eight years of internal phosphorus loading and changes in the sediment phosphorus profile of Lake Søbygaard. *Hydrobiologia*, **253**, 345–56.

Søndergaard, M., Borch, N. H. and Riemann, B. (2000). Dynamics of biodegradable DOC produced by freshwater plankton communities. *Aquatic Microbial Ecology*, **23**, 73–83.

Sorokin, Yu. I. (1999). *Aquatic Microbial Ecology: A Textbook for Students in Environmental Sciences*. Leiden: Backhuys.

(2002). Dynamics of inorganic phosphorus in pelagic communities of Sea of Okhotsk. *Journal of Plankton Research*, **24**, 1253–63.

Sorokin, Yu. I. and Sorokin, P. Yu. (2002). Microplankton and primary production in the Sea of Okhotsk in summer 1994. *Journal of Plankton Research*, **24**, 453–70.

Soto, D., Campos, H., Steffen, W., Parra, O. and Zuñiga, L. (1994). The Torres del Paine lake district (Chilean Patagonia): a case of potentially N-limited lakes and ponds. *Archiv für Hydrobiologie* (Suppl.) **99**, 181–97.

Sournia, A. (1978). *Phytoplankton Manual*. Paris: UNESCO.

Sournia, A., Chrétiennot-Dinet, M.-J. and Ricard, M. (1991). Marine plankton: how many species in the world oceans? *Journal of Plankton Research*, **13**, 1093–9.

Sousa, W. P. (1984). The role of disturbance in natural communities. *Annual Reviews of Ecology and Systematics*, **15**, 353–91.

Southwood, T. R. E. (1977). Habitat, the templet for ecological strategies. *Journal of Animal Ecology*, **46**, 337–65.

(1996). Natural communities: structure and dynamics. *Philosophical Transactions of the Royal Society of London B*, **351**, 1113–29.

Sparrow, F. K. (1960). *Aquatic Phycomycetes*, 2nd edn. Ann Arbor, MI: University of Michigan Press.

Spencer, C. N. and King, D. L. (1985). Role of light, carbon dioxide and nitrogen in the regulation of buoyancy, growth and bloom formation of *Anabaena flos-aquae*. *Journal of Plankton Research*, **11**, 283–96.

Spencer, M. and Warren, P. H. (1996). The effects of habitat size and productivity on food web structure in small aquatic microcosms. *Oikos*, **75**, 419–30.

Spencer, R. (1955). A marine bacteriophage. *Nature*, **175**, 690.

Spigel, R. H. and Imberger, J. (1987). Mixing processes relevant to phytoplankton dynamics in lakes. *New Zealand Journal of Marine and Freshwater Research*, **21**, 367–77.

Spijkerman, E. and Coesel, P. F. M. (1996a). Phosphorus uptake and growth kinetics of two planktonic desmid species. *European Journal of Phycology*, **31**, 53–60.

(1996b). Competition for phosphorus among planktonic desmid species in continuous-flow culture. *Journal of Phycology*, **32**, 939–48.

Spiller, S. C., Castelfranco, A. M. and Castelfranco, P. A. (1982). Effects of iron and oxygen on chlorophyll biosynthesis. I. *In vivo* observations on iron- and oxygen-deficient plants. *Plant Physiology*, **69**, 107–11.

Stadlemann, P., Moore, J. E. and Pickett, E. (1974). Primary production in relation to temperature structure, biomass concentration and light conditions at an inshore and offshore station in Lake Ontario. *Journal of the Fisheries Research Board of Canada*, **31**, 1215–32.

Starkwether, P. L., Gilbert, J. J. and Frost, T. M. (1979). Bacteria feeding by *Brachionus calyciflorus*: clearance and ingestion rates, behaviour and population dynamics. *Oecologia*, **44**, 26–30.

Steeg, P. F. T., Hanson, P. J. and Paerl, H. W. (1986). Growth-limiting quantities and accumulation of molybdenum in *Anabaena oscillaroides* (Cyanobacteria). *Hydrobiologia*, **140**, 143–7.

Steel, J. A. (1972). The application of fundamental limnological research in water supply and management. *Symposia of the Zoological Society of London*, **29**, 41–67.

(1973). Reservoir algal productivity. In *The Use of Mathematical Models in Water Pollution Control*, ed. A. James, pp. 107–35. Newcastle upon Tyne: University of Newcastle upon Tyne.

(1975). The management of Thames Valley Reservoirs. In *The Effects of Storage on Water Quality*, ed. R. E. Youngman, pp. 371–419. Medmenham: Water Research Centre.

Steel, J. A. and Duncan, A. (1999). Modelling the ecological aspects of bankside reservoirs and implications for management. *Hydrobiologia*, **395/396**, 133–47.

Steele, J. H. (1976). Patchiness. In *The Ecology of the Seas*, ed. D. H. Cushing and J. J. Walsh, pp. 98–115. Philadelphia, PA: W. B. Saunders.

Steemann Nielsen, E. (1952). The use of radioactive carbon (C^{14}) for measuring organic production in the sea. *Journal du Conseil pour l'exploration de la Mer*, **18(2)**, 117–40.

(1955). The interaction of photosynthesis and respiration and its importance for the determination of 14-C discrimination in photosynthesis. *Physiologia Plantarum*, **8**, 945–53.

Steemann Nielsen, E. and Jensen, Å.E. (1957). Primary oceanic production: the autotrophic production of organic matter in the oceans. *Galathea Reports*, **1**, 49–136.

Stein, J. R. (1973) *Handbook of Phycological Methods: Cultural Methods and Growth Measurements*. Cambridge: Cambridge University Press.

Stephenson, G. L., Hamilton, P., Kaushik, N. K., Robinson, J. B. and Solomon, K. R. (1984). Spatial distribution of phytoplankton in enclosures of three sizes. *Canadian Journal of Fisheries and Aquatic Sciences*, **41**, 1048–54.

(1989). The role of grazers in phytoplankton succession. In *Plankton Ecology*, ed. U. Sommer, pp. 107–70. Madison, WI: Brock-Springer.

(1993). *Daphnia* growth on varying quality of *Scenedesmus*: mineral limitation of zooplankton. *Ecology*, **74**, 2351–60.

Sterner, R. W. and Elser, J. J. (2002). *Ecological Stoichiometry: The Biology of Elements from Molecules to the Biosphere*. Princeton, NJ: Princeton University Press.

Sterner, R. W., Elser, J. J., Fee, E. J., Guildford, S. J. and Chrzanowski, T. H. (1997). The light : nutrient ratio in lakes: the balance of energy and materials affects ecosystem structure and process. *American Naturalist*, **150**, 663–84.

Stewart, K. M. (1976). Oxygen deficits, clarity and eutrophication in some Madison lakes. *Internationale Revue des gesamten Hydrobiologie*, **61**, 563–79.

Stewart, W. D. P. and Alexander, G. (1971). Phosphorus availability and nitrogenase activity in aquatic blue-green algae. *Freshwater Biology*, **1**, 389–404.

Stewart, W. D. P. and Lex, M. (1970). Nitrogenase activity in the blue-green alga *Plectonema boryanum* Strain 594. *Archiv für Mikrobiologie*, **73**, 250–60.

Stewart, W. D. P., Fitzgerald, G. P. and Burris, R. H. (1967). *In-situ* studies on N_2 fixation using the acetylene-reduction technique. *Proceedings of the*

National Academy of Sciences of the USA, **58**, 2071–8.

Stibor, H. (1992). Predator-induced life-history shifts in a freshwater cladoceran. *Oecologia*, 92, 162–5.

Stock, J. B., Ninfa, A. J. and Stock, A. J. (1989). Protein phosphorylation and regulation of adaptive responses in bacteria. *Microbiological Reviews*, **53**, 450–90.

Stocker, R. and Imberger, J. (2003). Horizontal transport and dispersion in the surface layer of a medium-sized lake. *Limnology and Oceanography*, **48**, 971–82.

Stockner, J. G. and Antia, N. J. (1986). Algal picoplankton from marine and freshwater ecosystems: a multidisciplinary perspective. *Canadian Journal of Fisheries and Aquatic Sciences*, **43**, 2472–503.

Stoermer, E. F. (1968). Near-shore phytoplankton populations in the Grand Haven, Michigan, vicinity during thermal bar conditions. In *Proceedings of the 11th Conference on Great Lakes Research*, pp. 137–50. Ann Arbor, MI: International Association for Great Lakes Research.

Stokes, G. G. (1851). On the effect of internal friction of fluids on the motion of pendulums. *Transactions of the Cambridge Philosophical Society*, **9** (II), 8–14.

Stoyneva, M. P. (1994). Shallows of the lower Danube as additional sources of potamoplankton. *Hydrobiologia*, **289**, 97–108.

Straile, D. (2000). Meteorological forcing of plankton dynamics in a large and deep continental lake. *Oecologia*, **122**, 44–50.

Straškraba, M. T., Jørgensen, S.-E. and Patten, B. C. (1999). Ecosystems emerging II. Dissipation. *Ecological Modelling*, **117**, 3–39.

Strathmann, R. R. (1967). Estimating the organic carbon content of phytoplankton from cell volume or plasma volume. *Limnology and Oceanography*, **12**, 411–18.

Strickler, J. R. (1977). Observations on swimming performances of planktonic copepods. *Limnology and Oceanography*, 22, 165–70.

Strickler, J. R. and Twombley, S. (1975). Reynolds' numbers, diapause and predatory copepods. *Verhandlungen der internationalen Vereinigung für theoretische und angewandte Limnologie*, **19**, 2943–50.

Stumm, W. and Morgan, J. P. (1981). *Aquatic Chemistry*, 2nd edn. New York: John Wiley.

(1996). *Aquatic Chemistry*, 3rd edn. New York: John Wiley.

Sugimura, Y. and Suzuki, Y. (1988). A high-temperature catalytic-oxidation method of non-volatile dissolved organic carbon in seawater by direct injection of liquid sample. *Marine Chemistry*, **24**, 105–31.

Sullivan, C. W. and Volcani, B. E. (1981). Silicon in the cellular metabolism of diatoms. In *Silicon and Siliceous Structures in Biological Systems*, ed. T. L. Simpson and B. E. Volcani, pp. 15–42. New York: Springer-Verlag.

Sültemeyer, D. (1998). Carbonic anhydrase in eukaryotic algae: characterisation, regulation and possible function during photosynthesis. *Canadian Journal of Botany*, **76**, 962–72.

Sültemeyer, D. F., Fock, H. P. and Canvin, D. T. (1991). Active uptake of inorganic carbon by *Chlamydomonas reinhardtii*: evidence of simultaneous transport of HCO_3^- and CO_2. *Canadian Journal of Botany*, **69**, 995–1002.

Sutcliffe, D. W., Carrick, T. R., Heron, J. *et al.* (1982). Long-term and seasonal changes in the chemical composition of precipitation and surface waters of lakes and tarns in the English Lake District. *Freshwater Biology*, **12**, 451–506.

Sverdrup, H. U. (1953). On conditions of vernal blooming of phytoplankton. *Journal du Conseil international pour l'exploration de la Mer*, **18**, 287–95.

Sverdrup, H. U., Johnson, M. W. and Fleming, R. H. (1942). *The Oceans: Their Physics, Chemistry and General Biology*. New York: Prentice Hall.

Swale, E. M. F. (1963). Notes on *Stephanodiscus hantzschii* Grun. in culture. *Archiv für Mikrobiologie*, **45**, 210–16

Swale, E. M. F. (1968). The phytoplankton of Oak Mere, Cheshire, 1963–1966. *British Phycological Bulletin*, **3**, 441–9.

Swift, D. G. (1980). Vitamins and phytoplankton growth. In *The Physiological Ecology of Phytoplankton*, ed. I. Morris, pp. 329–68. Oxford: Blackwell Scentific Publications.

Szeligiwicz, W. (1998). Phytoplankton blooms predictions: a new turn for Sverdrup's critical depth concept. *Polskie Archiwum Hydrobiologii*, **45**, 501–11.

Takamura, N., Otsuki, A., Aizaki, M. and Nojiri, Y. (1992). Phytoplankton species shift accompanied by transition from nitrogen dependence to phosphorus dependence of primary production in Lake Kasumigaura, Japan. *Archiv für Hydrobiologie*, **124**, 129–48.

Talling, J. F. (1951). The element of chance in pond populations. *The Naturalist, Hull*, October–December 1951, 157–70.

(1957a). Diurnal changes of stratification and photosynthesis in some tropical African waters. *Proceedings of the Royal Society of London B*, **147**, 57–83.

(1957b) Photosynthetic characteristics of some freshwater diatoms in relation to underwater radiation. *New Phytologist*, **56**, 29–50.

(1957c). The phytoplankton population as a compound photosynthetic system. *New Phytologist*, **56**, 133–49.

(1960). Self-shading effects in natural populations of a planktonic diatom. *Wetter und Leben*, **12**, 235–42.

(1965). The photosynthetic activity of phytoplankton in East African lakes. *Internationale Revue des gesamten Hydrobiologie*, **50**, 1–32.

(1966). Photosynthetic behaviour in stratified and unstratified lake populations of a planktonic diatom. *Journal of Ecology*, **54**, 99–127.

(1971). The underwater light climate as a controlling factor in the production ecology of freshwater phytoplankton. *Mitteilungen der internationalen Vereinigung für theoretische und angewandte Limnologie*, **19**, 214–43.

(1973). The application of some electro-chemical methods to the measurement of photosynthesis and respiration in freshwaters. *Freshwater Biology*, **3**, 335–62.

(1976). The depletion of carbon dioxide from lake water by phytoplankton. *Journal of Ecology*, **64**, 79–121.

(1984). Past and contemporary trends and attitudes in work on primary productivity. *Journal of Plankton Research*, **6**, 203–17.

(1986). The seasonality of phytoplankton in African lakes. *Hydrobiologia*, **138**, 139–60.

(1987). The phytoplankton of Lake Victoria (East Africa). *Ergebnisse der Limnologie*, **25**, 229–56.

Talling, J. F. and Heaney, S. I. (1988). Long-term changes in some English (Cumbrian) lakes subjected to increased nutrient inputs. In *Algae and their Aquatic Environments*, ed. F. E. Round, pp. 1–29. Bristol: Biopress.

Talling, J. F. and Talling, I. B. (1965). The chemical composition of African lake waters. *Internationale Revue des gesamten Hydrobiologie*, **50**, 421–63.

Talling, J. F., Wood, R. B., Prosser, M. V. and Baxter, R. M. (1973). The upper limit of photosynthetic productivity by phytoplankton: evidence from Ethiopian soda lakes. *Freshwater Biology*, **3**, 53–76.

Tallis, J. H. (1973). The terrestrialization of lake basins in North Cheshire, with special reference to a 'Schwingineer' structure. *Journal of Ecology*, **61**, 537–67

Tamiya, H., Iwamura, T., Shibata, K., Hase, E. and Nihei, T. (1953). Correlation between photosynthesis and light-independent metabolism in the growth of *Chlorella*. *Biochimica Biophysica Acta*, **12**, 23–40.

Tandeau de Marsac, N. (1977). Occurrence and nature of chromatic adaptation in Cyanobacteria. *Journal of Bacteriology*, **130**, 82–91.

Tang, E. P. Y. and Peters, R. H. (1995). The allometry of algal respiration. *Journal of Plankton Research*, **17**, 303–15.

Tansley, A. G. (1939). *The British Isles and Their Vegetation*. Cambridge: Cambridge University Press.

Temporetti, P. F. and Pedrozo, F. L. (2000). Phosphorus release rates from freshwater sediments affected by fish farming. *Aquaculture Research*, **31**, 447–55.

Tennekes, H. and Lumley, J. L. (1972). *A First Course in Turbulence*. Cambridge, MA: Massachussetts Institute of Technology Press.

Tett, P. and Barton, E. D. (1995). Why are there about 5000 species of phytoplankton in the sea? *Journal of Plankton Research*, **17**, 1693–704.

Tett, P., Heaney, S. I. and Droop, M. R. (1985). The Redfield ratio and phytoplankon growth rate. *Journal of the Marine Biological Association*, **65**, 487–504.

Thellier, M. (1970). An electrokinetic interpretation of the functioning of biological systems and its application to the study of mineral salt absorption. *Annals of Botany*, **34**, 983–1009.

Therriault, J.-C. and Platt, T. (1981). Environmental control of phytoplankton patchiness. *Canadian Journal of Fisheries and Aquatic Sciences*, **38**, 638–41.

Thienemann, A. (1918). Untersuchungen über die Beziehungen zwischen dem Sauerstoffgehalt des Wassers und der Zusammensetzung der Fauna in norddeutschen Seen. *Archiv für Hydrobiologie*, **12**, 1–65.

Thingstad, F., Heldal, M., Bratbak, G. and Dundas, I. (1993). Are viruses important partners in pelagic food webs? *Trends in Ecology and Evolution*, **8**, 209–13.

Thomas, E. A. (1949). Sprungschichtneigung in Zürichsee durch Strum. *Schweizerische Zeitschrift für Hydrobiolgie*, **11**, 527–45.

(1950). Auffällige biologische Folgen von Sprungschichtneigung in Zürichsee. *Schweizerische Zeitschrift für Hydrobiolgie*, **12**, 1–24.

Thomas, J. D. (1997). The role of dissolved organic matter, particularly free amino acids and humic substances, in freshwater ecosystems. *Freshwater Biology*, **38**, 1–36.

Thomas, R. H. and Walsby, A. E. (1985). Buoyancy regulation in a strain of *Microcystis*. *Journal of General Microbiology*, **131**, 799–809.

Thomasson, K. (1963). Araucanian Lakes. *Acta Phytogeographica Suecica*, 47, 1–139.

Thompson, J. M., Ferguson, A. J. D and Reynolds, C. S. (1982). Natural filtration rates of zooplankton in a closed system the derivation of a community grazing index. *Journal of Plankton Research*, **4**, 545–60.

Thorp, J. H. and Casper, A. F. (2002). Potential effects on zooplankton from species shifts in planktivorous mussels: a field experiment in the St Lawrence River. *Freshwater Biology*, **47**, 107–19.

Thouvenot, A., Richardot, M., Debroas, D. and Dévaux, J. (2003). Regulation of the abundance and growth of ciliates in a newly flooded reservoir. *Archiv für Hydrobiologie*, **157**, 185–93.

Tillmann, U. (2003). Kill and eat your predator: a winning strategy of the planktonic flagellate *Prymnesium parvum*. *Aquatic Microbial Ecology*, **32**, 73–84.

Tilman, D. (1977). Resource competition between planktonic algae: an experimental and theoretical approach. *Ecology*, **58**, 338–48.

(1987). On the meaning of competition and community structure. *Functional Ecology*, **1**, 304–15.

(1988). *Plant Strategies and the Dynamics and Structure of Plant Communities*. Princeton, NJ: Princeton University Press.

Tilman, D. and Kilham, S. S. (1976). Phosphate and silicate growth and uptake kinetics of the diatoms *Asterionella formosa* and *Cyclotella meneghiniana* in batch and semi-continuous culture. *Journal of Phycology*, **12**, 375–83.

Tilman, D., Kilham, S. S. and Kilham, P. (1982). Phytoplankton community ecology: the role of limiting nutrients. *Annual Reviews of Ecology and Systematics*, **13**, 349–72.

Tilzer, M. M. (1984). Estimation of phytoplankton loss rates from daily photosynthetic rates and observed biomass changes in Lake Constance. *Journal of Plankton Research*, **6**, 309–24.

Timms, R. M. and Moss, B. (1984). Prevention of growth of potentially dense phytoplankton populations by zooplankton grazing in the presence of zooplanktivorous fish in a shallow wetland ecosystem. *Limnology and Oceanography*, **29**, 472–86.

Titman, D. and Kilham, P. (1976). Sinking in freshwater phytoplankton: some ecological implications of cell nutrient status and physical mixing processes. *Limnology and Oceanography*, **21**, 409–17.

Tokeshi, M. (1997). Species co-existence and abundance: patterns and processes. In *Biodiversity: An Ecological Perspective*, ed. T. Abe, S. R. Levin and M. Higashi, pp. 35–55. New York: Springer-Verlag.

Tolland, H. G. (1977). *Destratification/Aeration in Reservoirs A Literature Review of the Techniques Used for Water Quality Management*, Technical Report No. 50. Medmenham: Water Research Centre.

Toms, I. P. (1987). Developments in London's water supply system. *Ergebnisse der Limnologie*, **28**, 149–67.

Torella, F. and Morita, R. Y. (1979). Evidence for a high incidence of bacteriophage particles in the waters of Yaquina Bay, Oregon: ecological and taxonomic implications. *Applied and Environmental Microbiology*, **37**, 774–8.

Tortell, P. D. (2000). Evolutionary and ecological perspectives in carbon acquisition in phytoplankton. *Limnology and Oceanography*, **45**, 744–50.

Tow, M. (1979). The occurrence of *Microcystis aeruginosa* in the bottom sediments of a shallow eutrophic pan. *Journal of the Limnological Society of Southern Africa*, **5**, 9–10.

Townsend, C. R., Winfield, I. J., Peirson, G. and Cryer, M. (1986). The response of young roach *Rutilus rutilus* to seasonal changes in the abundance of microcrustacean prey: a field determination of switching. *Oikos*, **46**, 372–8.

Tranvik, L. J. (1998). Degradation of dissolved organic matter in humic waters by bacteria. In *Aquatic Humic Substances*, ed. D. Hessen and L. J. Tranvik, pp. 259–83. Berlin: Springer-Verlag.

Tranvik, L. J. and Bertilsson, S. (2001). Contrasting effects of solar UV radiation on dissolved organic sources for bacterial growth. *Ecology Letters*, **4**, 458–63.

Trevisan, R. (1978). Noto sull'uso dei volumi algali per la stima della biomassa. *Rivista di Idrobiologia*, **17**, 345–58.

Trimbee, A. M. and Harris, G. P. (1983). Use of time-series analysis to demonstrate advection rates of different variables in a small lake. *Journal of Plankton Research*, **5**, 819–33.

Tsujimura, S. (2003). Application of the frequency of dividing cells technique to estimate the in-situ growth rate of *Microcystis* (Cyanobacteria). *Freshwater Biology*, **48**, 2009–24.

Turner, M. G., Baker, W. L., Peterson, C. J. and Peet, R. K. (1998). Factors influencing succession: lessons from large, infrequent natural disturbances. *Ecosystems*, **1**, 511–23.

Turpin, D. H. (1988). Physiological mechanisms in phytoplankton resource competition. In *Growth and Reproductive Strategies of Freshwater Phytoplankton*, ed. C. D. Sandgren, pp. 316–68. Cambridge: Cambridge University Press.

Tüxen, R. (1955). Das Systeme der nordwestdeutschen Pflanzengesellschaft. *Mitteilungen der Florist-soziologische Arbeitsgemeinschaft*, **5**, 1–119.

Tyler, J. E. and Smith, R. E. (1970). *Measurements of Spectral Irradiance Underwater*. New York: Gordon and Breach.

Tyler, P. A. (1996). Endemism in freshwater algae, with special reference to the Australian region. *Hydrobiologia*, **336**, 127–35.

Uhlmann, D. (1971). Influence of dilution, sinking and grazing on phytoplankton populations of hyperfertilised ponds and micro-ecosystems. *Mitteilungen der internationalen Vereinigung für theoretische und angewandte Limnologie*, **19**, 100–124.

Ulanowicz, R. E. (1986). *Growth and Development: Ecosystems Phenomenology*. New York: Springer-Verlag.

Upstill-Goddard, R. C., Watson, A. J., Liss, P. S. and Liddicoat, M. I. (1990). Gas transfer velocities in lakes measured with SF_6. *Tellus*, **42B**, 364–77.

Urbach, E., Scanlan, D. J., Distel, D. L., Waterbury, J. B. and Chisholm, S. W. (1998). Rapid diversification of marine picophytoplankton with dissimilar light-harvesting structures inferred from sequences of *Prochlorococcus* and *Synechococcus* (Cyanobacteria). *Journal of Molecular Evolution*, **46**, 188–201.

Utermöhl, H. (1931). Neue Wege in der quantitativen Erfassung des Planktons. *Verhandlungen der internationalen Vereinigung für theoretische und angewandte Limnologie*, **5**, 567–96.

Utkilen, H.-C., Oliver, R. L. and Walsby, A. E. (1985a). Buoyancy regulation in a red *Oscillatoria* unable to collapse gas vesicles by turgor pressure. *Archiv für Hydrobiologie*, **102**, 319–29.

Utkilen, H.-C., Skulberg, P. M. and Walsby, A. E. (1985b). Buoyancy regulation and chromatic adaptation in planktonic *Oscillatoria* species: alternative strategies for optimising light absorption in stratified lakes. *Archiv für Hydrobiologie*, **104**, 407–17.

Vadeboncoeur, Y., Vander Zanden, M. J. and Lodge, D. M. (2002). Putting the lake back together: reintegrating benthic pathways into lake food web models. *BioScience*, **52**, 44–54.

Vadstein, O., Olson, H., Reinertsen, H. and Jensen, A. (1993) The role of planktonic bacteria in lakes: sink and link. *Limnology and Oceanography*, **38**, 539–44.

Vander Zanden, M.J, Shuter, B. J., Lesster, N. and Rasmussen, J. B. (1999). Patterns of food chain length in lakes: a stable isotope study. *American Naturalist*, **154**, 406–16.

Vanderploeg, H. A. and Paffenhöfer, H. A. (1985). Modes of algal capture by the freshwater copepod *Diaptomus sicilis* and their relation to food-size selection. *Limnology and Oceanography*, **30**, 871–85.

Vanni, M. J., Luecke, C., Kitchell, J. F. and Magnusson, J. (1990). Effects of planktivorous fish mass mortality on the plankton community of Lake Mendota, Wisconsin: implications for biomanipulation. *Hydrobiologia*, **200/201**, 329–36.

Vaulot, D. (1995). The cell cycle of phytoplankton: coupling cell growth to population growth. In *The Molecular Ecology of Aquatic Microbes*, ed. I. Joint, pp. 303–32. Berlin: Springer-Verlag.

Vega-Palas, M. A., Flores, E. and Herrero, A. (1992). *NtcA*, a global nitrogen regulator from the cyanobacterium *Synechococcus* that belongs to the CRP family of bacterial regulators. *Molecular Microbiology*, **6**, 1853–9.

Velikova, V., Moncheva, S. and Petrova, D. (1999). Phytoplankton dynamics and red tides (1987–1997) in the Bulgarian Black Sea. *Oceanology*, **37**(3), 376–84. (English translation of original article in *Okeanologiya*.)

Venrick, E. L. (1990). Phytoplankton in an oligotrophic ocean: species structure and interannual variability. *Ecology*, **71**, 1547–63.

(1995). Scales of variability in a stable environment: phytoplankton in the central North Pacific. In *Ecological Time Series*, ed. T. M. Powell and J. H. Steele, pp. 150–80. New York: Chapman and Hall.

Ventelä, A.-M., Wiacowski, K., Moilanen, M. *et al.* (2002). The effect of small zooplankton on the microbial loop and edible algae during a cyanobacterial bloom. *Freshwater Biology*, **47**, 1807–19.

Verity, P. G., Robertson, C. Y., Tronzo, C. R. *et al.* (1992). Relationships between cell volume and carbon and nitrogen of marine photosynthetic nanoplankton. *Limnology and Oceanography*, **37**, 1434–46.

Vicente, E. and Miracle, M. R. (1988). Physicochemical and microbial stratification in a meromictic karstic lake in Spain. *Verhandlungen der internationalen Vereinigung für theoretische und angewandte Limnologie*, **23**, 522–9.

Vincent, W. F., Neale, P. J. and Richerson, P. J. (1984). Photoinhibition: algal responses to bright light during diel stratification and mixing in a tropical alpine lake. *Journal of Phycology*, **20**, 201–11.

Viner, A. and Kemp, L. (1983). The effect of vertical mixing on the phytoplankton of Lake Rotongaio (July 1979 – January 1981). *New Zealand Journal of Marine and Freshwater Science*, **17**, 402–22.

Volcani, B. E. (1981). Cell wall formation in diatoms: morphogenesis and biochemistry. In *Silicon and Siliceous Structures in Biological Systems*, ed. T. L. Simpson and B. E. Volcani, pp. 157–200. New York: Springer-Verlag.

Vollenweider, R. A. (1965). Calculation models of photosynthesis-depth curves and some implications regarding day rate estimates in primary production. *Memorie dell'Istituto italiano di Idrobiologia*, **18** (Suppl.), 425–57.

(1968). *Scientific Fundamentals of the Eutrophication of Lakes and Flowing Waters, with Particular Reference to Nitrogen and Phosphorus as Factors in Eutrophication*, Technical Report DAS/CSI/68.27. Paris: Organisation for Economic Co-Operation and Dvelopment.

(1974). *A Manual on Methods for Measuring Primary Production in Aquatic Environments*. Oxford: Blackwell Scientific Publications.

(1975). Input–output models with special reference to the phosphorus-loading concept in limnology. *Schweizerische Zeitschrift für Hydrologie*, **37**, 53–84.

(1976). Advances in defining critical load levels for phosphorus in lake eutrophication. *Memorie dell'Istituto italiano di Idrobiologia*, **33**, 53–83.

Vollenweider, R. A. and Kerekes, J. (1980). The loading concept as basis for controlling eutrophication philosophy and preliminary results of the OECD programme on eutrophication. *Progress in Water Technology*, **12**(2), 5–38.

Vollenweider, R. A., Munawar, M. and Stadelman, P. (1974). A comparative review of phytoplankton and primary production in the Laurentian Great Lakes. *Journal of the Fisheries Research Board of Canada*, **31**, 739–62.

Vollenweider, R. A., Montanari, G. and Rinaldi, A. (1995). Statistical inferences about the mucilage events in the Adriatic Sea, with special reference to recurrence patterns and claimed relationship to sun activity cycles. *Science of the Total Environment*, **165**, 213–24.

Votintsev, K. K. (1992). On the characteristics of the self-purification potential of Lake Baikal. *Sibirkskiy biologicheskiy Zhurnal*, **4**, 36–40.

Vrind-de Jong, E. W. de, Emburg, P. R. van and Vrind, J. P. M. de (1994). Mechanisms of calcification: *Emiliana huxleyi* as a model system. In *The Haptophyte Algae*, ed. J. C. Green and B. S. C. Leadbeater, pp. 149–166. Oxford: Oxford University Press.

Walker, B. H. (1992). Biodiversity and ecological redundancy. *Biological Conservation*, **6**, 18–23.

Walker, D. (1992). Excited leaves. *New Phytologist*, **121**, 325–45.

Wall, D. and Dale, B. (1968). Modern dinoflagellate cysts and evolution of the Peridiniales. *Micropalaeontology*, **14**, 265–304.

Wallis, S. G., Young, P. C. and Beven, K. (1989). Experimental investigation of an aggregated dead zone model for longitudinal transport in stream channels. *Proceedings of the Institute of Civil Engineers*, **87**, 1–22.

Walsby, A. E. (1971). The pressure relationships of gas vacuoles. *Proceedings of the Royal Society of London B*, **178**, 301–26.

(1978). The properties and buoyancy-providing role of gas vauoles in *Trichodesmium* Ehrenberg. *British Phycological Journal*, **13**, 103–16.

(1994). Gas vesicles. *Microbiological Reviews*, **58**, 94–144.

(2001). Determining the photosynthetic productivity of a stratified phytoplankton population. *Aquatic Sciences*, **63**, 18–43.

Walsby, A. E. and Bleything, A. (1988). The dimensions of cyanobacterial gas vesicles in relation to their efficiency in providing buoyancy. *Journal of General Microbiology*, **135**, 2635–45.

Walsby, A. E. and Klemer, A. R. (1974). The role of gas vacuoles in the microstratification of a population of *Oscillatoria agardhii* var. *isothrix* in Deming Lake, Minnesota. *Archiv für Hydrobiologie*, **74**, 375–92.

Walsby, A. E. and Reynolds, C. S. (1980). Sinking and floating. In *The Physiological Ecology of Phytoplankton*, ed. I. Morris, pp. 371–412. Oxford: Blackwell Scientific Publications.

Walsby, A. E. and Xypolyta, A. (1977). The form resistance of chitan fibres attached to the cells of *Thalassiosira fluviatilis* Hustedt. *British Phycological Journal*, **12**, 215–23.

Walsby, A. E., Utkilen, H. C. and Johnsen, I. J. (1983). Buoyancy changes in red-coloured *Oscillatoria agardhii* in Lake Gjersjoen, Norway. *Archiv für Hydrobiologie*, **97**, 18–38.

Walsby, A. E., Kinsman, R., Ibelings, B. W. and Reynolds, C. S. (1991). Highly buoyant colonies of the cyanobacterium *Anabaena lemmermannii* form persistent water blooms. *Archiv für Hydrobiologie*, **121**, 261–80.

Wardle, D. A., Bonner, K. I., Barker, G. M. *et al.* (1999). Plant removals in perennial grassland: vegetation dynamics, decomposers, soil biodiversity and ecosystem properties. *Ecological Monographs*, **69**, 535–68.

Wasmund, N. (1994). Phytoplankton periodicity in a eutrophic coastal water of the Baltic Sea. *Internationale Revue des gesamten Hydrobiologie*, **79**, 259–85.

Watson, A. J., Upstill-Goddard, R. C. and Liss, P. S. (1991). Air–sea exchange in rough and stormy seas measured by a dual-tracer technique. *Nature*, **349**, 145–7.

Watson, S. W. and Kalff, J. (1981). Relationships between nanoplankton and lake trophic status. *Canadian Journal of Fisheries and Aquatic Sciences*, **38**, 960–7.

Weiher, E. and Keddy, P. A. (1995). Assembly rules, null models and trait dispersion: new questions from old patterns. *Oikos*, **74**, 159–64.

Weiher, E., Clarke, D. P. and Keddy, P. A. (1998). Community assembly rules, morphological dispersion and the co-existence of plant species. *Oikos*, **81**, 309–22.

Weisse, T. (1994). Structure of microbial food webs in relation to the trophic state of lakes and fish grazing pressure: a key role of Cyanobacteria. In *Ecology and Human Impact on Lakes and Reservoirs in Minas Gerais with Special Reference to Future Development and Management Strategies*, ed. R. M. Pinto-Coelho, A. Giani and E. von Sperling, pp. 55–70. Belo Horizonte: Sociedade Editoria e Grafíca de Acão Communitaria.

  (2003). Pelagic microbes: protozoa and the microbial food web. In *The Lakes Handbook*, vol.1, ed. P. E. O'Sullivan and C. S. Reynolds, pp. 417–60. Oxford: Blackwell Science.

Weisse, T. and MacIsaac, E. (2000). Significance and fate of bacterial production in oligotrophic lakes in British Columbia. *Canadian Journal of Fisheries and Aquatic Science*, **57**, 96–105.

Welch, E. B. (1952) *Limnology*, 2nd edn. New York: McGraw-Hill.

Werner, D. (1977). Silicate metabolism. In *The Biology of Diatoms*, ed. D. Werner, pp. 110–49. Oxford: Blackwell Scientific Publications.

Wershaw, R. L. (2000). The study of humic substances: in search of a paradigm. In *Humic Substances: Versatile Components of Plants, Soil and Water*, ed. E. A. Ghabbour and G. Davies, pp. 1–7. Cambridge: Royal Society of Chemistry.

Wesenberg-Lund, C. (1904). *Studier over de dansk sersplankton*, vol. 1, *Tekst*. Copenhagen: Glydendanske Boghandel.

West, R. G. (1977). *Pleistocene Geology and Biology*, 2nd edn. London: Longman.

West, W. and West, G. S. (1909). The phytoplankton of the English Lake District. *The Naturalist* (in six parts), **626**, 1115–22; **627**, 134–41; **628**, 186–93; **629**, 260–7; **631**, 287–92; **632**, 323–31.

Wetzel, R. G. (1995). Death, detritus and energy flow in aquatic ecosystems. *Freshwater Biology*, **35**, 83–9.

Whipple, G. C. (1899). *The Microscopy of Drinking Water*. New York: John Wiley.

Whitton, B. A. (1975) Algae. In *River Ecology*, ed. B. A. Whitton, pp. 81–105. Oxford: Blackwell Scientific Publications.

Whitton, B. A. and Peat, A. (1969). On *Oscillatoria redekei* Van Goor. *Archiv für Mikrobiologie*, **68**, 362–76.

Wiackowski, K., Brett, M. T. and Goldman, C. R. (1994). Differential effects of zooplankton species on ciliate community structure. *Limnology and Oceanography*, **39**, 486–92.

Wildman, R., Loescher, J. H. and Winger, C. L. (1975). Development and germination of akinetes of *Aphanizomenon flos-aquae*. *Journal of Phycology*, **11**, 96–104.

Wilkinson, D. M. (1999). The disturbing history of intermediate disturbance. *Oikos*, **84**, 145–7.

Willén, E. (1976). A simplified method of phytoplankton counting. *British Phycological Journal*, **11**, 265–78.

Williams, D. B. (1975). On the concentrations of characterised dissolved organic compounds in seawater. In *The Micropalaeontology of Oceans*, ed. B. M. Funnell and W. R. Riedel, pp. 91–5. Cambridge: Cambridge University Press.

Williams, P. J. leB. (1970). Heterotrophic utilisation of dissolved organic compounds in the sea. I. Size distribution of population and relationship between respiration and incorporation of growth substances. *Journal of the Marine Biological Association of the United Kingdom*, **50**, 859–70.

  (1981). Incorporation of microheterotrophic processes into the classical paradigm of the planktonic food web. *Kieler meereforschungen Sonderheft*, **1**, 1–28.

(1995). Evidence for the seasonal accumulation of carbon-rich dissolved organic material: its scale in comparison with changes in particulate material and the consequential effect on net C : N assimilation ratios. *Marine Chemistry*, **51**, 17–29.

Williams, P. J. leB. and Lefevre, D. (1996). Algal ^{14}C uptake and total carbon metabolism. I. Models to account for the physiological processes of respiration and recycling. *Journal of Plankton Research*, **18**, 1941–59.

Wilsey, B. J. and Potvin, C. (2000). Biodiversity and ecosystem functioning: importance of species eveness in an old field. *Ecology*, **81**, 887–92.

Wilson, E. O. and Petr, F. M. (1986). *BioDiversity*. Washington, DC: National Academy Press.

Winkler, L. W. (1888). Die Bestimmung des im Wasser gelösten Sauerstoffs. *Berichte der deutsches chemische Gesellschaft*, **21**, 2843–54.

Wiseman, S. W. and Reynolds, C. S. (1981). Sinking rate and electrophoretic mobility of the freshwater diatom *Asterionella formosa*: an experimental investigation. *British Phycological Journal*, **16**, 357–61.

Wiseman, S. W., Jaworski, G. H. M. and Reynolds, C. S. (1983). Variability in the sinking rate of the freshwater diatom, *Asterionella formosa* Hass: the influence of the excess density of the colonies. *British Phycological Journal*, **18**, 425–32.

Witte, F., Goldschidt, T., Wanink, J. *et al.* (1992). The destruction of the endemic species flock: quantitative data on the decline of the haplochromine cichlids of Lake Victoria. *Environmental Biology of Fishes*, **34**, 1–28.

Woese, C. R. (1987). Bacterial evolution. *Microbiology Reviews*, **51**, 221–71.

Woese, C. R., Kandler, O. and Wheelis, M. L. (1990). Towards a natural system of organisms: proposal for the domains Archaea, Bacteria and Eucarya. *Proceedings of the National Academy of Sciences of the USA*, **87**, 4576–9.

Wolf-Gladrow, D. A. and Riebesell, U. (1997). Diffusion and reactions in the vicinity of plankton: a refined model for inorganic carbon transport. *Marine Chemistry*, **59**, 17–34.

Wolk, C. P. (1982). Heterocysts. In *The Biology of the Cyanobacteria*, ed. N. G. Carr and B. A. Whitton, pp. 359–86. Oxford: Blackwell Scientific Publications.

Wollast, R. (1974). The silica problem. In *The Sea*, vol. 5, ed. E. D. Goldberg, pp. 359–92. New York: John Wiley.

Woods, J. D. and Onken, R. (1983). Diurnal variation and primary production in the ocean: preliminary results of a Lagrangian ensemble model. *Journal of Plankton Research*, **4**, 735–56.

Wüest, A. and Lorke, A. (2003). Small-scale hydrodynamics in lakes. *Annual Reviews in Fluid Mechanics*, **35**, 273–412.

Yates, M. G. (1977). Physiological aspects of nitrogen fixation. In *Recent Developments in Nitrogen Fixation*, ed. W. Newton, J. R. Postgate and C. Rodriguez-Barrueco, pp. 219–70. London: Academic Press.

Yoshida, T., Gurung, T. B., Kagami, M. and Urabe, J. (2001). Contrasting effects of a cladoceran (*Daphnia galeata*) and a calanoid copepod (*Eodiaptomus japonicus*) on algal and microbial plankton in a Japanese Lake, Lake Biwa. *Oecologia*, **129**, 602–10.

Youngman, R. E. (1971). *Algal Monitoring of Water Supply Reservoirs and Rivers*, Technical Memorandum No. TM63. Medmenham: Water Research Association.

Yunes, J. S., Niencheski, L. F., Salomon, P. S. and Parise, M. (1998). Toxicity of Cyanobacteria in the Patos Lagoon Estuary, Brazil. *Verhandlungen der internationalen Vereinigung für theoretische und angewandte Limnologie*, **26**, 1796–2000.

Zacharias, O. (1898) Das Potamoplankton. *Zoologische Anzeiger*, **21**, 41–8.

Zehr, J. P. (1995). Nitrogen fixation in the sea: why only *Trichodesmium*? In *Molecular Ecology of Aquatic Microbes*, ed I. Joint, pp. 335–64. Berlin: Springer-Verlag.

Zhuravleva, V. I. and Matsekevich, Ye.S. (1974). Electrokinetic properties of *Microcystis aeruginosa*. *Hydrobiological Journal*, **10**(1), 55–7. (English translation of original paper in *Gidrobiologeskii Zhurnal.*)

Ziegler, S. and Benner, R. (2000). Effects of solar radiation on dissolved organic matter cycling in a tropical seagrass meadow. *Limnology and Oceanography*, **45**, 257–66.

Zimmermann, U. (1969) Ökologische und physiologische Untersuchungen an der planktonischen Blaualge *Oscillatoria rubescens* D.C. unter besonderen Berücksichtigung von Licht und Temperatur. *Schweizerische Zeitschrift für Hydrologie*, **31**, 1–58.

Zohary, T. and Pais-Madeira, A. M. (1990). Structural, physical and chemical characteristics of

Microcystis aeruginosa hyperscums from a hypertrophic lake. *Freshwater Biology*, **23**, 339–52.

Zohary, T. and Robarts, R. D. (1989). Diurnal mixed layers and the long-term dominance of *Microcystis aeruginosa*. *Journal of Plankton Research*, **11**, 25–48.

Zouni, A., Witt, H.-T., Kern, J., *et al.* (2001). Crystal structure of photosystem II from *Synechococcus* *elongatus* at 3.8 resolution. *Nature*, **409**, 739–43.

Zuniño, L. and Diaz, M. (2000). Autotrophic picoplankton along a trophic gradient in Andean–Patagonian lakes. *Verhandlungen der internationalen Vereinigung für theoretische und angewandte Limnologie*, **27**, 1895–99.

Index to lakes, rivers and seas

LAKES AND RESERVOIRS

Akan-Panke, Japan 43° 27′ N,
144° 06′ E 326

Ammersee, Germany 48° N, 11° 6′ E
329, 330 Table 7.3

Aralskoye More,
Kazakhstan/Uzbekistan 45° N,
60° E 318, 353

Arresø, Denmark 55° 59′ N, 12° 7′ E
344

Attersee, Austria 47° 50′ N, 13° 35′ E
329, 330 Table 7.3

Balaton, Hungary 46° 50′ N, 18° 42′ E
89, 345, 354

Balkhash, Ozero, Kazakhstan 46° N,
76° E 343, 344

Bangweolo, Zambia 11° S, 30° E 343

Bassenthwaite Lake, UK 54° 39′ N,
3° 12′ W 332 Table 7.5

Bautzen Reservoir, Germany
51° 11′ N, 14° 29′ E 403

Baykal, Ozero, Russia 53° N, 108° E
89, 322, 324, 390

Bayley-Willis, Lago, Argentina
40° 39′ S, 71° 43′ W 196

Biwa-Ko, Japan 35° N, 136° W 142,
219, 327

Blelham Tarn, UK 54° 24′ N, 2° 58′ W
82, 120, 245, 284, 332 Table 7.5,
333

Bodensee, Austria/Germany/
Switzerland 47° 40′ N, 9° 20′ E
48 Table 2.2, 71, 329, 330 Table
7.3

Brothers Water, UK 54° 21′ N,
2° 56′ W 332 Table 7.5

Budworth Mere, UK 53° 17′ N,
2° 31′ W 344

Buenos Aires/General Carrera, Lago,
Argentina/Chile 46° 35′ S, 72° W
322

Buttermere, UK 54° 33′ N, 3° 16′ W
332 Table 7.5

Cabora Bassa Dam, Mozambique
15° 38′ S, 33° 11′ E 205

Carioca, Lagoa, Brasil 19° 10′ S,
42° 1′ W 115

Carrilaufquen Chica, Argentina
41° 13′ S, 69° 26′ W 335

Carrilaufquen Grande, Argentina
41° 7′ S, 69° 28′ W 335

Castle Lake, California, USA
41° 14′ N, 122° 23′ W 167

Caviahue, Lago, Argentina 37° 54′ S,
70° 35′ W 425

Como, Lago di, Italy 45° 58′ N,
9° 15′ E 330, 331 Table 7.4

Coniston Water, UK 54° 20′ N,
3° 4′ W 331, 332 Table 7.5

Correntoso, Lago, Argentina
40° 24′ S, 71° 35′ W 336

Crater Lake, Oregon, USA 42° 56′ N,
122° 8′ W 111, 117 Table 3.2

Crose Mere, UK 52° 53′ N, 2° 50′ WE
109, 117 Table 3.2, 245, 339,
368, 373

Crummock Water, UK 54° 31′ N,
3° 18′ W 332 Table 7.5

Dead Sea, Israel/Jordan 31° 25′ N,
35° 29′ W 172

Derwent Water, UK 54° 34′ N, 3° 8′ W
331, 332 Table 7.5

Drontermeer, Netherlands 52° 16′ N,
5° 32′ E 345

Eglwys Nynydd, UK 51° 33′ N,
3° 44′ W 82, 87

Ennerdale Water, UK 54° 32′ N,
3° 23′ W 332 Table 7.5

Erie, Lake, Canada/USA 41° 45′ N,
81° W 290, 323, 324

Erken, Sjön, Sweden 59° 25′ N,
18° 15′ E 338

Espejo, Lago, Argentina 40° 36′ S,
71° 48′ W 336

Esrum Sø, Denmark 56° 00′ N,
12° 22′ E 338

Esthwaite Water, UK 54° 21′ N,
3° 0′ W 88, 108, 126, 295, 332,
332 Table 7.5, 370, 371, 390

Eyre, Lake, Australia 29° S, 137° E
343

Fairmont Lakes Reservoir,
Minnesota, USA 43° 16′ N,
93° 47′ W 416

Fonck, Lago, Argentina 41° 20′ S,
71° 44′ W 196

Garda, Lago di, Italy 45° 35′ N,
10° 25′ E 330, 331 Table 7.4

George, Lake, Uganda 0° 0′ N,
30° 10′ E 59, 105, 344

Grasmere, UK 54° 28′ N, 3° 1′ W 241,
332 Table 7.5, 333, 352

Great Bear Lake, Canada 66° N,
120° W 323

Great Slave Lake, Canada 62° N,
114° W 324

Hamilton Harbour, Lake Ontario,
Canada 43° 15′ N, 79° 51′ E 344

Hartbeespoort Dam, South Africa
25° 24′ S, 27° 30′ E 215

Hawes Water, UK 54° 31′ N, 2° 48′ W
331, 332 Table 7.5

Huron, Lake, Canada/USA 44° 30′ N,
82° W 323

Iseo, Lago d', Italy 45° 43′ N, 10° 4′ E
330, 331 Table 7.4

Issyk-kul, Ozero, Kyrgyzstan 43° N,
78° E 89, 325

Kaspiyskoye More, Azerbaijan/
Iran/Kazakhstan/Russia/
Turkmenistan 42° N,
51° E 322

Kasumigaura-Ko, Japan 36° 0′ N,
140° 26′ E 235, 345

Kilotes, Lake, Ethiopia 7° 5′ N,
38° 25′ E 113, 117 Table 3.2

Kinneret, Yam, Israel 32° 50′ N,
35° 30′ E 233, 327, 366

Kivu, Lac, D. R. Congo/Rwanda 2° S,
29° E 339, 341

Königsee, Germany 47° 40′ N,
12° 55′ E 329, 330 Table 7.3

Lacar, Lago, Argentina 40° 10′ S,
71° 30′ W 336

Ladozhskoye, Ozero, Russia 61° N,
32° E 89, 324

Lanao, Philippines 7° 52′ N,
124° 13′ E 233, 342, 354

Léman, Lac, France/Switzerland
46° 22′ N, 6° 33′ E 329, 330
Table 7.3

Llanquihue, Lago, Chile 41° 6′ S,
72° 48′ W 336

Loughrigg Tarn, UK 54° 26′ N,
3° 1′ W 332 Table 7.5

Lowes Water, UK 54° 34′ N, 3° 21′ W
332, 332 Table 7.5

Lugano, Lago di, Italy 45° 57′ N,
8° 58′ E 330, 331 Table 7.4

Maggiore, Lago, Italy/Switzerland
46° N, 8° 40′ E 330, 330 Table
7.3, 331 Table 7.4

Malham Tarn, UK 54° 6′ N, 2° 4′ W
129

Mälaren, Sweden 59° 18′ N, 17° 6′ E
198 Fig. 5.6

Malawi, Lake (formerly Lake Nyasa),
Malawi/Mozambique/Tanzania
12° S, 35° E 339, 340

Matano, Danau, Indonesia 2° 35′ S,
121° 23′ E 322

Memphrémagog, Lac, Canada/USA
45° 5′ N, 72° 16′ W 89

Mendota, Lake, USA 43° 21′ N,
89° 25′ W 289

Michigan, Lake, USA 43° 45′ N,
86° 30′ W 290, 323, 324, 390

Mikołajske, Jezioro, Poland 53° 10′ N,
21° 33′ E 298, 339

Millstätter See, Austria 46° 48′ N,
13° 33′ E 325, 350

Montezuma's Well, Arizona, USA
34° 37′ N, 111° 51′ W 233, 241

Murtensee, Switzerland, 46° 55′ N,
7° 5′ E 115

Mount Bold Reservoir, Australia,
35° 0′ S, 138° 48′ E 113, 117
Table 3.2

Nauel Huapi, Lago, Argentina
40° 50′ S, 71° 50′ W 336

Neagh, Lough, UK 54° 55′ N, 6° 30′ W
48 Table 2.2, 72, 345

Negra, Laguna, Chile 37° 49′ S,
70° 2′ W 113

Ness, Loch, UK 57° 16′ N, 4° 30′ W
126, 390

Neusiedlersee, Austria/Hungary
47° 47′ N, 16° 44′ E 377 Fig. 7.22

Nyos, Lake, Cameroon 6° 40′ N,
10° 13′ E 126

N'zigi (formerly Lake Albert), D. R.
Congo/Uganda 1° 50′ N, 31° E
339

Oak Mere, UK 53° 13′ N, 2° 39′ W
339, 345, 425

Okeechobee, Lake, Florida, USA
27° N, 81° W 413

Onezhskoye, Ozero, Russia 62° N,
36° E 89, 324, 325

Ontario , Lake, Canada/USA
43° 40′ N, 78° W 324

P. K. le Roux Reservoir, South Africa
30° 0′ S, 24° 44′ E 113, 117
Table 3.2

Ranco, Lago, Chile 40° 12′ S,
72° 25′ W 336

Rapel, Embalse de Chile 34° 2′ S,
71° 35′ S 233

Red Rock Tarn, Australia 38° 20′ S,
143° 30′ W 105

Rostherne Mere, UK 53° 21′ N,
2° 23′ W 119, 339, 366

Rotongaio, Lake, New Zealand
38° 40′ S, 176° 2′ E 121

Rydal Water, UK 54° 27′ N, 3° 0′ W
332 Table 7.5

Sagami-Ko, Japan 35° 24′ N, 139° 6′ E
198 Fig. 5.6

Schlachtensee, Germany 52° 28′ N,
13° 25′ E 410

Schöhsee, Germany 54° 10′ N,
10° 25′ E 338

Śniardwy, Poland 53° 45′ N, 21° 45′ E
344

Søbygaard, Denmark 56° 3′ N,
9° 40′ E 413

St James's Park Lake, UK 51° 30′ N,
0° 9′ E 345

Stechlinsee, Germany 53° 10′ N,
13° 3′ E 326, 337

Superior, Lake, Canada/USA
47° 30′ N, 89° W 117 Table 3.2,
323

Tahoe, Lake, California/Nevada, USA
39° N, 120° W 198 Fig. 5.6

Tai Hu, China 31° N, 120° E 344

Tanganyika, Lac, Burundi/ D. R.
Congo/Tanzania/Zambia 6° S,
30° E 263, 339, 340, 392

Tchad, Lac, Chad/Niger/Nigeria
12° 30′ N, 14° 30′ E 105, 343,
344

Thames Valley Reservoirs, UK
51° 30′ N, 0° 40′ W 139, 405,
415, 416

Thirlmere, UK 54° 32′ N, 3° 4′ W 332
Table 7.5

Titicaca, Lago, Bolivia/Peru 15° 40′ S,
69° 35′ W 342

Todos Los Santos, Lago, Chile
41° 5′ S, 72° 15′ W 336

Traful, Lago, Argentina 40° 36′ S,
71° 25′ W 336

Trummen, Sjön, Sweden 56° 52′ N,
14° 50′ E 413

Turkana (formerly Lake Rudolf),
Kenya/Ethiopia 4° N, 36° E
339

Ullswater, UK 54° 35′ N, 2° 53′ W
332, 332 Table 7.5

Valencia, Lago, Venezuela 10° 11′ N,
67° 40′ W 342

Veluwemeer, Netherlands 52° 55′ N,
5° 45′ E 345, 410

Victoria, Lake,
Kenya/Tanzania/Uganda 1° S,
33° E 339, 341

Vierwaldstättersee, Switzerland
47° N, 8° 20′ E 59, 329, 330
Table 7.3

Villarrica, Lago, Chile 39° 15′ S,
72° 35′ W 336

Volta Grande Reservoir, Brasil
20° 4′ S, 48° 13′ W 415

Vostok, Lake, Antarctica approx.
85° S, 50° W 322

Wahnbach Talsperre, Germany
50° 50′ N, 7° 8′ E 409

Walensee, Switzerland 47° 7′ N,
9° 16′ E 330 Table 7.3

Washington, Lake, Washington, USA
47° 35′ N, 122° 14′ W 408

Wast Water, UK 54° 26′ N, 3° 17′ W
332 Table 7.5

Wellington Reservoir, Australia
33° 20′ S, 116° 2′ E Fig 29

Windermere, UK 54° 20′ N, 2° 53′ W
80, 115, 117 Table 3.2, 119, 221,
225, 245, 247, 284, 289, 292,
332, 332 Table 7.5, 333, 354,
368, 401, 402

Winnipeg, Lake, Canada 53° S,
98° W 343

Wolderwijd, Netherlands 52° 29′ N,
5° 51′ E 345

Wörthersee, Austria 46° 37′ N,
14° 10′ E 325

Zürichsee, Switzerland 47° 12′ N,
8° 42′ E 59, 235, 330 Table 7.3,
366

RIVERS AND ESTUARIES

Angara, Russia 55° 50′ N, 110° 0′ E
322

Ashes Hollow, UK 52° 30′ N, 2° 50′ W
  48 Table 2.2
Danube, Romania/Russia 43° 15′ N,
  29° 35′ E 312
Dnepr, Ukraine 46° 30′ N, 32° 0′ E
  313
Dnestr, Moldova/Ukraine 46° 3′ N,
  30° 23′ E 312
Don, Ukraine 47° 11′ N, 39° 10′ E 313
Guadiana, at Mourão, Portugal
  38° 24′ N, 7° 22′ W 242
Mississippi, USA 29° 0′ N, 89° 20′ W
  126
Selenga, Russia 52° 15′ N, 106° 40′ E
  322
Severn, UK 51° 20′ N, 3° W 112, 242
Thames, at Reading, UK 51° 27′ N,
  0° 29′ W 48 Table 2.2
White Nile, Sudan/Uganda 15° 30′ N,
  32° 50′ E 339

OCEANS AND SEAS

Arctic Ocean
  Polar Basin 80° N, 180° W 306
Atlantic Ocean 111, 134
  Baltic Sea 55° N, 20° E 134, 310,
    312, 426
  Bay of Fundy 45° N, 66° W 42
  Benguela Current (Gabon) 0° N,
    0° E 134, 309
  Canaries Current 30° N, 15° W
    309
  Grand Banks of Newfoundland
    45° N, 52° W 310, 427

Gulf of Bothnia 62° N, 20° E 112,
  312
Gulf of Finland 60° N, 28° E 312
Gulf of Maine 43° N, 68° W 310,
  407
Gulf of Mexico 25° N, 90° W
Gulf Stream (Bahamas) 25° N,
  75° W
English Channel 50° N, 2° W 234,
  310, 311
Irish Sea – Clyde Estuary
  55° 50′ N, 5° 00′ W 112
Irish Sea – Severn Estuary
  51° 20′ N, 4° 00′ W 48 Table 2.2,
  117 Table 3.2
Labrador Sea 60° N, 55° W 306
Naragansett Bay (RI) 41° 23′ N,
  71° 24′ W 310
North Atlantic Drift Current
  45° N, 45° W 289, 306
North Atlantic Subtropical Gyre
  (Mauritania/Senegal) 15° N,
  20° W 309
North Atlantic Tropical Gyre 5° N,
  15° W 306
North Sea 56° N, 4° E 1, 310, 407,
  427
Norwegian Sea 70° N, 0° E 306,
  307
Øresund 56° 00′ N, 12° 50′ E
Oslofjord 59° N, 11° E 101, 312
Ria de Vigo 42° N, 9° W 311, 313
Sargasso Sea 30° N, 60° W 111, 117
  Table 3.2, 162
St Lawrence Estuary 49° N, 64° W
  134, 407

Trondheimsfjord 64° N, 9° E 311
Ullsfjord 71° N, 27° E 311
Mediterranean Sea
  Black Sea 43° N, 35° E 259, 312
  Golfo di Napoli 40° 45′ N, 14° 15′ E
    313
  Tyrrhenian Sea 40° N, 12° E
    313
Indian Ocean 111, 134, 162
  Andaman Sea 12° N, 96° E 309
  Arabian Sea 15° N, 65° E 309
  Arafura Sea 10° S, 135° E 310
  Red Sea 20° N, 38° E 40
Pacific Ocean
  ALOHA 22° 45′ N, 158° W
    305
  Baja California 25° N, 110° W
    407
  California Current (San Francisco)
    35° N, 115° W 309
  East China Sea 30° N, 125° E
    310
  Kuroshio Current 40° N, 150° E
    306, 307
  North Pacific Subtropical Gyre
    25° N, 160° W 133, 162, 304,
    354, 382
  Peru Current (Galapagos Islands)
    5° S, 90° W 134, 309
  Polar Front ~40° S 308
  Sea of Okhotsk 55° N, 90° W 134,
    310
  South Pacific Gyre 25° S, 140° W
    306
Southern Ocean 60° S, 30° E 168,
  306, 307

Index to genera and species of phytoplankton

Genera mentioned in the text are listed together with their systematic position (**Phylum** ORDER) and authorities for each species are cited. Synonyms are included where appropriate. Entries are suffixed 'M' or 'F' to indicate marine or freshwater occurrence, or 'M/F' for genera appearing in both. The bold entry for freshwater species (A, B, etc.) refers to the trait selected functional grouping of Reynolds et al. (2002) as summarised in Table 7.1.

Achnanthes (**Bacillariophyta** BACILLARIALES) M/F 7 Table 1.1

Achnanthes taeniata Grun. M 306, 312

Actinocyclus (**Bacillariophyta** BIDDULPHIALES) M/F 309, 312, 317, 353

Alexandrium (**Dinophyta** GONYAULACALES) M 233, 312, 313

Alexandrium tamarense (Lebour) Balech M 201, 311, 407

Amphidinium (**Dinophyta** GYMNODINIALES) M 7 Table 1.1

Amphisolenia (**Dinophyta** DINOPHYSIALES) M 203 Table 5.2, 234

Anabaena (**Cyanobacteria** NOSTOCALES) mostly F 6 Table 1.1, 26 Table 1.2, 58, 59, 65, 67, 81, 121, 129, 130, 165, 172, 205, 206, 213, 216, 226, 228, 229, 230, 232, 241, 249, 271, 293, 297, 328, 329, 330, 333, 340, 341, 342, 344, 366, 401, 404, 405

Anabaena circinalis Rabenh. ex Born. et Flah. F, **H1** 24, 26 Table 1.2, 57 Table 2.4, 339

Anabaena cylindrica Lemmermann F, **S$_N$**? 187

Anabaena flos-aquae (Lyngb.) Bréb. ex Born. et Flah. F, **H1** 54 Table 2.3, 157 Table 4.2, 165, 184 Table 5.0, 194, 195, 219 Table 5.4, 248, 320 Table 7.1, 333, 338

Anabaena lemmermannii Richter F, **H2** 58, 233, 312, 320 Table 7.1, 322, 332, 426

Anabaena minutissima Pridmore F, **S$_N$** 320 Table 7.1

Anabaena ovalisporum Forti F, **H2** 327

Anabaena planctonica Brunnth. F, **H2**? 338

Anabaena solitaria Kleb. F, **H2** 233, 332, 335, 336

Anabaena spiroides Kleb. F, **H1** 338

Anabaenopsis (**Cyanobacteria** NOSTOCALES) F 6 Table 1.1, 26 Table 1.2, 81, 205, 341, 344, 401

Anacystis (**Cyanobacteria** CHROOCOCCALES) F Genus defunct, part of *Synechococcus*

Anacystis nidulans, see also *Synechococcus* 183

Ankistrodesmus (**Chlorophyta** CHLOROCOCCALES) F 6 Table 1.1, 344, 425

Once large genus, many separated into other genera

Ankistrodesmus braunii 184 Table 5.0

Ankistrodesmus falcatus, see also *Monoraphidium contortum* 338

Ankyra (**Chlorophyta** CHLOROCOCCALES) F 6 Table 1.1, 82, 212, 218, 226, 228, 229, 230, 231, 232, 247, 274, 284, 285, 300, 320 Table 7.1, 333, 344, 359, 370

Ankyra judayi (G. M. Smith) Fott F, **X1** 20 Table 1.1, 26 Table 1.2, 219 Table 5.4, 277 Table 6.3, 284 Table 6.4, 285 Table 6.5, 338

Anthosphaera (**Haptophyta** COCCOLITHOPHORIDALES) M 312

Apedinella (**Chrysophyta** PEDINELLALES) M 7 Table 1.1

Apedinella spinifera (Throndsen) Throndsen M 35 Table 1.7

Aphanizomenon (**Cyanobacteria** NOSTOCALES) mostly F 6 Table 1.1, 26 Table 1.2, 59, 67, 81, 129, 165, 205, 216, 230, 233, 320 Table 7.1, 325, 330, 341, 401, 404, 405

Aphanizomenon flos-aquae Ralfs ex Born. et Flah. F, **H1** 24, 26 Table 1.2, 184 Table 5.0, 186, 219 Table 5.4, 312, 333, 338, 339, 426

Aphanizomenon gracile Lemm. F, probably **H2** 233

Aphanocapsa (**Cyanobacteria** CHROOCOCCALES) F Several species, all colonial; F, **K**; free cells are picoplanktic *Synechoccus*, **Z**

Aphanothece (**Cyanobacteria** CHROOCOCCALES) F, **K** 6 Table 1.1, 26 Table 1.2, 320 Table 7.1, 344

Arthrodesmus (**Chlorophyta** ZYGNEMATALES) F 6 Table 1.1

Arthrospira (**Cyanobacteria** OSCILLATORIALES) mostly F 6 Table 1.1, 26 Table 1.2, 320 Table 7.1

Asterionella (**Bacillariophyta** BACILLARIALES), M/F 7 Table 1.1, 23, 24, 33 Table 1.6, 54, 63, 64, 65, 66, 67, 80, 158, 172, 191, 192, 195, 196, 197, 201, 202, 206, 207, 208, 213, 218, 220, 221, 222, 224, 225, 226, 228, 231, 233, 242, 245, 246, 248, 249, 269, 270, 274, 280, 290, 292, 293, 294, 298, 299, 300, 325, 329, 333, 338, 339, 344, 365, 378

Asterionella formosa Hassall. F, **B** and **C** 20 Table 1.1, 26 Table 1.2, 29 Table 1.3, 31, 31 Table 1.5, 32, 52, 54 Table 2.3, 61 Table 2.5, 63, 66, 107, 113, 119, 129, 157 Table 4.2, 184 Table 5.1, 195, 197, 202, 219 Table 5.4, 220, 221 Table 5.5, 221 Table 5.6, 232, 245, 246, 247, 277 Table 6.3, 320 Table 7.1, 324, 326, 328, 330, 331, 338

Asterionella (**Bacillariophyta**) (cont.)
Asterionella japonica Cleve M 234, 311

Aulacoseira (**Bacillariophyta** BIDDULPHIALES) F 7 Table 1.1, 23, 28, 54, 62, 63, 130, 177, 213, 214, 243, 245, 249, 250, 325, 329, 336, 341, 342, 344, 415
Aulacoseira alpigena (Grun.) Krammer F, A (also recorded as *Melosira distans*) 203 Table 5.2
Aulacoseira ambigua (Grun.) Simonsen 232, 320 Table 7.1, 326
Aulacoseira baicalensis (K. I. Mey.) Simonsen F, A 322
Aulacoseira binderana Kütz. F, **B** 26 Table 1.2
Aulacoseira granulata (Ehrenb.) Simonsen F, **P** 6 Table 1.1, 29 Table 1.3, 123, 219 Table 5.4, 232, 320 Table 7.1, 324, 327, 328, 330, 336, 339, 340, 341, 373
Aulacoseira islandica (O. Müller) Simonsen F, **C** 225, 233, 320 Table 7.1, 322, 323, 324, 330
Aulacoseira solida (Eulenstein) Krammer F, **B** 225, 233, 320 Table 7.1, 322, 323, 324, 330
Aulacoseira subarctica (O. Müller) Haworth F, **B** 29 Table 1.3, 54 Table 2.3, 61 Table 2.5, 62, 129, 202, 225, 233, 247, 320 Table 7.1, 324, 325, 331, 333
Aureococcus (**Bacillariophyta** BIDDULPHIALES) M 407

Bacillaria (**Bacillariophyta** BIDDULPHIALES) M
Bacteriastrum (**Bacillariophyta** BIDDULPHIALES) M 203 Table 5.2, 311
Binuclearia (**Chlorophyta** ULOTRICHALES) F 233, 259
Bitrichia (**Chrysophyta** HIBBERDIALES) F 7 Table 1.1
Botryococcus (**Chlorophyta** CHLOROCOCCALES) F 6 Table 1.1, 54, 320 Table 7.1, 322, 339, 344
Botryococcus braunii Kützing F, **F** 331, 425

Calciopappus (**Haptophyta** COCCOLITHOPHORIDALES) M 312
Carteria (**Chlorophyta** VOLVOCALES) F 234, 311
Cerataulina (**Bacillariophyta** BIDDULPHIALES) M 7 Table 1.1, 234, 312, 317
Cerataulina bergonii H. Perag. M 312
Cerataulina pelagica (Cleve) Hendey M 311, 312
Ceratium (**Dinophyta** GONYAULACALES, M/F 7 Table 1.1, 13, 24, 60, 68, 88, 206, 213, 216, 218, 219, 228, 230, 231, 232, 233, 234, 249, 293, 295, 311, 320 Table 7.1, 329, 330, 331, 332, 333, 338, 339, 348, 366, 370, 371, 381, 385
Ceratium arcticum (Ehrenb.) Cleve M 306, 307
Ceratium azoricum Cleve M 307
Ceratium carriense Gourret M 307
Ceratium furca (Ehrenb.) M 307, 310, 311, 316
Ceratium fusus (Ehrenb.) M 234, 306, 307, 310, 311, 312, 313, 316
Ceratium hexacanthum Gourret M 307
Ceratium hirundinella (O. F. Müller) Dujardin F, L_M 6 Table 1.1, 20 Table 1.1, 83, 129, 184 Table 5.0, 205, 216, 219 Table 5.4, 229, 248, 325, 338, 339
Ceratium lineatum (Ehrenb.) M 307
Ceratium longipes (Bail.) Gran M 306, 307, 310, 312
Ceratium tripos (O. F. Müller) Nitzsch M 305, 307, 310, 311, 312, 313, 316
Chaetoceros (**Bacillariophyta** BIDDULPHIALES) mostly M 7 Table 1.1, 60, 306, 307, 310, 311, 312, 313, 314, 317, 353
Chaetoceros atlanticum Cleve M 308
Chaetoceros compressus Lauder M 234, 310, 311, 312
Chaetoceros convolutum Castracene M 307
Chaetoceros curvisetum Cleve M 313
Chaetoceros debilis Cleve M 309, 310, 311, 312

Chaetoceros diadema (Ehrenb.) Gran M 306, 310
Chaetoceros neglectus Karsten M 308
Chaetoceros radicans Schütt M 311
Chaetoceros socialis Lauder M 311, 312, 313
Chilomonas (**Cryptophyta** CRYPTOMONADALES) F 6 Table 1.1
Chlamydocapsa (**Chlorophyta** TETRASPORALES) F 293
Chlamydocapsa planctonica (W. and .G. S.West) Fott F, **F** (formerly known as *Gloeocystis planctonica*) 57 Table 2.4
Chlamydomonas (**Chlorophyta** VOLVOCALES) M/F 6 Table 1.1, 35 Table 1.7, 49, 78, 212, 232, 233, 312, 326, 333, 343, 352, 425
Chlamydomonas reinhardtii P. A. Dangeard F, **X1** 127, 157 Table 4.2
Chlorella (**Chlorophyta** CHLOROCOCCALES) F 32, 33 Table 1.6, 113, 127, 137, 148, 150, 182, 184 Table 5.0, 188, 196, 207, 208, 209, 212, 230, 231, 260, 298, 320 Table 7.1, 324, 333, 343, 344, 352, 357, 365, 370, 378, 425
Chlorella minutissima Fott et Nováková F, **X3** 203 Table 5.2, 332
Chlorella pyrenoidosa Chick F **X1** 20 Table 1.1, 26 Table 1.2, 157 Table 4.2, 424
Chlorella vulgaris Beijerinck F, **X1** (includes strains formerly known as *Chlorella pyrenoidosa*) 54 Table 2.3, 61 Table 2.5
Chlorobium (**Anoxyphotobacteria**) F 321 Table 7.1
Chlorobotrys (**Eustigmatophyta**) F 6 Table 1.1, 424
Chlorococcum (**Chlorophyta** CHLOROCOCCALES) F 54 Table 2.3, 61 Table 2.5, 352
Chlorogonium (**Chlorophyta** VOLVOCALES) F 326
Choricystis (**Chlorophyta** CHLOROCOCCALES) F 6 Table 1.1, 328
Chromatium (**Anoxyphotobacteria**) F 321 Table 7.1

Chromulina (**Chrysophyta**
 CHROMULINALES) F 7 Table 1.1,
 20 Table 1.1, 247, 251, 284, 300,
 328, 425
Chroococcus (**Cyanobacteria**
 CHROOCOCCALES) F 6 Table
 1.1, 26 Table 1.2, 171, 325, 327
Chroomonas (**Cryptophyta**
 CRYPTOMONADALES) M/F 6
 Table 1.1, 338
 Chroomonas salina (Wislouch)
 Butcher M 35 Table 1.7
Chrysochromulina (**Haptophyta**
 PRYMNESIALES) M/F 7 Table 1.1,
 13, 233, 234, 251, 312, 320 Table
 7.1, 323, 324, 327, 330, 331, 336,
 338, 343, 402
 Chrysochromulina herdlensis
 Leadbeater M 35 Table 1.7
 Chrysochromulina parva Lackey F,
 X2 20 Table 1.1, 338
 Chrysochromulina polylepis Manton
 et Parke M 312, 407
Chrysococcus (**Chrysophyta**
 CHROMULINALES) F 7 Table 1.1,
 20 Table 1.1, 251, 320 Table 7.1,
 332
Chrysolykos (**Chrysophyta**
 CHROMULINALES) F 7 Table 1.1,
 203 Table 5.2
Chrysosphaerella (**Chrysophyta**
 CHROMULINALES) F 7 Table 1.1,
 129, 213
Clathrocystis (**Anoxyphotobacteria**) F
 6 Table 1.1
Closterium (**Chlorophyta**
 ZYGNEMATALES) F 6 Table 1.1,
 23, 192, 340, 341, 344
 Closterium aciculare T.West
 F, **P** 20 Table 1.1, 26
 Table 1.2, 219 Table 5.4,
 232, 320 Table 7.1, 328,
 330, 339
 Closterium acutum Bréb. F, **P** 15 Fig
 1.1
Coccolithus (**Haptophyta**
 COCCOLITHOPHORIDALES) M 7
 Table 1.1
 Coccolithus pelagicus (Wallich)
 Schiller M 35 Table 1.7
Coccomyxa (**Chlorophyta**
 CHLOROCOCCALES) F 352
 Some planktic species are now in
 Coenocystis

Coelastrum (**Chlorophyta**
 CHLOROCOCCALES) F 320 Table
 7.1, 341, 344
 Coelastrum microporum Nägeli F, **J**
Coenochloris (**Chlorophyta**
 CHLOROCOCCALES) F 24,
 56, 82, 129, 203 Table 5.2,
 210, 228, 273, 293, 320
 Table 7.1, 325, 326, 340,
 359
 Coenochloris fottii (Hindák)
 Tsarenko F, **F** (formerly known
 as *Sphaerocystis schroeteri*) 57
 Table 2.4, 219 Table 5.4, 230,
 248, 331
Coenococcus (**Chlorophyta**
 CHLOROCOCCALES) F 231
Coenocystis (**Chlorophyta**
 CHLOROCOCCALES) F 203 Table
 5.2, 212
 Genus includes some species
 previously attributed to
 Coccomyxa
Corethron (**Bacillariophyta**
 BIDDULPHIALES) M 307, 311
 Corethron criophilum Castracene M
 306, 308
 Corethron hystrix Hensen M 307
Coscinodiscus (**Bacillariophyta**
 BIDDULPHIALES) mostly M 307,
 309, 310, 312, 317, 341
 Coscinodiscus subbulliens Jørgensen
 M 306
 Coscinodiscus wailseii Gran et Angst.
 M 50, 53
Cosmarium (**Chlorophyta**
 ZYGNEMATALES) F 6 Table 1.1,
 129, 202, 203 Table 5.2, 233,
 279, 320 Table 7.1, 331, 336,
 385, 424
 Cosmarium abbreviatum Raciborski
 F, **N** 195, 201, 247, 332
 Cosmarium contractum Kirchner F,
 N 332
 Cosmarium depressum (Nägeli)
 Lundell F, **N** 20 Table 1.1
Crucigena (**Chlorophyta**
 CHLOROCOCCALES) F 231
Cryptomonas (**Cryptophyta**
 CRYPTOMONADALES) M/F 6
 Table 1.1, 228, 229, 230, 247,
 260, 269, 274, 284, 300, 312, 320
 Table 7.1, 323, 324, 328, 330,
 335, 338, 359

Cryptomonas erosa Ehrenb. F, **Y** 334
Cryptomonas marssoni Skuja F, **Y**
 334
Cryptomonas ovata Ehrenb. F, **Y** 6
 Table 1.1, 20 Table 1.2, 184
 Table 5.0, 219 Table 5.4, 226,
 276, 277 Table 6.3, 334, 338,
 339
Cryptomonas profunda Butcher M
 35 Table 1.7
Cyanobium (**Cyanobacteria**
 CHROOCOCCALES), F 213, 269,
 327
 Picoplanktic **Z**, free cells of
 Cyanodictyon or *Synechocystis*
Cyanodictyon (**Cyanobacteria**
 CHROOCOCCALES) F 213, 269
 Two species, all colonial; F, **K?**;
 free cells are picoplanktic
 Cyanobium, **Z**
Cyanophora (**Glaucophyta**) F, **X1** 6
 Table 1.1, 11
Cyanothece (**Cyanobacteria**
 CHROOCOCCALES), F 269
 Relatively new genus
 accommodating larger species
 of *Synechococcus*, **X2?**
Cyathomonas (**Cryptophyta**
 CRYPTOMONADALES) M/F 159,
 425
Cyclococcolithus (**Haptophyta**
 COCCOLITHOPHORIDALES) M
 234
 Cyclococcolithus fragilis (Lohmann)
 Deflandre M 234
Cyclostephanos (**Bacillariophyta**
 BIDDULPHIALES) F 340, 341
Cyclotella (**Bacillariophyta**
 BIDDULPHIALES) M/F 7 Table
 1.1, 23, 63, 129, 195, 197, 201,
 202, 208, 225, 232, 249, 313,
 325, 326, 353
 Cyclotella bodanica Eulenstein F, **A**
 323, 329, 385
 Cyclotella caspia Grun. M 234, 312,
 313, 317
 Cyclotella comensis Grunow F, **A**
 203 Table 5.2, 320 Table 7.1,
 325, 331
 Cyclotella glomerata Bachm. F, **A**
 203 Table 5.2, 323, 325,
 329
 Cyclotella litoralis Lange et
 Syvertsen M 309, 317

Cyclotella (cont.)
  Cyclotella meneghiniana Kütz. F, **B**
    20 Table 1.1, 61 Table 2.5, 195,
    197
  Cyclotella nana Hustedt M 53
  Cyclotella praeterissima Lund F, **B**
    20 Table 1.1, 202, 247, 331
  Cyclotella radiosa (Grun.)
    Lemmermann F, A (formerly
    known as Cyclotella comta) 203
    Table 5.2, 233, 323, 326, 331
Cylindrocystis (**Chlorophyta**
  ZYGNEMATALES) F 424
Cylindrospermopsis (**Cyanobacteria**
  NOSTOCALES) F, S$_N$ 6 Table 1.1,
  26 Table 1.2, 205, 320 Table 7.1,
  340, 341, 342, 344, 366, 401
Cylindrospermopsis raciborskii
  (Woloszynska) Seenayya et
  Subba Raju F, S$_N$ 216, 384

Dactylococcopsis (**Cyanobacteria**
  CHROOCOCCALES), F 325
Detonula (**Bacillariophyta**
  BIDDULPHIALES) M 7 Table 1.1
  Detonula pumila (Castracene) Gran
    M 35 Table 1.7
Diacanthos (**Chlorophyta**
  CHLOROCOCCALES) F, **X3** 352
Diacronema (**Haptophyta**
  PAVLOVALES) M 7 Table 1.1
Diatoma (**Bacillariophyta**
  BACILLARIALES) F 7 Table 1.1,
  324
  Diatoma elongatum (Lyngb.) C.
    Agardh F, **C** 172, 338
Dictyosphaerium (**Chlorophyta**
  CHLOROCOCCALES) F 6 Table
  1.1, 210, 336
  Dictyosphaerium pulchellum
    H. C.Wood F, **F** 20 Table 1.1, 331
Dinobryon (**Chrysophyta**
  CHROMULINALES) F 7 Table 1.1,
  89, 129, 130, 210, 219 Table 5.4,
  228, 229, 231, 232, 320 Table
  7.1, 323, 325, 329, 330, 333, 334,
  336, 338, 385
  Dinobryon bavaricum Imhof F, **E**
    331
  Dinobryon cylindricum Imhof F, **E**
    203 Table 5.2, 322, 326, 331
  Dinobryon divergens Imhof F, **E** 20
    Table 1.1, 184 Table 5.0, 331,
    336, 338

Dinobryon sertularia Ehrenb. F, **E**
  331
Dinobryon sociale Ehrenb. F, **E** 157
  Table 4.2
Dinophysis (**Dinophyta**
  DINOPHYSIDALES) M 203 Table
  5.2, 309, 311, 313
  Dinophysis acuminata Clap. et Lach.
    M 310
Ditylum (**Bacillariophyta**
  BIDDULPHIALES) M 55, 130
  Ditylum brightwellii (West) Grun. M
    55, 130
Dunaliella (**Chlorophyta**
  VOLVOCALES) M 6 Table 1.1, 32,
  78, 234, 311
  Dunaliella tertiolecta Butcher M 35
    Table 1.7

Elakatothrix (**Chlorophyta**
  ULOTRICHALES) F 6 Table 1.1
Emiliana (**Haptophyta**
  COCCOLITHOPHORIDALES) M 7
  Table 1.1, 13, 203 Table 5.2, 311,
  312
  Emiliana huxleyi (Lohm.) Hay et
    Mohl. M 35 Table 1.7, 130, 234,
    307, 310, 313
Ethmodiscus (**Bacillariophyta**
  BIDDULPHIALES) M 50, 234
  Ethmodiscus rex (Rattray) Wiseman
    et Hendey M 50, 234
Euastrum (**Chlorophyta**
  ZYGNEMATALES) F 6 Table 1.1
Eucampia (**Bacillariophyta**
  BIDDULPHIALES) M 311, 312
  Eucampia cornuta (Cleve) Grun. M
    311
  Eucampia zodiacus Ehrenb. M
    312
Eudorina (**Chlorophyta** VOLVOCALES)
  F 6 Table 1.1, 24, 182, 211, 216,
  228, 229, 230, 232, 247, 269,
  270, 293, 299, 320 Table 7.1,
  338, 343
  Eudorina elegans Ehrenb. F, **G**? 26
    Table 1.2, 157 Table 4.2
  Eudorina unicocca G. M. Smith F, **G**
    26 Table 1.2, 57 Table 2.4, 184
    Table 5.0, 219 Table 5.4, 226,
    339
Euglena (**Euglenophyta**
  EUGLENALES) M/F 6 Table 1.1,
  343, 425

Euglena mutabilis F. Schmitz F, **W1**
  424, 425
Eunotia (**Bacillariophyta**
  BACILLARIALES) M/F 425
Eutetramorus (**Chlorophyta**
  CHLOROCOCCALES) F
  Redundant genus name – see
  Coenococcus
Eutreptia (**Euglenophyta**
  EUTREPTIALES) M 6 Table 1.1,
  234, 311, 312
Eutreptiella (**Euglenophyta**
  EUTREPTIALES) M 313
Exuviella (**Dinophyta**
  PROROCENTRALES) M 7 Table
  1.1, 307, 353
  Exuviella baltica Lohm. M 312

Florisphaera (**Haptophyta**
  COCCOLITHOPHORIDALES) M 7
  Table 1.1, 309
Fragilaria (**Bacillariophyta**
  BACILLARIALES) M/F 7 Table 1.1,
  24, 51, 63, 226, 228, 229, 232,
  233, 245, 249, 269, 300, 330, 338
  Fragilaria capucina Desmaz. F, **P** 6
    Table 1.1
  Fragilaria crotonensis Kitton F, **P** 6
    Table 1.1, 20 Table 1.1, 23, 29
    Table 1.3, 31 Table 1.5, 54 Table
    2.3, 57 Table 2.4, 61 Table 2.5,
    62, 63, 66, 129, 184 Table 5.0,
    219 Table 5.4, 232, 247, 320
    Table 7.1, 324, 333, 336, 339
  Fragilaria oceanica Cleve M 307,
    311
Fragilariopsis (**Bacillariophyta**
  BACILLARIALES) M 7 Table 1.1,
  307, 308
  Fragilariopsis cylindrus (Grun.)
    Krieger M 311
  Fragilariopsis doliolus Wallich M
    309
  Fragilariopsis nana (Steemann
    Nielsen) Paasche M 307

Geminella (**Chlorophyta**
  ULOTRICHALES) F, **T** 6 Table
  1.1, 233, 294, 320 Table 7.1
Gephyrocapsa (**Haptophyta**
  COCCOLITHOPHORIDALES) M 7
  Table 1.1, 203 Table 5.2
  Gephyrocapsa oceanica Kamptner M
    234

Glaucocystis (**Glaucophyta**) F 6 Table 1.1

Glenodinium (**Dinophyta PERIDINIALES**) F 7 Table 1.1, 13, 324, 334

Gloeocapsa (**Cyanobacteria CHROOCOCCALES**) F 325

Gloeotrichia (**Cyanobacteria NOSTOCALES**) F 6 Table 1.1, 26 Table 1.2, 67, 81, 129, 205, 213, 216, 338, 401

  Gloeotrichia echinulata (J. E. Smith) Richter F, **H2** 216, 320 Table 7.1, 338

Golenkinia (**Chlorophyta CHLOROCOCCALES**) F, **X1** 320 Table 7.1, 352

Gomphosphaeria (**Cyanobacteria CHROOCOCCALES**) F 6 Table 1.1, 26 Table 1.2, 81, 166, 205, 233, 330, 332, 338

  Main species *G. naegeliana* transferred to *Woronichinia*

Goniochloris (**Xanthophyta MISCHOCOCCALES**) F 6 Table 1.1

Gonium (**Chlorophyta VOLVOCALES**) F, **W1** 321 Table 7.1, 343

Gonyostomum (**Raphidophyta CHLOROMONADALES**) F 6 Table 1.1, 12, 321 Table 7.1

  Gonyostomum semen (Ehrenb.) Diesing F, **Q** 344

Gymnodinium (**Dinophyta GYMNODINIALES**) M/F 7 Table 1.1, 233, 311, 325, 326, 328, 336, 425

  Gymnodinium baicalense Authority not found F 322

  Gymnodinium catenatum Graham M 68, 233, 309, 408

  Gymnodinium helveticum Penard F 131, 324

  Gymnodinium sanguineum Hirasaka M 35 Table 1.6

  Gymnodinium simplex Lohm. M 35 Table 1.6

  Gymnodinium uberrimum (Allman) Kofoid et Swezey F 324

  Gymnodinium vitiligo Ballantine M 35 Table 1.6

Gyrodinium (**Dinophyta PERIDINIALES**) M 7 Table 1.1, 233, 234, 316

Gyrodinium uncatenatum Hulbert M 35 Table 1.6

Gyrosigma (**Bacillariophyta BACILLARIALES**) F, D 344

Halosphaera (**Prasinophyta, HALOSPHAERALES**) M 306

Hemiaulus (**Bacillariophyta BIDDULPHIALES**) M 166, 305, 385

  Hemiaulus hauckii Grun. M 311, 316, 317

Hemidinium (**Dinophyta PHYTODINIALES**) F 7 Table 1.1

Hemiselmis (**Cryptophyta CRYPTOMONADALES**) M 234

Heterocapsa (**Dinophyta PERIDINIALES**) M 233, 234, 312

  Heterocapsa triquetra (Ehrenb.) Stein M 310, 311, 312, 313, 316

Heterosigma (**Raphidophyta CHLOROMONADALES**) M

  Heterosigma carterae (Hulbert) Taylor M 35 Table 1.6

Histoneis (**Dinophyta DINOPHYSALES**) M 203 Table 5.2

Isochrysis (**Haptophyta PRYMNESIALES**) M 7 Table 1.1, 234

  Isochrysis galbana Parke M 35 Table 1.7

Karenia (**Dinophyta GYMNODINIALES**) M 233

  Karenia brevis (Davis) G. Hansen et Moestrup M 408

  Karenia mikimotoi (Miyake et Kominami) G. Hansen et Moestrup M 35 Table 1.6, 310, 311, 312, 316, 408

Katablepharis (**Cryptophyta CRYPTOMONADALES**) M 159

Katodinium (**Dinophyta GYMNODINIALES**) M/F 316

Kephyrion (**Chrysophyta CHROMULINALES**) F 7 Table 1.1

  Kephyrion littorale Lund F, **X1** 20 Table 1.1

Keratococcus (**Chlorophyta CHLOROCOCCALES**) F

  Keratococcus raphidioides Hansgirg F, **X1** 425

Kirchneriella (**Chlorophyta CHLOROCOCCALES**) F 6 Table 1.1, 336

Koliella (**Chlorophyta ULOTRICHALES**) F 6 Table 1.1, 320 Table 7.1

  Koliella longiseta (Vischer) Hindák F, **X3** 332

Lagerheimia (**Chlorophyta CHLOROCOCCALES**) F 344

  Lagerheimia genevensis (Chodat) Chodat F, **X1** 425

Lauderia (**Bacillariophyta BIDDULPHIALES**) M

  Lauderia annulata Cleve M 311

  Lauderia borealis Gran M 312

Lepocinclis (**Euglenophyta EUGLENALES**) F, **W1** 6 Table 1.1, 343, 425

Leptocylindrus (**Bacillariophyta BIDDULPHIALES**) M 203 Table 5.2, 312, 317

  Leptocylindrus danicus Cleve M 309, 310, 311

Limnothrix (**Cyanobacteria OSCILLATORIALES**) F 6 Table 1.1, 26 Table 1.2, 114, 191, 213, 401, 405

  Limnothrix redekei (van Goor) Meffert F, **S1** 206, 233, 320 Table 7.1, 345, 366, 416

Lingulodinium (**Dinophyta GONIAULACALES**) M 7 Table 1.1, 68, 311

  Lingulodinium polyedrum (Stein) Dodge M 234, 309, 316, 408

Lyngbya (**Cyanobacteria OSCILLATORIALES**) F 6 Table 1.1, 26 Table 1.2, 59, 325

  Lyngbya limnetica Lemm. F, **R** 211

Mallomonas (**Chrysophyta SYNURALES**) F 7 Table 1.1, 23, 129, 232, 293, 320 Table 7.1, 323

  Mallomonas akrokomos Ruttner in Pascher F, **X1** 333

  Mallomonas caudata Iwanoff F, **X2** 20 Table 1.1, 130, 203 Table 5.2, 331

Mantoniella (**Prasinophyta** CHLORODENDRALES) M 6 Table 1.1

Mantoniella squamata Manton et Parke M 35 Table 1.7

Melosira (**Bacillariophyta** BIDDULPHIALES) F 344

Melosira varians Agardh F, **D** 344

Merismopedia (**Cyanobacteria** CHROOCOCCALES) F, **L$_O$** 6 Table 1.1, 26 Table 1.2, 165, 171, 325, 326

Micractinium (**Chlorophyta** CHLOROCOCCALES) F, **X1** 352

Microcystis (**Cyanobacteria** CHROOCOCCALES) 6 Table 1.1, 24, 26 Table 1.2, 50, 55, 59, 67, 81, 82, 87, 113, 123, 129, 165, 166, 179, 184, 186, 190, 192, 205, 206, 208, 213, 214, 215, 226, 228, 229, 231, 232, 247, 249, 267, 269, 271, 272, 275, 279, 285, 292, 293, 297, 298, 299, 300, 320 Table 7.1, 327, 328, 339, 341, 344, 348, 353, 359, 366, 370, 381, 401, 403, 404, 405

Microcystis aeruginosa (Kütz.) emend. Elenkin F, **L$_M$**, **M** 26 Table 1.2, 54 Table 2.3, 57 Table 2.4, 58, 81, 129, 157 Table 4.2, 184 Table 5.0, 198, 209, 219, 219 Table 5.4, 333, 339, 345

Microcystis wesenbergii (Komárek) Komárek in Kondrat'eva 195, 219, 271

Micromonas (**Prasinophyta** CHLORODENDRALES) M 6 Table 1.1, 234, 312

Micromonas pusilla (Butcher) Manton et Parke M 35 Table 1.7

Monochrysis (**Chrysophyta** CHROMULINALES) M, F 332

Monodus (**Eustigmatophyta**) F 6 Table 1.1, 20 Table 1.1

Monodus subterraneus 184 Table 5.0

Monoraphidium (**Chlorophyta** CHLOROCOCCALES) F 6 Table 1.1, 23, 312, 320 Table 7.1, 324, 333, 338, 343, 344, 352

Monoraphidium contortum (Thuret) Komárková-Legnerová F, **X1** (taxon rationalises various species and subspecies of *Ankistrodesmus*) 26 Table 1.2

Monoraphidium griffithii (Berkeley) Komárková-Legnerová F, **X1** (taxon rationalises various species and subspecies of *Ankistrodesmus*)

Mougeotia (**Chlorophyta** ZYGNEMATALES) F, **T** 233, 320 Table 7.1, 330, 336

Nannochloris (**Chlorophyta** VOLVOCALES) M 6 Table 1.1, 234, 311, 425

Nannochloris ocelata Droop M 35 Table 1.7

Nephrodiella (**Xanthophyta** MISCHOCOCCALES) F 6 Table 1.1

Nephroselmis (**Prasinophyta** CHLORODENDRALES) M/F 6 Table 1.1

Nitzschia (**Bacillariophyta** BACILLARIALES) M/F 7 Table 1.1, 194, 234, 307, 308, 313, 340, 424

Nitzschia acicularis (Kütz.) W. Smith F 320 Table 7.1, 322, 342

Nitzschia bicapitata Lagerstedt M 309

Nitzschia closterium W. Smith M 308

Nodularia (**Cyanobacteria** NOSTOCALES) M/F 6 Table 1.1, 26 Table 1.2, 165, 205

Nodularia spumigenea Mertens ex Born et Flah. M/F **H1** 312, 426

Ochromonas (**Chrysophyta** CHROMULINALES) F, **X3** 7 Table 1.1, 131, 159, 251, 312, 323, 328, 330, 331, 425

Oocystis (**Chlorophyta** CHLOROCOCCALES) F 6 Table 1.1, 104, 233, 312, 324, 325, 326, 340

Oocystis lacustris Chodat F, **F** 129, 203 Table 5.2, 231, 320 Table 7.1

Ophiocytium (**Xanthophyta** MISCHOCOCCALES) F 6 Table 1.1

Ornithocercus (**Dinophyta** DINOPHYSIALES) M 203 Table 5.2, 234, 306, 314, 316

Oscillatoria (**Cyanobacteria** OSCILLATORIALES) F 115

Many planktic genera reclassified into *Limnothrix*, *Planktothrix* and *Tychonema*

Pandorina (**Chlorophyta** VOLVOCALES) F 6 Table 1.1, 232, 338, 343

Pandorina morum (O. F. Müller) Bory F, **G** 219 Table 5.4, 329

Paulschulzia (**Chlorophyta** TETRASPORALES) F 6 Table 1.1, 129

Paulschulzia pseudovolvox (Schulz) Skuja F, **F** 331

Pavlova (**Haptophyta** PAVLOVALES) M 7 Table 1.1, 234

Pavlova lutheri (Droop) Green M (formerly known as *Monochrysis lutheri*) 32, 35 Table 1.7

Pediastrum (**Chlorophyta** CHLOROCOCCALES) F 6 Table 1.1, 51, 192, 211, 320 Table 7.1, 324, 339, 341, 344

Pediastrum boryanum (Turpin) Meneghin F, **J** 20 Table 1.1

Pediastrum duplex Meyen F, **J** 67

Pedinella (**Chrysophyta** PEDINELLALES) F 7 Table 1.1

Pedinomonas (**Prasinophyta** PEDINOMONADALES) F 6 Table 1.1

Pelagococcus (**Chrysophyta** PEDINELLALES) M 7 Table 1.1, 35 Table 1.7

Pelagomonas (**Chrsophyta** PEDINELLALES) M 7 Table 1.1, 305, 308

Pelodictyon (**Anoxyphotobacteria**) F 6 Table 1.1

Peridinium (**Dinophyta** PERIDINIALES) M/F 7 Table 1.1, 13, 68, 157 Table 4.2, 165, 213, 233, 307, 311, 326, 327, 342, 366, 385

Peridinium aciculiferum Lemmermann F, **W2** 326

Peridinium bipes Stein F, **L$_M$**? 336

Peridinium cinctum (O. F. Müller) Ehrenb. F, **L$_M$** 20 Table 1.1, 219 Table 5.4, 338

Peridinium depressum Bail. M 306, 307

Peridinium faroense Paulsen M 310

Peridinium gatunense Nygaard F, L₀ 205, 218, 327

Peridinium inconspicuum Lemmermann F, L₀ 332, 336, 338

Peridinium lomnickii Wooszyska F, **Y** 320 Table 7.1, 334

Peridinium ovatum Pouchet M 306

Peridinium pallidum Ostenfeld M 306

Peridinium umbonatum Stein F, L₀ 232, 338, 425

Peridinium volzii Lemmermann F, L₀ 336

Peridinium willei Huitfeldt-Kaas F, L₀ 232, 320 Table 7.1, 325, 331, 332, 336

Phacotus (**Chlorophyta** VOLVOCALES) F 6 Table 1.1, 172, 343

Phacus (**Euglenophyta** EUGLENALES) F 6 Table 1.1, 343

Phacus longicauda (Ehrenb.) Dujardin F, **W1** 15 Fig 1.2

Phaeocystis (**Haptophyta**, PRYMNESIALES) M 7 Table 1.1, 13, 24, 234, 271, 306, 308, 311, 385, 407

Phaeocystis antarctica Karsten M 308

Phaeocystis pouchetii (Hariot) Lagerheim M 35 Table 1.7, 312

Phaeodactylum (**Bacillariophyta** BACILLARIALES) 177

Phaeodactylum tricornutum Bohlin M 28

Plagioselmis (**Cryptophyta** CRYPTOMONADALES) F 6 Table 1.1, 218, 225, 230, 232, 233, 284, 320 Table 7.1, 323, 324, 327, 330, 331, 336, 338, 343, 425

Plagioselmis nannoplanctica (Skuja) Novarino et al. F, **X2** (formerly known as *Rhodomonas minuta*) 20 Table 1.1, 172, 219 Table 5.4, 277 Table 6.3, 338

Planktolyngbya (**Cyanobacteria** OSCILLATORIALES) F 59, 115, 342

Planktolyngbya limnetica (Lemm.) Komárková-Legnerová et Cronberg F, **R** 193

Planktothrix (**Cyanobacteria** OSCILLATORIALES) F 6 Table 1.1, 23, 26 Table 1.2, 58, 83, 114, 191, 207, 208, 209, 213, 225, 228, 241, 259, 269, 271, 275, 350, 358, 365, 370, 381, 401, 405

Planktothrix agardhii (Gom.) Anagn. et Kom. F, **S1** (formerly known as *Oscillatoria agardhii*) 54 Table 2.3, 115, 157 Table 4.2, 184 Table 5.0, 186, 190, 191, 192, 193, 206, 207, 231, 233, 320 Table 7.1, 324, 325, 333, 339, 345, 366, 378, 408, 410, 416, 422

Planktothrix mougeotii (Bory ex Gom.) Anagn. et Kom. F, **R** (formerly known as *Oscillatoria agardhii var. isothrix*) 26 Table 1.2, 33, 59, 219 Table 5.4, 226, 233, 320 Table 7.1, 332

Planktothrix prolifica (Gom.) Anagn. et Kom. F, **R** (formerly known as *Oscillatoria prolifica*) 59, 193

Planktothrix rubescens (Gom.) Anagn. et Kom. F, **R** (formerly known as *Oscillatoria rubescens*) 59, 115, 193, 211, 233, 234, 320 Table 7.1, 325, 328, 330, 346, 366, 409

Platymonas (**Prasinophyta** CHLORODENDRALES) M 311

Plectonema (**Cyanobacteria** OSCILLATORIALES) F 166

Porosira (**Bacillariophyta** BIDDULPHIALES) M 310, 311

Porosira glacialis (Grun.) Jørgensen M 310, 311

Prochlorococcus (**Cyanobacteria** PROCHLORALES) M 6 Table 1.1, 11, 26 Table 1.2, 203 Table 5.2, 213, 305, 306, 316, 322

Prochloron (**Cyanobacteria** PROCHLORALES) M 6 Table 1.1, 11, 26 Table 1.2

Prochlorothrix (**Cyanobacteria** PROCHLORALES) F, **T?** 6 Table 1.1, 11, 26 Table 1.2

Prorocentrum (**Dinophyta** PROROCENTRALES) M 7 Table 1.1, 233, 234, 305, 311, 313

Prorocentrum balticum (Lohm.) Loeblich M 311

Prorocentrum micans Ehrenb. M 312, 313

Prorocentrum minimum (Pavillard) Schiller M 312

Protoperidinium (**Dinophyta** PROROCENTRALES) M 313

Prymnesium (**Haptophyta** PRYMNESIALES) M 7 Table 1.1, 35 Table 1.7, 251, 270, 353, 402, 407

Prymnesium parvum N. Carter M 313, 316

Pseudanabaena (**Cyanobacteria** OSCILLATORIALES) F 6 Table 1.1, 26 Table 1.2, 233, 320 Table 7.1, 345

Pseudanabaena limnetica (Lemm.) Komárek F, **S1** 416

Pseudonitzschia (**Bacillariophyta** BACILLARIALES) M 307, 311, 312

Pseudonitzschia delicatissima (P. T. Cleve) M (formerly known as *Nitzschia delicatissima*) 307, 311, 312

Pseudopedinella (**Chrysophyta** PEDINELLALES) M/F 7 Table 1.1, 312, 332

Pseudopedinella pyriformis N. Carter M 35 Table 1.7

Pseudosphaerocystis (**Chlorophyta** TETRASPORALES) F 6 Table 1.1, 56, 129, 211, 293, 320 Table 7.1, 324

Pseudosphaerocystis lacustris (Lemm.) Novák F, **F** (formerly known as *Gemellicystis neglecta*) 57 Table 2.4, 231, 248, 331

Pyramimonas (**Prasinophyta** PYRAMIMONADALES) M 308, 313

Pyrenomonas (**Cryptophyta**, CRYPTOMONADALES) M 6 Table 1.1

Pyrenomonas salina (Wislouch) Santore M

Pyrocystis (**Dinophyta** GONIAULACALES) M 13, 234, 305, 316

Pyrocystis noctiluca Murray M 55

Pyrodinium (**Dinophyta** GONIAULACALES) M 316, 408

Pyrodinium bahamense (Böhm) Steidinger et al. M 316, 408

Radiococcus (**Chlorophyta**
CHLOROCOCCALES) F
Radiococcus planctonicus Lund F, **F**
(formerly known as *Coenococcus*
planctonicus) 248, 331
Rhizosolenia (**Bacillariophyta**
BIDDULPIALES) M 7 Table 1.1,
203 Table 5.2, 234, 305, 306,
307, 313
Rhizosolenia alata Brightw. M 130,
307, 308, 310, 311, 312
Rhizosolenia calcar-avis Schultze M
311, 313, 316, 317
Rhizosolenia delicatula Cleve M 309,
310, 311, 312
Rhizosolenia fragilissima Bergon M
312
Rhizosolenia hebetata Bail. M 306,
307, 308, 310
Rhizosolenia setigera Brightw. M 60
Rhizosolenia stolterforthii H. Perag.
M 309
Rhizosolenia styliformis Brightw. M
305, 306, 307, 310, 311, 316, 317
Rhodomonas (**Cryptophyta**,
CRYPTOMONADALES) M/F 6
Table 1.1, 212, 234, 260, 311,
312, 325
Rhodomonas lens Pascher et
Ruttner F 35 Table 1.7
Richiella (**Cyanobacteria**,
OSCILLATORIALES) M 166, 305

Scenedesmus (**Chlorophyta**,
CHLOROCOCCALES) F 6 Table
1.1, 67, 182, 195, 211, 269, 270,
320 Table 7.1, 324, 339, 341,
344
"*Scenedesmus quadricauda* (Turpin)
Brébisson' F, **J** Oft-recorded
species is no longer regarded as
an entity and is being
subdivided (John et al., 2002)
20 Table 1.1, 26 Table 1.2, 60,
157 Table 4.2, 195, 277 Table
6.3, 424
Scherfellia (**Prasinophyta**
CHLORODENDRALES) F 6 Table
1.1
Scourfieldia (**Prasinophyta**
SCOURFIELDIALES) M 6 Table
1.1, 425
Scrippsiella (**Dinophyta**
PERIDINIALES) M 312

Scrippsiella trochoidea (Stein)
Loeblich M 310, 311, 312, 313,
316
Skeletonema (**Bacillariophyta**
BIDDULPIALES) M 7 Table 1.1,
312, 317
Skeletonema costatum Grev. M 33,
130, 234, 306, 309, 311, 312, 313
Snowella (**Cyanobacteria**
CHROOCOCCALES) F 6 Table
1.1, 24, 26 Table 1.2
Sphaerella (**Chlorophyta**
VOLVOCALES) F
(Most species transferred to
Haematococcus) 425
Sphaerocauum (**Cyanobacteria**
CHROOCOCCALES) F, **M** 320
Table 7.1
Sphaerocystis (**Chlorophyta**
CHLOROCOCCALES) F
(some former species now in
Coenochloris) 82, 203 Table 5.2
Sphaerocystis schroeteri Chodat F, **F**
194
Spirulina (**Cyanobacteria**
OSCILLATORIALES) M 6 Table
1.1, 26 Table 1.2, 59, 320 Table
7.1, 341, 404
Spondylosium (**Chlorophyta**
ZYGMEMATALES) F, **P** 6 Table
1.1, 273
Staurastrum (**Chlorophyta**
ZYGMEMATALES) F 6 Table 1.1,
24, 51, 129, 202, 232, 326, 336,
338, 340, 415
Staurastrum brevispinum West F, **P**
57 Table 2.4
Staurastrum chaetoceras (Schröder)
G. M. Sm. F, **P** 201
Staurastrum cf. *cingulum* (W. et
G. S. West) F, **P** 247
Staurastrum dorsidentiferum W. et
G. S. West F, **P** 328
Staurastrum longiradiatum W. et
G. S. West F, **P?** 219
Staurastrum pingue Teiling F, **P** 20
Table 1.1, 26 Table 1.2, 195, 201,
219 Table 5.4, 228, 229, 320
Table 7.1, 333
Staurodesmus (**Chlorophyta**
ZYGMEMATALES) F 6
Table 1.1, 129, 203 Table 5.2,
233, 293, 320 Table 7.1, 325,
331, 385

Staurodesmus triangularis
(Lagerheim) Teiling F, **N** 336
Stephanodiscus (**Bacillariophyta**
BIDDULPIALES) F 7 Table 1.1,
23, 53, 54, 63, 148, 214, 233,
243, 245, 249, 279, 329, 344
Stephanodiscus binderanus (Kütz.) W.
Krieg. F, **B** (formerly known as
Melosira binderana) 198, 322,
324
Stephanodiscus hantzschii Grun. F, **D**
6 Table 1.1, 20 Table 1.1, 29
Table 1.3, 31 Table 1.5, 184
Table 5.0, 320 Table 7.1, 338
Stephanodiscus minutulus Cleve et
Moller F, **B** (formerly known as
S. astraea var. *minutula*) 333,
338
Stephanodiscus rotula (Kütz.)
Hendey F, **C** (formerly known as
S. astraea) 6 Table 1.1, 20 Table
1.1, 29 Table 1.3, 31 Table 1.5,
50, 52, 54 Table 2.3, 61 Table
2.5, 66, 232, 320 Table 7.1, 338,
339
Stichococcus (**Chlorophyta**
ULOTRICHALES) F, **X1** 6 Table
1.1
Surirella (**Bacillariophyta**
BACILLARIALES) F 341, 344
Synechococcus (**Cyanobacteria**
CHROOCOCCALES) M/F 6 Table
1.1, 11, 20 Table 1.1, 26 Table
1.2, 97, 155, 165, 181, 183, 184
Table 5.0, 186, 203 Table 5.2,
269, 277 Table 6.3, 305, 306,
320 Table 7.1, 326
Picoplanktic **Z**, free cells of
Aphanocapsa. Larger species in
Cyanothece
Synechocystis (**Cyanobacteria**
CHROOCOCCALES) M/F 6 Table
1.1, 26 Table 1.2, 269
Several species, all colonial; F, **K?**;
free cells are picoplanktic
Cyanobium, **Z**
Synechocystis limnetica Popovskaya
322
Synedra (**Bacillariophyta**
BACILLARIALES) F 7 Table 1.1,
62, 192, 233, 328, 344
Synedra acus 54 Table 2.3, 320
Table 7.1, 338
Synedra filiformis Grun. F, **D**

Synedra ulna (Nitzsch) Ehrenb. F, **B** 20 Table 1.1, 322

Synura (**Chrysophyta** SYNURALES) F 7 Table 1.1, 231, 232, 320 Table 7.1, 321 Table 7.1, 343

Synura petersenii Korshikov F, **W1** 130

Synura uvella Ehrenb. F, **E** 331

Tabellaria (**Bacillariophyta** BACILLARIALES) F 7 Table 1.1, 293, 320 Table 7.1, 325, 329, 330

Tabellaria fenestrata (Lyngb.) Kütz. F, **P?** 233, 323

Tabellaria flocculosa (Roth.) Kütz. F, **N** 29 Table 1.3, 54 Table 2.3, 233, 323, 332, 333

Tabellaria flocculosa var. *asterionelloides* (Grun.) Knuds. F, **N** 6 Table 1.1, 20 Table 1.1, 61 Table 2.5, 64, 184 Table 5.0, 247

Tetraedron (**Chlorophyta**, CHLOROCOCCALES) F 324

Tetrastrum (**Chlorophyta**, CHLOROCOCCALES) F 6 Table 1.1, 51

Thalassionema (**Bacillariophyta** BIDDULPHIALES) M 307, 311

Thalassionema nitzschioides (Grun.) Mereschkowsky M 307, 309, 310, 311

Thalassiosira (**Bacillariophyta** BIDDULPHIALES) M 7 Table 1.1, 60, 67, 271, 307, 311, 314, 317

Thalassiosira baltica Grun. M 312

Thalassiosira decipiens (Grun.) Jørgensen M 311

Thalassiosira fluviatilis (syn. *T. weissflogii*) q.v.

Thalassiosira hyalina Grun. M 311

Thalassiosira nordenskioeldii Cleve M 234, 307, 310, 311, 312

Thalassiosira pseudonana Hasle et Heindal M 35 Table 1.7

Thalassiosira punctigera (Castracene) Hasle M 130

Thalassiosira subtilis Ostenf. M 309

Thalassiosira weissflogii (Grun.) Fryxell et Hasle M (formerly known as *Thalassiosira fluviatilis*) 35 Table 1.7, 54 Table 2.3, 60, 130

Thalassiothrix (**Bacillariophyta** BIDDULPHIALES) M 307

Thalassiothrix longissima Cleve et Grun. M 307

Thiocystis (**Anoxyphotobacteria**) F 6 Table 1.1

Thiopedia (**Anoxyphotobacteria**) F 6 Table 1.1

Trachelomonas (**Euglenophyta** EUGLENALES) F 6 Table 1.1

Trachelomonas hispida (Perty) Stein F, **X2** 15 Fig 1.2

Trachelomonas volvocina Ehrenb. F, **W2** 321 Table 7.1

Treubaria (**Chlorophyta** CHLOROCOCCALES) F 352

Tribonema (**Xanthophyta** TRIBONEMATALES) F 6 Table 1.1, 12, 320 Table 7.1, 324

Trichodesmium (**Cyanobacteria** OSCILLATORIALES) M 6 Table 1.1, 26 Table 1.2, 81, 166, 169, 203 Table 5.2, 234, 305, 306, 316, 385

Trichodesmium thiebautii Gomont M 58, 166

Tychonema (**Cyanobacteria** OSCILLATORIALES) F 6 Table 1.1, 26 Table 1.2

Tychonema bourrellyi F, **S1** 115

Umbellosphaera (**Haptophyta** COCCOLITHOPHORIDALES) M 7 Table 1.1, 203 Table 5.2, 305, 306, 309, 333

Uroglena (**Chrysophyta** CHROMULINALES) F 7 Table 1.1, 24, 83, 129, 130, 203 Table 5.2, 213, 228, 229, 231, 292, 320 Table 7.1, 323, 325, 328, 330

Uroglena americana Calkins F, **U** 326, 328

Uroglena lindii Bourrelly F, **U** 20 Table 1.2

Urosolenia (**Bacillariophyta** BIDDULPHIALES) F 7 Table 1.1, 129, 232, 320 Table 7.1, 328, 336, 385

Uroslenia eriensis (H. L. Smith) Round et Crawford F, **A** 203 Table 5.2, 323, 331, 335

Volvox (**Chlorophyta** VOLVOCALES) F 6 Table 1.1, 12, 68, 83, 179, 205, 211, 216, 285, 292, 294, 320 Table 7.1, 343

Volvox aureus Ehrenb. F, **G** 26 Table 1.2, 157 Table 4.2, 182, 184 Table 5.0

Volvox globator Linnaeus F, **G** 20 Table 1.2

Westella (**Chlorophyta** CHLOROCOCCALES) F 352

Willea (**Chlorophyta** CHLOROCOCCALES) F 203 Table 5.2

Woloszynskia (**Dinophyta** GYMNODINIALES) F, **X2** 7 Table 1.1

Woronichinia (**Cyanobacteria** CHROOCOCCALES) F 6 Table 1.1, 24, 26 Table 1.2, 81, 166, 233, 272, 320 Table 7.1, 325, 332, 338, 401

Woronichinia naegeliana (Unger) Elenkin F, **Lₒ** (formerly *Gomphosphaeria naegeliana* and *Coelosphaerium naegelianum*) 248

Xanthidium (**Chlorophyta** ZYGNEMATALES) F, **N** 6 Table 1.1

Index to genera and species of organisms other than phytoplankton

Abramis Actinopterygii, Cypriniformes 422

Acanthometra Rhizopoda, Radiolaria 252 Table 6.1

Acartia Crustacea, Calanoidea 261, 262
Acartia clausi 262
Acartia longiremis 261

Acineta Ciliophora, Suctoria 252 Table 6.1

Actinophrys Rhizopoda, Heliozoa 252 Table 6.1

Aeromonas Bacteria, Vibrionaceae 162

Agrostis Angiospermae, Glumiflorae 335

Alburnus Actinopterygii, Cypriniformes 422

Alosa Actinopterygii, Clupeiformes 290
Alosa pseudoharengus 290

Apherusa Crustacea, Malacostraca 255 Table 6.1

Arcella Rhizopoda, Foraminifera 252 Table 6.1

Argulus Crustacea, Branchiura 255 Table 6.1

Artemia Crustacea, Anostraca 254 Table 6.1, 269

Asellus Crustacea, Isopoda 391

Asplanchna Rotatoria, Monogononta 253 Table 6.1, 281
Asplanchna priodonta 281

Asterias Echinodermata, Asteroidea 257 Table 6.1

Asterocaelum Rhizopoda, Amoebina 252 Table 6.1, 294

Aurelia Coelenterata, Scyphozoa 253 Table 6.1

Australocedrus Gymnospermae, Coniferae 335

Bacillus Bacteria, Bacillales 162

Balanus Crustacea, Cirripedia 255 Table 6.1

Beroë Coelenterata, Nuda 253 Table 6.1

Bicosoeca Zoomastigophora 252 Table 6.1, 259

Blastocladiella Fungi, Blastocladiales 293

Bodo Zoomastigophora 252 Table 6.1, 259

Boeckella Crustacea, Calanoidea 255 Table 6.1, 336
Boeckella gracilipes 336

Bosmina Crustacea, Cladocera 254 Table 6.1, 266 Table 6.2, 282, 287
Bosmina longirostris 266 Table 6.2

Brachionus, Rotatoria, Monogononta 253 Table 6.1, 259, 266 Table 6.2, 281
Brachionus calyciflorus 266 Table 6.2, 281

Bythotrephes Crustacea, Cladocera 254 Table 6.1, 261, 290

Calanus Crustacea, Calanoidea 255 Table 6.1, 261, 428
Calanus finmarchius 261

Canthocamptus Harpacticoidea 255 Table 6.1

Carchesium Ciliophora, Peritricha 252 Table 6.1

Carcinus Crustacea, Malacostraca 256 Table 6.1

Centropages Crustacea, Calanoidea 255 Table 6.1, 261, 262, 282
Centropages hamatus 261
Centropages typicus 262

Cephalodella Rotatoria, Monogononta 292

Ceratophyllum Angiospermae, Ranales 419

Ceriodaphnia Crustacea, Cladocera 254 Table 6.1, 261, 277, 287

Cetorhinus Selachii, Euselachii 263

Chaetonotus Gastrotricha 253 Table 6.1

Chaoborus Arthropoda, Diptera 256 Table 6.1, 269

Chara Chlorophyta, Charales 419

Chirocephalus Crustacea, Anostraca 254 Table 6.1

Chironomus Arthropoda, Diptera 422
Chironomus anthracinus 422
Chironomus plumosus 422

Chydorus Crustacea, Cladocera 254 Table 6.1, 261, 266 Table 6.2, 279, 287
Chydorus sphaericus 261, 266 Table 6.2, 279

Chytridium Fungi, Chytridiales 293

Ciona Urochorda, Ascidacea 258 Table 6.1

Clavelina Urochorda, Ascidacea 258 Table 6.1

Clio Mollusca, Gastropoda 256 Table 6.1

Clupea Actinopterygii, Clupeiformes 261, 263, 427
Clupea harengus 261, 427

Coleps Ciliophora, Spirotricha 259

Colpoda Ciliophora, Holotricha 252 Table 6.1

Conochilus Rotatoria, Monogononta 253 Table 6.1

Convoluta Platyhelminthes, Turbellaria 253 Table 6.1

Coregonus Actinopterygii, Clupeiformes 289
Coregonus artedii 289

Cortaderia Angiospermae, Glumiflorae 335

Craspedacusta Coelenterata, Trachylina 253 Table 6.1

Cyanea Coelenterata, Scyphozoa 253 Table 6.1

Cyclops Crustacea, Cyclopoidea 260
Cyclops vicinus 260

Cyprinus Actinopterygii, Cypriniformes 422
Cyprinus carpio 422

Cypris Crustacea, Ostracoda 254 Table 6.1

Cytophaga Bacteria, Cytophagales 142

Daphnia Crustacea, Cladocera 86, 221, 254 Table 6.1, 261, 265, 266, 266 Table 6.2, 267, 269, 270, 274, 276, 277, 277 Table 6.3, 279, 280, 282, 284, 285, 286, 287, 289, 290, 299, 359, 390, 392, 404, 409, 418, 421, 422
Daphnia cucullata 261, 287

Daphnia galeata 221, 261, 266, 266 Table 6.2, 267, 269, 276, 284, 287, 289

Daphnia hyalina 261, 266 Table 6.2

Daphnia lumholtzi 261

Daphnia magna 261, 266, 277, 421, 422

Daphnia pulex 266, 270

Daphnia pulicaria 261, 277, 289, 290, 418

Daphnia schoedleri 266

Diaphanosoma Crustacea, Cladocera 254 Table 6.1, 409, 421

Diaptomus Crustacea, Calanoidea 266 Table 6.2, 389

Diaptomus oregonensis 266 Table 6.2

Diastylis Crustacea, Malacostraca 255 Table 6.1

Difflugia Rhizopoda, Foraminifera 252 Table 6.1

Dinophysis Dinophyta 252 Table 6.1

Doliolum Urochorda, Thaliacea 258 Table 6.1

Dreissena Mollusca, Lamellibranchiata 256 Table 6.1, 323

Dreissena polymorpha 323

Echinus Echinodermata, Echinoidea 257 Table 6.1

Ensis Mollusca, Lamellibranchiata 256 Table 6.1

Eodiaptomus Crustacea, Calanoidea 287

Eodiaptomus japonicus 287

Epistylis Ciliophora, Peritricha 252 Table 6.1

Escherichia Bacteria, Entobacteriacae 49, 155, 158, 181

Escherichia coli 49, 155, 158, 181

Esox Actinopterygii, Mesichthyes 422

Esox lucius 422

Eudiaptomus Crustacea, Calanoidea 255 Table 6.1, 266 Table 6.2, 279, 280, 286, 390

Eudiaptomus gracilis 266 Table 6.2, 280, 390

Euphausia Crustacea, Malacostraca 256 Table 6.1, 263

Euphausia frigida 263

Euphausia tricantha 263

Euplotes Ciliophora, Spirotricha 252 Table 6.1, 269

Eurydice Crustacea, Malacostraca 255 Table 6.1

Eurytemora Crustacea, Calanoidea 255 Table 6.1

Evadne Crustacea, Cladocera 254 Table 6.1, 262

Festuca Angiospermae, Glumiflorae 335

Filinia Rotatoria, Monogononta 253 Table 6.1, 281

Filinia longiseta 281

Fitzroya Gymnospermae, Coniferae 335

Flavobacterium Bacteria 142, 162

Flustra Ectoprocta 257 Table 6.1

Fulica Aves, Gruiformes

Fulica atra 422

Gadus Actinopterygii, Gadiformes

Gadus morhua 427

Gammarus Crustacea, Amphipoda 389, 391

Gastrosaccus Crustacea, Malacostraca 255 Table 6.1

Gigantocypris Crustacea, Ostracoda 254 Table 6.1

Globigerina Rhizopoda, Foraminifera 252 Table 6.1

Halteria Ciliophora, Spirotricha 252 Table 6.1

Hertwigia Rotatoria, Monogononta 292

Holopedium Crustacea, Cladocera 254 Table 6.1

Holothuria Echinodermata, Holothuroidea 258 Table 6.1

Hydra Coelenterata, Hydrida 253 Table 6.1

Hydractinia Coelenterata, Anthomedusae 252 Table 6.1

Isoetes Pteridophyta, Isoetales 419

Kellicottia Rotatoria, Monogononta 253 Table 6.1, 281

Kellicottia longispina 281

Keratella Rotatoria, Monogononta 253 Table 6.1, 259, 281, 287

Keratella cochlearis 259, 281

Keratella quadrata 259, 281

Lates Actinopterygii, Perciformes 263, 264, 341

Lates niloticus 341

Lepomis Actinopterygii, Perciformes

Lepomis cyanellus 422

Leptodiaptomus Crustacea, Calanoidea 409

Leptodora Crustacea, Cladocera 254 Table 6.1, 261

Leptomysis Crustacea, Malacostraca 255 Table 6.1

Limacina Mollusca, Gastropoda 256 Table 6.1

Limnocnida Coelenterata, Trachylina 253 Table 6.1

Limnothrissa Actinopterygii, Clupeiformes 263, 264, 392

Limnothrissa miodon 263, 264, 392

Littorella Angiospermae, Campanales 419

Lobelia Angiospermae, Campanales 419

Loligo Mollusca, Cephalopoda 256 Table 6.1, 264

Lolium perenne Angiospermae, Glumiflorae 383

Macrohectopus Crustacea, Malacostraca 255 Table 6.1

Megalogrammus Actinopterygii, Gadiformes 427

Megalogrammus aeglifinus 427

Mesocyclops Crustacea, Cyclopoidea 254 Table 6.1, 260

Mesocyclops leuckarti 260

Mesodinium Ciliophora, Rhabdophorina 68

Metopus Ciliophora, Spirotricha 252 Table 6.1

Microstomum Platyhelminthes, Turbellaria 253 Table 6.1

Moina Crustacea, Cladocera 254 Table 6.1

Monas Zoomastigophora 252 Table 6.1, 259

Monosiga Zoomastigophora 252 Table 6.1, 259

Morone Actinopterygii, Perciformes 428

Mulinum Angiospermae, Umbellales 335

Myriophyllum Angiospermae, Myrtales 419, 422

Mysis Crustacea, Malacostraca 255 Table 6.1

Najas Angiospermae, Najadales 419

Nassula Ciliophora, Holotricha 252 Table 6.1, 259

Nebaliopsis Crustacea, Malacostraca 255 Table 6.1

Nitella Chlorophyta, Charales 419

Noctiluca Dinophyta 251, 252 Table 6.1

Nothofagus Angiospermae, Fagales 335

Notholca Rotatoria, Monogonta 253 Table 6.1

Notonecta Arthropoda, Hemiptera 263

Nyctiphanes Crustacea, Malacostraca 256 Table 6.1

Obelia Coelenterata, Leptomedusae 252 Table 6.1

Oikopleura Urochorda, Larvacea 258 Table 6.1, 259

Oithona Crustacea, Cyclopoidea 254 Table 6.1

Ophiura Echinodermata, Ophiuroidea 257 Table 6.1

Oreochromis Actinopterygii, Perciformes 341

Oreochromis niloticus 341

Ostrea Mollusca, Lamellibranchiata 256 Table 6.1

Oxyrrhis Dinophyta 251, 252 Table 6.1, 270

Palinurus Crustacea, Malacostraca 256 Table 6.1

Patella Mollusca, Gastropoda 256 Table 6.1

Pelagia Coelenterata, Scyphozoa 253 Table 6.1

Pelagonemertes Nemertea 253 Table 6.1

Pelagothuria Echinodermata, Holothuroidea 258 Table 6.1

Pelomyxa Rhizopoda, Amoebina 252 Table 6.1

Penilia Crustacea, Cladocera 262

Peranema Zoomastigophora 252 Table 6.1

Perca Actinopterygii, Perciformes 289, 290

Perca flavescens 290

Perca fluviatilis 289

Petromyzon Agnatha, Cyclostomata 290

Petromyzon marinus 290

Phoronis Phoronidea 257 Table 6.1

Phragmites Angiospermae, Glumiflorae 419

Physalia Coelenterata, Siphonophora 253 Table 6.1

Pleurobrachia Coelenterata, Tentaculata 253 Table 6.1

Pleuronema Ciliophora, Holotricha 252 Table 6.1, 259

Plumularia Coelenterata, Leptomedusae 252 Table 6.1

Podochytrium Fungi, Chytridiales 293

Podon Crustacea, Cladocera 254 Table 6.1, 262

Polyarthra Rotatoria, Monogonta 259, 281, 287

Pontomyia Arthropoda, Diptera 256 Table 6.1

Potamogeton Angiospermae, Najadales 419, 420

Prorodon Ciliophora, Holotricha 252 Table 6.1, 259

Protoperidinium Dinophyta 252 Table 6.1

Pseudopileum Fungi, Chytridiales 293

Pyrosoma Urochorda, Thaliacea 258 Table 6.1

Quercus Angiospermae, Fagales 331, 359

Rhizophydium Fungi, Chytridiales 293, 294, 295

Rhizophydium planktonicum emend 293

Rhizosiphon Fungi, Chytridiales 293

Rozella Fungi, Chytridiales 294

Rutilus Actinopterygii, Cypriniformes 278, 422

Rutilus rutilus 278, 422

Sagitta Chaetognatha 256 Table 6.1

Salmo 389

Salmo trutta 389

Salpa Urochorda, Thaliacea 258 Table 6.1

Salpingooeca Zoomastigophora 252 Table 6.1

Salvelinus Actinopterygii, Clupeiformes 290

Salvelinus namaycush 290

Schoenoplectus Angiospermae, Cyperales 419

Scomber Actinopterygii, Perciformes 261

Scomber scombrus 261

Sialis Arthropoda, Megaloptera 256 Table 6.1

Sida Crustacea, Cladocera 254 Table 6.1, 421

Simocephalus Crustacea, Cladocera 254 Table 6.1, 261, 421

Sphagnum Bryophyta, Sphagnales 424

Sphingopyxis Bacteria 142

Sphyraena Actinopterygii, Perciformes 428

Squilla Crustacea, Malacostraca 255 Table 6.1, 263

Stentor Ciliophora, Spirotricha 252 Table 6.1

Stolothrissa Actinopterygii, Clupeiformes 263, 264, 392

Stolothrissa tanganicae 263, 392

Strobilidium Ciliophora, Spirotricha 252 Table 6.1, 259

Strombidium Ciliophora, Spirotricha 252 Table 6.1, 259, 340

Synchaeta Rotatoria, Monogonta 253 Table 6.1, 259

Temora Crustacea, Calanoidea 255 Table 6.1, 261, 282

Temora longicornis 261

Thiobacillus Bacteria, Chromatiaceae 162

Thiobacillus denitrificans 162

Thunnus Actinopterygii, Perciformes 428

Tintinnidium Ciliophora, Spirotricha 252 Table 6.1, 259

Tomopteris Annelida, Polychaeta 253 Table 6.1, 263

Trichocerca Rotatoria, Monogonta 253 Table 6.1

Tropocyclops Crustacea, Cyclopoidea 254 Table 6.1

Typha Angiospermae, Typhales 419

Vampyrella Rhizopoda, Amoebina 294

Velella Coelenterata, Siphonophora 253 Table 6.1

Vibrio Bacteria, Vibrionaceae 158, 162

Vorticella Ciliophora, Peritricha 252 Table 6.1

Zostera Angiospermae, Najadales 420

Zygorhizidium Fungi, Chytridiales 293, 294

Zygorhizidium affluens 294

Zygorhizidium planktonicum 293

General index

absolute viscosity, 44
accessory pigments, 10, 12, 13, 25, 95, 115, 193
acetylene-reduction assay of nitrogenase activity, 164
'acid rain', 423
acidification of natural waters, 423; reversal by adding phosphorus, 425; reversal by liming, 425
Actinopterygii, 258 Table 6.1
adaptive traits in phytoplankton, 207
ADP (adenosine diphosphate), 99
advective patchiness, 87
aestivation, 59
aeolian deposition, 170
affinity adaptation of nutrient uptake, 157, 194, 202, 203
airborne remote sensing, 134
akinetes, 216, 249
alder flies (Megaloptera), 256 Table 6.1
alewife (*Alosa pseudoharengus*), 290
alkaline phosphatase, 155, 159
alkalinity, 124
allogenic vs. autogenic drivers of change in species composition, 360
alternative steady states in shallow lakes, 343, 382, 417, 421; forward and reverse switching, 421, 423; influence of fish, 421; role of zooplankton, 421
aluminium, 424
amino acids, 123
amoebae, 252 Table 6.1, 259
amphipods, 263
anatoxins, 403
annelids, 253 Table 6.1
anoxygenic photosynthesis, 5
anoxyphotobacteria, 5, 6 Table 1.1
anthropogenic impacts, on food webs, 290, 426; on pelagic systems, 395
antioxidants, in photosynthesis, 123
apatite, 151
apoptosis, 297
aquo polymers, 39
archaeans, 5, 93

Arrhenius coefficient, 107
Arrhenius temperature scale, 186
arrow worms (chaetognaths), 256 Table 6.1, 263
artificial neural networks (ANN), 235
ascendency, 303, 355, 357
ascidaceans, 258 Table 6.1
'ascidian tadpole', 258 Table 6.1
ascorbate, 123
ash content of phytoplankton, 25, 26 Table 1.2, 27
assembly rules for phyoplankton communities, 362, 362 Table 7.8
atelomictic lakes, 74, 342
atomic-absorption spectroscopy, 167
atmospheric carbon-dioxide invasion in, 127
ATP (adenosine triphosphate), 5, 95, 99; phosphorylation by, 148, 149
Auftrieb, 3
Aufwuchs, 419, 420
autopoesis; (*see also* 'self-organisation' of communities), 355
auxospore, 60, 64
Avogadro number, 95
avoidance reactions, 122

Bacillariales (pennate diatoms), 7 Table 1.1, 13
bacillariophytes; (*see also* diatoms); density of, 53, 55; deposition of silica, 182, 245; evolution of, 13; growth limitation through silicon deficiency, 196; intervention in silicon cycling by, 174; mixed-depth threshold for growth, 80, 244; pigmentation variability in, 115; primary evolutionary strategies of, 234; silicon content of, 28, 29 Table 1.3, 31 Table 1.5, 53, 220; silicon requirements of, 27, 173, 197; silicon uptake by, 28, 174; structure of, 13
Bacteria, 2, 5, 140, 158, 164, 170, 264, 277 Table 6.3, 388, 392
bacterial pathogens, 292, 295

bacteriochlorophylls, 5, 6 Table 1.1
bacteriophages, 295
bacterioplankton, 2, 140, 141; concentrations and daily production rates, 141 Table 3.4; definition of, 2
bacterivorous phytoplankton, 131, 141
bacterivory in algae, 159
barium, 167
barnacles, 255 Table 6.1
barracuda (*Sphyraena*), 428
basking shark (*Cetorhinus*), 263
bicarbonate transport, 128
bicosoecids, 251
Biddulphiales (centric diatoms), 7 Table 1.1, 13
bioassays, 169
biochemical oxygen demand (BOD), 297
biodiversity, 368
biologically available phosphorus (BAP), 153, 154 Table 4.1, 155, 398
biomanipulation, 263, 416
biomineral reinforcement of cell structures, 27
bioturbation, 250, 413
birds, 1, 76, 84
bivalves (lamellibranchs), 243, 256 Table 6.1, 290
bleak (*Alburnus*), 422
Blelham Enclosures, 217, 220, 226, 245, 274, 284, 284 Table 6.4, 299, 357, 366, 370, 373, 374
bloom, 56; in the Kattegat, 1988, 312
'bloom-forming' Cyanobacteria, 401
blue-green algal blooms, 401
bodonids, 251
boron, 28
bosminids, 261, 279
bottom deposits, sampling the superficial layer of, 220
bottom–up and top–down processes, 250, 287, 288, 416
boundary layer, around a plankter, 147, 148
branchiopods, 254 Table 6.1, 260
bream (*Abramis*), 422, 423

brown tides, 407

bryophytes, 11

bryozoans, 353

buoyancy regulation, through ionic balance, 55; in Cyanobacteria, 56, 58

buoyant forces in water masses, 72

burgundy blood alga, 115

C_4 carbon fixers, 98

C-, S- and R- primary evolutionary strategies, 209, 210 Table 5.3, 212, 231, 233, 234, 298, 315; ecological traits of C strategists, 352, 357, 370; ecological traits of S strategists, 359

CS, CR, RS intermediate strategies, 210

CSR triangle, 210, 232, 315, 363, 364, 396, 408

calanoids, 254 Table 6.1, 260, 261, 279

calcium, 28; algal requirement for, 171; ionic concentrations in the sea, 171; in lakes, 171

calcium bicarbonate and DIC, 171

calcium carbonate in phytoplankton exoskeletons, 13, 25, 53

calcium carbonate scales, 25

calcium hardness, 171

Calvin cycle, 94, 95, 96, 98, 99, 100

CAMP receptor protein (CRP), 128, 181

capacity, environmental carrying; of atmospheric carbon flux, 136; of available silicon, 202; of the nutrient resources, 152, 194; of PAR income, 138; of phosphorus availability, 159; of primary production,, 131, 134, 136, 138; as set by mixed depth, 138; as set by transparency, 117 Table 3.2

carapace gape of cladocerans, 260, 261, 280

carbamylation, 99

carbohydrate, 53

carbon; algal content of, 28, 30, 32, 33, 33 Table 1.6, 35 Table 1.7, 36, 146; availability to photosynthetic organisms, 124, 125; bacterial content of; concentration mechanism, 128; external subsidies of, to pelagic systems, 390; fixation in the plankton, 93;

limitation of photosynthesis by, 127, 130; limitation of phytoplankton growth by, 30; metabolic turnover of, 127; pelagic animal content of, 281; recycling of, in aquatic systems, 391; requirement of for algal cell doubling, 146, 188; sources of, 3

carbon dioxide; atmospheric content of, 14; concentrations in water, 124; depletion in photosynthesis, 96, 98, 126; as a factor regulating species composition, 13; flux from atmosphere to water, 127, 132; supersaturating concentrations of, 126; uptake by phytoplankton, 127

carbonic anhydrase, 130

carboxylation, 94, 99, 127

carboxylic acid, 100

carotenoids, 25, 95, 123

carp (Cyprinus carpio), 422, 423

cell growth cycle, 179

cell assembly, regulation of, 181

cell division, 180; in chlorophytes, 182; in diatoms, 180, 182; in dinoflagellates, 180; in prokaryotes, 181; light efficiency of (r/I), 190, 206

cell division rates; in culture, 183; as a function of algal morphology, 183; as a function of temperature, 186; in relation to fluctuating light intensity, 193; in relation to light income, 206; in relation to persistent low light intensity, 192, 207; in relation to nitrogen deficiency, 195; in relation to nutrient deficiency, 194, 200, 203, 204; in relation to nutrient stoichiometry, 200; in situ, 219 Table 5.4; in relation to phosphorus deficiency, 194; in relation to photoperiod, 189; in relation to resource interaction, 197; in relation to resource supply, 188; maxima at 20 °C, 184 Table 5.1

cell quota, definition, 31; of phosphorus, 31

cell wall, 24, 27

centropomids, 263

cephalopods, 1, 18, 256 Table 6.1

chaetognaths, 256 Table 6.1, 263

Chaetophorales, 12

chain formation, 60

chaoborines, 256 Table 6.1

chaotic behaviour, 396

Charales, 12

chelates, 168, 169

chemiluminescence, 167

chitin, 60

chloride, algal requirement for, 172

chlorine, 28

Chlorobiaceae, 6 Table 1.1, 193

Chlorococcales, 6 Table 1.1, 12

Chlorodendrales, 6 Table 1.1, 11

Chloromonadales, 6 Table 1.1

chlorophyll a, occurrence in algae, 5, 6 Table 1.1, 7 Table 1.1, 12, 13, 25, 26 Table 1.2, 95; algal content of, 33, 35 Table 1.7, 36, 37, 113; as a proxy of algal biomass, 34; relative to cell carbon, 114, 125

chlorophyll b, occurrence in algae, 6 Table 1.1, 11, 95

chlorophyll c, occurrence in algae, 6 Table 1.1, 7 Table 1.1, 12, 13

chlorophyll concentration supported in relation to TP, 399

chlorophyll synthesis, 168

Chlorophyta; classification of, 6 Table 1.1; DIC requirements of, 123, 129; evolution of, 11; osmotrophy in, 131

chloroplast, 146

choanoflagellates, 13, 251

Chromatiaceae, 6 Table 1.1, 193

chromatic adaptation, 115

chromatographic analysis, 167

chromophores, 123

chromosomes, 179

Chromulinales, 6 Table 1.1

Chroococcales, 6 Table 1.1, 26 Table 1.2

chrysolaminarin, 7 Table 1.1

Chrysophyta; classification of, 6 Table 1.1, 13; DIC requirements of, 129; distribution and calcium hardness, 171; evolution of, 11, 12; mixotrophy in, 131; pigmentation variability in, 115; silicon requirements, 27; storage products in, 25; vitamin requirements of, 170

chrysose, 7 Table 1.1

chydorids, 261, 279

chytrids, 66, 292
ciliates, 252 Table 6.1, 259, 260, 261, 353, 354
ciliophorans, 252 Table 6.1, 259
cirripedes, 255 Table 6.1
cisco (*Coregonus artedii*), 289
cladocerans, 254 Table 6.1, 260, 261, 354
Cladophorales, 12
climate change, 431
clupeoids, 263, 389
coastal waters, phytoplankton of, 233, 310
cobalt, 28, 167
coccoliths, 7 Table 1.1, 13, 25, 130
coccolithophorids, evolution of, 13, 14; morphology of, 7 Table 1.1, 25
coelenterates, 252 Table 6.1
Coleochaetales, 12
community assembly in the plankton, 302
community ecology of phytoplankton, 302, 350
compensation point (where photosynthetic gains and maintenance losses are balanced), 16, 116, 118, 120
competition, definition of, 152
competitive exclusion principle (of Hardin), 203, 369
compositional change in communities rate of, 367
condensates of assimilation, 6 Table 1.1, 7 Table 1.1, 25, 26 Table 1.2, 189
constancy, 304
Continuous Plankton Recorder (CPR), 307
contractile vacuoles, 25
convection, 52
coot (*Fulica atra*), 422, 423
copepods, 254 Table 6.1, 259, 260
copper, 28; algal requirement for, 167; toxicity of, 167
copper sulphate, as an algicide, 167, 416
coregonids, 389
corethrines, 256 Table 6.1
Coriolis force, 17, 42
crabs (decapods), 256 Table 6.1, 263
Craspedophyceae, 13
crepuscular habitats, 192
critical depth model (of Sverdrup), 119, 120

critical patch size, 88, 89
crustaceans, 259; classification of, 254 Table 6.1
Cryptomonadales, 6 Table 1.1; osmotrophy in, 131; photadaptation in, 115
cryptophytes, evolution of, 12; morphology of, 6 Table 1.1, 25
ctenophores (sea combs and sea gooseberries), 253 Table 6.1, 263
cultural eutrophication, definition of, 397
cultures, 167
cyanelles, 6 Table 1.1, 11
Cyanobacteria; buoyancy in, 56, 58, 67, 81, 124; cytology of, 25; classification of, 6 Table 1.1, 10, 26 Table 1.2; in deep chlorophyll layers, 193; DIC requirements of, 130; distribution and calcium hardness, 171; enhanced abundance with eutrophication, 402; evolution of, 10, 11; gas vacuoles of, 24, 56; habits of, 10; in relation to low DIN, 196; nitrogen fixation in, 164, 169, 398; in relation to N:P ratio, 198; overwintering of, 405; pH tolerance in, 124, 425; photoprotection in, 123; pico-microplanktic transformation, 269; production of siderophores, 169; storage products in, 25; structure of photosynthetic apparatus, 96; toxicity in, 56, 402, 403; 'warning' concentrations, 1229
cyanobacterial blooms, 401
cyanophages, 295
cyanophycin, 85
cyclopoids, 254 Table 6.1, 260, 278
cylindrical curves, 81
cysts, of dinoflagellates, 215
cytochrome, 96, 167, 168
cytoplasm, 25

DAPI (DNA-specific stain), 220
DCMU, 66
DNA, 140, 146
DNA:cell carbon ratio, as an index of DNA replication, 220
daphniids, features of, 261; habitats of, 261
decapods, 256 Table 6.1, 263

decomposition rates of algae, 297
deep 'chlorophyll maxima' (DCM), 59, 83, 115, 192, 328
density; of air, 46; of carbohydrates, 53; of diatoms, 53, 55; of metabolic oils, 53; of mucilage, 55; of nucleic acids, 53; of phytoplankton, 51, 53, 54 Table 2.3, 55; of polyphosphate bodies, 53; of proteins, 53; salinity dependence of, 40; of silica, 53; temperature dependence of, 40; of water, 39 Table 2.1, 40
density gradients in water columns, 72, 74, 75, 79, 204
density stratification, 73, 75
desmids, 6 Table 1.1, 12, 25; DIC requirements of, 129
detergents as a source of P, 398
detritus, 419
diadinoxanthin–diatoxanthin reaction, 123
diatoms; (*see also* bacillariophytes), 13; deposition of silica, 182, 245; DIC requirements of, 129; evolution of, 27, 234; growth limitation though silicon deficiency, 196; intervention in silicon cycling, 174; mixed-depth threshold for growth, 80, 244; morphology of, 27; photoprotection in, 123; settling velocities of, 52, 66, 67, 245; silicon content of, 28, 29 Table 1.3, 31 Table 1.5, 32, 53, 220; silicon requirements of, 27, 173, 197; uptake of silicon, 174; vitamin requirements of, 170
diatoms, centric, 6 Table 1.1, 25
diatoms, pennate, 7 Table 1.1
diatoxanthin, 6 Table 1.1, 12
dilution functions, 240
dimethyl sulphide (DMS), 173
dimethyl sulphonopropionate (DMSP), 173, 307; algal osmoregulation by, 173; DMS–DMSP metabolism, 173
dinitrogen reductase, 164
dinoflagellates; adaptive radiation in, 13, 14; DIC requirements of, 129, 130; evolution of, 12, 13, 16; mixotrophy in, 131; photosynthesis in, 123;

'swimming' velocities, 68; vitamin requirements of, 170

Dinophyta, 7 Table 1.1; dinophyte zooplankters, 252 Table 6.1

dipterans, 263

disentrainment of phytoplankton, 75, 123

disequilibrium explanations of species diversity, 369; consumer effects, 371; intermediate disturbances, 372, 374, 384; pathogenic effects, 371; resource-based competition, 371

dispersal of plankton, 353

dissipation of turbulence, 47

dissipative ecological unit (DEU), 351, 355, 382

dissolution of diatom frustules, 174

dissolved humic matter (DHM), 142, 390, 424

dissolved inorganic carbon (DIC), 125; uptake in phytoplankton, 128

dissolved inorganic nitrogen (DIN), 163

dissolved organic carbon (DOC), 93, 139, 140, 141, 142, 261, 264, 388; DOC produced by phytoplankton, 100, 124; DOC as a source of DIC, 126

dissolved organic matter (DOM), 142

dissolved organic nitrogen (DON), 163

disturbance, 304, 372

diversity indices, 366

doliolids, 258 Table 6.1, 262

droop model of nutrient uptake, 150

dry masses of phytoplankton, 25, 26 Table 1.2; in relation to volume, 25

ducks, 423

Dugdale model of nutrient uptake, 150

dugongs, 427

dust deposition (aeolian deposition), 170

echinoderms, 257 Table 6.1

ecological efficiency of energy transfer, 261

ecological stoichiometry, 199, 262

ectoprocts, 257 Table 6.1

eddy spectrum, 17, 18, 38, 42, 48, 251

El Niño, 305, 309, 382

electivity of predators, 278

electrophoretic mobility, 67

endemism among phytoplankton, 353

endoplasmic reticulum, 25

endosymbiosis, 11, 13, 15, 305

energy flow in pelagic systems, 387, 388; relevance to structure, 393, 427

engineer species, 382, 419

entrainability, 17

entrainment of phytoplankton, 38, 39, 67, 69, 74, 75, 204, 243

entropy, 355

environmental grain, 47

environmental heterogeneity, 78

epilimnion, 75

epilithic algae, 342

epiphytes, 285

equitability (evenness), 366

ethylene diamine tetra-acetic acid (EDTA), 168

Eubacteria, 5

Euglenales, 6 Table 1.1

euglenoids, 12, 25

euglenophytes; classification, 6 Table 1.1; evolution of, 11, 12; osmotrophy in, 131; storage products in, 25

eukaryotes, 11; early evolution of, 11, 12, 13

euphausids, 256 Table 6.1, 263

euphotic zone, 116

euplankton, 2

Eustigmatophyta, 6 Table 1.1, 12

Eutreptiales, 6 Table 1.1

eutrophic food webs, 140

eutrophic lakes, phytoplankton of, 232

eutrophication; of coastal waters, 426; of lakes, 397; of seas, 425; of Windermere, 333; reversal of, 408

evaporative heat losses from water bodies, 41

evenness (or equitability), 366

excretion, of excess photosynthate, 123

excretory products of zooplankton, 287

exergy, 303, 356; fluxes of, 357, 375

exoskeletons, 25, 27

exponential phase of population growth, 183

extracellular production, 100, 140, 239

eyespots, 12

factor limitation of capacity, 169

faecal pellets, 288

ferredoxin, 96, 165, 168

Fick's diffusion laws, 127, 147

Ficoll, 51

filter-feeders, 260, 261, 262

filter-feeding; on detrital POC, 277; impacts of, 264, 270, 274; intervention of planktivores, 277, 289; limiting food concentrations for, 265, 275, 277; measurement of, 265; nutritional aspects, 270; rates of, 265, 266, 274; in relation to food availability, 267; in relation to temperature, 274; saturating food concentrations for, 274, 276; size selection, 267; in turbid water, 280

filtering setae of cladocerans, 260, 261

fish, 1, 18

fishing industry, 427

fish-rearing ponds, 418

fish-supportive capacity of pelagic production, 389

flagellates, 6 Table 1.1; vertical distribution of, 83

flagellum, propulsion, 49

Flexibacteria, 5

flood-plain lakes, 352

flow cytometry, 140

fluorescence, 122; as a surrogate of biomass, 122; as a surrogate of phyletic composition, 122

fluorometry, 52, 132

flushing, 240

fluvial 'dead zones', 242

fluvioglacial deposits, 337

foamlines, 85, 86

food chains of the pelagic, 261; energetic dissipation in, 291; interactive linkage strength, 291; length of, 291

food location by chemoreception, 280

food requirements, of calanoids, 262
    of cladocerans, 262; of pelagic fish, 389
food, zooplankton-supportive capacity of, 262
food-web control nodes, 394
foraging in calanoids, 279; in relation to pelagic energy flows, 388; saturating food availability, 280; superiority over filter-feeding, 279, 286, 394
foraminiferans, 252 Table 6.1, 259
form resistance of phytoplankton (to sinking), 50, 60; coefficient of, 51; as contributed by chain formation, 60; as provided by protuberances and spines, 60; of stellate colonies, 63
friction velocity; (see also turbulent velocity), 46, 48 Table 2.2
frictional drag coefficients, 41
frustules; (see also diatoms, siliceous cell walls), 27, 32, 174
fucoxanthin, 6 Table 1.1, 7 Table 1.1, 12, 13
fulvic acids, 168
fungal epidemics, 293
fungi, 2; parasitic on algae, 292

Gaia principle, 173
gas vacuoles, 56
gas vesicles, 57, 58
gastropods, 256 Table 6.1
gastrotrichs, 253 Table 6.1
Gelbstoff, 111
generation time, 178, 188
germination of resting stages, 214, 229, 250
gill rakers, 263, 389
Gilvin, 111
glaciations, 336
Glaucophyta, 6 Table 1.1
global warming, 14
glucose, 100
glutathione, 123
glycerol, 52
glycogen, 6 Table 1.1, 25, 26 Table 1.2, 59, 98, 189
glycolate excretion, 100, 239
glycolic acid, 100, 123
Gonyaulacales, 7 Table 1.1
gonyaulacoids, 13

grazing losses of phytoplankton, 220
green tides, 407
greenhouse gases, oceanic absorption of, 428
gross primary production (GPP), 141
growth rate; (see also replication rate), 178
growth-regulating nucleotides, 158
growth and reproductive strategies of freshwater phytoplankton, 208; correspondence with morphology, 211; C-, S- and R- primary evolutionary strategies, 209, 233, 315; CS, CR, RS intermediate strategies, 210; SS strategy, 211, 234; violent, patient and explerent strategies, 211
guanosine bipyrophosphate (ppGpp), 158, 181, 214
Gulf Stream (North Atlantic Drift Current), 289, 306
Gymnoceratia, 216, 229
Gymnodiniales, 7 Table 1.1
gymnodinioids, 13

HNLC oceans (high nitrogen, low chlorophyll), 308
habitat filtration, 360
habitat templates, 315, 348, 394
haddock (Megalogrammus aeglifinus), 427
Haeckel, E., 3
Halobacteria, 5
Haptonema, 13
haptophytes; classification of, 7 Table 1.1; evolution of, 11, 13; morphology of, 24; pigmentation variability in, 115; vitamin requirements of, 170
harmful algal blooms (HAB), 312, 401
heat flux, solar, 72
heleoplankton, definition of, 2
hemipterans, 263
Henry's law, 125, 126
Hensen, V., 3
herbivory on phytoplankton, 250
herring (Clupea harengus), 261, 263, 427
heterocysts (heterocytes), 6 Table 1.1, 26 Table 1.2, 164, 196, 230
heterotrophy, 12, 126, 141
Hibberdiales, 7 Table 1.1

high-nutrient, iron-deficient areas of the ocean, 170
Hill–Bendall model of photosynthesis, 94
holopedids, 261
holotrichs, 252 Table 6.1, 259
horizontal patchiness of phytoplankton, 84, 88
human metabolites, as a source of P, 398
humic acids, 168
humic lakes, plankton of, 12, 326
hydrogen, algal content of, 28, 32, 33 Table 1.6
hydroxyapatite, 129
hydrozoans, 252 Table 6.1
hyperparasitism, 294
hyperscum, 215
hypersensitiviy of hosts to parasitic attack, 294
hypnozygotes, 215
hypolimnion, 75

ice cover, 306, 307, 308, 322
information theory, 366
intermediate disturbance hypothesis (IDH), 373, 377
invasion by species 'new to the locality', 384
internal phosphorus loading, 412
ionic regulation, 55
iron, 167, 335, 398, 424; algal requirement for, 168; cell contents of, 28, 168
    complexes with organic carbon (DOFE), 170; limitation of nitrogen fixation, 169; limiting concentrations, 168; sources of, 170; symptoms of cell deficiency, 168; uptake by algae, 168
iron-binding chelates, 168
IRONEX oceanic fertilisation, 170, 308
island biogeography, 352
isopycny, 19, 72, 82

jellyfish, 2
Jenkin mud sampler, 249

kairomones, 269
karyokinesis, 179
kataglacial lakes, phytoplankton of, 232
Kelvin–Helmholtz instability, 79

keystone species, 382, 394, 427
kinematic viscosity, 44, 48
KISS model of critical patch size, 88
Kolmogorov eddy scale, 42, 45
krill (euphausids), 256 Table 6.1, 263

lake trout (*Salvelinus namaycush*), 290
lakes as a source of CO_2, 390
lamellibranchs, 256 Table 6.1
laminar flow, 44
Langmuir circulations, 85, 85
larvaceans, 258 Table 6.1, 259
larvae, planktic: actinotrocha, 257
    Table 6.1; amphiblastulae, 252
    Table 6.1; appendicularia, 258
    Table 6.1; auricularia, 257 Table
    6.1; bipinnaria, 285 Table 6.5;
    cyphonautes, 257 Table 6.1;
    cypris, 255 Table 6.1; megalopa,
    256 Table 6.1; nauplii larvae, 259;
    paralarvae, 264; phyllosoma, 256
    Table 6.1; pilidium, 253 Table 6.1;
    pluteus, 257 Table 6.1;
    trochophore, 256 Table 6.1;
    trochosphere, 253 Table 6.1;
    veliger, 256 Table 6.1; zoea, 256
    Table 6.1
latent heat of evaporation, 41
Laurentian great lakes, 290
leptodorids, 261
leucosin, 6 Table 1.1
licensing factors, 181
Liebig's law of the minimum,
    152
light (*see also* photosynthetically
    active radiation), 95; absorption
    by water, 138; attenuation
    (extinction) of, underwater, 103,
    109, 110, 113, 116; effect of cloud
    on, 108; light underwater, 108,
    110; surface incidence of, 108;
    surface reflectance, 108
light harvesting by photosynthetic
    organisms, 94, 95, 123
light-harvesting complex (LHC);
    areal concentrations of, 137;
    mechanics of, 96, 97, 192;
    numbers of, 113; structure of, 95,
    96
light adaptation in phytoplankton,
    113, 114, 120
light-dependent growth in
    phytoplankton, 206
light:nutrient ratio in lakes, 199

light-saturation of photosynthesis,
    103, 106 Table 3.1, 122, 136
limiting capacity, 152
limiting factors, 151
limnoplankton,, 2
lipid, 6 Table 1.1, 7 Table 1.1, 54;
    accumulation, 54
lipo-polysaccharides, 403
Lloyd's crowding index, 82
lobsters (decapods), 256 Table 6.1,
    263
Lorenzian attractors, 396
loss rates of phytoplankton, 139,
    239; aggregated, 297; due to cell
    death, 240, 296; due to
    consumption by animals, 240,
    243, 278, 280, 285, 286; due to
    downstream transport, 239, 242;
    due to parasitism, 240, 292, 295;
    due to sedimentation, 240, 243;
    due to washout, 239, 240, 241;
    from suspension, 70, 76; seasonal
    variations, 297, 299
losses, distinction between
    physiological and demographic
    processes, 239
luxury uptake of nutrients, 31, 151,
    161, 194
Lycopodium spores, as a model of
    suspended phytoplankton, 76, 80,
    245

mackerel (*Scomber scombrus*), 261
macroecology, 380
macroinvertebrates (of macrophyte
    beds), 419
macrophytes, 342, 391, 395;
    architecture of, 418; effects of
    eutrophication on, 420; leaf area
    index in, 420; as refuges for
    potamoplankton, 243
macrophytoplankton, size definition
    of, 5
macroplanktic herbivores, 262
macroplanktic predators, 261
macrothricids, 261, 279
magnesium, 28; algal requirement
    for, 172; availability of, 172
maintenance requirement of
    resource, 189
mammals, 2, 18
malacostracans, 255 Table 6.1
manatees, 427
Mandala, of Margalef, 314, 408

manganese, 28, 167, 424; algal
    requirement for, 167
mannitol, 6 Table 1.1
mantis shrimps, 263
marine snow, 66, 166, 248, 249, 430
marl lakes, 129
mass mortalities of phytoplankton,
    214
match-mismatch hypothesis, 290
maturation-promoting factors (MPF),
    181
mechanical energy loss in water
    columns, 47
medusae, 252 Table 6.1
megalopterans, 256 Table 6.1, 263
megaplankton, 263
Mehler reaction, 100, 123
membrane transport systems, 148
memory in community assembly,
    366, 380, 384
meromictic lakes, 74, 263
meromixis, 74, 340
meroplankton, 2, 243
mesophytoplankton, size definition
    of, 5
mesotrophic lakes, phytoplankton
    of, 233
mesozooplankton: of freshwaters,
    260; of the sea, 261
metalimnion, 75
methanogens, 5
Michaelis–Menten kinetics, 150
microalgae (μ-algae), 2, 4
microbial food web, 140, 259, 261,
    318
microbial 'loop', 140
microcystins, 403, 404
micronutrients, inorganic, 166;
    organic, 170
microphytoplankton, size definition
    of, 5
microplanktic protists, 251
microplanktic metazoans, 259
microstratification, 81
microzooplankton, 170, 251, 259,
    261, 287, 392
migration of phytoplankton,
    vertical, 83
Milankovitch cycle, 431
minimal communities, 351
minimum cell quota, of a given
    nutrient, 150
Mischococcales, 6 Table 1.1
mitochondria, 25

mitosis, 180

mixing times, 75, 76, 117

mixolimnion, 340

mixotrophy in phytoplankton, 131, 159

models simulating phytoplankton growth and performance, 234

Mollusca, 256 Table 6.1

molybdate-reactive phosphorus (MRP), 155, 399

molybdenum, 28, 167, 335, 398; algal requirement for, 167

monimolimnion, 340

Monin–Obukhov length, 72, 79, 119, 121, 243

Monod equation, 150

monosccharides, 100

Monosigales, 13

monosilicic acid, 28, 173, 182

monovalent : divalent cation ratio, 172

morphological adaptations of phytoplankton, 19, 23, 24, 48, 53, 120

motility, advantages of for phytoplankters, 205, 206

mucilage, presence of in plankters, 271; as a defence against digestion, 273; as a defence against grazers, 270, 273; as a defence against heavy metals, 273; as a defence against oxygen, 273; for nutrient sequestration, 272; production as a buoyancy aid, 55, 67; properties of, 24, 55, 271; relative volume of, 57 Table 2.4; as a response to nutrient deficiency, 271; for self-regulation, 272; for streamlining, 272

mucilaginous phytoplankters, 24

mucilaginous threads, 67

Müller, J., 3

mycoplankton, 2, 2

mycosporine-like amino acids, 123

mysids, 255 Table 6.1, 263

myxomycetes, 295

NADP (nicotinamide adenine dinucleotide phosphate), 94, 96; reduction of, 165

nannoplankton, 2

nanocytes, 215

nanoflagellates, 259, 280, 287

nanophytoplankton size definition of, 5

nekton, 1, 2, 18, 263

nematodes, 253 Table 6.1

net primary production (NPP), 134; by global domains, 135 Table 3.3

netplankton, 4

niche, 303, 354

Nile perch (Lates), 341

Nile tilapia (Oreochromis), 341

nitrate reductase, 163; synthesis of, 168

nitrilotriacetic acid (NTA), 168

nitrogen; algal content of, 28, 30, 32, 33, 33 Table 1.6, 35 Table 1.7, 36, 161; availability to phytoplankton, 162, 163; cell quotas of, 164; external concentrations favouring dinitrogen fixers, 196; external concentrations half-saturating phytoplankton growth, 196; N-regulated capacity of Patagonian lakes, 162, 196, 335; potential chlorophyll yield, 164; requirement of for cell doubling, 146, 188, 195; sources of, 162; uptake rates of algal cells, 163

nitrogen fixation, 164; in Cyanobacteria, 164, 398; DIN sensitivity of, 165, 196; dependence upon energy, 165; dependence upon phosphorus, 165; redox sensitivity of, 164; requirement for trace metals, 165

nitrogenase, 168

North American Great Lakes, 89

North Atlantic Oscillation (NAO), 289

northern cod (Gadus morhua), 427

Nostocales, 6 Table 1.1, 26 Table 1.2

nucleus, 25; nuclear division (karyokinesis), 179

nucleic acids, 27, 53, 179

nutrients, 145; availability to phytoplankton, 145, 146; cell quotas, 150, 199; demand versus supply, 145, 188; flux rates to cells, 147; intracellular sensitivity and control, 149; limitation, 151, 152; uptake by phytoplankton, 145, 146, 148, 150; uptake, role of motion, 147

nutrient-induced fluorescent transients (NIFT) indicative of P deficiency, 159

nutrient ratios, in algal tissue, 28

nutrient regeneration, 287

ocean; heat exchanges, 41; renewal time, 39; total area of, 39; total volume of, 39; water circulation in the, 41, 42

oceanic fronts, 307, 308

oceanic provinces, 318

Oedogoniales, 12

oils (as storage product), 6 Table 1.1, 7 Table 1.1, 25; effect on cell densities, 53, 54

oligotrophic lakes, 388; phytoplankton of, 232

oligotrophic food webs, 140

open oceans, phytoplankton of, 233

operons, 149, 181

opossum shrimps (mysids), 255 Table 6.1

organic composition of phytoplankton, 28

orthophosphoric acid, 153

Oscillatoriales, 6 Table 1.1, 26 Table 1.2

osmoregulation, 25

osmotrophy, 131

outgassing of CO_2 from lakes and rivers, 126

overwintering of Cyanobacteria, 214, 405

oxygen, algal content of, 28, 32, 33 Table 1.6

pH, 124, 423

pH–CO_2–bicarbonate system in water, 125

pH sensitivity of phosphate solubility, 153

packaging effect on pigment distribution, 192

paradox of the plankton (of Hutchinson), 368, 379

paralarvae, 256 Table 6.1

paramylon (or paramylum), 6 Table 1.1, 25, 98, 189

parasites of phytoplankton, 292, 294

parasitic fungi, 66

particulate organic carbon (POC), 126, 279, 390, 393

particulate organic matter (POM), 390

patchiness of phytoplankton, 78, 84, 88

Pavlovales, 7 Table 1.1

Péclet number (*Pe*), 147

Pedinellales, 7 Table 1.1

Pedinomonadales, 6 Table 1.1, 11

penetrative convection, 72

perch (*Perca fluviatilis*), 289

perennation of phytoplankton, 249

Peridiniales, 7 Table 1.1

peridinians, 32

peridinin, 13

peritrichs, 252 Table 6.1

Permian extinctions, 14

phaeophytes, 12

phaeophytin, 96

phagotrophy, 12; in algae, as a means of nutrient sequestration, 159

phoronids, 257 Table 6.1

phosphatase production in algae, 159

phosphate binding in soils, 398

phosphoglycolate, 99

phosphoglycolic acid, 123

phosphorus, algal content of, 28, 31, 32, 33, 33 Table 1.6, 36, 151; affinity adaptation, 157; availability to phytoplankton, 398, 399; deficiency in cells, 158; external concentrations half-saturating phytoplankton growth, 195; as a factor limiting growth rate of phytoplankton, 158; fractions in water, 154 Table 4.1; minimum requirements of algae for, 151; potential chlorophyll yield, 160; quotas supporting replication, 158; requirement of for cell doubling, 146, 188, 194; sources of, 151; storage adaptation, 157; uptake rates of algal cells, 150, 155, 157 Table 4.2; velocity adaptation, 157

phosphorus, recycling of, 412

phosphorus 'release' from sediments, 413

photadaptation, to low light doses, 120, 193, 194

photoautotrophy, 5, 94, 141

photodamage, 124

photodegradation (of DOM), 143

photoheterotrophy, 5

photoinhibition, 104, 117, 121, 122, 124, 194, 246

photons, 95

photon flux density (PFD), 95, 123

photon interception and absorption, 95

photooxidation, 121

photoprotection, 104, 121, 122, 194

photorespiration, 99, 123, 139, 239

photosynthesis, as a basis of aquatic production,, 5, 93; biochemistry of, 94; carbon limitation of, 127, 130; carbon oxidation, 100; carbon sourcing in, 124, 125; condensates of, 25; electron transport, 94, 95; hexose generation, 94, 95; Hill–Bendall model of, 94; integration through depth, 104, 119, 132; integration through time, 132; light efficiency of (*P/ I*), 103, 105; light harvesting in, 94, 95; light saturation of, 103, 106 Table 3.1, 122, 136; measurement, 101, 105, 107; models of, 104, 110; net of respiration, 107, 132; oxygen generation, 94; productive yields, 133, 139; quantum efficiency of, 95, 98, 100; reduction of CO_2, 94, 96, 98, 146; reduction of water, 94; regulation by light, 101, 102, 108, 120, 121; in relation to depth, 103; surface depression of, 104

photosynthetic inhibitors, 66

photosynthetic pigments, 5, 6 Table 1.1, 7 Table 1.1, 11, 12, 25, 26 Table 1.2, 123

photosynthetic quotient (*PQ*), 100, 105, 107, 163, 190

photosynthetic rates, Tab 3.1, 104, 106 Table 3.1; temperature dependence of, 106

photosynthetic yield to planktic food webs, 139

photosynthetically active radiation (PAR); (*see also* light), 16, 95, 108; carbon yield, 100; energy equivalence, 95, 100

photosystem I, 94, 96, 122, 165

photosystem II, 94, 95, 122, 123

phototrophy, 12

phototrophic bacteria, 5

phycobilins, 6 Table 1.1, 10, 12, 25, 26 Table 1.2, 168

phycobiliproteins, 115

phycobilisomes, 10, 95

phycocyanin, 6 Table 1.1, 26 Table 1.2, 115, 193

phycoerythrin, 6 Table 1.1, 26 Table 1.2, 115, 193

phycomycetes, 292

phycoviruses, 295

phyllopods, 260, 279

Phytodiniales, 7 Table 1.1

phytoplankters, 3

phytoplankton, definition of, 2, 3, 36; functional classification of, 4; phyletic classification of, 4, 5, 36; general characteristics of, 16; numbers of species of, 4; size, 19, 20 Table 1.1, 34, 35 Table 1.7, 36, 53; shape, 19, 20 Table 1.1, 22, 23, 35 Table 1.7, 50; form resistance, 50, 53, 60, 243; morphological adaptation, 19, 23, 24, 48, 53, 120; turbulent embedding of, 48, 49; 'swimming', 49; effect of viscosity, 49; settling velocities of, 39, 49, 50, 52, 61 Table 2.5, 67, 68, 75, 77; as a function of size, 53; as a function of density, 53; as a function of chain formation, 62; as a function of cylindrical elongation, 62; as a function of colony formation, 63; stellate colonies, 63; settling velocities as a function of cell vitality, 65, 66; entrainment of, 38, 39, 67, 69, 74, 75, 77; vertical distribution of, 80, 83; horizontal patchiness of, 84, 88; dry masses of, 25, 26 Table 1.2; densities of, 51, 53, 54 Table 2.3; organic composition of, 28; elemental composition, 24, 25, 28, 35 Table 1.7, 36, 146; photosynthesis in, 93; light adaptation in, 113, 114, 120; uptake of carbon in, 127, 128; bacterivory in, 131; excretion in, 123; nutrient requirements of, 146; nutrient uptake, 19; availability of nitrogen to, 164; nitrogen uptake rates of, 163; availability of phosphorus to, 153, 154, 154 Table 4.1, 398, 399

phytoplankton, definition of (*cont.*)
phosphorus uptake rates of, 150, 155, 157 Table 4.2; phosphorus-constrained growth of, 158; cell phosphorus quotas, 161; major-ion requirements of, 171; as trophic-level indicators, 129; growth and replication of, 178; growth rates of, (*see also* cell division rates), 183; as a function of nutrient availability, 151; light-dependent growth in, 206; areas projected by, 192; adaptive traits of, 207; growth and reproductive strategies, 208; strategies for survival of resource exhaustion, 211; loss from suspension, 70, 76; losses to grazers, 139, 141, 220, 250; counter-grazing adaptations, 269; effects of zooplankton selection on, 287; beneficial effects of zooplankton, 287; selection by performance, 225; seasonality of selective mechanisms, 231; eutrophic species of, 232; oligotrophic species, 203 Table 5.2, 232, 233; of shallow lakes, 233; of coastal waters, 233; of open oceans, 233; of rivers, 242; entrainment criteria for, 243; sedimentary fluxes of, 247, 249, 318; parasites of, 292, 294; parasitic infections of, 66; perennation of, 249; regenerative strategies of, 249; trait separation of, 319; community assembly in, 350; succession in the sea, 313; biomass levels in the sea, 313, 388; biomass levels in inland waters, 345; dispersal mechanisms, 353; of acidic waters, 423, 424; of humic lakes, 326; methods to control abundance, 414

phytoplankton assemblages, communities and structure, 302; of high-latitude seas, 306, 308; of the North Pacific, 304; of the South Pacific, 306; of oceanic upwellings, 308; of continental shelves, 309, 310; of inshore waters, 317; of large oligotrophic lakes, 319; of small and medium

oligotrophic lakes, 319; of subarctic lakes, 328; of alpine lakes, 329; of the English Lakes, 330; of Auracanian lakes, 335; of kataglacial lakes, 336; of the African Great Lakes, 339; of Lake Titicaca, 342; of shallow lakes, 342; in relation to supportive capacity, 346; overview of mechanisms, 383

phytoplankton population dynamics in natural environments, 217; estimating *in-situ* rates of growth, 217; episodic outbursts of rapid increase, 217; observed increase rates, 217; frequency of dividing cells (FDC), 218; frequency of nuclear division, 220; from resource depletion, 220; spring increase in a temperate lake, 221; effect of inoculum size, 223, 225, 229; effect of nutrient availability, 223; effect of temperature, 223, 226; effect of underwater light field, 223, 226; effect of resource diminution, 223; effect of lake enrichment on timing of maxima, 224; species selection by performance, 225; exploitative efficiency, 225; comparison of species-specific performances, 226; effect of light-exposure thresholds, 226; in relation to phosphorus availability, 228; in relation to pH, 228; in relation to N:P ratio, 230; in relation to transparency, 230; in relation to carbon dioxide, 231; in relation to seasonality, 231, 232, 234; quantification in a Blelham Enclosure, 221 Table 5.5, 221 Table 5.6; as influenced by simultaneous loss rates, 298; species selection by performance, 298; effect of inoculum size, 299; sinking losses., 299; simulation models of, 234

phytosociology, 318; application to the pelagic, 319

Phytotelmata, 292, 352

picophytoplankton, size definition of, 5; settling velocities of, 68; in oligotrophic oceans, 234

plagioclimax, 370

Planck's constant, 95

plasmalemma, 24, 25, 146

planktivory, 392, 395

plankton, definition of, 1, 2

plastids, 11, 12, 13, 25, 123

plastoquinone (PQ), 96, 122

plate tectonics, 309

polder lakes, 366

polychaetes, 253 Table 6.1, 263

polymerase chain reaction (PCR), 140

polymictic lakes, 74

polynia, 306

polyphemids, 261

polyphosphate bodies, 53

polysaccharides, 98

Porifera (sponges), 27

post-upwelling relaxation, 309

potamoplankton, 2, 242

potassium, 28; algal requirement for, 172

power, 303, 355

Prasinophyta, 6 Table 1.1, 11

Prasiolales, 12

prawns (decapods), 256 Table 6.1

Prochlorales, 6 Table 1.1, 26 Table 1.2

prochlorobacteria, 6 Table 1.1, 11, 26 Table 1.2

programmed cell death, 297

projected area of algal shape, 192

prokaryotes, 25

Prorocentrales, 7 Table 1.1

PROTECH phytoplankton growth model, 236, 377, 383, 415

proteins, 27, 53

protistans, 2, 11, 27, 251

Protobacteria (α-), 142

Protobacteria (γ-), 142

protomonadids, 251

proton motive forces, 148

protoplasm, 25, 30

protozoans, parasitic on algae, 294

Prymnesiales, 7 Table 1.1

purple non-sulphur bacteria, 5

pycnocline, 75, 79, 245, 340

Pyramimonadales, 6 Table 1.1, 11

pyrosomans, 262

Q_{10}, of photosynthetic rate, 107; of cell division, 186

quanta, 95

quantum efficiency of photosynthesis, 95

quantum yield of photosynthesis, 98, 100
quinones, 96

r-, K- and w- selection, 209, 212, 231, 314
radiolarians, 3, 27, 252 Table 6.1, 259
rails, 423
Raphidomonadales, 6 Table 1.1
Raphidophyta, 6 Table 1.1, 12
raptorial feeders, planktic, 260, 278
recruitment of larval fish, 289
recycling of phosphorus, 412
recycling of resources in the pelagic, 288, 296
Redfield stoichiometric ratio, 32, 33, 33 Table 1.6, 37, 188, 198
red tides, 407
redox-sensitivity of phosphate solubility, 153; of nitrogen speciation, 162; of nitrogen fixation, 164; to toxicity of metal micronutrients, 167
relocation of phosphorus capacity, away from plankton, 411
replication rate, 178
reptiles, 1, 18
resilience (elasticity) of communities to recover from forcing, 304, 380, 383
resistance of communities to forcing, 304, 380
resource-based competition, 197, 371
resource segregation, 205
respiration; basal rate, 108, 115, 132, 190; measurement of, 102; 'dark' and 'light', 102, 189; as a balance to excess photosynthesis, 139, 239, 388
resting stages, 214, 249, 250
restoration of lakes by phosphorus-load reduction, 409; by sediment removal,, 413; using clay minerals, 413; sensitivity of systems to treatment, 413
resuspension of sedimented material, 79, 250
retention time (hydraulic), 240, 241
Reynolds number, of water flow, 45, 50, 243; of particle motion, 49, 50
rhizopods, 252 Table 6.1, 292
rhodophytes, 11, 12

Richardson numbers, 73
Rift Valley lakes, 339; food webs of, 263
river bed roughness, 46
rivet theory (of Ehrlich and Ehrlich), 382
RNA, 140, 146, 199, 262
roach (Rutilus rutilus), 278, 392, 422
rotifers, 253 Table 6.1, 259, 292, 354
RUBISCO (ribulose 1,5-biphosphate carboxylase), 94, 95, 98, 99, 127, 128; as an oxidase, 99, 123, 140
RuBP (ribulose1,5-biphosphate), 94, 98, 99, 123

salmonids, 389
salps, 258 Table 6.1, 261, 262
sampling strategies for phytoplankton, 89
saprophytes, 297
sarcodines, 259
satellite-borne remote sensing, 131, 134
saxitoxins, 403, 407
scales, calcareous, 7 Table 1.1, 25; siliceous, 25
scum formation, 402, 404
Scourfieldiales, 6 Table 1.1
sea bass (Morone), 428
sea birds, 428
sea combs (ctenophores), 263
sea gooseberries (ctenophores), 253 Table 6.1, 263
sea lamprey (Petromyzon marinus), 290
sea otters, 427
sea sawdust, 81, 166
sea urchin barrens, 426
seals, 427, 428
seasonality in phytoplankton performances, 231, 232, 234
seaweeds, 12
Secchi disk, 116, 118
secondary sewage treatment, 399
sediment deposition, 249
sediment diagenesis, 250
sediment focusing, 250
sediment semi-fluid surface layer, 249
sediment traps, 220, 246
sedimentary flux, 284, 388, 390, 430; of phytoplankton, 247, 249, 318

sedimented material, resuspension of, 250
seiching, 340
selective planktic feeding, 278; impacts of, 280
'self-organisation' of communities, 355
seston, definition of, 2
settlement of diatom frustules, 175
settling flux, 220
shallow lakes, characteristics of, 319, 342, 391; alternative steady states in, 343, 417; phytoplankton of, 233
Shannon–Weaver index, 366, 378
sharks, 427
shear, 148, 250, 413
shear velocity; (see also turbulent velocity), 46, 48 Table 2.2
shelf waters, 117 Table 3.2, 135, 309, 430
Sherwood Number (Sh), 148
shortwave electromagnetic radiation, 72
siderophores, 169
sidids, 260
sieve effect, 112
silica, 28; opaline, 174; skeletal, 174
siliceous cell walls, 25, 27, 53
silicification of diatoms, 13, 174
silicon, 28; algal content of, 33 Table 1.6, 173; cycling by diatoms, 174; deposition, in diatoms, 182, 245; external SRSi concentrations half-saturating diatom growth, 197; relative to carbon, 174; requirements of diatoms, 173, 197; silicon limitation of diatoms, 196; silicon requirements of Synurophyceae, 173; solution from sinking of diatoms, 174; sources, 27, 173; uptake by diatoms, 173
sinking behaviour, of phytoplankton, 39, 49, 50, 52, 243; of colonial aggregates, 63; of cylindrical shapes, 50, 62; of oblate spheroids, 50; of prolate spheroids, 50; in relation to mixed-layer depth, 70
relevance to nutrient uptake, 147; of sculpted models of algae, 50, 51; spherical shapes, 50; of stellate colonies, 63

sinking behaviour (cont.); of teardrop shapes, 51; vital regulation of, 52, 65, 66, 123, 246
sinking-rate determination, 51
size-efficiency hypothesis, 278
size spectra of phytoplankton assemblages, 348
slime moulds, 295
sodium, 28; algal requirement for, 172; cyanobacterial requirement for; , 172
soil chemistry, 398
SOIREE oceanic iron fertilisation, 170, 308
solar energy income, 41; as a function of latitude, 41
solar constant, 41, 95
solar energy income, 108
solar heat flux, 72
soluble reactive silicon (SRSi), 173
species composition in marine plankton, 302; in limnetic plankton, 318; in relation to light deficiency, 345; in relation to nutrient deficiencies, 345
species richness of phytoplankton assemblages, 354, 364, 366, 367; a-, β-, ?-diversity, 380, 394; relevance to community functioning, 382; in relation to successional maturation, 384
species selection, in oligotrophic systems, 203; in rivers, 242
specific heat, of water, 16, 41
spirotrichs, 252 Table 6.1, 259
squid (cephalopods), 256 Table 6.1, 264
stability in phytoplankton communities, 304, 381
standing crop, 223
starch, 6 Table 1.1, 7 Table 1.1, 25, 98, 189
statoblasts, 257 Table 6.1, 353
stoichiometric lake metabolism model, 401
stoichiometric oxygen demand of decomposition, 296
stoichiometry, as an ecological driver, 199, 287
Stokes' Equation, 49, 51, 243
stolothrissids, 389
stoneworts, 419
storage adaptation of nutrient uptake, 157, 194, 202

storage products of phytoplankton, 6 Table 1.1, 7 Table 1.1, 25, 26 Table 1.2
stratification of the water column, 73, 75, 79, 204
straw, anti-algal effects of, 405
structural viscosity, 67
structured granules, 25
succession, in terrestrial plant communities, 204; in the plankton, 303, 359, 370, 384
succession rate, 366
successional climax, 304, 359, 365
sulphate, algal requirement for, 172
sulphur, algal content of, 28, 32, 33 Table 1.6
sulphur-reducing bacteria, 5, 193
supportive capacity, 346
surface avoidance by flagellates, 83
surface-area-to-volume ratio in phytoplankton, 185
surface mixed layer, 47, 204; mixed-layer depth, 48 Table 2.2, 71, 74, 75, 79, 80, 204, 244; mixed depth, in relation to light penetration, 117, 119, 138, 244; relative to phytoplankton sinking rates, 244; artificial enhancement of, 414
suspension of planktic organisms, 1, 17, 38
Synurales, 7 Table 1.1
synurophycaens, 25
'Swimming' velocities in phytoplankton, 49, 68, 83, 205
'Telescoping' of algae at the surface, 402, 404

tench (Tinca), 423
terrestrial sources of POM, POC, 390
Tetrasporales, 6 Table 1.1, 12
thalliaceans, 258 Table 6.1, 261, 262
thermal bar, 89, 324
thermal stratification, 73, 75, 290
thermocline, 74, 75
thylakoids, 10, 12, 25, 96, 146; assembly, 168
tides, 42
tidal mixing, 42, 48
total iron (TFe), 169
total phosphorus (TP), 155, 399
toxic algae, 312, 401, 407
toxic metals, 167; redox sensitivity of, 167

trace elements, 167
trait selection, 362
trait-separated functional groups of freshwater phyoplankton, 319, 320 Table 7.1, 346, 347 Table 7.6, 394
Tribonematales, 6 Table 1.1
tripton, definition of, 2
trishydroxymethyl-aminomethane (tris), 168
trophic cascades, 263, 288
trophic states, 400; separation criteria, 401 Table 8.1
tuna (Thunnus), 428
tunicates, 258 Table 6.1, 261, 262
turbellarians, 253 Table 6.1
turbulence, 42, 46; in confined channels, 46; associated with moving particles, 49
turbulent dissipation, 47, 71; rate of, 47, 48, 48 Table 2.2
turbulent embedding of phytoplankton, 48, 49
turbulent extent, 69
turbulent intensity, 74, 75
turbulent kinetic energy, 47
turbulent velocity, 39, 45, 46, 48 Table 2.2, 69; fluctuations of, 45, 46
turgor pressure, 57, 59
tychoplankton, 1, 2

ubiquity of phytoplankton, 353
Ulotrichales, 6 Table 1.1, 12
ultraplankton, 2, 4
Ulvales, 12
ultraviolet radiation, protection from, 123
underwater light fields, 108, 110, 111, 119
upwellings, 308
urea, as a source of nitrogen available to phytoplankton, 163
Utermöhl counting technique, 179

vacuoles, 25
vanadium, 28, 167, 335, 398
Várzeas (flood-plain lakes), 352
vascular plants, 11
velocity adaptation of nutrient uptake, 157, 194, 202
velocity gradients through water columns, 47, 80, 83

vertical migration of phytoplankton, 205
viruses, 2, 5, 292, 295
viscosity of water, 39 Table 2.1, 40
vitamins, 170; microalgal benefits from, 170; B$_{12}$ requirements of phytoplankton, 170
Vollenweider–OECD model, 399, 400, 408, 410
volume; of the ice caps, 39; of inland waters, 39; of the sea, 39
Volvocales, 6 Table 1.1, 12, 25; DIC requirements of, 130
vorticellids, 259

Water, physical properties of, 16, 39; et seq., 39 Table 2.1; molecular structure, 39
density of, 39 Table 2.1, 40; viscosity of, 39 Table 2.1, 40;

viscous behaviour of, 44, 49; specific heat of, 41
water, expansion coefficient of, 72
water movements, 17, 38, 39, 41, 47, 77
water blooms, 56, 81, 401
Wedderburn number, 74, 75, 243
WETSTEM electron microscopy, 67
whales, 263, 428
'white water' events, 13
wind action, 42; stress applied to water, 46, 69; velocity, 46

X-ray microanalysis, 167
xanthophylls, 12, 13, 25, 95; , 115, 123
Xanthophyta, 12, 25; classification of, 6 Table 1.1

yellow perch (Perca flavescens), 290

young-of-the-year (YOY) fish, 289

zeaxanthin–violoxanthin reaction, 123
zebra mussel (Dreissenia polymorpha), 290, 323
zeta potential, 67
zinc, 28, 167
zoobenthos, 392, 395
zoomastigophorans, 251, 252 Table 6.1
zooplankton, 2, 2, 251; phyletic classification of, 252 Table 6.1; to, 258 Table 6.1; anti-predator defences, 269; growth rates of, 274; competitive interactions among calanoids and cladocerans, 286; zooplankton and optimal foraging by fish, 392, 394
zoospores, 66
Zygnematales, 6 Table 1.1, 12